Analyzing Network Data in Biology and Medicine
An Interdisciplinary Textbook for Biological, Medical, and Computational Scientists

The increased and widespread availability of large network data resources in recent years has resulted in a growing need for effective methods for their analysis. The challenge is to detect patterns that provide a better understanding of the data. However, this is not a straightforward task because of the size of the datasets and the computer power required for the analysis. The solution is to devise methods for approximately answering the questions posed and these methods will vary depending on the datasets under scrutiny. This cutting-edge text introduces biological concepts and biotechnologies producing the data, graph and network theory, cluster analysis and machine learning, before discussing the thought processes and creativity involved in the analysis of large-scale biological and medical datasets, using a wide range of real-life examples. Bringing together leading experts, this text provides an ideal introduction to and insight into the interdisciplinary field of network data analysis in biomedicine.

Nataša Pržulj is Professor of Biomedical Data Science at University College London (UCL) and an ICREA Research Professor at Barcelona Supercomputing Center. She has been an elected academician of The Academy of Europe, Academia Europaea, since 2017 and is a Fellow of the British Computer Society (BCS). She is recognized for designing methods to mine large real-world molecular network datasets and for extending and using machine learning methods for integration of heterogeneous biomedical and molecular data, applied to advancing biological and medical knowledge. She received two prestigious European Research Council (ERC) research grants, Starting (2012–2017) and Consolidator (2018–2023), and USA National Science Foundation (NSF) grants among others. She is a recipient of the BCS Roger Needham Award for 2014. She was previously an Associate Professor (Reader, 2012–2016) and Assistant Professor (Lecturer, 2009–2012) in the Department of Computing at Imperial College London and an Assistant Professor in the Computer Science Department at University of California Irvine (2005–2009). She obtained a PhD in Computer Science from University of Toronto in 2005.

Analyzing Network Data in Biology and Medicine

An Interdisciplinary Textbook for Biological, Medical, and Computational Scientists

Edited and authored by

NATAŠA PRŽULJ

*Professor of Biomedical Data Science, Computer Science Department,
University College London
ICREA Research Professor at Barcelona Supercomputing Center*

CAMBRIDGE
UNIVERSITY PRESS

University Printing House, Cambridge CB2 8BS, United Kingdom

One Liberty Plaza, 20th Floor, New York, NY 10006, USA

477 Williamstown Road, Port Melbourne, VIC 3207, Australia

314-321, 3rd Floor, Plot 3, Splendor Forum, Jasola District Centre,
New Delhi – 110025, India

79 Anson Road, #06–04/06, Singapore 079906

Cambridge University Press is part of the University of Cambridge.

It furthers the University's mission by disseminating knowledge in the pursuit of
education, learning, and research at the highest international levels of excellence.

www.cambridge.org
Information on this title: www.cambridge.org/bionetworks
DOI: 10.1017/9781108377706

© Cambridge University Press 2019

This publication is in copyright. Subject to statutory exception
and to the provisions of relevant collective licensing agreements,
no reproduction of any part may take place without the written
permission of Cambridge University Press.

First published 2019

Printed and bound in Great Britain by Clays Ltd, Elcograf S.p.A.

A catalogue record for this publication is available from the British Library.

Library of Congress Cataloging-in-Publication Data
Names: Pržulj, Nataša, editor.
Title: Analyzing network data in biology and medicine : an interdisciplinary
   textbook for biological, medical and computational scientists / edited by
   Nataša Pržulj, University College London.
Description: Cambridge, United Kingdom ; New York, NY : Cambridge University
   Press, 2019. | Includes bibliographical references.
Identifiers: LCCN 2018034214 | ISBN 9781108432238 (hardback : alk. paper)
Subjects: LCSH: Medical informatics–Data processing. | Bioinformatics.
Classification: LCC R858 .A469 2019 | DDC 610.285–dc23
   LC record available at https://lccn.loc.gov/2018034214

ISBN 978-1-108-43223-8 Paperback

Additional resources for this publication at www.cambridge.org/bionetworks

Cambridge University Press has no responsibility for the persistence or accuracy
of URLs for external or third-party internet websites referred to in this publication
and does not guarantee that any content on such websites is, or will remain,
accurate or appropriate.

To my loving family: Cvita, Bogdan, Nina, Sofia, and Laurentino.
And to my best friend, Vesna.

Contents

List of Contributors   page ix
Preface   xiii

1   **From Genetic Data to Medicine: From DNA Samples to Disease Risk Prediction in Personalized Genetic Tests**   1
LUIS G. LEAL, ROK KOŠIR, AND NATAŠA PRŽULJ

2   **Epigenetic Data and Disease**   63
RODRIGO GONZALEZ-BARRIOS, MARISOL SALGADO-ALBARRÁN, NICOLÁS ALCARAZ, CRISTIAN ARRIAGA-CANON, LISSANIA GUERRA-CALDERAS, LAURA CONTRERAS-ESPINOSA, AND ERNESTO SOTO-REYES

3   **Introduction to Graph and Network Theory**   111
THOMAS GAUDELET AND NATAŠA PRŽULJ

4   **Protein–Protein Interaction Data, their Quality, and Major Public Databases**   151
ANNE-CHRISTIN HAUSCHILD, CHIARA PASTRELLO, MAX KOTLYAR, AND IGOR JURISICA

5   **Graphlets in Network Science and Computational Biology**   193
KHALIQUE NEWAZ AND TIJANA MILENKOVIĆ

6   **Unsupervised Learning: Cluster Analysis**   241
RICHARD RÖTTGER

7   **Machine Learning for Data Integration in Cancer Precision Medicine: Matrix Factorization Approaches**   286
NOËL MALOD-DOGNIN, SAM F. L. WINDELS, AND NATAŠA PRŽULJ

8   **Machine Learning for Biomarker Discovery: Significant Pattern Mining**   313
FELIPE LLINARES-LÓPEZ AND KARSTEN BORGWARDT

9   **Network Alignment**   369
NOËL MALOD-DOGNIN AND NATAŠA PRŽULJ

10   **Network Medicine**   414
PISANU BUPHAMALAI, MICHAEL CALDERA, FELIX MÜLLER, AND JÖRG MENCHE

11   **Elucidating Genotype-to-Phenotype Relationships via Analyses of Human Tissue Interactomes**   459
IDAN HEKSELMAN, MORAN SHARON, OMER BASHA, AND ESTI YEGER-LOTEM

12 **Network Neuroscience** 490
ALBERTO CACCIOLA, ALESSANDRO MUSCOLONI, AND CARLO VITTORIO CANNISTRACI

13 **Cytoscape: A Tool for Analyzing and Visualizing Network Data** 533
JOHN H. MORRIS

14 **Analysis of the Signatures of Cancer Stem Cells in Malignant Tumors Using Protein Interactomes and the STRING Database** 593
KREŠIMIR PAVELIĆ, MARKO KLOBUČAR, DOLORES KUZELJ, NATAŠA PRŽULJ, SANDRA KRALJEVIĆ PAVELIĆ

Index   621

Contributors

Nicolás Alcaraz
The Bioinformatics Centre Section for RNA and Computational Biology, University of Copenhagen, Copenhagen, Denmark

Cristian Arriaga-Canon
CONACyT-Instituto Nacional de Cancerología, Mexico

Omer Basha
Department of Clinical Biochemistry and Pharmacology, Faculty of Health Sciences, Ben-Gurion University of the Negev, Beer-Sheva, Israel

Karsten Borgwardt
Machine Learning and Computational Biology Lab, Department of Biosystems Science and Engineering, Basel, ETH Zurich, Switzerland
Swiss Institute of Bioinformatics, Basel, Switzerland

Pisanu Buphamalai
CeMM Research Center for Molecular Medicine of the Austrian Academy of Sciences, Vienna, Austria

Alberto Cacciola
Biomedical Cybernetics Group, Biotechnology Center (BIOTEC), Center for Molecular and Cellular Bioengineering (CMCB), Center for Systems Biology Dresden (CSBD), Department of Physics, Technische Universität Dresden, Dresden, Germany
Brain bio-inspired computing (BBC) lab, IRCCS Centro Neurolesi "Bonino Pulejo," Messina, Italy, Department of Biomedical, Dental Sciences and Morphological and Functional Images, University of Messina, Italy

Michael Caldera
CeMM Research Center for Molecular Medicine of the Austrian Academy of Sciences, Vienna, Austria

Carlo Vittorio Cannistraci
Biomedical Cybernetics Group, Biotechnology Center (BIOTEC), Center for Molecular and Cellular Bioengineering (CMCB), Center for Systems Biology Dresden (CSBD), Department of Physics, Technische Universität Dresden, Dresden, Germany
Brain bio-inspired computing (BBC) lab, IRCCS Centro Neurolesi "Bonino Pulejo," Messina, Italy

Laura Contreras-Espinosa
Universidad Nacional Autonoma de Mexico (UNAM), Mexico

Thomas Gaudelet
Department of Computer Science, University College London, London, UK

Rodrigo González-Barrios
Instituto Nacional de Cancerología, Mexico

LIST OF CONTRIBUTORS

Lissania Guerra-Calderas
Instituto Nacional de Cancerología, Mexico

Anne-Christin Hauschild
Krembil Research Institute, Toronto Western Hospital, Toronto, Canada, Department of Pharmacogenetics Research, Center for Addiction and Mental Health, Toronto, Canada

Idan Hekselman
Department of Clinical Biochemistry & Pharmacology, Faculty of Health Sciences, Ben-Gurion University of the Negev, Beer-Sheva, Israel

Igor Jurisica
Krembil Research Institute, Toronto Western Hospital, Toronto, Canada
University of Toronto, Toronto, Canada

Marko Klobučar
University of Rijeka, Department of Biotechnology, Centre for High-Throughput Technologies, Rijeka, Croatia

Rok Košir
Institute of Biochemistry, Faculty of Medicine, University of Ljubljana
BIA Separations CRO, Labena Ltd, Ljubljana, Slovenia

Max Kotlyar
Krembil Research Institute, Toronto Western Hospital, Toronto, Canada

Sandra Kraljević Pavelić
University of Rijeka, Department of Biotechnology, Centre for High-Throughput Technologies, Rijeka, Croatia

Dolores Kuzelj
University of Rijeka, Department of Biotechnology, Centre for High-Throughput Technologies, Rijeka, Croatia

Luis G. Leal
Department of Life Sciences, Imperial College London, UK
Supported by a President's PhD Scholarship from Imperial College London

Felipe Llinares-López
Machine Learning and Computational Biology Lab, Department of Biosystems Science and Engineering, Basel, ETH Zurich, Switzerland
Swiss Institute of Bioinformatics, Basel, Switzerland

Noël Malod-Dognin
Department of Computer Science, University College London, London, UK

Jörg Menche
CeMM Research Center for Molecular Medicine of the Austrian Academy of Sciences, Vienna, Austria

Tijana Milenković
Department of Computer Science and Engineering, Eck Institute for Global Health, and Interdisciplinary Center for Network Science and Applications (iCeNSA), University of Notre Dame, Notre Dame, Indiana, USA

John H. Morris
Department of Pharmaceutical Chemistry, University of California San Francisco, USA

Felix Müller
CeMM Research Center for Molecular Medicine of the Austrian Academy of Sciences, Vienna, Austria

Alessandro Muscoloni
Biomedical Cybernetics Group, Biotechnology Center (BIOTEC), Center for Molecular and Cellular Bioengineering (CMCB), Center for Systems Biology Dresden (CSBD), Department of Physics, Technische Universität Dresden, Dresden, Germany

Khalique Newaz
Department of Computer Science and Engineering, Eck Institute for Global Health, and Interdisciplinary Center for Network Science and Applications (iCeNSA), University of Notre Dame, Notre Dame, Indiana, USA

Chiara Pastrello
Krembil Research Institute, Toronto Western Hospital, Toronto, Canada

Krešimir Pavelić
Juraj Dobrila University of Pula, Pula, Croatia

Nataša Pržulj
ICREA Research Professor at Barcelona Supercomputing Center, Barcelona, Spain; Professor of Biomedical Data Science at Computer Science Department, University College London, London, UK

Richard Röttger
Department of Mathematics and Computer Science, University of Southern Denmark, Odense, Denmark

Marisol Salgado-Albarrán
Instituto Nacional de Cancerología, Mexico

Moran Sharon
Department of Clinical Biochemistry and Pharmacology, Faculty of Health Sciences, Ben-Gurion University of the Negev, Beer-Sheva, Israel

Ernesto Soto-Reyes
Natural Science Department, Universidad Autónoma Metropolitana-Cuajimalpa (UAM-C), Mexico

Sam F. L. Windels
Department of Computer Science, University College London, London, UK

Esti Yeger-Lotem
Department of Clinical Biochemistry and Pharmacology, Faculty of Health Sciences, Ben-Gurion University of the Negev, Beer-Sheva, Israel National Institute for Biotechnology in the Negev, Ben-Gurion University of the Negev, Beer-Sheva, Israel

Preface

We are witnessing tremendous changes in the world around us. Technological advances are impacting our lives and increasing our ability to measure things. They are yielding an astounding harvest of data about all aspects of life that form large systems of diverse interconnected entities. We are beginning to utilize the data systems to improve our understanding of the world and find solutions to some of the foremost challenges.

One such challenge is to better understand biological phenomena and apply the newly acquired understanding to improve medical treatments and outcomes. Even at the level of a cell, we are far from fully understanding the processes that we measure by genomic, epigenomic, transcriptomic, proteomic, metabolomic, metagenomic, and other "omic" data. All these different data types measure different aspects of the functioning of a cell. As these observational data grow, it is increasingly harder to analyze them and understand what they are telling us about the cell, not only due to their sizes, but also their complexities. It is not only the biology that we need to understand, which is being measured, but also the ways to abstract these complex data systems by using mathematical models that make the data amenable to computational analyses. In addition, we need to comprehend the computational challenges coming from the theory of computing, which teach us about the problems that we can efficiently and exactly solve by using computers, and about those that we cannot. Furthermore, we need to put all this biology, mathematics, and computing jointly in use by the medical sciences if we are to contribute to personalizing treatments and improving our health.

This textbook provides a resource for training upper level undergraduate students, graduate students, and researchers in this multidisciplinary area. The goal is to enable them to understand these complex issues and undertake independent research in this exciting, emerging field. The textbook presents the material in a way understandable to researchers of diverse backgrounds. Exercises are provided at the end of each chapter to put the learned material into practice. The solutions to exercises are also provided for lecturers on www.cambridge.org/bionetworks.

The textbook material is carefully chosen to start from basics and lead to more advanced concepts in a succession of chapters that build on the previous ones. The book first introduces the complex genomic and epigenomic data related to diseases and risk prediction along with the main machine learning, bioinformatics and other methods used in this domain (Chapters 1 and 2). Then it introduces the widely adopted mathematical models of graphs (networks) and the basic theory needed to understand the tools constructed for analyzing complex omics network data (Chapter 3). A very important and widely studied omics network is that of physical interactions between proteins in a cell. Hence, the biotechnologies producing these data are surveyed in Chapter 4, the quality of the data is discussed and major public databases containing the data are introduced. An introduction into methods for advanced analyses of these data is given in Chapter 5.

The textbook proceeds with the basics of machine learning commonly used to analyze network data. First, it introduces a key methodology of unsupervised learning, cluster analysis (Chapter 6) and the applications of it in this interdisciplinary area. Then it proceeds with the basics of machine learning for data integration (Chapter 7) and advanced topics in machine learning for biomarker discovery (Chapter 8).

Just as aligning genetic sequences has revolutionized our biological and medical understanding, aligning molecular networks is expected to have similar groundbreaking impacts. This important topic is addressed and network alignment methods introduced in Chapter 9. The field of network medicine is introduced in Chapter 10. Methodology for elucidating genotype-to-phenotype relationships via analyses of human tissue-specific interactomes is presented in Chapter 11. Another important interconnected network is that of neurons in our brain. The basics of network neuroscience are presented in Chapter 12. Finally, a description of how the material presented in the textbook can be put to practice by using a major software package for analyzing network data, Cytoscape, and a major protein interaction database, STRING, are presented in the last two chapters.

I hope you will find this textbook a good resource for getting you started with doing research in this exciting and inspiring multidisciplinary area. I wish you enjoyable learning!

Nataša Pržulj

1 From Genetic Data to Medicine: From DNA Samples to Disease Risk Prediction in Personalized Genetic Tests

Luis G. Leal, Rok Košir, and Nataša Pržulj

1.1 Background

The completion of the 1000 Genomes Project has given us a comprehensive insight into the variability of the human genome. On average, a typical human genome will differ from the reference genome in 4.1 to 5 million sites, the majority of which (86% on average) consist of single nucleotide polymorphisms (SNPs) [1]. SNPs are defined as locations in the DNA sequence where at least two different nucleotides appear in the human population [2]. They have been the focus of many studies, as their presence may have functional consequences: They may affect the transcription factor binding affinity, the mRNA transcript stability, and could produce changes in the amino acid sequences of proteins [3, 4]. These functional changes have effects on the predisposition of individuals to diseases, or the efficacy of drugs on patients.

Given that functional changes could increase predisposition to diseases, SNPs are used as genetic markers to identify genes associated with diseases. According to Szelinger et al. [5], a gene's function may be altered by SNPs at different levels. There are silent, or non-functional SNPs, which do not interfere with the functions of genes, SNPs which increase the risk of a disease, and SNPs having strong functional effects upon disease development (Mendelian disorders); however, only some hundreds of them are likely to contribute to disease risk [1, 6].

Detecting these genetic alterations is fundamental to understanding the development of diseases. With the advent of SNP microarrays, searching for inherited genomic variants was enabled for the first time and it boosted the relationship between computational methodologies and biological understanding [2]. It was not until the rise in next generation sequencing (NGS) and the increase in the density of SNP microarrays, that the SNP identification and genotyping tasks could be executed in mass. Both technologies have shifted the amount of data generated from single SNP studies to whole-genome analyses of multiple individuals at the same time (e.g., the Cancer Genome Atlas Project,[1] the NHLBI Exome Sequencing Project,[2] and the 1000 Genomes Project[3]) [7]. Accurate computational approaches are, however, needed to elucidate heterogeneous disorders from these raw data.

Genome-wide association studies (GWAS) are ideal for detecting novel SNP-disease associations, because disease predisposition can be closely related to the presence of genetic variants. A large number of susceptible loci for common complex diseases (e.g., heart disease, diabetes, obesity, hypertension, cancer) have been found in recent GWAS [8]. For example, genome-wide approaches are important to uncover multiple genetic alterations occurring in cancer development [9]. Different types of cancer, including breast cancer [10] and lung cancer [11], have been studied using this approach. Also, thanks to simultaneous genotyping of SNPs, we have broadened the understanding of diabetes [12], coronary artery disease [13], and hypertension [14], to name a few.

The number of published GWAS increases every year for a wide range of complex traits and different websites gather the data generated from these studies. The full catalog of GWAS is administrated by the National Human Genome Research Institute and the European Bioinformatics Institute (NHGRI-EBI).[4] Other public databases with relevant information are the Single Nucleotide Polymorphism database (dbSNP),[5] the Human Gene Mutation Datbase (HGMD)[6] and the Catalogue of Somatic Mutations in Cancer (COSMIC).[7]

A major purpose of GWAS is to formulate a predictive model based on SNPs for disease diagnostics. GWAS were conceived with the hope of revealing the genetic causes of complex diseases, in the same way that single SNPs driving Mendelian diseases (e.g., cystic fibrosis, hemophilia A, muscular dystrophy) were identified in the past with other approaches [15]. To date, the vast majority of the variants identified by GWAS explain only a fraction of disease heritability, in part because complex diseases have been shown to be the result of multiple interacting SNPs, also known as gene–gene epistatic interactions, and environmental factors [16, 17].

Genetic studies of complex diseases have been approached from two perspectives [18]. First, it is hypothesized that cumulative effects of common variants (i.e., SNPs with allele frequencies higher than 5% in the population) result in a complex disease, which is a focus of many GWAS. Second, it is hypothesized that low frequency variants (0.5%–5%) and rare variants (<0.5%) can have large effects resulting in a complex disease [18, 19]. Thus, the allele's frequency in the population and its effect size on

---

[1] http://cancergenome.nih.gov
[2] http://evs.gs.washington.edu/EVS
[3] www.1000genomes.org
[4] www.ebi.ac.uk/gwas
[5] www.ncbi.nlm.nih.gov/snp
[6] www.hgmd.cf.ac.uk
[7] http://cancer.sanger.ac.uk/cosmic

the disease are crucial not only to identify the origin of complex diseases, but also to determine the technology (e.g., rare variants can only be determined by NGS, not by microarrays) and the sample size in genetic studies (e.g., large samples are needed to find significant rare variants) [2]. While the amount of studies focused on the effects of low frequency and rare variants is still limited, there are already encouraging results coming from these studies. For example, rare variants associated with osteoporosis, type 2 diabetes, Alzheimer's disease, risk of heart attack, as well as several variants associated with lipid metabolism have been identified [20]. With the ever increasing number of genome projects worldwide based on sequencing we can expect these numbers to rise in the near future. For further information on rare and low frequency variants please refer to an excellent review by Bomba et al. [21].

Traditional univariate statistical methods are used to identify single SNP-disease associations [10, 11, 22]. The association tests examine each SNP independently for association to the disease by means of logistic regression models or contingency table methods when the trait is qualitative (e.g., case/control phenotype), or by means of analysis of variance (ANOVA) when the trait is quantitative (e.g., artery thickness) [3]. Even though these strategies are adequate to study single SNPs, detecting complex genetic architectures demands more sophisticated data-mining approaches [23]. Thus, new algorithms capable of discovering complex multigenic SNPs are being developed for mining data from GWAS studies [24, 25].

Thanks to the completion of the Human Genome Project, the technological advances to genotype SNPs and the detection of markers associated with complex traits via GWAS, new opportunities have appeared for the clinical translation of these discoveries to personalized medicine. In this way, genetic tests have enabled the confirmation or prediction of specific disorders by identifying changes in the chromosomes, DNA sequence, or gene products of individuals [26]. Genetic testing has grown to cover a wide range of variants, including variants associated with adult disease onset, drug dosage, and adverse reactions [27]. As it was envisioned some years ago, the accelerated improvement in genome sequencing techniques has brought GWAS results a step closer to the personal benefit of patients.

Personalized genetic tests (PGTs) have revolutionized our perception of healthcare services under the promise of accurate prediction of disease risk. PGTs are founded on the synergistic relationship between technological advances, medical knowledge, and computational methods, translating the best of them for the benefit of patients. Currently, PGTs can be indicated by health providers, but they also can be accessed through direct-to-consumer (DTC) providers. The DTC genetic testing is offered worldwide via the Internet by various companies; typically, after sending a saliva sample, the consumers receive a report detailing if they carry specific mutations which may increase the disease risk. The idea of DTC services came to life with the availability of GWAS data from different populations; however, the predictive ability of the genetic risk models is a concern [28], especially when inappropriate reference populations are used and the non-genetic factors are omitted [29].

The purpose of this chapter is to summarize a foremost component of PGTs: the methods to transform the raw data from genotyping technologies into disease risk predictions. Because accuracy in risk assessment is essential for personalized medicine, we emphasize the current state and perspectives of the algorithms for SNP identification, as well as the main approaches for predicting SNPs causative of disease. In parallel, we discuss how these components have been implemented in the PGTs market by DTC companies, hence providing the reader with a global picture of the science behind disease risk prediction.

This chapter is structured as follows. First, we introduce the health-related genetic tests and list some companies offering personalized genetic services, including their locations, prices, and types of services they offer. Then, we outline the main platforms for SNP genotyping, along with the algorithms designed for detecting SNPs from their output data. Next, we survey the techniques to predict single-SNP-disease and multiple-SNP-disease associations. We discuss some predictive genetic risk models in DTC services and the factors affecting these approaches. Finally, we discuss perspectives and give recommendations for the improvements of algorithms in personalized genetic testing.

Box 1.1 contains a glossary of terms used in this chapters.

Box 1.1: Glossary of biological concepts

This box presents brief definitions of the biological terms used in the book. Most of these definitions have been adapted from the Genetic Home Reference Glossary.[a]

- **Allele**: Allele represent one of two or more versions of the same gene. Each individual inherits two alleles, one from each parent.
- **Allele frequency**: The measure of an allele's relative frequency (percentage) in a population.
- **Alternative splicing**: The usage of different exons that are all part of the initial transcript, to form the mature mRNA, which will be translated into a protein. Alternative splicing results in the generation of related, but different, proteins from one gene.
- **Coding region/sequence (CDS)**: Represent the region of DNA that will be transcribed into a mature messenger RNA (mRNA) and translated into the amino acid sequence of a protein.
- **Common variants**: Alternative forms of a gene, which are present with a minor allele frequency (MAF) higher than 5%.
- **Contiguous SNPs**: SNPs lying next to each other on the DNA strand.
- **Copy number variants (CNVs)**: A type of structural variation where a section of DNA is present in two or more copies instead of only one.
- **Duplication**: A type of mutation, where a portion of a gene, a whole gene, several genes, or larger regions of the chromosome are copied and are present in duplicate amounts.

(cont.)

---

[a] http://ghr.nlm.nih.gov/glossary

- **Effect size**: Contribution of a SNP to the genetic component (i.e., heritability) of the disease. This is usually the odds ratio reported in GWAS for the SNP [30, 31].
- **Exons**: Exons represent portions of the DNA sequence of a gene that are transcribed into mRNA and are translated into proteins.
- **Gene**: Genes are the basic physical and functional units of heredity made out of DNA. They make instructions on how to make proteins. The human genome is composed of approximately 19,000 genes [32].
- **Gene–gene epistatic interactions (epistatsis)**: A condition in which the expression of one gene is affected by the expression of one or more independently inherited genes. For example, when the expression of gene B depends on the expression of gene A, then the expression of gene B will not occur if gene A is not expressed. In such a case, gene A is said to be epistatic to gene B.
- **Genotype**: Represents all of the alleles an individual inherited from parents. It can also refer to two specific alleles of a particular gene. At the genomic level, each SNP can have two alleles (e.g., allele *A* and allele *a*); hence, a SNP is linked to one of three possible genotypes, e.g., *AA*, *Aa*, or *aa*.
- **Haplotype**: Describes a combination of alleles or a set of SNPs that are found on the same chromosome and tend to be inherited together. The International HapMap Project collects information of haplotypes.
- **Heritability component**: The heritability component of a disease is the proportion of phenotypic variability in the population explained by genetic factors [24].
- **Heterozygous**: Contrary of the homozygous: an individual inherits two different alleles from parents.
- **Homozygous**: When an individual receives the same alleles from parents, he/she is said to be homozygous.
- **Insertions/deletions (INDELs)**: Types of genetic variation involving the addition (insertion) or loss (deletion) of smaller (single nucleotide) or larger pieces of the DNA strand from a part of a chromosome.
- **Introns**: Introns are portions of the DNA molecule that are transcribed into mRNA, but are not translated into proteins.
- **Inversion**: A type of mutation in which a smaller or larger segment of the DNA molecule is broken away, inverted from end to end and re-inserted back into the chromosome.
- **Linkage disequilibrium (LD)**: Indicates that alleles are physically close to one another on the DNA strand. They occur together more often than accounted by chance alone.
- **Loci**: Particular sites on a chromosome.
- **Minor allele frequency (MAF)**: Refers to the frequency of the least abundant (minor) allele of a SNP in a population.

(cont.)

> **Box 1.1: Glossary of biological concepts (cont.)**
>
> - **Rare variants**: Alternative forms of a gene, which are present with a minor allele frequency (MAF) of less than 1%.
> - **SNPs (rSNPs)**: Single nucleotide polymorphisms involve a variation in one single base pair at a specific location in the genome. They represent the main type of single nucleotide variants present in the human genome. SNPs differ from SNVs in that their variation in the population is known. A variation can be said to be a SNP if it is present in at least 1% of the population.
> - **Single nucleotide variations (SNV)**: In NGS sequence analysis, variations in a single nucleotide are referred to as SNV, since their population frequency is not known.
> - **Structural variants (SV)**: Represent different types of genomic alternations, including duplications, inversions, insertions, deletions etc. To be qualified as SV, the affected region of the DNA has to be 1 kb or larger in size.
> - **Untranslated regions (UTRs)**: UTRs represent regions of DNA on either side of the coding regions (CDS) that are not translated into the amino acid sequence of a protein.

1.2 Genetic Tests in Healthcare

Genetic tests are predominantly used to determine whether a patient's DNA sequence has alterations that may result in chromosomal, monogenic, or complex disorders (see Box 1.2) [26, 33]. These alterations in specific genes or chromosomes are important for healthcare in different contexts; for example, they may be responsible for inherited disorders, or they could affect the sensitivity of individuals to a drug therapy. Therefore, types of PGTs have been formulated for a range of applications (see Section 1.2.1) and a number of specialized PGT providers has increased around the world (see Section 1.2.2).

1.2.1 Types of Genetic Tests

While a wide variety of PGTs are available for non-health concerns, including paternity, siblingship, forensic testing, and ancestry, we are interested in health-related genetic tests. Most of the health-related genetic tests evaluate if the patient carries a specific genetic mutation that may increase the disease risk, or a physical trait (Box 1.2). Hence, the test may reveal specific mutations in the DNA, effectiveness of drugs, possibility of drug side effects, or the influence of genetic variants on physical traits [26].

> **Box 1.2: Types of genetic disorders and PGTs**
>
> - **Chromosomal disorders**: Abnormalities such as extra copies, or missing parts of one chromosome.
> - **Monogenic disorders or Mendelian diseases**: Mutations in one gene that arise in a severe disorder. The alteration may be linked to one or both alleles, and a person carrying the mutation may have the disorder's symptoms or not (healthy carrier).
> - **Complex genetic disorders**: The joint effect of alterations in many genes, lifestyle and environmental factors.
> - **Predictive genetic tests**: Detect gene mutations that increase the risk of developing a disorder in adult life. They are thought to be performed in individuals without disease symptoms.
> - **Diagnostic genetic tests**: They are thought to be performed in individuals who show disease symptoms. They may confirm the physician's diagnosis and help choose the right treatment.
> - **Carrier tests**: These tests find single mutated alleles in asymptomatic individuals. The patient does not show signs of the disease, but their children are at risk of having the genetic condition.
> - **Pharmacogenomic tests**: Tests specially designed to evaluate the sensitivity to drug therapy in a patient. They target SNPs associated to drug dosage and risk of adverse effects.

Among the types of health-related PGTs preseted in Box 1.2, we focus on the predictive genetic tests. The results of these tests predict the risk of onset of a particular disease, which depends on the patient's genetic profile and the methodology used to assess the risk. Still the current methodologies do not consider other non-genetic factors of importance (e.g., environmental factors, lifestyle), so the results are highly inaccurate [29]. The probabilistic nature inherent to predictive genetic tests has opened opportunities for improvement, as discussed in Sections 1.4.3 and 1.5.

1.2.2 Genetic Tests Providers

Typically, there are two ways to access the genetic screening services. If a genetic disorder is suspected, a physician orders the test from a laboratory; the laboratory sends the reports back to the healthcare provider and the physician counsels the patient in the interpretation of the results. On the other hand, any person can order a DTC genetic test straight from private companies [34]. The consumer receives a kit to collect a sample of saliva and returns the sample to the company. After the DNA is isolated from the sample and the screening is completed, the reports are sent back to the consumer, or posted online. Despite the variety of tests covered, most of the reports are only for informational purposes, the consumer does not receive a diagnosis and in most cases the companies do not supply medical counselling [28]. Table 1.1 shows

Table 1.1: Some examples of companies offering PGTs for health purposes (prices as per year 2016)

Company	Technologies	Health services	URL
23andme	Illumina Human Omni Express – 24 format chip	• *Health reports*: Personalized information about how the individuals' genetics influences susceptibility to diseases. Reports include inherited conditions, drug responses, genetic risk factors and genetic traits. • The price of a personalized saliva-collection kit is 200 USD. These DTC services are available to customers in the USA, Canada, Denmark, Finland, Ireland, Sweeden, the Netherlands, and the United Kingdom.	www.23andme.com
Counsyl	• Array based ge-notyping. • NGS genotyping test with Illumina HiSeq 2000.	• *Family prep screen*: A screening test for parents. It predicts if parents carry genetic diseases that may be inherited to their descendants. • *Informed pregnancy screen*: A screening test for pregnant women. It predicts if the baby could suffer any chromosomal conditions, resulting in birth defects. • *Inherited cancer screen*: A screening test to detect genes associated with cancer and the chances of suffering different types of cancer. • Costs fluctuate between 150 and 300 USD for individuals with insurance in the USA. The service can be accessed through physician order.	www.counsyl.com
Gene-by-Gene	• Illumina Omni Express array • Illumina MiSeq	• Among the services offered by the company we find: WES[a] (1.095 USD), GWAS studies (199 USD) and WGS[b] (10.395 USD). These DTC services are available in the USA through saliva sample collection.	www.genebygene.com
FullGenomes	Illumina's HiSeq X	• WES (775 USD) and WGS at different sequencing depths: 30× (1,850 USD), 10× (745 USD), 4× (375 USD) and 2× (250 USD). DTC available in USA	www.fullgenomes.com
GenePlanet	Microarrays and sequencing platforms	• *Personal DNA analysis* (560 USD): The service includes information on susceptibility to 20 diseases, predicted response to 6 drugs and 14 traits. This DTC company is based in Slovenia	www.geneplanet.com/

[a] Whole-exome sequencing
[b] Whole-genome sequencing

some examples of genetic tests for health purposes that can be accessed either through physicians or directly from DTC companies.

The National Institute of Health administers the Genetic Testing Registry (GTR) [35].[8] This database enhances access to details of health-related genetic tests and laboratories worldwide. Although the laboratories voluntarily submit the information and the GTR does not include DTC tests, the database has standardized information for over 32,000 tests from 45 laboratories. The test-specific information is integrated with other NCBI databases in the domains of genomic sequence, sequence variation, genotype-phenotype relationships, and medical literature.

1.3 Common Technologies and Algorithms for SNPs Identification

A number of different technologies are used to assay DNA samples for genetic variants in PGTs. The advent of sequencing technologies has broadened the landscape of variant detection, including SNPs, INDELs, and structural variants. However, the non-sequencing technologies are still crucial for pinpointing specific SNPs and genotyping them in individuals, at low cost. This progress has simultaneously prompted advances in the algorithms for inferring potential genotypes from the raw data (Figure 1.1). The aim of this section is to summarize two common technologies for identifying SNPs, namely microarrays and NGS, and the resulting algorithms that are being used in response to the platforms' evolution (Sections 1.3.1 and 1.3.2).

1.3.1 Microarrays

Microarray technologies provide different alternatives for exploring whole genomes, including gene differential expression identification, copy number estimation, and genotyping [36, 37]. In genotyping, the SNP arrays determine the genotypes of individuals by measuring their relative allele intensities [36]. The first whole-genome sampling method for SNP genotyping was developed by Affymetrix in 2003 [9]. Since then, new generation microarrays have decreased the cost of this technology, improved the coverage and allowed for high throughput genotyping in GWAS [38].

Two main microarray platforms used for the genotyping of SNPs are the Affymetrix GeneChip and the Illumina Bead Array. Despite differences in the physical design and SNP content, both platforms have led to the discovery of hundreds of SNPs related to both complex traits and diseases [39].

1.3.1.1 Affymetrix SNP Microarrays
The Affymetrix SNP microarrays consist of a printed-array format that is produced in parallel by photolithographic manufacturing (see Figure 1.2(a)). For every SNP on the array there are two probes present, each one specific for one SNP allele (see Boxes 1.1 and 1.3 for definitions of biological and technical methods). After fragmenting,

---

[8] www.ncbi.nlm.nih.gov/gtr/

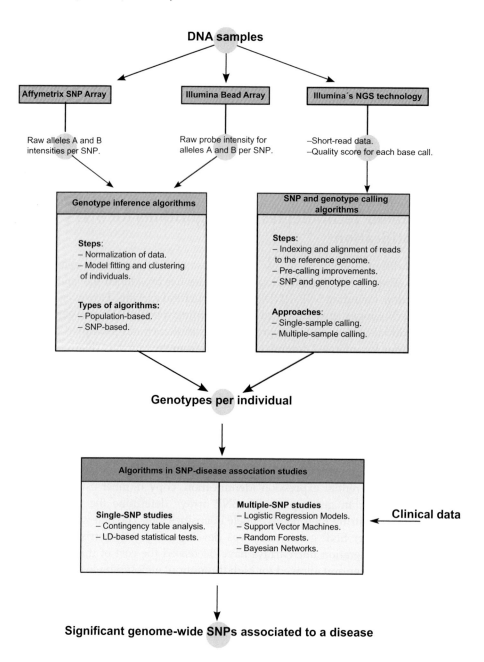

Figure 1.1: Workflow of the technologies and algorithms in the discovery of SNP-disease associations.

fluorescence marking and hybridizing of the patient's DNA to the array, the array is scanned and the fluorescence signals (i.e., intensities) are measured. In the initial versions of the Affymetrix GeneChip genotyping microarrays, SNP were detected with the use of five probes that perfectly matched the targeted SNP (perfect match

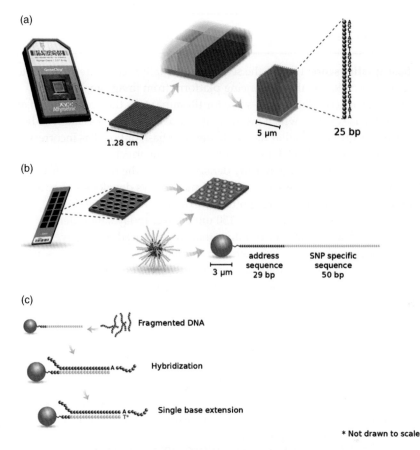

Figure 1.2: Schematic representation of high density microarrays. (a) Affymetrix GeneChip microarrays are composed of 25 bp long oligonucleotides which are synthesized by photolithography directly on a glass surface. Each array consists of hundreds of thousands of 5×5 μm sized square blocks that harbor millions of copies of the same oligonucleotide. The position of each oligonucleotide spot on the arrays is predetermined and known. (b) Illumina's BeadArrays consist of silica beads (3 μm in size) that are covered with hundreds of thousands of copies of a specific oligonucleotide. The beads randomly assemble at a uniform spacing of approximately 5.7 μm in microwells etched out of planar silica slides. In order to determine the position of each bead, the oligonucleotide is composed of two parts: a 50 bp sequence specific to the target SNP and a 29 bp address, which allows unambiguous identification of the oligonucleotide. (c) Illumina's single base extensions procedure. Fragmented DNA is hybridized to the genotyping array. After the un-hybridized DNA is washed away (not shown) a labeled terminating nucleotide is incorporated. The extended nucleotide is subsequently stained to amplify the signal and scanned with the BeadArray reader (not shown).

probes) and five probes with a single base mismatch in the middle of the probe (mismatch probes). Perfect match and mismatch probes were used to overcome the problem of unspecific binding of DNA fragments to probes. In new arrays, only perfect match probes are used.

> **Box 1.3: Technical concepts on microarrays and NGS**
>
> - **Base quality score**: During the sequencing, quality scores are assigned to each base called by the sequencing platform from image analysis. The quality score (i.e. Phred or Q-score for Illumina) tells the probability of an error in base calling, which means that a base is more or less likely to be correct. A Q10 means that the prediction of 1 base out of 10 is incorrect, while Q20 means that 1 base call out of 100 is incorrect.
> - **High-density arrays**: Microarray density refers to the number of features (probes) present on the array. The first microarrays developed were low density arrays. Probes were spotted onto a glass microscope slide, creating features between 100 to 150 µm in size. Today's high-density arrays, such as Affymetrix and Illumina, are produced using novel technologies which enable generation of smaller features with sizes of 5 µm or less. This also enables having more features per array (>106).
> - **Library**: Depending on the NGS method used (WGS, WES, amplicon sequencing, RNA-seq) a library is composed of fragmented nucleic acids (DNA or RNA) with added, platform specific, adapters at each end.
> - **Polymerase chain reaction (PCR)**: The PCR is a method used to amplify DNA sequences. It is capable of producing several billion copies (amplicons) of a target sequence from a small amount of sample. The method employs temperature cycling where two specific short oligonucleotides bind to DNA and then DNA polymerase amplifies the DNA segment between the two oligonucleotides. In each cycle, the amount of the target sequence doubles.
> - **Probes**: Single stranded sequences of DNA/RNA used to detect complementary sequences in samples (cDNA, RNA, DNA).
> - **Reads**: In NGS instruments, a read refers to the sequence of A, T, C, and G nucleotide bases that make up a DNA or RNA molecule that was sequenced. NGS instruments enable sequencing of many millions of different reads in a single run.
> - **Reference genome**: The representative nucleotide sequence database of a species. This is put together after the whole genome of a species has been sequenced. The reference sequence is constantly updated to fill in the sequence gaps that were missing.
> - **Reversible terminator**: Nucleotide bases (A,C,T,G) in which the 3'-OH position on the ribose sugar is reversibly blocked, preventing the addition of the next nucleotide by DNA polymerase.
> - **RNA sequencing (RNA-seq)**: Refers to NGS methods used to determine the sequence of each RNA molecule of an organism.
> - **Sequencing depth (coverage)**: Refers to the number of times a particular nucleotide (or short sequence) is read during an NGS sequencing process.
>
> (cont.)

- **Sequencing throughput**: Sequencing throughput per run refers to the number of base pairs a specific NGS machine can read in one run. This amount is, however, not equal to the length of the DNA sequence obtained. For example, to sequence a human genome (size 3 Gbp), a NGS machine with a throughput of 3 Gbp per run is not enough, because for one sequencing the depth needs to be considered. Thus, to reach a 30× depth we would need at least 30 runs to sequence the whole genome. Optionally, we can use a NGS machine with a throughput $\geq$ 90 Gbp to complete the sequence in one run.
- **Short reads**: The number of nucleotides an NGS platform is capable of sequencing in a single run is much shorter than what is attained by Sanger sequencing. For this reason they are defined as short reads.
- **Targeted sequencing**: Refers to NGS methods used to determine the DNA sequence of a subset of genes or regions of the genome of an organism. Targeted exome sequencing is one example of targeted sequencing where we sequence a subset of genes of interest.
- **Whole-exome sequencing (WES)**: Refers to NGS methods used to determine the complete DNA sequence of an organism's protein coding genome.
- **Whole-genome sequencing (WGS)**: Refers to NGS methods used to determine the complete DNA sequence of an organism's genome.

The Affymetrix platforms have been in constant growth after the release of the first SNP genotyping array, which contained only 1,494 SNPs [22]. Technical improvements resulted in the 10K, 100K, and 500K SNP versions with 11,555, 116,204, and 500,568 SNPs assayed in the chips respectively [22]. The latest Affymetrix genome-wide array (Affymetrix Genome-Wide Human SNP Array 6.0.) is capable of determining more than 906,600 SNP and consists of 6.8 million 5 x 5 μm spots each containing more than 1 million copies of a 25 base pair (bp) oligonucleotide probe.[9]

1.3.1.2 Illumina SNP BeadChips

Contrary to the Affymetrix microarrays, the Illumina Bead Array Technology is based on silica beads that randomly assemble onto a glass/silica slide etched with an array of millions of small holes (Figure 1.2(b)(c)). Each bead is covered with many hundreds of thousands of copies of a specific 79 bp long oligonucleotide. This oligonucleotide is composed of a 23 bp long address sequence, needed for determination of bead location on the array and a 50 bp long SNP specific sequence.[10] The SNP specific sequence terminates one base prior to the investigated SNP. After hybridization of unlabeled sample DNA, a single-base extension is carried out on the array which incorporates a fluorescence labeled nucleotide. The Bead Array is scanned and the fluorescence

---

[9] http://media.affymetrix.com/support/downloads/package_inserts/
[10] www.illumina.com

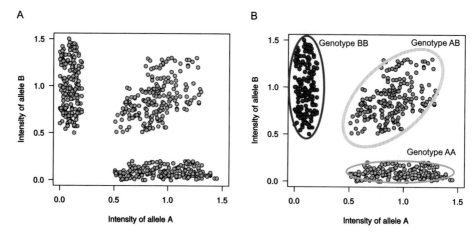

Figure 1.3: Genotyping of individuals by clustering microarray SNP data. (a) Allele intensities for a single SNP. Each point represent a sample, or an individual. (b) Clusters of genotypes.

intensity is measured. Consequently, one SNP allele measurement is retrieved in each bead [40].

The Bead Array Technology enables generation of higher density arrays compared to printed, or spotted arrays. The Illumina family of genome-wide SNP arrays covers several different BeadChip arrays with the largest of them (HumanOmni5-4 Bead-Chip) interrogating over 4,200,000 markers where each SNP is measured with at least 15× redundancy (i.e., at least 15 measurements per SNP for each DNA sample).

1.3.1.3 Algorithms for Genotyping

In parallel with the technical advances in microarrays producing SNP data, a number of methodologies for analyzing the data have been proposed. Most of the algorithms preprocess the raw probe data through quantile normalization, fit a model to the normalized data and then apply a clustering method to assign genotypes to individuals (Figure 1.3) [8]. Table 1.2 summarizes some algorithms for identifying SNPs from microarray data. Li et al. [7] group these algorithms into population-based and SNP-based algorithms.

Population-based algorithms, also known as between-sample models: These types of algorithms analyse simultaneously the data of all the individuals (i.e., samples) taking one SNP at a time [7]. The algorithm forms a cluster for each possible genotype. Thus, three clusters of individuals are obtained for the genotypes of alleles A and B and the individual's genotype is determined from cluster membership.

For example, GStram is a population-based algorithm for SNP and CNV genotyping in GWAS studies [38]. The method has been tested with data from Illumina BeadArray genotyping technology. It transforms the normalized intensities into allele frequencies, estimates a probability density function (PDF) from the allele frequencies and uses the peaks in the PDF to identify cluster membership for each SNP.

Table 1.2: Representative algorithms for SNP microarray data analysis

Algorithm	Type	Approach	Other features/platform
RLMM [45]	Population-based algorithm	Supervised learning algorithm based on a linear model with Mahalanobis distance classifier.	It uses training data to learn the intensities for each genotype group at each SNP. (Affymetrix)
BRLMM [46]	Population-based algorithm	Based on the RLMM algorithm.	It adds a Bayesian step to consider the variability of low allele frequencies. (Affymetrix)
CRLMM [47]	Population-based algorithm	Based on the BRLMM algorithm.	It adds a pre-processing step to remove artefacts. It uses likelihood ratios as uncertainty measures in the genotype calling. (Affymetrix)
MA-SNP [8]	Population-based algorithm	Linear model for single arrays.	It decomposes probe intensities into biological signals and noise. It provides a confidence measure for the genotype calling. (Affymetrix)
GStram [38]	Population-based algorithm	Heuristic comparison of probability density functions for allele frequencies.	Allows the discovery of CNVs in close linkage disequilibrium with SNPs. (Illumina)
Genotype calling algorithm implemented in the program Illuminus [40]	Population-based algorithm	Bivariate mixture model of truncated t-distributions.	It does not rely on training data. It defines a quality metric for the stability of genotype calls. (Illumina)
GenoSNP [44]	SNP-based algorithm	Mixture of Student t-distributions.	The method is independent of the sample size. It does not rely on parameters from the population. (Illumina)
M3 [7]	SNP-based and population-based algorithm	Gaussian mixture model.	Improves the accuracy in the calling of rare variants. (Illumina)
optiCall [48]	Mixture of SNP-based and population-based algorithm	Mixture of t-distributions.	It uses within and across sample intensity data to call rare and low frequent SNPs. Less stringent SNP quality control is needed prior GWAS. (Illumina)
zCall [49]	Post-processing tool after using a SNP- or population-based algorithm	Linear regression model.	Designed to call rare variants on the genotyping results of other algorithms. (Illumina)

Another population-based algorithm designed for the Illumina microarrays is proposed by Teo et al. [40]. This algorithm fits a mixture model for the normalized intensities, finds the model parameters by using an expectation–maximization (EM) framework and assigns genotypes conditional to clusters (see Box 1.4).

The main limitation of population-based algorithms is their dependence on the sample size. The adequate sample size is mainly a function of the minor allele frequency (MAF) in the model, so as showed by [41], special care should be taken when assigning genotypes for SNPs with low MAF [42]. A comparison of genotyping algorithms [43] showed that at least 100 samples are needed to estimate the model parameters and reduce the miscalls. In particular, 100 individuals are needed if < MAF 10% and 10,000 if MAF $\sim$ 1% [44]. In addition, the algorithms require that

---

Box 1.4: Mixture models and the expectation–maximization (EM) algorithm

In general, the population of individuals consists of three subpopulations given by the genotype classes *AA*, *Aa*, or *aa*. Each individual is assigned to a subpopulation by an unknown cluster membership. This variability between individuals leads to the finite mixture models which allow to estimate the proportion of the subpopulations and the cluster membership. Thus, the finite mixture model combines three probability density functions to approximate the distribution of SNP intensities in the overall population. Frequently, the cluster membership determination is performed under an EM framework described below [50]:

1. Fix the number of subpopulations. It is usually three, as only three possible genotypes are analyzed, but it could be extended to capture outliers in a null class [44].
2. Define the distributions for each subpopulation (e.g., a bivariate mixture model using truncated *t*-distributions [40]).
3. Give a starting guess of the component membership.
4. Asses the relative frequencies, mean intensities of each subpopulation and other parameters in the model (e.g., location parameter, variance-covariance matrix, mixture proportions [40].
5. Asses the probability (p_{ij}) that individual i belongs to subpopulation j by using the Bayes' theorem (step E).
6. Replace the component membership with p_{ij} and obtain a estimation of the relative frequencies, mean intensities of each subpopulation and other parameters in the model (step M).
7. Repeat steps 5 to 6 until convergence.

The EM approach can be seen as the calibration of the model parameters conditional on the assigned genotypes (step M), and the assignment of the genotypes to SNP intensity data conditional on the cluster features (step E) [40].

non-biological differences, such as the background noise, to be minimized among samples [36].

SNP-based algorithms (also known as within-sample models): This class of algorithms infers genotypes at the individual level. Instead of genotyping one SNP at a time, they genotype all the SNPs of every individual at a time. For example, in the GenoSNP algorithm for Illumina, the SNPs are called simultaneously from the measurements within a sample [44]. Thus, clustering is done on the pair of allele-specific intensities for each beadpool separately. First, the SNP intensities are modelled by a four-component mixture model of t-distributions, where each mixture corresponds to one of the three genotypes and a null class. Then, an EM-based algorithm computes the expected parameters maximizing the expected log-likelihood of the data. The genotype with the maximum probability conditional to the parameters is assigned to each SNP [44].

Globally, the calling results have discrepancies when multiple algorithms are tested simultaneously [43, 51, 52]. The work of Ritchie et al. [43] compares four methods (GenCall, Illuminus, GenoSNP, and CRLMM) using GWAS data of multiple sclerosis and data from the HapMap project [53]. For large sample sizes (> 50 individuals), CRLMM showed higher accuracy, followed by GenoSNP, Illuminus, and GenCall. Although all of them had variations in the calling of low MAFs, GenoSNP and Illuminus outperformed the other methods. These findings suggested that the SNP-based algorithms deal better than population-based algorithms when low MAF are genotyped. In a recent study, however, Lemieux et al. [51] also tested the performance of four genotyping tools (GenCall, GenoSNP, optiCall, and zCall) in 10,520 unique samples from the Montreal Heart Institute Cohort, and the 1000 Genomes Project data was used as gold standard [6]. While all the tools showed the same level of performance calling common variants, the performance decreased for rare variants. GenCall, the proprietary method from Illumina, has the higher concordance rate for rare variants and zCall outperformed other tools when considering low misclassification rates. In this case, the SNP-based algorithm, GenoSNP, did not outperform other tools, proving that methods' accuracies depend on the experiment and that it is not straightforward to recommend a unique method for genotyping tasks [51].

1.3.2 Next Generation Sequencing

In the last four decades, since Sanger's seminal publication in 1977 [54], the field of DNA sequencing has seen constant development. The first major success was the publication of the Human Genome Project in 2001 with the use of fluorescently labeled Sanger sequencing. But just five years after that, another major leap in sequencing technology occurred, the development of massively parallel sequencing, or next generation sequencing (NGS)[55].

The main advantage of NGS is the large sequencing throughput per run, which has increased from around 80 kilo bp in the 96 well Sanger sequencers to several hundred giga bp in today's NGS platforms (Figure 1.4). In contrast to the Human Genome Project, which took 13 years to complete at a cost of nearly 3 billion USD,

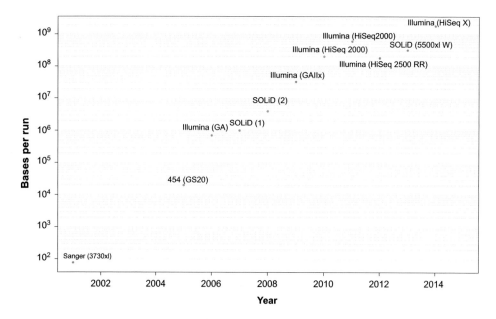

Figure 1.4: Number of bases sequenced per run and the year of releasing of the platform. Among all the platforms released every year, this plot shows the platforms with the maximum number of bases sequenced per run. Data consulted in January 2016. Data available from: https://flxlexblog.wordpress.com/2014/06/11/developments-in-next-generation-sequencing-june-2014-edition/

today a human genome can be completed within a week for around 1,000 USD,[11] with the increase in throughput of NGS platforms, significant changes in data analysis pipelines were also introduced. The main challenges were related to the enormous amount of data generated per run, the analysis of short read lengths and differences in error profiles compared to Sanger sequencing [55].

Several different NGS platforms were developed in the years following 2005, the main competitors being Roche (454 GS Junior, 454 FLX+), Life Technologies (SOLiD, Ion Proton, Ion Torent), and Illumina (MiSeq, NextSeq, HiSeq). Despite specific differences between these platforms [56], all massively parallel approaches have four things in common: (1) A fast and simple library preparation (in comparison to Sanger sequencing) which includes ligation of adapters to the fragmented DNA. (2) Fragment amplification with the use of PCR, to produce millions of fragment copies (needed for signal detection). (3) Sequencing reactions occur in a series of repeating steps, whereby a nucleotide is incorporated and determined at each step. (4) DNA fragments can be sequenced from both sides [55].

In addition to NGS platforms mentioned above, which work based on sequencing by synthesis (like Illumina) or sequencing by ligation (ABI SOLiD), single-molecule sequencing platforms have also been developed. One such platform currently available is the Pacific Biosciences RSII. The main advantage of a single-molecule approach

---

[11] www.nature.com/news/is-the-1-000-genome-for-real-1.14530

is the read length is substantially longer compared to above mentioned platforms and can exceed 40,000 bp. These long read lengths are especially important because they enable assembly of long continuous sequence stretches even in large genomes such as human, which is not possible with other approaches. The current downside of the single-molecule sequencing is, however, the cost, which is still substantially higher compared to sequencing by synthesis [57].

The following sections examine the features of NGS platforms and the SNP calling algorithms in the analysis of NGS data. A more detailed look into Illumina's sequencing technology will be presented below, since their technology has seen widespread use in research, direct-to-consumer and clinical settings.

1.3.2.1 The Illumina NGS Platform

Illumina's success can be attributed to several factors. One of the more important ones is the relatively short time in which the company was able to increase throughput per instrument run from only 1 Gb in the Solexa 1G machine to 600 Gb in the HiSeq 2000 series. With the later system, it is possible to sequence six human genomes in a matter of just 11 days [58]. The company also offers several sequencing systems ranging from low throughput options in the MiniSeq and MiSeq series over medium throughput options in the NextSeq series to high throughput options with the HiSeq and HiSeq X series.

Illumina's sequencing technology is referred to as sequencing by synthesis (SBS), where each nucleotide is determined at the time of incorporation into the emerging DNA strand. However, as with any NGS sequencing protocol, regardless of the platform used, the first step is the preparation of the sequencing library (Figure 1.5). The library workflow in Illumina is similar to other technologies and includes fragmentation of isolated DNA to the appropriate size, followed by ligation of Illumina specific adapters to the fragments. Once the library's concentration is determined, an exact amount is denatured (single-stranded DNA (ssDNA) fragments) and transferred to a flow cell, which is composed of a flag glass with eight microfluidic channels (Figure 1.5 (a), (b)) (The number of channels depends on the Illumina system used). The surface of the channel is covered with covalently linked adaptors that are complementary to the ssDNA ligated adaptors. After the ssDNA fragments have hybridized to the flow cell oligos (Figure 1.5 (c)), the oligos are used as primers to synthesise the second strand of the fragment [58, 59].

Because Illumina's technology is not sensitive enough to determine incorporation of one single nucleotide, the fragments bound to the flow cell must first be amplified. This is achieved through a process called bridge amplification (Figure 1.5 (d)), which amplifies the initial fragment up to about 1,000 copies (Figure 1.5 (e)). The result of bridge amplification is a high number of clusters, which can reach numbers of up to 180million per single lane. The final step of bridge amplification is generation of ssDNA fragments by removing one strand of the double-stranded DNA (dsDNA) fragment with the use of a cleavage site on the surface oligos (Figure 1.5 (f)). At this point, clusters of fragments are ready to be sequenced one nucleotide at a time with the SBS approach. Each sequencing cycle is composed of several steps, which include (Figure 1.5 (g)): (1) Addition of fluorescently labeled nucleotides to the flow cell. Each nucleotide is labeled with a specific dye and acts as a reversible

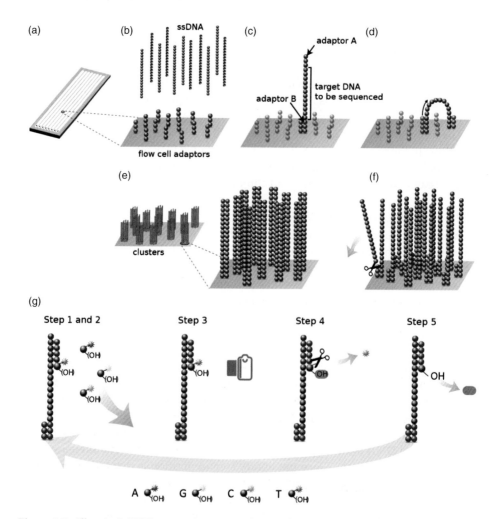

Figure 1.5: Illumina's NGS sequencing protocol. DNA to be sequenced is randomly fragmented and ligated with specific adaptors at both ends. Once denatured, single stranded DNA fragments are put onto Illumina's flow cell (a) where they hybridize to oligonucleotides (which are complementary to the added adapters) present on the flow cell surface (b,c). Solid phase bridge amplification is carried out (d) which produces several million dense clusters composed of identical double stranded DNA fragments (e). Enzymatic cleavage (f) produces single stranded DNA fragments ready for sequencing by synthesis (g).

terminator. (2) One nucleotide is incorporated by DNA polymerase, unincorporated nucleotides are washed away. (3) A detailed image of the flow cell is captured. (4) Fluorescent groups are cleaved of the nucleotides. (5) 3'-OH groups are deblocked allowing another cycle to commence [58, 59].

It is important to note that for all NGS sequencing platforms the input of good quality DNA is important and the initial steps of sample preparation and DNA extraction are critical to achieve high quality sequencing results.

1.3.2.2 Algorithms for SNP Calling and Genotyping

Next generation sequencing (NGS) technologies have expanded the amount of data available and a number of algorithms to identify SNPs have been published in recent years [18] (see Table 1.3). The first common step of these algorithms is to index the reference genome using data structures, mainly, hash tables and suffix trees (see Box 1.5) [60, 61]. Thus, the reference genome and the reads are assigned a set of indices to efficiently organize them in the memory. After the indexing, aligners based on either the Smith–Waterman algorithm [62] or Needleman–Wunsch algorithm [63] are used to align the reads to the sequence genome. Some aligners also include local realignment and recalibration steps specifically designed to improve the variant detection (SNP calling) (see Box 1.6) [61].

Table 1.3: Representative algorithms for SNP and genotype calling in NGS data

Algorithm	Approach	Features
MAQ [64]	Bayesian	This model incorporates SNP calling on diploid samples. It estimates the error probability of each alignment and introduces quality scores to derive genotype calls.
SOAPsnp [65]	Bayesian	This is a program in the Short Oligonucleotide Analysis Package (SOAP). It uses a compression index to accelerate the indexing of sequences.
VarScan2 [66]	Heuristic	This is an analysis tool for WES data. It is specially designed for the detection of CNVs and somatic mutations across tumor samples. It relies on heuristic thresholds for quality data (*e.g.*, coverage) to determine the genotype of each SNP.
seqEM [67]	Bayesian	This is a Bayes classifier for genotype calling. It applies the EM algorithm to maximize the data likelihood given the genotype frequencies.
Atlas-SNP2 [68]	Bayesian	This is a computational tool specialized in recognizing sequencing errors. A Bayesian model estimates the sequencing error for each allele.
GATK [69]	Bayesian	This is a suite of tools for DNA sequence analysis. It handles single sample, multiple samples and low coverage data. It realigns reads to minimize the number of mismatches.
SAMtools [70]	Bayesian	This software implements algorithms for the analysis of alignments in SAM format. The samtools `mpileup` and `bcftools` routines execute the calling based on the likelihood of the observed data for each genotype.
MAFsnp [71]	Probabilistic	This model introduces a likelihood-based statistic. It provides p-values for calling SNPs and avoids posterior filtering steps.

> **Box 1.5: Indexing of the reference genome**
>
> Prior the alignment of the reads to the reference genome, most of the aligners index the reference genome based on hash tables and suffix trees [60]. Hash tables are data structures that store short fragments of the query sequence (e.g., reference genome sequence) called k-mers. They are obtained by a mapping function which splits the original query sequence and assigns indices in an array or seed index table. Subsequently, the algorithm searches the k-mers in a second sequence (e.g., read sequences) to provide a set of preliminary short seed matches. The seeds are extended to allow full completion of the alignment, including insertions, deletions, and gaps [72].
>
> On the other hand, suffix trees are data structures that represent all the suffixes for a given string (e.g., reference genome sequence). The suffixes are all the possible substrings, which include the last letter of the full string; for example, for string ACG, the suffixes are G, CG, and ACG. Thus, the suffix tree contains paths of nodes and edges storing these suffixes. The edges are labeled with concatenated letters and the nodes contain the letter positions in the main string. Once the tree is constructed, it allows to query a second sequence (e.g., read sequences) by finding the matching path [60]. As it is impractical to store the suffix trees in memory even for short reference genomes, algorithms have improved to process compressed data structures [65, 72, 73].

Once the reads are aligned to the reference genome, the algorithms search along the aligned reads for sequence variations (SNP calling), and assign genotypes to the individuals (genotype calling) [74]. The Bayesian framework is the preferred strategy because it allows the calling of potential variants in regions of low sequencing depth, it also provides a measure of confidence for the inferred genotypes [71, 69]. In this approach, the sequencing reads overlapping a nucleotide position are examined and the likelihood values are assessed for each of the three possible genotypes. The genotype with the highest posterior probability is assigned to its respective SNP [61] (see details of the Bayesian framework and a SNP genotyping algorithm in Box 1.7).

To describe in more detail the SNP and genotype calling processes, we will review the Genome Analysis Toolkit (GATK) [69]. This software contains tools for DNA sequence analyses that gained popularity after being applied in The Cancer Genome Atlas,[12] and the 1000 Genomes Project [1]. It performs a three-step variant discovery process, which includes a Bayesian model for SNP and genotype calling followed by variant filtering [75]. First, the variants are called per single sample by UnifiedGenotyper or HaplotypeCaller internal algorithms. UnifiedGenotyper is a simple genotyper that works with the classic Bayesian framework described

---

[12] http://cancergenome.nih.gov

> **Box 1.6: Alignment and post-alignment steps**
>
> **Needleman–Wunsch (N–W) and Smith–Waterman (S–W) alignment algorithms:** These algorithms fall in the category of dynamic programming algorithms, which consist of splitting the general problem in smaller pieces, finding their solutions and putting them together to find the optimal solution. They are based on the concept that along the optimal alignment, some partial sub-alignments can be found. Therefore, they divide the full sequence into small pieces, perform pair-wise comparisons of nucleotides, score them according to a scoring system for matches, mismatches, and INDELs, perform optimal alignment of these pieces and reconstruct the optimal alignment from them. While the N–W algorithm finds an optimal global alignment of two sequences, the S–W algorithm finds an optimal local alignment of two sequences by comparing segments of any possible length and finding the one that maximizes the alignment score [63, 62].
>
> **Alignment improvement:** Before performing the SNP calling, some alignment artefacts must be removed. One of these artefacts corresponds to wrongly aligned reads that may be erroneously assumed as SNPs. As misaligned reads increase the number of false-positive SNPs, reads should be locally realigned especially near to INDELs [69]. The realignment step can also be followed by a correction of the base quality scores (see Box 1.3). Some algorithms use the quality score (e.g., SOAPsnp [65]) as an input in the SNP calling functions, so prior the calling they estimate a mismatch rate in the base calling and use that estimation to recalibrate the raw scores.

in Box 1.7. This genotyper checks locus by locus, using the aligned reads and the quality scores of each base to assess the genotype likelihoods. Due to its sensibility to alignment errors, UnifiedGenotyper has been deprecated in favor of HaplotypeCaller algorithm. This new algorithm identifies active regions in which substantial variations occur between the sample and the genome. Thus, instead of walking along locus like UnifiedGenotyper, it walks along regions that are more likely to show variations and omit regions identical to the reference. The algorithm produces a set of possible variants and estimates the likelihoods of observing a given read at each allele (per-read likelihoods) (see Box 1.8). This information is used in the second step where genotypes are assigned to the samples. To improve the sensitivity of the genotype calls, the authors recommend a joint genotyping of all the samples simultaneously (cohort-wide analysis). The genotype calling follows the Bayes' theorem by assessing the likelihood of each possible genotype, using the per-read likelihoods as evidence. Subsequently, in the last step, the variants are refined depending on the requirements of each project. The user can specify, among other things, the alleles of interest for genotyping and the minimum base quality score (i.e., Phred score) to filter out low quality called variants [75].

Box 1.7: SNP and genotype calling

SNP calling (variant calling): The process of identifying the positions where the aligned reads show variation of one base or more relative to the genome of reference [76].

Genotype calling: The process of assigning a genotype to an individual in the position where a SNP was identified [76].

Bayesian framework for genotype and SNP calling: This statistical framework applies Bayes' theorem to assess the posterior probability, $p(G \mid E)$, that an individual has genotype G given evidence E (i.e., the read data at a specific sequence position):

$$p(G \mid E) = \frac{p(G)p(E \mid G)}{p(E)}. \tag{1.1}$$

In Equation 1.1, the prior probability of observing the genotype, $p(G)$, is constant, therefore, genotype $\check{G}$ with the highest $p(G \mid E)$ is computed by Equation 1.2:

$$\check{G} = \arg\max_{G}\{p(G)p(E \mid G)\}, \tag{1.2}$$

where $p(E \mid G)$ can be seen as the rescaled quality scores of the base, and $p(G)$ is the probability a priori of the genotype. Here, $p(G)$ may be based on information from external databases such as dbSNP. Generally, $p(G \mid E)$ provides the statistical uncertainty for the genotype calling and this is used to separate high confidence calls from low confidence calls in downstream analyses [76, 77].

Simple genotype walker algorithm: To assess the posterior probability of each genotype, a genotype walker algorithm makes use of the following equations [69]:

$$p(E \mid G) = \prod_{b \in \mathbf{B}} \left(\frac{1}{2}p(b \mid A_1) + \frac{1}{2}p(b \mid A_2)\right). \tag{1.3}$$

$$p(b \mid A) = \begin{cases} \frac{e}{3} & \text{if } b \neq A \\ 1 - e & \text{if } b = A \end{cases}. \tag{1.4}$$

Equation 1.3 describes the posterior probability of evidence E given genotype G, where b is a base in the pile of reads aligned to the target locus. Also, it is assumed that genotype G has two alleles A_1 and A_2. Equation 1.4 describes the probability of observing base b given allele A, where e is the scaled base quality score.

(cont.)

Then, the algorithm proceeds as follows:

Algorithm 1.1 Simple genotype walker (adapted from [69])

Input: List of genotypes **G**; list of alleles, **A**, per genotype; list of bases, **B**, covering the target locus (given the pileup of reads aligned to the locus); base quality scores.

Output: Genotype call

1: for $G \in$ **G** do
2:   for $b \in$ **B** do
3:     for $A \in$ **A** do
4:       Compute Equation 1.4
5:       Compute Equation 1.3
6:     Compute Equation 1.2
7: Choose the genotype with the highest posterior probability

---

The GATK is an example of how the SNPs and genotype calling steps use prior probabilities as measures of confidence [71]. Indeed, most algorithms take advantage of these measures to exclude poor called variants, but exceptions exist. For instance, Hu et al. [71] designed a method (MAFsnp) that transforms the SNP calling problem into a hypothesis testing problem and produces p-values for each candidate locus. Other algorithms rely on heuristics to determine the genotype, as is the case of VarScan2 [66]. For the SNP calling, VarScan2 considers the following heuristic: a SNP is called in a locus if and only if this is supported by at least two independent reads and a minimum allele frequency of 8%. Thus, it counts the bases observed in the reads at a given position and identifies a SNP if there is any variant base fulfilling the heuristic. In addition, only samples with minimum base quality Q20 (defined in Box 1.3) and 3× minimum coverage thresholds should be analyzed by VarScan2. The genotype calling is performed in a similar way: a sample is called homozygous (same alleles) if supported by 75% or more of the reads at a position; otherwise it is called heterozygous (two different alleles) [66].

Overall, heuristic strategies to call genotypes do not provide measures of uncertainty for each SNP/genotype called, many true heterozygous genotypes are not called (i.e., high false-negative rates) and the algorithms are affected by low sequencing depth [76]. The probabilistic methods, therefore, have been adopted for advanced studies, allowing the users to handle multiple samples simultaneously, minimizing the loss of information [70] (see Table 1.3). For a full list of algorithms on variant identification, variant annotation and visualization using NGS data, we refer the reader to the review articles by Mielczarek et al. [61] and Pabinger et al. [18].

> **Box 1.8: SNP and genotype calling with HaplotypeCaller**
>
> This is a variant and genotype caller from GATK. Its routine is summarized as follows:[a]
>
> 1. Search the genome for regions showing variation relative to the genome of reference. In this step, an activity score is calculated per each position on the genome as the probability of containing a variant. Then, the activity scores are smoothed and the regions encompassing local maxima scores are labeled as activity regions.
> 2. For each region, reassemble the reads to discover potential haplotypes. A De Bruijn graph (defined below) is constructed for the reference sequence in the active region, the reads are mapped against the graph and the most mapped paths in the graph are selected as potential haplotypes.
> 3. Align each potential haplotype against a reference haplotype. The reads are realigned to their most likely haplotypes using the S–W algorithm. Variation sites are identified in the aligned reads and labeled as potential variants.
> 4. Evaluate the potential variants in the aligned haplotypes. The base quality scores are used to calculate the likelihood of observing a read given the haplotype. Then, these likelihoods are aggregated to find the likelihood of a read given the potential variant (per-read likelihoods).
> 5. Call genotypes for the potential variants. The per-read likelihood values are used as evidence to calculate the posterior probability of a given genotype. The genotype with the highest posterior probability is assigned to the potential variant.
>
> **De Bruijn graph**: A directed graph showing the overlapping between sequences. To construct a De Bruijn graph, the sequence is split into substrings of a given size k (k-mers, nodes in the graph). A directed edge is drawn between two k-mers, A and B, if the last $k-1$ nucleotides of A overlap the first $k-1$ nucleotides of B [78].
>
> ---
> [a] www.broadinstitute.org/gatk/

1.3.3 Pros and Cons of Microarrays and NGS

Both microarray and NGS technologies have unique advantages and disadvantages for genotyping, depending on the research objective. A summary of these technological differences affecting the SNP identification is presented in Table 1.4.

The principal difference between NGS and microarrays is the coverage. The potential of NGS to interrogate millions of bases has moved forward this technology as preferred for SNPs detection. Whereas all polymorphisms may be detected in a single sample by NGS, the microarrays are constrained to a predefined number of probes. In some scenarios, however, NGS is costly or unnecessary, but the interest may reside on

Table 1.4: Main differences between microarrays and NGS technologies for SNP/genotype calling

Microarrays	NGS
Interrogate 4 million SNPs per sample.	Interrogate 3.2 billion bases of the human genome.
Tolerate lower quality and quantity DNA samples.	High quality DNA is needed in the samples. High sequencing depth is also a contributing factor for SNP calling [87].
Offer high resolution and can detect unknown CNVs. Tends to include more common variants than rare SNPs.	Both common and rare SNPs are potentially detected [52, 61].
Allow detection of SNPs on specific regions of the genome that can be predisposed to disease.	Allow detection of SNPs across the whole genome giving a comprehensive view of the variations.
Less expensive than NGS.	The costs are decreasing but it is still an expensive technology for multiple sample studies [18].

targeted genotyping for specific genes by the use of customized microarrays, which are also attractive for the study of new species and specific genomic regions.

Price and quality are also determinant variables to choosing an NGS technology. Whilst WGS covers almost 100% of the genome, generates a huge amount of data, and is a costly technique, WES focuses on capturing approximately 1% of the genome that codes for proteins and therefore, it can be performed at a lower cost [18]. Nevertheless, WGS has identified a large number of mutations, has higher power for CNVs detection and better sequencing quality than WES [79]. These findings have confirmed that WGS can capture exons from protein-coding genes that would be missed by WES. The slightly superiority of WGS over WES, added to an expected decreasing in its cost, will mark the preferences between both approaches in the near future.

Two components to NGS should be taken into account before choosing this technology. First, the risk of false SNP calling due to structural variations. If SNPs are the focus of genotyping tasks, other variations between the reference genome and the sequenced sample, such as CNVs, INDELs, duplications and inversions, may cause incorrect mapping of the short reads to the reference genome, and thereby, the accuracy in the detection of SNPs decreases [74]. Second, low sequencing depth makes it hard to differentiate between sequencing errors and true SNPs because the algorithms require a redundant number of reads aligned to each base [4]. While a 30-fold depth per sample increases the genotyping accuracy (> 99%) and has been successful in research of Mendelian disorders in small individuals cohorts, it is hardly achievable in large-scale genotyping for complex disorders due to the sequencing costs [80]. Studies including hundreds and thousands of samples (e.g., the 1000 Genomes Project (Boxes 1.9 and 1.10)) can be accomplished with low-coverage WGS, where a 2–6-fold depth is suggested to preserve the power of variant calling even for low frequency alleles [80, 81].

Box 1.9: The 1000 Genomes Project

The 1000 Genomes Project is an international collaborative work to create an extensive catalogue on genetic variation in humans.[a] The full project was divided into four phases (Pilot [82], Phase 1 [6], Phase 2, and Phase 3 [1]) to comprise the genotyping of 2,504 individuals from 26 different populations in Africa, East Asia, Europe, South Asia, and America, by using high-density SNP microarrays, targeted exome sequencing and low-coverage WGS.

During the project phases, multiple algorithms were used depending on the sequencing technology and the institution that collected the samples. In the last phase, for example, most of the alignment of the WGS data was performed with the Burrows–Wheeler Alignment tool [83] followed by local realignment of reads around INDELs and recalibration of base quality scores using GATK [69]. For the variant calling stage, the institutions processed the data with SNPTools [84], HaplotypeCaller (see Box 1.8), Atlas2 [85], etc. More than 84.7 million of SNPs were found, most of them (99%) with frequencies of at least 1% [1].

Since the launching of the project in 2008, the methods' advancements have broadened the spectrum of genetic variations to be characterized. The extensive genotype data generated in Phase 3 covered not only bi-allelic SNPs, but also multi-allelic SNPs and structural variants. These data are now available for researchers, allowing them to combine the data with their own and impute unknown genotypes in their samples, saving money and time. Additionally, regions of the genome affecting cellular phenotypes can be mapped, boosting studies on heritability, natural selection and population structure, to name a few.

---
[a] www.1000genomes.org

Box 1.10: Explosion of genome projects

Although the 1000 Genomes Project may have been the first of its kind, the continuing fall of sequencing cost and increase in sequencing throughput has in recent years led to an explosion of genome projects. Their common goal is a significant increase in the amount of samples being sequenced; from several tens or hundreds of thousands to millions. Below we mention only a few projects, some of which have recently ended and offer valuable data to the community while others are still ongoing.

(cont.)

- Deep Sequencing of 10,000 Genomes (US): In August 2016, a team from J. Craig Venter Institute and Human Longevity Inc. published an article in PNAS reporting analysis of 10,545 human genomes at 3,040 coverage. The depth of sequencing enabled them to uncover over 8,000 previously unknown variants as well as parts of the genome that are not part of the main build of the hg38 reference genome. Human Longevity offers a browser-based search of their genomic database for researchers [86].
- Faroe Genome Project (Fargen): This project aims to sequence genomes of inhabitants of the Faroe Islands, a North Atlantic archipelago with a population size of 49,000 inhabitants. The population is presumed to be the most genetically homogenous population in the North Atlantic and will therefore provide valuable information on both monogenic and complex disorders.[a]
- The 100,000 Genomes Project (UK): Started in 2012 by the prime minister and funded by the Department of Health its goal is to sequence 100,000 whole genomes from about 70,000 NHS patients and their families with rare diseases and patients with cancer by 2017. In December 2016, 16,171 genomes had already been sequenced. One of the main aims of the project is to enable the use of genomics as part of routine care in the NHS.[b]
- The 1 million Genome Projects (USA and China): Both the US and China announced their precision medicine initiatives in 2015 and 2016 respectively. In the US, a $215 million initiative aims to collect data on health records and genomic sequences from 1 million participants to learn about the interplay between genetics, lifestyle, and the environment. The goal is to combine data from ongoing projects such as the Million Veterans Project that has already collected 343,000 DNA samples from former soldiers as well as data from new studies funded by NIH. Similarly, the Chinese Academy of Sciences' goal is to put China on the forefront of precision medicine by 2030 with an estimated $9.2 billion in funding. A variety of studies including metabolic disease, breast cancer, gut cancer, and other conditions will be studied in addition to looking at genetic differences between China's southern, central, and northern populations.[c] [d] [e]

---

[a] www.fargen.fo/fo/um-fargen/the-fargen-infrastructure-en/
[b] www.genomicsengland.co.uk/the-100000-genomes-project/
[c] www.nextbigfuture.com/2016/06/chinas-92-billion-precision-medicine.html
[d] www.technologyreview.com/s/534591/us-to-develop-dna-study-of-one-million-people/
[e] www.nature.com/news/china-s-bid-to-be-a-dna-superpower-1.20121

Some other factors to discern between technologies are the input quality and the quantity of samples. High-throughput methods for sequencing genomes still require not only high quality DNA, but also large quantities of sample. For example, in the discovery of malaria vaccine agents, the patient samples are hardly sequenced due to low parasite DNA and high human DNA contents [87]. Instead, the microarrays are preferred because they can tolerate parasite DNA of low quantity and quality from human samples, plus they can be technically customized to include new loci in the array [87].

1.4 Algorithms to Predict SNP-Disease Association

A huge volume of genotyping data is being produced by microarray and NGS technologies to be further exploited in GWAS. Classical statistical analysis can be used to understand the genetic association between a single SNP and a disease trait from these data, especially when the SNP has a large effect size on the trait (i.e., common variants). However, prediction of genetic associations for complex disorders is not achievable through classical methods due to: (i) non-genetic factors and (ii) multiple SNPs having a minor effect in the disease susceptibility, but acting synergistically [24]. Also, as the number of genotyped SNPs is in the hundreds of thousands in small samples of individuals, more sophisticated data mining methods are required to acquire predictive power. In this chapter we will describe the classical single-SNP and multi-SNP methods for GWAS (Sections 1.4.1 and 1.4.2, respectively). Then, we will discuss some methodological considerations on the genetic risk models used by DTC companies (Section 1.4.3).

1.4.1 Single-SNP Association Studies

The simplest single-SNP association studies aim to find variants conferring direct disease susceptibility in case-control designs [88]. Statistical methods for analyzing common variants compare the frequency of SNP alleles and genotypes in two groups of individuals. The first group shows a specific trait, generally a disease-related trait (e.g., individuals diagnosed with a disease, or cases); the second group does not show the trait (e.g., individuals not diagnosed with the disease, or controls). Generally, both groups are drawn from a homogeneous population with similar ethnicity to assure a matching in their genetic background. For each single SNP, the frequency data for either the genotype or the allele is presented in a contingency table. A basic test of association, namely the chi-square test, is used to evaluate the risk of disease with respect to the allele or genotype (Box 1.11).

Other studies take the linkage disequilibrium (LD) as a source of information to show genotypic differences between case and control samples [89]. In these, the data is compared through statistical tests, such as the Hotelling's T^2 test [90, 91, 92], the sum of squared score test [93] and the LD contrast test [94]. In contrast with the contingency table analysis methods, the LD-based methods try to find indirect associations between SNPs in LD and a true causal variant (Box 1.12).

Box 1.11: Contingency table analysis

The genotyping data of n_d cases and n_c controls, for a single SNP with alleles A and a, is the baseline to perform a contingency table analysis. When the three genotypes are analyzed, a full genotype table is constructed [88, 95] as follows:

Group	AA	Aa	aa
Cases	$f_{1,1}$	$f_{1,2}$	$f_{1,3}$
Controls	$f_{2,1}$	$f_{2,2}$	$f_{2,3}$

From this table, a range of models of penetrance (i.e., risk of disease) may be studied:

Recessive model: The penetrance increases only with genotype aa.

Group	AA + Aa	aa
Cases	$f_{1,1} + f_{1,2}$	$f_{1,3}$
Controls	$f_{2,1} + f_{2,2}$	$f_{2,3}$

Dominant model: The penetrance increases with either genotype Aa or genotype aa.

Group	AA	Aa + aa
Cases	$f_{1,1}$	$f_{1,2} + f_{1,3}$
Controls	$f_{2,1}$	$f_{2,2} + f_{2,3}$

A chi-square statistic is used to evaluate the deviation between the observed frequencies ($f_{i,j}$) and the expected frequency ($E_{i,j}$) across the cells of the tables (Equations 1.5, 1.6). Here, H_0 is the hypothesis of no association between rows and columns, and thus, no association of the SNP with the disease.

$$X = \sum_{i=1}^{2} \sum_{j=1}^{J} \frac{(f_{i,j} - E_{i,j})^2}{E_{i,j}}. \tag{1.5}$$

$$E_{i,j} = \frac{n_d n_c}{(n_d + n_c)}. \tag{1.6}$$

The distribution of X is χ^2 with 2 degrees of freedom (d.f.) for the general genetic model and 1 d.f. for the models of penetrance [96].

> **Box 1.12: Hotelling's T^2 test**
>
> The Hotelling T^2 statistic is used to reveal LD between a group of SNPs and an SNP causative of disease. Considering a design of n_d cases and n_c controls genotyped in n_s markers, the ith individual in the disease group has the following genotype coding, $X_{i,s}$ [91]:
>
> $$X_{i,s} = \begin{cases} 1, & \text{if } G_{i,s} = AA \\ 0, & \text{if } G_{i,s} = Aa \\ -1, & \text{if } G_{i,s} = aa \end{cases}$$
>
> where X is the indicator variable and $G_{i,s}$ is the genotype observed in individual i at the sth SNP. An indicator variable Y is defined in the same way for the individuals from the control group. Then, the individual profiles, $X_i = (X_{i,1}, \ldots, X_{i,n_s})$, are used to assess the vector of means ($\bar{X}$) and the covariance matrix (Σ) of the indicator variables (Equations 1.7, 1.8, 1.9).
>
> $$\bar{X}_s = \frac{1}{n_d} \sum_{i=1}^{n_d} X_{i,s}. \tag{1.7}$$
>
> $$\bar{X} = (\bar{X}_1, \ldots, \bar{X}_{n_s})^T. \tag{1.8}$$
>
> $$\Sigma = \frac{1}{n_d + n_c - 2} \left(\sum_{i=1}^{n_d} (X_i - \bar{X})(X_i - \bar{X})^T + \sum_{i=1}^{n_c} (Y_i - \bar{Y})(Y_i - \bar{Y})^T \right). \tag{1.9}$$
>
> The T^2 statistic, defined in Equation 1.10 has a χ^2 distribution with n_s degrees of freedom (d.f.):
>
> $$T^2 = \frac{n_d n_c}{n_d + n_c} (\bar{X} - \bar{Y})^T \Sigma^{-1} (\bar{X} - \bar{Y}). \tag{1.10}$$
>
> This test can be referred to as a test to compare the means of coded genotypes between case and control groups [97]. Here, H_0 is the hypothesis of no LD between the SNPs tested and a SNP causative of disease, versus H_a: "At least one SNP showing LD with a SNP causative of disease" [91].

1.4.2 Multi-SNP Association Studies

Methods for multi-SNP association studies range from machine-learning algorithms to exhaustive and stochastic search approaches [16]. These methods aim to model the joint effect of clusters of SNPs on the disease, while performing a supervised classification of individuals into two classes, disease (case), or no disease (control). The input predictor variables correspond to non-genetic factors (e.g, age, area of residence, poisonous agents exposure) and genetic factors or SNPs, whose effect is thought to be mutually correlated either in the genomic neighbourhood, or in genes sharing the disease pathways [19].

In this section, we will review four methods widely used for GWAS, namely logistic regression models (LRMs), support vector machines (SVMs), random forests (RFs) and Bayesian networks (BNs) (see Table 1.5). However, several other methods have been used in GWAS, including dimension-reduction techniques (e.g., principal component-based logistic regression) [98], linear mixed models for longitudinal cohorts [99], algorithms for clustering SNPs [100, 101] and exhaustive search algorithms for epistatic interactions [102].

1.4.2.1 Logistic Regression Models

The logistic regression model (LRM) is the standard method for multiple-SNP analyses with binary outcomes [22, 10, 11]. LRMs estimate the probability of having the disease in agreement with the individuals' genotypes, while controlling the covariates' effect (e.g., gender, population stratification, age). In the LRMs, the independent variables (SNP genotypes) explain the odds of the dependent binary variable (case/control) by means of a logit function (Box 1.13). Adjusted p-values are obtained for each SNP to identify their genome-wide significance and odds ratios are assessed to quantify an SNP's effect on a trait [103].

LRMs provide measures to rank SNVs (e.g., p-values) and the option of making inference for the genetic and non-genetic effect on the trait. A potential problem is, however, that the power to detect meaningful associations decreases in settings of small number of cases. LRMs have the risk of false-positive findings due to the presence of substructures in the data (e.g., correlations between loci, also known as linkage disequilibrium (LD),[13] inflate the statistics) [104].

Despite of these drawbacks, regression approaches (including logistic, linear and LASSO-penalised models [106]) are well-established methods for GWAS and have resulted in the discovery of more than 24,000 SNP-trait associations summarized in the GWAS catalogue [107]. This compilation of results covers numerous genomic loci of complex diseases in ancestrally diverse populations. For example, variants associated with type 2 diabetes in the glycated hemoglobin have been reported in European, African, East Asian, and South Asian cohorts [108]. Other interesting examples are the GWAS of late-onset Alzheimer's disease; SNPs associations found in caucasian individuals are evident in cohorts of African-Americans, Japanese, and Israeli-Arabs too [109]. Hence, regression models are preferred for metha analysis and transethnic studies but also to facilitate replication of results in independent samples.

1.4.2.2 Support Vector Machines (SVMs)

Support vector machines generate a maximum separation between case and control individuals by means of a kernel function. Briefly, SVMs receive as input the SNP genotypes for each individual, define a separating hyperplane and discriminate the data points between two classes in the feature space (Box 1.14) [110, 111, 112].

---

[13] Linkage disequilibrium means that alleles are physically close to one another on the DNA strand. They occur together more often than accounted by chance alone.

Table 1.5: Pros and cons of multi-SNP methods for GWAS

Method	Pros	Cons	References
Logistic regression models	• Significant genome-wide SNPs are detected and prioritized for subsequent analyses. • The covariates effect is controlled.	• Ignore the LD by assuming the independence of variables (i.e., SNPs). • Large cohorts of patients are needed to find significant p-values. • Risk of type I error.	[10, 11, 22, 116, 117, 118, 119]
Support vector machines	• Handle high-dimensional data, allowing the input of more features (SNPs) than samples. • Do not require previous knowledge of SNPs linked to a disease. • Able to learn non-linear decision boundaries. • Robust to noise in SNPs and less prone to overfitting.	• Sensitive to the choice of kernels. • Sensitive to parameter tuning steps. • Not direct feature selection.	[24, 111, 113, 115, 120, 121]
Random forests	• The predictive model is built without assumptions about the genotype–phenotype link. • Almost free of parameters to be tuned. • Provide variable importance measures (VIMs) to rank SNPs.	• The performance decreases when the number of SNPs increases. • High dimensional data have an impact in the detection of epistatic interactions. • LD affects negatively the outcome by hiding the joint effects of SNPs. • The VIMs lack measures of uncertainty. • Hardly detect SNPs with small effect size on the disease.	[23, 122, 123, 124, 125, 126, 127]
Bayesian networks	• Allow to make statements on the patterns of SNP-SNP and SNP-disease relationships. • Incorporate prior knowledge (e.g., distances between genes) to reduce the number of possible directed acyclic graphs (DAG).	• Provide little information on which variables (e.g., SNPs) are important. • It is not known which scoring criteria perform the best when the model is learned. • Scalability issues.	[25, 128, 129, 130, 131, 132, 133, 134]

Box 1.13: Summary of LRMs

Many multiple-SNP association tests use a regression setting by fitting a LRM. Given n independent observations of disease status (Y) and genotype code (X) for the individuals, the LRM tests any association between disease and genotype. The disease status for the ith individual is $Y_i = 1$ if diagnosed with the disease (case), otherwise $Y_i = 0$ (control). Considering a single SNP with alleles A and a, the genotype G_i can take different coding schemes [103]:

- Linear coding: $X_i = \begin{cases} 0, & \text{if } G_i = AA \\ 1, & \text{if } G_i = Aa \\ 2, & \text{if } G_i = aa \end{cases}$

- Dominant effect coding: $X_i = \begin{cases} 0, & \text{if } G_i = AA \\ 1, & \text{if } G_i = Aa \text{ or } G_i = aa \end{cases}$

- Recessive effect coding: $X_i = \begin{cases} 0, & \text{if } G_i = AA \text{ or } G_i = Aa \\ 1, & \text{if } G_i = aa \end{cases}$

An LRM makes use of the logit function, which links the expected value of the disease status (p_i) with the linear effects of the genotypes (Equation 1.11). This function also describes the odds of disease, defined as the ratio between the probability of disease (p_i) and the probability of non-disease ($1 - p_i$).

$$\text{logit}(p_i) = \frac{p_i}{1 - p_i} = Odds. \tag{1.11}$$

The simplest LRM is written in terms of a single genomic variable X_1 (i.e., a single SNP) in Equation 1.12. Then, a test of association for the SNP is formulated on the parameter β_1 and a p-value is calculated from the test. The likelihood ratio test, Wald test, and score test are traditionally used to evaluate $H_0 : \beta_1 = 0$ vs $H_a : \beta_1 \neq 0$ [89]

$$\text{logit}(p(Y_i = 1 \mid X_{1_i})) = \beta_0 + \beta_1 X_{1,i}. \tag{1.12}$$

Multiple-SNP association analyses expand the model to include more SNPs, X_s ($s = 1, \ldots, n_s$) (Equation 1.13). Under this scenario, n_s tests of association are performed jointly on the parameters $\beta = (\beta_1, \ldots, \beta_{n_s})$; thus, one p-value per SNP is obtained.

$$\text{logit}(p(Y_i = 1 \mid X_{1,i}, \ldots, X_{n_s,i})) = \beta_0 + \sum_{s=1}^{n_s} \beta_s X_{s,i}. \tag{1.13}$$

Similarly, LRM can combine non-genomic variables ($W_j : j = 1, \ldots, J$), such as clinical factors, and information across SNPs in an additive model (Equation 1.14). When the number of SNPs is higher than the number of individuals, shrinkage approaches (i.e., penalized LASSO regression) are fitted to data [105].

(cont.)

> **Box 1.13: Summary of LRMs (cont.)**
>
> $$\text{logit}(p(Y_i = 1 \mid \boldsymbol{X}, \boldsymbol{W}) = \beta_0 + \sum_{s=1}^{n_s} \beta_s X_{s,i} + \sum_{j=1}^{J} \gamma_j W_{j,i}. \quad (1.14)$$
>
> The *p*-values are subsequently adjusted to control the type I error by using, in the simplest way, a Bonferroni correction [103].
> **Type I error**: Also known as the false-positive rate, or the significance level. This is the probability of rejecting H_0 when it is true. In other words, it is the proportion of false positives that the researcher will tolerate in the study [96]. Here, a false positive means a SNP erroneously predicted to be associated with the disease.

Mittag *et al.* [24] reported an SVM for Parkinson's disease and Type 1 diabetes. The method integrates a Gaussian radial basis function kernel followed by an extensive grid search for optimization parameters. An important conclusion of their work is that the disease risk prediction improves when including SNPs with a small effect size in the models. Thus, the authors recommend the inclusion of rare and common variants (effect size between 0.6 to 1.5) to obtain feasible predictions.

Other SVMs approaches have had success for type 2 diabetes [113], rheumatoid arthritis [114] and type 1 diabetes [115], but the performance decreases when the disease has moderated or low heritability [24], as is the case with Parkinson's disease. For this disease unknown non-genetic factors are thought to affect the risk profiling, specially, in late onset cases.

1.4.2.3 Random Forests (RFs)
Random forest methods consist of classification and regression tree classifiers (CARTs), which are aggregated to predict a SNP-disease association. CARTs provide easy rules to be interpreted visually in a tree-shaped graph, where nodes can be either internal (fork in a branch) indicating a decision rule, or external (leaf) indicating the class label [136]. By moving through the CART from its root to an output node, it is possible to classify a new individual. RFs are constructed by growing a specified number of CARTs based upon boostrapping of the data; then, a consensus classifier is learnt from the collection of CARTs (Box 1.15).

Random forests are popular in this field, because they provide measures to rank and filter SNPs in follow-up studies [23]. The variable importance measures (VIMs), such as the Gini importance and the permutation importance [126], are thought to capture the causal effect of the SNP, weighting its contribution to the trait. Recent works on RFs have proposed novel strategies for feature selection (step 3(*i*) in Box 1.15) and improvements in VIMs [122, 123, 124, 125] . For example, a tree-based ensemble method called trees inside trees (T-Trees), performs feature selection from blocks of contiguous SNPs in LD. Then, instead of the traditional univariate split functions inside a node, a new tree is built into the node by using the SNPs at each block [124].

> **Box 1.14: Summary of SVMs**
>
> The SVMs analyze input data with labels (case or control) and find the optimal separating hyperplane, i.e, the hyperplane that maximizes the distance, or margin between the closest points of each class (see Figure 1.6).
>
> The input data are the genotyping profiles of dimension D for N individuals: $X = (x_1, \ldots, x_N)$ with $x_i \in \Re^D$. The class labels are $Y = (y_1, \ldots, y_N)$ with $y_i \in \{-1, 1\}$ [111]. The optimal separating hyperplane is given by:
>
> $$w^T x - b = 0. \tag{1.15}$$
>
> This is the plane equation where $w \in \Re^D$ is the weight vector, and b is a threshold. Then, the class assignment for the individuals is:
>
> $$g(x_i) = sign(w^T x_i - b). \tag{1.16}$$
>
> To maximize the margin M between data points, the objective function is given by equation:
>
> $$\arg\max_{w \in \Re^D, b \in \Re} M(w, b) = \frac{1}{\|w\|} \min_{i=1,\ldots,N} \{y_i(w^T x_i - b)\}. \tag{1.17}$$
>
> After constraining w and b, the objective function is reduced to Equation 1.18. Hence, the hyperplane that gives the maximum margin is found by minimizing $\|w\|$ for which Lagrange multipliers are used to solve the optimization problem [112, 135].
>
> $$\arg\min_{w \in \Re^D, b \in \Re} w^T w, \text{ constrained to } \min_{i=1,\ldots,N} \{y_i(w^T x_i - b)\} = 1. \tag{1.18}$$
>
> Despite the fact that classification problems are hardly solved by a linear classifier, the points can still be transformed into a higher dimensional feature space to make them linear separable. The points can be transformed by a kernel function, for example, the Gaussian kernel function [111] (Equation 1.19):
>
> $$K(x_i, x_j) = \exp\left(-\frac{\|x_i - x_j\|^2}{2\sigma^2}\right). \tag{1.19}$$

Considering the hundreds of thousands of markers in cohorts of thousands of individuals, an important challenge for RFs is the high-dimensionality of GWAS data. Although the data can be split up in small chunks to be analyzed independently, the interaction effects can be lost in the process and still many SNPs are uninformative or redundant. Therefore, efficient implementations of RF are being specifically developed for large-scale GWAS [126].

Overall, RF approaches have shown higher accuracy than standard LRMs because of their ability to separate informative from irrelevant SNPs [126, 127]. Applications to hypertensive heart disease [123], Parkinson's disease and Alzheimer data [122] have also proved the usefulness of RFs (Table 1.5).

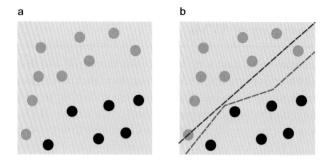

Figure 1.6: A classification problem is addressed by SVMs: (a) class labels for disease (black) and no disease (orange) are separated by using (b) a linear function (blue line) and a non-linear function (green line).

1.4.2.4 Bayesian Networks (BNs)

Bayesian networks (BNs) aim to find the pattern of multiple SNP associations (i.e., epistatic interactions) for a given disease. This approach constructs directed acyclic graphs (DAG) whose nodes correspond to discrete random variables (e.g., disease status, SNPs) and the edges represent the dependences between variables under the joint probability distribution of the graph [133] (Box 1.16). The goal is to find the SNP nodes that are parents of the disease status node in a two-layer structure (Figure 1.7); thus, to explore causal relationships between variants in the first layer and a disease in the second layer.

Applications of BNs to GWAS data have used score-and-search algorithms to learn the graph and aimed to improve the scoring criterion, which is a bottleneck of this approach [25, 134, 129]. For instance, Han et al. [25] defined a new scoring function to capture the complexity of a network and to weight how well it fits the data. This approach outperformed SVMs when fitting data for Alzheimer and autism diseases. Overall, BNs hold promise for revealing the SNPs affecting the trait [128], but current methods are still affected by a large number of noisy SNPs that may be included as predictors. In addition, there is no consensus in which scoring function achieves the best performance in the learned model [130].

1.4.3 Predictive genetic risk models in DTC services

The vast number of GWAS published to date has broadened our understanding of the SNPs behind common diseases. In parallel, DTC companies have made use of research studies and public GWAS data to develop their own predictive risk models, which generally underestimate the complexity of SNP interactions. As information on the data sources and algorithms used by DTC companies to fit their models is restricted, in this subsection we will not review their models, but we will cover some generalities on the procedures, methodological considerations, and factors that may be affecting their predictive abilities.

Reports on genetic risks include quantitative estimations of the association between a set of SNPs and a condition (Box 1.17). A basic methodology to assess these indicators is as follows [141]:

> **Box 1.15: The RF growing algorithm**
>
> CARTs are classifiers representing predictor variables (i.e., SNPs) by nodes and a sequence of predictions by paths. Thus, a CART can be seen as the set of SNPs that segregates the individuals in the phenotypes of interest [137]. As single CARTs are biased to single sets of SNPs, the RF algorithms create and ensemble thousands of CARTs into a consensus predictor. The standard RF growing algorithm has the following methodology (adapted from [135, 136, 137]):
>
> 1. Split the data points (i.e., individuals) into a group of training data (two-thirds of the original data) and a group of testing data (one-third of the original data).
> 2. Draw a sample with replacement from the group training data. This kind of sample, known as a bootstrap sample, must has the same size of the training data and its data points may be repeated.
> 3. Grow a CART with the bootstrap sample: (i) At each node, randomly pick a small number of SNPs (random feature selection). (ii) Select the SNP that better discriminates the individuals in the bootstrap sample by maximizing a measure of node purity (e.g., Gini index, see below). In this way, each node is split sequentially resulting in two child nodes. This step is repeated until the largest possible CART is grown. (iii) Estimate the prediction error for the CART using the testing data.
> 4. Repeat steps 2 and 3 to grow a given number of CARTs.
> 5. Use an aggregate voting strategy over the CARTs' predictions to classify new instances. The majority vote over all predictions is used to classify the testing data and assess the RF prediction accuracy.
>
> **Node labels**: The nodes in a CART are traditionally labeled with a decision rule, which is univariate binary test for the variable X of the form: $X < c$, where c is the cut-point that divides the individuals into two disjoint sets [124].
> **Gini importance**: The Gini importance for an SNP in the RF, ΔI_G, is an indicator of the SNP relevance to split the group of individuals into two classes [138]. This measure is the aggregation of the Gini impurities, $\Delta i(\tau, T)$, at each node τ, where SNP s was selected over all T RFs:
>
> $$\Delta I_G(s) = \sum_T \sum_\tau \Delta i(\tau, T). \qquad (1.20)$$
>
> The Gini impurity, $\Delta i(\tau, T)$, is calculated with the following equations:
>
> $$i(\tau) = 1 - p_d^2 - p_c^2 \qquad (1.21)$$
>
> $$\Delta i(\tau) = i(\tau) - p_1 i(\tau_1) - p_2 i(\tau_2), \qquad (1.22)$$
>
> where p_d and p_c are the fractions of individuals from classes "disease" and "control" out of the total number of individuals n at node τ: $p_d = \frac{n_d}{n}$ and $p_c = \frac{n_c}{n}$. Likewise, p_1 and p_2 are the fractions of individuals in the child nodes, τ_1 and τ_2: $p_1 = \frac{n_1}{n}$ and $p_2 = \frac{n_2}{n}$. Thus, $\Delta i(\tau)$ is the decrease in the Gini impurity after splitting n individuals in node τ to the child nodes τ_1 and τ_2 [138].

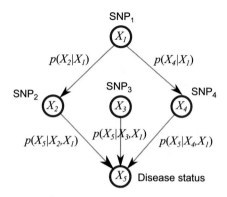

Figure 1.7: DAG representing the causal relationships between SNPs and a disease.

Box 1.16: Summary of BNs

A Bayesian network is a graphical description of the probabilistic relationships between predictor and response variables [139]. In the GWAS context, the DAG consists of random variables $X = \{X_1, \ldots, X_N\}$ inside a set of nodes V (e.g., SNP nodes and disease status nodes), plus a set of edges E describing the joint probability distribution of the network. Consequently, a BN has a DAG component showing the network structure ($G = \{V, E\}$) (Figure 1.7) and a probability component (P) describing the dependences between variables [130, 132]:

$$p(X_1, \ldots, X_N) = \prod_{i=1}^{N} p(X_i \mid R_i). \tag{1.23}$$

The P component satisfies the local Markov assumption, which establishes that, each variable X_i (node) is conditionally independent of their non-descendant variables, given its parents (R_i) [131]. In Figure 1.7, the P component is illustrated by the conditional probabilities on directed edges; for example, the DAG can be used to calculate $p(X_5 \mid X_2, X_1)$, the conditional probability that an individual has the disease given that his/her genotype profile has risky alleles in SNP_1 and SNP_2.

BNs are commonly learned by two types of algorithms [140]:

Score-and-search algorithms: This family of algorithms evaluates the fit of a DAG given the data by using a scoring criterion, for example, the posterior probability of the DAG. The method returns the DAG that maximizes the score as follows:

1. Start the search from a baseline DAG (e.g., an empty DAG).
2. Transform the DAG: add, withdraw, or reverse an edge from a list of possible transformations.

(cont.)

3. Select the transformation that improves the score while ensuring that the graph remains acyclic.
4. Repeat the previous two steps until no longer improvement in the score is achieved.

Constraint-based algorithms: This family of algorithms build a DAG constrained to a test of conditional independence among variables (e.g., partial correlation):

1. Test the conditional independence among all pairs of variables.
2. Find all pairs of variables dependant of each other and propagate undirected edges on the neighbourhood of each node. This undirected graph is assumed to contain the skeleton of the true graph.
3. Eliminate indirect dependences from the skeleton and determine directions of edges avoiding the creation of directed cycles.

Conditional independence: Let three random variables be X, Y, and Z. If $P(X \mid Y, Z) = P(X \mid Y)$ we say that X is conditionally independent of Z given Y [131].

---

1. Find studies on SNP-disease associations from peer-reviewed literature and public databases.
2. Select studies meeting quality and reporting criteria, and form an initial variant set. Each SNP must be reported with an estimate of relative effect (e.g., odds ratio (OR), Box 1.17). The variant set may be further modified to match the consumer origin (i.e., population ethnicity).
3. Assess the risk indicator. For example, if RR (Box 1.17) is used as an indicator for a disease based on three SNPs:

   - Estimate the RR for each genotype from the OR reported in GWAS by using Equation 1.27. The assessment may assume that the prevalence of the condition and the genotype frequencies in the population are known.
   - Asses the overall RR for the consumer given his/her genotyping profile. Similar to the GCI (Equation 1.28), this may be calculated as the product of all RR over the three SNPs observed in the consumer [142].

4. Integrate non-genetic risks. If other factors are consistently associated with the disease (e.g., smoking is a well-known factor for lung cancer), integrate their relative effects into the risk model.

Direct-to-consumer reports are questioned in the scientific community due to the variability in the risk interpretation among companies [144, 145]. A number of reasons can be inferred on the revision of the above four-step methodology [146]. First, SNPs selected from GWAS generally do not overlap between companies. Except for SNPs with high effect on the condition, companies' methodologies vary in the inclusion of SNPs with low effects to improve their models. Moreover, the assessed risk may change when new disease-associated SNP are added to the model, so a consumer may be reclassified from low to high risk depending on the state-of-art of GWAS. Second,

> **Box 1.17: Risk indicators**
>
> The risk indicators are measures of association between a genetic variant and the disease. Common risk indicators are (adapted from [141]):
>
> **Absolute risk** (*AR*): The probability of disease, *D*, in the population with genotype *G*.
>
> $$AR = p(D \mid G). \tag{1.24}$$
>
> **Odds ratio** (*OR*): The ratio of odds of disease in the population with genotype, *G*, to the population without the genotype (G^-). This is the risk indicator reported in GWAS:
>
> $$OR = \frac{Odds(D \mid G)}{Odds(D \mid G^-)} = \frac{\frac{p(D \mid G)}{1 - p(D \mid G)}}{\frac{p(D \mid G^-)}{1 - p(D \mid G^-)}}. \tag{1.25}$$
>
> **Relative risk** (*RR*): The probability of disease in the population with the genotype, divided by the probability of disease in the population without the genotype.
>
> $$RR = \frac{p(D \mid G)}{p(D \mid G^-)}. \tag{1.26}$$
>
> *RR* is estimated from *OR* by [143]:
>
> $$RR = \frac{OR}{1 - p(D \mid G^-) + p(D \mid G^-) \cdot OR}. \tag{1.27}$$
>
> For example, $RR = 2$, means that someone carrying the genotype is two times more likely to develop the disease than someone without the genotype.
>
> **Genetic Composite Index (GCI)**: This index was used by Navigenics, a former DTC company, to estimate the combined effect of a set of genotypes. The GCI is obtained by a multiplicative model for an individual with genotype profile ($G_1, \ldots, G_{n_s}$):
>
> $$GCI(G_1, \ldots, G_{n_s}) = \prod_{s=1}^{n_s} RR_s. \tag{1.28}$$

the reference population from which the risk data was obtained do not match the consumer's origin. Historical comparisons of DTC reports have illustrated that differences in the reference-population variables (e.g., the incidence of disease in the population without the SNP) have an impact on the values of predicted risks [29, 147, 146]. For instance, if the reference GWAS included different follow-up times for a disease cohort, then it is likely that the reference-population variables are biased, so they will produce different estimates of population risk [29]. Third, and a more relevant reason of variability between results, is the use of different models. Kalf *et al.* [29] simulated genotype data and used published models of Navigenics, deCODEme (two former DTC companies) and a 23andMe model from 2007.[14] Their results revealed that models for the first two companies appeared to overestimate risks, but in the absence of complete knowledge on the SNPs affecting disease heritability, it was impossible to define which model best captured the risk.

The sources of variability mentioned here bring us to a frequent concern regarding DTC reports: limited predictive power in their models. Predictive ability of genetic tests is limited because common diseases have a heavy non-genetic component of lifestyle and environmental factors. Although disease risks based on genetic markers could be accurate, they comprise only a fraction of the overall disease susceptibility, and unknown gene–environment interactions adding to the overall risk are being omitted [146]. In addition, the heritability of some diseases is low, such as in Parkinson's disease, which detract the credibility of tests based exclusively on genetic markers. Another reason why the prediction of disease risk can be inaccurate is that, even when the heritability is high, the SNP effect size may be small [24]. Therefore, SNPs are playing specific roles whose joint effect is still too complex to be unveiled by the existing algorithms.

Despite criticisms of DTC services, the growing market of DTC companies has attracted hundreds of thousands of customers since late 2007, and with them the formation of large databases of genetic information that may be exploited for medical research. For example, 23andMe, a company with over one million genotyped customers,[15] conducts collaborative work with universities to construct new machine-learning methods for human phenotype prediction. They have merged data from Parkinson's disease patient cohorts, including samples drawn from their customer base, to propose disease classification models based on LRMs [148, 149]. The models yielded high classification accuracy (area under curve > 0.9) by integrating non-genetic factors (age, sex, family history, testing of sense of smell) and a genetic risk score from 30 genetic risk factors. Other improvements in learning methods from 23andMe scientists are available on their webpage.[16]

1.5 Perspectives and Recommendations

With continuing advances of sequencing technology and reductions in sequencing cost the field of personalized genomics is expanding with exponential speed. NGS methods are seeing increased use in hospital laboratories where targeted, whole exome and whole genome sequencing are used to facilitate diagnosis, therapeutic decisions, and disease predictions [150]. With our increasing knowledge of diseases and genome variation novel opportunities will also arise for DTC companies. However, the current approach of the majority of DTC companies, to provide individuals with likelihoods

---

[14] http://stanford.edu/class/gene210/files/readings/23-01_WhitePaper.pdf
[15] http://blog.23andme.com/news/one-in-a-million/
[16] http://blog.23andme.com/23andme-research/

for disease development based on SNPs alone and without expert counselling, has raised well founded doubts from both academic and clinical institutions [151]. The American College of Medical Genetics and Genomics in 2015 revised their position of DTC testing, cautioning consumers about minimum requirements when ordering PGTs [152]. In addition, regulatory frameworks have been set-up not only in the United States, but also in several European countries pertaining to DTC use [153]. The success of DTC companies in the future will therefore probably depend on the incorporation of medical experts and genetic counselors, as well as an issuing of clinically accurate and relevant data, with the use of novel computational approaches.

Novel approaches are not only relevant for DTC companies, but also for clinical applications of PGTs. It is clear that common diseases, which are on the rise in developing countries, cannot be explained simply by the presence of common SNPs (usually the target of research in GWAS). While rare variants were thought to be the missing link in the unexplained heritability, it seems that they only have a modest-to-small effect on variation of complex traits [154]. The challenge of novel algorithms will therefore likely be the ability to *integrate* various sources of data. For instance, integration of patient genomic data with their clinical, health history and lifestyle data could lead to better disease prognosis and personalized treatment. Small steps towards this have already been introduced with the so called pathology supported genetic testing [155]. While this is still done at a small scale, with only a dozen SNPs taken into account per patient, the true challenge of future algorithms will be the ability to integrate high throughput data from genomics, proteomics, and metabolomics.

We have pointed out that the identification of SNPs and their functional consequences on diseases lays the basis of personalized medicine and opens opportunities for research in data-mining. Below we highlight some points to be considered in the future:

- Clearly, our understanding on how SNPs affect protein sequences, mRNA and transcription factors can not be expanded by using only genetic data. Methods to merge and exploit large amounts of information (e.g., WGS, clinical data, omics data) would clarify the mechanisms of common diseases. Current strategies (e.g., LRMs, SVMs, BNs, RFs) are not able to combine diverse dataset types while controlling simultaneously for noise, data over-fitting, and high dimensionality. Implementation of these methods has also proved that some of them are not suited for genetic data. For example, when algorithms rely on permutation procedures for bootstrapping or cross-validation [136], the computational capacity is limited to a few hundred SNPs (e.g., permutation procedures in RFs [156]).

    Hence, data fusion frameworks must be envisioned to decode the multi-type interacting entities that underlie biological functions and diseases. For example, network-based methods [157], kernel-based methods [158], and non-negative matrix factorization [159, 160] might be utilized to produce SNP-disease predictions that incorporate multiple types of data. In addition, data fusion frameworks have the potential to mine the non-genetic influence on disease risk by integrating environmental factors and lifestyle variables in the analyses, which

ultimately trigger the disease. These external variables have led to better discrimination of cases than did models based exclusively on genotypes [161], so the prediction may be improved if we ensure that the interaction between genes, environment, and lifestyle are included in the model.
- Methods should be able to formulate polygenic models of disease susceptibility, including variants across the spectrum of effect sizes and all allele frequencies [15, 3]. We have, therefore, three critical factors affecting the selection of genetic variants in the models: Number of SNPs, effect size, and allele frequency. First, parsimonious models with a small number of SNPs, although preferred for clinical diagnostic, hardly explain the whole heritability. In contrast, if more SNPs are included in the model, larger sample sizes will be needed to find significant associations [136]. Second, when SNPs' effect size are small, they do not result in genome-wide significance by using standard single-SNP analyses [105]; even multi-SNP methods have not unveiled all the true associations for SNPs when effect sizes drops. Third, as commented by Visscher *et al.* [15], the portfolio of risk alleles is likely to be different for each affected individual in the population, thus, to identify the causative loci and the SNPs in strong LD with the causative loci, a mix of common, low frequency, and rare variants should be considered. Summarizing, we are demanding methods not only to select a set of potential SNPs for different allele frequencies and effect sizes, but also to perform well with small cohorts of patients.
- Machine-learning approaches are needed for predicting the influence of epistatic SNP–SNP interactions on disease susceptibility [135, 162]. Beyond the additive effect size by each SNP alone, it is thought that complex epistatic interactions could have a higher variance that may be used to explain disease susceptibility [163, 164]. Interestingly, the heritability of some disorders is explained by only a fraction of the SNP-disease associations discovered so far; consequently, it has also been postulated that the "missing heritability" could be explained by epistatic effects (a postulation that is controverted given the difficulties in determining the significance of those interactions [162, 163]). Whether or not epistasis could explain this gap between additive contribution of SNPs and the known heritability, the advent of more GWASs have evidently increased the number of theoretical interactions between SNPs, whose joint impact can only be disclosed with novel algorithms.
- With regard to the SNP/genotype calling procedures, more research is needed to improve the measures of confidence for the inferred SNP/genotypes [76]. The inclusion of these measures as additional data in downstream GWAS could reduce the uncertainty in disease risk prediction. Similarly, algorithms for indexing of reads and reference genomes are being updated to keep up with the technological advances of sequencing platforms [61], warranting more research in the mean time.

As algorithmic solutions surge, the researchers have to contemplate a number of decision factors prior the SNP identification and SNP-disease prediction. We finalize this review by giving some recommendations for choices of algorithms:

- In the SNP/genotype calling step, the sample size is a relevant variable when microarray technologies are used. Although many discrepancies have been observed in the performance of genotyping algorithms, small samples should be handled with algorithms that do not rely on training data, or population parameters (e.g., GenoSNP [44]). If the focus is on rare variants, then special approaches accounting for them should be used (e.g., M3 [7]). Likewise, when the data come from NGS technologies, the number of samples guides the algorithm selection, as not all of them admit multiple samples. Low sequencing depth and precalling realignment steps should also be judged to reduce mismatches (e.g., GATK, SAMtools).
- Concerning the SNP-disease association studies, there is not a method suitable for handling the problems of high-dimensional data and addressing all the aims of GWAS simultaneously. Among the methods surveyed here, SVMs are the only approach dealing with large input sets of individuals and SNPs. Other machine-learning methods (BNs and RFs) do not succeed with high-dimensional data, and LRMs have the risk of error type I in the statistical tests performed on thousands of SNPs. Sadly SVMs' outputs offer fewer opportunities of analysis in comparison with other methods. For example, RFs rank the SNPs by using VIMs for follow-up studies. Another strength of RFs is their ability to reduce the starting number of SNP to yield a small subset which might be used as input to another method [136]. Regarding BNs, patterns of interaction between multiple SNP and diseases can be extracted from the graph component, making them attractive for epistasis search. Finally, LRMs, the preferred approach for GWAS of case-control designs, offer a measure of SNP effect size while allowing the adjustment of non-genetic components in the model [3].

Classical methods surveyed here are adequate for the analysis of fairly straightforward cases involving limited information, but emerging levels of medical data from PGTs are now on the horizon and they demand novel integrative approaches to yield advancements in this field.

1.6 Exercises

1.1 Define SNP and why it is important.

1.2 Select the two major technologies that mostly influenced genome-wide detection of SNPs:

- PCR
- Microarrays
- Next generation sequencing
- Sanger sequencing

1.3 Imagine you are a doctor and your patient is not responding to the treatment. What type of genetic testing would you prescribe for them?

- Mendelian disorder testing
- Chromosomal disorder testing

- Complex genetic disorder testing
- Pharmacogenomic testing
- Monogenic disorder testing

1.4 Name three differences between Affymetrix spotted microarrays and Illumina BeadChip microarrays.

1.5 Name three features that massively parallel sequencing approaches have in common.

1.6 Order the steps in Illumina sequencing from start to finish:

- Bridge amplification
- Sequencing by synthesis
- DNA fragmentation
- DNA isolation
- Ligation of adapters
- Data analysis

1.7 Several genome projects of larger scale (compared to the 1000 Genomes Project) are currently under way. Some have already been mentioned here but many more exist. Find two additional genome projects of your interest.

1.8 Perform variant calling with the UnifiedGenotyper and HaplotypeCaller algorithms from GATK tools. Use your own sequencing and reference genome data, or the example data provided in the GATK documentation.

1.9 Generate a synthetic genotyping dataset of 100 individuals ($n_d = 50$ cases, $n_c = 50$ controls) and one single SNP. Assume a multiplicative model with genotype relative risk, r, for the causal allele a as follows:

Group	AA	Aa	aa
Cases	$(1-p_s)^2/t$	$2rp_s(1-p_s)/t$	$r^2 p_s^2/t$
Controls	$(1-p_s)^2$	$2p_s(1-p_s)$	p_s^2

where p_s is the allele frequency in the population and $t = (1-p_s)^2 + 2rp_s(1-p_s) + r^2 p_s^2$. Assume $p_s = 0.1$ and $r = 3$.

1.10 Perform a contingency table analysis with the genotyping data of Exercise 1.9. Use a chi-square statistic to evaluate the hypothesis of no association of the SNP with the disease. Assume a recessive model of penetrance. Is the SNP associated with the disease at $\alpha = 0.05$ (level of significance)? Perform the same analysis but change to a dominant model of penetrance. Explain the differences in the results.

1.11 Fit LRMs for a dataset of 1,000 SNPs, 50 cases, and 50 controls, and evaluate the association of each SNP with the disease as follows:

(a) Generate data for two groups of SNPs: (i) 990 SNPs with genotype relative risk $r = 1$. Thus, 990 SNPs without association with the disease. (ii) 10 SNPs with genotype relative risk $r = 1.5$. Thus, 10 SNPs associated with the disease. Assume that the allele frequency in both SNP groups follows an uniform distribution ($P_s \sim \text{Unif}(0.1, 0.9)$).

(b) Using a dominant coding scheme, fit a LRM for each SNPs and obtain the adjusted p-values from the association tests.
(c) Evaluate the sensitivity in detecting SNPs associated with the disease (level of significance: 0.05).
(d) Increase the sample size to 500 cases and 500 controls. Did the sensitivity improve?
(e) Increase the sample size to 500 cases and 500 controls, and the relative risk to $r = 3$ for the SNPs associated with the disease. Comment the disadvantages of LRMs to find significant associations.

1.12 Take the genetic dataset from Exercise 1.6e and define a SVM to classify the patients between cases and controls. Report the cross-validated accuracy in classification when (a) a linear kernel function and (b) a Gaussian kernel function are used in training and predicting.

1.13 Fit an RF for a genetic dataset consisting of 500 cases, 500 controls, and 7 SNPs described as follows:

SNP id.	Description	p_s	r
1	Low-frequency variant causative of disease	0.02	3
2	Common variant causative of disease	0.1	3
3 to 7	Random SNPs without association with the disease	0.2	1

Report the confusion matrix for the model, the classification error and prioritize the SNPs according to their average Gini importance.

1.14 Construct a BN for the dataset of Exercise 1.6. Show the network structure when using (a) a score-and-search algorithm and (b) a constraint-based algorithm.

1.15 The ORs of three SNPs and each possible genotype are presented in the following table:

Marker	AA	Aa	aa
SNP 1	1	1.5	1.9
SNP 2	1	2.1	2.5
SNP 3	1	1.8	2.2

Calculate the overall relative risk of disease for two individuals with genotypes $\{Aa, aa, AA\}$ and $\{Aa, AA, Aa\}$, respectively. Assume that the prevalence of disease in the control group is 8%.

Note: Solutions are available to instructors at www.cambridge.org/bionetworks.

1.7 Acknowledgments

This work was supported by the European Research Council (ERC) Starting Independent Researcher Grant 278212, the European Research Council (ERC) Consolidator Grant 770827, the President's PhD Scholarhip Scheme from Imperial College London, the Serbian Ministry of Education and Science Project III44006, the Slovenian Research Agency project J1-8155 and the awards to establish the Farr Institute of Health Informatics Research, London, from the Medical Research Council, Arthritis Research UK, British Heart Foundation, Cancer Research UK, Chief Scientist Office, Economic and Social Research Council, Engineering and Physical Sciences Research Council, National Institute for Health Research, National Institute for Social Care and Health Research, and Wellcome Trust (grant MR/K006584/1).

References

[1] The 1000 Genomes Project Consortium. A global reference for human genetic variation. *Nature*, 2015;526(7571):68–74. Available from: http://dx.doi.org/10.1038/nature1539310.1038/nature15393 http://www.nature.com/nature/journal/v526/n7571/abs/nature15393.html#supplementary-information.

[2] LaFramboise T. Single nucleotide polymorphism arrays: A decade of biological, computational and technological advances. *Nucleic Acids Research*, 2009;37(13):4181–4193.

[3] Bush WS, Moore JH. Chapter 11: Genome-wide association studies. *PLoS Computational Biology*, 2012;8(12):e1002822. Available from: http://journals.plos.org/ploscompbiol/article?id=10.1371/journal.pcbi.1002822.

[4] Yu X, Sun S. Comparing a few SNP calling algorithms using low-coverage sequencing data. *BMC Bioinformatics*, 2013;14(1):274. Available from: www.biomedcentral.com/1471-2105/14/274.

[5] Szelinger S, Pearson J, Craig D. Microarray-based genome-wide association studies using pooled DNA. In DiStefano J, ed., *Disease Gene Identification Methods and Protocols*. Human Press; 2011, pp. 107–124.

[6] The 1000 Genomes Project Consortium. An integrated map of genetic variation from 1,092 human genomes. *Nature*, 2012;491(7422):56–65. Available from: http://dx.doi.org/10.1038/nature11632 and www.nature.com/nature/journal/v491/n7422/abs/nature11632.html#supplementary-information.

[7] Li G, Gelernter J, Kranzler HR, Zhao H. M3: an improved SNP calling algorithm for Illumina BeadArray data. *Bioinformatics*, 2012;28(3):358–365. Available from: http://bioinformatics.oxfordjournals.org/cgi/doi/10.1093/bioinformatics/btr673.

[8] Wen Y, Li M, Fu WJ. MA-SNP: A new genotype calling method for oligonucleotide SNP arrays modeling the batch effect with a normal mixture model. *Statistical Applications in Genetics and Molecular Biology*, 2011;10(1):1–23. Available from: www.degruyter.com/view/j/sagmb.2011.10.issue-1/sagmb.2011.10.1.1698/sagmb.2011.10.1.1698.xml.

[9] Mao X, Young BD, Lu YJ. The application of single nucleotide polymorphism microarrays in cancer research. *Current Genomics*, 2007;8(4):219–228.

[10] Purrington KS, Slager S, Eccles D, et al. Genome-wide association study identifies 25 known breast cancer susceptibility loci as risk factors for triple-negative breast cancer. *Carcinogenesis*, 2014;35(5):1012–1019. Available from: www.carcin.oxfordjournals.org/cgi/doi/10.1093/carcin/bgt404.

[11] Wang LE, Gorlova OY, Ying J, et al. Genome-wide association study reveals novel genetic determinants of DNA repair capacity in lung cancer. *Cancer Research*, 2013;73(1):256–264. Available from: www.ncbi.nlm.nih.gov/pmc/articles/PMC3537906/

[12] Billings LK, Florez J. The genetics of type 2 diabetes: What have we learned from GWAS? *Annals of the New York Academy of Sciences*, 2010;1212(1):59–77.

[13] Lee JY, Lee BS, Shin DJ, et al. A genome-wide association study of a coronary artery disease risk variant. *Journal of Human Genetics*, 2013;58(3):120–126. Available from: www.nature.com/doifinder/10.1038/jhg.2012.124.

[14] Hastie CE, Padmanabhan S, Dominiczak AF. Genome-wide association studies of hypertension: light at the end of the tunnel. *International Journal of Hypertension*, 2010;2010:509581. Available from: www.ncbi.nlm.nih.gov/pmc/articles/PMC2958365/

[15] Visscher PM, Brown MA, Mccarthy MI, Yang J. Five years of GWAS discovery. *The American Journal of Human Genetics*, 2012;90(1):7–24. Available from: http://dx.doi.org/10.1016/j.ajhg.2011.11.029.

[16] Jing PJ, Shen HB. MACOED: A multi-objective ant colony optimization algorithm for SNP epistasis detection in genome-wide association studies. *Bioinformatics*, 2015;31(5):634–641. Available from: http://bioinformatics.oxfordjournals.org/cgi/doi/10.1093/bioinformatics/btu702.

[17] Arkin Y, Rahmani E, Kleber ME, et al. EPIQ: Efficient detection of SNP-SNP epistatic interactions for quantitative traits. *Bioinformatics*, 2014;30(12):19–25.

[18] Pabinger S, Dander A, Fischer M, et al. A survey of tools for variant analysis of next-generation genome sequencing data. *Briefings in Bioinformatics*, 2014;15(2):256–278. Available from: http://bib.oxfordjournals.org/cgi/doi/10.1093/bib/bbs086.

[19] Wang YT, Sung PY, Lin PL, Yu YW, Chung RH. A multi-SNP association test for complex diseases incorporating an optimal P-value threshold algorithm in nuclear families. *BMC Genomics*, 2015;16(1):381. Available from: www.biomedcentral.com/1471-2164/16/381.

[20] Walter K, Min JL, Huang J, et al. The UK10K project identifies rare variants in health and disease. *Nature*, 2015; 526(7571):82–90. Available from: www.nature.com/doifinder/10.1038/nature14962.

[21] Bomba L, Walter K, Soranzo N. The impact of rare and low-frequency genetic variants in common disease. *Genome Biology*, 2017; 18(1):77. Available from: http://genomebiology.biomedcentral.com/articles/10.1186/s13059-017-1212-4.

[22] Ueyama C, Horibe H, Yamase Y, et al. Association of FURIN and ZPR1 polymorphisms with metabolic syndrome. *Biomedical Reports*, 2015;3(5):

641–647. Available from: www.spandidos-publications.com/10.3892/br.2015.484.

[23] Winham SJ, Colby CL, Freimuth RR, et al. SNP interaction detection with Random Forests in high-dimensional genetic data. *BMC Bioinformatics*, 2012;13(1):164. Available from: https://bmcbioinformatics.biomedcentral.com/articles/10.1186/1471-2105-13-164.

[24] Mittag F, Büchel F, Saad M, et al. Use of support vector machines for disease risk prediction in genome-wide association studies: Concerns and opportunities. *Human Mutation*, 2012;33(12):1708–1718.

[25] Han B, Chen Xw, Talebizadeh Z, Xu H. Genetic studies of complex human diseases: characterizing SNP-disease associations using Bayesian networks. *BMC Systems Biology*, 2012;6 (Suppl 3):S14. Available from: www.ncbi.nlm.nih.gov/pmc/articles/PMC3524021/.

[26] Sequeiros J, Paneque M, Guimarães B, et al. The wide variation of definitions of genetic testing in international recommendations, guidelines and reports. *Journal of Community Genetics*, 2012;3(2):113–124.

[27] Katsanis SH, Katsanis N. Molecular genetic testing and the future of clinical genomics. *Nature Reviews Genetics*, 2013;14(6):415–26. Available from: http://dx.doi.org/10.1038/nrg3493.

[28] Covolo L, Rubinelli S, Ceretti E, Gelatti U. Internet-based direct-to-consumer genetic testing : A systematic review. *Journal of Medical Internet Research*, 2015;17(12):e279.

[29] Kalf RRJ, Mihaescu R, Kundu S, et al. Variations in predicted risks in personal genome testing for common complex diseases. *Genetics in Medicine*, 2014;16(1):85–91. Available from: http://dx.doi.org/10.1038/gim.2013.80.

[30] Stringer S, Wray NR, Kahn RS, Derks EM. Underestimated effect sizes in GWAS: fundamental limitations of single SNP analysis for dichotomous phenotypes. *PLoS One*, 2011;6(11):e27964. Available from: http://journals.plos.org/plosone/article?id=10.1371/journal.pone.0027964.

[31] Park JH, Gail MH, Weinberg CR, et al. Distribution of allele frequencies and effect sizes and their interrelationships for common genetic susceptibility variants. *Proceedings of the National Academy of Sciences of the United States of America*, 2011;108(44):18026–18031. Available from: www.ncbi.nlm.nih.gov/pmc/articles/PMC3207674/.

[32] Ezkurdia I, Juan D, Rodriguez JM, et al. Multiple evidence strands suggest that there may be as few as 19,000 human protein-coding genes. *Human Molecular Genetics*, 2014;23(22):5866–5878. Available from: www.ncbi.nlm.nih.gov/pubmed/?term=24939910.

[33] Council of Europe. *Genetic Tests for Health Purposes*. Alsace Media Science; 2012.

[34] Roberts JS, Ostergren J. Direct-to-consumer genetic testing and personal genomics services: A review of recent empirical studies. *Current Genetic Medicine Reports*, 2013;1(3):182–200.

[35] Rubinstein WS, Maglott DR, Lee JM, et al. The NIH genetic testing registry: A new, centralized database of genetic tests to enable access to comprehensive information and improve transparency. *Nucleic Acids Research*, 2013; 41(Database issue):D925–35. Available from: http://nar.oxfordjournals.org/content/41/D1/D925.long.

[36] Yang J, Zhang W, Wu B. A note on statistical method for genotype calling of high-throughput SNP arrays. *Journal of Applied Statistics*, 2013;40(6):1372–1381.

[37] Govindarajan R, Duraiyan J, Kaliyappan K, Palanisamy M. Microarray and its applications. *Journal of Pharmacy and Bioallied Sciences*, 2012;4(6):310. Available from: www.jpbsonline.org/text.asp?2012/4/6/310/100283.

[38] Alonso A, Marsal S, Tortosa R, Canela-Xandri O, Julià A. GStream: Improving SNP and CNV coverage on genome-wide association studies. *PLoS One*, 2013;8(7):e68822. Available from: http://dx.plos.org/10.1371/journal.pone.0068822.

[39] Distefano JK, Taverna DM. Technological issues and experimental design of gene association studies. *Methods in Molecular Biology*, 2011;700:3–16. Available from: www.ncbi.nlm.nih.gov/pubmed/21204023.

[40] Teo YY, Inouye M, Small KS, et al. A genotype calling algorithm for the Illumina BeadArray platform. *Bioinformatics*. 2007;23(20):2741–6. Available from: http://europepmc.org/articles/PMC2666488.

[41] Kang SJ, Finch SJ, Haynes C, Gordon D. Quantifying the percent increase in minimum sample size for SNP genotyping errors in genetic model-based association studies. *Human Heredity*, 2004;58(3–4):139–144. Available from: www.karger.com/Article/FullText/83540.

[42] Powers S, Gopalakrishnan S, Tintle N. Assessing the impact of non-differential genotyping errors on rare variant tests of association. *Human Heredity*, 2011;72(3):153–160. Available from: www.ncbi.nlm.nih.gov/pmc/articles/PMC3214826/.

[43] Ritchie ME, Liu R, Carvalho BS, Irizarry RA. Comparing genotyping algorithms for Illumina's Infinium whole-genome SNP BeadChips. *BMC Bioinformatics*, 2011;12(1):68. Available from: http://bmcbioinformatics.biomedcentral.com/articles/10.1186/1471-2105-12-68.

[44] Giannoulatou E, Yau C, Colella S, Ragoussis J, Holmes CC. GenoSNP: A variational Bayes within-sample SNP genotyping algorithm that does not require a reference population. *Bioinformatics*, 2008;24(19):2209–2214. Available from: www.ncbi.nlm.nih.gov/pubmed/18653518.

[45] Rabbee N, Speed TP. A genotype calling algorithm for affymetrix SNP arrays. *Bioinformatics*, 2006;22(1):7–12. Available from: http://dx.doi.org/10.1093/bioinformatics/bti741.

[46] Affymetrix. *BRLMM: An Improved Genotype Calling Method for the GeneChip® Human Mapping 500K Array Set*. Affymetrix; 2006. Available from: www.biostat.jhsph.edu/~iruczins/teaching/misc/gwas/papers/affymetrix2006.pdf.

[47] Carvalho B, Bengtsson H, Speed TP, Irizarry Ra. Exploration, normalization, and genotype calls of high-density oligonucleotide SNP array data. *Biostatistics*, 2007;8:485–499.

[48] Shah TS, Liu JZ, Floyd JAB, et al. optiCall: A robust genotype-calling algorithm for rare, low-frequency and common variants. *Bioinformatics*, 2012;28(12):1598–603. Available from: www.ncbi.nlm.nih.gov/pmc/articles/PMC3371828/.

[49] Goldstein JI, Crenshaw A, Carey J, et al. zCall: A rare variant caller for array-based genotyping: Genetics and population analysis. *Bioinformatics*, 2012;28(19):2543–2545. Available from: www.ncbi.nlm.nih.gov/pmc/articles/PMC3463112/.

[50] Schlattmann P. *Medical Applications of Finite Mixture Models*. Gail M, Krickeberg K, Samet J, Tsiatis A, Wong W, series eds. Springer; 2009. Available from: www.springerlink.com/index/10.1007/978-3-540-68651-4.

[51] Lemieux Perreault LP, Legault MA, Barhdadi A, et al. Comparison of genotype clustering tools with rare variants. *BMC Bioinformatics*, 2014;15(1):52. Available from: http://bmcbioinformatics.biomedcentral.com/articles/10.1186/1471-2105-15-52.

[52] Baross Á, Delaney AD, Li HI, et al. Assessment of algorithms for high throughput detection of genomic copy number variation in oligonucleotide microarray data. *BMC Bioinformatics*, 2007;8(1):368. Available from: www.biomedcentral.com/1471-2105/8/368.

[53] Frazer KA, Ballinger DG, Cox DR, et al. A second generation human haplotype map of over 3.1 million SNPs. *Nature*, 2007;449(7164):851–861. Available from: http://dx.doi.org/10.1038/nature06258.

[54] Sanger F, Nicklen S, Coulson AR. DNA sequencing with chain-terminating inhibitors. *Proceedings of the National Academy of Sciences of the United States of America*, 1977;74(12):5463–5467. Available from: www.ncbi.nlm.nih.gov/pmc/articles/PMC431765/.

[55] Mardis ER. A decade's perspective on DNA sequencing technology. *Nature*, 2011;470(7333):198–203. Available from: http://dx.doi.org/10.1038/nature09796.

[56] Metzker ML. Sequencing technologies: The next generation. *Nature Reviews Genetics*, 2010;11(1):31–46. Available from: http://dx.doi.org/10.1038/nrg2626.

[57] Mardis ER. DNA sequencing technologies: 2006–2016. *Nature Protocols*, 2017;12(2):213–218. Available from: www.nature.com/doifinder/10.1038/nprot.2016.182.

[58] Mardis ER. Next-generation sequencing platforms. *Annual Review of Analytical Chemistry*, 2013;6:287–303. Available from: www.annualreviews.org/doi/abs/10.1146/annurev-anchem-062012-092628.

[59] Buermans HPJ, den Dunnen JT. Next generation sequencing technology: Advances and applications. *Biochimica et Biophysica Acta*,

2014;1842(10):1932–1941. Available from: www.sciencedirect.com/science/article/pii/S092544391400180X.

[60] Li H, Homer N. A survey of sequence alignment algorithms for next-generation sequencing. *Briefings in Bioinformatics*, 2010;11(5):473–483. Available from: http://bib.oxfordjournals.org/content/11/5/473.abstract.

[61] Mielczarek M, Szyda J. Review of alignment and SNP calling algorithms for next-generation sequencing data. *Journal of Applied Genetics*, 2016;57(1): 71–79. Available from: http://link.springer.com/article/10.1007/s13353-015-0292-7.

[62] Smith TF, Waterman MS. Identification of common molecular subsequences. *Journal of Molecular Biology*, 1981;147(1):195–197. Available from: www.sciencedirect.com/science/article/pii/0022283681900875.

[63] Needleman SB, Wunsch CD. A general method applicable to the search for similarities in the amino acid sequence of two proteins. *Journal of Molecular Biology*, 1970;48(3):443–453. Available from: www.sciencedirect.com/science/article/pii/0022283670900574.

[64] Li H, Ruan J, Durbin R. Mapping short DNA sequencing reads and calling variants using mapping quality scores. *Genome Research*, 2008;18(11):1851–1858. Available from: www.ncbi.nlm.nih.gov/pmc/articles/PMC2577856/.

[65] Liu CM, Wong T, Wu E, et al. SOAP3: Ultra-fast GPU-based parallel alignment tool for short reads. *Bioinformatics*, 2012 mar;28(6):878–879. Available from: http://bioinformatics.oxfordjournals.org/content/28/6/878.long.

[66] Koboldt DC, Zhang Q, Larson DE, et al. VarScan 2 : Somatic mutation and copy number alteration discovery in cancer by exome sequencing. *Genome Research*, 2012;22(3):568–576.

[67] Martin ER, Kinnamon DD, Schmidt MA, Powell EH, Zuchner S, Morris RW. SeqEM: An adaptive genotype-calling approach for next-generation sequencing studies. *Bioinformatics*, 2010;26(22):2803–2810. Available from: https://academic.oup.com/bioinformatics/article/26/22/2803/227284.

[68] Shen Y, Wan Z, Coarfa C, et al. A SNP discovery method to assess variant allele probability from next-generation resequencing data. *Genome Research*, 2010;20(2):273–280.

[69] McKenna A, Hanna M, Banks E, et al. The Genome Analysis Toolkit: A MapReduce framework for analyzing next-generation DNA sequencing data. *Genome Research*, 2010;20(9):1297–1303. Available from: http://genome.cshlp.org/cgi/doi/10.1101/gr.107524.110.

[70] Li H, Handsaker B, Wysoker A, et al. The Sequence Alignment/Map format and SAMtools. *Bioinformatics*, 2009;25(16):2078–2079. Available from: http://bioinformatics.oxfordjournals.org/cgi/content/full/25/16/2078.

[71] Hu J, Li T, Xiu Z, Zhang H. MAFsnp: A multi-sample accurate and flexible SNP caller using next-generation sequencing data. *PLoS One*, 2015;10(8):e0135332. Available from: http://dx.plos.org/10.1371/journal.pone.0135332.

[72] Homer N, Merriman B, Nelson SF. BFAST: An alignment tool for large scale genome resequencing. *PLoS One*, 2009;4(11):e7767. Available from: http://journals.plos.org/plosone/article?id=10.1371/journal.pone.0007767.

[73] Li R, Yu C, Li Y, et al. SOAP2: An improved ultrafast tool for short read alignment. *Bioinformatics*, 2009;25(15):1966–1967. Available from: http://bioinformatics.oxfordjournals.org/content/25/15/1966.full.pdf.

[74] Sasaki E, Sugino RP, Innan H. The Linkage Method: A novel approach for SNP detection and haplotype reconstruction from a single diploid individual using next-generation sequence data. *Molecular Biology and Evolution*, 2013;30(9): 2187–2196. Available from: http://mbe.oxfordjournals.org/cgi/doi/10.1093/molbev/mst103.

[75] Van der Auwera G. HC overview: How the HaplotypeCaller works. Broad Institute; 2014. Available from: http://gatkforums.broadinstitute.org/discussion/4148/hc-overview-how-the-haplotypecaller-works.

[76] Nielsen R, Paul JS, Albrechtsen A, Song YS. Genotype and SNP calling from next-generation sequencing data. *Nature Reviews Genetics*, 2011;12(6): 443–451. Available from: www.ncbi.nlm.nih.gov/pmc/articles/PMC3593722/.

[77] Altmann A, Weber P, Bader D, et al. A beginners guide to SNP calling from high-throughput DNA-sequencing data. *Human Genetics*, 2012;131(10):1541–1554. Available from: www.ncbi.nlm.nih.gov/pubmed/22886560.

[78] Compeau PEC, Pevzner PA, Tesler G. How to apply de Bruijn graphs to genome assembly. *Nature Biotechnology*, 2011;29(11):987–991. Available from: http://dx.doi.org/10.1038/nbt.2023.

[79] Belkadi A, Bolze A, Itan Y, et al. Whole-genome sequencing is more powerful than whole-exome sequencing for detecting exome variants. *Proceedings of the National Academy of Sciences of the United States of America*, 2015;112(17):5473–8. Available from: www.pnas.org/content/112/17/5473.abstract.

[80] Li Y, Sidore C, Kang HM, Boehnke M, Abecasis GR. Low-coverage sequencing: implications for design of complex trait association studies. *Genome Research*, 2011;21(6):940–951. Available from: www.ncbi.nlm.nih.gov/pmc/articles/PMC3106327/.

[81] Li H. A statistical framework for SNP calling, mutation discovery, association mapping and population genetical parameter estimation from sequencing data. *Bioinformatics*, 2011;27(21):2987–2993. Available from: http://bioinformatics.oxfordjournals.org/cgi/doi/10.1093/bioinformatics/btr509.

[82] The 1000 Genomes Project Consortium. A map of human genome variation from population-scale sequencing. *Nature*, 2010;467(7319):1061–1073. Available from: http://dx.doi.org/10.1038/nature09534.

[83] Li H, Durbin R. Fast and accurate short read alignment with Burrows–Wheeler transform. *Bioinformatics*, 2009;25(14):1754–1760. Available from: http://bioinformatics.oxfordjournals.org/content/25/14/1754.long.

[84] Wang Y, Lu J, Yu J, Gibbs RA, Yu F. An integrative variant analysis pipeline for accurate genotype/haplotype inference in population NGS data. *Genome Research*, 2013;23(5):833–842. Available from: http://genome.cshlp.org/content/23/5/833.long.

[85] Challis D, Yu J, Evani US, et al. An integrative variant analysis suite for whole exome next-generation sequencing data. *BMC Bioinformatics*, 2012;13(1):8. Available from: http://bmcbioinformatics.biomedcentral.com/articles/10.1186/1471-2105-13-8.

[86] Telenti A, Pierce LCT, Biggs WH, et al. Deep sequencing of 10,000 human genomes. *Proceedings of the National Academy of Sciences of the United States of America*, 2016;113(42):11901–11906. Available from: www.ncbi.nlm.nih.gov/pmc/articles/PMC5081584.

[87] Jacob CG, Tan JC, Miller BA, et al. A microarray platform and novel SNP calling algorithm to evaluate *Plasmodium falciparum* field samples of low DNA quantity. *BMC Genomics*, 2014;15:719. Available from: www.ncbi.nlm.nih.gov/pmc/articles/PMC4153902/.

[88] Lewis CM, Knight J. Introduction to genetic association studies. *Cold Spring Harbor Protocols*, 2012;2012(3):297–306. Available from: http://cshprotocols.cshlp.org/content/2012/3/pdb.top068163.long.

[89] Han F, Pan W. Powerful multi-marker association tests: unifying genomic distance-based regression and logistic regression. *Genetic Epidemiology*, 2010;34(7):680–688. Available from: www.ncbi.nlm.nih.gov/pmc/articles/PMC3345567/.

[90] Fan RZ, Knapp M. Genome association studies of complex diseases by case-control designs. *American Journal of Human Genetics*, 2003;72(4):850–868. Available from: www.cell.com/ajhg/fulltext/S0002-9297(07)60608-9?code=ajhg-site.

[91] Xiong M, Zhao J, Boerwinkle E. Generalized T^2 test for genome association studies. *American Journal of Human Genetics*, 2002;70(5):1257–1268. Available from: www.ncbi.nlm.nih.gov/pmc/articles/PMC447600/.

[92] Chen L, Zhong M, Chen WV, Amos CI, Fan R. A genome-wide association scan for rheumatoid arthritis data by Hotelling's T2 tests. *BMC Proceedings*, 2009;3 Suppl 7:S6. Available from: www.ncbi.nlm.nih.gov/pmc/articles/PMC2795960/.

[93] Zaykin DV, Meng Z, Ehm MG. Contrasting linkage-disequilibrium patterns between cases and controls as a novel association-mapping method. *American Journal of Human Genetics*, 2006;78(5):737–746. Available from: www.sciencedirect.com/science/article/pii/S0002929707638099.

[94] Pan W. Asymptotic tests of association with multiple SNPs in linkage disequilibrium. *Genetic Epidemiology*, 2009;33(6):497–507. Available from: http://doi.wiley.com/10.1002/gepi.20402.

[95] Lewis CM. Genetic association studies: Design, analysis and interpretation. *Briefings in Bioinformatics*, 2002;3(2):146–153.

[96] Clarke GM, Anderson CA, Pettersson FH, et al. Basic statistical analysis in genetic case-control studies. *Nature Protocols*, 2011;6(2):121–133. Available from: www.ncbi.nlm.nih.gov/pmc/articles/PMC3154648/.

[97] Wang X, Zhang S, Sha Q. A new association test to test multiple-marker association. *Genetic Epidemiology*, 2009;33(2):164–171. Available from: www.ncbi.nlm.nih.gov/pmc/articles/PMC3572742/.

[98] Yi H. Comparison of dimension reduction-based logistic regression models forcase-control genome-wide association study: principalcomponents analysis vs. partial least squares. *Journal of Biomedical Research*, 2015;29(April):298–307. Available from: www.jbr-pub.org/ch/reader/view_abstract.aspx?file_no=JBR150404&flag=1.

[99] Sikorska K, Montazeri NM, Uitterlinden A, et al. GWAS with longitudinal phenotypes: performance of approximate procedures. *European Journal of Human Genetics*, 2015;(January):1–8. Available from: www.nature.com/doifinder/10.1038/ejhg.2015.1.

[100] Yan B, Wang S, Jia H, Liu X, Wang X. An efficient weighted tag SNP-set analytical method in genome-wide association studies. *BMC Genetics*, 2015;16(1):25. Available from: www.biomedcentral.com/1471-2156/16/25.

[101] Wang C, Kao WH, Hsiao CK. Using hamming distance as information for SNP-sets clustering and testing in disease association studies. *PLoS One*, 2015;10(8):e0135918. Available from: http://dx.plos.org/10.1371/journal.pone.0135918.

[102] Wan X, Yang C, Yang Q, et al. BOOST: A fast approach to detecting gene-gene interactions in genome-wide case-control studies. *The American Journal of Human Genetics*, 2010;87(3):325–340. Available from: http://linkinghub.elsevier.com/retrieve/pii/S0002929710003782.

[103] Leblanc M, Goldman B, Kooperberg C. Methods for SNP regression analysis in clinical studies. In Crowley J, Hoering A, eds., *Handbook of Statistics in Clinical Oncology*. 3rd edn. Chapman and Hall/CRC; 2012. p. 657.

[104] Thomas A, Abel HJ, Di Y, et al. Effect of linkage disequilibrium on the identification of functional variants. *Genetic Epidemiology*, 2011;35 Suppl 1:S115–119. Available from: www.ncbi.nlm.nih.gov/pmc/articles/PMC3248791/.

[105] Szymczak S, Biernacka JM, Cordell HJ, et al. Machine learning in genome-wide association studies. *Genetic Epidemiology*, 2009;33 Suppl 1:S51–57. Available from: www.ncbi.nlm.nih.gov/pubmed/?term=19924717.

[106] Papachristou C, Ober C, Abney M. A LASSO penalized regression approach for genome-wide association analyses using related individuals: application to the Genetic Analysis Workshop 19 simulated data. *BMC Proceedings*, 2016;10(S7):53. Available from: http://bmcproc.biomedcentral.com/articles/10.1186/s12919-016-0034-9.

[107] MacArthur J, Bowler E, Cerezo M, et al. The new NHGRI-EBI Catalog of published genome-wide association studies (GWAS Catalog). *Nucleic Acids Research*, 2017;45(D1):D896–D901. Available from: https://academic.oup.com/nar/article-lookup/doi/10.1093/nar/gkw1133.

[108] Wheeler E, Leong A, Liu CT, et al. Impact of common genetic determinants of Hemoglobin A1c on type 2 diabetes risk and diagnosis in ancestrally diverse populations: A transethnic genome-wide meta-analysis. *PLoS Medicine*, 2017;14(9):e1002383. Available from: www.pubmedcentral.nih.gov/articlerender.fcgi?artid=PMC5595282.

[109] Jun GR, Chung J, Mez J, et al. Transethnic genome-wide scan identifies novel Alzheimer's disease loci. *Alzheimer's & Dementia: The Journal of the Alzheimer's Association*, 2017 jul;13(7):727–738. Available from: www.ncbi.nlm.nih.gov/pubmed/28183528 and www.pubmedcentral.nih.gov/articlerender.fcgi?artid=PMC5496797.

[110] Roshan U, Chikkagoudar S, Wei Z, Wang K, Hakonarson H. Ranking causal variants and associated regions in genome-wide association studies by the support vector machine and random forest. *Nucleic Acids Research*, 2011;39(9):e62. Available from: http://nar.oxfordjournals.org/content/39/9/e62.short.

[111] Brænne I. Machine learning methods for genome-wide association data. PhD thesis, University of Lubeck; 2011. Available from: www.zhb.uni-luebeck.de/epubs/ediss1086.pdf.

[112] Burges CJC. A tutorial on support vector machines for pattern recognition. *Data Mining and Knowledge Discover*, 1998;2:121–167.

[113] Ban HJ, Heo JY, Oh KS, Park KJ. Identification of type 2 diabetes-associated combination of SNPs using support vector machine. *BMC Genetics*, 2010;11(1):26. Available from: http://bmcgenet.biomedcentral.com/articles/10.1186/1471-2156-11-26.

[114] Negi S, Juyal G, Senapati S, et al. A genome-wide association study reveals ARL15, a novel non-HLA susceptibility gene for rheumatoid arthritis in North Indians. *Arthritis and Rheumatism*, 2013;65(12):3026–3035. Available from: www.ncbi.nlm.nih.gov/pubmed/23918589.

[115] Wei Z, Wang K, Qu HQ, Zhang H, Bradfield J, Kim C, et al. From disease association to risk assessment: an optimistic view from genome-wide association studies on type 1 diabetes. *PLoS Genetics*, 2009;5(10):e1000678. Available from: http://journals.plos.org/plosgenetics/article?id=10.1371/journal.pgen.1000678.

[116] Liu JZ, van Sommeren S, Huang H, et al. Association analyses identify 38 susceptibility loci for inflammatory bowel disease and highlight shared genetic risk across populations. *Nature Genetics*, 2015;47(9):979–986. Available from: www.nature.com/ng/journal/v47/n9/full/ng.3359.html?WT.ec_id=NG-201509&spMailingID=49416217&spUserID=MzcwMzk3NDU5ODMS1&spJobID=744114307&spReportId=NzQ0MTE0MzA3S0.

[117] Chang YC, Liu PH, Yu YH, et al. Validation of type 2 diabetes risk variants identified by genome-wide association studies in Han Chinese population: A replication study and meta-analysis. *PLoS One*, 2014;9(4):e95045. Available from: www.ncbi.nlm.nih.gov/pmc/articles/PMC3988150/.

[118] Bradfield JP, Qu HQ, Wang K, et al. A genome-wide meta-analysis of six type 1 diabetes cohorts identifies multiple associated loci. *PLoS Genetics*, 2011;7(9):e1002293. Available from: www.ncbi.nlm.nih.gov/pmc/articles/PMC3183083/.

[119] Stahl EA, Raychaudhuri S, Remmers EF, et al. Genome-wide association study meta-analysis identifies seven new rheumatoid arthritis risk loci. *Nature Genetics*, 2010;42(6):508–514. Available from: www.ncbi.nlm.nih.gov/pmc/articles/PMC4243840/.

[120] Kim J, Sohn I, Kim DDH, Jung SH. SNP selection in genome-wide association studies via penalized support vector machine with MAX test. *Computational and Mathematical Methods in Medicine*, 2013;2013:340678. Available from: www.ncbi.nlm.nih.gov/pmc/articles/PMC3794570/.

[121] Chen SH, Sun J, Dimitrov L, et al. A support vector machine approach for detecting gene-gene interaction. *Genetic Epidemiology*, 2008;32(2):152–167. Available from: www.ncbi.nlm.nih.gov/pubmed/17968988.

[122] Nguyen TT, Huang J, Wu Q, Nguyen T, Li M. Genome-wide association data classification and SNPs selection using two-stage quality-based Random Forests. *BMC Genomics*, 2015;16 (Suppl 2):S5. Available from: www.biomedcentral.com/qc/1471-2164/16/S2/S5.

[123] Yang W, Charles Gu C. Random forest fishing: A novel approach to identifying organic group of risk factors in genome-wide association studies. *European Journal of Human Genetics*, 2014;22(2):254–259. Available from: http://dx.doi.org/10.1038/ejhg.2013.109.

[124] Botta V, Louppe G, Geurts P, Wehenkel L. Exploiting SNP correlations within random forest for genome-wide association studies. *PLoS One*, 2014;9(4):e93379. Available from: http://dx.plos.org/10.1371/journal.pone.0093379.

[125] Wei C, Lu Q. GWGGI: Software for genome-wide gene-gene interaction analysis. *BMC Genetics*, 2014;15(1):101. Available from: http://bmcgenet.biomedcentral.com/articles/10.1186/s12863-014-0101-z.

[126] Schwarz DF, König IR, Ziegler A. On safari to random jungle: A fast implementation of random forests for high-dimensional data. *Bioinformatics*, 2010;26(14):1752–1758. Available from: http://bioinformatics.oxfordjournals.org/content/26/14/1752.long.

[127] Jiang R, Tang W, Wu X, Fu W. A random forest approach to the detection of epistatic interactions in case-control studies. *BMC Bioinformatics*, 2009;10 Suppl 1(1):S65. Available from: http://bmcbioinformatics.biomedcentral.com/articles/10.1186/1471-2105-10-S1-S65.

[128] Zhang Y. A novel bayesian graphical model for genome-wide multi-SNP association mapping. *Genetic Epidemiology*, 2012;36(1):36–47. Available from: www.ncbi.nlm.nih.gov/pmc/articles/PMC3337957/.

[129] Jiang X, Neapolitan RE, Barmada MM, Visweswaran S. Learning genetic epistasis using Bayesian network scoring criteria. *BMC Bioinformatics*, 2011;12(1):89. Available from: www.biomedcentral.com/1471-2105/12/89.

[130] Jiang X, Neapolitan RE, Barmada MM, Visweswaran S. Learning genetic epistasis using Bayesian network scoring criteria. *BMC Bioinformatics*, 2011;12:89. Available from: www.ncbi.nlm.nih.gov/pmc/articles/PMC3080825/.

[131] Han B, Chen XW. bNEAT: A Bayesian network method for detecting epistatic interactions in genome-wide association studies. *BMC Genomics*, 2011;12 Suppl 2(2):S9. Available from: http://bmcgenomics.biomedcentral.com/articles/10.1186/1471-2164-12-S2-S9.

[132] Jiang X, Barmada MM, Visweswaran S. Identifying genetic interactions in genome-wide data using Bayesian networks. *Genetic Epidemiology*, 2010;34(6):575–581. Available from: www.ncbi.nlm.nih.gov/pmc/articles/PMC3931553/.

[133] Verzilli CJ, Stallard N, Whittaker JC. Bayesian graphical models for genomewide association studies. *American Journal of Human Genetics*, 2006;79(1):100–112. Available from: www.ncbi.nlm.nih.gov/entrez/query.fcgi?cmd=Retrieve&db=PubMed&dopt=Citation&list_uids=16773569.

[134] Zhang Y, Liu JS. Bayesian inference of epistatic interactions in case-control studies. *Nature Genetics*, 2007;39(9):1167–1173.

[135] Koo CLC, Liew MMJ, Mohamad MS, Salleh AHM. A review for detecting gene-gene interactions using machine learning methods in genetic epidemiology. *BioMed Research International*, 2013;2013:13. Available from: www.ncbi.nlm.nih.gov/pmc/articles/PMC3818807 and www.hindawi.com/journals/bmri/2013/432375/abs/.

[136] Ziegler A, DeStefano AL, König IR, et al. Data mining, neural nets, trees–problems 2 and 3 of Genetic Analysis Workshop 15. *Genetic Epidemiology*, 2007;31 Suppl 1:S51–60. Available from: www.ncbi.nlm.nih.gov/pubmed/18046765.

[137] Niel C, Sinoquet C, Dina C, Rocheleau G. A survey about methods dedicated to epistasis detection. *Frontiers in Genetics*, 2015;6:285. Available from: http://journal.frontiersin.org/article/10.3389/fgene.2015.00285/abstract.

[138] Menze BH, Kelm BM, Masuch R, et al. A comparison of random forest and its Gini importance with standard chemometric methods for the feature selection and classification of spectral data. *BMC Bioinformatics*, 2009;10(1):213. Available from: http://bmcbioinformatics.biomedcentral.com/articles/10.1186/1471-2105-10-213.

[139] Szymczak S, Biernacka JM, Cordell HJ, et al. Machine learning in genome-wide association studies. *Genetic Epidemiology*, 2009;33(Suppl. 1):51–57.

[140] Perrier E, Imoto S, Miyano S. Finding optimal Bayesian network given a super-structure. *Journal of Machine Learning*, 2008;9:2251–2286. Available from: http://jmlr.csail.mit.edu/papers/volume9/perrier08a/perrier08a.pdf.

[141] Stack CB, Gharani N, Gordon ES, Schmidlen T, Christman MF, Keller MA. Genetic risk estimation in the Coriell Personalized Medicine Collaborative. *Genetics in Medicine*, 2011;13(2):131–139. Available from: http://dx.doi.org/10.1097/GIM.0b013e318201164c.

[142] Bellcross CA, Page PZ, Meaney-Delman D. Direct-to-consumer personal genome testing and cancer risk prediction. *Cancer Journal*, 2012;18(4):293–302. Available from: www.ncbi.nlm.nih.gov/pubmed/22846729.

[143] Zhang J, Yu KF. What's the relative risk? *JAMA*, 1998;280(19):1690. Available from: http://jama.jamanetwork.com/article.aspx?articleid=188182.

[144] Bloss CS, Darst BF, Topol EJ, Schork NJ. Direct-to-consumer personalized genomic testing. *Human Molecular Genetics*, 2011;20(R2):R132–141. Available from: http://hmg.oxfordjournals.org/content/early/2011/08/09/hmg.ddr349.abstract.

[145] Aiyar L, Shuman C, Hayeems R, et al. Risk estimates for complex disorders: comparing personal genome testing and family history. *Genetics in Medicine*, 2014;16(3):231–237. Available from: http://dx.doi.org/10.1038/gim.2013.115.

[146] Swan M. Multigenic condition risk assessment in direct-to-consumer genomic services. *Genetics in Medicine*, 2010;12(5):279–288. Available from: www.ncbi.nlm.nih.gov/pubmed/20474084.

[147] Imai K, Kricka LJ, Fortina P. Concordance study of 3 direct-to-consumer genetic-testing services. *Clinical Chemistry*, 2011;57(3):518–21. Available from: www.ncbi.nlm.nih.gov/pubmed/21159896.

[148] Do CB, Tung JY, Dorfman E, et al. Web-based genome-wide association study identifies two novel loci and a substantial genetic component for Parkinson's disease. *PLoS Genetics*, 2011;7(6):e1002141. Available from: http://journals.plos.org/plosgenetics/article?id=10.1371/journal.pgen.1002141.

[149] Nalls MA, McLean CY, Rick J, et al. Diagnosis of Parkinson's disease on the basis of clinical and genetic classification: A population-based modelling study. *The Lancet Neurology*, 2015;14(10):1002–1009. Available from: http://www.thelancet.com/article/S1474442215001787/fulltext.

[150] Wilson BJ, Nicholls SG. The Human Genome Project, and recent advances in personalized genomics. *Risk Management and Healthcare Policy*, 2015;8:9–20. Available from: www.ncbi.nlm.nih.gov/articles/PMC4337712/.

[151] Hawkins AK, Ho A. Genetic counseling and the ethical issues around direct to consumer genetic testing. *Journal of Genetic Counseling*, 2012;21(3):367–373. Available from: www.ncbi.nlm.nih.gov/pubmed/22290190.

[152] American College of Medical Genetics and Genomics Board of Directors. Direct-to-consumer genetic testing: A revised position statement of the American College of Medical Genetics and Genomics. *Genetics in Medicine*. 2016;18(2):207–208. Available from: http://dx.doi.org/10.1038/gim.2015.190.

[153] Borry P, van Hellemondt RE, Sprumont D, et al. Legislation on direct-to-consumer genetic testing in seven European countries. *European*

Journal of Human Genetics, 2012;20(7):715–721. Available from: www.ncbi.nlm.nih.gov/pmc/articlesPMC3376265/.

[154] Auer PL, Lettre G. Rare variant association studies: considerations, challenges and opportunities. *Genome Medicine*, 2015;7(1):16. Available from: www.ncbi.nlm.nih.gov/pmc/articles/PMC4337325/.

[155] Kotze MJ, van Velden DP, Botha K, et al. Pathology-supported genetic testing directed at shared disease pathways for optimized health in later life. *Personalized Medicine*, 2013;10(5):497–507. Available from: www.futuremedicine.com/doi/abs/10.2217/pme.13.43.

[156] Nicodemus KK, Malley JD, Strobl C, Ziegler A. The behaviour of random forest permutation-based variable importance measures under predictor correlation. *BMC Bioinformatics*, 2010;11(1):110. Available from: http://bmcbioinformatics.biomedcentral.com/articles/10.1186/1471-2105-11-110.

[157] Dutkowski J, Kramer M, Surma MA, et al. A gene ontology inferred from molecular networks. *Nature Biotechnology*, 2013;31(1):38–45. Available from: http://dx.doi.org/10.1038/nbt.2463.

[158] Wang Y, Chen S, Deng N, Wang Y. Drug repositioning by kernel-based integration of molecular structure, molecular activity, and phenotype data. *PLoS One*, 2013;8(11):e78518. Available from: http://journals.plos.org/plosone/article?id=10.1371/journal.pone.0078518.

[159] Gligorijević V, Janjić V, Pržulj N. Integration of molecular network data reconstructs gene ontology. *Bioinformatics*, 2014;30(17):i594–600. Available from: www.ncbi.nlm.nih.gov/pmc/articles/PMC4230235/.

[160] Gligorijević V, Pržulj N. Methods for biological data integration: Perspectives and challenges. *Journal of The Royal Society Interface*, 2015;12(112):20150571. Available from: http://rsif.royalsocietypublishing.org/lookup/doi/10.1098/rsif.2015.0571.

[161] Talmud PJ, Hingorani AD, Cooper JA, et al. Utility of genetic and non-genetic risk factors in prediction of type 2 diabetes: Whitehall II prospective cohort study. *BMJ (Clinical research ed)*, 2010;340(jan14):b4838. Available from: www.bmj.com/content/340/bmj.b4838.

[162] Wei WH, Hemani G, Haley CS. Detecting epistasis in human complex traits. *Nature Reviews Genetics*, 2014;15(11):722–733. Available from: http://dx.doi.org/10.1038/nrg3747.

[163] Cordell HJ. Detecting gene-gene interactions that underlie human diseases. *Nature Reviews Genetics*, 2009;10(6):392–404. Available from: http://dx.doi.org/10.1038/nrg2579.

[164] Phillips PC. Epistasis: The essential role of gene interactions in the structure and evolution of genetic systems. *Nature Reviews Genetics*, 2008;9(11):855–67. Available from: www.ncbi.nlm.nih.gov/pmc/articles/PMC2689140/.

2 Epigenetic Data and Disease

Rodrigo González-Barrios, Marisol Salgado-Albarrán, Nicolás Alcaraz, Cristian Arriaga-Canon, Lissania Guerra-Calderas, Laura Contreras-Espinosa, and Ernesto Soto-Reyes

2.1 Background

Genetic material carries the information for every process necessary for life. Environmental exposure to mutagens can alter genetic information throughout lifetime and together with genetic predisposition could generate diseases in an organism. However, genetic information is not the only mechanism underlying the transgenerational transmission and environment influence on human variation and disease. The relatively new field of epigenetics has given a new perspective to the origin of diseases. The term "epigenetics" was first defined by Waddington in 1939 as "The random interactions between genes and their products, which result in a phenotype" [1]. Nowadays, epigenetics is mostly defined as heritable changes that regulate gene expression and chromosome architecture independent of any DNA sequence. In recent years, it has been emphasized that epigenetic components dictate and coordinate gene expression. The human body has more than 100 distinct cell types, which have essentially the same genome but contain a unique epigenome that serves to instruct specific gene expression programs present within each cell type. Epigenetic modifications are highly dynamic and can be altered throughout aging and environmental exposure. There are many types of epigenetic modifications, which include DNA methylation and posttranslational histone modifications such as methylation, acetylation, phosphorylation, ubiquitylation, and sumolyation. Also, non-coding RNAs play a key role in the regulation of epigenetic processes. Altogether, epigenetic mechanisms are essential to many cellular functions, and when dysregulated, major adverse health and behavioral effects occur.

Epigenetic aberrations could explain the origin and prevalence of certain diseases. The most studied relationship between epigenetics and disease has been cancer [2], but aberrant epigenetic patterns go beyond oncology, stretching to a variety of biomedical fields including imprinting, metabolism, neurology, immunology, development, cardiovascular disease, etc. [2, 3, 4, 5, 6]. Researchers worldwide understand the importance of epigenetic mechanisms in disease biology, leading to the origin of many techniques to study different levels of epigenetic information, regarding DNA methylation, histone post-translational modification, chromatin associated proteins, chromatin remodelers, and non-coding RNAs. With the introduction of whole genome sequencing and epigenomics, the knowledge of epigenetic mechanisms and its implications in cellular processes rapidly increased and became one of the most flourishing areas in biology and medicine, leading to great scale international efforts to characterize the epigenome from different samples and diseases. However, this field is still in its infancy and further research is still needed to understand the epigenetic machinery and its biological function. In this chapter, we will review different mechanisms from a bioinformatics perspective, experimental approaches and computational methods to study epigenetic data (see Box 2.1 for glossary of terms).

2.2 DNA Methylation and its Role in Genome Regulation

Among the most widely studied epigenetic mechanisms is DNA methylation, a covalent modification that mainly occurs at the fifth carbon position of the cytosine, thus forming the 5-methylcytosine (5mC). The molecular machinery related to 5mC can be divided into three components that can establish, read, and remove this mark: (1) The three active DNA methyltransferases (DNMT1, DNMT3a and 3b) which catalyze and maintain cytosine methylation patterns; (2) the methyl binding proteins (MBPs), that read this mark and carry out different effector mechanisms like gene silencing; and (3) the DNA demethylases (TET1, 2 and 3) that actively "erase" 5mC through a series of oxidations and glycosylations (see Table 2.1). In normal cells, genome wide studies demonstrated that 70 to 80% of the CpG sites are methylated, predominantly in repetitive genomic regions, including satellite repeats, viral sequences, LINEs, and SINEs (long and short interspersed transposable elements) and gene bodies [7, 8, 9, 10].

Paradoxically, the regions of the genome that contains most of CpG dinucleotides are usually not methylated. These regions are known as "CpG islands" (CGI) and were first described in 1987 by Gardiner and Frommer. They are short genomic regions (around 200–1000bp) that are defined by an increase in cytosine and guanine greater than 50%. CGI are particularly associated with gene promoters and regulatory regions, where approximately 70% of annotated gene promoters are associated with CGI, making this the most common promoter type element in the vertebrate genome [11, 12]. Virtually all housekeeping genes, as well as a proportion of tissue specific genes and developmental regulator genes are associated with CGI [13, 14]. There is evidence that a large class of CGIs are located remotely from annotated transcription start sites (TSS), moreover, some show evidence for promoter function, although an increasing number of exceptions are being identified suggesting other functions related to this CGIs [15].

Methylation at CGI in promoters is associated with transcriptional repression by interfering with transcription factor binding; nevertheless, repression seems to occur largely indirectly, via recruitment of MBPs that induce chromatin remodeling. However, the strength of repression mediated by 5mC depends on the local concentration of CpGs within the promoter. Through DNA methylation mapping by massive

> **Box 2.1: Glossary**
>
> **Chromatin conformation capture (5C, HiC):** A technique used to profile all chromatin interactions in specific regions of the genome by the hybridization of a mixture of DNA primers to chromosome conformation capture (3C) templates followed by high-throughput sequencing.
> **Cis-acting:** Regions of non-coding DNA, which regulate transcription within the same chromosome.
> **CpG Island:** Genomic regions with a minimum of 200 bp, with a G+C content greater than 50% and observed/expected CpG ratio above 60%.
> **Enhancer:** A cis-acting regulatory sequence that markedly increases expression of a neighboring gene. Enhancers are typically capable of operating over considerable distances (sometimes ~50 kb) upstream or downstream of the gene and in either orientations.
> **Epigenomics:** Is the systematic analysis of the global state of gene expression modulated by epigenetic processes such as DNA methylation, posttranslational modifications of histones non-coding RNA and the organization of chromatin inside the nucleus.
> **Euchromatin:** Less densely packed or open chromatin that is often associated with active transcription.
> **Global hypomethylation:** Loss of DNA methylation across the genome that commonly occurs in cancer cells.
> **Heterochromatin:** Tightly packed form of chromatin that lack high number of genes and is commonly constituted by repetitive sequences in the genome, which is associated with inactive transcription and serves as a structural element of the chromosome.
> **Hi-C contact matrix:** Matrix which displays all chromatin interactions found within a genomic range. First, the genome is partitioned into bins of fixed size. Then, a contact matrix is generated, where every cell corresponds to the frequency of contacts between the associated pair of loci. The frequency or number of interactions is then converted into color signal.
> **Local hypermethylation:** Gain of methylation that occurs at specific regulatory regions that alters the normal state of transcription in diseases like cancer.

paralleled sequencing it was reported that there are three classes of promoters based on CpG ratio, CG content, and length of the CpG rich region: High CpG promoters, intermediate CpG promoters and low CpG promoters (HCP, ICP, and LCP respectively) [16]. HCP and most ICP remain largely hypomethylated and transcriptionally active, these types of promoters are associated with housekeeping and some tissue specific genes [16]. In contrast, CGI with LCP are predominantly methylated, and such hypermethylation does not affects gene expression, which indicates that repression

Table 2.1: DNA modifications, function, and writer enzymes

DNA Modification	Name	Function
5mC	5-methylcitosyne	Repression, alternative expression
5hmC	5-hydroxymethylcytosine	Transcription regulation
5fmC	5-formylcytosine	Unknown
5cmC	5-carboxylcytosine	Unknown
Human enzymes	**Gene name**	**Function**
DNMT1	DNA methyltransferase 1	DNA methylation maintenance
DNMT2	DNA methyltransferase 2	Low DNA activity, methylates tRNA
DNMT3A	DNA methyltransferase 3A	De novo methylation
DNMT3B	DNA methyltransferase 3B	De novo methylation
DNMT3L	DNA methyltransferase 3L	Cofactor of De novo methyltransferases activity
TET1	Ten-eleven-translocation protein 1	DNA oxydase/demethylation
TET2	Ten-eleven-translocation protein 2	DNA oxydase/demethylation
TET3	Ten-eleven-translocation protein 3	DNA oxydase/demethylation

by DNA methylation requires high 5mC density, and such density of 5mC is also necessary for the repressive function of MBPs.

DNA methylation is a major epigenetic mechanism that determines cellular outcome and response to environmental signals, by dynamically regulating gene expression and chromatin formation. Due to its importance in development and cell differentiation, specific patterns of 5mC are established among cell types. This pattern of methylation is not static and can be altered in response to environmental stressors. Specific cell patterns are maintained by molecular mechanisms that keep "epigenetic memory" and maintain the cellular identity across its lifetime. However, altered DNA methylation is frequently observed in diseases, like cancer, compared with corresponding healthy cells.

Comprehensively studying the profiles of different healthy individuals and tissue types enables us to estimate the variance of a particular CpG site or of regions such as promoters. Reference data sets are now being created in consortia such as Blueprint, the International Human Epigenome Consortium (IHEC) and Roadmap using high-resolution technologies, which will be discussed later. Focusing on normal tissue types, these joint efforts aim freely to release reference datasets of integrated epigenomic profiles of stem cells, as well as developmental and somatic tissue types to the research community. Concordantly, filtering for *loci* that are unstable in DNA methylation between individuals excludes unsuitable CpG sites before biomarker candidate selection approaches. Systematic screening of reference data sets obtained from different individuals will enable us to identify and exclude variable CpG sites and regions, facilitating future biomarker selections.

2.2.1 DNA Demethylation and its Role in Genomic Profiles

Until recently, it was believed that DNA methylation was an irreversible epigenetic mark because it is a covalent modification. Now, it is known that DNA methylation is a dynamic process because different processes can remove it. DNA demethylation can occur passively by the reduction or absence in the enzymatic activity of the DNMTs, or actively, mediated by enzymes with the ability to remove the methyl group from the cytosines [17].

In 2009, it was discovered the first enzyme capable of erasing the 5mC, by an oxidation process [18]. These enzymes belong to the family of dioxigenases known as TET (ten-eleven-translocation proteins), composed of three members (TET1, TET2, and TET3) [19]. TET enzymes have the ability to convert de 5mC to 5-hydroxymethylcytosine (5hmC), mediating an active DNA demethylation process. Also, the 5hmC can be further catalyzed to 5-formylcytosine (5fC), presenting the highest level of oxidation with 5-carboxylcytosine (5caC). 5caC can be efficiently removed by thymine DNA glycosylase (TDG)-mediated bases excision repair.

The distribution of 5hmC depends on multiple factors such as cell type, cell differentiation and response to the environment. These questions have been addressed from a genome wide analysis perspective. An example of the importance of these enzymes in diseases is TET2. TET2 is one of the most frequently mutated genes in hematopoietic malignancies as an early event in cancer [20].

2.2.2 Different Experimental Strategies for the DNA Methylation Analysis

DNA methylation has been actively studied for the past four decades, using different approaches according to the available technologies (Figure 2.1). Initially global 5mC levels were studied with different approaches like: High-performance liquid chromatography (HPLC) or luminometric methylation assay (LUMA) [21, 22]. However, these methods only screen general levels of 5mC regardless of its genomic patterns. Other methods use methyl-sensitive restriction endonucleases; nevertheless, these analyses were restricted to the sequence of the enzymes and show digestion biases. Bisulfite conversion promptly became a gold standard technique. With this method, methylation research acquired the possibility to do aimed studies for specific sites by methyl specific polymerase chain reaction (MS-PCR), or to search for methylation patterns from specific regions to whole genome mapping by combining sequencing to bisulfite converted sequences (see Section 2.3).

Alternatively, region specific or global 5mC status can be assessed by inmunoprecipitation of methylated sequences (MeDIP). This method is an adaptation of the chromatin immunoprecipitation (ChIP) protocol for DNA and uses an antibody against 5-methylcytosine. However, some biases come from MeDIP method, because it cannot establish CpG methylation patterns due to the resolution based on the size of immunoprecipitated DNA fragments (~200 to 500 bp), and the amount of enrichment of methylated DNA depends on the abundance of CpGs in a given sequence and the capacity of the antibody to detect such levels.

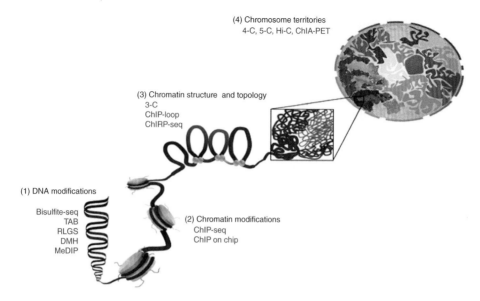

Figure 2.1: Overview of epigenetic modifications and different methods of analysis (1) Analysis of DNA covalent modifications, that includes and restriction analysis (RLGS, etc.), bisulfite sequencing and modifications for 5hmC analysis (TAB seq), and methylation immunoprecipitation (MeDIP) approach oxidation levels. (2) Chromatin modifications, which include histone post-translational modifications and other chromatin binding complexes. (3) Chromatin structure and topology, massive profile analysis of one vs one interactions of proteins associated to chromatin or ncRNAs that show 3D architecture of the chromatin regions and determine its transcriptional or structural function, methods commonly used for this are 3C, ChIP-loop and ChIP-seq. (4) Chromosome territories.

Also, the introduction of microarrays technology for methylation also became of importance to quickly establish epigenetic patterns for specific regions, like CpG islands located in promoter regions of selected genes. By combining this technology with epigenetic methods like sodium bisulfite, this method has helped the discovery of different roles of DNA methylation in the genome and its translation to applied approaches in biomedical research.

From the discovery of DNA structure to the publication of the Human Genome Project, massive genome sequencing approaches have advanced in a very important way. This was possible thanks to the discovery of next generation sequencing (NGS) methods. This technology is significantly cheaper, faster, more accurate and more reliable than the ones used in the past, thus allowing the development of clinical approaches (see Box 2.2) These approaches have also allowed the development of interdisciplinary projects such as The International Genome Sample Resource (IGSR; www.internationalgenome.org). This consortium represents the largest open collection of human variation data [23]. Most of current epigenomic data come from NGS, however, some data is still obtained by microarrays (ChIP on chip) design for specific regions, like gene regulators or CGI (Figure 2.2). Taken together, the application of different approaches to study DNA methylation has provided tools for researchers to understand its implication in many biological functions at a

> **Box 2.2: Scope and limitations of genomic experimental strategies**
>
> **DNA microarrays:** Developed in the 1980s, this genomic strategy is based on the hybridization of fluorophore labeled DNA to a solid surface. The number of DNA molecules bound to the surface can be determined by the intensity of the signal emitted by the fluorophore. The limitations of this technique are the number of probes printed in the array.
>
> **NanoString:** Similar to microarrays or quantitative PCR, this technology is based on fluorophore-labeled probes that target a gene of interest. It provides high resolution, less than one copy per cell, and fidelity. The limitation of this assay is the low number of gene targets (around 800, which are below microarrays).
>
> **Short-read NGS:** This methodology relies upon sequencing by ligation (SBL) or sequencing by synthesis (SBS). The SBL approach is based on a fluorophore-labeled probe that is hybridized to a DNA sample and ligated to an adjacent oligonucleotide, which together are used for image capture. The SBS identifies nucleotide addition by employing a polymerase and a signal such as a fluorophore or a change in ionic concentration. These two experimental strategies are based on solid surface DNA amplification.
>
> **Sequencing by synthesis (CRT):** CRT assay employs a terminator molecule similar to those used in Sanger sequencing, where the ribose 3′-OH group is blocked, preventing elongation. The fluorophore and the blocking group can be removed before starting a new cycle.
>
> **Sequencing by synthesis (SNA, 454 Ion Torrent):** This technology is based on the incorporation of a dNTP, which works as a single signal, into an elongating strand. In this technique, it is not necessary to employ blocking dNTPs. The 454 pyrosequencing was the first NGS technology.
>
> **Single-molecule long-read sequencing (PacBio and ONT):** One of the most used platforms for long-read sequencing is the single-molecule real-time (SMRT) sequencing method developed by Pacific Biosciences (PacBio). This equipment employs a flow cell containing thousands of individual wells where the polymerases travel along the DNA template.

genome-wide level and obtain new insights on the mechanisms underlying genome regulation, and the implication of changes in methylation patterns in diseases.

2.2.3 Processing and Analysis Methods and Tools for DNA Methylation Data from Bisulfite Based Assays

2.2.3.1 Bisulfite Conversion

Bisulfite conversion is one of the most accepted methods to determine the methylation state in a genomic region, using different experimental strategies such as bisulfite

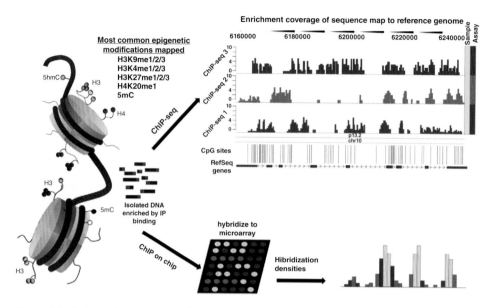

Figure 2.2: Epigenetic mapping performed by NGS and microarrays. Enrichment coverage example taken from WashU EpiGenome Browser viewer.

sequencing (BS-seq) and methylation sensitive PCR (MS-PCR). Moreover, bisulfite conversion is unable to differentiate 5mC from 5hmC, because both protect cytosine from its conversion to uracil [24]. New methods have been developed to distinguish and profile 5mC and 5hmC, using specific antibodies (MeDIP) or TET assisted bisulfite sequencing (TAB-seq).

In general, bisulfite conversion method generates a chemical conversion of unmethylated cytosines (C) to thymines (T). Therefore, the analysis aims for counting and mapping the number of C to T conversions and quantifying the proportion of methylation per base. We can carry this out by identifying C to T conversions in the aligned reads and dividing the number of Cs by the sum of Ts and Cs for each cytosine in the genome. However, bisulfite treatment has a 95% conversion efficiency rate, and base-calling quality is not constant and could change within the same read and between sequencing runs. Hence, bisulfite libraries are susceptible to errors and biases that could generate miscalled bases that could be counted as C–T conversions erroneously. To avoid this, attention must be made to initial quality control, trimming, and suitable alignment of bisulfite libraries.

Performing quality control of the data is important to avoid mis-mapping events and incorrect methylation calls. Therefore, it is important to check the base quality, which represents the level of confidence in the base calls, in order to avoid miscalled bases that can be counted as C–T conversion erroneously. Such basic checks can be performed via fastQC software (www.bioinformatics.babraham.ac.uk/projects/fastqc/). In addition, evaluating, and reducing potential sequence contamination or adapter contamination from the libraries is a common first approach. Base composition and CG-content plots will be useful for this purpose. It is common for BS-seq experiments performed in mammals, to observe an average of cytosine content around 1–2% of the

Table 2.2: Alignment methods for bisulfite based experiments

Aligner	Webpage	Reference
MethylCoder	https://github.com/brentp/methylcode.	[26]
BS-seeker2	http://pellegrini.mcdb.ucla.edu/BS_Seeker2/	[27]
Bison	http://sourceforge.net/projects/dna-bison/	[28]
BSMAP	https://sourceforge.net/projects/bsmapper/	[29]
VALiBS	https://github.com/wwwyxder/valibs	[30]

complete sequence length. Also, the CG content observed in BS-seq libraries peaks around 30%. This will certainly change among different tissues and cell types or species with different methylation rates. However, the rate of C and CG content helps to identify adapter contamination by observing spikes increase the occurrence of C at the later cycles, or an increase of the CG profile to more than 40%. Such errors can be fixed by trimming the sequence file. This can be achieved using trimming programs such as Trim Galore (www.bioinformatics.babraham.ac.uk/projects/trim_galore/).

After pre-alignment quality control and processing is done, the next step is the alignment. For this purpose, the BS-seq methods rely on modification of know short read alignment methods, and the conversion of the genome that is studied to and *in silico* bisulfite converted sequence. For example Bismark employs Bowtie (or Bowtie 2) and *in silico* C–T conversion of reads and genomes [25], some methods that use this strategy are cited in Table 2.2. After completing the alignment and methylation calling, there is no need for further quality control, and analysis of methylation proportions and differential methylation regions (DMRs) analysis can start. We will discuss differential methylation analysis later in the chapter.

2.2.3.2 *Methylation Microarrays*

As an alternative to the bisulfite-based mass sequencing, microarrays emerge as another promising method for the understanding of DNA methylation in cell biology and diseases. This experimental approach has decreased in costs and increased in coverage. The simplicity of the generation of specific panels for the analysis of discrete regions of the genome is also an advantage (for further discussion of microarray methods and applications see Chapter 1). There are different types of array-based assessments of methylation, but here we will focus on Illumina's array assays from Illumina 27K and 450K platforms (see Box 2.3).

Illumina's microarray technology has been adapted for the study of DNA methylation from sequences converted by bisulfite. Where the types of beads are quantified as the average signal representing methylated and non-methylated alleles, which are computed as β-values. The β-value is defined as the ratio of the methylated probe intensity to the overall intensity that summarized methylated and unmethylated probe intensities. Following the notation used by Illumina methylation assay [12], the β-value for an interrogated CpG site is defined as:

$$\beta = \frac{\text{Max}(M,0)}{\text{Max}(M,0) + (U,0) + 100}. \tag{2.1}$$

> **Box 2.3: Microarrays platforms and different probe types**
>
> Illumina methylation 27K platform contains 27,578 CpG loci that targets 14,000 genes. It covers around 2 CpG in regions within 1Kb upstream to 0.5 kb downstream of genes TSS. Biologically is a rather weak coverage of methylation of promoter regions. Although this platform was phased out in 2010, many datasets are held in public databases like Gene Expression Omnibus (GEO) that can be used for analysis and data mining. Human methylation 450K platform contains approximately 485,577 CPGs sites covering 99% RefSeq genes and 96% of CpG islands outside coding regions. These assays are configured with conjugated oligonucleotides in beads to measure specific target sequences. The Illumina 450 chip has two probe designs Infinium I and II. Infinium I targets each CpG with two beads labeled by the same dye. These probes contain from 0 to 10 CpG sites, designed either methylated or unmethylated to match the bisulfite converted sequence at the target sites. The type II probes employ one bead type labeled with two different dye colors for unmethylated (red) and methylated (green) CpGs. The Ilumina 27K only has Infinium I probes, while the Ilumina 450K platform contains both types of probes.

Here U and M represent the intensities measured in a specific region by the unmethylated and methylated probes, respectively. Any negative value will be reset to zero, in order to avoid negative values after background adjustment. A constant offset (100) is added to the denominator to regularize the β-value when both U and M probe intensities are low. A β-value statistic results in a number from 0 to 1, or 1 to 100%. Where a β-value of zero represents a site in which all copies were completely unmethylated, and a β-value of 1 indicates a CpG site that is fully methylated in all copies. Illumina has developed their Genome-Studio software [31], for basic data analysis and other purposes.

Quality control is essential to evaluate the data quality of the samples and to avoid miscalled and false positive errors. For this reason, Illumina's array includes several control probes to determine the data quality. In order to analyze the quality and to detect samples that behave poorly in the arrays, diagnostic plots of the control probes are generated using different tools, such as the Genome-Studio software [31], or the HumMethQCReport R-package [32]. Also, other packages available in R and Bioconductor, which are based on the use of P values can be employed for quality control measures; such as IMA, the pre-processing and analysis pipeline, Minfi [31, 33, 34, 35], available at http://bioconductor.org/packages/2.12/bioc/vignettes/minfi/inst/doc/minfi.pdf and www.bioconductor.org/packages/2.12/bioc/html/methylumi.html.

Note: Bisulfite conversion may present incomplete conversion, where some unmethylated Cs are not converted to Ts. Causing false positive results. For mammals and other species non-CpG methylation are almost not present in somatic tissues, hence, we can calculate the conversion rate by using the percentage of non-CpG methylation. The closer the conversion rate is to 100%, the higher the quality of the experiment. Typical values for a good quality of bisulfite conversion will be higher than 99.5%. This will not apply in organisms such as plants, which commonly present non-CpG methylation [36, 37].

Microarray Normalization and Batch Effect Correction The purpose of normalization is to remove sources of experimental artefacts, random noise, and systematic technical biases introduced during the assay; in addition to biases between samples of different batches [38]. There are two different types of normalization: Between array and within array normalization. The first type removes technical artefacts between samples on different arrays; the second type corrects intensity related dye biases [39]. There are many normalization and data processing algorithms available for 450K arrays (Table 2.3). The peak adjustment is initially proposed to correct probe I and II bias. Therefore, many researchers use the *M*-value, which uses methylation *M*-value and rescales design II probe data to the peak positions of design I probes:

$$M = \log_2 \frac{\text{Max}(M,0)}{\text{Max}(U,0)}. \tag{2.2}$$

It is a normalization method within a sample, which is implemented in R package IMA. A normalized *M*-value near zero signifies a semimethylated locus, a positive *M*-value indicates mere CpGs are methylated than unmethylated, whereas a negative *M*-value represents the opposite interpretation. However, because DNA methylation is not distributed in a balanced manner in the genome, a bias is generated in the distribution of the log-ratio of methylation. This bias is dependent on the levels of methylation found in a particular tissue. Such imbalance is created by the non-random distribution of CpG sites throughout the genome. For example CGI, which are the

Table 2.3: 450k array data processing algorithms/methods

Method/Package	Utility	Reference
Genome Studio	Raw data processing, background control.	[31]
Subset quantile normalization	Probe I and II bias correction, color bias adjustment across sample normalization.	[33]
Subset quantile within normalization	Probe I and II bias correction, color bias adjustment.	[40]
Methylumi/lumi	Raw data processing, background control, color bias adjustment across sample normalization.	[41]
Beta mixture quantile dilation	Probe I and II bias correction.	[42]
Ilumina methylation analyser	Probe I and II bias correction, across sample normalization.	[34]

densest CpG regions are normally not methylated, whereas the opposite behavior is observed in the less dense CpG regions in human cells. It is also important to note that this is common behavior of normal cells. Likewise, this behavior in DNA methylation is reversed in diseases such as cancer. Because of this and other reasons, there is still a lack of consensus regarding which is the optimal approach for normalization of methylation data.

Differential Methylation Region Analysis Once processing of the data has been performed, the most common goal of DNA methylation profiling is to find differentially methylated CpGs (DMC) between two groups of samples. For this purpose, the average β-values from microarray and methylation ratio from BS-seq data are equivalent, thus the same statistic can be applied. An advantage of methylation sequencing against microarrays is the direct count of methylated and unmethylated cytosines, which is much more accurate than the signal intensity of the probes. This is because DNA methylation occurs in discrete regions of the genome, called differentially methylated regions (DMR), where such DMR clusters have associated specific functions in the regulation of the genome. The search for DMR helps us to understand the biological significance of the DNA methylation. For this reason, several algorithms that identify DMR have been developed for BS-seq and microarrays data. In summary, these methods need two steps: First they identify DMCs through Fisher or chi square test. Afterwards, these methods put together the CpGs with the similar statistics into a DMR according to the defined threshold of distance and DMC statistics. Some algorithms such as BSmooth conduct a smoothing step before DMC detection, which smooth out outlier CpGs and utilize CpGs with a low coverage. Some of the methods employed for DMC and DMR analysis are summarized in Table 2.4.

The study of DNA methylation profiles is a growing field in the understanding of diseases, whose implications is both in the understanding of genomic and cell regulation, as well as in the development of diseases. The methylation profiles serve to recognize specific characteristics of the disease as well as possible new treatment or biomarkers that can contribute to modern medicine. Many statistical methods can be used for DMC or DMR detection, however, they have performed very differently. More tests need to be developed and deployed for clinical use and better understanding of DNA methylation in a genome scale.

Table 2.4: Examples of DMC and DMR detection methods

Method/Package	Usage	Algorithm	Reference
Methylsig	DMC/DMR	Beta binominal	[43]
MOABS	DMC/DMR	Beta-binominal hierarchical model	[44]
Radmeth	DMC/DMR	Beta-binominal	[45]
methylKit	DMR	Logistic regression	[46]
Bump hunting	DMR	Linear regression	[47]
BSmooth	DMR	Smooth t test	[48]
Biseq	DMR	Smooth/beta regression	[49]

2.3 The Post-Translational Modifications of Histones

DNA is packaged in a complex with proteins and RNA known as chromatin. The fundamental unit of this complex is the nucleosome, that is composed of 165 base pairs (bp) rolled around an octamer of histones (H2A, H2B, H3, and H4, a couple from each) [50, 51]. Chromatin can be arranged in a more compact structure, that has a diameter of 30 nm, which is known as solenoid. This structure is stabilized by the union of the histone H1. At this level the formation of heterochromatin alters the expression of multiple genes [52].

The modifications in the chromatin architecture are regulated by different epigenetic mechanisms, including DNA methylation, non-coding RNA, ATP-dependent complexes, histone variants, and histone post-translational modifications. This latter happens mainly in the histone amino-terminal region and has distinct effects on chromatin state depending on the modification and the amino acid modified. They can induce relaxation and/or compaction of the chromatin, affecting the accessibility to the DNA sequence. Among the biochemical changes in histones are: ADP-ribosylation, lysine and arginine methylation, serine and threonine phosphorylation, ubiquitination, SUMOylation and lysine acetylation [50, 53, 54, 55]. These post-translational modifications act as a code, known as the "histone code," that tells the genome of the cell which genes to repress or activate, whether promoters or enhancers are active or not, or whether a sequence is a structural region or parasitic sequence that must be compacted. Among the most studied activation marks are H3K4me3, H3K9ac and H3K27ac, H3K36me3. These are recognized by different elements associated with the genetic transcription. Moreover, histone marks such as H3K9me3 and H3K27me3 are known to recruit protein complexes associated with genetic repression and chromatin compaction [50, 53, 54, 55]. The most used technique to make profiles and mapping of how histones are found in the genome is the chromatin immunoprecipitation assay (ChIP). In the next subsection we will describe this method and how the analysis is performed and which computational tools are available.

2.3.1 Experimental Evaluation of Post-Translational Modifications of Histones

To evaluate post-translational modifications among the entire genome, the preferred method is chromatin immunoprecipitation (ChIP). A classic ChIP assay evaluates the presence of a protein or histone modification in a specific region of the genome. It starts from a sheared cross-linked chromatin (the complex of proteins bound to the DNA sequence) incubated with an antibody (Ab) that binds to a specific histone modification, so the Ab will only recognize its target epitope. Then, the Ab–chromatin complex is isolated from the rest of the chromatin and the DNA contained in the Ab–chromatin complex is purified [56]. Finally, using specific PCR primers, the target region is amplified by PCR [57].

Currently, ChIP experiments are coupled to next generation sequencing technologies (ChIP-seq) to evaluate the presence of a post-translational modification, not only in a specific gene, but along the entire genome. DNA fragments from ChIP-seq are sequenced as reads, which are then mapped onto the reference genome, and the genomic regions that are significantly enriched for ChIP reads, compared with input reads, are detected as peaks [58, 59].

ChIP-seq experiments can be used to evaluate several targets, from DNA-bound proteins, post-translational modifications of histones, among others. Depending on the type of target, the bioinformatics approach will vary. Usually, there are three modes of protein-DNA interactions:

- **Sharp mode:** Certain proteins bind to specific sequences in the genome and produce a highly localized signal; for instance, transcription factors and cofactors.
- **Broad mode:** This mode of interaction with DNA is characterized by proteins associated with large genomic domains; for example, histone modifications involved in elongation of transcription or heterochromatin.
- **Mixed mode:** Includes proteins that with differential distribution along the genome, from sharp to broad mode depending on the genomic region, one example is RNA Pol II.

Depending on the mode of interaction between a protein and the DNA, some considerations are needed in the experimental design. For proteins displaying sharp signals, 10–14 millions of reads (M) are recommended; meanwhile, for broad signals a higher number of reads is suggested (20–40 M) [60].

Also, controls are important for a well-designed study. Each ChIP condition should have an input sample as control. The input sample consists of a sheared cross-linked chromatin obtained in the same conditions as the immunoprecipitated sample, but it is not incubated with any Ab. This control serves to correct biases during the data analysis.

2.3.2 ChIP-seq Data Analysis

A ChIP-seq experiment produces millions of reads. Once the raw data is obtained, an appropriate analysis is needed. Here, we describe the basic steps to analyze a ChIP-seq experiment [60, 61, 62, 63].

1. **Quality metrics**: Raw data obtained from sequencing platforms (FASTQ files) is evaluated to discard low quality reads. This step also includes trimming of barcodes and low-quality nucleotides. The preferred tool to evaluate read quality is FastQC [61, 64].
2. **Reads mapping**: Once the quality of the reads is evaluated, they are aligned to a reference genome using mapping tools, producing a BAM file. This step usually allows for a small number of mismatches in the alignment [65].
3. **Background evaluation**: Given that a ChIP-seq assay uses an Ab to recognize a specific protein–DNA complex (the signal), it can also bind non-specifically to a region (noise). Most of a ChIP-seq library is noise rather than signal (80–90% of reads). Thus, a method to establish a threshold to better identify signal peaks from background is needed [60].
4. **Peak calling**: A peak is a region of the genome where multiple reads have mapped. Peak calling tools have been developed to better identify true peaks.

Table 2.5: List of peak calling tools to analyse ChIP-seq data

Name	Webpage	Reference
MACS	https://github.com/taoliu/MACS/tree/macs_v1	[66]
SICER	http://home.gwu.edu/approximatelywpeng/Software.htm	[67]
PeakSeq	https://github.com/gersteinlab/PeakSeq	[68]

Table 2.6: List of differential binding analysis tools for ChIP-seq data

Name	Webpage	Reference
ChIPComp	https://bioconductor.org/packages/release/bioc/html/ChIPComp.html	[69]
ChIPDiff	http://cmb.gis.a-star.edu.sg/ChIPSeq/paperChIPDiff.htm	[70]
DBChIP	http://bioconductor.org/packages/release/bioc/html/DBChIP.html	[71]
DESeq	http://bioconductor.org/packages/release/bioc/html/DESeq.html	[72]
diffReps	https://github.com/shenlab-sinai/diffreps	[73]
EdgeR	http://bioconductor.org/packages/release/bioc/html/edgeR.html	[74]
GFOLD	https://bitbucket.org/feeldead/gfold	[75]
JAMM	https://github.com/mahmoudibrahim/JAMM	[76]
MAnorm	http://bcb.dfci.harvard.edu/~gcyuan/MAnorm/MAnorm.htm	[77]
ODIN	www.regulatory-genomics.org/odin-2/download-installation/	[78]
PePr	https://github.com/shawnzhangyx/PePr	[79]
RSEG	http://smithlabresearch.org/software/rseg/	[80]
THOR	www.regulatory-genomics.org/thor-2/download-installation/	[81]

There is no better method for peak calling and multiple tools have been developed. Some of these tools are shown in Table 2.5

5. **Differential binding analysis**: After obtaining peak sets, the identification of regions with differential protein binding patterns between conditions is important. For this end, normalization is performed [61]. Some of the differential binding analysis tools are listed in Table 2.6.

6. **Integrative analyses**: Further analysis can be performed by integrating ChIP-seq data to other type of experiments such as RNA-seq, Hi-C, and DNA methylation. Also, ChIP-seq data can be used to identify protein binding sites of transcription factors [61].

In summary, transcriptional regulation of a gene can be influenced by post-translational modifications of histones. ChIP-seq is an antibody-based technique to evaluate the presence of a protein or a histone modification along the entire genome. ChIP-seq experiments produce millions of sequences, named "reads," which are further analyzed to obtain enrichment peaks and which can be used to identify differential binding of proteins among conditions using different bioinformatics tools.

The ChIP-seq has opened a new window of opportunities in biomedicine. This allows a global perspective to search for novel regulators of transcription, which, viewed in a locus specific manner, would have been a challenge to predict. There are many examples of the contribution of this method to the knowledge of diseases, allowing the proposal of possible new targets for therapies for multiple diseases. Using the ChIP-seq approach, in solid tumors derived from breast cancer patients it was shown the differential estrogen receptor binding and H3K27me3 histone mark has been associated with poor clinical outcome [82].

2.4 Higher Order Chromatin Organization

Each cell in the human body contains the same DNA sequence. Different cell types in an organism are the result of different expression profiles that are presented in each of them. Where the position of these genes within the nucleus is will affect their transcription capacity. Therefore, one of the current efforts in epigenetics is to know how the tridimentionality of the nucleus influences gene regulation. In this regard, the eukaryotic genome is organized in a hierarchical and three-dimensional fashion. On the first level exist double stranded DNA, which is then packaged with histones to form nucleosomes, then nucleosome contacts form fibers or clutches, which can form dynamic long distance loops [83]. Some of them are established by architectural and regulatory proteins to give rise to structural landmarks, named domains [84, 85, 86]. Chromatin domains are comprised of clusters of genes with similar patterns of expression. For example, actively transcribed or inactive domains. The interaction among domains with similar characteristics form compartments, and the fusion of compartments in the same chromosome form chromosome territories [87].

In this context, the observation that chromosomes contain chromatin loops, led to the idea that these are separated by regulatory sequences. In this hypothesis, domains have well-defined borders marked by specific characteristics. These boundaries, also named insulators, represent one of the components that contribute to chromatin domain formation and maintenance of a specific configuration [88, 89].

Another component involved in the establishment of three-dimensional organization are enhancers. Enhancers are control sequences, that interact with promoters to regulate gene expression, they are usually located far from the promoter [90]. Transcription factors bind to enhancers or promoters to establish chromatin loops, which allow these distant regulatory sequences to be brought into close spatial proximity with its target gene [91]. It should be noted, that the activity of insulators or enhancers depends on the cell type and stage of development; meaning that cells from an adult kidney may have different patterns of enhancer/insulator activity than brain cells from embryos.

Some experimental methods have been developed to study enhancer/promoter interactions and the three-dimensional conformation of chromatin. The aim of this section is to show some of the most common experimental techniques used to study higher-order chromatin organization and the bioinformatics methods to analyze these data.

2.4.1 Technologies to Study Chromatin Conformation

The study of long-range chromatin interactions has increased after the development of the chromosome conformation capture (3C) technique [92]. The development of 3C-derived techniques made possible a genome-wide study of chromatin conformation. These methods include 4C, 5C, chromatin interaction analysis with paired-end-tag sequencing (ChIA-PET), and Hi-C (genome-wide 3C).

The 3C and 4C techniques require choosing one locus and evaluating an interaction with another locus (3C) or genome wide (4C). The 5C method probes multiple loci genome wide, while Hi-C allows a genome-wide analysis of chromatin interactions. In sum, 3C can be seen as a one-vs-one experiment, 4C as one-vs-all, 5C as many-vs-many and Hi-C as all-vs-all. All these approaches produce a DNA fragment composed of the two interacting sequences, depending on its tri-dimensional localization within the nucleus. (See Box 2.3.)

2.4.1.1 The 3C, 4C, 5C, and ChIA-PET Technologies

In a 3C experiment, the template is a pool of DNA fragments that reflect the interaction between two genomic loci and which can be detected by PCR using primer pair combinations [93, 94]. The abundance of each DNA fragment is a measurement of the frequency with which the two loci interact [95]. Thus, the 3C method can be used to capture and quantify only one physical interaction between a gene and a distant element, both in cis and in trans. On the contrary, a 4C experiment evaluates the DNA contacts made across the genome by a given genomic site of interest [96]. The 4C technique has been applied to demonstrate that individual gene loci can be engaged in many long-range DNA contacts with loci elsewhere on the same chromosome and on other chromosomes [97, 98]. The 5C method can be used to generate dense interaction maps that cover most or all potential interactions between all fragments of any genomic region. Dense interaction maps can provide a global overview of the conformation of a given genomic region [95]. Finally, ChIA-PET is a combination of 3C and ChIP that allows the genome-wide identification of potential interacting loci bound by a given protein [99].

2.4.1.2 The Hi-C Technology

A Hi-C experiment produces hundreds of millions of reads; these reads are analyzed and generate genome-wide maps showing interactions between genomic loci. A Hi-C experiment consists of these steps: (1) Cross-linking cells, (2) digestion of the DNA with a restriction enzyme, (3) mark the restriction sites with biotin, (4) ligation of the cross-linked fragments, (5) shearing the resulting DNA and pulling down the fragments with biotin, and (6) sequencing the pulled down fragments using paired-end reads. This procedure produces a genome-wide sequencing library that provides a proxy for measuring the three-dimensional distances among all possible locus pairs in the genome [100].

Box 2.4: 3C-based approaches to study chromatin conformation

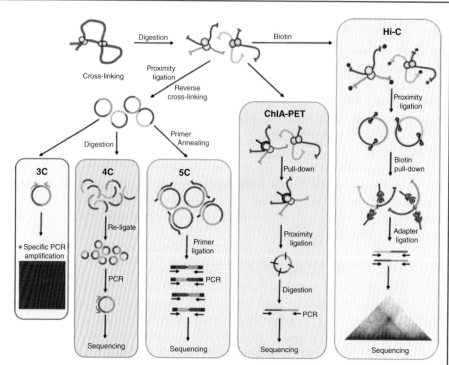

Detecting DNA fragments that preferentially interact together based on their proximity in the 3D space are the basis of all chromosome conformation capture (3C) technologies, including Hi-C (a high-throughput derivative of 3C). The first step of most 3C-based methods involves cross-linking of cells to capture chromatin organization, followed by fragmentation of the chromatin by digestion with a restriction enzyme or by sonication. The digestion is pieced back together by proximity-ligation of adjacent DNA ends.
After reverse cross-linking, different approaches can be used to identify the chromatin interactions. During 3C, 4C, and 5C, the ends are simply re-ligated to generate circular ligation products. For Hi-C, restriction ends are first marked with biotin to identify sites that were cut and re-ligated. During 3C, primers are designed upstream of specific cut sites to measure the frequency of given junctions by (qPCR) one at a time. In 4C, the 3C library is further digested with an enzyme that cuts more frequently and fragments are circularized by ligation. Primers designed against a region of interest are then used to simultaneously amplify all other fragments contacting it, and the products are sequenced. In the 5C approach, primer sequences overlapping restriction fragment ends are ligated only when the two ends are immediately adjacent, then products are amplified and sequenced. In the Hi-C method the restriction fragment ends are labeled using biotin, ligated products are enriched using streptavidin pull-down after and interactions are interrogated in a genome-wide all-versus-all manner. This method could potentially be combined with an enrichment step, either using sequence-specific probes or with an antibody against a protein of interest (ChIA-PET).

2.4.2 Bioinformatic Methods of Hi-C Analysis

A Hi-C experiment generates a huge amount of information, which needs to be analyzed to identify those chromatin interactions that have a functional impact on the cell from those "random" interactions, also known as non-functional. Thus, the bioinformatics methods available to analyze Hi-C data are mainly focused in the normalization and statistical analysis. Nevertheless, in the next subsection we briefly review the steps required in a typical Hi-C experiment.

2.4.3 Mapping and Filtering

The first step in the analysis of Hi-C data is the alignment of reads to a reference genome. There are four methods for alignment of reads obtained from a Hi-C experiment: (1) Pre-truncation, (2) iterative mapping, (3) allow split alignments, and (4) split if not mapped. For a more detailed explanation of the alignment methods see reference [101].

After the mapping step, a filtration of reads is needed. Usually, filters applied are: (1) Mismatches, (2) read quality, and (3) uniqueness of reads. Then, reads that passed the previous filters are classified to identify the "informative" ones. These filters are: (1) Strand filters and (2) distance filters [101, 102].

2.4.4 Normalization

The input data for normalization is a raw contact matrix. There are three normalization methods [101]: (1) Explicit-factor correction, (2) matrix balancing, and (3) joint correction.

- **Explicit-factor correction**: These methods need an a priori knowledge of bias factors (GC content, mappability, and fragment length) and corrects by modeling the probability of observing an interaction between two sequences [101].
- **Matrix balancing**: These methods correct for all bias causing factors without modeling them. They assume that in case of no bias, then loci would produce equal number of reads [101].
- **Joint correction**: These methods correct considering the one-dimensional distance between the two sequences, given that regions adjacent to each other in one dimension cannot be far way in the three-dimensional space, and this does not necessarily mean they form a functional loop [101].

Some of the tools that can be employed for Hi-C data normalization are shown in Table 2.7.

2.4.5 Statistical Analysis

Once biases are eliminated during normalization, the identification of statistically significant interactions between genomic regions is important. The number of interactions between two regions depends on the distance between them; there is random looping that needs to be evaluated when assigning statistical significance to the contacts. Many tools are available for Hi-C data analysis; some of them are enlisted in

Table 2.7: List of Hi-C data normalization tools

Name	Programming language	Website	Reference
Hi-Corrector	ANSI C	http://zhoulab.usc.edu/Hi-Corrector/	[103]
Hi-Five	Phyton	http://taylorlab.org/software/hifive/	[104]
HiCNorm	R	www.people.fas.harvard.edu/junliu/HiCNorm/	[105]

Table 2.8. The methods developed to assess the significance of an interaction in a Hi-C experiment are described below [101].

2.4.6 Visualization of Hi-C Data

There are generally three ways of visualizing Hi-C data: (1) Square heat maps, (2) circular plots, and (3) genome browsers. To visualize large-scale data, such as complete chromosomes or the genome, a square heat map or a circular plot are the selected options. On the contrary, to evaluate local data corresponding to a specific region of the genome, a genome browser is the appropriate option [101, 102, 117, 118]. Browsers of Hi-C data are summarized in Table 2.9.

Square Heat Maps and Circular Plots Contact matrices are square heat maps where the contact count is transformed into colour. Every Hi-C heat map is a diagonal in the middle, showing adjacent loci and they can represent the entire genome or only one chromosome. Circular plots can be used to visualize intra-chromosomal interactions, where the contacts are show by arcs connecting distal loci [118].

Genome Browsers Genome browsers are useful for the inspection of a small region of the genome. Usually, the DNA sequence is shown horizontally and the Hi-C data is added in parallel to the DNA sequence. They allow the visualization of Hi-C data by local arc tracks, similar to circular plots [118].

2.4.7 Topological Associated Domain Identification from Hi-C Data

One of the purposes of epigenetics is the identification of regulatory domains, which have been characterized by specific histone marks and transcription factors. Topological associated domains (TADs) are densely interacting regions that can be identified in a contact map as interacting squares. TADs are of importance in the tri-dimensional organization of chromatin, thus many methods have been developed to study TADs from Hi-C experiments (Table 2.10).

The importance of data generated by experiments that capture chromatin interaction in the clinics is a growing topic in the field of epigenetics. One of the applications of techniques such as Hi-C is the prediction of the emergence of chromosomal

Table 2.8: List of Hi-C data analysis tools

Name	Programming language	Website	Reference
diffHic	R	https://bioconductor.org/packages/release/bioc/html/diffHic.html	[106]
Fit-Hi-C	R	https://bioconductor.org/packages/release/bioc/html/FitHiC.html	[107]
Gothic	R	https://bioconductor.org/packages/release/bioc/html/GOTHiC.html	[108]
HiC-Inspector*	R	https://github.com/HiC-inspector/HiC-inspector	[109]
HiC-Pro*	Phyton, R	http://github.com/nservant/HiC-Pro	[110]
HiCdat*	R	https://github.com/MWSchmid/HiCdat	[111]
Hiclib*	Phyton	https://bitbucket.org/mirnylab/hiclib	[112]
HiCUP*	Perl, R	www.bioinformatics.babraham.ac.uk/projects/hicup/	[113]
HIPPIE*	Phyton	wanglab.pcbi.upenn.edu/hippie	[114]
HOMER	Perl	http://homer.ucsd.edu/homer/interactions/	[115]
Juicer*	Java	https://github.com/theaidenlab/juicer	[116]

*Tool that provides the complete workflow to analyze Hi-C data.

Table 2.9: List of genome browsers to visualize Hi-C data

Name	Webpage	Reference
3D Genome Browser	http://promoter.bx.psu.edu/hi-c/	[119]
Epigenome Browser	http://epigenomegateway.wustl.edu/browser/	[120]
HiBrowse	http://hyperbrowser.uio.no/3d	[121]
Juicebox	www.aidenlab.org/juicebox/	[122]
My5C	http://my5c.umassmed.edu/welcome/welcome.php	[123]

Table 2.10: List of TAD identification tools from Hi-C experiments

Name	Website	Reference
DI-HMM	https://github.com/gcyuan/diHMM	[70]
HiCseg	https://cran.r-project.org/web/packages/HiCseg/index.html	[124]
TADbit	https://github.com/3DGenomes/TADbit	[125]
TADtree	http://compbio.cs.brown.edu/projects/tadtree/	[126]

translocations. It is currently believed that for a translocation to occur, there must be proximity between two chromosomes at the chromosomal territory level, this is known as the "proximity effect." Proximity effects have been described in analysis of cancer-causing translocations. An example widely studied is the one involving ABL and BCR genes. Genesis of the BCR–ABL hybrid gene appears to be a fundamental step in the development of chronic myeloid leukemia. These genes have been found in close proximity in normal hematopoietic cells and its Hi-C contact frequency is significantly higher compared with other loci in the same nucleus [127, 128, 129, 130, 131]. Suggesting the predisposition to generate relapses in these patients. Similar results were found in Burkitt's lymphoma. In this case, the gene MYC (8q24) translocate in ~90% of the cases with IGH (14q21), and less frequently with IGL (22q11) and IGK (2p11). In such patients, locus proximity correlated to the observed incidence of translocation [132].

2.5 Long Non-Coding RNAs, Novel Molecular Regulators

Non-coding RNAs (ncRNAs) have been known for decades, with classic examples such as the ribosomal, transfer, small nucleolar RNAs, among others [133]. With the development of genome-wide sequencing and other high throughput technologies the presence of thousands of novel transcripts have been revealed and reported by the ENCODE consortium [133]. ncRNAs are now suggested to be functional products and not transcription noise. Of these novel transcripts, many are considered to be long non-coding RNA (lncRNA). However, we still do not know the function of the great majority of these transcripts. Although RNA polymerase II can generate spurious transcripts by non-specific binding, current evidence has shown that these ncRNAs are involved in many biological processes with specific expression patterns and regulation

[134, 135, 136, 137]. For this reason it is very important to elucidate the functions of this kind of non-coding transcripts in regulation of gene expression and diseases.

In general, lncRNAs are defined as transcripts longer than 200 base pairs, lacking open reading frames (ORF). The majority of lncRNAs have in addition the structure called CAP at the 5' end, the nonetheless, some exceptions have been observed. Likewise, as mRNAs, lncRNAs can be polyadenilated, suffer splicing, and may have several isoforms [138]. Expression analysis based RNA-seq performed on cell lines and different human tissues have suggested a higher amount of lncRNAs in the human genome than protein-coding genes [139]. The main features of lncRNAs are: (1) that they are synthetized in low scale, (2) they are located in both nucleus and cytoplasm, and (3) they are expressed in a tissue specific manner [137]. Depending on their genomic location and the sense in which they are expressed, lncRNAs can be divided into intergenic, intragenic, bidirectional, enhancers, as well as sense and antisense.

Experimental evidence has demonstrated that lncRNAs may be involved in different biological processes such as DNA methylation [140], genome imprinting [141], nuclear compartmentalization [142], gene to gene interactions [143], and regulation of chromatin structure [144], among others [145].

The first long non-coding RNA was described in 1988. It was identified as a transcript that was expressed during liver development in mice and it does not generate a protein product in vivo [146]. Actually, a high number of lncRNA have been identified through chromatin modifications associated with active promoters such as H3K4me3, and epigenetic modifications associated with transcription elongation as H3K36me3 [138]. Interestingly, many of these RNAs are transcribed relatively close to protein coding genes, occurring mostly in an antisense manner [147]. Moreover, genomic context does not necessarily explain the function or evolution of these genes [138, 148, 149]. Besides, in an attempt to understand lncRNAs and their cellular functions, they have been classified in four archetypes of molecular mechanisms: Signal, decoy, guide, and scaffold [150]. Despite what it is known about lncRNAs, there is a great debate on how to classify them [151, 145]

In summary, our knowledge of lncRNA is very limited and we are still unaware of the function of the great majority of these transcripts, therefore, it has increasingly become an exciting field of study in biology. Yet, their implications in biological mechanisms are leading to a high research interest which is additionally enhanced by the discovery of its possible implications in many diseases, such as cancer and maybe future therapeutic targets [152].

2.5.1 The Implications of lncRNAs in Precision Medicine

Conventional treatments are designed to treat sickness in terms of the average response of patients, assuming that one average approach fits all individuals. However, what is successful for some might not work for other patients. In order to improve the effectiveness of treatments, an emerging approach for disease treatment and prevention has gained attention, which is currently known as precision medicine. Precision medicine is based in the knowledge of the individual variability observed in genes, epigenetic profiles, environment, and lifestyle for each person [153]. The concept is sustained by the assumption that genetic or molecular aberrations can be observed as a cause or contributor of a disease. Therefore, genomic studies and related information can shed light on various questions associated with the health and disease of an individual. Genomic approaches such as DNA sequence variation, expression profiles, proteomics, and metabolomics have become valuable tools for precise disease management and prediction. The information from an individual genome sequence and the associated expressed biomarker are essential to establish risk checkpoints and achieve precision therapies [154].

Actually, numerous reports have demonstrated the use of genomic data and expression signatures as clinical prognostic factors in cancer and other diseases. Most of these studies are focused on coding genes and regulatory regions to identify different types of biomarkers and in some cases establish companion diagnostics that will positively impact the outcome for patients [155]. Diverse reports have observed aberrant expression patterns of lncRNAs associated with numerous diseases including cardiovascular disease, Alzheimer', fragile X syndrome, and cancer. Therefore, lncRNAs have become novel targets with potential for therapeutics, prognoses, and diagnostics of diseases [156].

In order to understand the implications and the biology of these transcripts, several computational tools have been generated that help in their study. In the next section, we will discuss some of these tools, as well as the analysis strategies that can be used.

2.5.2 Bioinformatic Tools for lncRNAs Analysis

The study of the transcriptome represents a new challenge for the modern medical area, since it breaks with the traditional analysis of only some coding genes [157]. In a conceptual manner, the transcriptome is defined as the set of total transcripts present in a cell, at a particular time in its life cycle [158, 159]. The current tools of NGS allow us to obtain information about the transcriptome of a specific sample, and in consequence we can know the expression levels of all the transcripts. This implies that the amount of data obtained is too large and requires sophisticated bioinformatics and statistical tools for processing. Despite this, advances in bioinformatics have developed various *in silico* methodologies [160, 161].

On the other hand, NGS methods such as RNA-seq, show some advantages over methodologies such as expression microarrays. In the sense that RNA-seq scans the whole transcriptome and is not limited to the detection of sequences or transcripts that are already known, therefore, this tool can be used to discover new isoforms, gene fusions, and unknown lncRNAs. Thus, these observations have shown the great potential of RNA-seq for the search of new transcripts that can be of coding and noncoding nature [162].

Once the expression information of an RNA-seq has been obtained, it is important to be clear about what you want to know or search. In this sense, there are two main ways to analyze the results obtained: (2) Work with already known annotated transcripts [163], or look for new transcripts that have not been characterized or annotated

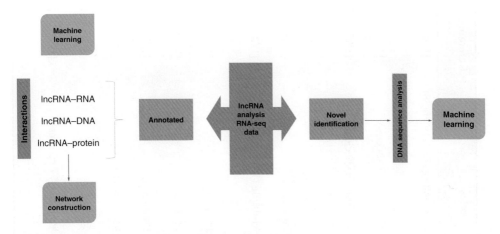

Figure 2.3: New methodological approaches for the bioinformatic analysis of lncRNAs. From RNA-seq data, it is possible to identify the expression of already annotated transcripts, in order to perform interaction analysis among some transcripts of interest, and determine the possible functions of the RNAs of interest, or their relationship with a specific cellular process.

[163, 164, 165] (Figure 2.3). In the specific case of lncRNA, they are not fully characterized, so the field of work in annotated and non-annotated transcripts is extensive [163].

2.5.3 Analysis of Annotated lncRNAs

When it is desired to work with lncRNA that have already been previously annotated, there are different strategies for the analysis of these, all bioinformatics analysis of them begins with the alignment of the transcriptome [165]. For this, it is important to know that the reference files for the alignment will depend on each source. GENCODE is a source that has alignment files for the complete genome, or only the subset belonging to the lncRNA can be used [166]. Ensembl, on the other hand, has coding and non-coding transcripts in independent files [167]. The databases also provide, in addition to these files, information of a structural, functional, or expression type (see Table 2.11).

Once the alignment of the reads has been done, the expression quantification process is carried out, those analyses can be analyzed by using fragments per kilobase of transcript per million mapped reads (FPKM), reads per kilobase of transcript per million mapped reads (RPKM), or transcripts per kilobase million (TPM). Once transcript quantification is done a differential expression analysis is desirable, because it is used to determine which will be the lncRNA of interest to study. These analyses allow us to compare the levels of expression between different conditions (normal vs disease, experimental vs control, stages of development, etc.) [178]. An example could be the determination of the level expression of a transcript under normal and cancer conditions, in order to determine which transcripts are up- or down-regulated. All of the above analysis can be carried out with packages such as R, or with the use of tools in Galaxy, depending on the user's experience [178, 179, 180].

Table 2.11: LncRNAs and databases

Database	Description	Disease-association data	Number of lncRNA	Species	Reference
LNCipedia 5.0	A database for annotated human lncRNA transcripts sequences, structures, and available literature.	No	120,353	Human (hg37/hg19 Hg38)	[168]
NONCODEV5.0	A database for annotated ncRNA (except tRNAs and rRNAs).	Yes	548,640	Multiple (17 species)	[169]
lncRNAtor	A database for annotated lncRNA sequences, expression profiles, protein interactions, and phylogenetic conservation.	No	31,725	Multiple (6 species)	[170]
TANRIC	An interactive platform for research and exploration of lncRNAs in cancer.	Yes	~13,000	Human	[171]
ChIPBase v2.0	An integrative database of transcription regulation and expression profiles of lncRNAs.	Yes	38,293	Multiple (10 species)	[172]
lnc2Cancer	A curated database for cancer-associated lncRNAs.	Yes	666	Human	[173]
lncRNADisease	A database for lncRNA and disease associaton.	Yes	1,564	Human	[174]
lncRNome	A database for human lncRNA transcripts.	No	~170,000	Human	[175]
lncRNAdb v2.0	A database for eukaryotic lncRNA structural information, gene expression data, genomic context, and phylogenetic conservation.	No	>200	Multiple (68 species)	[176]
EVLncRNAs	An integrative and curated database for annotated and experimentally validated human lncRNAs.	Yes	1,543	Multiple (77 species)	[177]

From the results of quantification and differential expression analysis, other types of analysis can be performed, which allow us to understand relationships between the expression of the transcripts and the conditions analyzed [181]. In this sense, there are three main types of analysis: Those that relate the expression of lncRNA with a particular pathology; those that associate the levels of lncRNA with the expression of proteins; or those that associate the levels of lncRNA with the expression of other no-coding transcripts such as miRNA [182, 183, 184]. This is called correlation studies, and they are based mainly on the Pearson correlation coefficient, which is calculated with operations between matrices [184]. The result of this process is the construction of a network that allows analyzing the relationship between the species and/or the study condition [185].

Furthermore, to carry out a deeper functional study about lncRNA functions, other types of tools capable of predicting the interaction between molecules have been developed. In this case, we can determine the possible interaction between lncRNA–ncRNA–miRNA [184, 186, 187], lncRNA–protein. [188, 189] or lncRNA–DNA [190] (Figure 2.4). Basically, the reason why those algorithms were designed is to analyze the binding sites that each of the biological molecules has, and the probability that the molecules of interest can interact with each other. These tools use for their construction, information that has been previously corroborated experimentally, or that is available in the literature (text manning), but its represents a limitation of analysis [188, 189]. However, there are reports of tools capable to thermodynamically predict some interactions between nucleic acids molecules (ncRNA [187] and DNA [190]), thus the complementary use of these tools can be a way to resolve the lack of reported information on interactions between nucleic acids.

As a consequence, all of the above criteria lead us to obtain much more complete information about the possible function of each lncRNA, or even the cellular processes in which it participates. For example, if an lncRNA is able to interact with chromatin remodeling complex, or transcription factors, among others regulatory elements, with this information we can get a general landscape of the lncRNA function. Actually, there exist two main methodologies for functional analysis of lncRNAs: Networks construction [183, 184, 188], and machine learning [191]. The algorithms designed by machine learning have taken on a lot of importance, since they provide reliable results, and they are capable of working with much more information. Currently, there are groups working on the improvement of network construction algorithms. The algorithms differ from each other by automation and training, in addition to being conditioned by the inputs accepted by each of the matrices that are used. In this sense, there is a lot of literature that describes in detail each of the algorithms and the matrices used for each of them.

By working on the transcripts already annotated, it can be clearly seen that the focus of these analyses is guided to be the functional characterization of each of them (Figure 2.5). In Epigenetics, it is important to describe the role that each lncRNAs play as epigenetic regulators and describe the pathways by which they could act. In this sense, the tools described above allow us to measure the relationship that exists between the expression of some proteins that participate in epigenetic processes, or the association with other non-coding transcripts and even begin to discern the possible mechanisms of action. Thus, the lncRNAs display a more tissue-specific expression

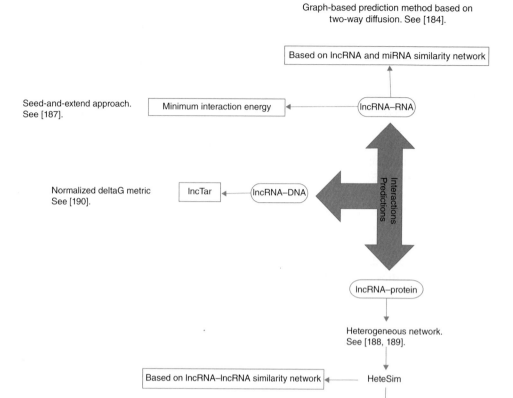

Figure 2.4: Algorithms designed to predict RNA interactions. Examples of algorithms used to predict lncRNA interactions with different types of RNAs, its can be predict interactions with DNA and proteins too. Among them, the figure shows the algorithms based on the construction of networks (HeteSim) and in the calculation of some parameters that characterize the physical interactions, such as the minimum interaction energy.

pattern even in pathological conditions or stages of development, for this reason this kind of transcripts can be useful indicators of disease status in patients this makes them attractive molecules for the development of new algorithms that allow us to develop new tools to determine which of the diseases are specific to diseases and can be used to modern therapies [192].

2.5.4 Analysis of Unannotated lncRNAs

The bioinformatics tools to discover new transcripts have faced several challenges along their development, among them are the inherent characteristics of the lncRNA, such as the little evolutionary conservation, and splicing sites that differ with the sites of canonic splicing sites of mRNAs [164, 165]. Despite this, the first tools designed based their pipeline on the analysis of mRNAs. Currently, there are tools and pipelines

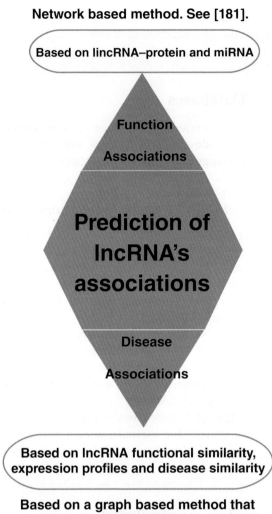

Figure 2.5: Algorithms used to predict the association of lncRNAs with diseases. Examples of algorithms used to predict the association of lncRNAs based on diseases. Some of these methodologies depend strictly on the data already reported in the literature, as is the case of the associations of lncRNAs with diseases. However, the characteristic shared by many of the pipelines is that they are based on previous networks of similarity of the lncRNAs in terms of function, structure, expression profile, or association to diseases. Another important input that these methodologies receive is the information coming from the co-expression studies.

designed for the discovery of new lncRNA transcripts, mainly based on the identification of splicing sites, genomic sequences, and de novo assemblies [164]. In this sense, the implementation of methods based on machine learning have been proven to have better results than previously designed tools [164]. lncRNAs are transcripts difficult to characterize for their low abundance, for this reason more efforts need to be made in this area in order to get the all compendium of non-coding transcripts that form part of the human transcriptomes.

2.6 Epigenetic Databases

Here we provide a brief overview of the most relevant projects, consortia, and databases containing large-scale epigenetic datasets, which have been used to provide novel insights into a wide range of disease-focused studies.

2.6.1 Encyclopedia of DNA Elements in the Human Genome

The Encyclopedia of DNA elements (ENCODE, www.encodeproject.org) is an international collaborative effort funded by the National Human Genome Research Institute (NHGRI) in the US. Its main objective is to map and characterize all the functional and regulatory regions within the genome [193]. Though initially intended for humans, the ENCODE project has expanded to include other model organisms such as mouse [194], fruit fly [195], and worm [196].

The current version of ENCODE (as of February 2017) contains information of more than 6,500 samples which span different types of immortalized cell lines, tissues, and stem cells. Types of epigenetic information included are DNA methylation (MeDIP-seq), histone mark enrichment (ChIP-seq), open chromatin regions (DNase-seq), and non-coding transcripts (RNA-seq).

The rich datasets in ENCODE have already helped us to understand the mechanisms of disease development. For example, studies using ENCODE data have shown that 90% of variants linked to disease risk fall outside protein-coding regions in the DNA.

2.6.2 The Roadmap Epigenomics Project

The aim of the NIH Roadmap Epigenomics project (www.roadmapepigenomics.org/) is to produce and make publicly available human epigenomic data from various stem cell types and ex-vivo tissues [197]. Their vision is to provide a genome-wide landscape of epigenetic variation that can take place during normal human development and disease. Also, a major aim is to provide reference epigenomes, to which Version 9 of the Roadmap currently includes 2,804 genome-wide datasets of which 1,821 are histone modifications datasets, 360 DNase datasets, 277 DNA methylation, and 166 RNA-seq datasets. Release 9 also includes 111 reference epigenomes, each containing a set of five core histone modifications (H3K4me3, H3K4me1, H3K27me3, H3K9me3, and H3K36me3).

Analysis of the Roadmap data has led to elucidation on epigenetic mechanisms behind Alzheimer's disease and non-coding SNPs in autoimmune diseases [4, 198]. Another study revealed that the mutation density in cancer genomes can be predicted by epigenetic features such as chromatin accessibility and modification, derived from the most likely cell of origin of the corresponding tumor [199].

2.6.3 Functional Annotation of the Mammalian Genome

The Functional Annotation of the Mammalian Genome (FANTOM, http://fantom.gsc.riken.jp) consortium is an international research effort that attempts to annotate and understand all the elements encompassing the complex transcriptional regulatory network of the mammalian genome. In phase 5 (FANTOM5) of the project, by using capped analysis of gene expression (CAGE) followed by next generation sequencing [200, 201], researchers of FANTOM systematically profiled the sets of transcripts, transcription factors, promoters, and enhancers active in the majority of mammalian primary cell types and a series of cancer cell lines, and tissues.

Datasets of FANTOM5 have led to important discoveries, such as that enhancers are the first to be transcribed, followed by transcription factors and finally non-transcription factor genes [202]. Also, it was shown that enhancers produce bidirectional, exosome-sensitive unspliced RNAs, which so far are the strongest predictors of enhancer activity [203].

FANTOM5 data has been used to identify new pan-cancer biomarkers by comparing the transcriptional profiles with RNA-seq datasets from The Cancer Genome Atlas [204, 205]. Enhancer RNAs were shown to be upregulated in cancer, where also promoters overlap with repetitive elements that are also upregulated in cancer.

2.6.4 BLUEPRINT Epigenome

Project BLUEPRINT is a European funded effort with the objective to obtain and make available epigenomes from all types of blood cells [206]. Samples in their data repository include primary cells from healthy patients, as well as patients suffering from blood-bases diseases such as type I diabetes and hematopoietic neoplasia and autoimmune diseases among others. Types of assays include DNA methylation (WGBS), histone modifications (ChIP-seq), chromatin accessibility (DNase-seq), and RNA expression (RNA-seq).

2.6.5 The International Human Epigenome Consortium

The International Human Epigenome Consortium (IHEC, http://ihec-epigenomes.org/) is an organization that coordinates the productions of reference epigenomes for multiple cell types relevant to health and disease research [206]. Their goal is to map 1,000 epigenomes by 2020. The IHEC data portal collects epigenetic information totaling 8,753 (February 2017) datasets produced by all its members, these include ENCODE, NIH Roadmap [207], the European Union BLUEPRINT Project, the Canadian Epigenetics Environment, and Health Research Consortium (CEEHRC), and the German Epigenome Programme DEEP, among others (Table 2.12).

Table 2.12: A non-exhaustive list of epigenetic focused databases

Name	Type of data	Platforms	Description
MethylomeDB	DNA methylation	Methyl-MAPS	Provides genome-wide methylation profiles for human and mouse brains.
NGSmethDB	DNA methylation	Whole genome bisulfite sequencing	Contains methylation profiles for different human cell lines, primary tissues, pathological biopsies and autopsies curated mainly from NCBI GEO [20] and the Roadmap project.
MethBase	DNA methylation	Whole genome bisulfite sequencing and reduced representation bisulfite sequencing	Contains hundreds of annotated methylomes for human, mouse and other model organisms.
4DGenome	Chromatin interactions	3C, 4C, 5C, Hi-C, Capture-C, ChIA-PET, IM-Pet	Contains thousands of chromatin interactions for human and other model organisms.
EpiFactors	Genes and gene products corresponding to epigenetic factors.	N/A	Contains information about genes and gene products that code into epigenetic factors such as histones, histone modifiers, DNA methyltransferases, etc.
DNAmod	DNA chemical modifications	N/A	Contains information about all types of covalent DNA modifications such as 5mC and 5-hydroxymethylcytosine and several others, found in many model species.
NONCODE	Long non-coding RNAs	N/A	Collects and annotates information about lncRNAs from human and other model organisms. Provides information about conservation and the relationships between lncRNAs and diseases

2.7 Conclusion and Final Remarks

The epigenome is an important mechanism that regulates which parts of the genome are condensed and inactive to transcription and which are accessible and therefore transcriptionally active. Hence, epigenetic modifications are a major driver of development and determine and maintain cellular identity. The epigenome suffers different aberrations in diseases, where such aberrations have been related to many biological alterations related to the aetiology and the clinical outcome for patients. Different fundamental epigenetic mechanisms that regulate gene expression: Including DNA methylation, histone modifications, chromatin remodeling proteins, associated proteins, lncRNAs, and master regulators of the chromatin have been observed in association with diseases. Here we summarized some of the different epigenetic components and their importance in different pathologies, and present an overview of profiling methods and technologies that have rapidly matured in the last decade with examples of some of the epigenetic databases that exist.

There is currently a worldwide effort to understand and determine the precise human epigenome in healthy and diseased tissues, in order to help establish profiles that predict disease outcome in terms of patient prognosis and treatment response and also the possibility of epigenetic specific therapies, which could improve the life expectancy of patients. The comparison of epigenetic profiles and the integration of data from different studies from healthy and diseased tissue will allow the generation of predictive models for disease outcomes using epigenetic biomarkers and signatures. However, many key questions in the field remain unanswered, regarding the rules that command the epigenetic code, and the functions of the epigenetic components in the regulation process. Do we know all the players in epigenetics? It is clear that further effort in research towards completing the epigenetic map are still needed, therefore detailed epigenomic analysis are needed to create the maps in healthy and diseased tissues. The advances in this field will be of great importance in promising research of epigenetic therapies.

2.8 Exercises

2.1 How is epigenetics defined and why is it important in cellular mechanisms?
2.2 What is the importance of epigenetics in medicine?
2.3 Select the methodologies for DNA methylation analysis.

- LUMA
- ChIP
- RT-PCR
- MeDIP
- MS-PCR
- Western Blot
- Bisulfite conversion

2.4 You did a trial of bisulfite conversion and sequencing in a sample and you are ready to analyze them. What is the order of the steps you must follow for the analysis? And what computer tools could you use in the steps of your pipeline?

2.5 Explore epigenetic patterns changes of RUNX3, TIMP3, and GAPDH genes between cancer and non-neoplastic cell lines. We will use primary T helper cells and compared the two-leukaemia derivate cell line K562.

(a) Search for RNAseq, H3K4me3 (for promoters), H3K9me3, and H3K9me3 for repressive marks. Explain the state of these genes in these cell lines.
(b) Obtain the possible regulatory region of the gene by H3K4me3 enrichment and search and observe if it falls in a CpG island site.
(c) Use different databases to observe epigenome data (http://egg2.wustl.edu/roadmap/web_portal/ or https://genome.ucsc.edu/ to get data).

2.6 Let's suppose that in your laboratory you need to analyze the expression levels of a gene regulatory CpG island, in brain tissue samples let's call that gene Batman. However, you notice that when you treat the cells you study with an inhibitor, there are fewer transcript expressions compared to the control (vehicle). By performing a bisulfite and sequence analysis, you realize that in both treated and untreated cells, they have the same pattern in sequencing (as shown in Figure 2.6):

Please answer the following questions:

(a) If the expression profile of the transcript changes in cells treated with inhibitor compared to the control, how do you explain that by methylation sequencing there are no changes in cytosines in CpG context?
(b) Which experimental technique would you use to test your response?

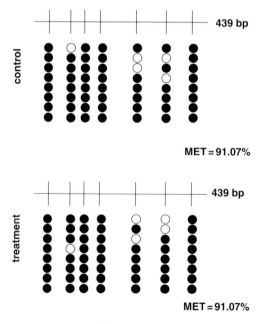

Figure 2.6

2.7 Suppose that you have a research project where objective is to analyze the expression of the lncRNA (W) in the cell line A with respect to the cell line B. The main objective is to determine if there is a significant differential expression between cell lines (A vs B). According to the text, please propose a pipeline (flow diagram) to determine the differential expression between cell lines, is important that you can distinguish between the control cell line and the problem cell line. Assume which part of RNA-seq data is needed for both cell lines.

    Note: Your results must have statistical significance. For this reason, how much RNA-seq data is there supposed to be, at least, for each cell line?

2.8 Access to the Washu Epigenome Browser webpage (http://epigenomegateway.wustl.edu/browser/), select the human hg19 genome and select all the public hubs, then load the tracks from the Encyclopedia of DNA Elements and answer the next questions:

(a) Look for the *GAPDH* gene (transcript variant2) and indicate the beginning and the end of the gene.
(b) Load the tracks for the following histone post-translational modifications in the K562 cell line: H3K4me3, H3K9me3 and H3K27me3.
(c) Evaluate the histone post-translational modification marks listed above along the *GAPDH* human gene. Then indicate which is the leading histone modification mark observed at *GAPDH* promoter.
(d) Investigate the biological function of the histone modification mark found at *GAPDH* promoter and indicate if the gene is transcriptionally active or repressed.

2.9 Describe at least two mechanisms by which lncRNAs can carry out their regulatory functions in the cell.

2.10 What methodology would you need to measure the chromatin interaction of a specific locus with the complete genome?

2.11 The different chromatin conformational capture methodologies answer different experimental questions regarding the interaction of the genome in the nucleus. What kind of experimental questions do these techniques answer?

Note: Solutions are available to instructors at www.cambridge.org/bionetworks.

2.9 Acknowledgements

We thank Consejo Nacional de Ciencia y Tecnología (CONACyT) 182997 and 284748 and FOSISS 0261181 for the support in the present work.

References

[1] Waddington CH. The epigenotype. 1942. *International Journal of Epidemiology* 2012;41:10–13.

[2] Jones PA, Baylin SB. The fundamental role of epigenetic events in cancer. *Nature Reviews Genetics* 2002;3:415–428.

[3] Kular, Kular S. Epigenetics applied to psychiatry: Clinical opportunities and future challenges. *Psychiatry and Clinical Neurosciences.* 2018;72(4):195–211. Doi:10.1111/pcn.12634.

[4] Farh KKH, Marson A, Zhu J, et al. Genetic and epigenetic fine mapping of causal autoimmune disease variants. *Nature*, 2015;518:337–343.

[5] Chen Z, Miaof, paterson AD, et al. Epigenomic profiling reveals an association between persistence of DNA methylation and metabolic memory in the DCCT/EDIC type 1 diabetes cohort. *Proceedings of the National Academy of Sciences of the USA* 2016;113:E3002–E3011.

[6] Soubry A. Epigenetics as a driver of developmental origins of health and disease: Did we forget the fathers? *BioEssays*, 2018;40:1700113.

[7] Doerfler, W. De novo methylation, long-term promoter silencing, methylation patterns in the human genome, and consequences of foreign DNA insertion. In Doerfler W, Böhm P, eds., *DNA Methylation: Basic Mechanism. Current Topics in Microbiology and Immunology*, vol. 301. Springer; 2006, pp. 125–175.

[8] Illingworth RS, Gruenewald-Schneider U, Webb S, et al. Orphan CpG islands identify numerous conserved promoters in the mammalian genome. *PLoS Genetics*, 2010;6, e1001134.

[9] Sadakierska-Chudy A, Kostrzewa RM, Filip M. A comprehensive view of the epigenetic landscape part I: DNA methylation, passive and active DNA demethylation pathways and histone variants. *Neurotoxicty Research* 2015;27:84–97.

[10] Li E, Zhang Y. DNA methylation in mammals. *Cold Spring Harbor Perspectives Biology* 2014;6:a019133.

[11] Meissner A, Mikkelsen TS, Gu H, et al. Genome-scale DNA methylation maps of pluripotent and differentiated cells. *Nature* 2008;454:766–770.

[12] Bird A, Taggart M, Frommer M, Miller OJ, Macleod D. A fraction of the mouse genome that is derived from islands of nonmethylated, CpG-rich DNA. *Cell* 1985;40:91–99.

[13] Larsen F, Gundersen G, Lopez R, Prydz H. CpG islands as gene markers in the human genome. *Genomics*, 1992;13:1095–1107.

[14] Dai W, Zeller C, Masrow N, et al. Promoter CpG island methylation of genes in key cancer pathways associates with clinical outcome in high-grade serous ovarian cancer. *Clinical Cancer Research*, 2013;19:5788–5797.

[15] Maunakea AK, Chepelev I, Cui K, Zhao K. Intragenic DNA methylation modulates alternative splicing by recruiting MeCP2 to promote exon recognition. *Cell Research*, 2013;23:1256–1269.

[16] Weber M, Hellmann I, Stadler MB, et al. Distribution, silencing potential and evolutionary impact of promoter DNA methylation in the human genome. *Nature Genetics*, 2007;39:457–466.

[17] Wu X, Zhang Y. TET-mediated active DNA demethylation: Mechanism, function and beyond. *Nature Reviews Genetics*, 2017;18:517–534.

[18] Tahiliani M, Kohk P, Shen Y, et al. Conversion of 5-methylcytosine to 5-hydroxymethylcytosine in mammalian DNA by MLL partner TET1. *Science*, 2009;324:930–935.

[19] Rawłuszko-Wieczorek AA, Siera A, Jagodziński PP. TET proteins in cancer: Current 'state of the art'. *Critical Reviews Oncology Hematology* 2015;93:425–436. Doi:10.1016/j.critrevonc.2015.07.008

[20] Inoue S, Lemonnier F, Mak TW. Roles of IDH1/2 and TET2 mutations in myeloid disorders. *International Journal of Hematology*, 2016;103:627–633.

[21] Karimi M, Johansson S, Stach D, et al. LUMA (LUminometric Methylation Assay): A high throughput method to the analysis of genomic DNA methylation. *Experimental Cell Research*, 2006;312:1989–1995.

[22] Feinberg AP, Vogelstein, B. Hypomethylation distinguishes genes of some human cancers from their normal counterparts. *Nature* 1983;301:89–92.

[23] Clarke L, Fairley S, Zheng-Bradley X, et al. The International Genome Sample Resource (IGSR): A worldwide collection of genome variation incorporating the 1000 Genomes Project data. *Nucleic Acids Research*, 2017;45:D854–D859.

[24] Booth MJ, Ost TW, Beraldi D, et al. Oxidative bisulfite sequencing of 5-methylcytosine and 5-hydroxymethylcytosine. *Nature Protocols*, 2013;8:1841–1851.

[25] Krueger F, Andrews SR. Bismark: A flexible aligner and methylation caller for Bisulfite-Seq applications. *Bioinformatics*, 2011;27:1571–1572.

[26] Pedersen B, Hsieh TF, Ibarra C, Fischer, RL. MethylCoder: Software pipeline for bisulfite-treated sequences. *Bioinformatics*, 2011;27:2435–2436.

[27] Harrison A, Parle-McDermott A. DNA methylation: A timeline of methods and applications. *Frontiers in Genetics*, 2011;2:74.

[28] Ryan DP, Ehninger D. Bison: Bisulfite alignment on nodes of a cluster. *BMC Bioinformatics*, 2014;15:337.

[29] Xi Y, Li W. BSMAP: Whole genome bisulfite sequence MAPping program. *BMC Bioinformatics*, 2009;10:232.

[30] Li M, Huang P, Yan X, et al. VAliBS: A visual aligner for bisulfite sequences. *BMC Bioinformatics*, 2017;18:410.

[31] Bibikova, M. et al. High density DNA methylation array with single CpG site resolution. *Genomics*, 2011;98:288–295.

[32] Mancuso FM, Montfort M, Carreras A, Alibés A, Roma G. HumMeth27 QCReport: An R package for quality control and primary analysis of Illumina Infinium methylation data. *BMC Research Notes*, 2011;4:546.

[33] Touleimat N, Tost J. Complete pipeline for Infinium®Human Methylation 450K BeadChip data processing using subset quantile normalization for accurate DNA methylation estimation. *Epigenomics*, 2012;4:325–341.

[34] Wang D, Yan L, Hu Q, et al. IMA: An R package for high-throughput analysis of Illumina's 450K Infinium methylation data. *Bioinformatics*, 2012;28:729–730.

[35] Marabita F, Almgren M, Lindholm ME, et al. An evaluation of analysis pipelines for DNA methylation profiling using the Illumina HumanMethylation450 BeadChip platform. *Epigenetics* 2013;8:333–346.

[36] Takeda S, Paszkowski J. DNA methylation and epigenetic inheritance during plant gametogenesis. *Chromosoma*, 2006;115:27–35.

[37] How-Kit A, Daunay A, Mazaleyrat N, et al. Accurate CpG and non-CpG cytosine methylation analysis by high throughput locus-specific pyrosequencing in plants. *Plant Molecular Biology*, 2015;88:471–485.

[38] Sun S, Huang YW, Yan PS, Huang TH, Lin S. Preprocessing differential methylation hybridization microarray data. *BioData Mining*, 2011;4:13.

[39] Siegmund KD. Statistical approaches for the analysis of DNA methylation microarray data. *Human Genetics*, 2011;129:585–595.

[40] Maksimovic J, Gordon L, Oshlack A. SWAN: Subset-quantile within array normalization for illumina infinium HumanMethylation450 BeadChips. *Genome Biology*, 2012;13:R44.

[41] Du P, Kibbe WA, Lin SM. lumi: A pipeline for processing Illumina microarray. *Bioinformatics*, 2008;24:1547–1548.

[42] Teschendorff AE, Marabita F, Lechner F, et al. A beta-mixture quantile normalization method for correcting probe design bias in Illumina Infinium 450 k DNA methylation data. *Bioinformatics*, 2013;29:189–196.

[43] Park Y, Figueroa ME, Rozek LS, Sartor MA. MethylSig: A whole genome DNA methylation analysis pipeline. *Bioinformatics*, 2014;30:2414–2422.

[44] Sun D, Xi Y, Rodriguez B, et al. MOABS: model based analysis of bisulfite sequencing data. *Genome Biology*, 2014;15:R38.

[45] Dolzhenko E, Smith AD. Using beta-binomial regression for high-precision differential methylation analysis in multifactor whole-genome bisulfite sequencing experiments. *BMC Bioinformatics*, 2014;15:215.

[46] Akalin A, Kormaksson M, Li S, et al. methylKit: A comprehensive R package for the analysis of genome-wide DNA methylation profiles. *Genome Biology*, 2012;13:R87.

[47] Jaffe AE, Murakami P, Lee H, et al. Bump hunting to identify differentially methylated regions in epigenetic epidemiology studies. *International Journal of Epidemiology*, 2012;41:200–209.

[48] Hansen KD, Langmead, B, Irizarry, R. A. BSmooth: From whole genome bisulfite sequencing reads to differentially methylated regions. *Genome Biology*, 2012;13:R83.

[49] Hebestreit K, Dugas M, Klein HU. Detection of significantly differentially methylated regions in targeted bisulfite sequencing data. *Bioinformatics*, 2013;29:1647–1653.

[50] Felsenfeld G, Groudine M. Controlling the double helix. Nature 421, 448–453. 2003;

[51] Travers AA, Vaillant C, Arneodo A, Muskhelishvili G. DNA structure, nucleosome placement and chromatin remodelling: A perspective. *Biochemical Society Transactions*, 2012;40:335–340.

[52] Luger K, Hansen JC. Nucleosome and chromatin fiber dynamics. *Current Opinion in Structural. Biology*, 2005;15:188–196.

[53] Bannister AJ, Schneider R, Kouzarides T. Histone methylation: Dynamic or static? *Cell*, 2002;109:801–806.

[54] Bannister AJ, Kouzarides T. Regulation of chromatin by histone modifications. *Cell Research*, 2011;21:381–395.

[55] Peserico A, Simone C. Physical and functional HAT/HDAC interplay regulates protein acetylation balance. *Journal of Biomedicine and Biotechnology*, 2011;2011:371832.

[56] Gilmour DS, Lis JT. In vivo interactions of RNA polymerase II with genes of *Drosophila melanogaster*. *Molecular and Cellular Biology*, 1985;5:2009–2018.

[57] Mukhopadhyay A, Deplancke B, Walhout AJM, Tissenbaum HA. Chromatin immunoprecipitation (ChIP) coupled to detection by quantitative real-time PCR to study transcription factor binding to DNA in *Caenorhabditis elegans*. *Nature Protocols*, 2008;3:698–709.

[58] Robertson G, Hirst M, Bainbridge M, et al. Genome-wide profiles of STAT1 DNA association using chromatin immunoprecipitation and massively parallel sequencing. *Nature Methods*, 2007;4:651–657.

[59] Johnson DS, Mortazavi A, Myers RM, Wold B. Genome-wide mapping of in vivo protein-DNA Interactions. *Science*, 2007;316:1497–1502.

[60] Landt SG, Marinov GK, Kundaje A, et al. ChIP-seq guidelines and practices of the ENCODE and modENCODE consortia. *Genome Research*, 2012;22:1813–1831.

[61] Nakato R, Shirahige K. Recent advances in ChIP-seq analysis: From quality management to whole-genome annotation. *Briefings in Bioinformatics*. 2017;18:279–290. DOI:10.1093/bib/bbw023.

[62] ENCODE. Histone ChIP-seq Data Standards and Processing Pipeline. Current information for Histone ChIP-seq is available at www.encodeproject.org/chip-seq/histone/.

[63] Epigenie. Guide: Getting started with ChIP-seq. Available at https://epigenie.com/guide-getting-started-with-chip-seq/.

[64] Babraham Bioinformatics. FastQC: A quality control tool for high throughput sequence data. Available at www.bioinformatics.babraham.ac.uk/projects/fastqc/.

[65] Langmead B, Trapnell C, Pop M, Salzberg SL. Ultrafast and memory-efficient alignment of short DNA sequences to the human genome. *Genome Biology*, 2009;10:R25.

[66] Zhang Y, Liu T, Meyer CA, et al. Model-based Analysis of ChIP-Seq (MACS). *Genome Biology*, 2008;9:R137.

[67] Zang C, Schones DE, Zeng C, et al. A clustering approach for identification of enriched domains from histone modification ChIP-Seq data. *Bioinformatics*, 2009;25:1952–1958.

[68] Rozowsky J, Euskirchen G, Auerbach RK. et al. PeakSeq enables systematic scoring of ChIP-seq experiments relative to controls. *Nature Biotechnology*, 2009;27:66–75.

[69] Chen L, Wang C, Qin ZS, Wu H. A novel statistical method for quantitative comparison of multiple ChIP-seq datasets. *Bioinformatics*, 2015;31:1889–1896.

[70] Xu H, Wei CL, Lin F, Sung, WK. An HMM approach to genome-wide identification of differential histone modification sites from ChIP-seq data. *Bioinformatics*, 2008;24:2344–2349.

[71] Liang K, Kele S. Detecting differential binding of transcription factors with ChIP-seq. *Bioinformatics*, 2012;28:121–122.

[72] Anders S, Huber W. Differential expression analysis for sequence count data. *Genome Biology*, 2010;11:R106.

[73] Shen L, Shao NY, Liu X, et al. diffReps: Detecting differential chromatin modification sites from ChIPseq data with biological replicates. *PLoS One*, 2013;8:e65598.

[74] Robinson MD, McCarthy DJ, Smyth GK. edgeR: A bioconductor package for differential expression analysis of digital gene expression data. *Bioinformatics*, 2010;26:139–140.

[75] Feng J, Meyer CA, Wang Q, et al. GFOLD: A generalized fold change for ranking differentially expressed genes from RNA-seq data. *Bioinformatics*, 2012;28:2782–2788.

[76] Ibrahim MM, Lacadie SA, Ohler U. JAMM: A peak finder for joint analysis of NGS replicates. *Bioinformatics*, 2015;31:48–55.

[77] Shao Z, Zhang Y, Yuan GC, Orkin SH, Waxman DJ. MAnorm: A robust model for quantitative comparison of ChIP-Seq data sets. *Genome Biology*, 2012;13:R16.

[78] Allhoff M, Seré K, Chauvistré H, et al. Detecting differential peaks in ChIP-seq signals with ODIN. *Bioinformatics*, 2014;30:3467–3475.

[79] Zhang Y, Lin YH, Johnson TD, Rozek LS, Sartor MA. PePr: A peak-calling prioritization pipeline to identify consistent or differential peaks from replicated ChIP-Seq data. *Bioinformatics*, 2014;30:2568–2575.

[80] Song Q, Smith AD. Identifying dispersed epigenomic domains from ChIP-Seq data. *Bioinformatics*, 2011;27:870–871.

[81] Allhoff M, Seré K, Pires JF, Zenke M, Costa IG. Differential peak calling of ChIP-seq signals with replicates with THOR. *Nucleic Acids Research*, 2016;44:e153.

[82] Ross-InnesCS, Stark R, Teschendorff AE, et al. Differential oestrogen receptor binding is associated with clinical outcome in breast cancer. *Nature*, 2012;481:389–393.

[83] Allis CD, Jenuwein T. The molecular hallmarks of epigenetic control. *Nature Reviews Genetics*, 2016;17:487–500.

[84] Dixon JR, Selvaraj S, Yue F, et al. Topological domains in mammalian genomes identified by analysis of chromatin interactions. *Nature*, 2012;485(7398):376–380.

[85] Splinter E, Heath H, Kooren J, et al. CTCF mediates long-range chromatin looping and local histone modification in the β-globin locus. *Genes & Development*, 2006;20:2349–2354.

[86] Ong C, Corces VG. CTCF: An architectural protein bridging genome topology and function. *Nature Reviews Genetics*, 2014;15:234–246.

[87] Lichter P, Cremer T, Borden J, Manuelidis L, Ward DC. Delineation of individual human chromosomes in metaphase and interphase cells by in situ suppression hybridization using recombinant DNA libraries. *Human Genetics*, 1988;80:224–234.

[88] Gaszner M, Felsenfeld G. Insulators: Exploiting transcriptional and epigenetic mechanisms. *Nature Reviews Genetics*, 2006;7:703–713.

[89] Burgess-Beusse B, Farrell C, Gaszner M, et al. The insulation of genes from external enhancers and silencing chromatin. *Proceedings of the National Academy of Sciences of the USA*, 2002;99(Suppl 4):16433–16437.

[90] Calo E, Wysocka J. Modification of enhancer chromatin: What, how, and why? *Molecular Cell*, 2013;49:825–837.

[91] Bulger M, Groudine M. Enhancers: The abundance and function of regulatory sequences beyond promoters. *Developmental Biology*, 2010;339:250–257.

[92] Dekker J. Capturing chromosome conformation. *Science*, 2002;295:1306–1311.

[93] de Wit E, de Laat W. A decade of 3C technologies: Insights into nuclear organization. *Genes & Development*, 2012;26:11–24.

[94] Dekker J. The three 'C' s of chromosome conformation capture: Controls, controls, controls. *Nature Methods*, 2006;3:17–21.

[95] Dostie J, Richmond TA, Arnaou RA, et al. Chromosome Conformation Capture Carbon Copy (5C): A massively parallel solution for mapping interactions between genomic elements. *Genome Research*, 2006;16:1299–1309.

[96] van de Werken HJG, de Vree PJ, Splinter E, et al. 4C technology: Protocols and data analysis. *Methods in Enzymology*, 2012;513:89–112.

[97] Wu C, Allis CD. *Nucleosomes, Histones & Chromatin*. Elsevier;2012.

[98] Simonis M, Klous P, Splinter E, et al. Nuclear organization of active and inactive chromatin domains uncovered by chromosome conformation capture-on-chip (4C). *Nature Genetics*, 2006;38:1348–1354.

[99] Fullwood MJ, Liu MH, Pan YF, et al. An oestrogen-receptor–bound human chromatin interactome. *Nature*, 2009;462:58–64.

[100] Forcato M, Nicoletti C, Pal K, et al. Comparison of computational methods for Hi-C data analysis. *Nat. Methods*, 2017;14:679–685.

[101] Ay F, Noble WS. Analysis methods for studying the 3D architecture of the genome. *Genome Biology*, 2015;16:183.

[102] Lajoie BR, Dekker J, Kaplan N. The hitchhiker's guide to Hi-C analysis: Practical guidelines. *Methods* 2015;72:65–75.

[103] Li W, Gong K, Li Q, Alber F, Zhou XJ. Hi-Corrector: A fast, scalable and memory efficient package for normalizing large-scale Hi-C data. *Bioinformatics*, 2015;31(6):960–962. DOI:10.1093/bioinformatics/btu747.

[104] Sauria ME, Phillips-Cremins JE, Corces VG, Taylor J. HiFive: A tool suite for easy and efficient HiC and 5C data analysis. *Genome Biology*, 2015;16:237.

[105] Hu M, Deng K, Selvaraj S, et al. HiCNorm: Removing biases in Hi-C data via Poisson regression. *Bioinformatics*, 2012;28:3131–3133.

[106] Lun ATL, Smyth GK. diffHic: A bioconductor package to detect differential genomic interactions in Hi-C data. *BMC Bioinformatics*, 2015;16:258.

[107] Ay F, Bailey TL, Noble WS. Statistical confidence estimation for Hi-C data reveals regulatory chromatin contacts. *Genome Research*, 2014;24:999–1011.

[108] Mifsud B, Martincorena I, Darbo E, et al. GOTHiC, a probabilistic model to resolve complex biases and to identify real interactions in Hi-C data. *PLoS One*, 2017;12:e0174744.

[109] Castellano G, Le Dily F, Pulido AH, Beato M, Roma G. HiC-inspector: A toolkit for high-throughput chromosome capture data. bioRxiv preprint available at www.biorxiv.org/content/biorxiv/early/2015/06/18/020636.full.pdf. 2015.

[110] Servant N, Varoquaux N, Lajoie BR, et al. HiC-Pro: An optimized and flexible pipeline for Hi-C data processing. *Genome Biology*, 2015;16:259.

[111] Schmid MW, Grob S, Grossniklaus U. HiCdat: A fast and easy-to-use Hi-C data analysis tool. *BMC Bioinformatics*, 2015;16:277.

[112] Imakaev M. Documentation for the Hi-C data analysis library by Leonid Mirny lab: HiC correction library 0.9 documentation. 2013. Available online at https://mirnylab.bitbucket.io/hiclib/.

[113] Wingett S, Ewels P, Furlan-Magaril M, et al. HiCUP: Pipeline for mapping and processing Hi-C data. *F1000Research*, 2015;4:310.

[114] Hwang YC, Lin CF, Valladares O, et al. HIPPIE: A high-throughput identification pipeline for promoter interacting enhancer elements. *Bioinformatics*, 2015; 31:1290–1292. DOI:10.1093/bioinformatics/btu801.

[115] Homer (webpage): Software for motif discovery and next-gen sequencing analysis. Available online at http://homer.ucsd.edu/homer/interactions/.

[116] Durand NC, Shamim, MS, Machol I, et al. Juicer provides a one-click system for analyzing loop-resolution Hi-C experiments. *Cell Systems*, 2016;3: 95–98.

[117] Mora A, Sandve GK, Gabrielsen OS, Eskeland R. In the loop: Promoter–enhancer interactions and bioinformatics. *Briefings in Bioinformatics*, 2015;17:980–995. DOI:10.1093/bib/bbv097.

[118] Yardımcı GG, Nobl WS. Software tools for visualizing Hi-C data. *Genome Biology*, 2017;18:26.

[119] Wang Y, Zhang B, Zhang L, et al. The 3D Genome Browser: a web-based browser for visualizing 3D genome organization and long-range chromatin interactions. bioRxiv preprint available at www.biorxiv.org/content/early/2017/02/27/112268.

[120] Zhou X, Lowdon RF, Li D, et al. Exploring long-range genome interactions using the WashU Epigenome Browser. *Nature Methods*, 2013;10:375–376.

[121] Paulsen J, Sandve GK, Gundersen S, et al. HiBrowse: Multi-purpose statistical analysis of genome-wide chromatin 3D organization. *Bioinformatics*, 2014;30:1620–1622.

[122] Durand NC. Robinson JT, Shamim MS, et al. Juicebox provides a visualization system for Hi-C contact maps with unlimited zoom. *Cell Systems*, 2016;3:99–101.

[123] Lajoie BR, van Berkum NL, Sanyal A, Dekker J. My5C: Web tools for chromosome conformation capture studies. *Nat. Methods* 2009;6:690–691.

[124] Levy-Leduc C, Delattre M, Mary-Huard T, Robin S. Two-dimensional segmentation for analyzing Hi-C data. *Bioinformatics*, 2014;30:i386–i392.

[125] Serra F, Baù D, Goodstadt M, et al. Automatic analysis and 3D-modelling of Hi-C data using TADbit reveals structural features of the fly chromatin colors. *PLOS Computational Biology*, 2017;13:e1005665.

[126] Weinreb C, Raphael BJ. Identification of hierarchical chromatin domains. *Bioinformatics* 2016;32:1601–1609.

[127] Lukášová E, Kozubek S, Kozube M, et al. Localisation and distance between ABL and BCR genes in interphase nuclei of bone marrow cells of control donors and patients with chronic myeloid leukaemia. Human *Genetics*, 1997;100:525–535.

[128] Neves H, Ramos C, da Silva MG, Parreira A, Parreira L. The nuclear topography of ABL, BCR, PML, and RARalpha genes: Evidence for gene proximity in specific phases of the cell cycle and stages of hematopoietic differentiation. *Blood*, 1999;93:1197–1207.

[129] Bártová E, Kozubek S, Kozube M, et al. The influence of the cell cycle, differentiation and irradiation on the nuclear location of the *abl*, *bcr* and *c-myc* genes in human leukemic cells. *Leukemia Research*, 2000;24:233–241.

[130] Kozubek S, Lukášová E, Marečková A, et al. The topological organization of chromosomes 9 and 22 in cell nuclei has a determinative role in the induction of t(9,22) translocations and in the pathogenesis of t(9,22) leukemias. *Chromosoma*, 1999;108:426–435.

[131] Engreitz JM, Agarwala V, Mirny LA. Three-dimensional genome architecture influences partner selection for chromosomal translocations in human disease. *PLoS One*, 2012;7:e44196.

[132] Roix JJ, McQueen PG, Munson PJ, Parada LA, Misteli T. Spatial proximity of translocation-prone gene loci in human lymphomas. *Nature Genetics*, 2003;34:287–291.

[133] Djebali S, Davis CA, Merkel A, et al. Landscape of transcription in human cells. *Nature*, 2012;489:101–108.

[134] Froberg JE, Yang L, Lee JT. Guided by RNAs: X-inactivation as a model for lncRNA function. *Journal of Molecular Biology*, 2013;425:3698–3706.

[135] Lee JT. Lessons from X-chromosome inactivation: Long ncRNA as guides and tethers to the epigenome. *Genes & Development*, 2009;23:1831–1842.

[136] Mercer TR, Dinger ME, Mattick JS. Long non-coding RNAs: Insights into functions. *Nature Reviews Genetics*, 2009;10:155–159.

[137] Derrien T, Johnson R, Bussotti G, et al. The GENCODE v7 catalog of human long noncoding RNAs: Analysis of their gene structure, evolution, and expression. *Genome Research*, 2012;22:1775–1789.

[138] Guttman M, Amit I, Garber M, et al. Chromatin signature reveals over a thousand highly conserved large non-coding RNAs in mammals. *Nature*, 2009;458:223–227.

[139] Iyer MK, Niknafs YS, Malik R, et al. The landscape of long noncoding RNAs in the human transcriptome. *Nature Genetics*, 2015;47:199–208.

[140] Devaux Y, Zangrando J, Schroen B, et al. Long noncoding RNAs in cardiac development and ageing. *Nature Reviews Cardiology*, 2015;12:415–425.

[141] Quinn JJ, Chang HY. Unique features of long non-coding RNA biogenesis and function. *Nature Reviews Genetics*, 2016;17:47–62.

[142] Schmitz KM, Mayer C, Postepska A, Grummt I. Interaction of noncoding RNA with the rDNA promoter mediates recruitment of DNMT3b and silencing of rRNA genes. *Genes & Development*, 2010;24:2264–2269.

[143] Mancini-Dinardo D, Steele SJS, Levorse JM, Ingram RS, Tilghman SM. Elongation of the Kcnq1ot1 transcript is required for genomic imprinting of neighboring genes. *Genes & Development*, 2006;20:1268–1282.

[144] Mao YS, Sunwoo H, Zhang B, Spector DL. Direct visualization of the cotranscriptional assembly of a nuclear body by noncoding RNAs. *Nature Cell Biology*, 2011;13:95–101.

[145] Hacisuleyman E, Goff LA, Trapnell C, et al. Topological organization of multichromosomal regions by the long intergenic noncoding RNA Firre. *Nature Structural & Molecular Biology*, 2014;21:198–206.

[146] Tsai MC, Manor O, Wan Y, et al. Long noncoding RNA as modular scaffold of histone modification complexes. *Science*, 2010;329:689–693.

[147] Pachnis V, Brannan CI, Tilghman SM. The structure and expression of a novel gene activated in early mouse embryogenesis. *The EMBO Journal*, 1988;7:673–681.

[148] Rinn JL, Kertesz M, Wang JK, et al. Functional demarcation of active and silent chromatin domains in human HOX loci by noncoding RNAs. *Cell*, 2007;129:1311–1323.

[149] He Y, Vogelstein B, Velculescu VE, Papadopoulos N, Kinzler KW. The antisense transcriptomes of human cells. *Science*, 2008;322:1855–1857.

[150] Katayama S, Tomaru Y, Kasukawa T, et al. Antisense transcription in the mammalian transcriptome. *Science*, 2005;309:1564–1566.

[151] Wang KC, Chang HY. Molecular mechanisms of long noncoding RNAs. *Molecular Cell*, 2011;43:904–914.

[152] Schmitt AM, Chang HY. Long noncoding RNAs in cancer pathways. *Cancer Cell*, 2016;29:452–463.

[153] Rubin, EH, Allen JD, Nowak JA, Bates SE. Developing precision medicine in a global world. *Clinical Cancer Research*, 2014;20:1419–1427.

[154] West M, Ginsburg GS, Huang AT, Nevins JR. Embracing the complexity of genomic data for personalized medicine. *Genome Research*, 2006;16: 559–566.

[155] Castaneda C, Nalley K, Mannion C et al. Clinical decision support systems for improving diagnostic accuracy and achieving precision medicine. *Journal of Clinical Bioinformatics*, 2015;5:4.

[156] Wahlestedt C. Targeting long non-coding RNA to therapeutically upregulate gene expression. *Nature Reviews Drug Discovery*, 2013;12:433–446.

[157] Wang L, Nie J, Sicotte H, et al. Measure transcript integrity using RNA-seq data. *BMC Bioinformatics*, 2016;17:58.

[158] Frith MC, Pheasant M, Mattick JS. Genomics: The amazing complexity of the human transcriptome. *European Journal of Human Genetics*, 2005;13, 894–897.

[159] Adams J. Transcriptome: Connecting the genome to gene function. *Nature Education*, 2008;1(1):195. See definition at www.nature.com/scitable/definition.

[160] Audoux J, Salson M, Grosset CF, et al. SimBA: A methodology and tools for evaluating the performance of RNA-Seq bioinformatic pipelines. *BMC Bioinformatics*, 2017;18:428.

[161] Khomtchouk BB, Hennessy JR, Wahlestedt C. MicroScope: ChIP-seq and RNAseq software analysis suite for gene expression heatmaps. *BMC Bioinformatics*, 2016;17:390.

[162] Kukurba KR, Montgomery SB. RNA Sequencing and Analysis. *Cold Spring Harbor Protocols*, 2015;2015:951–969.

[163] Yotsukura S, duVerle D, Hancock T, Natsume-Kitatani Y, Mamitsuka H. Computational recognition for long non-coding RNA (lncRNA): Software and databases. *Briefings in Bioinformatics*, 2017;18:9–27.

[164] Yu N, Yu Z, Pan Y. A deep learning method for lincRNA detection using auto-encoder algorithm. *BMC Bioinformatics*, 2017;18:511.

[165] Weirick T, Militello G, Müller R, et al. The identification and characterization of novel transcripts from RNA-seq data. *Briefings in Bioinformatics*, 2016;17:678–685.

[166] The GENCODE Project: Encyclopedia of genes and gene variants available online at https://www.gencodegenes.org; 2018.

[167] Accessing Ensembl data/FTP download available online at www.ensembl.org/info/data/index/html. 2018.

[168] Volders PJ, Verheggen K, Menschaert G, et al. An update on LNCipedia: A database for annotated human lncRNA sequences. *Nucleic Acids Research*, 2015;43:4363–4364.

[169] Fang S, Zhang, Jin LL, Guo C, et al. NONCODEV5: A comprehensive annotation database for long non-coding RNAs. *Nucleic Acids Research*, 2018;46:D308–D314.

[170] Park C, Yu N, Choi I, Kim W, Lee S. lncRNAtor: A comprehensive resource for functional investigation of long non-coding RNAs. *Bioinformatics*, 2014;30:2480–2485.

[171] Li J, Han L, Roebuck P, et al. TANRIC: An interactive open platform to explore the function of lncRNAs in cancer. *Cancer Research*, 2015;75:3728–3737.

[172] Zhou KR, Liu S, Wen-Ju Sun WJ, et al. ChIPBase v2.0: Decoding transcriptional regulatory networks of noncoding RNAs and protein-coding genes from ChIP-seq data. *Nucleic Acids Research*, 2017;45:D43–D50.

[173] Ning S, Zhang J, Wang P, et al. Lnc2Cancer: A manually curated database of experimentally supported lncRNAs associated with various human cancers. *Nucleic Acids Research*, 2016;44:D980–D985.

[174] Chen G, Wang Z, Wang DQ, et al. LncRNADisease: A database for long-non-coding RNA-associated diseases. *Nucleic Acids Research*, 2013;41:D983–D986.

[175] Bhartiya D, Pal K, Ghosh S, et al. lncRNome: A comprehensive knowledgebase of human long noncoding RNAs. *Database*, 2013;2013:bat034.

[176] Quek XC, Thomson DW, Maag JLV, et al. lncRNAdb v2.0: Expanding the reference database for functional long noncoding RNAs. *Nucleic Acids Research*, 2015;43:D168–D173.

[177] Zhou B, Zhao H, Yu J, et al. EVLncRNAs: A manually curated database for long non-coding RNAs validated by low-throughput experiments. *Nucleic Acids Research*, 2018;46:D100–D105.

[178] Conesa A, Madrigal P, Tarazona S, et al. A survey of best practices for RNA-seq data analysis. *Genome Biology*, 2016;17:13.

[179] LoVerso PR, Cui F. A computational pipeline for cross-species analysis of RNA-seq data using R and bioconductor. *Bioinformatics and Biology Insights*, 2015;9:165–174.

[180] Galaxy Project Platform. Available online at http://usegalaxy.org.018.

[181] Liu K, Beck D, Thoms JAI, et al. Annotating function to differentially expressed LincRNAs in myelodysplastic syndrome using a network-based method. *Bioinformatics*, 2017;33:2622–2630.

[182] Chen X. KATZLDA: KATZ measure for the lncRNA-disease association prediction. *Science Reports*, 2015;5:16840.

[183] Wang Z, Fang H, Tang NL, Deng M. VCNet: Vector-based gene co-expression network construction and its application to RNA-seq data. *Bioinformatics*, 2017;33:2173–2181.

[184] Huang YA, Chan KCC, You ZH. Constructing prediction models from expression profiles for large scale lncRNA-miRNA interaction profiling. *Bioinformatics*. 2017;34:812–819. DOI:10.1093/bioinformatics/btx672.

[185] Chen X, Yan CC, Zhang X, You ZH. Long non-coding RNAs and complex diseases: From experimental results to computational models. *Briefings in Bioinformatics*, 2017;18:558–576.

[186] Yan Y, Huang SY. RRDB: A comprehensive and nonredundant benchmark for RNARNA docking and scoring. *Bioinformatics*, 2017;34:453–458. DOI:10.1093/bioinformatics/btx615.

[187] Fukunaga T, Hamada M. RIblast: An ultrafast RNA–RNA interaction prediction system based on a seed-and-extension approach. *Bioinformatics*, 2017;33:2666–2674.

[188] Zheng X, Wang Y, Tian K, et al. Fusing multiple protein-protein similarity networks to effectively predict lncRNA-protein interactions. *BMC Bioinformatics*, 2017;18:420.

[189] Xiao Y, Zhang J, Deng L. Prediction of lncRNA-protein interactions using HeteSim scores based on heterogeneous networks. *Science Reports*, 2017;7:3664.

[190] Li J, Ma W, Zeng P, et al. LncTar: A tool for predicting the RNA targets of long noncoding RNAs. *Briefings in Bioinformatics*, 2015;16:806–812.

[191] Navarin N, Costa F. An efficient graph kernel method for non-coding RNA functional prediction. *Bioinformatics*, 2017;33:2642–2650.

[192] Ulitsky I, Bartel DP. lincRNAs: Genomics, evolution, and mechanisms. *Cell* 2013;154:26–46.

[193] The ENCODE (ENCyclopedia Of DNA Elements) Project. *Science*, 2004;306:636–640.

[194] Mouse ENCODE Consortium, Stamatoyannopoulos JA, Snyder M, et al. An encyclopedia of mouse DNA elements (Mouse ENCODE). *Genome Biology*, 2012;13:418.

[195] The modENCODE Consortium, Roy S, Ernst J, et al. Identification of functional elements and regulatory circuits by *Drosophila* modENCODE. *Science*, 2010;330:1787–1797.

[196] Gerstein MB, Lu ZJ, Van Nostrand EL, et al. Integrative analysis of the *Caenorhabditis elegans* genome by the modENCODE project. *Science*, 2010;330:1775–1787.

[197] Roadmap Epigenomics Consortium, Kundaje A, Meuleman W, et al. Integrative analysis of 111 reference human epigenomes. *Nature*, 2015;518:317–330.

[198] Gjoneska E, Pfenning AR, Mathys H, et al. Conserved epigenomic signals in mice and humans reveal immune basis of Alzheimer's disease. Nature, 2015;518:365–369.

[199] Polak P, Karlić R, Koren A, et al. Cell-of-origin chromatin organization shapes the mutational landscape of cancer. Nature, 2015;518:360–364.

[200] Kodzius R Kojima M, Nishiyori H, et al. CAGE: Cap analysis of gene expression. *Nature Methods*, 2006;3:211–222.

[201] Takahashi H, Kato S, Murata M, Carninci P. CAGE (cap analysis of gene expression): a protocol for the detection of promoter and transcriptional networks. *Methods in Molecular Biology*, 2012;786:181–200.

[202] Arner E, Daub CO, Vitting-Seerup K, et al. Transcribed enhancers lead waves of coordinated transcription in transitioning mammalian cells. Science, 2015;347:1010–1014.

[203] Andersson R, Gebhard C, Miguel-Escalada I, et al. An atlas of active enhancers across human cell types and tissues. *Nature*, 2014;507:455–461.

[204] Kaczkowsk, B, Tanaka Y, Kawaji H, et al. Transcriptome analysis of recurrently deregulated genes across multiple cancers identifies new pan-cancer biomarkers. *Cancer Research*, 2016;76:216–226.

[205] Weinstein JN, Collisson EA, Mills GB, et al. The Cancer Genome Atlas Pan-Cancer analysis project. *Nature Genetics*, 2013;45:1113–1120.

[206] Bujold D, de Lima Morais DA, Carol Gauthier, et al. The International Human Epigenome Consortium Data Portal. *Cell Systems*, 2016;3:496–499.e2.

[207] Adams D, Altucci L, Antonarakis SE, et al. BLUEPRINT to decode the epigenetic signature written in blood. *Nature Biotechnology*, 2012;30:224–226.

3 Introduction to Graph and Network Theory

Thomas Gaudelet and Nataša Pržulj

3.1 Motivation

Hardly any entity exists in absolute isolation of others. It is usually a part of a larger system of greater complexity. This remains true at any scale. For instance, stars and planets interact via different forces of attraction and repulsion that shape the universe. Species interact tightly through predator–prey relationships to form an often fragile ecosystem. Neurons in a brain interact to carry out complex tasks ranging from limb motion to information processing. Studying single entities in isolation of others is crucial to our understanding. However, it is also essential to study their roles as part of wider systems of interacting components.

The organization of such systems is far from random and reflects the function of the system. Understanding the organizational principles of the systems can shed light on the impact of changes on the system or on its environment. A question could be, for instance, to investigate what happens if a species in a food network was to become extinct; the balance of the predator–prey interactions and competition for food could be seriously impacted. Similarly, careful consideration needs to be taken when trying to (re)introduce a species to a given environment not to upset the fragile equilibrium of the local food network.

Graphs (networks) are means of representing entities and their interactions, while abstracting much of the data about them. These representations are needed to develop tools to investigate the organization of complex systems. Everyday examples of network representations of real systems include maps of public transport networks, which connect various stations, and the World Wide Web. We formally introduce graph theory in Section 3.3.

In biology, many systems can naturally be modeled by networks. Here, we briefly introduce a number of biological networks that we will use to illustrate the theory. Some of those networks are discussed in detail in Chapters 4 and 10. Table 3.1 summarizes these networks.

Protein–protein interaction (PPI) networks represent physical interactions between proteins (detailed in Chapter 4). They are generally represented with *simple graphs*, introduced in Section 3.3.

Table 3.1: Existing databases for protein–protein interaction (PPI) networks, transcriptional regulatory networks (TR), metabolic networks (M), cell signaling networks (CS), genetic interaction networks (GI), disease-gene networks (DG), drug-target networks (DT), and protein structure networks (PS).

Databases	Url	Networks	Reference
KEGG	www.genome.jp/kegg	M; TR	[14]
GeneDB	www.genedb.org	M	[15]
BioCyc, EcoCyc, MetaCyc, HumanCyc	biocyc.org	M; TR	[16, 17, 18]
MetaTIGER	www.bioinformatics.leeds.ac.uk/metatiger	M	[19]
ERGO (previously known as WIT)	ergo.integratedgenomics.com	M	[20]
GeneNet	wwwmgs.bionet.nsc.ru/mgs/systems/genenet	TR	[21]
Reactome	reactome.org	TR	[22, 23]
RegulonDB	regulondb.ccg.unam.mx	TR	[24]
JASPAR	jaspar.cgb.ki.se	TR	[25]
Phospho.ELM	phospho.elm.eu.org	TR	[26]
NetPhorest	netphorest.info	TR	[27]
PHOSIDA	141.61.102.18/phosida/index.aspx	TR	[28]
TRANSPATH	www.biobase.de/transpath	TR; CS	[29]
SMPDB	smpdb.ca	TR; DG	[30]
BioGRID	thebiogrid.org	PPI; GI	[31]
HPRD	www.hprd.org	PPI	[32]
SGD	www.yeastgenome.org	PPI	[33]
IntAct	www.ebi.ac.uk/intact	PPI	[34]
HPID	www.hpid.org	PPI	[35]
DroID	www.droidb.org	PPI	[36]
MIPS	mips.gsf.de	PPI	[37]
DIP	dip.doe-mbi.ucla.edu/dip/Main.cgi	PPI	[38]
MINT	mint.bio.uniroma2.it	PPI	[39]
IID (previously known as I2D/OPHID)	ophid.utoronto.ca/ophidv2.204	PPI	[40]
STRING	string-db.org	PPI	[41]
MiST	mistdb.com	CS	[42]
SynlethDB	histone.sce.ntu.edu.sg/SynLethDB	GI	[43]
COXPRESdb	coxpresdb.jp	COEX	[44]
GeneFriends	genefriends.org	COEX	[45]
OMIM	www.omim.org	DG	[46]
DrugBank	www.drugbank.ca	DT	[47]
MATADOR	matador.embl.de	DT	[48]
CTD	ctdbase.org	DT	[49]
PDB	www.wwpdb.org	PS	[50]

Transcriptional regulation networks [1, 2, 3] model gene expression regulation. Specific genes code for proteins that regulate the expression of other genes; such a protein is called a *transcription factor*. A transcriptional regulation network is a simplified representation of this phenomenon, where gene X is connected to gene Y if the protein product of X controls the expression of Y. Such a relation is *asymmetric*, as X controls the expression of Y, but Y does not influence the expression of X. Systems having asymmetric interactions are typically represented by *directed graphs*, introduced in Section 3.3. In this representation of transcriptional regulation, a shortcoming is that there is no difference between repression and enhancement of gene expression. This can be remedied by using *weighted graphs* to represent these phenomena, introduced in Section 3.3.

Cell signaling networks [4, 5, 6] model the complex communication processes within a cell, or between a cell and the extracellular environment. A cell signaling network captures the sequence of transient interactions between proteins that transduce a signal from the cell membrane to the nucleus. As a result, transcription happens in the nucleus enabling the cell to react to the received environmental signal. A signaling pathway corresponds to an ordered sequence of signal transduction reactions, a cascade of reversible chemical modifications of proteins that enable the modified proteins to interact with other proteins and hence transduce the signal. All signaling pathways of a cell form its signaling network. Cell signaling networks are modeled with *directed* graphs (defined in Section 3.3).

Metabolic networks [7, 8, 9] model the metabolism of a cell. The set of all metabolic reactions in a cell forms a metabolic network. A metabolic reaction transforms one metabolite (small molecule) into another and is catalyzed by an enzyme (protein). In short, a metabolic reaction involves at least two metabolites and an enzyme. The interactions are asymmetric, as the biochemical reactions involved are typically irreversible. There exist various network representations of metabolic reactions. For instance, *bipartite graphs* (defined in Section 3.3) can be used to represent them as follows. Links exist only between metabolites and enzymes (and not between two metabolites, or between two enzymes) such that a metabolite is linked to the enzymes catalyzing a reaction in which it participates. That is, an enzyme is linked to metabolites that are inputs and outputs of the reaction that it catalyzes. Also, there exist metabolite-centric and enzyme-centric network representations. In the metabolite-centric representation, only metabolites and their interactions are modeled by the metabolic network; metabolites are connected if there is an enzyme that mediates the reaction that transforms one metabolite into the other. In the enzyme-centric representation, only enzymes and interactions between them are modeled; two enzymes are connected if they catalyze reactions involving the same metabolite. The choice of a metabolic network representation depends on the problem being studied. For instance, if it is the enzymes that one is interested in analyzing, then enzyme-centric representation should be used.

Gene co-expression networks [10] represent the correlation of the expression of genes over time. Two genes are connected if their expression is significantly correlated over time. Gene co-expression networks are typically modeled by undirected graphs.

Genetic interaction networks [11, 12] link genes that genetically interact. A *genetic interaction* is the phenomenon when the effects of one gene are modified by one or more other genes [13]. They are usually measured as follows. When a single gene is mutated, we observe the growth of the mutated cell (called fitness); we measure the growth of the mutated cell as the percentage of the growth of the wild type (non-mutated) cell. Then we mutate a pair of genes, A and B, in the same cell (double mutant) and measure the cell's growth. We expect the growth of the double mutant to be a function of the growths of the two single mutants, i.e., of the cell with only gene A mutated and the cell with only gene B mutated. However, when the growth of the double mutant is significantly different from the expected combined growths of the single mutants, then we say that genes A and B genetically interact. These genetic interactions are negative if the growth of the double mutant is smaller than expected from the combined growths of the single mutants, with the extreme case being "synthetic lethality" meaning that the double mutant cell dies while single mutants live. They are positive if the growth of the double mutant is larger than expected from the combined growth of the single mutants. Genetic interaction networks are generally modeled by undirected graphs.

Biological networks also include *neuronal networks* modeling the wiring of neurons a brain [51], *disease-gene networks* [52] modeling the genes affected by the same diseases, *drug-target networks* [53] modeling the proteins that are targeted by the same drugs, and *protein structure networks* modeling the spatial organization of amino acid residues of a protein's three-dimensional structure [50].

This chapter is structured as follows: First we introduce notions from linear algebra and algorithmic complexity that are needed to describe algorithms introduced subsequently. Then, we introduce notions of graph and network theory. Graphs and networks refer to the same mathematical formalism. A few practice exercises are given at the end of the chapter.

3.2 Background

In this section, we introduce the mathematical and algorithmic notions that will be used throughout this chapter.

3.2.1 Mathematical Background

We start with the linear algebra background necessary to navigate this Chapter.

A *set* is a collection of unique elements. Braces are used to denote a set, for instance, $A = \{1, 2, 3, 4\}$, or $B = \{a, d, z\}$ are sets. The empty set is denoted by the symbol $\emptyset$, the set containing all positive integers is denoted by $\mathbb{N}$, and the set $\mathbb{R}$ denotes the set of all real numbers. The cardinality of a set is the number of elements it contains and is denoted by $|.|$, for instance, $|A| = 4$ and $|B| = 3$ in the above examples. A set is *finite* if it contains a finite number of elements. Both $\mathbb{N}$ and $\mathbb{R}$ are *infinite* sets. An *ordered set* is a set where the elements are ordered. A *subset* A of set S, is a set of elements included in S, $A \subseteq S$.

A *matrix* is a two-dimensional ordered set of elements arranged in rows and columns. The size of a matrix is given by its number of rows m and its number of columns n, denoted by $m \times n$. If M is a matrix of size $m \times n$ containing real numbers, we write $M \in \mathbb{R}^{m \times n}$. In the following, we only consider real matrices. The entry of M located on the ith row and j^{th} column is denoted by M_{ij}. A *diagonal matrix* is a square matrix in which all off-diagonal entries are zero.

Row vectors and *column vectors* are special cases of a matrix where $m = 1$ and $n = 1$, respectively. They are generally denoted with bold lower case letters. Henceforth, we simply refer to column vectors of dimension $m \times 1$ as *vectors* of dimension m.

The examples below illustrate a matrix $M_1 \in \mathbb{R}^{2 \times 3}$, a square matrix $M_2 \in \mathbb{R}^{3 \times 3}$, and a column vector $\mathbf{v} \in \mathbb{R}^3 (= \mathbb{R}^{3 \times 1})$:

$$M_1 = \begin{pmatrix} 1 & 2 & 3 \\ 4 & 5 & 6 \end{pmatrix}, \quad M_2 = \begin{pmatrix} 1 & 2 & 3 \\ 4 & 5 & 6 \\ 7 & 8 & 9 \end{pmatrix}, \quad \mathbf{v} = \begin{pmatrix} 1 \\ 2 \\ 3 \end{pmatrix}. \tag{3.1}$$

3.2.1.1 Matrix Operations

The *transpose* of a matrix $M \in \mathbb{R}^{m \times n}$ is denoted by M^T and is the matrix where the rows of M^T correspond to the columns of M, and vice versa. We have $M^T \in \mathbb{R}^{n \times m}$. The entry in the ith row and j^{th} column of M^T corresponds to the entry in the jth row and i^{th} column of M, i.e. $\forall (i,j) \in [1,n] \times [1,m] : M^T_{ij} = M_{ji}$. The transpose of a column vector is a row vector and vice versa. The transposes of the above example matrices, M_1, M_2, and $\mathbf{v}$, are

$$M_1^T = \begin{pmatrix} 1 & 4 \\ 2 & 5 \\ 3 & 6 \end{pmatrix}, \quad M_2^T = \begin{pmatrix} 1 & 4 & 7 \\ 2 & 5 & 8 \\ 3 & 6 & 9 \end{pmatrix}, \quad \mathbf{v}^T = \begin{pmatrix} 1 & 2 & 3 \end{pmatrix}. \tag{3.2}$$

The sum of two matrices L and M in $\mathbb{R}^{m \times n}$ is done by summing entry-wise the elements of the two matrices, i.e., if we denote the sum by $S = L + M$, then $S_{ij} = L_{ij} + M_{ij}$.

For two matrices $L \in \mathbb{R}^{m \times n}$ and $M \in \mathbb{R}^{p \times q}$, the matrix product $P = LM$ is defined if and only if $n = q$. We have $P \in \mathbb{R}^{m \times q}$ and the entries of the product are $P_{ij} = \sum_{k=1}^{n} L_{ik} M_{kj}$. The product of two vectors $\mathbf{l}$ and $\mathbf{m}$, of same dimension, is called the *dot product* and denoted by $\mathbf{p} = \mathbf{l} \cdot \mathbf{m}$. It is equivalent to the matrix product as $\mathbf{l} \cdot \mathbf{m} = \mathbf{l}^T \mathbf{m}$. The existence of the product LM does not imply the existence of the product ML and even in cases where both are defined, we do not generally have $ML = LM$.

3.2.1.2 Special Matrices

A *null matrix* is a matrix where all entries are zero. A *square* matrix is a matrix that has the same number of rows and columns, i.e., $m = n$. Henceforth, we only consider square matrices.

An *identity matrix* is a diagonal matrix where all the diagonal entries are equal to one. The identity matrix of dimension $n \times n$ is denoted by I_n, or I if the dimension is clear from the context. For any matrix $M \in \mathbb{R}^{m \times n}$, we have $MI_n = M$ and $I_m M = M$.

A matrix M is *invertible* if there exists a matrix L such that $LM = ML = I$. The matrix L is then called the *inverse* of M and is denoted by M^{-1}. Clearly $\left(M^{-1}\right)^{-1} = M$, i.e., the inverse of M^{-1} is M.

A matrix M is *diagonalizable* if and only if there exists two matrices D and P such that D is diagonal, P is invertible, and $M = P^{-1} D P$.

A matrix M is *symmetric* if and only if $M^T = M$, a matrix is *antisymmetric* if and only if $M^T = -M$. An important property used in spectral graph theory (Section 3.3.7) is that if a matrix is symmetric, then it is diagonalizable.

3.2.1.3 Sets of Vectors
The vectors considered in the following definitions are assumed to have the same, arbitrary dimension.

Vector **l** is a *linear combination* of n vectors $\mathbf{v}_1, \mathbf{v}_2, \cdots, \mathbf{v}_n$ if $\mathbf{l} = \lambda_1 \mathbf{v}_1 + \lambda_2 \mathbf{v}_2 + \cdots + \lambda_n \mathbf{v}_n$ where $\{\lambda_1, \lambda_2, \ldots, \lambda_n\} \in \mathbb{R}^n$ are n scalars. A *vector space* is a collection of vectors, which may be added together and multiplied ("scaled") by numbers, called scalars. The *span* of a *vector space* is the set of all linear combinations of its elements.

A set of non-null vectors $\{\mathbf{v}_1, \mathbf{v}_2, \ldots, \mathbf{v}_n\}$ is said to be *linearly independent* if it is not possible to express any of the vectors as a linear combination of the others. In other words, the set of non-null vectors, $\{\mathbf{v}_1, \mathbf{v}_2, \ldots, \mathbf{v}_n\}$, is linearly independent if and only if $\lambda_1 \mathbf{v}_1 + \lambda_2 \mathbf{v}_2 + \cdots + \lambda_n \mathbf{v}_n = 0 \Rightarrow \lambda_1 = \lambda_2 = \cdots = \lambda_n = 0$.

A set of vectors in a vector space V is called a *basis*, or a set of basis vectors, if the vectors are linearly independent and every vector in the vector space is a linear combination of this set. In more general terms, a basis is a linearly independent spanning set.

A *norm* on a vector space $\mathcal{V}$, is a function $n : \mathcal{V} \to \mathbb{R}^+$ that assigns to each vector a positive length and verifies $n(\mathbf{v}) = 0 \Rightarrow \mathbf{v} = \mathbf{0}$. A vector space $\mathcal{V}$ defined with a norm $n_\mathcal{V}$ is called a *normed vector space*. A *unit vector* in such a space is a vector $\mathbf{v}$ such that $n_\mathcal{V}(\mathbf{v}) = 1$.

Two vectors **u** and **v** are *orthogonal* if they are perpendicular to each other, i.e., if their dot product is zero, $\mathbf{u} \cdot \mathbf{v} = 0$. It is easy to prove that a set of orthogonal vectors is linearly independent. A set of vectors is *orthonormal* if every vector is a unit vector and the vectors are mutually orthogonal. A square matrix Q is orthogonal if all its columns (or equivalently rows) are orthogonal to each other. A matrix Q is *orthonormal* if it is orthogonal and $Q^T Q = I$, which means that all its columns (or equivalently rows) are unit vectors.

3.2.1.4 Matrix Spectral Decomposition
A vector **v** of dimension n is an *eigenvector* of a matrix $M \in \mathbb{R}^{n \times n}$ if and only if there exist a scalar λ, termed *eigenvalue*, such that $M\mathbf{v} = \lambda \mathbf{v}$. The set of eigenvalues of a matrix is called its *spectrum*. The *geometric multiplicity* of an eigenvalue corresponds to the number of linearly independent eigenvectors associated with the eigenvalue. If the sum of the geometric multiplicities of the spectrum of matrix $M \in \mathbb{R}^{n \times n}$ is equal to n then the matrix is diagonalizable. We denote by $\lambda_1, \lambda_2, \ldots, \lambda_n$ the list of the eigenvalues of M with corresponding eigenvectors $\mathbf{v}_1, \mathbf{v}_2, \ldots, \mathbf{v}_n$. Note that some eigenvalues can be identical. Then we have $M = P \Lambda P^{-1}$, where Λ is the diagonal matrix with i^{th} diagonal entry corresponding to λ_i and P is the matrix where the ith column is $\mathbf{v}_i$. A crucial property of real symmetric matrices for our purposes, is that it is always possible to find a set of orthonormal eigenvectors, such that any real symmetric matrix S has a spectral decomposition of the form $S = Q \Lambda Q^T$, where Q is an orthonormal matrix.

This concludes the short introduction of linear algebra. We refer the reader to textbooks such as [54, 55] for more complete introductions to the subject area.

3.2.2 Computational Complexity

Here we present a high-level introduction to some of the concepts of computational complexity necessary to understand a part of the discussion in the following sections.

Computational complexity is a branch of theoretical computer science that focuses on classifying computational problems according to their inherent difficulty and relating those classes to each other. The complexity of a problem is captured by the complexity of the best algorithm for solving it. The complexity of an algorithm is characterized by the amount of computational resources it requires for solving an instance of the problem on an input of size n.

A Turing machine [56, 57, 58] is a mathematical model of an abstract computer which uses a set of predefined operations and rules. A problem's complexity is usually computed by considering the time required by a Turing machine to solve it with a set of suitable elementary operations. Elementary operations correspond to operators in a computer, such as addition, multiplication, or comparison. An algorithm's complexity is measured by the number of elementary operations it needs to perform to solve the problem. The complexity of an algorithm varies depending on its input, e.g., for a sorting algorithm, its running time will be different depending on the size of the input, or if the input is already sorted or not. Thus, to get a consistent measure, an algorithm is generally characterized by its worst-case complexity that measures the resources (running time, memory) an algorithm requires in the worst case. The asymptotic, "*big-O*" notation describes the limiting behavior of a function when the argument tends towards a particular value or infinity. It is used to describe the complexity of an algorithm. The letter "*O*" is used because the growth rate of a function is called the *order of the function*. For instance, an algorithm that requires $7n^2 + 3n$ operations for an input of size n has a complexity of $O(n^2)$, an algorithm that requires $5n^3 + 3n^2 + 100n + 1200$ has a complexity of $O(n^3)$, etc. So "big-O" provides the order of an upper bound to the growth rate of the function. Formally, for arbitrary functions $f(\cdot)$ and $g(\cdot)$, we have $f(x) = O(g(x))$ if and only if there exists a positive real number $c \in \mathbb{R}$ and a real number x_0 such that for all $x \geq x_0, f(x) \leq cg(x)$.

Computational problems are divided in different complexity classes, such as P, NP, NP-complete, and NP-hard, amongst others [59, 60, 61]. P stands for decision problems that can be solved on a deterministic sequential machine in the amount of time that is *polynomial* in the size of the input. Intuitively, computational problems are in P if they are solvable in polynomial time, i.e., have a polynomial (in the size of the input) upper bound on the number of operations they require. NP stands for decision problems whose solutions can be *verified* in polynomial time in the size of the input, or equivalently, whose solution can be found in polynomial time on a *non-deterministic* machine (more than one action can be performed at a given situation). The set of NP problems include the problems for which any solution is verifiable in polynomial time, but for which finding a potential solution can be costly ($O(2^n)$ for instance). The set of P problems is a subset of the set of NP problems. *NP-hard* problems are those computational problems that are *at least as hard as the hardest problem in NP*. More formally, a problem H is NP-hard if every problem L in NP can be reduced in polynomial time to H. Finally, *NP-complete* problems are those that are both NP and NP-hard.

The notions introduced in this section will be used in the following sections to characterize the hardness of the computational problems.

3.3 Graph Theory

3.3.1 Definitions

A graph, or a network, G, is defined by a set of vertices V (also called vertices, or points) and a set of edges E (also called arcs, or links) corresponding to pairs of vertices and representing interactions between vertices. A graph is formally denoted by the pair $G = (V, E)$, with $E \subseteq V \times V$. We also use the notations $(V(G), E(G))$ and (V_G, E_G) to denote unambiguously the vertex set and edge set of graph G.

An *edge* $e \in E$ corresponds to a pair of vertices $(u, v) \in V^2$. Edge $e = (u, v)$ (also denoted by uv for brevity) is said to *join* the vertices u and v, which, in turn, are called the *ends* or *endnodes* of the edge e. An edge is *incident* to both its ends and the vertices joined by an edge are called *adjacent*. If both ends of an edge correspond to the same vertex, this edge is called a *loop*. An edge can be *undirected* or *directed*. An undirected edge represents a symmetric interaction between vertices, for instance, a physical contact between two proteins in a protein–protein interaction (PPI) network. A directed edge represents an asymmetric interaction, for instance a predator–prey relationship between two species in a food network. Such edges are represented by arrows and a directed edge originates at the *source vertex* and points to the *target vertex*. If two vertices are joined by two or more edges, then we say that there is a *multi-edge* or *multiple edge* between those two vertices.

A graph is *undirected* if all its edges are undirected. A graph is *simple* if it is undirected, contains no loops, and no multi-edge. For instance, graph G of Figure 3.1 is simple. There is a wealth of theoretical results for simple graphs that can be found in any textbook dedicated to graph theory [62], illustrating their importance. In biology, PPI networks are typically modeled by simple graphs.

A graph is *directed* if all of its edges are directed. Graph H in Figure 3.1 is an example of a directed graph. In nature, food webs with directed predator–prey relationships are modeled by directed graphs. The same is true for transcriptional regulation networks.

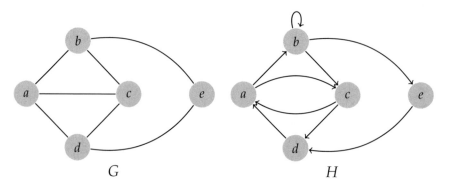

Figure 3.1: Examples illustrating: A simple graph G; a directed graph H.

Undirected and directed are the most common graphs and will be used to introduce various notions in the subsequent sections. However, in some instances, they are not sufficient to describe a given system (e.g., metabolic networks discussed in Section 3.1). Hence, we will discuss other types of graphs in Section 3.3.4.

3.3.2 Degree and Neighborhood

The *degree* of vertex u corresponds to the number of edges incident to u and is denoted by $d(u)$. The *neighborhood* of vertex u is the set of all vertices adjacent to u, formally defined as $N(u) = \{v \in V : uv \in E\}$. Consider graph G of Figure 3.2: vertex b (in red) has degree 3 ($d(b) = 3$) and the vertices in its neighbourhood are circled in red. If the edges are directed, each vertex u has an *indegree* and an *outdegree*. The indegree of u corresponds to the number of edges having u as target vertex and the outdegree of u is the number of edges having source vertex u. In graph H of Figure 3.2, these notions are illustrated on vertex d: the indegree of d is 2 and the outdegree is 1, illustrated by red and blue edges, respectively.

The following two lemmas introduce basic results of graph theory. The proofs are given to illustrate how to conduct proofs.

Lemma 3.1 *The sum of the degrees of all vertices of an undirected graph is equal to the number of edges in the graph times two. That is, for an undirected graph, $G = (V, E)$, $\sum_{v \in V} d(v) = 2|E|$.*

Proof To prove this result, it suffices to remark that each edge contributes 1 to the degree of each of its ends. Thus, when summing over the degree of all vertices, each edge is counted twice. □

Lemma 3.2 *In an undirected graph G, the number of vertices of odd degree is even.*

Proof A way to prove this, is to partition the set of vertices V in two sets: The set of vertices of even degree V_{even} and the set of vertices of odd degree V_{odd} such that $V = V_{even} \cup V_{odd}$. Jointly with the previous lemma, we have that

$$2|E| = \sum_{v \in V} d(v) = \sum_{v \in V_{even}} d(v) + \sum_{v \in V_{odd}} d(v).$$

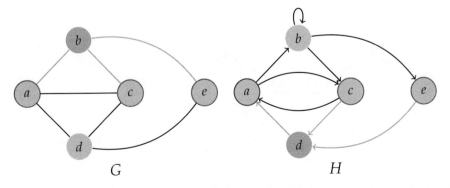

Figure 3.2: Examples illustrating neighborhood and degree notions in undirected graph G and directed graph H.

The term on the left hand side is clearly even and by definition the sum over V_{even} is also even. Thus, $\sum_{v \in V_{odd}} d(v)$ is also even. Since $d(v)$ is odd for any vertex $v \in V_{odd}$, $|V_{odd}|$ must be even. □

3.3.3 Subgraphs and Connectedness

A *subgraph* H of graph G contains a subset of vertices of G and a subset of edges connecting those vertices. Formally, if $V(H) \subseteq V(G)$ and $E(H) \subseteq E(G)$, then H is a subgraph of G. If H contains all edges from G that connect its vertices, then H is *induced*, or *node-induced*. Consider graph $G = (V, E)$ and set of vertices $A \subseteq V$, then $G[A]$ denotes the subgraph *induced* (or *node-induced*) by A on G. Its vertex set is the set A and its edge set includes all the edges in E that have both ends in A. In contrast, *edge-induced* graph $G[E']$ is the graph with edge set $E' \subseteq E$ and vertex set containing all vertices of V connected by edges in E'.

A subgraph H of G such that $V(H) = V(G)$ is called a *spanning subgraph* of G. A graph is *complete* if it contains all possible edges between its vertices. A subgraph of G that is complete is called a *clique*.

Figure 3.3 illustrates these concepts. In graph G, red denotes a subgraph which is neither a spanning subgraph nor an induced subgraph. It is not spanning because it does not contain vertex b. It is not induced, because it does not contain ed edge. The red subgraph in graph H is a spanning subgraph, as it contains all vertices. The red subgraph of graph I is induced.

Two graphs G and H are *isomorphic* if there exists a function $f : V_G \mapsto V_H$ such that $uv \in E(G)$ if and only if $f(u)f(v) \in E(H)$. The function f is a *bijection*, which means that f is a one-to-one mapping of the vertices of G to the vertices of H. We say that G *contains a copy* of H if G has a subgraph isomorphic to H. Figure 3.4 gives an example of two isomorphic graphs.

Finding approximately isomorphic mappings between biological networks is a recurrent problem that is detailed in Chapter 9.

The generalization of the graph isomorphism problem is to find if a graph contains a copy of another graph. This problem is called the *subgraph isomorphism problem* and has long been known to be NP-complete [63, 59]. The clique problem is a special case of the subgraph isomorphism problem. It refers to searching for the largest clique in graph G. Detection of cliques in molecular networks has been used to identify groups

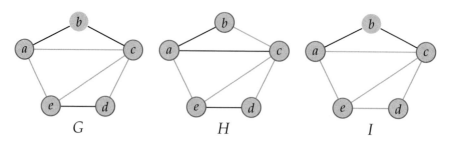

Figure 3.3: Examples illustrating in red in graph G a subgraph, in graph H a spanning subgraph, and in graph I an induced subgraph for the same graph.

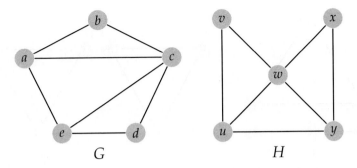

Figure 3.4: Examples of isomorphic graphs G and H. The isomorphic function maps the vertices a, b, c, d, and e of graph G to the vertices u, v, w, x, and y of graph H, respectively.

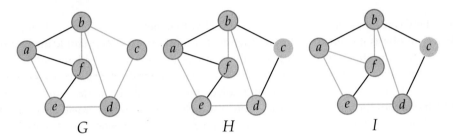

Figure 3.5: Examples illustrating: a–f-walk in graph G, an a–f-path in graph H, and a cycle in graph I. In graph G, edge bd is used twice to complete the walk.

of consistently co-expressed genes [64, 65] and to match three-dimensional structures of molecules [66, 67].

A *walk* in an undirected graph G is a sequence of vertices of V such that consecutive vertices are adjacent. A walk $w = u_0 \ldots u_n \in V^{n+1}$ is *closed* if $u_0 = u_n$. The length of a walk w corresponds to the number of edges it contains; here w is of length n. A u–v-walk, where $(u,v) \in V^2$, is a walk that starts at vertex u and ends at vertex v. The shortest u–v-walk is a u–v-walk of minimum length (minimum number of edges, or minimum sum of edge weights in weighted graphs). If a walk never uses an edge more than once, it is called a *trail*. A closed trail is called a *tour*. If a walk never visits any vertex in V more than once, it is called a *path*. Analogous to the shortest walk, the shortest path is a path of minimum length. There is an u–v-path in G if and only if there is an u–v-walk. This implies that a shortest u–v-walk is a shortest u–v-path in unweighted graphs. A closed walk that never visits a vertex more than once is called a *cycle*. A graph that contains at least a cycle is *cyclic*, and a graph that contains no cycles is *acyclic*. Figure 3.5 illustrates in red examples of a walk (G), a path (H), and a cycle (I).

Similar notions are defined for directed graphs with the difference being that the direction of the edges need to be followed, i.e., a walk $w = u_0 \ldots u_n \in V^{n+1}$ is defined in a directed graph if for any consecutive pair of vertices there is a directed edge with source u_{i-1} and target u_i, with $i \in [1, n]$.

The length of the shortest path beween two nodes u and v in a graph is called the *distance* between the nodes and is denoted by $\delta(u, v)$. It is infinity if there is no

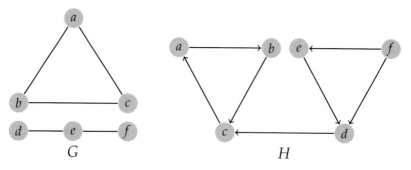

Figure 3.6: Graph G has two connected components: $\{a,b,c\}$ and $\{d,e,f\}$. Graph H is weakly connected and has one strongly connected component, $\{a,b,c\}$.

path between the nodes. It is trivial to show that this metric is well defined on undirected graphs as it is positive definite, symmetric, and satisfies the triangle inequality. Computing the shortest path between two vertices is a classical problem and several algorithms have been developed. Perhaps the most famous is Dijkstra's algorithm developed in 1959 [68]. The algorithm solves the *single-source shortest path problem*, which searches for the shortest paths from a source node of graph G to all other nodes of G. The pseudocode for Dijkstra's algorithm is given in Exercise 3.6. The exercise asks that you prove that its complexity is $O\left(|V|^2\right)$.

Two vertices are *connected* if there exist a walk from one vertex to the other. Formally, for any $u, v \in V$, $u \sim v \Leftrightarrow \exists\, u\text{–}v\text{-walk}$ in G. The subgraphs of graph G formed by connected vertices are called *components* of G, or *connected components* of G. For instance, graph G in Figure 3.6 has two connected components. A graph is *connected* if it has a single component. In a directed graph, the notion is extended to *weak and strong connectedness*. The former corresponds to the notion of connectedness on the underlying undirected graph, i.e., the graph obtained from the directed graph by removing the directions of edges. Two vertices u and v are *strongly connected* if there exists a directed path from u to v and a directed path from v to u, i.e., if there exists at least one directed cycle containing both vertices. The subgraphs of a graph in which all vertices are strongly connected are called *strongly connected components*. Similarly, the subgraphs of a graph in which all vertices are weakly connected are called *weakly connected components*. Graph H in Figure 3.6 illustrates these concepts.

3.3.4 Types of Graphs

A *multi-graph* is a graph containing at least one multiple edge (see graph G in Figure 3.7, which has two edges between nodes c and d). In biology, multiple edges, and thus multi-graphs, are used when one wants to overlay different types of interactions on the same set of entities. This means that two entities can interact in different ways, e.g., by a physical and by a genetic interaction between proteins.

A *mixed* graph contains both directed and undirected edges. Similar to multi-graphs, mixed graphs are used to represent various types of interactions between the same entities. For instance, a mixed network could represent undirected protein

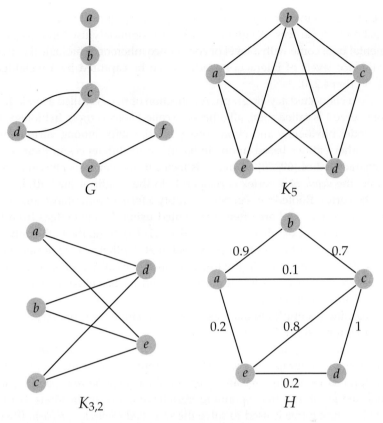

Figure 3.7: Examples illustrating an undirected multi-graph G, the complete graph K_5, the bipartite graph $K_{3,2}$, and a weighted graph H.

physical interactions together with directed transcriptional regulation interactions between proteins that regulate expression of other proteins. We use genes and their protein products interchangeably and represent them by the same vertices.

Recall that in a complete graph, each pair of vertices is adjacent. A simple graph that is complete is denoted by K_n, where n represents the number of vertices in the graph. K_5 is illustrated in Figure 3.7.

A graph $G(V,E)$ is *bipartite* if we can divide its vertex set V into two non-overlapping sets A and B, such that $V = A \cup B$ and every edge e in E connects a vertex from A to a vertex from B. A bipartite graph that has all possible edges between its partitions is called *complete bipartite graph* and is denoted by $K_{m,n}$, where $m = |A|$, $n = |B|$ (and $E(K_{m,n}) = A \times B$). As mentioned in Section 3.1, bipartite graphs are used to model metabolic networks. Bipartite graph, $K_{3,2}$, with $A = \{a,b,c\}$ and $B = \{d,e\}$ is illustrated in Figure 3.7.

A *weighted* graph has a weight, or score, given to each edge. Formally, a weighted graph is defined by its vertex and edge sets (V and E), as well as a set of possible weights Ω and a function $\omega : E \mapsto \Omega$ associating a weight to each edge (illustrated with graph H in Figure 3.7). Weighted graphs can be used to include the probability of existence of a given interaction (in this case, $\Omega = [0, 1]$). It could also be used to mark distances between vertices. Weighted graphs are commonly used in biology, as most experimental data come with a level of confidence inherent to biological experiments. The confidence levels of interactions (edges) can be captured by a weighted graph (e.g., see Section 4.2.3).

A *tree* is a connected acyclic graph. A collection of trees is called a *forest*. In a tree, a vertex of degree 1 is called a *leaf*, all others are called *inner vertices*. In biology, trees are used to model phylogeny and evolutionary relationships among various biological species or other entities based upon similarities and differences in their physical or genetic characteristics [69, 70]. A tree T is *rooted* if one vertex is chosen as a root. In such a tree, the *depth* of a vertex corresponds to the length of the path between the root and the vertex. Rooted trees generally imply a form of inheritance and order from the root to the leaves and are often represented using directed edges from the root. For a given vertex u at depth $n \in \mathbb{N}$, all vertices visited along the path connecting the root to u are *ancestors* of u and the vertex visited at depth $n - 1$ is the *parent* of u. The neighbors of u at depth $n + 1$ in the tree are called the *children* of u. A *binary tree* is a rooted tree where each vertex has at most two children. See illustrations in Figure 3.8. Rooted trees are used in a wide variety of applications, ranging from data structures to machine learning (examples in machine learning include decision trees, classification trees, random forests) [71, 72, 73].

If a spanning subgraph of graph G is a tree, it is called a *spanning tree* of G. In Figure 3.3 the red subgraph in graph H is a spanning tree of the graph. Every connected graph has a spanning tree and every minimally connected spanning subgraph of a connected graph is a tree. Spanning trees have many applications. For instance, an optimal spanning tree is used to solve the so-called *connector problem*: If we have a set of cities which we want to connect by a system of railway lines and if we know the cost of constructing lines between pairs of cities, how do we construct the system so that the cost is minimal.

We introduce two more types of graphs that are becoming increasingly used in the literature. The first one is *hypergraphs* [74], which are defined by a set of vertices V and a set of edges E, called *hyperedges*. A hyperedge can connect an arbitrary number (greater than zero) of vertices. Figure 3.9 gives an illustration of a hypergraph with $V = \{a, b, c, d, e, f, g\}$ and $E = \{\{a, b, c\}, \{b, c\}, \{d\}, \{c, e, f\}\}$. In biology, the need for hypergraphs arises as interactions are not simply binary in many systems. For instance, a protein complex is not able to perform a function unless all of the

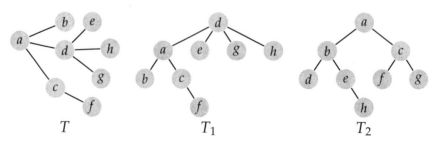

Figure 3.8: Examples illustrating tree T, the tree T rooted at vertex d (T_1), and binary tree T_2.

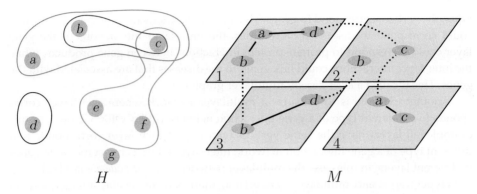

Figure 3.9: Examples illustrating a hypergraph, H, and a multilayer graph, M. The hypergraph has seven vertices and four hyperedges. The multilayer graph has four layers, G_1, G_2, G_3, and G_4. Each layer corresponds to a graph: $G_1 = (V_1 = \{a^1, b^1, d^1\}, E_1 = \{a^1b^1, a^1d^1\})$, $G_2 = (V_2 = \{b^2, c^2\}, E_2 = \emptyset)$, $G_3 = (V_3 = \{b^3, d^3\}, E_3 = \{b^3d^3\})$, and $G_4 = (V_4 = \{a^4, c^4\}, E_4 = \{a^4c^4\})$. Finally, the set of inter-layer edges corresponds to $\{b^1b^3, d^1c^2, b^2d^3, c^2a^4\}$, where the superscripts denote the layers.

constituent proteins are present. A network representation of protein complexes has proteins as vertices and a hyperedge contains all proteins that are part of a specific complex. When using a standard binary graph to represent such interactions, the information about which proteins form a complex can be blurred. For instance, imagine that two proteins can form a complex by themselves, but can also form another complex by interacting with a third protein; in a binary representation, the two complexes would not be distinguishable as the three proteins would form a triangle, while they would be precisely represented by two distinct hyperedges within a hypergraph representation. Hence, hypergraphs have gained interest in computational biology [75].

The second type of networks are *multilayer networks*. They are used to represent systems with heterogeneous entities, or heterogeneous types of interactions. A multilayer network is a collection of networks, called *layers*, with the potential addition of inter-layer edges, i.e., edges connecting a vertex from a layer to a vertex from another layer. There are various formal definitions of a multilayer network [76, 77], in the following we present the one from Boccaletti *et al.* (2014) [77]. A multilayer network M is defined by a pair $M = (\mathcal{G}, \mathcal{C})$, where $\mathcal{G}$ denotes the set of layers of M and $\mathcal{C}$ corresponds to its set of inter-layer edges. If M has n layers, we denote them by $\mathcal{G} = \{G_1, \ldots, G_n\}$, where G_i corresponds to layer i of M and $G_i = (V_i, E_i)$ (recall that a layer is a graph). An inter-layer edge can connect any pair of vertices in any two layers (i.e., $\mathcal{C} \subseteq \cup_{i,j \in [1,n]: i \neq j} V_i \times V_j$). Note that the vertex sets of the layers are not necessarily disjoint, thus, to avoid confusion, a superscript denoting the index of the layer is used, such that $V_i = \{v_1^i, v_2^i, \ldots, v_{p_i}^i\}$, where $p_i = |V_i|$. For instance, in Figure 3.9, M has four layers, G_1, G_2, G_3, and G_4.

Multilayer networks can be used to model disease–gene systems; for instance, a multilayer network is constructed with two layers corresponding to two data types: Diseases and genes. Hence, in one layer vertices correspond to diseases while in the other, the vertices correspond to genes. The intra-layer interactions of the disease layer could correspond to disease comorbidities, the intra-layer interactions of the gene layer could correspond to protein–protein interactions between gene products, while the inter-layer interactions could link genes to the diseases that are associated with the genes. This is called a *layer-disjoint* multilayer graph [76].

Another example is to consider a multilayer network where each layer corresponds to a different type of a gene interaction network (PPI, COEX, GI, etc.). In this example, all layers have the same vertices (genes), but each layer corresponds to a different type of a gene interaction network; inter-layer edges connect the same genes in different layers. In this case, the multilayer network is called *node-aligned* [76].

Hypergraphs and multilayer networks as models of molecular interaction data have increased the flexibility and the accuracy with which we can model these systems-level data. However, the tools for extracting new biological information from these models are still lacking and are a subject to active research.

3.3.5 Classic Graph Theory Problems

We give an overview of three classical problems in graph theory that found applications in computational biology. The first two deal with finding special paths in a graph, an *Eulerian circuit* and a *Hamiltonian path*. The last one deals with finding a particular mapping between two sets of vertices, a *matching*.

3.3.5.1 Eulerian Circuit

An *Eulerian circuit* in a graph G is a closed tour that uses every edge exactly once and visits all the vertices (possibly more than once). A graph is *Eulerian* if it contains an Eulerian circuit.

The name comes from a Swiss mathematician, Leonhard Euler, who in 1736 was the first to investigate the existence of such a tour in a graph. This is also regarded as the birth of graph theory. In his seminal paper, he showed that there was no Eulerian circuit in the graph formed by the bridges of the town of Königsberg (today's Kaliningrad). His insight was to represent the town as a graph, where each bridge corresponds to an edge and land masses that the bridges connect as vertices. The resulting graph is shown in Figure 3.10.

The theorem that resulted from this study gives a necessary and sufficient condition to verify if a graph is Eulerian.

Theorem 3.3 *An undirected graph G is Eulerian if and only if it is connected and has no vertex of odd degree.*

Proof Assume that G has an Eulerian circuit, c. Then, G is connected, as it contains a tour, c, that connects all vertices. Each time a vertex $v \in V$ is visited in c, two of its incident edges are traversed. Since c visits all vertices of G and traverses each edge of G exactly once, the vertices of G have even degree.

Conversely, let G be a connected graph with no vertices of odd degree. Consider a trail $t = u_0 \ldots u_n$ in G (recall that a trail can visit a vertex multiple times) and suppose that it is a longest trail in G.

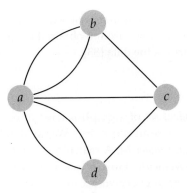

Figure 3.10: A graph model of 1736 Königsberg. Each edge corresponds to a bridge and each vertex represents a land mass that the bridges connect.

Suppose that there is an edge $e = uv$ not traversed by t. Either exactly one of the vertices, u or v, belongs to t, or neither of the vertices belong to t.

1. Assume that one vertex, say u, is in t. I.e. $\exists i \in [0, n] : u_i = u$. Then the trail $\tilde{t} = vu_i u_{i+1} \ldots u_n u_1 \ldots u_i$ is a longer trail than t which gives a contradiction.
2. Assume both vertices u and v do not appear in tour t. Since graph G is connected, there exists a path p connecting u_0 to u in G. Since $u_0 \in t$ and $u \notin t$, there is an edge along the path p with only one end in t, which brings us back to the first case.

Thus, we know that t is a trail using all edges of G. Since G is connected, t visits all vertices.

Suppose now that t is not closed, i.e., $u_0 \neq u_n$. Let j be the number of times t visits vertex u_n. Then, t contains $2j + 1$ edges incident to u_n. Since, by hypothesis, vertex u_n has an even degree, there is, at least one edge incident to u_n that is not traversed by t, which means one could extend t. Or, by assumption, t is the longest trail in G, thus necessarily t is a closed trail, i.e., a tour, and $u_0 = u_n$.

Therefore, t is an Euler circuit and G is Eulerian, which concludes the proof. □

3.3.5.2 Hamiltonian Paths

A path of a graph G that contains all vertices of G is called a *Hamiltonian path*. A *Hamiltonian cycle* of a graph G is a cycle that contains every vertex of G. A graph G containing a Hamiltonian cycle is *Hamiltonian*. (The name comes from William Rowan Hamilton who created the icosian game of finding a Hamiltonial cycle along the edges of a dodecahedron.) In contrast to Eulerian graphs, there is no known necessary and sufficient condition for a graph G to be Hamiltonian. Furthermore, answering whether there exists a Hamiltonian cycle in a graph is an NP-complete problem [59, 63].

Eulerian and Hamiltonian paths have been used for the analysis of DNA fragment reads. The problem consists in assembling a genome from DNA reads [78, 79, 80]. It can be solved either by looking for a Hamiltonian path in a graph built from the reads (called a read overlap graph), or by searching for an Eulerian path in a particular graph called de Bruijn graph, also constructed from the data [81, 82]. As Eulerian problems are tractable, they should have computational advantage. However, Li *et al.* showed that this depends on the technology used to obtain the data, notably the length of the reads and the coverage (size) of the data [83].

3.3.5.3 Matching

A *matching*, or an *independent set* of a graph is a set of edges that have no common vertices (i.e., no two edges share an end-point). We say that the two ends of an edge in matching M of G are *matched under M*. A matching M in $G = (V, E)$ *saturates* a vertex $u \in V$ if there is an edge in M incident to u; if true, the vertex u is said to be *M-saturated*, otherwise u is *M-unsaturated*. If every vertex in V is M-saturated, the matching M is *perfect*. Furthermore, if G has no matching $\tilde{M}$ such that $|\tilde{M}| > |M|$, then M is a *maximum matching*.

Let M be a matching in graph G. An *M-alternating path* in G is a path whose edges are alternatively in M and not in M. An *M-augmenting path* is an M-alternating path whose origin and end vertices are M-unsaturated.

A classical problem solved by finding a matching corresponds to the *personnel assignment* problem, it goes: In a given company, there are n workers $w_1, w_2, \ldots, w_n$ available for n jobs $j_1, j_2, \ldots, j_n$. Each worker is qualified for one or more of these jobs. Is there an optimal job assignment, one worker per job, based on each worker's qualifications?

To solve this problem, the idea is to first construct a bipartite graph G with partition (W, J) of the vertex set, where $W = \{w_1, w_2, \ldots, w_n\}$ represents the workers and $J = \{j_1, j_2, \ldots, j_n\}$ represents the jobs. An edge connects a worker to a job if the worker is qualified for that particular job. The problem is to find out if graph G has a perfect matching. An important result on which the algorithm is based is Hall's theorem, which states that *if G is a bipartite graph, with partition (W,J), then G contains a matching that saturates every vertex in W if and only if for any subset S of W, the number of elements in the neighbor set of S has at least as many elements as S*, i.e., $|N(S)| \geq |S|$. The neighbor set of a set S is defined as the set containing all the neighbors of the set S in a graph G and is denoted by $N(S)$. This result is used by the *Hungarian algorithm* that solves the personal assignment problem. The Hungarian algorithm has the computational complexity of $O(n^3)$ and has been used to solve other network-related problems, e.g., network alignment [84] (detailed in Chapter 9).

A wealth of problems is tackled by graph theoretic approaches that have application in bioinformatics, too many to list here. For instance, the survey by Zhang *et al.* gives an overview of applications in precision oncology [85].

3.3.6 Data Structures and Search Algorithms for Graphs

3.3.6.1 Data Structures

Intuitively, data structures are ways to represent data in a computer. Two most commonly used data structures to represent graphs are adjacency matrix and adjacency list. The choice of which to use depends on the graph and the task at hand, detailed below.

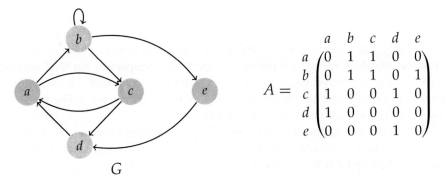

Figure 3.11: An illustration of graph G and its adjacency matrix A. The 1s represent the existence of a directed edge from the vertex in a row to the vertex in the column, 0s represent the absence of an edge.

For a simple graph G containing n vertices, indexed from 1 to n, the adjacency matrix is denoted by A, with $A \in \{0, 1\}^{n \times n}$. Each row/column index represents a vertex of the graph and the entries of A are such that:

$$A_{i,j} = \begin{cases} 1, & \text{if there is an edge between vertex } i \text{ and vertex } j \text{ in } G, \\ 0, & \text{otherwise.} \end{cases} \quad (3.3)$$

If the graph is directed, then A has entries:

$$A_{i,j} = \begin{cases} 1, & \text{if there is an edge from vertex } i \text{ to vertex } j \text{ in } G, \\ 0, & \text{otherwise.} \end{cases} \quad (3.4)$$

Figure 3.11 presents an adjacency matrix of a directed graph.

If a graph is weighted, the entries of the adjacency matrix correspond to the weights of the edges.

For undirected graphs, weighted or not, the adjacency matrix is symmetric, i.e., $A = A^T$, which is a property used in spectral graph theory that we will discuss in Section 3.3.7. Also, we note that for an undirected graph, the sum of the entries on a row, or column, corresponds to the degree of the vertex. For a directed graph, the sum of a row's entries is the outdegree of the vertex, whilst the sum of a column's entries is the indegree of the vertex. Similarly, for a weighted graph, we get a *weighted degree*, also called the *strength of a node*.

Storing a graph represented by its adjacency matrix requires $O(n^2)$ memory. However, this can often be reduced. For instance, for undirected graphs as the adjacency matrix is symmetric, storing only the entries above (or below) the diagonal is sufficient. The same can be done for a directed graph with no loops and multiple edges, by defining the adjacency matrix A such that:

$$A_{i,j} = \begin{cases} 1, & \text{if there is an edge from vertex } i \text{ to vertex } j \text{ in } G, \\ -1, & \text{if there is an edge from vertex } j \text{ to vertex } i \text{ in } G, \\ 0, & \text{otherwise.} \end{cases} \quad (3.5)$$

A is then antisymmetric and storing only the entries above (or below) the diagonal is sufficient.

Adjacency matrix structure is preferred if a graph is dense, which means that the number of edges is high (of the order $O(n^2)$) with respect to the total number of possible edges (which is equal to $\binom{n}{2} = \frac{n(n-1)}{2}$ for a simple graph). If the graph is sparse, in the sense that it has $O(n)$ or fewer edges, the adjacency list defined below is a more space efficient structure to store a graph.

An adjacency list is a list of vertices and all of their adjacent vertices, as illustrated in Figure 3.12. To store an adjacency list, one needs $O(m+n)$ memory, where m is the number of edges and n is the number of vertices in the graph.

3.3.6.2 Graph Search Algorithms

Visiting nodes and edges of a graph is used in many applications. For instance, we may want to know how far from a node of interest are all other nodes of the graph. Hence, efficient algorithms to explore and search a graph are essential to investigate its structure. We introduce two main algorithms for exploring and searching graphs: Breadth first search (BFS) and depth first search (DFS).

BFS is illustrated in Figure 3.13. It starts from vertex v (vertex 1 in the example of Figure 3.13) and visits all of its neighbors (vertices 2, 3, and 4 in Figure 3.13). Then, it moves to the first of the neighbors of v (vertex 2 in Figure 3.13) and visits all of its neighors (vertices 5 and 6 in Figure 3.13). Then, it goes to the second neighbor

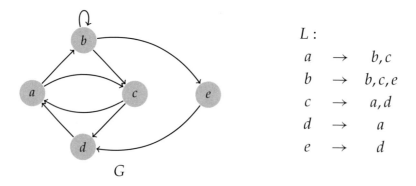

Figure 3.12: Graph G and its adjacency list, L.

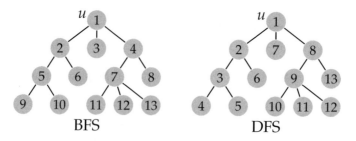

Figure 3.13: Illustrations of BFS and DFS. Numbers in vertices represent the order in which vertices are examined.

of v (vertex 3 in Figure 3.13) and visits all of its neighbors (none in Figure 3.13). It repeats this until it visits all neighbors of all neighbors of v. And the process repeats in this "breadth first" manner until there are no unexplored vertices. If a graph is disconnected, the algorithm needs to be ran from starting vertices belonging to different connected components of the graph, as the algorithm visits only the vertices connected to the starting vertex. BFS can be used to solve single-source shortest path problems as an alternative to Dijkstra's algorithm in an unweighted network. BFS can also be used to compute the connected components of a graph, or to find a cycle, if one exists.

DFS is similar to BFS in that it explores all vertices of G, but it explores the neighbors of a vertex in a depth first manner. An illustration is given in Figure 3.13. DFS can be used to find if a graph is connected, to find its connected components, to find paths and cycles, and make a topological sort of a directed graph (i.e., a linear ordering of the vertices of graph $G = (V, E)$ such that for each edge uv in E, u appears before v in the ordering).

Both algorithms have the same running time complexity and space complexity. When using and adjacency list to represent graph $G = (V, E)$, DFS and BFS have time complexity $O(|V| + |E|)$ and space complexity $O(|V|)$.

3.3.7 Spectral Graph Theory

Spectral graph theory uses the wealth of algebraic tools to investigate matrix representations of graphs. It mostly studies graphs through the eigenvalues and eigenvectors of their matrix representations. It has found many applications: in chemistry, where eigenvalues are used to quantify the stability of molecules [86]; in theoretical physics, for instance, to minimize energies of Hamiltonians systems in quantum mechanics [87]; in machine learning, e.g., in image processing [88, 89, 90]. Spectral graph methods are especially useful in the study of simple graphs as their adjacency matrices are symmetric. As mentioned in Section 3.2.1, any symmetric matrix has a spectral decomposition in $\mathbb{R}$. Henceforth, we only consider simple graphs.

The Laplacian matrix L of G is defined by $L = D - A$, where A is the adjacency matrix of G and D is the degree matrix for the graph. The degree matrix for $G = (V, E)$ is the diagonal matrix D such that

$$D_{ij} = \begin{cases} d(v_i) & \text{if } i = j, \\ 0 & \text{otherwise.} \end{cases} \tag{3.6}$$

Since A is symmetric, so is L. Two other types of Laplacian matrices are often found in the literature [91, 92]:

- The *normalized Laplacian matrix* is defined as $\mathcal{L} = I - D^{-\frac{1}{2}} A D^{-\frac{1}{2}}$, with the convention $D_{i,i}^{-1} = 0$ if $d(v_i) = 0$.
- The *random-walk normalized Laplacian matrix*, defined as $\mathcal{L}_{rw} = I - D^{-1} A$.

For an undirected graph G on n vertices, different Laplacian matrices are all symmetric and semi-positive, meaning that they have n non-negative eigenvalues.

Henceforth, we consider only normalized Laplacian matrices. We denote by $\lambda_0, \lambda_1, \ldots, \lambda_{n-1}$ the eigenvalues of graph $G = (V, E)$ with $|V| = n$ and assume $\lambda_0 \leq \lambda_1 \leq \cdots \leq \lambda_{n-1}$. We introduce basic properties of the spectrum of a graph (introduced in Section 3.2.1) underlining the link between graph structure and its normalized Laplacian eigenvalues. For an in-depth exposition, we refer the reader to Chung's textbook [91].

We start by remarking that 0 is always an eigenvalue of $\mathcal{L}$ (as eigenvector $\mathbf{v} = D^{\frac{1}{2}}\mathbf{u}$, where $\mathbf{u} = (1, 1, \ldots, 1)$, is always a solution of $\mathcal{L}\mathbf{u} = 0$). Furthermore, the geometric multiplicity of the eigenvalue 0 gives the number of connected components of graph G, i.e., if $\lambda_i = 0$ and $\lambda_{i+1} \neq 0$ then G has exactly i connected components and the spectrum of G corresponds the union of the spectra of its connected components.

There is a number of results on bounds for the eigenvalues of graphs. We present a few of the most classical ones:

- For any $0 \leq i \leq n-1$, $0 \leq \lambda_i \leq 2$.
- If $\lambda_{n-1} = 2$, the graph G is bipartite.
- For any graph G, $\sum_i \lambda_i \leq n$, with equality holding if G has no isolated vertices.
- For $n \geq 2$, $\lambda_1 \leq \frac{n}{n-1}$, with equality holding if G is complete.
- For a graph with no isolated vertices, $\lambda_{n-1} \geq \frac{n}{n-1}$.
- For a graph that is not complete, $\lambda_1 \leq 1$.

Spectral decompositions of graphs are used for various problems, for instance to find random walks [88, 93, 94, 91], or to visualize a graph using its space embedding based on eigenvectors [95, 91]. One can also define distances between two graphs on n vertices using their spectra [96]. A common application of spectral graph theory for image processing is spectral clustering [88]. Clustering problems are introduced in detail in Chapter 6. Clustering deals with partitioning a set of vertices based on a measure of their similarity. For instance, in an image, clustering could result in identifying different objects based on image and pixel properties. An image can be modeled as a weighted graph, with pixels corresponding to vertices and edges between neighboring pixels weighted according to a chosen measure of pixel similarity. Spectral clustering is then used to identify different components or objects present in the picture based on the Laplacian representation of the graph model of the image.

3.4 Network Measures

In this section, we introduce network measures commonly used to analyze real world networks. Some are also outlined in Chapters 10 and 12. Network measures are used to give insight into the structure of a network that can shed new light onto the underlying biology. We introduce basic measures and basic random graph models that are used to understand the structural properties of real world networks. The more complex measures are covered in Chapter 5.

3.4.1 Network Properties

Measures of network structure, also called network properties, can historically and roughly be divided into *global* and *local* ones. In general, if they involve the full network, then they are global; if they involve only a node and its neighborhood, then they are local.

The simplest local property of a node is its degree. It has been shown that proteins of high degree in a protein–protein interaction (PPI) network have been linked to diseases, such as cancer [52, 97]. Similarly, degrees of proteins in PPI networks has been correlated to their essentiality, in the sense that mutations affecting them are lethal to the cell [98]. However, in higher confidence PPI networks of yeast (see Chapter 4 for details on confidence of PPIs), this correlation was shown not to hold [99]. This demonstrates an inability of such a simple measure of network structure to capture biological signal. Hence, more sophisticated measures have been developed (also detailed in Chapter 5).

The *degree distribution* $P(k)$ of a network measures the percentage of nodes in the network having degree k. Many molecular networks have degree distributions that follow a *power law*, $P(k) \sim k^{-\gamma}$, where the *degree exponent* $\gamma > 0$ [100, 101]. In a log–log plot, a power-law distribution corresponds to a straight line with slope $-\gamma$. Networks with a power-law degree distribution are called *scale-free*. The name comes from the fact that power laws have the same functional forms at all scales. This means that the power law $P(k)$ remains unchanged when rescaling variable k as it satisfies: $P(\lambda k) = \lambda^{-\gamma} P(k)$. In scale-free networks, most nodes are of degree 1, a much smaller percentage of nodes is of degree 2, etc. But there exists a small number of nodes of high degree, called *hubs*. Figure 3.14 illustrates the degree distribution of the PPI network of fruit fly. The *cumulative degree distribution*, denoted by $P_c(k)$, gives the probability that a node has the degree smaller than k.

The *density* of a network measures the number of edges in a graph as a fraction of the maximum possible number of edges:

$$\text{density} = \frac{|E|}{\binom{|V|}{2}} = \frac{|E|}{\frac{|V|(|V|-1)}{2}} = \frac{2|E|}{|V|(|V|-1)} \qquad (3.7)$$

If this density is small, i.e. if $|E| = O\left(|V|^k\right)$ with $k < 2$, the graph is said to be *sparse*. If the density is close to 1, i.e., $|E| = O\left(|V|^2\right)$, then the graph is *dense*. Most biological networks are sparse. Network density has been linked to robustness of networks to random perturbations [102]. Intuitively, the denser the network, the less likely it is to

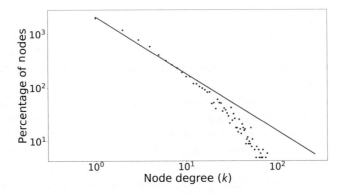

Figure 3.14: Degree distribution of fruit fly PPI network (obtained from BioGrid database [31] in April 2017). The power-law $\gamma = 1.07$ is also presented.

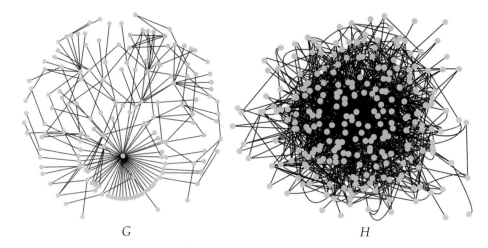

Figure 3.15: Illustration of two networks G and H with densities of 0.014 and 0.033, respectively. The layouts are generated with Cytoscape software package for network analysis [103] detailed in Chapter 13.

be disconnected upon removal of a randomly chosen node. However, increased density increases biological complexity and cost. Hence, molecular networks are sparse, but wired in ways that make them robust (detailed below). Figure 3.15 illustrates two networks with different densities.

To measure how "far spread" the nodes of a graph are, we use the *average path length* $\bar{\delta}$ of the graph and the *diameter D* of the graph. One computes the shortest path lengths δ (cf. Section 3.3.3) between all pairs of nodes and averages them to obtain the average path length. The maximum path length over all pairs of nodes is the network diameter. Formally, for graph $G = (V, E)$ on n nodes, such that $V = \{v_1, \ldots, v_n\}$, we have

$$\bar{\delta} = \frac{1}{n(n-1)} \sum_{i=1}^{n} \sum_{j=1}^{n} \delta(v_i, v_j), \tag{3.8}$$

$$D = \max_{i,j} \delta(v_i, v_j). \tag{3.9}$$

Clearly, $\bar{\delta} < D$. If two nodes u and v are not connected in a graph, $\delta(u,v) = \infty$. In practice, if u and v are not connected, we set $\delta(u,v) = D+1$, where D is the maximum of the diameters of the connected components of the graph. Graph G in Figure 3.16 has diameter 4, with the longest path being between nodes a and e, and average path length of $\bar{\delta} = 1.93$.

Graph $G = (V, E)$ is said to be *small-world* if every pair of nodes is connected by a relatively short path, generally if $D = O(\log(|V|))$, or smaller. The small-world property is frequent in real world networks. The idea originated from Frigyes Karinthy in 1929, who hypothesized that anything in the world, and particularly any two individuals, are on average six connections from each other. John Guare used this idea in his famous play titled *Six Degrees of Separation*. In the context of social networks, the idea was studied by Gurevitch in 1961 [104], and Travers and Milgram in 1967 [105], the latter through social experiments. The property has since been uncovered in other networks, including biological ones [106].

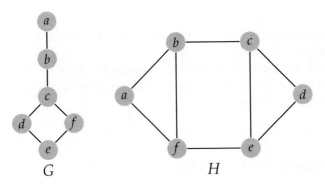

Figure 3.16: Two graphs to illustrate network properties. G has diameter of 4, average path length of 1.93, and average clustering coefficient of 0. Graph H has $D = 3$, $\bar{\delta} = 1.13$, and $\overline{C} = 0.55$. Furthermore, in H, node c has degree centrality 3, closeness centrality 0.71, betweenness centrality 2, eigenvector centrality 0.44, eccentricity centrality 2, and subgraph centrality 8.87. Finally nodes c and f have matching index of 0.5 whilst a and c have matching index of 0.25.

Another network measure is the *average clustering coefficient*, which evaluates if a network has subsets of densely connected nodes. For node u, the clustering coefficient is defined as $C_u = \frac{2e}{k(k-1)}$, where k is the degree of u and e is the number of edges between the neighbors of u. For instance, all nodes of graph G in Figure 3.16 have clustering coefficient of 0, since none of their neighbors are connected. On the other hand, in graph H on the same figure we have $C_b = C_c = C_e = C_f = \frac{2}{3(3-1)} = 0.33$ and $C_a = C_d = \frac{2}{2(2-1)} = 1$. This measure is similar to the density, introduced earlier, but this time we measure the density in the neighborhood of a node: It measures the ratio of the number of edges in the neighborhood of a node to the number of all possible edges in the neighborhood. The clustering coefficient can also be thought of as the ratio of the number of *triangles* (complete graph K_3) containing node u to the number of triplets of nodes containing node u. The clustering coefficient takes values between 0 and 1, 0 if no edges connect any pair of nodes in the neighborhood of the node, and 1 if the neighborhood forms a complete graph. The average clustering coefficient of a network is defined as the average of the clustering coefficients of its nodes, $\overline{C} = \frac{1}{n}\sum_{u \in V} C_u$. In Figure 3.16, graph G has average clustering coefficient $\overline{C} = 0$, but graph H has $\overline{C} = \frac{1+1+\frac{1}{3}+\frac{1}{3}+\frac{1}{3}+\frac{1}{3}}{6} = 0.55$. For a given network, the closer $\overline{C}$ is to 1, the more its nodes tend to form clusters. Another property is the clustering spectrum, $C(k)$, defined as the average clustering coefficient over all nodes of degree k. Biological networks have high average clustering coefficients when compared to random models (random network models are presented in Section 3.4.2), which suggests a modular organization of biological systems, i.e., the organization of their entities into groups, or functional subunits that are sparsely interconnected [107, 108, 109].

In a network, if nodes of high degree tend to be connected to each other, the network is said to be *assortative*. This is the case in social networks [101, 110]. In contrast, if nodes of high degree preferentially connect to nodes of low degree, then the network is called *disassortative*. Molecular networks are mostly disassortative, with hubs connecting nodes of low degree [110]. The level of assortativity can be measured in various ways:

- Compute a joint probability distribution, $P(k_i, k_j) = P(k_i)P(k_j)$, for pairs of degrees, (k_i, k_j), where $P(k_i)$ is the probability of a node in a network to have degree k_i.
- Compute the *assortativity coefficient* r. For graph $G = (V, E)$, denoting $|E| = m$ and for each edge, $e \in E$, denoting the degree of its ends by $d_{e,0}$ and $d_{e,1}$, Newman [110] defines r as

$$r = \frac{\frac{1}{m}\sum_{e \in E} d_{e,0} d_{e,1} - \left(\frac{1}{2m}\sum_{e \in E}[d_{e,0} + d_{e,1}]\right)^2}{\frac{1}{2m}\sum_{e \in E}\left(d_{e,0}^2 + d_{e,1}^2\right) - \left(\frac{1}{2m}\sum_{e \in E}[d_{e,0} + d_{e,1}]\right)^2}. \quad (3.10)$$

This is the Pearson's correlation coefficient of the degrees at either ends of an edge, $-1 \leq r \leq 1$.
- Compute the *average neighbor degree distribution*, d_{davg}, defined for graph $G = (V, E)$ as

$$d_{davg}(k) = \frac{1}{|N_k|} \sum_{v \in N_k} \bar{d}_v, \quad (3.11)$$

where N_k is the set of nodes in G with degree k and $\bar{d}_v$ denotes the average degree of the neighbors of v, defined as $\bar{d}_v = \frac{1}{d(v)} \sum_{u \in N(v)} \omega(uv) d(u)$, with $\omega(\cdot)$ being the weight of the edge (1 in an unweighted network).

There is a variety of centrality measures that have been introduced to measure the "significance" of a node in a network. We introduce the most common ones.

The *degree centrality* of a node is simply its degree. If a network is directed, recall that both the in- and outdegrees need to be taken into consideration. The underlying assumption is that important nodes of a network are involved in a high number of interactions. It has been shown that molecular networks are robust to random mutations, i.e., a mutation to a randomly chosen gene will not be lethal for the cell. This has been attributed to the fact that since molecular networks are scale-free, by selecting a node at random we are likely to select a node of low degree, the mutation of which is not likely to spread negative effects throughout the network (since the node has very few neighbors through which the perturbation could spread). However, molecular networks are fragile to targeted attacks, mutations in genes that are highly linked (hubs), as disfunction of such genes is likely to disconnect the network [111, 112].

The *closeness centrality* measures how closely a node is connected to all other nodes. For node u in graph $G = (V, E)$, the closeness centrality is computed as $C_c(u) = \frac{1}{\sum_{v \in V, v \neq u} \delta(u,v)}$. Its normalized version representing the average lengths of shortest paths instead of their sum is $\frac{n-1}{\sum_{v \in V, v \neq u} \delta(u,v)}$, where $n = |V|$. The closeness centrality has been used to highlight the existence of a core-periphery structure in metabolic networks, with some nodes being central (close to all other nodes) and others peripheral (distant from most other nodes) [113]. It was also used to identify central metabolites, study the evolution of a metabolic network's organisation, and compare metabolic networks across species [114, 115].

The *betweenness centrality* of a node measures how many shortest paths between pairs of other nodes in the network go through the node. In graph $G = (V, E)$, consider three distinct nodes $u, v, w \in V$. Let $\sigma_{u,v}$ be the number of shortest paths connecting u

and v and $\sigma_{u,v}(w)$ the number of shortest u–v-paths going through w. The betweenness centrality of node $w \in V$ is defined as $\sum_{(u,v)\in(V)^2, u\neq v\neq w} \frac{\sigma_{u,v}(w)}{\sigma_{u,v}}$. Nodes with high betweenness centrality are *bottlenecks* in the network. Bottleneck proteins of the yeast PPI network and of directed biological networks, such as regulatory networks, are often essential [116, 117].

The *eigenvector centrality* identifies nodes that are connected to important nodes as measured by a relative score associated to each node. The rationale behind it is that nodes are not equal, some are more important than others, and a connection to an important node should weight more than to a low ranked (unimportant) node. Hence, the eigenvector centrality of a node is a function of the eigenvector centrality of its neighbors. It is defined by using spectral graph theory and the spectrum of a graph (Section 3.3.7). Consider graph $G = (V, E)$ such that $V = \{v_1, v_1, \ldots v_n\}$ and A is its adjacency matrix. Let $\sigma(A)$ denote the spectrum of A and $\tilde{\lambda}$ be its eigenvalue of largest absolute value, $\tilde{\lambda} = \mathrm{argmax}_{\lambda \in \sigma(A)} |\lambda|$. The eigenvector centrality is defined by using the normalized eigenvector associated to $\tilde{\lambda}$, i.e. $x \in \mathbb{R}^n$ such that $Ax = \tilde{\lambda} x$ and $\|x\| = 1$. The eigenvector centrality of a given node v_i is the ith coordinate of eigenvector x. This measure has been used to rank webpages, to identify network hubs in PPI networks and re-examine their essentiality [111], to predict synthetic lethal interactions from PPI networks [118], to identify gene–disease associations in literature mined gene interaction networks [119], etc.

The *eccentricity* of a node is the maximum distance from it to any other node in the network, along a shortest path. Hence, low eccentricity of a node means that it is on the "boundary" of a network, while high eccentricity means that it is "deep inside" the network. The *eccentricity centrality* of a node is defined as the inverse of its eccentricity. The eccentricity centrality was used to study the yeast PPI network, however, it was not able to distinguish essential from non-essential proteins [120].

The *subgraph centrality* measures the participation of a node in all subgraphs in a network; smaller subgraphs are given more weight than larger ones in this measure to make the measure appropriate for characterizing network motifs (small recurring subgraphs of a network, detailed below) [121]. For graph $G = (V, E)$ with $V = \{v_1, \ldots, v_n\}$ and adjacency matrix A, the subgraph centrality of node $v_j \in V$ is defined as $\sum_{k=0}^{\infty} \frac{(A^k)_{i,i}}{k!}$. The measure has been used to study the molecular structure of drug compounds [122], to compute the degree of folding in protein chains [123], to identify essential proteins [111], etc.

The last measure we discuss is slightly different, as it gives a similarity score of two nodes within a network. It is called the *matching index*. The rationale behind this measure is that two genes that have similar function do not necessarily need to be adjacent in a molecular network. Thus, the matching index of two nodes, u and v, is measured as the overlap between their neighborhood node sets, $N(u)$ and $N(v)$, respectively. It corresponds to the *Jaccard index* of those two sets defined as $\frac{|N(u) \cap N(v)|}{|N(u) \cup N(v)|}$, i.e, the ratio of the number of common neighbors and the number of all neighbors. The matching index can be used to cluster nodes of a network, i.e., group together nodes that have a high matching index, roughly speaking. It has been used for predicting connectivity in primate cortical networks [124].

Thus far, we have introduced the most commonly used centrality measures. It is relatively easy to compute most of them and a number of software packages are available, for instance the python package NetworkX [125]. There exist other types of centrality measures in network science that are not discussed in this chapter, for instance, see Paladugu et al. [118].

More sophisticated measures include *network motifs* and *graphlets*. The former are small subgraphs that occur in a network at frequencies statistically significantly higher than expected at random according to a certain random model [126, 127] (see Section 3.4.2 for classical random models). A number of studies have investigated the motifs of biological networks, e.g., in signal transduction and gene regulatory networks [128, 129], finding that they are functional. Also, Barabasi et al. [108] highlighted a correlation between specific motifs and optimised biological functions. However, it has been shown that motifs heavily depend on the choice of the random network model used in their identification [130]. Since network comparison is computationally intractable due to NP-completeness of the underlying subgraph isomorphism problem, it is difficult to find a well-fitting random graph model that is to be used to meaningfully identify network motifs from a particular network data. Hence, graphlets have been introduced, which are independent of any random graph model. Graphlets are small, connected, non-isomorphic, induced subgraphs of larger networks, introduced by Pržulj [131, 132]. A wealth of tools to analyze real-world networks have been built upon graphlets (detailed in Chapter 5).

3.4.2 Network Models

The study of network models aims at understanding the structure of real world networks. The applications are multiple and include studying the evolutionary principles of biological systems organization, investigating dynamic processes on networks, such as diseases spreading [133, 134], predicting interactions or de-noising of biological networks [135]. We present some of the most frequently used network models.

3.4.2.1 *Erdős–Renyi Random Graphs*

The earliest random model was introduced by Erdős and Rényi [136]. There are two closely related ways to generate an Erdős–Rényi (ER) random graph on n nodes and m edges. The first one is to pick pairs of nodes uniformly at random to form the m edges. The second is to, for each possible pair of vertices, add an edge connecting the pair with probability p, with $p = \frac{m}{\binom{n}{2}}$.

ER graphs are well studied and they have many proven properties, such as the degree distribution following a Poisson law, low clustering, and a small diameter ($O(\log(n))$). It has also been proven that when $m \geq \frac{n}{2}$, one giant connected component containing $O(n)$ appears. For more details, see Bollobas' books [137, 138].

An extension of this model is the *generalized random graph model (ER-DD)* [101] in which the random network is generated as for ER graphs, but the degree distribution is forced to match the degree distribution of a real-world network that we are trying to model. To generate an ER-DD graph on n nodes, we first associate to a node a number of "stubs" that corresponds to its degree. We add edges between pairs of nodes chosen uniformly at random if each node has at least one stub available. When an edge is added, both its ends lose a stub. We proceed until all m edges are added. Note that due to randomness of selecting pairs of nodes, this procedure may not terminate, as it may encounter impossible constraints for adding new edges. In that case, it needs to be restarted until it produces a model network.

3.4.2.2 Scale-free Networks

Barabasi and Albert introduced a network model of preferential attachment that produces networks with power-law degree distributions [139]. The idea is to start from a small seed network $G_0 = (V_0, E_0)$ and to add nodes based on the *"rich get richer"* principle, hence the probability $p(v)$ that a node u added at time t is connected to an existent node v is directly proportional to the current degree of the existing node $\frac{d(v)}{\sum_{w \in V_t} d(w)}$, with $d(v)$ denoting the degree of node v and V_t being the set of nodes of the graph at time t. These networks are good models of the World Wide Web.

Another scale-free model is the one *with gene duplication and divergence (SF–GD)* [140]. Similar to the previous model, theses graphs are generated iteratively from a small seed network (containing at least an edge). Each iteration can be decomposed in two phases: duplication and divergence. During the former, an existing node u is chosen at random and is duplicated, i.e., node u' is added to the network and is connected to all the neighbors of u; also, an edge between u and u' is added with predefined probability p. In the divergence step, for each node connected to u and u', one of the two edges (to u or u') is chosen and removed with probability q, also set by the user. This network model was introduced to model the phenomena of gene duplication and divergence through evolution.

3.4.2.3 Geometric Networks

Geometric graphs model spatial relationships between objects.

In a *geometric random graph* (GEO) [141], nodes are distributed in a metric space uniformly at random and they are connected by edges if they are close enough (within radius r) in space according to a chosen distance norm. The radius r is typically chosen so that the network has a desired density. These models are well fitting wireless ad-hoc networks. In these networks, people carrying walkie-talkies (hand-held, portable, two-way radio transceivers) are represented by nodes. Edges exist between two nodes if the people modeled by the nodes are close enough in space so that they can receive and transmit messages between each other by their walkie-talkies.

A biologically inspired variation of this model is the *geometric graph with gene duplications and mutations (GEO-GD)* [142]. Instead of distributing the nodes uniformly at random in space, we start with a small graph and add nodes to mimic gene duplications and mutations. A node is chosen randomly and duplicated. Then the duplicate is placed uniformly at random within distance $2r$ of the original node, where r is the radius parameter introduced earlier. The duplication process iterates until the number of nodes reaches the number specified by the user. The nodes are then connected if they are within radius r. This model is good for modeling PPI networks [142].

3.4.2.4 Stickiness Index Based Networks

The idea is to assign a "stickiness" to a protein proportional to its degree. Then nodes are connected with probability proportional to the product of there stickiness indices [143]. The model is used to examine the structure of PPI networks [143].

There are many other graph families and network models, too many to describe here. We detailed only the most commonly used ones. Note that choosing a well-fitting network model is essential to define the significance of network motifs. They are also used for algorithmic development, since an NP-hard problem may be tractable on a particular graph family, so an efficient algorithm for that particular graph family could be developed. Most models are also defined for directed and other types of networks [139, 144].

3.5 Summary

In this chapter, we introduced various notions from graph and network theory used to analyze real-world biological networks. We covered the basics of linear algebra and computational complexity necessary for understanding the rest of the chapter. Then, we introduced the formal definition of a graph with some key concepts and properties from graph theory, such as neighborhood, subgraphs, and connectedness. Various types of graphs were discussed to show the range of models that can be used to represent real-world networks. A few classical problems of graph theory were discussed, linking them to applications in computational biology. Two widely used data structures to store and manipulate graphs in a computer were also introduced. Basics of spectral graph theory that are used in analyses of biological networks were also given. We covered various measures of nodes and networks and their use to analyze biological network data. Finally, we outlined some of the commonly used random graph models.

3.6 Exercises

3.1 Prove the following: In a tree, any two vertices are connected by a unique path.
3.2 Prove that if two graphs $G = (V_G, E_G)$ and $H = (V_H, E_H)$ are isomorphic then for any subgraph S_G of G there exists a subgraph S_H of H such that S_G and S_H are isomorphic.
3.3 Are the two graphs G and H Figure 3.17 isomorphic? Explain.

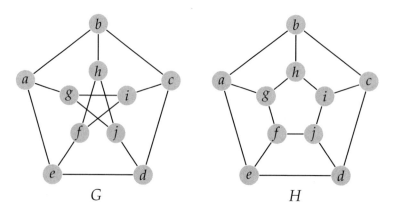

Figure 3.17: Graphs for in Exercise 3.6

3.4 Dijkstra's algorithm
The pseudocode for Dijkstra's algorithm is given below. What is its computational time complexity? Give a proof.

Algorithm 3.1 Dijkstra's algorithm

Algorithm takes graph $G = (V, E)$ and source vertex u as inputs and returns the distances δ from u to all vertices in V

1: **function** DIJKSTRA(G, u)
2:     $current \leftarrow u$
3:     $\delta(u) \leftarrow 0$
4:     $Q \leftarrow V$ ▷ Put all vertices in unvisited set
5:     **for** $v \in V \setminus \{u\}$ **do**
6:        $\delta(v) \leftarrow \infty$
7:     **while** $Q \neq \emptyset$ **do**
8:        remove $current$ from Q
9:        **for** $v \in N(current)$ **do**   ▷ Neighborhood of $current$
10:           $d \leftarrow \delta(u, current) + \omega(current, v)$
11:           **if** $d < \delta(u, v)$ **then**   ▷ If d smaller, replace stored value
12:             $\delta(u, v) \leftarrow d$
13:        $current \leftarrow \operatorname{argmim}_{v \in Q} \delta(v)$ ▷ Move to the vertex with smallest δ in Q
14:        **if** $\delta(current) \leftarrow \infty$ **then**
15:           **return** δ ▷ Terminate if smallest distance is infinity
16:     **return** δ

3.5 Connector problem. The algorithm typically used to find a spanning tree of a graph G and solve the connector problem mentioned in Section 3.3.4 is called Kruskal's algorithm. Its pseudocode is detailed below.

Algorithm 3.2 Kruskal's algorithm

Algorithm takes weighted graph $G = (V_G, E_G, \Omega, \omega)$ and returns a spanning tree T of minimal weight

1: **function** KRUSKAL(G)
2:     $E_T \leftarrow \{\operatorname{argmin}_{e \in E_G}\}$
3:     $S \leftarrow E_G \setminus E_T$
4:     **while** 1 **do**
5:        Pick $e \in E_G \setminus E_T$ such that the edge-induced graph of G
6:        on $E_T \cup e$ is acyclic and $\omega(e)$ is as small as possible.
7:        **if** e cannot be found **then**
8:           **return** $T = G[E_T]$
9:        **else**
10:           add e to E_T

Extend the Kruskal's algorithm to construct a minimal weight spanning tree of a given weighted graph $G = (V, E, \Omega, \omega)$ with the additional requirement that some edges need to be included in the edge set of the spanning tree.

3.6 Show that graph $G = (V, E)$ is bipartite if and only if it contains no odd cycle.

3.7 Let $N = (V, E)$ be a network with $|V| = n$ and $(u, v) \in V^2$ be two adjacent vertices of N. Show that we have:

$$\frac{n-1}{cc(u)} = \frac{n-1}{cc(v)} + n_> - n_<, \qquad (3.12)$$

where $cc(.)$ denotes the closeness centrality of a vertex defined with the shortest path distance measure $\delta(.,.)$ and $n_< = |\{w \in V : \delta(w, u) < \delta(w, v)\}|$ and $n_> = |\{w \in V : \delta(w, u) > \delta(w, v)\}|$.

3.8 Consider the undirected graph with vertex set $V = \{1, 2 \ldots, n\}$ and edge set $E = \{(i, i+1), \forall i \in V \setminus \{n\}\}$. The graph corresponds to a one-dimensional axis with vertices at all integers between 1 and n. Compute the betweenness centrality for each vertex i.

3.9 Consider a connected network with a power-law degree distribution such that the probability that a vertex has degree k is given by $p(k) = \lambda k^{-\gamma}$, with $\gamma > 2$.

(a) Compute λ as a function of γ.
(b) Compute the percentage P_K of vertices of degree K or greater.
(c) Compute the percentage of edges E_K that connect to at least one vertex of degree K or greater.

Hint: Note that $\sum_{k=x}^{\infty} p(k)$ can be approximated by $\int_x^{\infty} p(k)dk$.

Note: Solutions are available to instructors at www.cambridge.org/bionetworks.

3.7 Acknowledgments

This work was supported by the European Research Council (ERC) Starting Independent Researcher Grant 278212, the European Research Council (ERC) Consolidator Grant 770827, the Serbian Ministry of Education and Science Project III44006, the Slovenian Research Agency project J1-8155 and the awards to establish the Farr Institute of Health Informatics Research, London, from the Medical Research Council, Arthritis Research UK, British Heart Foundation, Cancer Research UK, Chief Scientist Office, Economic and Social Research Council, Engineering and Physical Sciences Research Council, National Institute for Health Research, National Institute for Social Care and Health Research, and Wellcome Trust (grant MR/K006584/1).

References

[1] Blais A, Dynlacht BD. Constructing transcriptional regulatory networks. *Genes & Development*, 2005;19:1499–1511.

[2] Yamaguchi-Shinozaki K, and Shinozaki K, Transcriptional regulatory networks in cellular responses and tolerance to dehydration and cold stresses. *Annual Review of Plant Biology*, 2006;57(1):781–803.

[3] Lee TI. Transcriptional regulatory networks in saccharomyces cerevisiae. *Science*, 2002;298(5594):799–804.

[4] Holland JH. Signaling networks, *Complexity*, 2002;7(2):34–45.

[5] Eungdamrong NJ, Iyengar R. Modeling cell signaling networks, *Biology of the Cell*, 2004;96(5):355–362.

[6] Samaga R, Klamt S. Modeling approaches for qualitative and semi-quantitative analysis of cellular signaling networks. *Cell Communication and Signaling*, 2013;11(1):43.

[7] Jeong H, Tombor B, Albert R, Oltvai ZN, Barabasi AL. The large-scale organization of metabolic networks. *Nature*, 2000;407(6804):651–654.

[8] Ma H, and Zeng AP. Reconstruction of metabolic networks from genome data and analysis of their global structure for various organisms. *Bioinformatics*, 2003;19(2):270–277.

[9] Ma HW, Zhao XM, Yuan YJ, Zeng AP. Decomposition of metabolic network into functional modules based on the global connectivity structure of reaction graph. *Bioinformatics*, 2004;20(12):1870–1876.

[10] Stuart JM. A Gene-coexpression network for global discovery of conserved genetic modules. *Science*, 2003;302(5643):249–255.

[11] Tong AH, Lesage G, Bader GD, et al. Global mapping of the yeast genetic interaction network. *Science*. 2004;(303):808–813.

[12] Costanzo M, VanderSluis B, Koch EN, et al. A global genetic interaction network maps a wiring diagram of cellular function. *Science*, 2016;353(6306):aaf1420.

[13] Mani R, Onge RPS, Hartman JL, Giaever G, and Roth FR. Defining genetic interaction. *Proceedings of the National Academy of Sciences*, 2008;105(9):3461–3466.

[14] Kanehisa M, Goto M, Furumichi S, Tanabe M, Hirakawa M. KEGG for representation and analysis of molecular networks involving diseases and drugs. *Nucleic Acids Researh*, 2010;38(Database issue):D355–D360.

[15] Hertz-Fowler C. GeneDB: A resource for prokaryotic and eukaryotic organisms. *Nucleic Acids Research*, 2004;32(Database issue):339D–D343.

[16] Weaver D, Gama-Castro S, Muñiz-Rascado L, et al. The EcoCyc database. *EcoSal Plus*, 2014;(6):1.

[17] Romero P, Karp P. PseudoCyc, a pathway-genome database for pseudomonas aeruginosa. *Journal of Molecular Microbiology and Biotechnology*, 2003;5(4):230–239.

[18] Caspi R, Billington RL, Ferrer L, et al. The MetaCyc database of metabolic pathways and enzymes and the BioCyc collection of pathway/genome databases. *Nucleic Acids Research*, 2016;44(D1):D471–D480.

[19] Whitaker JW, Letunic I, McConkey GA, Westhead DR. metaTIGER: A metabolic evolution resource. *Nucleic Acids Research*, 2009;37(Suppl):1D531–8.

[20] Overbeek R, Larsen N, Walunas T, et al. The ErgoTM genome analysis and discovery system. *Nucleic Acids Research*, 2003;31:164–171.

[21] Ananko EA, Podkolodny NL, Stepanenko IL, et al. GeneNet: A database on structure and functional organisation of gene networks. *Nucleic Acids Research*, 2002;30(1):398–401.

[22] Croft D, Fabregat Mundo A, Haw R, et al. The Reactome pathway knowledgebase. *Nucleic Acids Research*, 2014:42, (D1): D472–D477.

[23] Fabregat A, Sidiropoulos K, Garapati P, et al. The Reactome pathway knowledgebase. *Nucleic Acids Research*, 2016;44, (D1): D481–D487.

[24] Gama-Castro S, Salgado H, Peralta-Gil M, et al. RegulonDB version 7.0: Transcriptional regulation of Escherichia coli K-12 integrated within genetic sensory response units (gensor units). *Nucleic Acids Research*, 2011;39(Suppl.):D98–D105.

[25] Sandelin A, JASPAR: An open-access database for eukaryotic transcription factor binding profiles. *Nucleic Acids Research*, 2004;32(Database issue):91D–D94.

[26] Diella F., Cameron S, Gemünd C, et al. Phospho.ELM: a database of experimentally verified phosphorylation sites in eukaryotic proteins. *BMC bioinformatics*, 2004;5(1):79.

[27] Miller ML, Jensen LJ, Diella F, et al. Linear motif atlas for phosphorylation-dependent signaling. *Science Signaling*, 2008;1(35):ra2.

[28] Gnad F, Ren S, Cox J, et al. PHOSIDA (phosphorylation site database): Management, structural and evolutionary investigation, and prediction of phosphosites. *Genome Biology*, 2007;8(11)R250.

[29] Choi C, Krull M, Kel A, et al. TRANSPATH: A high quality database focused on signal transduction. *Comparative and Functional Genomics*, 2004;5(2):163–168.

[30] Frolkis A, Knox C, Lim E, et al. SMPDB: The small molecule pathway database. *Nucleic Acids Research*, 2009;38(Suppl. 1):D480–D487.

[31] Chatr-Aryamontri A, Oughtred R, Boucher L, et al. The BioGRID interaction database: 2017 update. *Nucleic Acids Research*, 2017;45(D1):D369–D379.

[32] Keshava Prasad TS, Goel R, Kandasamy K, et al. Human protein reference database - 2009 update. *Nucleic Acids Research*, 2009;37(Suppl. 1):D767–D772.

[33] Cherry JM, Hong EL, Amundsen C, et al. Saccharomyces genome database: the genomics resource of budding yeast. *Nucleic Acids Research*, 2012;40(D1):D700–D705.

[34] Kerrien S, Alam-Faruque Y, Aranda B, et al. IntAct: Open source resource for molecular interaction data. *Nucleic Acids Research*, 2007;35(Suppl. 1):D561–D565.

[35] Han K, Park B, Kim H, Hong J, Park J. HPID: The human protein interaction database. *Bioinformatics*, 2004;20(15):2466–2470.

[36] Yu J, Pacifico S, Liu G, Finley Jr. RL. DroID: the drosophila interactions database, a comprehensive resource for annotated gene and protein interactions. *BMC Genomics*, 2008;9(1):461.

[37] Mewes HW, Frishman D, Güldener U, et al. MIPS: a database for genomes and protein sequences. *Nucleic acids research*, 2002;30(1):31–4.

[38] Xenarios I, DIP: The database of interacting proteins. *Nucleic Acids Research*, 2000;28(1):289–291.

[39] Licata L, Briganti L, Peluso D, et al. MINT, the molecular interaction database: 2012 update. *Nucleic Acids Research*, 2012;40(D1):2006–2008.

[40] Brown KR, Jurisica I. Online predicted human interaction database. *Bioinformatics*, 2005;21(9):2076–2082.

[41] Szklarczyk D, Franceschini A, Wyder S, et al. STRING v10: Protein-protein interaction networks, integrated over the tree of life. *Nucleic Acids Research*, 2015;43(D1):D447–D452.

[42] Ulrich LE, Zhulin IB. MiST: a microbial signal transduction database. *Nucleic Acids Research*, 2007;35(Suppl. 1):D386–D390.

[43] Guo L, Liu H, Zheng J. SynLethDB: Synthetic lethality database toward discovery of selective and sensitive anticancer drug targets. *Nucleic Acids Research*, 2016;44(D1):D1011–D1017.

[44] Obayashi T, Kinoshita K. COXPRESdb: A database to compare gene coexpression in seven model animals. *Nucleic Acids Research*, 2011;39(Suppl. 1):D1016–D1022.

[45] van Dam S, Craig T, de Magalhães JP. GeneFriends: A human RNA-seq-based gene and transcript co-expression database. *Nucleic Acids Research*, 2015;43(Database issue):D1124–D1132.

[46] Online Mendelian Inheritance in Man, OMIM®. McKusick-Nathans Institute of Genetic Medicine, Johns Hopkins University (Baltimore, MD). 2018. Available online at https://omin.org/.

[47] Law V, Knox C, Djoumbou Y, et al. DrugBank 4.0: Shedding new light on drug metabolism. *Nucleic Acids Research*, 2014;42(D1):D1091–D1097.

[48] Gunther S, Kuhn M, Dunkel M, et al. SuperTarget and Matador: Resources for exploring drug-target relationships. *Nucleic Acids Research*, 2008;36(Suppl. 1):D919–D922.

[49] Mattingly CJ, Colby GT, Forrest JN, Boyer JL. The comparative toxicogenomics database (CTD). *Environmental Health Perspectives*, 2003;111(6):793.

[50] Burley SK, Berman HM, Kleywegt GJ, et al. Protein data bank (PDB): The single global macromolecular structure archive. In Wlodawer A, Dauter Z, Jaskolski M, eds., *Protein Crystallography. Methods in Molecular Biology*. Humana Press 2017, vol. 1607, pp. 627–641.

[51] Lettvin L, Maturana H, McCulloch W, What the frog's eye tells the frog's brain. *Proceedings of the IRE*, 1959;47(11):1940–1951.

[52] Emmert-Streib F, Tripathi S, de Matos R, Simoes R, Hawwa AF, Dehmer M, The human disease network. *Systems Biomedicine*, 2013;1(1):20–28.

[53] Yildirim MA, Goh KI, Cusick ME, Barabási AL, Vidal M. Drug–target network. *Nature Biotechnology*, 2007;25(10):1119–1126.

[54] Lang S. *Linear Algebra*, undergraduate Texts in Mathematics series. Springer; 1987, vol. 42.

[55] Strang G. *Introduction to Linear Algebra*, Wellesley-Cambridge Press; 1993, vol. 3.

[56] Turing AM. On computable numbers, with an application to the entscheidungsproblem. *Proceedings of the London Mathematical Society*, 1937;s2-42(1):230–265.

[57] ——, On computable numbers, with an application to the entscheidungsproblem. a correction. *Proceedings of the London Mathematical Society*, 1938;s2-43(1):544–546.

[58] Hopcroft JE, Motwani R, Ullman JD. *Introduction to Automata Theory, Languages, and Computation*, 2nd edn Addison-Wesley; 2001, vol. 32, no. 1.

[59] Garye MR, Johnson DS. *Computers and Intractability: A Guide to the Theory of NP-Completeness* Mathematical Sciences W.H. Freeman;1990, vol. 24, no. 1.

[60] Arora S, Barak B. *Computational Complexity: A Modern Approach*. Cambridge University Press;2009.

[61] Sipser M. *Introduction to the Theory of Computation*. Cengage Learning;2012.

[62] Bondy JA, Murty US. *Graph Theory with Applications*. Macmillan;1976.

[63] Karp RM. Reducibility among combinatorial problems. In Junger M, Liebling TM, Naddef D, eds., *50 Years of Integer Programming 1958-2008: From the Early Years to the State-of-the-Art*. Springer;2010, 219–241.

[64] Ben-Dor A, Shamir R, Yakhini Z, Clustering gene expression patterns. *Journal of Computational Biology*, 1999;6(3-4):281–297.

[65] Shi Z, Derow CK, Zhang B. Co-expression module analysis reveals biological processes, genomic gain, and regulatory mechanisms associated with breast cancer progression. *BMC Systems Biology*, 2010;4(1):74.

[66] Gardiner EJ, Artymiuk PJ, Willett P. Clique-detection algorithms for matching three-dimensional molecular structures. *Journal of Molecular Graphics and Modelling*. 1997;15(4):245–253.

[67] Samudrala R, Moult J. A graph-theoretic algorithm for comparative modeling of protein structure. *Journal of Molecular Biology*, 1998;279(1):287–302.

[68] Dijkstra EW. A note on two problems in connexion with graphs. *Numerische Mathematik*, 1959;1(1):269–271.

[69] Saitou N, Nei M. The neighbor-joining method: a new method for reconstructing phylogenetic trees. *Molecular biology and evolution*, 1987;4(4):406–425.

[70] Huelsenbeck JP, Ronquist F. Mrbayes: Bayesian inference of phylogenetic trees. *Bioinformatics*, 2001;17(8):754–755.

[71] Praagman J. Classification and regression trees. *European Journal of Operational Research*, 1985;19(1):144.

[72] Rokach L, Maimon O. *Data Mining with Decision Trees*, World Scientific;2008.

[73] Ho TK. Random decision forests. In *Proceedings of 3rd International Conference on Document Analysis and Recognition*, vol. 1. IEEE Computer Society Press;pp. 278–282.

[74] Berge C. *Graphs and Hypergraphs*. North-Holland Publishing Company;1973.

[75] Klamt S, Haus UU, Theis T. Hypergraphs and cellular networks. *PLOS Computational Biology*, 2009;5(5):e1000385.

[76] Kivelä M, Arenas A, Barthelemy B, et al. Multilayer networks. *Journal of Complex Networks*, 2014;2(3)203–271.

[77] Boccaletti S, Bianconi G, Criado R, et al. The structure and dynamics of multilayer networks. *Physics Reports*, 2014;544(1)1–122.

[78] Schatz MC, Delcher AL, Salzberg SL. Assembly of large genomes using second-generation sequencing. 2010;20(9):1165–1173.

[79] Flicek P, Birney E. Sense from sequence reads: methods for alignment and assembly. *Nature Methods*, 2010;7(6):479–479.

[80] Miller JR, Koren S, Sutton G. Assembly algorithm for next-generation sequencing data. *Genomics*, 2010;95(6):315–327.

[81] Idury R, Waterman MS. A new algorithm for DNA sequence assembly. *Journal of Computational Biology*, 1995;2(2):291–306.

[82] Pevzner PA, Tang H, Waterman MS. An Eulerian path approach to DNA fragment assembly. *Proceedings of the National Academy of Sciences*, 2001;98(17):9748–9753.

[83] Li Z, Chen Y, Mu D, et al. Comparison of the two major classes of assembly algorithms: Overlap-layout-consensus and de-bruijn-graph. *Briefings in Functional Genomics*, 2012;11(1):25–37.

[84] Milenković T, Ng WL, Hayes W, Pržulj N. Optimal network alignment with graphlet degree vectors. *Cancer Informatics*, 2010;9:121.

[85] Zhang W, Chien J, Yong J, Kuang R. Network-based machine learning and graph theory algorithms for precision oncology. *npj Precision Oncology*, 2017;1(1):25.

[86] Biggs NL, Hoare MJ. The sextet construction for cubic graphs. *Combinatorica*, 1983;3(2):153–165.

[87] Kottos T, Smilansky U. Periodic orbit theory and spectral statistics for quantum graphs. *Annals of Physics*, 1999;274:76–124.

[88] Uw S, Ng AY, Jordan MI, Weiss Y. On spectral clustering: Analysis and an algorithm. In Dietterich TG, Becker S, Ghahramani Z, eds., *Advances in Neural Information Processing Systems 14*. MIT Press; 2001, pp. 849–856.

[89] Zelnik-manor L, Perona P. Self-tuning spectral clustering. In Saul LK, Weiss Y, Boltou L, eds., *Advances in Neural Information Processing Systems 17*. MIT Press; vol. 2 2004, pp. 1601–1608.

[90] Hammond DK, Vandergheynst P, Gribonval R. Wavelets on graphs via spectral graph theory. *Applied and Computational Harmonic Analysis*, 2011;30(2): 129–150.

[91] Chung F. *Lectures on Spectral Graph Theory*. American Mathematical Society; 2001.

[92] Aldous D, Fill JA. Reversible Markov Chains and Random Walks on Graphs. Unfinished monograph, recompiled 2014, available at www.stat.berkeley.edu/users/aldous/Rwa/book.pdf; 2002, pp. 1–516.

[93] Seidel R. On the all-pairs-shortest-path problem. In *Proceedings of the Twenty-Fourth Annual ACM Symposium on Theory of Computing, STOC '92*, ACM;1992, pp. 745–749.

[94] Lovász L. Random walks on graphs. *Combinatorics, Paul Erdos is Eighty*, 1993;2:1–46.

[95] Koren Y. Drawing graphs by eigenvectors: theory and practice. *Computers and Mathematics with Applications*, 2005;49(11-12):1867–1888.

[96] Gu J, Hua B, Liu S. Spectral distances on graphs. *Discrete Applied Mathematics*, 2015;190–191(5):56–74.

[97] Jonsson PF, Bates PA. Global topological features of cancer proteins in the human interactome. *Bioinformatics*, 2006;22(18):2291–2297.

[98] Jeong H, Mason SP, Barabási SP, Oltvai ZN. Lethality and centrality in protein networks. *Nature*, 2001;411(6833):41–42.

[99] Coulomb S, Bauer M, Bernard D, Marsolier-Kergoat MC. Gene essentiality and the topology of protein interaction networks. *Proceedings of the Royal Society B: Biological Sciences*, 2005;272(1573):1721–1725.

[100] Albert R. Scale-free networks in cell Biology. *Journal of Cell Science*, 2005;118(21):4947–4957.

[101] Newman M. *Networks: An Introduction*. Oxford University Press; 2010.

[102] Leclerc RD. Survival of the sparsest: Robust gene networks are parsimonious. *Molecular Systems Biology*, 2008;4:213.

[103] Shannon P, Markiel A, Owen O, et al. Cytoscape: a software environment for integrated models of biomolecular interaction networks. *Genome Research*, 2003;13:2498–2504.

[104] Gurevitch M. The social structure of acquaintanceship networks. PhD Thesis, Massachusetts Institute of Technology; 1961.

[105] Milgram S. The small world problem. *Psychology Today*, 1967;2(1):60–67.

[106] Newman MEJ, Watts DJ, Barabasi AL. *The Structure and Dynamics of Networks*. Princeton University press; 2006.

[107] Ravasz E. Hierarchical organization of modularity in metabolic networks. *Science*, 2002;297(5586):1551–1555.

[108] Barabási AL, Gulbahce N, Loscalzo J. Network medicine: A network-based approach to human disease. *Nature Reviews Genetics*, 2011;12(1):56–68.

[109] Clune J, Mouret JB, Lipson H. The evolutionary origins of modularity. in *Procceedings of the Royal Society B*, 2013;280(1755):20122863.

[110] Newman MEJ. Assortative mixing in networks. *Physical Review Letters*, 2002;89:208701.

[111] Zotenko E, Mestre J, O'Leary DP, Przytycka TM. Why do hubs in the yeast protein interaction network tend to be essential: Reexamining the connection

between the network topology and essentiality. *PLoS Computational Biology*, 2008;4(8):e1000140.

[112] Levy SF, Siegal ML. Network hubs buffer environmental variation in *Saccharomyces cerevisiae*. *PLoS Biology*, 2008;6(11):2588–2604.

[113] Da Silva MR, Ma H, Zeng AP. Centrality, network capacity, and modularity as parameters to analyze the core-periphery structure in metabolic networks. *Proceedings of the IEEE*, 2008;96(8):1411–1420.

[114] Ma HW, and Zeng AP. The connectivity structure, giant strong component and centrality of metabolic networks. *Bioinformatics*, 2003;19(11):1423–1430.

[115] Mazurie A, Bonchev D, Schwikowski B, Buck GA. Evolution of metabolic network organization. *BMC Systems Biology*, 2010;4:59.

[116] Joy MP, Brock A, Ingber DE, Huang S. High-betweenness proteins in the yeast protein interaction network. *Journal of Biomedicine & Biotechnology*, 2005;2005(2):96–103.

[117] Yu H, Kim PM, Sprecher E, Trifonov V, Gerstein M. The importance of bottlenecks in protein networks: correlation with gene essentiality and expression dynamics. *PLoS Computational Biology*, 2007;3(4):e59.

[118] Paladugu SR, Zhao S, Ray A, Raval A. Mining protein networks for synthetic genetic interactions. *BMC Bioinformatics*, 2008;9(1):426.

[119] Özgür A, Vu T, Erkan G, Radev DR. Identifying gene-disease associations using centrality on a literature mined gene-interaction network. *Bioinformatics*, 2008;24(13):i277–i285.

[120] Wuchty S, Stadler PF. Centers of complex networks. *Journal of Theoretical Biology*, 2003;223(1):45–53.

[121] Estrada E, Rodríguez-Velázquez JA. Subgraph centrality in complex networks. *Physical Review E*, 2005;71(5):056103.

[122] Estrada E, Uriarte E. Recent advances on the role of topological indices in drug discovery research. *Current Medicinal Chemistry*, 2001;8(13):1573–1588.

[123] Estrada E. Characterization of the folding degree of proteins. *Bioinformatics*, 2002;18(5):697–704.

[124] Costa LDF, Kaiser M, Hilgetag CC. Predicting the connectivity of primate cortical networks from topological and spatial node properties. *BMC Systems Biology*, 2007;1(1):16.

[125] Hagberg A, Schult D, Swart P. Exploring network structure, dynamics, and function using NetworkX. *SciPy 2008: Proceedings of the 7th Python in Science Conference*, 2008;2:11–15.

[126] Milo R, Shen-Orr S, Itzkovitz S, et al. Network motifs: simple building blocks of complex networks. *Science*, 2002;298(5594):824–827.

[127] Alon U. Network motifs: theory and experimental approaches. *Nature Reviews. Genetics*, 2007;8(6):450–61.

[128] Shen-Orr SS, Milo R, Mangan S, Alon U. Network motifs in the transcriptional regulation network of Escherichia coli. *Nature Genetics*, 2002;31(1):64–68.

[129] Yeger-Lotem E, Sattath S, Kashtan N, et al. Network motifs in integrated cellular networks of transcription-regulation and protein-protein interaction. *Proceedings of the National Academy of Sciences*, 2004;101(16):5934–5939.

[130] Artzy-Randrup Y, Fleishman SJ, Ben-Tal N, Stone L. Comment on "Network motifs: simple building blocks of complex networks" and "Superfamilies of evolved and designed networks." 2004;305(5687):1107; author reply 1107.

[131] Pržulj N, Corneil DG, Jurisica I. Modeling interactome: Scale-free or geometric? *Bioinformatics*, 2004;20(18)3508–3515.

[132] Pržulj N. Biological network comparison using graphlet degree distribution. *Bioinformatics*, 2007;23(2):e177–e183.

[133] Brockmann D, and Helbing D. The hidden geometry of complex, network-driven contagion phenomena. *Science*, 2013;342(6164):1337–1342.

[134] Taylor D, Klimm F, Harrington H, et al. Topological data analysis of contagion maps for examining spreading processes on networks. *Nature Communications*, 2015;6:7723.

[135] Kuchaiev O, Rašajski M, Higham DJ, Pržulj N. Geometric de-noising of protein-protein interaction networks. *PLoS Computational Biology*, 2009;5(8):e1000454.

[136] Erdős P, Rényi A. On the strength of connectedness of a random graph. *Acta Mathematica Academiae Scientiarum Hungarica*, 1964;12(1-2):261–267.

[137] Bollobás B. *Modern Graph Theory*, ser. Graduate Texts in Mathematics. vol. 184 Springer New York; 1998.

[138] Bollobas B. *Extremal Graph Theory*. Dover Publications, 2004.

[139] Barabási AL, Albert R. Emergence of scaling in random networks. *Science*, 1999;268(5439):509–512.

[140] Vazquez A, Flammini A, Maritan A, Vespignani A. Modeling of protein interaction networks. *ComPlexUs*, 2003;138–44.

[141] Penrose M. *Random Geometric Graphs*, Oxford Studies in Probability, vol. 5. Oxford University Press; 2003.

[142] Przulj N, Kuchaiev O, Stevanovic A, Hayes W. Geometric evolutionary dynamics of protein interaction networks. *Pacific Symposium on Biocomputing*, 2010;15:178–189.

[143] Pržulj N, Higham DJ. Modelling proteinprotein interaction networks via a stickiness index. *Journal of The Royal Society Interface*, 2006;3(10):711–716.

[144] Sarajlić A, Malod-Dognin N, Yaveroğlu ON, Pržulj N. Graphlet-based characterization of directed networks. *Scientific Reports*, 2016;6:35098.

4 Protein–Protein Interaction Data, their Quality, and Major Public Databases

Anne-Christin Hauschild, Chiara Pastrello, Max Kotlyar, and Igor Jurisica

4.1 Protein–Protein Interactions: Introduction and Motivation

Proteins carry out most cellular activities, such as growth and DNA repair, and form building blocks for most cellular structures and functions. A majority of these tasks are accomplished through stable and transient protein–protein interactions (PPIs). Each protein's function requires highly specific interactions, characterized by binding partners, location, affinity, and duration of interaction. Proteins are encoded in the genome of each cell in an organism. However, complex processes during differentiation of cells, such as epigenetic modifications of the genome, alternative splicing and post-translational modification (PTM), lead to cell type specific regulation of gene expression and abundance of corresponding proteins. All these mechanisms alter the quantity of available proteins, therefore regulating the type of possible interactions in cell-specific sets. There are multiple possibilities to define which interactions should be included in a network of interest. On the one hand, not all possible direct pairwise interactions are biologically relevant: some may not occur in a physiological setting (e.g., due to different subcellular localizations) or may be very weak. On the other hand, indirect protein interactions, for example involving proteins in the same complex but not in direct contact, can be biologically important, and could be considered for inclusion in an interactome. The interactome is the entire set of interactions of a specific entity – for example an organism, an organ or a cell type [1]. Moreover, cell type specific factors such as alternative splicing and post-translational modifications or aberrant changes like protein misfolding or mutations can produce protein variants with distinct interactions [2, 3].

Many applications prove that PPI information provides important support for many biological and medical studies. Given an interactome, specific analysis can yield

> **Box 4.1: Notation**
>
> | | |
> |---|---|
> | BioGRID | Biological General Repository for Interaction Datasets |
> | GO | Gene Ontology |
> | GPCR | G protein-coupled receptor |
> | HT | high-throughput |
> | HUPO | Human Proteome Organization; Montreal, QC, Canada |
> | ICGC | International Cancer Genome Consortium |
> | ID | identifier |
> | IID | Integrated Interactions Database |
> | IMEx | International Molecular Exchange |
> | LT | low-throughput |
> | MCODE | Molecular Complex Detection algorithm |
> | MPIDB | Microbial Protein Interaction DataBase |
> | PathDIP | Pathway Data Integration Portal |
> | PPI | protein–protein interaction |
> | PSI | Proteomics Standards Initiative |
> | PSI-MI | Proteomics Standards Initiative Molecular Interaction |
> | PTM | post-translational modification |
> | TCGA | The Cancer Genome Atlas |

scientific insights into diseases and treatment mechanisms, for instance, by identifying relevant proteins and protein pairs. For example, the modeling of signaling cascades and molecular processes in healthy and diseased organisms can increase the understanding of disease mechanisms and foster the identification and characterization of clinically relevant prognostic signatures [4, 5, 6, 7, 8]. Furthermore, this understanding aids the study of drug mechanisms and the development of potential new drug targets. Thus, it is essential to identify functional roles of proteins, but also the context, including cell type, condition, and parameters of interactions, such as, binding affinity, dynamics, etc.

Unfortunately, obtaining information about context remains challenging as there are many conditions with differing interactions. Consequently, interactome studies often focus on the primary task of identifying interactions, and frequently have some limitations on the types of proteins and interactions they investigate. One of the most interesting applications of PPIs is the analysis of diseases. Gene mutations or epigenetic modifications can lead to gain or loss of PPIs and changed signaling cascades. Clearly, this may result in severe signaling anomalies, frequently leading to diseases such as cancer [2, 9, 10, 11, 12]. Annotating interactions with these changes across diverse phenotypes will provide an infrastructure to study, model and fathom complex diseases.

Despite significant advances in technologies, identifying and fully characterizing all interactions in an organism remains a difficult challenge. Many proteins remain without known interactions, so-called orphans, and the understanding of the interaction context (e.g., phenotype, tissue, localization, strength, etc.) remains limited [13]. In fact, less than 40% of the human interactome is currently known, assuming a total of about 650,000 human interactions [14], which is a lower bound not considering splice- and mutation-specific variations. Potential reasons are the large diversity in depth of our understanding of interactions for specific protein classes [13], such as GPCRs or membrane proteins [15], due to the inherent difficulty to study them, but also the limited number of suitable biological assays [16], see Section 4.2. As a result, current interactomes rarely include interactions that happen only under certain conditions or at specific times (transient interactions) due to the experimental difficulties to detect such type of interactions.

In the next sections we describe biological assays and computational methods for PPI discovery and annotation, discuss the errors and biases in resulting PPI networks, their annotation, analysis and visualization, and applications of these networks in biology. Figure 4.1 illustrates the steps that have led to making thousands of PPIs available through public databases, and subsequent analysis and visualization of these PPI networks. At first, experimental methods are used to detect interactions, which are then published in scientific journals. In order to properly characterize, annotate, and ensure easy accessibility to this data, various PPI databases have been developed. The derived PPI networks can be helpful in numerous workflows in molecular biology research, such as understanding of gene function [17], disease mechanisms [18], or drug mechanisms of action [19]. However, inappropriate network analysis may lead to incorrect conclusions [20]. Effective use of PPI networks requires an understanding of

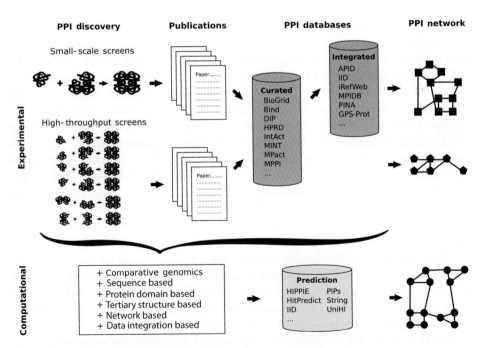

Figure 4.1: General overview illustrating the sequence of main steps that enabled the accumulation and storage of information required to construct and analyze PPI networks. These main steps include: experimental PPI detection and publication, curation of this information into databases, extending the information by computational predictions, and finally the creation, analysis, and visualization of PPI networks. Created in Inkscape (https://inkscape.org).

how the networks were identified, the types of interactions and biases/errors that they contain, and the use of network analysis methods that are appropriate for the given task. This chapter will provide the reader with the required knowledge to understand and evaluate the available PPI data, and highlights methods for appropriate network analysis, considering the most frequent workflows.

4.2 Experimental Detection and Computational Prediction of PPIs

PPIs can be detected by a wide variety of bioassays, and can also be predicted by computational methods. The need for different methods has arisen from the complexity of the problem; each individual method is limited in the types of interactions it can identify, has specific biases and error rates, as well as cost and throughput [21].

4.2.1 Experimental Methods

Experimental PPI detection methods can be characterized by: The *number of interactions* detected in an experiment (high vs low-throughput), the *types of proteins* involved (e.g., membrane, soluble), whether the proteins are *modified*, the *types of interactions* detected (e.g., direct, indirect), and the *settings* where interactions are detected. Methods with different characteristics will result in networks with very different properties.

Specific detection methods identify different **numbers of interactions**: Small-scale screens, so called low-throughput methods (LT), typically identify in the order of tens of interactions, while high-throughput (HT) screens identify hundreds to thousands of interactions. Both have advantages and disadvantages, and these are listed in Table 4.1. Until the late 1990s all detection methods were small-scale screens. These screens include methods such as affinity chromatography, affinity precipitation, dosage lethality, biochemical assays, and synthetic lethality [22]. Studies based on small-scale screens often employ a combination of these methods to identify interactions [23], and consequently, their results are often considered reliable [14, 24, 25]. HT-screens have brought about the possibility of identifying a large fraction of all PPIs in human and other eukaryotic species. However, they have not replaced small-scale screens due to their higher cost and complexity, as well as concerns about their accuracy.

Each detection method has limitations on the **types of proteins** involved in the interactions it can identify – key limiting factors include the subcellular localization of interactions and interaction stability. Commonly used methods such as yeast two-hybrid do not detect interactions of membrane proteins while some methods (e.g., membrane yeast two-hybrid, mammalian protein–protein interaction trap, mammalian membrane two-hybrid) only detect interactions involving membrane proteins, or only extracellular interactions (e.g., avidity-based extracellular interaction screen) [21]. Interaction stability is another important limiting factor; all methods can detect stable interactions, but transient interactions [26], which can be easily disrupted, may be missed by some methods (e.g., luminescence-based mammalian interactome mapping, affinity purification-mass spectrometry) [21].

Table 4.1: Advantages and drawbacks of experimental methods

	Low-throughput	High-throughput
Advantages	• reliable results	• identify many interactions simultaneously
Disadvantages	• only identify few interactions simultaneously	• higher cost • larger complexity • questionable accuracy
Examples	• affinity chromatography • affinity precipitation • dosage lethality • biochemical assays • synthetic lethality	• yeast two-hybrid • (mammalian) membrane yeast two-hybrid • mammalian protein–protein interaction trap • avidity-based extracellular interaction screen • luminescence-based mammalian interactome mapping • affinity purication-mass spectrometry

Another important distinction between detection methods is related to the **type of interactions**. The most common difference is whether reported interactions are direct or indirect. Many methods, such as yeast two-hybrid and mammalian membrane two-hybrid, detect direct binding between pairs of proteins, though they may not always distinguish between direct binding and close proximity [27]. Other common methods, including affinity purification-mass spectrometry and proximity-dependent biotin identification coupled to mass spectrometry, can identify complexes involving more than two proteins, but do not report which pairs of proteins are in direct contact. Protein pairs in the same complex but not in direct contact are referred to as indirect interactions. Identification of direct and indirect interactions (large complexes) are important for the understanding of interaction networks. Some methods such as fluorescence resonance energy transfer can provide additional information about interactions including their cellular location, and the equilibrium between the bound and unbound state of proteins [28].

A drawback of most methods is that they cannot detect interactions in an entirely natural **setting**. Most methods, including yeast two-hybrid, luminescence-based mammalian interactome mapping, and fluorescence resonance energy transfer, require modifications to candidate interacting proteins – potentially interfering with interactions. Some common methods, including yeast two-hybrid and membrane yeast two-hybrid, are carried out in yeast cells – an environment that may be very different from the species of the tested proteins. Some methods require cell-lysis (e.g., affinity purification mass spectrometry) and some frequently involve over-expression of tested proteins (e.g., yeast two-hybrid).

4.2.2 Computational Methods

Numerous computational methods have been developed to predict PPIs, relying on machine learning, statistical, and graph-theoretical approaches. These methods can be divided into five categories, based on the evidence they use to predict interactions: *Genomic conservation* in different species, *protein sequence, protein domains, protein tertiary structure, topology of interaction networks*. More recent methods *integrate* diverse types of evidence to further increase proteome coverage and reduce false discovery rates.

PPI prediction methods based on **comparative genomics** include conservation of gene neighbourhood, gene co-occurrence, and gene fusion [29]. Conservation of gene neighborhood identifies pairs of genes that occur in close proximity in several genomes [30, 31]. Gene co-occurrence identifies gene pairs that are either present or absent together in multiple genomes [32]. Gene fusion identifies gene pairs that occur as a single gene in some species [33]. Gene pairs identified by these methods are often functionally related and encode interacting proteins.

Protein sequence information has been used for PPI prediction in two ways: either by learning mappings from sequence to interaction, or by assessing similarity to known interacting protein pairs. Several studies have predicted interactions directly from protein sequence [34, 35, 36, 37, 38] using support vector classifiers. Inputs to the classifiers have included physiochemical characteristics of residues and frequencies of residue combinations. Other approaches use sequence homology to predict interactions. The paralogous verification method (PVM) predicts that two proteins interact if their sequences are highly similar (homologous) to a known pair of interacting proteins in the same species [23]. An alternative of this approach considers homologous pairs of known interacting proteins in other species [39, 40, 41].

Protein domains are conserved units of protein structure that can function independently from the rest of the protein. Some PPIs occur through binding between pairs of domains [42, 43]. Consequently, several methods predict interactions based on the presence of certain domains on candidate interacting proteins [44, 45, 46, 47]. These methods first identify domains that may facilitate interactions by searching for overrepresented domain pairs among known interacting protein pairs; then these domains are used to predict new interactions.

Tertiary structure has been used to predict PPIs mainly through homology [48, 49]. To predict whether two proteins interact, some methods assess whether their primary and tertiary structures are homologous to interacting proteins that form a complex with a known three-dimensional structure [48, 49, 50].

Several methods predict PPIs based on the **topology of experimentally detected PPI networks** [51, 52, 53, 54, 55]. These methods are based on the observation that PPI networks have densely connected local neighbourhoods [52]: Two proteins that interact with each other, tend to share many interaction partners. Consequently, two proteins can be predicted to interact if they are known to share many interaction partners.

Some of the most successful PPI prediction methods **integrate** diverse interaction evidence [13, 55, 56, 57, 58, 59, 60]. Several studies have shown that a combination of evidence improves performance by 50% over any individual evidence type [55, 57, 58]. Evidence types frequently used in integrative methods include gene coexpression, similarity of protein function, localization or process, domain complementarity, orthologous interactions, and network topology. Approaches for predicting interactions from a combination of evidence have included naive Bayes [55, 56, 58, 59], decision trees [58], logistic regression [58, 61], random forest [58, 62], support vector machines [57, 58], and association mining [13].

4.2.3 Errors and Challenges

4.2.3.1 Error Rates

Understanding of the biases and errors in PPI networks is important for their analysis and application. All datasets of experimentally detected PPIs include false positives – protein pairs that do not interact [25, 63]. The false positive rate is the percentage of detections or predictions that are true. Such protein pairs may be caused by two sources of error: experiments reporting false positive interactions and curation errors, where experimental results are incorrectly recorded.

Both sources of error are difficult to measure. False positive rates of interaction detection studies vary widely, depending on the detection method, experimental procedure, and the approach used for assessing false positive rate. For example, false positive rates of several HT-studies, determined by averaging estimates from different assessment approaches, ranged from 35% to 83% [63]. The second source of false positives – curation errors – was investigated by Cusick et al. [25]. The authors re-curated subsets of interactions from major PPI databases, and found that up to 45% were not supported by the original publications. While we can estimate false positives based on existing knowledge, it is important to keep in mind that all experimental methods and databases have an unknown percentage of false negatives. Consequently, also computational methods based on this protein–protein interaction data will be biased towards the false negative rates of the experimental techniques.

The International Molecular Exchange (IMEx) consortium is a collaboration among major primary PPI databases, founded to create a standard for literature curation of protein interaction in order to build a non-redundant high-quality data repository [64]. IMEx developed mechanisms based on improved standards, quality-control measures and external references, to curate internal and integrated databases, to minimize curation error and eliminate errors potentially introduced when remapping identifiers of third-party resources [65].

A common strategy for obtaining high-confidence PPI datasets is to select interactions reported by multiple studies. While this can reduce false positives from both detection and curation, it substantially increases false negative error (i.e., existing PPIs not present). An important limitation of this approach is that relatively few interactions are reported by multiple experimental studies (e.g., about 15% of human experimentally detected PPIs) and many excluded interactions are correct, but may be specific to the tissue, phenotype, or detection method. This tradeoff between the false positive rate of a PPI dataset and the number of true interactions it includes, also occurs within individual PPI studies. If a study reduces its false positive rate it will likely increase its false negative rate (i.e., many true interactions will be undetected).

An assessment of five HT detection methods by Braun *et al.* found that the false positive rate of each method could be reduced to less than 5% but the false negative rate was 65%–80% [66]. Taking the union of interactions can increase the false positive rate and reduce the false negative rate. Braun *et al.* showed that taking the union of five HT methods increased the false positive rate to 10% and reduced the false negative rate to 41% [66]. In comparison, the computational PPI prediction method FpClass achieved lower error rates on the same test set: a false positive rate of 1% and a false negative rate of 10% [13]. While performance of prediction and detection methods varies widely across datasets, in general, combining both approaches can be an effective strategy for interactome mapping [67].

Individual PPI detection and prediction methods have strengths and pitfalls. Therefore, the union of the results produced by a diverse set of algorithms will reduce the false negative rate, but may lead to increased false positive rate. Annotating PPIs with confidence measures, scoring the methods used to detect or predict the interaction, can help to select or to filter out interactions at a specific confidence level [68], tailoring the data to specific task and analysis.

4.2.3.2 Biases

Biases in PPI datasets lead to under-representation of certain proteins or types of interactions. Sometimes PPI studies are specifically designed to detect certain types of interactions (e.g., interactions of membrane proteins), and their observed bias has no negative implications. However, many biases that are unintentional (or remain unknown) can lead to inappropriate analysis and in turn to incorrect conclusions.

Biases in PPI datasets have multiple causes: Limitations of PPI detection and prediction methods, restrictions of commonly used experimental procedures, research biases, and other factors. Some PPI detection methods are ineffective for proteins with low expression or particular subcellular localizations, or for interactions that are transient [21]. Interaction prediction methods may have similar limitations because they are often trained on experimentally detected PPI datasets containing these biases. Naturally, machine learning and data mining-based prediction methods trained on biased experimental data will provide biased prediction. Prediction methods also have their own biases; for example, predictions based on orthologous interactions will fail to identify interactions of species-specific proteins. Experimental procedures of PPI studies may lead to biases as well. For example, interactions are typically detected in either yeast or cancer cell lines – interactions requiring different cell types may not be found. A third, and possibly the biggest source of biases in PPI data, is research bias: the tendency of researchers to focus on particular proteins, often ones that have been previously studied [69]. Previously studied proteins may be easier or less expensive to work with, or they may have a known phenotypic role, motivating further research, leaving out so called interactome orphans [13].

Biases in PPI datasets can lead to inaccurate estimates of interactome properties, and may render networks less effective. Currently known PPI datasets have been used to estimate characteristics of interactomes such as their size and topological properties (e.g., degree distribution). These estimates are less accurate (or completely incorrect) when PPI datasets are deficient for certain types of proteins or interactions.

Applications of PPI networks such as prediction of disease genes or drug discovery, often assume that a protein's role is related to its centrality or proximity to other genes with known roles. However, factors such as research bias may artificially link a protein's centrality to its importance in disease – known disease proteins are likely to be extensively interactive. Furthermore, the described biases strongly influence PPI detection or prediction methods and thereby can lead to incorrect conclusions about the structure of PPI networks, and thus ineffective strategies to evaluate and compare the quality of PPI data and networks. Finally, the issue might affect the assessments of detection (and prediction) methods, with respect to the understanding of biological networks and their applications in research.

4.3 Challenges of Data Integration

Integrating datasets can provide a more complete view of a fragmented domain from which insights are more readily inferred. However, a misguided integration may distort our view of the target interactome, and amplify existing biases. Frequently made available to the public as a database accessible over the Internet, these integrated datasets have become commonplace and indispensable, particularly in biology and medicine. In order to analyze the entire interactome, datasets generated with different methods need to be curated and integrated, and made available in a standard format. The IMEx consortium, for instance, was founded to create such a standard for literature curation of protein interaction and a non-redundant data set available in a single search interface [64]. Another community standard for data representation in proteomics is defined by the Proteomics Standards Initiative (PSI) [70]. Similar to the IMEx, it aims to facilitate data comparison, exchange and verification. Integration is also key to obtain a more reliable interactome, as shown in [71], but it faces several challenges due to data heterogeneity.

4.3.1 Heterogeneity in Biological Data

Heterogeneity is the main obstacle in biological data integration. There are different sources of heterogeneity:

Molecular: Experiments can describe physical interactions among proteins, as well as functional interactions between proteins and other entities (e.g., DNA, RNA, metabolites, drugs, etc.). Integration requires proper annotation of all the interactions as well as a clear categorization of the type of molecules present in the set or in the network.

Experimental: As described in the previous section, diverse techniques are available to identify PPIs, each one of them with different properties and outputs. Moreover, even when using the same technique, heterogeneity is present between different experiments or even replicas of the same experiment. These can result from the sample itself as, for instance, variations in the state of the cells (e.g. circadian rhythm), differences in instruments or assays used, the sensitivity of the technology and samples selected as control.

Integration needs to take into account this variability and normalize results from different experiments to make them comparable.

Condition specific: One cannot study all possible interactions of a complex organism at once. For this reason, each experiment is set in a specific environment (e.g., a cell line), compartment (e.g., membrane), condition (e.g., drug treatment), or protein status (e.g., mutation). Consequently, the integration of such data needs to take into account all of these conditions and each interaction needs to be viewed in the detailed context it was discovered. It should be clear that the data integrating different sources can not entirely represent the true interactome of an organism, without taking conditions into account. For example, integration of interactions detected in a diseased tissue and its normal counterpart will create a dataset very useful to study the molecular causes of the diseases. However, in order to create a more complete dataset for the tissue, the phenotype specific subsets should be considered for meaningful analysis. It is then important to carefully curate the original conditions when trying to integrate data to obtain a more complete interactome.

Nomenclature: There are few databases that aim at addressing nomenclature heterogeneity assigning a specific identifier (ID) to each protein or fragment of thereof. The more generic and most frequently used ones are Uniprot,[1] Entrez,[2] and Ensembl.[3] Moreover, researchers can use the protein or gene names to identify the molecules. Integration of such variety of IDs requires careful mapping, and sometimes it cannot completely reduce ambiguity, in particular in one-to-many mappings or when names and not IDs are used in the annotation [72].

4.4 Protein–Protein Interaction Databases

Over the years, large-scale and small-scale experiments, literature mining, and predictions identified many PPIs. To store this information a number of independent public databases have been developed, and are constantly updated in order to provide the most up-to-date and the most annotated interactomes.

A database is a structured dataset that manages data and access to it. Data are organized according to a *schema*, which defines the structure of the underlying dataset stored in the database, and a list of the constraints the database will enforce on data access or modification.

Classically, the existing databases are characterized into three different classes, curated databases, prediction databases, and integrated databases (that will be further described in the next section). In order to cope with the increasingly data-driven nature of modern biology and medicine, and growing demands for large and comprehensive data resources, integrated databases have become an indispensable infrastructure in bioinformatics workflows.

However, similar to data integration, database integration faces several challenges that can be described by *variety*, or the presence of the same data in diverse representations, *velocity*, being data produced at increasing rates, and *veracity*, considering varying quality of databases [73].

---

[1] Uniprot (www.uniprot.org)
[2] Entrez (www.ncbi.nlm.nih.gov/protein)
[3] Ensembl (www.ensembl.org)

Variety: The form or structure of the data. Although each repository follows a certain standard, data-overlap is limited and the majority of overlapping information is lacking a generalized structure. Further, there are challenges for example with respect to the unique identification of proteins or details about evidence, scores and data provenance. In combination, the described issues lead to challenges when designing generalized databases, without knowing ahead of time what "forms" of data will be encountered. Even if the form of the data is fully or partially known ahead of time, if we fail to determine a method to exploit the structure in the data it is still effectively *unstructured* or *semi-structured*.

Velocity: The rate at which the data is generated and becomes available or obsolete for a given purpose. The available information about PPIs is constantly increasing and besides the differences between PPI databases, with each update interactions may be added, verified, or removed.

Veracity: Any uncertainty, ambiguity, incompleteness, conflicting conclusions, and even fraudulent or erroneous results that may raise doubts as to the reliability of a dataset for a given purpose. Therefore, we refer to the *veracity* of data when talking about the trust or a lack thereof in the accuracy of data when it is used as evidence to support or reject a hypothesis.

In summary, PPI data integration creates a foundation for the network analysis and interpretation, and poses a significant challenge in order to ensure quality and provenance.

4.4.1 Curated Databases

Over the last decades there has been a great effort to build biological databases and resources for PPI data, determined in thousands of experimental studies in various biological systems, and reported in thousands of publications. The curated databases are built solely on manually curated, experimentally validated interactions, both small-scale and high-throughput methods. State-of-the-art examples include the BIOlogical General Repository for Interaction Datasets (BioGRID) [74] or IntAct [75]. Table 4.2 shows a selection of commonly used primary databases.

Table 4.2: Curated databases

Database	Website	Number of Interactions	Organisms	Version	Reference
BioGrid	thebiogrid.org	1,418,871	66	3.4.146	[74]
Bind	inactive	58,266	10	2003	[76]
DIP	dip.doe-mbi.ucla.edu	81,731	834	2004	[77]
HPRD	www.hprd.org	41,327	human	Release 9	[78]
MIntAct	ebi.ac.uk/intact	718,180	>275	4.2.6	[75]
MINT	mint.bio.uniroma2.it	235,635	~30	2012	[79]
MIPS	mips.helmholtz-muenchen.de/proj/ppi	1,800	10	2005	[80]

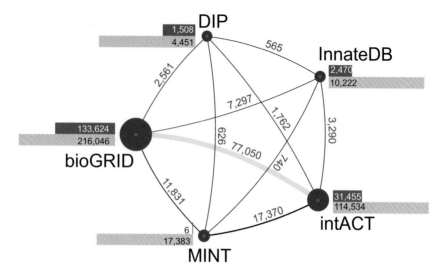

Figure 4.2: Visualization of overlap between different primary databases. Each circle represents an individual database, the bars indicate the total number (light blue) and the number of unique PPI records (dark blue) in the database. The level of agreement on recorded interactions between two databases is represented by the edge thickness and color. Created in NAViGaTOR 3.0 [81], exported in SVG, and finalized using Adobe Illustrator.

However, the existing primary databases partially contain redundant but mostly complementary information, depicted in Figure 4.2. Therefore, several international consortia published standards, such as the HUPO (Human Proteome Organization; Montreal, QC, Canada), Proteomics Standards Initiative Molecular Interaction (PSI-MI) controlled vocabulary and data structure [70]. Together with the continuing efforts of the IMEx Consortium [64], these initiatives made important contributions toward creating standards, and providing integrated access to PPI data from multiple data sources into larger PPI networks [82, 83]. Although the available aggregated data is useful, data sources are only partially consolidated and subject to many outstanding issues. Particular challenges are, for instance, referencing gene and protein identifiers across databases, varying nomenclature of the organism investigated, and varying conventions for the representation of multi-protein complexes. Other obstacles arise when combining raw unprocessed data from high-throughput studies with high-confidence data, without a general agreement between databases on which of these are best used for redistribution [82].

4.4.2 Prediction Databases

An important problem of current interactomes in human and model organisms are missing interactions. Integrated curated databases comprise about 350,000 human PPIs. However, at best, only around 40% of all protein interactions have been detected by at least one experiment and only about 15% are verified by two independent publications, see Figure 4.3. Furthermore, the interactomes of many other organisms

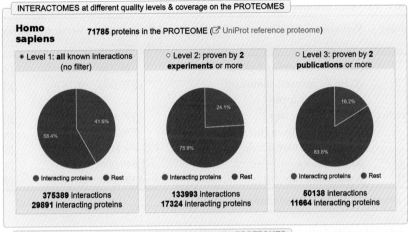
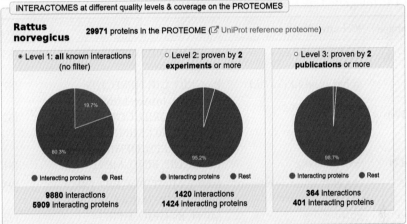

Figure 4.3: Illustration of varying coverage of interactomes across diverse species. Even well-studied organisms like *Homo sapiens* have interactions for less than half of all known proteins (~43%). For other less-analyzed species like *Rattus norvegicus*, hardly any interactions are experimentally validated. The figure comprises two screen captures, for human and rat, generated by the APID data server (Version: 13.July.2018), available at http://apid.dep.usal.es [83].

are not known at all, or remain very sparse. As described in Section 4.2.2, diverse computational algorithms have been used to predict PPIs aiming at complementing experimentally obtained data and achieving a more complete interactome. Based on these algorithms, a set of different prediction databases have been created. Table 4.3 shows a selection of state-of-the-art prediction databases. A more comprehensive list of all current PPI databases and analysis tools is available at: https://omictools.com/ppis-category.

4.4.2.1 Integrated Databases

In order to solve the described challenges, integrated databases were created, which combine multiple curated or predicted databases and occasionally new experimental

Table 4.3: Prediction databases

Database	Website	Number of Interactions	Version	Reference
BCI	califano.c2b2.columbia.edu	21,156	2009	[84]
HIPPIE	cbdm-01.zdv.uni-mainz.de/mschaefer/hippie/	>270,000	2016	[85]
HitPredict	hintdb.hgc.jp/htp	398,696	2015	[86]
IID	ophid.utoronto.ca/iid	726,363	2017–07	[87]
PIPs	compbio.dundee.ac.uk/www-pips	498–79,441[a]	2009	[88]
String	string-db.org	~1,840,000	2007	[89]
UniHI	www.unihi.org	~350,000	2014	[90]

[a] *Note:* Depending on the selected confidence score.

Table 4.4: Integrated curated databases

Database	Website	Resources	Reference
APID	apid.dep.usal.es	BioGRID, DIP, HPRD, IntAct, MINT	[83]
IID	ophid.utoronto.ca/iid	BioGRID, DIP, HPRD, I2D, InnateDB, IntAct, MINT	[87]
iRefWeb	wodaklab.org	BIND, BioGRID, CORUM, DIP, IntAct, HPRD, MINT, MIPS (MPact, MPPI), OPHID	[82]
MPIDB	www.jcvi.org	IntAct, DIP, BIND, MINT	[91]
PINA	cbg.garvan.unsw.edu.au	IntAct, MINT, BioGRID, DIP, HPRD, MIPS (MPact, MPPI)	[93]
GPS-Prot	gpsprot.org	MINT, BioGRID, HPRD Human – HIV interactome	[92]

data. These combined data sources tremendously increased the completeness and usability of the PPI data within or even across multiple species. Similar to primary and predicted databases, there are two different types of integrated databases, those that focus on summarizing curated data and those that integrate both curated and predicted databases.

Examples for the **integrated curated databases** include iRefWeb [82] or Integrated Interaction Database (IID) [87]. Other databases focus on specific species like the Microbial Protein Interaction Database (MPIDB) [91] or focus on the host–pathogen interactome (GPS-Prot) [92]. Table 4.4 provides further examples of state-of-the-art integrated databases.

It is clear that PPI predictions can expand the known interactome and, even with limitations described in Section 4.2.3, lead interactomes closer to completeness. For this reason, integration of experimental and predicted PPIs is the key to provide researchers a more comprehensive and higher quality data for interactome-based analyses (described in Section 4.6). Nevertheless, many databases tend to focus solely on either experimentally validated or predicted interactions.

Table 4.5: Integrated curated and prediction databases

Database	Website	#Interactions[a]	Organisms	Vers.	Ref.
IID	ophid.utoronto.ca/iid	E: 543,488 O: 645,208 C: 726,363 T: 1,783,963	Human, mouse, rat, Fly, worm, yeast	2017-07	[87]
InnateDB	innatedb.com	E: 367,478 C: 462,421 T: 829,899	Human, mouse, bovine	03-2017	[94]
HPIDB 2.0	agbase.msstate.edu/hpi/main.html	T: 55,505	Host: 55 (human 97%) Pathogen: 523	March 14, 2017	[95]
PAIR	cls.zju.edu.cn/pair	E: 5,990 C: 145,494	A. thaliana	2011	[96]
PrePPI	technology.sbkb.org/portal/page/350/	E: 199,863 T: ~60,000 for yeast and ~370,000 for human [b]	Human, yeast	2013	[97]

[a] E: Experimental detection, O: Orthologous, C: Computational prediction, T: Total.
[b] High confidence PPIs (Probability >0.5); Low confidence PPIs (probability >~0.1) ~2,000,000

The **integrated curated and prediction databases** represent a type of databases that combine both experimental and predicted interactions, examples are listed in Table 4.5. The integration is effective in reducing the number of false negative results, as reviewed in [98], albeit at the price of increasing false positives. The integration of orthologous PPIs further extends the coverage of the interactome space, especially for model organisms, whose interactomes are heavily lacking interactions (a key problem since these organisms are widely used in many research fields). Therefore, network analyses based on combined interactomes are now an integral part of systems biology research [11]. One example is finding network motifs in PPI networks of integrated databases. Network motifs are network topological characteristics defined as overrepresented small connected subgraphs in networks that can be considered as an interaction pattern and therefore as an essential functional unit in the organization of modules in networks [99, 100].

4.4.3 PPI Context Annotation

Availability of complete interactomes is essential in systems biology research, but it is important to be aware that it includes interactions relevant across different conditions. To understand molecular mechanisms requires to know when and where are PPIs functional, and have a good model of resulting spatio-temporal signaling cascades they form. PPIs can be annotated at different functional levels, such as pathways, tissues, subcellular localization, and disease associations. Few databases provide PPI-specific annotations, and if they do, it is mostly for human.

Examples of annotation databases can be found in Table 4.6.

Table 4.6: Annotation databases

Database	Website	Type of annotation	Organisms	Reference
ComPPI	comppi.linkgroup.hu	Subcellular localization	Yeast, worm, fly, human	[101]
IID	ophid.utoronto.ca/iid	Tissue	Human, mouse, rat, fly, worm, yeast	[87]
TissueNet	netbio.bgu.ac.il/tissuenet	Tissue	Human	[102]
SPECTRA	alpha.dmi.unict.it/spectra/	Tissue and disease	Human	[103]
HIPPIE	cbdm-01.zdv.uni-mainz.de/ mschaefer/hippie/	Disease	Human	[85]
GPS-Prot	gpsprot.org	Disease	Human – HIV	[92]

4.4.3.1 *Subcellular Localization*

Proteins are located in one or more cellular compartments. In fact, compartmentalization is one mechanism of molecular regulation. For example, TP53 functions depend on its localization: If DNA damage occurs, TP53 is imported into the nucleus, forms a tetramer and activates DNA repair pathways. TP53 is the most frequently mutated gene in cancer, but when it is not mutated, loss of TP53 activity is associated with its exclusion from the nucleus, either because of cytoplasmic sequestration, or hyperactive nuclear export [104]. It is obvious that if two proteins are located in different compartments, they cannot interact; unfortunately, subcellular localization data remain incomplete, redundant, and often poorly structured, and a big part of the proteome has only computationally predicted localization [101]. Yet, it is fundamental to annotate PPIs with their subcellular localization, especially when analyzing molecular processes that span across compartments.

4.4.3.2 *Tissue Annotation*

Tissue-specific subnetworks have fewer interactions, and are more fragmented than the entire interactome, but they are biologically more relevant [105]. It has been shown that tissues with similar functions tend to share a high number of interactions, and that tissue-specific networks change substantially during growth and differentiation, highlighting the need for properly annotating interactions with the tissue or the developmental stage being studied [106]. Interestingly, proteins related to hereditary diseases tend to have tissue-specific PPIs (occurring exclusively in the tissues related to the pathology), a perfect example of how tissue annotation can be useful in identifying disease molecular mechanisms [107].

4.4.3.3 *Disease*

Proteins can be altered in diseases due to changes in their sequence (mutations), epigenetic modifications, post-translational modification, and hence in their gene expression. It has been shown that mutations that remove part of a protein mimic the disappearance of such protein from the interactome, while mutations that alter specific domains of a protein impair fewer interactions [108]. It is also known that mutations on interactors of disease related proteins lead to similar disease phenotypes, probably due to their functional relationship. It is clear that PPI networks can be used to explore the differences between healthy and disease states – to prioritize gene candidates for further functional studies, to identify prognostic markers, or to pinpoint the most relevant candidates for drug targets [109, 110, 111, 112, 113, 114].

4.5 Protein–Protein Interaction Networks and their Properties

In order to investigate the entire interactome, the data is represented as a network, where each protein is represented as a node, and each interaction between proteins as an edge. Identifying and analyzing these networks of interactions enables a better understanding of the molecular mechanisms, and thus different phenotypes. Numerous studies have shown how PPI networks enhance many applications, for instance, prediction of gene function, identification of disease genes and drug discovery, see Section 4.6 for further details. In the following sections we provide a brief overview of notations, concepts, and properties of networks, both formal and informal.

4.5.1 Short Introduction to Networks

A biological network is an abstract conceptual model of a definite set of entities (e.g., proteins in a specific complex, pathway, or organism), and the relationships between those entities (i.e., interactions among proteins). In computer science and mathematics, a network is often referred to as a *graph*. Formally, the set of entities we seek to represent are the set of vertices V in a graph $G(V, E)$, where the set of edges E defines pair-wise relationships between members of V.

4.5.2 Network Construction

Informal notations for depicting a graph $G(V, E)$ resemble the ball-and-stick model used to convey chemical structures, where a line (edge) connects two vertices (nodes or molecules). In the construction of a graph or network model, an edge between vertices can be directed or undirected, and may have specific weight or other annotation. If an edge is undirected, then the order of the pair $(u, v) \in E$ is not significant, whereas if the edge is directed then we say the relationship between u and v represented by the edge (u, v) is from the *source* at $u \in V$ to the *destination* at $v \in U$. If we model the transition of one chemical compound A into another compound B the relationship is directional. Experimentally detected PPIs are usually undirected, except when the edge describes a functional interaction between two proteins, it is more often directed (an enzyme that catalyzes a certain reaction, for instance a kinase catalyzing phosphorylation of a protein). Edges can also be weighted, for instance by the physical strength of the interaction or the certainty that a predicted interaction exists.

More details about network construction and graph and network theory is discussed in Chapter 3, while a thorough evaluation of biological network alignment and analysis techniques in general will be given in Chapters 9 and 13. The following paragraphs will introduce measures to evaluate PPI networks.

4.5.3 Properties of PPI Networks

4.5.3.1 Degree and Betweenness Centrality

Node centrality is a crucial analysis for identifying nodes that play important roles in a graph. The most frequently used node centrality measures are degree and betweenness centrality.

The degree of a vertex v is defined as the number of edges e touching a node v:

$$D_v = \sum_i e_{vi}, \quad i \in V(G). \tag{4.1}$$

In PPI networks, proteins with high degree – sometimes referred to as hubs – have been described to be essential to cellular functions in yeast [115], leading to the hypothesis that disease genes in human could be hubs as well. In lung cancer, up-regulated genes are indeed highly connected [116], and genes frequently implicated in different types of human cancers have twice the amount of connection compared to non-cancerous proteins [117]. A more comprehensive study considering diseases other than cancer highlighted that disease-related proteins tend to interact more among each other [118], they tend to be expressed in the same tissue and they have similar functionality, but they are not hubs in the interactome. A possible explanation for such different results, beside the wider inclusion of diseases, is that disease-causing proteins can have higher degree because they are more studied [119] – an example of how the biases described in previous sections can affect the results of analyses performed using interactomes.

Betweenness centrality uses global information of the connectivity of the whole graph. The betweenness centrality of v is calculated as the number of shortest paths between pairs of nodes in graph G that pass through v:

$$BC_v = \sum_{i \neq j} \frac{s_{ij}(v)}{s_{ij}}, \tag{4.2}$$

where s_{ij} is the number of shortest paths (described in Section 4.5.3.4) between vertices i and j, and $s_{ij}(v)$ is the number of such paths that pass through v [120]. Nodes with high betweenness centrality are also called bottlenecks and in regulatory networks they tend to correlate with essentiality of genes [20].

4.5.3.2 Articulation Points

An articulation point is a node whose removal disconnects the graph and for this reason it is often considered critically important in a network [121]. In PPI networks they have been associated with proteins essential for survival [122].

4.5.3.3 Graph Density

It is a measure based on the number of connections in a graph and is defined as:

$$\text{Den} = \frac{2|E|}{|V|(|V|-1)}, \tag{4.3}$$

where $|V|$ is the number of vertices in the graph and $|E|$ is the number of edges. If density is close to 1, then the graph is densely connected, while if the density is close to 0 the graph is sparsely connected. PPI networks are sparse (Table 4.7 shows the number of edges, nodes, and density of six networks built on six different species interactomes), mainly because selection should favor parsimonious networks, imposing a fundamental design constraint in the evolutionary drive [123]. However, the interactome is largely incomplete and shows huge difference in local density, i.e., the network surrounding well studied proteins tends to be dense, while areas around barely studied proteins with few interactions, tend to be rather sparse. Proteins without any interactions are so called orphans [13]. Moreover, not all proteins can interact with each other all the time, and multiple levels of regulation of the molecular processes, combined with tissue-specific expression and localization, contribute to the regulation. It has been demonstrated that each single gene is controlled by a limited number of other genes, which is small compared to the total gene content of an organism [124].

4.5.3.4 Distance

A path is a finite or infinite sequence of consecutive edges in a graph connecting a set of vertices which are all distinct from one another. The length of a path is defined by the number of edges in such sequence. The path with minimal length between two vertices is a *shortest path*, and defines the *distance* between the two vertices. The longest distance between any pair of vertices in a graph defines the graph's *diameter*. Path lengths and diameters of biological networks are small compared to the network size. This is usually referred to as a small world property [125]. This feature is thought to indicate that information can be transmitted quite efficiently in biological networks [126]. Moreover, the comparison of the distribution of shortest paths in several graphs can highlight possible mechanisms underlining the molecular differences among such graphs. For example, it has been shown that cancer networks tend to have shorter paths than normal networks, suggesting the creation of fast and efficient shortcuts during carcinogenesis [127]. Another example is the temporal analysis of

Table 4.7: Number of edges, nodes and density for the interactome of the six listed species (experimental interactions extracted from IID http://ophid.utoronto.ca/iid).

Species	Edges	Nodes	Density
Homo sapiens	268,024	163,55	0.002004
Mus musculus	30,537	9,005	0.000753
Rattus norvegicus	5,697	2,984	0.00128
Drosophila melanogaster	58,723	10,240	0.00112
Caenorhabditis elegans	13,948	5,103	0.001071
Saccharomyces cerevisiae	145,472	6,255	0.007437

biological networks: Wong et al. found meaningful temporal changes between different networks specific to different stages of cancer progression [113].

4.5.3.5 Clustering Coefficient

PPI networks can contain densely connected subnetworks, known to be functional modules that include proteins performing similar functions or involved in related biological tasks. The detection and understanding of the biological significance of these subnetworks need to be thoroughly investigated [128]. One basic measure of the tendency of a graph to group into modules is the clustering coefficient. If v is a vertex with degree D_v and there are λ_v edges between the D_v neighbours of v, then the local clustering coefficient is:

$$CC_v = \frac{2\lambda_v}{D_v(D_v - 1)}. \tag{4.4}$$

The closer this value is to 1, the more likely the network includes clusters. Due to the modularity nature of biological processes, it is not surprising to find a higher average clustering coefficient in biological networks than in random networks as defined by the Erdős–Rényi model [19, 129]. An Erdős–Rényi random network consists of a fixed set of vertices and a fixed number of edges of equal likelihood. A limitation of clustering coefficients is the inability to indicate vertices that are present in different biological modules [130].

4.5.3.6 Cliques

Cliques are complete subgraphs, where all the vertices are fully connected (i.e., there is an edge between any two vertices in the clique) and are considered the most basic clusters. Cliques are models representing protein complexes, and groups of proteins whose molecular functions are tightly related [131]. As mentioned before, cliques identified in disease-related PPI networks have been found to be associated to the disease phenotype [132]. However, a full clique is a stringent structure, as in a biological system one or few edges are likely missing from the clique, either due to biological parsimony or to data incompleteness. Moreover, if edges are selected based on specific conditions prior to clique analyses (for example, considering only edges present in a specific disease being studied), the number of edges composing the clique can be further reduced. A solution is to analyze quasi-cliques – cliques missing a limited number of edges – even if this type of analysis requires complex computation. More details can be found in the comprehensive review by Wu and Hao [133].

4.5.3.7 Other Properties

Beyond the described network properties, many other structural features such as motives exist. They are described in Chapter 3 on graphs and network theory and Chapter 5 on graphlets in network science and computational biology.

4.5.4 PPI Network Annotations and Visualization

The graph representation provides a powerful, global overview of the interactome. However, the visualization of large PPI networks often results in an incomprehensible

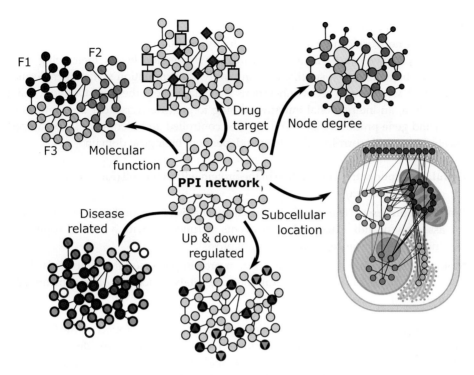

Figure 4.4: The annotation of PPI networks using different sources of information enables to highlight potential relations of important genes or proteins. The illustration exemplifies different types of annotations, namely molecular function (blue, green, and yellow, each represents different GeneOntology molecular functions), drug targets (blue square and yellow diamond represent targets of different drugs), node degree (size and colour of each node correspond to the degree of each node), subcellular location (proteins are positioned according to their location in the cell, membrane, cytoplasm, nucleus, mitochondria, etc.), up- and down-regulation (up-regulated and down-regulated proteins are represented by red up-pointing and green down-pointing triangles, respectively), and disease-related proteins (nodes outlined with shades of red according to the strength of the association). Networks created in NAViGaTOR 3.0 [81] exported in SVG, and finalized using Inkscape (https://inkscape.org).

hairball, not directly interpretable. Therefore, it may be difficult to extract useful information from un-annotated interactomes. The integration of interactome with additional information about the interactions and proteins may help to organize and clarify the networks as well as highlight the key attributes [68]. Figure 4.4 depicts a few examples for commonly used network annotations. These annotations for nodes and edges can be included in the original data files, retrieved from text files, or imported from PPI or other databases.

Similarly to interactions (as described in Section 4.4.3), proteins can be annotated with different types of identifiers for different purposes. The appearance of nodes and edges can be set based on this annotation information, which can increase the understanding of the network, see Figure 4.4 for examples. Next, we will present a few examples for resources that provide valuable information to help in gaining further insight into molecular mechanisms the network or a subnetwork represents.

4.5.4.1 Qualitative Annotations

Nodes in a network can be annotated with any qualitative or quantitative data. For example, proteins can be annotated with their expression levels, their being drug targets or belonging to a specific protein family, their mutation status or molecular function. A widely used set of annotations is available from the Gene Ontology (GO) Consortium, an international initiative aiming to provide consistent descriptions of genes and gene products [68]. GO uses three controlled and hierarchically organized vocabularies, each corresponding to an independent biological domain:

Cellular component: The location of the protein within the cell (for instance, nucleus or mitochondrion).

Biological process: The specific biological mechanism, accomplished by an ordered assembly of individual molecular events or functions with a defined beginning and end (for instance, glycolysis or apoptosis).

Molecular function: The purpose a certain protein possesses at the molecular level e.g., in a biological process (examples are catalytic activity or transmembrane transport).

A more detailed description of definitions and policies can be found on the GO website http://geneontology.org. The GO project was intended to provide information for large sets of genes and gene products, and strategies have been developed to make the manual and automated access as efficient and effective as possible. This comprehensive and easily accessible gene ontology data is particularly suitable for the annotation of PPI networks. In addition, GO annotations enable enrichment analysis to identify significantly over- or under-represented GO terms (e.g., molecular functions) in a given protein list or a PPI subnetwork.

The most commonly used ontology for this purpose is the biological process ontology. As depicted in Figure 4.4 (upper left), modules of proteins that are part of the same process can be identified, providing a functional interpretation of the network [68]. Qualitative annotations can be visualised through qualitative visual attributes, such as color, size, outline, and shape. Examples for this annotation are further described in Section 4.6 and in [15, 134, 135].

4.5.4.2 Quantitative Annotations

Besides the mentioned qualitative annotations a number of quantitative measures can be used to annotate proteins and interactions in a network. For instance, popular measures used to annotate and visualize proteins in a network are experimentally determined, including differential gene or protein expression, abundance or mutation status. Other interesting continuous annotations arise from epigenetics like the percentage of promoter methylation and the number of post-translational modifications. Furthermore, interactions represented by edges in the network can be annotated by confidence of the interaction or co-expression of genes defined by correlation. In addition, both nodes and edges can be annotated by network measures such as betweenness or centrality, as described in Section 4.5.3. Quantitative measures can be visualized through quantitative visual attributes such as size (nodes) or thickness (edges), transparency (both), or shades of colors (both) [136, 137, 138, 139].

4.6 Applications of PPI Network Analysis

Over the last two decades, the field of PPI network analysis has been continuously expanded and specialized towards diverse applications. In order to highlight the value and importance of the field, this section describes possible questions from diverging areas of systems biology and related network analyses workflows.

4.6.1 Identification of Disease-associated Genes

Many diseases are the result of a complex combination of malfunctioning proteins and other gene products involved in one or multiple molecular processes. Therefore, we can assume that disease-associated genes are interacting with each other and likely to be strongly connected in a PPI network. It follows that the analyses of PPI networks, in combination with known disease proteins, can be used to find novel associations between genes and diseases.

For instance, many clinical studies analyze the expression of genes in patients and compare these to the expression level of normal controls. These studies typically result in a list of genes ranked by the strength of diseases association. In order to confirm and prioritize this disease association the proximity (e.g., many direct interactions) of the potential candidates to known disease proteins is considered.

A variety of network-based methods have been developed to identify and characterize potential novel disease-associated genes and gene products. The first approaches, so-called **linkage methods**, assumed that the direct interaction partners of a protein were likely to be associated with the same disease phenotype [140]. Later, it was shown that topological attributes (e.g., hubs) enriched in disease–gene connecting subnetworks are very likely disease candidates [141]. Subsequently, *local information* was included in the prioritization process, analyzing the neighborhood of known disease genes, such as the presence on shortest paths between known disease genes [136, 141, 142]. Disease **module-based methods** assume that all cellular components that belong to the same topological or functional module have a high likelihood of being involved in the same disease. Topological or functional disease modules are defined as a combination of network components (nodes and edges) that influence cellular functions and thereby their malfunction can result in particular disease phenotypes [19]. Module-based methods are focused on identifying these disease modules – constructing the interactome of the condition of interest and identifying the subnetworks that contain the highest amount of disease associated genes [19, 143, 144].

More recent methods, such as **diffusion-based methods**, incorporated the complete *global network topology* into the analysis. The aim is to identify the paths that are closest to the known disease genes, assuming that genes in close overall proximity to known disease genes are more likely to be involved in the disease as well. Random walkers start at a known disease protein and diffuse along the interactome. Thus, nodes in closest proximity to the known disease genes will be most often visited by the random walkers, and proteins that interact with several disease proteins will have a higher weight. It has been shown that global approaches, such as diffusion-based, generally perform better than local (linkage) or module-based methods. However, the integration of both global and local approaches into a consensus method outperforms both single approaches [145]. HotNet is a variant of the described approach and based on the concept of heat diffusion – the mutated source genes are considered "hot" and their heat disperses to the surroundings [146]. A newer version of the algorithm incorporates the impact of heat directionality in order to decipher which rare alterations are relevant driver, rather than passenger, mutations [147]. In 2017, Porta-Pardo *et al.* conducted a comprehensive evaluation of these algorithms, with respect to the detection of cancer drivers [148].

4.6.2 Improvement of Gene Signatures

A signature is a set of molecules linked to a phenotype of interest (for example, diagnosis, prognosis, response to treatment). While many molecular signatures have been identified for diverse diseases, very few have been successfully translated into clinical practice. Lack of reproducibility and variable performance on independent test sets are the main reasons, which are mostly caused by batch effects, molecular heterogeneity between samples and within populations, and the small sample sizes [149].

It has been proposed that many signatures perform equally well because of biological redundancy [150]. In this case, integration of all the signatures available for a certain disease can provide insight into the common molecular mechanisms underlying such signatures. Beside the obvious identification of intersection, it is possible to discover that some proteins from different signatures interact to form complexes or pathways, or the genes present in different signatures share common regulators (e.g., a transcription factor). Finally, network topology analyses can be used to predict signatures instead of expanding or annotating them [151]. A recent study shows that prediction models for disease classification built upon novel network based multi-gene features are considerably more stable and provide valuable insights into disease mechanisms [143].

4.6.3 Prediction of Drug Targets

A significant amount of research has been directed to discover new drugs in a variety of diseases, but the success of new treatments remains relatively low. This is in part due to a limited understanding of the effects of a compound on the entire organism, especially adverse effects [152]. Interaction networks provide a necessary infrastructure to model and analyze a drug's mechanism of action, identify putative targets, and study possible off-target effects. Drug effect on proteins can be retrieved from specialized databases, as described in Section 4.5.4. Predictions performed using networks integrating PPI and drug-target data can lead to drug repurposing – the identification of effectiveness of a drug to treat a disease different from the one for which it is currently used, potentially saving money and time [153].

4.6.4 Annotation of Protein Functions

Functional annotation of proteins is a fundamental challenge in computational biology. Over the last decade, the growing availability of PPIs for many model organisms has stimulated the development of network-based approaches to annotate functions for poorly characterized proteins [154]. These methods are based on the assumption that interacting proteins are likely components of the same biological process, and the function of an unannotated protein can be deduced when the function of its binding partners is known [134]. Various methods have been developed to address this problem, starting with direct approaches, such as, neighborhood-based that commonly consider the number of edges connected to annotated proteins. For example, Schwikowski et al. predicted the function of an unannotated protein to be the most common one among its neighbours [155], while Hishigaki et al. developed a method based on chi-square statistics [135]. More frequently, module-based approaches utilize graph theoretic, network topology or network clustering based algorithms, and propagate function within dense sub-networks [154]. For instance, the molecular complex detection algorithm (MCODE) assigns weights based on the core clustering coefficient, which is increasing in heavily interconnected graph regions [156]. Another tool called NetworkBlast, utilizes maximum likelihood-based scoring of subnetworks to detect molecular complexes [157].

Most recent integration approaches, such as INGA, combine PPI network with domain architecture search and sequence similarity of proteins. The improved prediction accuracy of this strategy mostly results from the synergy effect when uniting the strengths and weaknesses of the different sources of information [158]. Sharan *et al.* provide a comprehensive evaluation of other available methods, such as, graph theoretic, Markov random fields, or graph clustering [122, 154, 159, 160].

4.7 Integrative Computational Biology Workflow

Typical applications for integrative PPI network analysis are clinical studies or basic research, that result in a set of genes or proteins of interest with respect to a disease or biological mechanism. For instance, a set of genes that is significantly over-expressed in patients suffering from cancer. In order to confirm and prioritize this list of genes, for example according to their disease association, a variety of network-based analyses can be applied. This section summarizes the steps required to carry out such a PPI network analysis as depicted in Figure 4.5, based on the knowledge acquired in this chapter. Finally, we will utilize this general workflow and describe in detail the application in a typical example such as a set of cancer-related genes. Our aim is to identify novel genes and proteins that are potentially involved in the molecular processes associated with cancer.

1. **Selection of omics data:** Depending on the area of interest and the type of study, the up-front analysis of e.g., gene expression or protein abundance results in a set of most relevant proteins. This set will be used to extract the corresponding subnetworks from one or several of the previously described PPI databases, see Section 4.4.
2. **Annotation:** The network is then annotated with data pertinent to the subject of the study. This can be the previously utilized data (level of expression or up-/down-regulation) or we can use data extracted from annotation

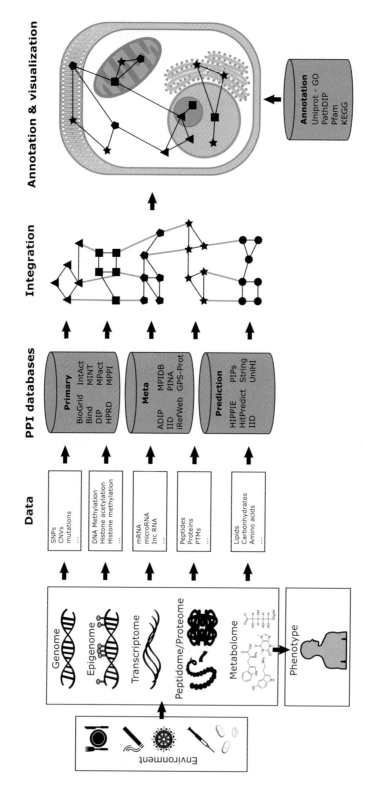

Figure 4.5: General overview of the sequence of steps required to construct and analyze a PPI network. Created in Inkscape (https://inkscape.org).

Table 4.8: Selected genes and frequency of mutation in cancer. Genes are listed with the name reported in tumor portal and, in brackets, with the official gene name. Frequency shows the percentage of samples across different tumors where mutations in the listed gene were identified.

Gene	Frequency	Gene	Frequency
TP53	36%	NF1	4%
PIK3CA	14%	EGFR	4%
PTEN	7%	ATM	4%
KRAS	7%	PIK3R1	3%
APC	6%	BRAF	3%
MLL3 (KMT2C)	6%	CDKN2A	3%
FAT1	6%	SETD2	3%
MLL2 (KMT2D)	5%	CREBBP	3%
ARID1A	5%	FBXW7	3%
VHL	4%	SPEN	3%
PBRM1	4%	MTOR	3%

databases such as the molecular function in gene ontology (as described in Section 4.4.3).

3. **Visualization:** Finally, the data can be visualized by various software tools, including NAViGaTOR [81] or Cytoscape (described in detail in Chapter 13).
4. **Evaluation:** As described previously, this is a crucial part of the PPI network analysis, since a proper visualization can highlight interesting features and reveal important insights of the selected network and its subject.

In the following, we will illustrate this workflow on a practical example:

Step 1: The starting point will be a set of genes collected from tumor portal, a database that reports mutation frequencies identified by analyzing data from 4,742 samples spanning 21 tumor types[4] [161]. We selected genes that are reported as mutated in at least 3% of all cancer samples. The resulting 22 genes (as analyzed on January 27, 2017) are listed in Table 4.8. We first collect PPIs for genes in Table 4.8 from BioGRID (the largest curated database, as shown in Table 4.2) and IID (the largest database including both experimental and predicted PPIs, as listed in Table 4.5). Due to the structure of the IID database, it offers the possibility to build three different networks, containing only experimental, with only predicted, and comprising the entire set of PPIs. To demonstrate the differences in resulting networks when choosing either of these options, we create and evaluate a PPI network for all three options, respectively.

Step 2: In each network we calculate degree and betweenness centrality and identify articulation points. Degree and betweenness for each network are shown in Figure 4.6. As shown, our query genes have in general higher degree and betweenness centrality than the average of the entire network (in line with what was mentioned in Section 4.5.3). In particular, in the degree distribution we can notice that, when using

---

[4] (www.tumorportal.org)

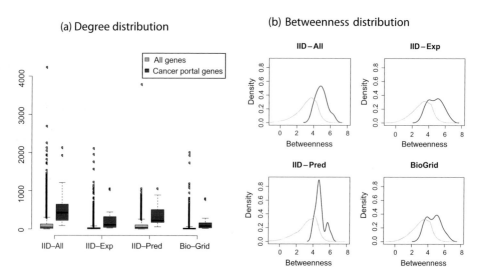

Figure 4.6: Degree (a) and betweenness centrality (b) distribution for four networks. In black are the values for the entire network while in red are the values for the query genes.

integrated data (in this example, the entire IID–IID–All), the degree of our query genes is the highest – meaning that we detect the highest number of interactors of our genes of interest. Note that, if we would use only experimental PPIs (from IID only experimental or from BioGrid) we would lose most of the interactions (as visible in Figure 4.6a), resulting in a much smaller network. However, if we build our network using predicted PPIs (from IID only predictions), the degree of our query proteins would be closer to the degree the query proteins have in the integrated network.

Step 3: Both degree and betweenness can be determined for each node and visually annotated in the network using NAViGaTOR[5] 3.0 [81]. Subsequently, we annotate the nodes in the network using the mutation information in COSMIC[6] v.75 [162].

Step 4: COSMIC is a database curating somatic mutations in genes across 43 different tissue types from genome-wide LT and HT studies, such as TCGA and ICGC. We extracted the list of COSMIC genes that show a mutation rate of at least 5% in at least one of the 42 cancer tissues. Thereafter, for each gene in our network we determine the median of mutation rate in all cancer tissues and annotate the corresponding node. Finally, we count the number of interactors in each network that have a median mutation rate across cancers higher than zero. This reveals that the number of mutated interactors is 444 in the BioGRID network, 594 in IID with only experimental data, 849 in IID with only predicted interactions and 1,040 in the entire IID. As expected, the integrated approach leads to the identification not only of a higher number of interactors but also of a higher number of cancer-related interactors (that is, genes that are frequently mutated in cancer), shedding light on the possible connection among genes linked to cancer.

---

[5] Network Analysis, VIsualization, and GrAphing TORonto (http://ophid.utoronto.ca/navigator/)
[6] Catalogue Of Somatic Mutations In Cancer, http://cancer.sanger.ac.uk/cosmic

To investigate if the genes frequently mutated in cancer that are connected through PPIs in our network are also part of the same molecular processes, a pathway enrichment analysis is performed. One example for pathway enrichment platforms is pathDIP,[7] it integrates information from 20 different pathway databases, with literature curated pathways with physical evidence as well as predicted biologically relevant protein-pathway associations [163]. We select all genes in each network that have a median Cosmic mutation rate higher than zero and use pathDIP to extract literature curated pathways that are present with high confidence. As shown in Table 4.9, the number of enriched pathways decreases with the increase in the number of genes tested, indicating a higher presence of genes with common features in the biggest set (that is, the entire IID set).

Looking at the top 10 pathways in each set, we notice that the sets from experimental data (BioGRID and experimental only IID) are not completely overlapping but both highlight cancer-specific pathways, such as EGFR, IGF1R, PDGF, and VEGF signaling, that are normally active during fetal development and abnormally activated in cancer [164]. Cancer research discovered that a growing number of genes and signaling pathways active during embryogenesis have been shown to be re-activated in cancer, and to regulate tumor development and progression [165]. Key developmental signaling pathways (e.g., Wnt, Hedgehog, and Notch pathways) are frequently deregulated in cancer and participate in all stages of tumor progression [166]. Interestingly, sets originated from IID – Pred and the entire IID present developmental biology and axon guidance as the most enriched pathways. The class of genes related to axon guidance (i.e., semaphorins, slits, netrins, and ephrins) comprises important regulators of neuronal migration and positioning during embryonic development, which have also been implicated in many carcinogenic stages, such as cell growth, survival, invasion and angiogenesis [167].

The two sets of pathways are clearly related and linked to the overlap between development and cancer; however, the results show very specific and focused pathways. In contrast, the larger lists of interactors provide a more extensive idea of the broader mechanisms involved in cancer.

4.8 Closing Remarks

Different experimental technologies have been developed to identify protein–protein interactions (PPIs). However, the majority are time consuming and cost-intensive. Over the last years computational prediction methods have been designed to successfully fill the remaining gaps. All types of interactions are accumulated in large databases with different focus and content. In general they can be categorized into three groups: (1) **curated** for experimental data, (2) **prediction** for predicted interactions and (3) **integrated** databases summarizing the information, from different curated or prediction databases.

Integrated databases contain the most complete information about PPI networks and therefore yield the best resources for further network analysis. The annotation

---

[7] Pathway Data Integration Portal (http://ophid.utoronto.ca/pathDIP)

Table 4.9: Top ten enriched pathways, for BioGRID (a), experimentally validated PPIs in IID (b), predicted PPIs in IID (c), all PPIs in IID (d). Pathways are enriched using pathDIP. The number of enriched pathways for that network is shown in brackets. The presented Q-values are Bonferroni corrected.

(a)

BioGrid (327)	
Pathway name	Q-value
EGFR1	2.35E-31
Pathways in cancer	1.21E-26
Signaling pathways in glioblastoma	1.31E-22
Diseases of signal transduction	2.7E-19
Developmental biology	2.26E-17
Thyroid hormone signaling pathway	6.96E-17
Signaling by PDGF	4.51E-16
Interleukin-3, 5 and GM-CSF signaling	9.39E-16
Signaling by ERBB4	2.37E-15
IGF1R signaling cascade	1.95E-15

(b)

IID – Experimental (303)	
Pathway name	Q-value
EGFR1	1.14E-29
Pathways in cancer	1.54E-24
Signaling pathways in glioblastoma	9.41E-24
Developmental Biology	2.95E-23
IGF1R signaling cascade	1.44E-19
IRS-related events triggered by IGF1R	1.44E-19
Signaling by Type 1 insulin-like growth Factor 1 receptor (IGF1R)	1.44E-19
Insulin receptor signaling cascade	7.79E-19
VEGFA-VEGFR2 pathway	9.58E-19
Signaling by VEGF	7.14E-19

(c)

IID – Pred (307)	
Pathway name	Q-value
Developmental biology	2.09E-35
Axon guidance	3.9E-32
Pathways in cancer	1.42E-26
Signaling by VEGF	1.08E-24
Signaling by NGF	1.39E-23
EGFR1	1.83E-23
VEGFA-VEGFR2 pathway	3.26E-23
Focal adhesion	1.12E-22
Focal adhesion	2.55E-22
IGF1R signaling cascade	2.28E-22

(d)

IID – All (271)	
Pathway name	Q-value
Developmental biology	9.75E-39
Axon guidance	1.57E-34
Pathways in cancer	1.88E-24
Focal adhesion	5.18E-24
Signaling by VEGF	1.19E-22
Focal adhesion	1.72E-21
VEGFA-VEGFR2 Pathway	2.3E-21
Signaling by NGF	3.45E-21
IGF1R signaling cascade	6.62E-21
IRS-related events triggered by IGF1R	6.62E-21

of PPIs by protein function, localization, or disease involvement, can lead to further insights on biomedical processes and disease mechanisms. This enables a large amount of applications such as characterizing disease genes and signatures, or predicting drug targets and protein function. In summary, PPI data and its analysis is a major resource in biomedical research today and can provide useful insights into biological phenomena and potential treatment targets for medicine.

4.9 Exercises

4.1 What are the steps that led to the creation of data necessary for modern PPI network analysis?

4.2 Why are current PPI networks incomplete and why is this a problem for PPI network analysis?

4.3 Review the section on experimental methods.

   (a) What are the two general terms for experimental techniques to analyze PPIs.
   (b) What are the main differences between those methods?
   (c) Name two examples for both types of experimental methods.

4.4 Name two different methods for computational prediction of PPIs, describe the algorithm in your own words, and compare them in terms of their advantages, disadvantages and biases.

4.5 Which initiatives aim at the integration of PPI data?

4.6 Describe in your own words the four main challenges of data heterogeneity.

4.7 Why is it impossible to study all interactions of a complex organism at once?

4.8 How is the problem of nomenclature heterogeneity addressed?

4.9 Name the three challenges in database integration. Give a short description how this relates to biomedical data.

4.10 What are the four main types of PPI databases? What are current examples?

4.11 What is the main problem of curated databases?

4.12 Name advantages of integrated databases.

4.13 Assume you are working on a study that acquired experimental datasets of liver cancer in rats and humans. Which PPI database would you use to compare the PPI data and network of the two species? Explain why you chose this or these databases.

4.14 Describe in your own words, what a network is.

4.15 What is an articulation point?

4.16 Graph density

   (a) How is graph density calculated?
   (b) What does it mean, if a graph shows a density of 0.98?
   (c) Is this a common value for a PPI network?

4.17 What is a clique?

4.18 What is the main goal of PPI network annotation?

4.19 What are the three branches of gene ontology?

4.20 What is the main difference between quantitative and qualitative annotations? Name examples.

4.21 What are the four most common steps in network analyses?

4.22 Search for TP53 in IID (http://ophid.utoronto.ca/iid) and collect all interactors with interactions of TP53 across tissues. How many interaction partners do you find for experimental, computational predictions and both? (Hint you can use the parameters of the IID query.)

4.23 Let's focus on the AHSG protein.

(a) Please collect all interactors (experimental and predicted) of AHSG across tissues using IID (http://ophid.utoronto.ca/iid). How many protein interaction partners do you find?

(b) Repeat the IID query of AHSG restricted to experimentally validated interactions occurring in liver tissue. What happens if you search for interactions in heart tissue?

(c) Select the PPI network containing all interactions for AHSG and its interaction partners in liver (including orthologous). Note: You can retrieve the interactions among partners of query proteins automatically when activating the parameter during the IID query.

(d) Annotate the proteins in the network with their gene symbols and their node degree. Can you find gene ontology annotation for the proteins, e.g., molecular function?

Note: Some visualization tools enable automated calculation of note degree and gene ontology annotation, for instance NAViGaTOR 3.0[8] and Cytoscape (with the aid of plug-ins like GOlorize[9]).

(e) Visualize the acquired AHSG network. Depict the gene symbols as node labels and their node degree as size of the node. Can you colour the nodes according to their gene ontology annotation?

Note: Solutions are available to instructors at www.cambridge.org/bionetworks.

4.10 Acknowledgments

IJ was supported in part by funding from Natural Sciences Research Council (NSERC #203475), Canada Foundation for Innovation (CFI #225404, #30865), Canada Research Chair Program (CRC #203373, #225404), Ontario Research Fund (RDI #34876), Ontario Research Fund (GL2-01-030), IBM and Ian Lawson van Toch Fund. The funders had no role in study design, data collection, and analysis, decision to publish, or preparation of the manuscript.

References

[1] Cusick ME, Klitgord N, Vidal M, Hill DE. Interactome: Gateway into systems biology. *Human Molecular Genetics*, 2005;14(suppl 2):R171–R181.

[2] Petschnigg J, Groisman B, Kotlyar M, et al. The mammalian membrane two-hybrid assay (mamth) for probing membrane-protein interactions in human cells. *Nature Methods*, 2014;11(5):585–592.

---

[8] NAViGaTOR (http://ophid.utoronto.ca/navigator/)
[9] GOlorize (http://apps.cytoscape.org/apps/golorize)

[3] Yang X, Coulombe-Huntington J, Kang S, et al. Widespread expansion of protein interaction capabilities by alternative splicing. *Cell* 2016;164(4):805–817.

[4] Arkin MR, Wells, JA. Small-molecule inhibitors of protein-protein interactions: progressing towards the dream. *Nature Reviews Drug Discovery*, 2004;3(4):301–317.

[5] Ruffner H, Bauer A, Bouwmeester T. Human protein–protein interaction networks and the value for drug discovery. *Drug Discovery Today*, 2007;12(17-18):709–716.

[6] Chua, H. N, Wong, L. Increasing the reliability of protein interactomes. *Drug Discovery Today*, 2008;13(15-16):652–658.

[7] Arkin MR, Tang Y, Wells JA. Small-molecule inhibitors of protein–protein interactions: Progressing toward the reality. *Chemistry & Biology*, 2014;21(9):1102–1114.

[8] Szilagyi A, Nussinov R, Csermely P. Allo-network drugs: Extension of the allosteric drug concept to protein-protein interaction and signaling networks. *Current Topics in Medicinal Chemistry*, 2013;13(1):64–77.

[9] Petschnigg J, Kotlyar M, Blair L, et al. Systematic identification of oncogenic egfr interaction partners. *Journal of Molecular Biology*, 2017;429(2):280–294.

[10] Savas S, Geraci J, Jurisica I, Liu G. A comprehensive catalogue of functional genetic variations in the EGFR pathway: Protein–protein interaction analysis reveals novel genes and polymorphisms important for cancer research. *International Journal of Cancer*, 2009;125(6):1257–1265.

[11] Mosca R, Pons T, Ceol A, Valencia A, Aloy P. Towards a detailed atlas of protein–protein interactions. *Current Opinion in Structural Biology*, 2013;23(6):929–940.

[12] Porta-Pardo E, Hrabe T, Godzik A. Cancer3d: Understanding cancer mutations through protein structures. *Nucleic Acids Research*, 2014;43(D1):D968–D973.

[13] Kotlyar M, Pastrello C, Pivetta F, et al. In silico prediction of physical protein interactions and characterization of interactome orphans. *Nature Methods*, 2015;12(1):79–84.

[14] Stumpf MP, Thorne T, de Silva E, et al. Estimating the size of the human interactome. *Proceedings of the National Academy of Sciences USA*, 2008;105(19):6959–6964.

[15] Sokolina K, Kittanakom S, Snider, J, et al. Systematic protein–protein interaction mapping for clinically relevant human GPCRS. *Molecular Systems Biology*, 2017;13(918):1–19.

[16] Edwards AM, Isserlin R, Bader GD, et al. Too many roads not taken. *Nature* 2011;470(7333):163–165.

[17] Vazquez A, Flammini A, Maritan A, Vespignan, A. Global protein function prediction from protein-protein interaction networks. *Nature Biotechnology*, 2003;21(6):697–700.

[18] Lage K. Proteinprotein interactions and genetic diseases: The interactome. *Biochimica et Biophysica Acta: Molecular Basis of Disease*, 2014;1842(10):1971–1980.

[19] Barabasi AL, Gulbahce N, Loscalzo, J. Network medicine: A network-based approach to human disease. *Nature Reviews Genetics*, 2011;12(1):56–68.

[20] Przytycka TM, Singh M, Slonim DK. Toward the dynamic interactome: It's about time. *Briefings in Bioinformatics*, 2010;11(1):15.

[21] Snider J, Kotlyar M, Saraon P, et al. Fundamentals of protein interaction network mapping, *Molecular Systems Biology*, 2015;11(12):848.

[22] Breitkreutz BJ, Stark C, Tyers M. The GRID: The General Repository for Interaction Datasets. *Genome Biology*, 2003;4(3):R23.

[23] Deane CM, Salwinski L, Xenarios I, Eisenberg D. Protein interactions: Two methods for assessment of the reliability of high throughput observations. *Molecular Cell Proteomics*, 2002;1(5):349–356.

[24] von Mering C, Krause R, Snel B, et al. Comparative assessment of large-scale data sets of protein-protein interactions. *Nature*, 2002;417(6887): 399–403.

[25] Cusick ME, Yu H, Smolyar A, et al. Literature-curated protein interaction datasets. *Nature Methods*, 2009;6(1):39–46.

[26] Perkins JR, Diboun I, Dessailly BH, Lees JG, Orengo C. Transient protein-protein interactions: Structural, functional, and network properties. *Structure*, 2010;18(10):1233–1243.

[27] Fields S, Song O. A novel genetic system to detect protein–protein interactions. *Nature*, 1989;340(6230):245–246.

[28] Ma L, Yang F, Zheng J. Application of fluorescence resonance energy transfer in protein studies. *Journal of Molecular Structure*, 2014;1077:87–100.

[29] Valencia, A, Pazos, F. Computational methods for the prediction of protein interactions. *Current Opinions in Structural Biology*, 2002;12(3):368–373.

[30] Dandekar T, Snel B, Huynen M, Bork P. Conservation of gene order: A fingerprint of proteins that physically interact. *Trends in Biochemical Sciences*, 1998;23(9):324–328.

[31] Overbeek R, Fonstein M, D'Souza M, Pusch GD, Maltsev N. Use of contiguity on the chromosome to predict functional coupling. *Silico Biology*, 1999;1(2):93–108.

[32] Pellegrini M, Marcotte EM, Thompson MJ, Eisenberg D, Yeates TO. Assigning protein functions by comparative genome analysis: protein phylogenetic profiles. *Proceedings Of The National Academy of Sciences USA*, 1999;96(8):4285–4288.

[33] Enright AJ, Iliopoulos I, Kyrpides NC, Ouzounis CA. Protein interaction maps for complete genomes based on gene fusion events. *Nature*, 1999;402(6757):86–90.

[34] Bock JR, Gough DA. Predicting protein–protein interactions from primary structure. Bioinformatics 2001;17(5):455–460.

[35] Martin S, Roe D, Faulon JL. Predicting protein-protein interactions using signature products. *Bioinformatics*, 2005;21(2):218–226.

[36] Shen J, Zhang J, Luo X, et al. Predicting protein-protein interactions based only on sequences information. *Proceedings of the National Academy of Sciences USA*, 2007;104(11):4337–4341.

[37] Guo Y, Yu L, Wen Z, Li M. Using support vector machine combined with auto covariance to predict protein-protein interactions from protein sequences. *Nucleic Acids Research*, 2008;36(9):3025–3030.

[38] Yu CY, Chou LC, Chang DT. Predicting protein-protein interactions in unbalanced data using the primary structure of proteins. *BMC Bioinformatics*, 2010;11:167.

[39] Walhout AJ, Sordella R, Lu X, et al. Protein interaction mapping in C. elegans using proteins involved in vulval development. *Science*, 2000;287(5450):116–122.

[40] Brown KR, Jurisica I. Unequal evolutionary conservation of human protein interactions in interologous networks. *Genome Biology*, 2007;8(5):R95.

[41] Huang TW, Lin CY, Kao CY. Reconstruction of human protein interolog network using evolutionary conserved network. *BMC Bioinformatics*, 2007;8:152.

[42] Pawson T, Gish GD, Nash P. SH2 domains, interaction modules and cellular wiring. *Trends in Cell Biology*, 2001;11(12):504–511.

[43] Pawson, T, Nash, P. Assembly of cell regulatory systems through protein interaction domains, *Science* 2003;300(5618):445–452.

[44] Sprinzak E, Margalit H. Correlated sequence-signatures as markers of protein–protein interaction. *Journal of Molecular Biology*, 2001;311(4):681–692.

[45] Wojcik J, Boneca IG, Legrain, P. Prediction, assessment and validation of protein interaction maps in bacteria. *Journal of Molecular Biology*, 2002;323(4):763–770.

[46] Deng M, Mehta S, Sun F, Chen T. Inferring domain-domain interactions from protein–protein interactions. *Genome Research*, 2002;12(10):1540–1548.

[47] Kim I, Liu Y, Zhao H. Bayesian methods for predicting interacting protein pairs using domain information. *Biometrics*, 2007;63(3):824–833.

[48] Aloy P, Russell RB. Interrogating protein interaction networks through structural biology. *Proceedings of the National Academy of Sciences* USA, 2002;99(9):5896–5901.

[49] Lu L, Arakaki AK, Lu H, Skolnick J. Multimeric threading-based prediction of protein-protein interactions on a genomic scale: Application to the *Saccharomyces cerevisiae* proteome. *Genome Research*, 2003;13(6A):1146–1154.

[50] Zhang QC, Petrey D, Deng L, et al. Structure-based prediction of protein–protein interactions on a genome-wide scale. *Nature*, 2012;490(7421):556–560.

[51] Saito R, Suzuki H, Hayashizaki Y. Interaction generality: A measurement to assess the reliability of a protein–protein interaction, *Nucleic Acids Research*, 2002;30(5):1163–1168.

[52] Goldberg DS, Roth, FP. Assessing experimentally derived interactions in a small world. *Proceedings of the National Academy of Sciences USA*, 2003;100(8):4372–4376.

[53] Bader JS. Greedily building protein networks with confidence. *Bioinformatics*, 2003;19(15):1869–1874.

[54] Saito R, Suzuki H, Hayashizaki Y. Construction of reliable protein–protein interaction networks with a new interaction generality measure. *Bioinformatics*, 2003;19(6):756–763.

[55] Scott MS, Barton GJ. Probabilistic prediction and ranking of human protein-protein interactions. *BMC Bioinformatics*, 2007;8:239.

[56] Jansen R, Yu H, Greenbaum D, Kluger Y, et al. A Bayesian networks approach for predicting protein–protein interactions from genomic data. *Science*, 2003;302(5644):449–453.

[57] Ben-Hur A, Noble WS. Kernel methods for predicting protein-protein interactions. *Bioinformatics*, 2005;21(Suppl 1):i38–i46.

[58] Qi Y, Bar-Joseph Z, Klein-Seetharaman J. Evaluation of different biological data and computational classification methods for use in protein interaction prediction. *Proteins*, 2006;63(3):490–500.

[59] Rhodes DR, Tomlins SA, Varambally S, et al. Probabilistic model of the human protein-protein interaction network. *Nature Biotechnology*, 2005:23(8):951–959.

[60] Elefsinioti A, Sarac OS, Hegele, A, et al. Large-scale de novo prediction of physical protein–protein association. *Molecular & Cellular Proteomics*,2011;10(11):M111.010629.

[61] Bader JS, Chaudhuri A, Rothberg JM, Chant J. Gaining confidence in high-throughput protein interaction networks. *Nature Biotechnology*, 2004;22(1):78–85.

[62] Xenarios I, Salwinski L, Duan X. J, et al. DIP, the Database of Interacting Proteins: A research tool for studying cellular networks of protein interactions. *Nucleic Acids Research*, 2002;30(1):303–305.

[63] Hart GT, Ramani AK, Marcotte EM. How complete are current yeast and human protein-interaction networks?, *Genome Biology*, 2006;7(11):120.

[64] Orchard S, Kerrien S, Abbani S, et al. Protein interaction data curation: The international molecular exchange (imex) consortium. *Nature Methods*, 2012;9(4):345–350.

[65] Orchard S, Kerrien S, Jones P, et al. Submit your interaction data the imex way. *Proteomics*, 2007;7(S1):28–34.

[66] Braun P, Tasan M, Dreze M, et al. An experimentally derived confidence score for binary protein-protein interactions. *Nature Methods*, 2009;6(1):91–97.

[67] Schwartz AS, Yu J, Gardenour KR, Finley RLJ, Ideker T. Cost-effective strategies for completing the interactome. *Nature Methods*, 2009;6(1):55–61.

[68] Koh GC, Porras P, Aranda B, Hermjakob H, Orchard SE. Analyzing protein–protein interaction networks. *Journal of Proteome Research*, 2012;11(4):2014–2031.

[69] Edwards AM, Kus B, Jansen R, et al. Bridging structural biology and genomics: Assessing protein interaction data with known complexes. *Trends in Genetics*, 2002;18(10):529–536.

[70] Hermjakob H, Montecchi-Palazzi L, Bader G, et al. The HUPO PSI's molecular interaction formata community standard for the representation of protein interaction data. *Nature Biotechnology*, 2004;22(2):177–183.

[71] Karagoz K, Sevimoglu T, Arga KY. Integration of multiple biological features yields high confidence human protein interactome. *Journal of Theoretical Biology*, 2016;403:85–96.

[72] Fundel K, Zimmer R. Gene and protein nomenclature in public databases. *BMC Bioinformatics*, 2006;7(1);372.

[73] Laney D. 3D data management: Controlling data volume, variety and velocity. META Group Inc. Technical Report, file number 949, 2001:1–4.

[74] Stark C, Breitkreutz BJ, Reguly T, et al. Biogrid: A general repository for interaction datasets. *Nucleic Acids Research*, 2006;34(suppl 1):D535–D539.

[75] Orchard S, Ammari M, Aranda B, et al. The mintact project intact as a common curation platform for 11 molecular interaction databases. *Nucleic Acids Research*, 2014;42(D1):D358–D363.

[76] Bader GD, Betel D, Hogue CW. Bind: The biomolecular interaction network database. *Nucleic Acids Research*, 2003;31(1):248–250.

[77] Salwinski L, Miller CS, Smith AJ, et al. The database of interacting proteins: 2004 update. *Nucleic Acids Research*, 2004;32(suppl 1):D449–D451.

[78] Prasad TK, Goel R, KandasamyK, et al. Human protein reference database: 2009 update. *Nucleic Acids Research*, 2009;37(suppl 1):D767–D772.

[79] Licata L, Briganti L, Peluso D, et al. Mint, the molecular interaction database: 2012 update, *Nucleic Acids Research*, 2012;40(D1):D857–D861.

[80] Pagel P, Kovac S, Oesterheld M, et al. The mips mammalian protein–protein interaction database. *Bioinformatics*, 2005;21(6):832–834.

[81] Brown KR, Otasek D, Ali M, et al. Navigator: Network analysis, visualization and graphing toronto. *Bioinformatics*, 2009;25(24):3327–3329.

[82] Turinsky AL, Razick S, Turner B, Donaldson IM, Wodak SJ. Interaction databases on the same page. *Nature Biotechnology*, 2011;29(5):391–393.

[83] Alonso-Lopez D, Gutierrez MA, Lopes KP, et al. Apid interactomes: Providing proteome-based interactomes with controlled quality for multiple species and derived networks. *Nucleic Acids Research*, 2016;44(W1):W529–W535.

[84] Lefebvre C, Lim WK, Basso K, Dalla Favera R, Califano A. A context-specific network of protein-DNA and protein-protein interactions reveals new regulatory motifs in human B cells. In Ideker T, Bafna V, eds., *Systems Biology and Computational Proteomics. Lecture Notes in Computer Science*, vol. 4532. Springer;2007, pp. 42–56.

[85] Alanis-Lobato G, Andrade-Navarro, MA, Schaefer, MH. (2016):Hippie v2. 0: Enhancing meaningfulness and reliability of protein–protein interaction networks. *Nucleic Acid Research*, 45(D1):D408–D414.

[86] Lopez Y, Nakai K, Patil A. Hitpredict version 4: comprehensive reliability scoring of physical protein–protein interactions from more than 100 species. *Database*, 2015;2015:bav117.

[87] Kotlyar M, Pastrello C, Sheahan N, Jurisica I. Integrated interactions database: Tissue-specific view of the human and model organism interactomes. *Nucleic Acids Research*, 2016;44(D1):D536.

[88] McDowall MD, Scott MS, Barton GJ. Pips: Human protein–protein interaction prediction database, *Nucleic Acids Research*, 2009;37(suppl 1):D651–D656.

[89] Szklarczyk D, Morris JH, Cook H, et al. The string database in 2017: Quality-controlled protein–protein association networks, made broadly accessible. *Nucleic Acids Research*, 2016;45(D1):D362–D368.

[90] Kalathur RKR, Pinto JP, Hernandez-Prieto MA, et al. UniHI 7: An enhanced database for retrieval and interactive analysis of human molecular interaction networks. *Nucleic Acids Research*, 2014;42(D1):D408–D414.

[91] Goll J, Rajagopala SV, Shiau SC, et al. MPIDB: The microbial protein interaction database. *Bioinformatics*, 2008;24(15):1743–1744.

[92] Fahey ME, Bennett MJ, Mahon C, et al. Gps-prot: A web-based visualization platform for integrating host-pathogen interaction data. *BMC Bioinformatics*, 2011;12(1):298.

[93] Cowley MJP, Kassahn M, Waddell KS, et al. Pina v2.0: Mining interactome modules. *Nucleic Acids Research*, 2012;40(D1):D862–D865.

[94] Breuer K, Foroushani AK, Laird MR, et al. InnateDB: Systems biology of innate immunity and beyondrecent updates and continuing curation. *Nucleic Acids Research*, 2012;41(D1):D1228–D1233.

[95] Ammari MG, Gresham CR, McCarthy FM, Nanduri B. Hpidb 2.0: A curated database for host–pathogen interactions. *Database*, 2016;2016:baw103.

[96] Lin M, Shen X, Chen X. Pair: The predicted arabidopsis interactome resource. *Nucleic Acids Research*, 2011;39(suppl 1):D1134–D1140.

[97] Zhang QC, Petrey D, Garzon JI, Deng L, Honig B. Preppi: A structure-informed database of protein–protein interactions. *Nucleic Acids Research*, 2012;41(D1):D828–D833.

[98] Wetie AGN, Sokolowska I, Woods, AG, Roy U, Deinhardt K, Darie CC. Protein–protein interactions: Switch from classical methods to proteomics and bioinformatics-based approaches. *Cellular and Molecular Life Sciences*, 2014;71(2):205–228.

[99] Yeger-Lotem E, Sattath S, Kashtan N, et al. Network motifs in integrated cellular networks of transcription–regulation and protein–protein interaction. *Proceedings of the National Academy of Sciences USA*, 2004;101(16):5934–5939.

[100] Wu J, Vallenius T, Ovaska K, Westermarck J, Makela TP, Hautaniemi S. Integrated network analysis platform for protein–protein interactions. *Nature Methods*, 2009;6(1):75–77.

[101] Veres DV, Gyurko DM, Thaler B, et al. Comppi: A cellular compartment-specific database for protein–protein interaction network analysis. *Nucleic Acids Research*, 2015;43(D1):D485–D493.

[102] Basha O, Barshir R, Sharon M, et al. The tissuenet v. 2 database: A quantitative view of protein-protein interactions across human tissues. *Nucleic Acids Research*, 2017;45(D1):D427–D431.

[103] Micale G, Ferro A, Pulvirenti A, Giugno R. SPECTRA: An integrated knowledge base for comparing tissue and tumor-specific PPI networks in human. In Pellegrini M, Magi A, Iliopoulos CS, eds., *Repetitive Structures in Biological Sequences: Algorithms and Applications, Frontiers in Bioengineering and Biotechnology*. Frontiers Media;2016; p. 59.

[104] O'Brate A, Giannakakou P. The importance of p53 location: Nuclear or cytoplasmic zip code? *Drug Resistance Updates*, 2003;6(6):313–322.

[105] Lopes TJ, Schaefer M, Shoemaker J, et al. Tissue-specific subnetworks and characteristics of publicly available human protein interaction databases. *Bioinformatics*, 2011;27(17):2414–2421.

[106] Liu W, Wang J, Wang T, Xie H. Construction and analyses of human large-scale tissue specific networks. *PloS One*, 2014;9(12):e115074.

[107] Yeger-Lotem E, Sharan R. Human protein interaction networks across tissues and diseases. *Frontiers in Genetics*, 2015;6:257.

[108] Zhong Q, Simonis N, Li QR, et al. Edgetic perturbation models of human inherited disorders. *Molecular Systems Biology*,2009;5(1):321.

[109] Schuster-Bockler B, Bateman A. Protein interactions in human genetic diseases. *Genome Biology*, 2008;9(1):R9.

[110] Santin AD, Zhan F, Bellone S, et al. Gene expression profiles in primary ovarian serous papillary tumors and normal ovarian epithelium: Identification of candidate molecular markers for ovarian cancer diagnosis and therapy. *International Journal of Cancer*, 2004;112(1):14–25.

[111] List M, Hauschild AC, Tan Q, et al. Classification of breast cancer subtypes by combining gene expression and DNA methylation data. *Journal of Integrative Bioinformatics*, 2014;11(2):1–14.

[112] Srivas R, Shen J, Yang C, et al. A network of conserved synthetic lethal interactions for exploration of precision cancer therapy. *Molecular Cell*, 2016;63(3):514–525.

[113] Wong SW, Pastrello C, Kotlyar M, Faloutsos C, Jurisica I. Modeling tumor progression via the comparison of stage-specific graphs. *Methods*, 2018;132:34–41.

[114] Fortney K, Griesman J, Kotlyar M, et al. Prioritizing therapeutics for lung cancer: An integrative meta-analysis of cancer gene signatures and chemogenomic data. *PLoS Computational Biology*, 2015;11(3):e1004068.

[115] Jeong H, Mason SP, Barabasi AL, Oltvai ZN. Lethality and centrality in protein networks. *Nature*, 2001;411(6833):41–42.

[116] Wachi S, Yoneda K, Wu R. Interactome-transcriptome analysis reveals the high centrality of genes differentially expressed in lung cancer tissues. *Bioinformatics*, 2005;21(23):4205.

[117] Jonsson PF, Bates PA. Global topological features of cancer proteins in the human interactome. *Bioinformatics*, 2006;22(18):2291.

[118] Goh KI, Cusick, ME, Valle D, Childs B, Vidal M, Barabasi AL. The human disease network. *Proceedings of the National Academy of Sciences*, 2007;104(21):8685–8690.

[119] Ideker T, Sharan R. Protein networks in disease. *Genome Research*, 2008;18(4):644–652.

[120] Djebbari A, Ali, M, Otasek D, et al. Navigator: Large scalable and interactive navigation and analysis of large graphs. *Internet Mathematics*, 2011;7(4):314–347.

[121] Tian L, Bashan A, Shi DN, Liu YY. Articulation points in complex networks, *Nature Communications*, 2017;8(1):14223.

[122] Przulj N, Wigle DA, Jurisica I. Functional topology in a network of protein interactions. *Bioinformatics*, 2004;20(3):340–348.

[123] Leclerc RD. Survival of the sparsest: Robust gene networks are parsimonious. *Molecular Systems Biology*, 2008;4(1):213.

[124] Bailly-Bechet M, Braunstein A, Pagnani A, Weigt M, Zecchina R. Inference of sparse combinatorial-control networks from gene-expression data: a message passing approach. *BMC Bioinformatics*, 2010;11(1):355.

[125] Maslov S, Sneppen K. Specificity and stability in topology of protein networks. *Science*, 2002;296(5569):910–913.

[126] Yook SH, Oltvai ZN, Barabasi AL. Functional and topological characterization of protein interaction networks. *Proteomics*, 2004;4(4):928–942.

[127] Wong SW, Cercone N, Jurisica I. Comparative network analysis via differential graphlet communities. *Proteomics*, 2015;15(2–3):608–617.

[128] Hartwell LH, Hopfield JJ, Leibler S, Murray AW. From molecular to modular cell biology. *Nature*, 1999;402:C47–C52.

[129] Barabasi AL, Oltvai ZN. Network Biology: understanding the cell's functional organization. *Nature Reviews Genetics*, 2004;5(2):101–113.

[130] Tadaka S, Kinoshita K. Ncmine: Core-peripheral based functional module detection using near-clique mining. *Bioinformatics*, 2016;btw488.

[131] Yang L, Tang X. Protein-protein interactions prediction based on iterative clique extension with gene ontology filtering. *The Scientific World Journal*, 2014;2014:523634.

[132] Yang L, Zhao X, Tang X. Predicting disease-related proteins based on clique backbone in protein-protein interaction network. *International Journal of Biological Sciences*, 2014;10(7):677–688.

[133] Wu Q, Hao JK. A review on algorithms for maximum clique problems. *European Journal of Operational Research*, 2015;242(3):693–709.

[134] Lin C, Jiang D, Zhang A. Prediction of protein function using common neighbors in protein-protein interaction networks. In *Sixth IEEE Symposium on BioInformatics and BioEngineering (BIBE'06)*. IEEE;2006, pp. 251–260.

[135] Hishigaki H, Nakai K, Ono T, Tanigami A, Takagi T. Assessment of prediction accuracy of protein function from protein–protein interaction data. *Yeast*, 2001;18(6):523–531.

[136] Fortney K, Kotlyar M, Jurisica, I. Inferring the functions of longevity genes with modular subnetwork biomarkers of *Caenorhabditis elegans* aging. *Genome Biology*, 2010;11(2):R13.

[137] Li YH, Tavallaee G, Tokar T, et al. Identification of synovial fluid microrna signature in knee osteoarthritis: Differentiating early- and late-stage knee osteoarthritis. *Osteoarthritis and Cartilage*, 2016;24(9):1577–1586.

[138] Zhang Y, Xie J, Yang J, Fennell A, Zhang C, Ma Q. Qubic: A bioconductor package for qualitative biclustering analysis of gene co-expression data. *Bioinformatics*, 2017;33(3):450–452.

[139] Zeng Y, Zhang L, Zhu W, et al. Quantitative proteomics and integrative network analysis identified novel genes and pathways related to osteoporosis. *Journal of Proteomics*, 2016;142:45–52.

[140] Oti M, Snel B, Huynen MA, Brunner HG. Predicting disease genes using protein–protein interactions. *Journal of Medical Genetics*, 2006;43(8): 691–698.

[141] Dezso Z, Nikolsky Y, Nikolskaya T, et al. Identifying disease-specific genes based on their topological significance in protein networks. *BMC Systems Biology*, 2009;3(1):36.

[142] Segal E, Shapira M, Regev A, et al. Module networks: Identifying regulatory modules and their condition-specific regulators from gene expression data, *Nature Genetics*, 2003;34(2):166–176.

[143] Alcaraz N, List M, Batra R, et al. De novo pathway-based biomarker identification. *Nucleic Acids Research*, 2017;45(16):e151.

[144] Batra R, Alcaraz N, Gitzhofer K, et al. On the performance of de novo pathway enrichment. *Systems Biology and Applications*, 2017;3:1.

[145] Navlakha S, Kingsford C. The power of protein interaction networks for associating genes with diseases. *Bioinformatics*, 2010;26(8):1057–1063.

[146] Vandin F, Upfal E, Raphael BJ. Algorithms for detecting significantly mutated pathways in cancer. *Journal of Computational Biology*, 2011;18(3):507–522.

[147] Killock D. Genetics: Hotnet2 – see the wood for the trees. *Nature Reviews Clinical Oncology*, 2015;12(2):66–66.

[148] Porta-Pardo E, Kamburov A, Tamborero D, et al. Comparison of algorithms for the detection of cancer drivers at subgene resolution. *Nature Methods*, 2017;14(8):782–787.

[149] Sung J, Wang Y, Chandrasekaran S, Witten DM, Price ND. Molecular signatures from omics data: From chaos to consensus. *Biotechnology Journal*, 2012;7(8):946–957.

[150] Statnikov A, Aliferis CF. Analysis and computational dissection of molecular signature multiplicity. *PloS Computational Biology*, 2010;6(5):e1000790.

[151] Cun Y, Frohlich H. Biomarker gene signature discovery integrating network knowledge. *Biology*, 2012;1(1):5–17.

[152] Azuaje F. Drug interaction networks: An introduction to translational and clinical applications. *Cardiovascular Research*, 2012;97(4):631–641.

[153] Nosengo N. Can you teach old drugs new tricks? *Nature*, 2016;534(7607):314–316.

[154] Sharan R, Ulitsky I, Shamir R. Network-based prediction of protein function. *Molecular Systems Biology*, 2007;3(1):88.

[155] Schwikowski B, Uetz P, Fields S. A network of protein–protein interactions in yeast. *Nature Biotechnology*, 2000;18(12):1257–1261.

[156] Bader GD, Hogue CW. An automated method for finding molecular complexes in large protein interaction networks. *BMC Bioinformatics*, 2003;4(1):2.

[157] Sharan R, Ideker, T, Kelley B, Shamir R, Karp RM. Identification of protein complexes by comparative analysis of yeast and bacterial protein interaction data. *Journal of Computational Biology*, 2005;12(6):835–846.

[158] Piovesan D, Giollo M, Leonardi E, Ferrari C, Tosatto SC. Inga: Protein function prediction combining interaction networks, domain assignments and sequence similarity. *Nucleic Acids Research*, 2015;43(W1):W134–W140.

[159] Nabieva E, Jim K, Agarwal A, Chazelle, B, Singh M. Whole-proteome prediction of protein function via graph-theoretic analysis of interaction maps. *Bioinformatics*, 2005;21(suppl 1):i302–i310.

[160] Deng M, Tu Z, Sun F, Chen T. Mapping gene ontology to proteins based on protein–protein interaction data. *Bioinformatics*, 2004;20(6):895–902.

[161] Lawrence MS, Stojanov P, Mermel CH, et al. Discovery and saturation analysis of cancer genes across 21 tumour types. *Nature*, 2014;505(7484):495–501.

[162] Forbes SA, Beare D, Boutselakis H, et al. COSMIC: Somatic cancer genetics at high-resolution. *Nucleic Acids Research*,2017;45(D1):D777.

[163] Rahmati S, Abovsky M, Pastrello C, Jurisica I. pathdip: An annotated resource for known and predicted human gene pathway associations and pathway enrichment analysis. *Nucleic Acids Research*, 2017;45(D1):D419.

[164] Hopfner M, Schuppan D, Scherubl H. Growth factor receptors and related signaling pathways as targets for novel treatment strategies of hepatocellular cancer. *World Journal of Gastroenterology*, 2008;14(1):1.

[165] Enfield KS, Rowbotham DA, Holly A, et al. Abstract a21: Mir-106a and mir-106b affect growth and metastasis of lung adenocarcinoma. *Cancer Research*, 2016;76(6Suppl):A21.

[166] Aiello NM, Stanger, BZ. Echoes of the embryo: using the developmental biology toolkit to study cancer. *Disease Models & Mechanisms*, 2016;9(2):105–114.

[167] Mehlen P, Delloye-Bourgeois C, Chedotal A. Novel roles for slits and netrins: Axon guidance cues as anticancer targets? *Nature Reviews Cancer*, 2011;11(3):188–197.

5 Graphlets in Network Science and Computational Biology

Khalique Newaz and Tijana Milenković

5.1 Introduction

Networks (or graphs), which consist of a set of nodes and a set of edges, can be used to elegantly model and analyze many real-world phenomena. Examples are the Internet, Facebook, stock markets, disease spread, brain, or cell [1, 2]. Network-based analyses of real-world phenomena are important. As a naive illustration, both graphite and diamond are composed of the same elements, namely carbon atoms. However, they have very different patterns of connections (links, interactions, or edges) between the atoms. Clearly, the individual elements (the atoms) alone would fail to distinguish between the very different properties of graphite and diamond (graphite being soft and dark, diamond being hard and clear). Yet, their interaction patterns, i.e., their networks, can easily do so. Of all types of real-world networks, this chapter primarily focuses on biological networks (BNs). Nonetheless, the discussed work is applicable to any network type.

In BNs, nodes are biomolecules, and edges are interactions between the biomolecules. Typically, by "biomolecules," one assumes genes/proteins. For example, a prominent type of BNs is a protein–protein interaction (PPI) network, in which nodes are proteins and edges correspond to physical bindings between the proteins [3]. Other examples of BNs whose nodes are genes/proteins include gene regulatory and co-expression networks, where nodes are genes and edges represent functional interactions between the genes (Chapter 3). Also, we note that other types of BNs exist in which nodes are different biomolecules than genes/proteins (Chapter 3). For example, in a protein structure network (PSN) [4], or contact map [5], nodes are amino acids of the given protein and edges correspond to proximity of the amino acids in the protein's 3-dimensional structure. PSNs are different than PPI (as well as gene regulatory and co-expression) networks in the sense that a PSN, i.e., an entire protein structure network, is a node of a PPI network. That is, PSNs and PPI networks are networks at different scales. In the rest of the chapter, unless explicitly stated otherwise, we will assume that nodes in BNs are genes/proteins. Also, henceforth, we will use terms "gene" and "protein" interchangeably, as is often done in the literature, given that proteins are "simply" products of genes.

Computational advances in bioinformatics, such as analyses of genomic sequence or gene expression data, have revolutionized our biological understanding [6, 7, 8]. Yet, proteins function by interacting with each other, and this is what BNs model (unlike sequence or expression data). Thus, network science-based data analyses, i.e., BN research, have already furthered and will continue to further our biological understanding compared to the non-network sequence or expression data analyses. For example, analyses of the human PPI network have already yielded novel knowledge about human aging that cannot be obtained via the non-network-based analyses (Section 5.4.2).

Network science and BN research are interdisciplinary fields, which draw on many scientific branches, including graph theory from mathematics, data mining, information visualization, or graphical models from computer science, inferential modeling from statistics, statistical mechanics from physics, etc. In this chapter, we focus on the *graph theoretic* and *data mining* aspects of network science and BN research. Intuitively, since cellular function is often encoded in the topology (or structure) of BNs, the goal is to link network topological patterns to the observed functional (including phenotypic) behavior. In the above example, the goal would be to identify network patterns that can distinguish between graphite that is soft and dark and diamond that is hard and clear. In more detail, the goal is to extract functional knowledge from well-studied network regions and transfer that knowledge to poorly studied (functionally uncharacterized) regions that have similar network patterns as the well-studied regions. This is important, especially in biology, because experiments for determining protein function are long and costly. Since proteins aggregate to perform functions, and since BNs model these aggregations, computational prediction of protein function via BN analyses can suggest candidates for experimental validation and consequently save time and money.

To efficiently capture and compare network topological patterns of interest, many *network properties* (also called *measures of network structure* or *topological descriptors/features*) have been proposed (Chapter 3) [1, 9, 10, 11]. Among them, *graphlets* [12] have been shown to capture extremely well detailed topological characteristics of complex real-world networks, where intuitively graphlets are small subgraphs of a network, such as a cycle (e.g., triangle or square), a path, a clique (complete graph), or any other network pattern (Figure 5.1 (a)). Consequently, graphlets have been used as a basis for sensitive measures of network [11, 12, 14, 15] and node (or edge) [16, 17] similarities. These in turn have been used to develop state-of-the-art algorithms for many computational problems, such as network comparison [15, 18, 19, 20, 21] and alignment [22, 23, 24, 25, 26, 27], link prediction [28], node centrality computation [29], and clustering [17, 30], as well as for various application problems, such as studying human aging [31, 32], cancer [33, 34], other diseases [35], or pathogenicity [17, 29], quantifying protein-ligand binding affinity [19, 37], or classifying protein structures [36, 19, 37]. Importantly, graphlets have been shown to be superior to many other measures of network topology, such as network motifs [10, 30, 38], random walks [29, 39], PageRank-like [32, 40] and spectral graph theoretic [41, 42] "topological signatures," and various centrality measures [29, 31].

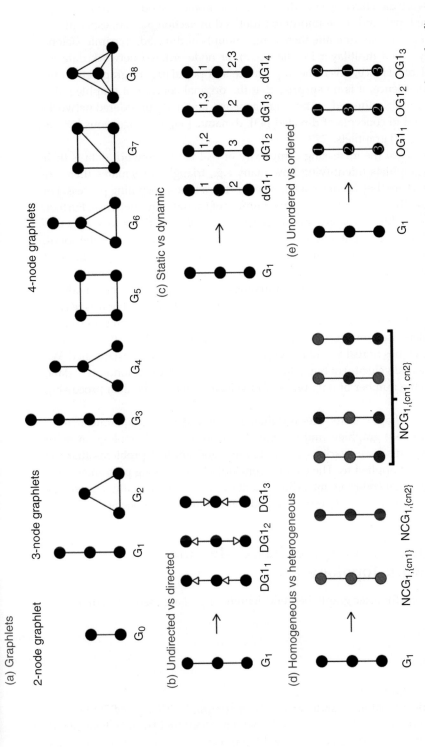

Figure 5.1: An illustration of different types of graphlets. Panel (a) shows all 2–4-node original (undirected, static, homogeneous, and unordered) graphlets. The remaining panels show, for a representative original graphlet, namely a 3-node path, all possible corresponding: (b) directed graphlets, (c) dynamic graphlets containing up to three events, (d) heterogeneous graphlets containing up to two node colors, and (e) ordered graphlets. Note that in panel (d), with the exhaustive approach for enumerating all possible heterogeneous graphlets corresponding to a 3-node path, given two colors, there are six heterogeneous graphlets, shown on the right of the arrow, each accounting for both which colors are present in the graphlet and which node position has which color. On the other hand, with the approach by Gu et al. [13] that is discussed in this chapter, there are three possible colored graphlets, denoted by $NCG_{1,\{cn1\}}$, $NCG_{1,\{cn2\}}$, and $NCG_{1,\{cn1,cn2\}}$, each accounting only for which colors are present in the graphlet, ignoring the node-specific color information. Consequently, with the latter approach, the last four graphlets on the right of the arrow, which all have the same two colors present in them, will be treated as the same heterogeneous graphlet.

195

The complexity of real-world network data has been increasing. Namely, while traditional network data have been undirected, static, homogeneous (single node type and single edge type), and node-unordered (defined in Section 5.2.1.5), especially in the computational biology domain, increasing amounts of directed, dynamic, heterogeneous (partly called multilayer in Chapter 3), or node-ordered network data are becoming available. Hence, given the importance and popularity of graphlets in the field of network science, it is no surprise that the original concept of graphlets that allow for analyses of undirected, static, homogeneous, and node-unordered networks has been extended to concepts of directed [43], dynamic [30], heterogeneous [13], or node-ordered [19, 37] graphlets.

Another aspect of the increasing complexity of real-world network data is their size. Counting graphlets (identifying how many e.g., triangles or squares there are and potentially where they occur) in a large network is a time-consuming process. For example, suppose that we want to count the number of triangles in a network. Further, suppose that we do this naively by checking each combination of three nodes in a network, i.e., if a three-node combination results in a triangle, we include it in the count. For a network with only 10 nodes, this process would require us to check 120 different three-node combinations, which increases to 161,700 three-node combinations for a network with 100 nodes. That is, with just a 10 times increase in the number of nodes in a network, the possible number of triangles increases by $\sim 1,000$ times. For this reason, several recent efforts have focused on speeding up the graphlet counting process, to allow for graphlet-based analyses of large networks [44, 45, 46, 47, 48, 49]. Note that the above complexities of real-world networks are related to the well established 3Vs (volume, variety, and velocity) of big data, where volume refers to the amount of data, variety refers to the number of data types, and velocity refers to the data processing speed.

The rest of the chapter introduces (original as well as directed, dynamic, heterogeneous, and ordered) graphlets and graphlet-based measures of topology in more detail, and then it discusses various computational and applied problems that the measures have been applied to. The chapter concludes by reviewing prominent software tools for graphlet counting, including recent tools for faster counting, as well as implementations of a number of graphlet-based computational approaches.

5.2 Graphlets and Graphlet-based Measures of Network Topology

Please see Chapter 3 for basic graph theoretic definitions that are used throughout this chapter.

5.2.1 Graphlets

5.2.1.1 Original Graphlets

Recall from Chapter 3 that in an undirected network or graph $G(V, E)$, or G for brevity, V is a node set and E is an edge set, where E contains unordered pairs of elements of V. Note that in this chapter, we only deal with graphs that do not contain any loops (i.e., no node is linked to itself) or multi-edges (i.e., any two nodes are linked by at most one edge).

Recall that intuitively graphlets are small subgraphs of a network [12], such as a cycle (e.g., triangle or square), a path, or a clique (complete graph) (Figure 5.1 (a)). More formally, graphlets are defined as connected non-isomorphic induced subgraphs of an undirected graph (all of these concepts are defined in Chapter 3). Graphlets are different from network motifs [38, 50, 51, 52], since the former must be induced subgraphs while the latter are partial subgraphs. An additional difference between the two is that for a subgraph (e.g., triangle or square) to be identified as a motif, it is required that the count of how many times the subgraph appears in a real-world network is significantly higher than the count of how many times the subgraph appears in random model networks (Chapter 3). However, it is unclear which random network model fits real-world networks the best and should thus be used for motif identification; using an inappropriate model may identify as significantly over-represented those subgraphs that otherwise would not have been identified as motifs [53]. On the other hand, graphlet counting is only done in a real-world network and does not depend on any random network model. Thus, graphlet counting avoids the above problem that network motif identification is prone to.

Figure 5.1 (a) illustrates all possible graphlets containing two to four nodes. The number of graphlets containing k nodes increases super-exponentially with k. The computational complexity of counting graphlets on k nodes in a network with n nodes is $\mathcal{O}(n^k)$. For these reasons, i.e., because counting of large graphlets in large networks is time-consuming, in practice, graphlets on up to five nodes have typically been studied. Yet, given short diameters of real-world networks due to their small-world nature (also known as the "six degrees of separation" phenomenon) [54], even using only up to 5-node graphlets typically still allows for exploring a large portion of network topology. Further, it has been shown that sometimes less is more, i.e., that using up to 4-node graphlets can be sufficient to correctly extract function from network topology [28, 30, 42]. This may be attributed to the noisiness of network data, the lack of meaningful functional signal beyond the neighborhood depth around a node that can be captured by 4-node graphlets, or dependencies and redundancies between the different graphlets that may blur the topological signal [15].

5.2.1.2 *Directed Graphlets*

Recall from Chapter 3 that in a directed graph $G(V, E)$, V is a node set and E is an edge set, where E contains ordered pairs of elements of V, i.e., $E \subseteq V \times V$.

Directed graphlets (Figure 5.1 (b)) [43] are connected non-isomorphic induced subgraphs of a directed graph, which do not contain any anti-parallel directed edge (i.e., for any two nodes u and v in a given subgraph, if the subgraph has a directed edge from u to v, then it cannot have a directed edge from v to u.)

Note that when generalizing undirected graphlets to directed graphlets, each undirected edge (u, v) of a given graphlet can be replaced by either a directed edge from u to v or a directed edge from v to u. Hence, a graphlet with k edges can potentially result in 2^k different directed graphlets, minus the isomorphic configurations. For example, the 3-node path undirected graphlet G_1, which has two undirected edges, results in three different directed graphlets: $2^2 = 4$ from replacing the two undirected edges with four directed ones, minus the only one isomorphic configuration of the directed graphlet $DG1_1$ (Figure 5.1). Consequently, the number of directed graphlets containing n nodes is at least as large as (and typically much larger than) the number of undirected graphlets on the same number of nodes. That is, as n increases, the number of different directed graphlets increases even more rapidly than the corresponding number of undirected graphlets (which is already super-exponential with respect to n). For instance, for $n = 2$, there is a single directed graphlet and a single undirected graphlet (an edge), while for $n = 4$, there are 34 directed graphlets and six undirected ones [43].

5.2.1.3 Dynamic Graphlets

Definition of a raw dynamic network. In this section, let $G(V, E)$ be a *dynamic network*, where V is the set of nodes and E is the set of *events* (temporal edges) that are associated with a start time and duration [55]. An event can be represented as a 4-tuple $(u, v, t_{start}, \sigma)$, where u and v are its endpoints, t_{start} is its starting time, and σ is its duration [30]. Thus, if we were to aggregate all events of a dynamic network into a static version of the network (by removing the temporal information from the events and then taking their union), each event would correspond to a unique edge in the aggregate static version of the dynamic network, whereas each static edge could correspond to multiple events. Note that this section considers undirected events, but most ideas can be extended to directed events as well. Also, note that not all dynamic network data contain explicit information about event duration. For example, when events are instantaneous, as is the case with e.g., email or sms communications, all events can be assumed to have the same negligible duration.

Definition of a snapshot-based representation of the raw dynamic network. Given a dynamic network as defined above, the network can be represented as a series of snapshots, where each snapshot aggregates the dynamic data (i.e., all events) observed during the corresponding time interval. Specifically, the entire time interval of the dynamic data is split into time windows of size t_w. For each window, a static network snapshot is generated such that an edge between nodes u and v will exist if and only if there are at least f events between u and v during the given window (Figure 5.2). More formally, a dynamic network $G(V, E)$ can be represented as a sequence of k network snapshots $\{G_0, G_1, \ldots, G_k\}$, where each snapshot $G_i = (V_i, E_i)$ is a static graph capturing network structure during time interval i (as described above), and $V_i \subseteq V$ and $E_i \subseteq E$ (where E_i is restricted to nodes in V_i).

Using static graphlets to analyze the snapshot-based representation of the raw dynamic network. Given the snapshot-based representation of the dynamic network, an obvious way to apply graphlet-based analyses to it is to count static graphlets (as defined above, in Subsection 5.2.1.1) within each snapshot, and then analyze the time-series of the results to gain insights into network structural changes with time [31]. However, this strategy ignores important relationships between different snapshots. To explicitly explore these relationships, the notion of dynamic graphlets was recently proposed [30], as follows.

Event time	Node 1	Node 2	Duration
January 2	s	u	2 minutes
January 5	u	t	5 minutes
February 1	v	t	1 minute
February 7	t	u	6 minutes
March 20	v	s	3 minutes
March 24	v	t	4 minutes
March 26	v	u	5 minutes

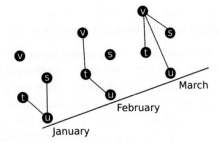

Figure 5.2: An illustration of how raw temporal data (left) is modeled as a dynamic network, which contains a sequence of static network snapshots, with each snapshot corresponding to a given time window (right). The raw data is phone call records data, where the nodes are people and the events denote communication between the people via phone calls. In this illustration, the time window size (t_w) is one month, and the minimum number of interactions that need to exist between two nodes within the given time window in order to connect the nodes (f) is one.

Using dynamic graphlets to analyze the raw dynamic network. Intuitively, static graphlets were extended by adding temporal dimension *on top* of the topological structure of a static graphlet. Specifically, the relative order of events is added onto the edges of a static graphlet. For example, a static graphlet that is a 3-node path captures two different interactions between three nodes. To answer *when* these interactions happen, dynamic graphlets need to be used (Figure 5.1 (c)).

To formalize the above intuition, we need to first introduce the notion of a Δt-time-respecting path (in the raw dynamic network $G(V, E)$ that contains information on duration of each event). A Δt-time-respecting path between two nodes is a sequence of events that connects the two nodes such that for any two consecutive events in the sequence, the start time of the later event and the end time of the earlier event are within Δt time of each other. A dynamic network $G(V, E)$ is then considered Δt-connected if there exists a Δt-time-respecting path between every pair of nodes in the network. Let $G'(V', E')$ be a *dynamic subgraph* of G with $V' \subseteq V$ and $E' \subseteq E$, where E' is restricted to nodes in V'. Then, a *dynamic graphlet* is an equivalence class of isomorphic Δt-connected dynamic subgraphs (these concepts are defined in Chapter 3); equivalence is taken with respect to the relative temporal order of events, regardless of the events' start times [30]. Because it is events' relative ordering that is considered and not their actual times, two connected dynamic subgraphs will correspond to the same dynamic graphlet if they are topologically equivalent and if their corresponding events occur in the same order. Figure 5.1 (c) illustrates the concept of dynamic graphlets. For additional illustrations and more formal definitions, see [30].

Using dynamic graphlets to analyze the snapshot-based representation of the raw dynamic network. The above definitions related to dynamic graphlets are provided in the context of a raw dynamic network and thus rely on event durations. When dealing with the snapshot-based representation of the raw dynamic network that does not contain any event duration information, dynamic graphlets can still be used. In this case, the only thing that one needs to do is to consider each edge between any pair of nodes from snapshot i of the dynamic network to be an event that is active from time i to time $i + 1$. Then, all of the above definitions that rely on event durations are applicable to the snapshot-based network representation.

On the number of static versus dynamic graphlets of the given size. Note that for a given dynamic graphlet with n nodes and k events, discarding the order of the events and removing duplicate events over the same edge results in a static graphlet with n nodes and $k' \leq k$ edges [30]. Each dynamic graphlet corresponds to a single static graphlet, while one static graphlet backbone can correspond to multiple dynamic graphlets. For example, the static backbone G_2, i.e., the triangle, corresponds to a single dynamic graphlet when $k = 3$, to 90 different dynamic graphlets when $k = 6$, and to 3,025 different dynamic graphlets when $k = 9$. To our knowledge, there is no closed form solution for computing the number of dynamic graphlets corresponding to the same graphlet backbone; these statistics for up to 5-node static backbones and up to $k = 10$ are available in Supplementary Table S1 in [30]. A reason why one static graphlet backbone can and typically does correspond to multiple dynamic graphlets is that for a given number of events k, as n increases, the number of different dynamic graphlets increases more rapidly than the corresponding number of static graphlets. For instance, given $k = 4$, for $n = 2$, there is a single dynamic graphlet and a single static graphlet (an edge), while for $n = 4$, there are 18 dynamic graphlets and six static graphlets [30].

5.2.1.4 Heterogeneous Graphlets

Traditional methods for analyzing BNs typically deal with a single homogeneous BN type. However, different BN types exist. BN data integration into a heterogeneous network (including what is called a multilayer network in Chapter 3 or a "network of networks" in the literature) that encompasses different node types or different edge types is expected to yield deeper insights into cellular functioning compared to traditional homogeneous BN analyses of individual data types in isolation.

Extending original (homogeneous) graphlets to their heterogeneous counterparts to allow for dealing with different node or edge types (colors) is a natural step. Given a heterogeneous network containing n nodes and c different node or edge types (also called colors), a naive (exhaustive) extension would track both which combinations of node (or edge) colors exist in a given graphlet as well as at which node (or edge) positions in the graphlet the colors occur. With such an approach, the computational complexity of the problem, namely both the enumeration of all possible heterogeneous graphlet types on up to n nodes (the space complexity) and counting of the heterogeneous graphlets in a network (the time complexity), increase exponentially with the number of colors [18]. For example, with the naive approach, for c different node colors, the number of heterogeneous graphlets increases exponentially with n as $c^n - I$, where I is the number of isomorphic configurations of the given heterogeneous graphlet [18].

A more computationally efficient colored graphlet approach was recently proposed, which only tracks which combinations of node (or edge) colors exist in a given graphlet but not at which node (or edge) positions in the graphlet the colors occur [13]. Consequently, with the new approach: (1) the number of possible colored graphlets and thus the computational space complexity is much smaller (though still exponential in terms of the number of colors) compared to the exhaustive approach, but (2) importantly, the computational time complexity of counting colored graphlets in a heterogeneous network is the same as that of counting original graphlets in a homogeneous network, unlike with the exhaustive approach. The rest of this section explains this recent more efficient colored graphlet approach (henceforth referred to simply as colored graphlets or heterogeneous graphlets).

First, for ease of explanation, we discuss *node-colored graphlets* (NCGs), and we do so via an illustration. Given k possible node colors, and given one of all possible combinations of the k colors, for a homogeneous graphlet G_i, a *node-colored graphlet* is the set of all distinct graphs that are isomorphic to G_i, while in each such graph, each node is colored with one of the colors from the color combination in question, and also, each color from the color combination in question has to be present in each such graph. Given k node colors, there are $2^k - 1$ possible node-colored graphlets. For example, let us assume that a heterogeneous network has nodes with two possible node colors: c_{n1} and c_{n2}. These two node colors have $2^2 - 1 = 3$ possible combinations: $\{c_{n1}\}$, $\{c_{n2}\}$, and $\{c_{n1}, c_{n2}\}$. As a result, for each homogeneous graphlet G_i, there are three possible node-colored graphlets: $NCG_{i,\{c_{n1}\}}$, $NCG_{i,\{c_{n2}\}}$, and $NCG_{i,\{c_{n1},c_{n2}\}}$, where $NCG_{i,\{c_{n1}\}}$ is a colored version of G_i that contains only c_{n1}-colored nodes, $NCG_{i,\{c_{n2}\}}$ contains only c_{n2}-colored nodes, and $NCG_{i,\{c_{n1},c_{n2}\}}$ contains both c_{n1}- and c_{n2}-colored nodes. Figure 5.1 (d) illustrates this for a homogeneous graphlet corresponding to a 3-node path [13].

This definition of node-colored graphlets is more efficient than the naive exhaustive enumerative definition is. For example, when $k = 2$, with the naive definition, there are six node-colored graphlets for a 3-node path, while with the above presented approach, there are only three of them (Figure 5.1 (d)). When $k = 3$, with the naive definition, there are 18 node-colored graphlets for a 3-node path, while with the above presented approach, there are only seven of them. Even with this more efficient approach, the number of node-colored graphlets (and thus the computational space complexity) increases drastically with the increase of k. Yet, this is not a major concern, because in practice, one may expect a relatively small value of k (e.g., one can study a heterogeneous network whose nodes are proteins, functions, diseases, and drugs with k value of only four) [13]. Further, the computational time complexity of searching for a given colored graphlet in a heterogeneous network remains the same as that of searching for its homogeneous equivalent. This is because the former involves: (1) counting in the heterogeneous network all graphlets, independent of their colors (which is the same as counting homogeneous graphlets in the network), and (2) for each of the homogeneous graphlets found in the network, simply determining which node colors appear in it and thus which node-colored graphlet the non-colored graphlet corresponds to. Step 1 is the time consuming part of the node-colored graphlet counting process, unlike step 2, which is trivial.

Second, analogous to node-colored graphlets, without going again through all details due to space constraints, *edge-colored graphlets* (ECGs) were defined [13].

Third, the above ideas can be combined into graphlets that have different node as well as edge colors [13]. A computationally simple option is to concatenate node- and edge-colored graphlet-based statistics. A more complex option is, for each node-colored graphlet, to vary its edge colors (or vice versa). While this would further increase the number of colored graphlets of interest, innovative approaches have been introduced for faster homogeneous graphlet counting (discussed in more detail in Section 5.5), even in networks with millions of nodes counting [46, 56], and similar directions can be pursued for heterogeneous graphlet counting.

5.2.1.5 Ordered Graphlets

In this section, let $G(V, E)$ be an undirected *node-ordered* (or simply, ordered) graph, with V as a node set and E as an edge set, where E contains unordered (i.e., undirected) pairs of elements of V, and each element of V is labeled with a unique number, where the node labels (numbers) represent a strict total ordering of nodes in the graph. For example, when a 3-dimensional structure of a protein is modeled using a PSN, the information about the position of amino acids in the sequence (or amino acid chain) of the protein can be imposed as the order onto nodes of the PSN.

Ordered graphlets (Figure 5.1 (e) and Figure 5.3) [19, 37] are connected non-isomorphic induced subgraphs of an undirected ordered graph, whose nodes acquire the *relative* node order of their originating graph.

Note that for a given ordered graphlet with n nodes and k edges, discarding the node order results in an unordered graphlet of the same size. Hence, more than one ordered graphlet can correspond to the same unordered graphlet. For example, all 3-node path ordered graphlets, which differ only in the relative ordering of their nodes, correspond to the same unordered 3-node path graphlet (Figure 5.1 (e)). In other words, the number of ordered graphlets on n nodes is always at least as large as (and typically much larger than) the number of unordered graphlets on the same number of nodes. That is, as n increases, the number of different ordered graphlets increases more quickly than the corresponding number of unordered graphlets. For instance, for $n = 2$, there is a single ordered graphlet and a single unordered graphlet, while for $n = 4$, there are 38 ordered graphlets and six unordered graphlets [37].

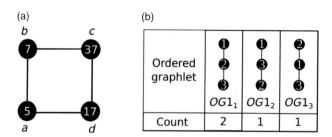

Figure 5.3: An illustration of a toy ordered network (a) and counts of all 3-node path ordered graphlets that exist in the network (b). For subgraph a–b–c, its raw node order 5–7–37 is converted to relative order 1–2–3, which corresponds to ordered graphlet $OG1_1$. The same holds for subgraph a–d–c. Hence, there are two occurrences of $OG1_1$ in the network. For subgraph b–c–d, its raw node order 7–37–17 is converted to relative order 1–3–2, which corresponds to ordered graphlet $OG1_2$. This is the only occurrence of $OG1_2$ in the network. Finally, for subgraph, b–a–d, its raw node order 7–5–17 is converted to relative order 2–1–3, which corresponds to ordered graphlet $OG1_3$. This is the only occurrence of $OG1_3$ in the network.

5.2.2 Graphlet-based Measures of Topological Position of Individual Nodes, Edges, or Non-edges

Henceforth, unless otherwise indicated, the focus is on original graphlets. However, all ideas are trivially applicable (and in many cases have already been applied) to directed, dynamic, heterogeneous, or ordered graphlets as well.

5.2.2.1 Graphlet Orbits

Before we can explain any graphlet-based measures, we need to define the notion of graphlet orbits. Recall that an isomorphism f from graph G to graph H is a bijection f of nodes of G to nodes of H such that vu is an edge of G if and only if $f(v)f(u)$ is an edge of H. An automorphism is an isomorphism from G to G. The automorphisms of G form a group denoted by Aut(G). If v is a node of G, then the automorphism orbit of v is Orb(v) = $\{u \in V | u = f(v)$ for some $f \in$ Aut(G)$\}$, where V is the set of nodes of G. Hence, end nodes of a G_1 belong to one orbit, whereas the middle node of a G_1 belongs to another orbit. See Figure 5.5 for an illustration. There are 15 orbits for 2–4-node graphlets (Figure 5.4(a)) and 73 orbits for 2–5-node graphlets [16].

5.2.2.2 Graphlet Degree Vector (GDV)

Many network science tasks require a way of summarizing the topology around a node in a network, i.e., quantifying the node's network neighborhood. Recall from Chapter 3 that a measure called the degree of a node, which is the number of the node's neighbors in the network, has been widely used for this purpose. Clearly, this measure is limited to the node's direct contacts in the network. However, useful information from the rest of the network is missed. So, a more advanced, graphlet-based measure of a node's network neighborhood was proposed, which generalizes the degree of node v, that counts the number of edges that v touches (where an edge is the only graphlet on two nodes, G_0 in Figure 5.1 (a) and in the top of Figure 5.4), into GDV of v, that counts the number of graphlets on up to n nodes (G_0, G_1, etc. in the top of Figure 5.4) that v touches [16]. Importantly, it is relevant to distinguish between a node touching, for example, a G_1 at an end node (one of the orbits of G_1) or at the middle node (the other orbit of G_1). Thus, GDV of v has 15 or 73 elements (depending on the considered graphlet size) counting how many orbits of each type touch v (v's degree is the first element). See Figure 5.6 for an illustration. The GDV captures v's connectivities up to its fourth neighborhood. Later in this chapter, we discuss what computational and applied tasks the GDV measure has been used for.

Assuming the existence of o orbits ($o = 15$ for 2–4-node graphlets and $o = 73$ for 2–5-node graphlets), GDVs of all n nodes in a network can be represented by an $n \times o$ matrix, which we call *GDV matrix*, where matrix entry (i, j) contains the information on how many times node i touches orbit j (Figure 5.7). This matrix will be crucial for explaining some concepts later on in this chapter.

Note that the original (undirected, static, homogeneous) GDV concept was extended to its directed [43], dynamic [30], heterogeneous [13], or ordered [19] counterpart.

Much of network research has traditionally focused on measuring network positions of *nodes*. Since a graph consists of both nodes *and* edges, more recently, the

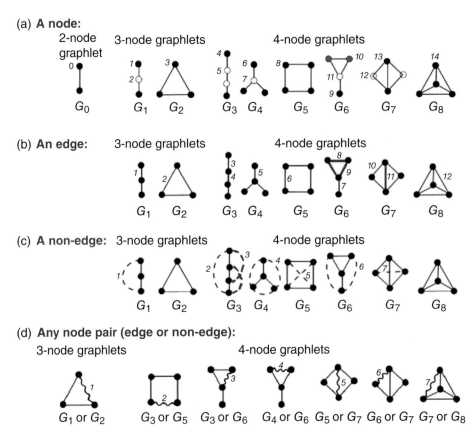

Figure 5.4: All topological positions (orbits) in up to 4-node graphlets of a node ((a); node shade), an edge ((b); solid line), a non-edge ((c); broken line), and any node pair, an edge or a non-edge ((d); wavy line) are shown. For example: (1) in graphlet G_3, the two end nodes are in node orbit 4, while the two middle nodes are in node orbit 5; (2) in G_3, the two outer edges are in edge orbit 3, while the middle edge is in edge orbit 4; (3) in G_3, the non-edge touching the end nodes is in non-edge orbit 2, while the two non-edges that touch the end nodes and the middle nodes are in non-edge orbit 3; (4) a node pair at node pair orbit 1 touches a G_2 at edge orbit 2, if it is an edge, or a G_1 at non-edge orbit 1, if it is a non-edge (hence, mutually exclusive edge orbit 2 and non-edge orbit 1 are reconciled into a common node pair orbit 1). There are 15 node, 12 edge, 7 non-edge, and 7 node pair orbits for up to 4-node graphlets. In a graphlet, different orbits are colored differently. All up to 5-node graphlets are used, but only up to 4-node graphlets are illustrated. There are 73 node, 68 edge, 49 non-edge, and 49 node pair orbits for up to 5-node graphlets. The figure was adapted from Hulovatyy *et al.* [28].

focus has also been extended to measuring network positions of edges [57, 58]. In the context of graphlets, analogous to the concept of GDV, the concept of *edge-GDV* was introduced, in order to count the number of graphlets that an *edge* touches at a given *edge orbit* (Figure 5.4(b)) [17]. Given the automorphism group of graph G, Aut(G), if vu is an edge of G, then the edge orbit of vu is $\text{Orb}_e(vu) = \{xy \in E | x = f(v) \text{ and } y = f(u) \text{ for some } f \in \text{Aut}(G)\}$, where E is the set of edges of G (Figure 5.4). (An alternative definition of edge orbits exists, with the same end result; see [17] for details.)

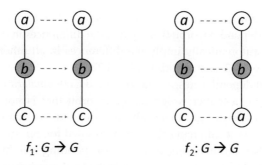

$f_1: G \to G$ $\qquad$ $f_2: G \to G$

Figure 5.5: All possible automorphisms of a toy network G. For ease of explanation, the toy networks just happens to be a simple 3-node path with nodes a, b, and c. There exist two automorphisms, denoted by $f_1 : G \to G$ and $f_2 : G \to G$. The former one maps nodes a, b, and c to nodes a, b, and c, respectively. The latter one maps nodes a, b, and c to nodes c, b, and a, respectively. Because a can be mapped to both a (by f_1) or c (by f_2), and because c can be mapped to both c (by f_1) or a (by f_2), Orb(a) = Orb(c) = {a,c}. Similarly, because b can be mapped only to b, Orb(b) = {b}. Consequently, there are two different orbits in G, as illustrated with different node colors.

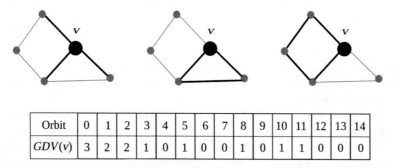

Orbit	0	1	2	3	4	5	6	7	8	9	10	11	12	13	14
GDV(v)	3	2	2	1	0	1	0	0	1	0	1	1	0	0	0

Figure 5.6: An illustration of the graphlet degree vector (GDV) of node v, for 2–4-node graphlets. Coordinates (or dimensions) of v's GDV count how many times v is touched by a particular orbit, such as an edge (the leftmost panel), a triangle (the middle panel), or a square (the rightmost panel). Hence, the degree (the leftmost panel) is generalized to the GDV. The GDV of v is presented in the table for orbits 0 to 14; v is touched three times by an edge (orbit 0), twice by an end node of graphlet G_1 (orbit 1), etc.

Analogously, the concept of *non-edge-GDV* was defined to count the number of graphlets that a *non-edge* touches at a given *non-edge orbit* (Figure 5.4(c)) [28].

Finally, mutually exclusive edge and non-edge orbits were reconciled by defining *node-pair-GDV* to count the number of graphlets that a *node pair*, an edge or a non-edge, touches at a given *node-pair orbit* (Figure 5.4(d)) [28].

5.2.2.3 GDV-similarity

Comparing GDVs of two nodes gives a measure of topological similarity of the nodes' extended network neighborhoods. Using general-purpose measures to compare raw GDVs may be inappropriate, as some node orbits are dependent on other node orbits (see below) [16]. For example, the differences in orbit 0 (i.e., in the degree) of two nodes will automatically imply the differences in all other orbits for these nodes, since all orbits contain, i.e., depend on, orbit 0. Similarly, the differences in orbit 3 (the triangle) of two nodes will automatically imply the differences in all other orbits of the two nodes that contain orbit 3, such as orbits 14 and 72.

To remove orbit dependencies, a measure called *GDV-similarity* was designed [16]. This measure assigns a higher weight w_i to an orbit that is not affected by many other orbits, and a lower weight to an orbit that depends on many other orbits. This way, the dependencies of different orbits are accounted for. Specifically, if an orbit i is affected by o_i number of other obits, then its corresponding weight w_i is computed as $w_i = 1 - \{(\log(o_i)/\log(n_o))\}$, where n_o represents the total number of different orbits for graphlets of up to size n. For example, when $n = 5$, i.e., for up to 5-node graphlets, there are $n_o = 73$ different node orbits. For more details on orbit dependencies and weights w_i, see [16].

Then, given two nodes u and v, their GDV-similarity is computed as follows [16]. Let u_i and v_i denote the ith coordinate of u's and v's respective GDVs. The distance $D_i(u,v)$ between the ith orbits of nodes u and v is: $D_i(u,v) = w_i \times (|\log(u_i + 1) - \log(v_i + 1)|)/\log(max\{u_i, v_i\} + 2)$. Given the $D_i(u,v)$ values for each orbit i (e.g., for all 73 orbits of the 2–5 node graphlets), the total distance $D(u,v)$ between nodes u and v is then: $D(u,v) = \sum_{i=0}^{72} D_i / \sum_{i=0}^{72} w_i$. The value of $D(u,v)$ lies in the interval $[0,1)$, where $D(u,v) = 0$ means that GDVs of nodes u and v are identical. Finally, GDV-similarity $S(u,v)$ between nodes u and v is: $S(u,v) = 1 - D(u,v)$. Now, the higher the GDV-similarity value, the higher the topological similarity between the two nodes. Clearly, GDV-similarity $S(u,v)$ compares two rows of the GDV matrix (defined above and illustrated in Figure 5.7).

Original GDV-similarity (as defined above) considers all of the orbits present in all graphlets on up to n nodes (i.e., it considers all 15 orbits, i.e., all 15 columns of the GDV matrix, that exist for 2–4 node graphlets, or all 73 orbits, i.e., all 73 columns of the GDV matrix, that exist for 2–5 node graphlets). However, it was recognized that for a given node, some of its orbit counts (i.e., some of the columns of the GDV matrix) are redundant, meaning that the counts of these redundant orbits can be derived from the counts of the non-redundant orbits [15]. Specifically, of the 15 orbits of all 2–4-node graphlets, 11 are non-redundant, and of the 73 orbits of all 2–5-node graphlets, 56 are non-redundant [15]. Hence, GDV-similarity can be computed by only considering counts for the non-redundant orbits. Specifically, one can first discard from the GDV matrix those columns that contain counts for the redundant orbits and then apply the above GDV-similarity definition to obtain $S(u,v)$ between nodes u and v using the resulting filtered GDV matrix. For a description of how to determine redundant orbits, see [15].

An alternative definition of GDV-similarity exists. Given the GDV matrix (whether the full matrix or the filtered one), one can first apply principal component analysis (PCA) to reduce the dimensionality of the GDV matrix, i.e., the number of its columns, by extracting only the most useful information from all of the columns. Then, one can use Euclidean distance (or some other distance metric) on the PCA-dimensionality-reduced GDV matrix, in order to capture GDV-similarity of two nodes [30]. PCA is a standard dimension reduction technique and more details can be found in [59]. The PCA-based definition of GDV-similarity is particularly

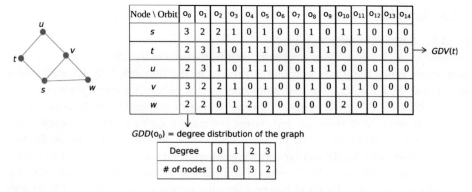

Figure 5.7: An illustration of the *GDV* matrix for 2–4-node graphlets (i.e., their 15 orbits) (upper right) of a toy graph containing five nodes (upper left). Clearly, the matrix dimension is 5×15 (ignoring the top row listing the orbits and the first column listing the nodes). Each row of the GDV matrix represents the GDV of the corresponding node. For example, the row colored in red represents the GDV of node t. By comparing any two rows, one obtains GDV-similarity between the corresponding two nodes. Each column of the GDV matrix contains information needed to compute the graphlet degree distribution (GDD) of the corresponding orbit. For example, the column colored in green contains information needed to compute the GDD of orbit 0, i.e., the degree distribution. The degree distribution of the illustrated network is as follows: three nodes have degree of two and two nodes have degree of three (bottom). Similar reasoning can be applied to any column i (corresponding to orbit i) to get the ith GDD, which counts how many nodes touch orbit i once, twice, three times, and so on. By correlating each pair of the 15 columns, one obtains the 15×15 GCM of the network.

useful for e.g., dynamic or heterogeneous graphlets, where it might be hard to derive orbit weights w_i and thus use the original GDV-similarity measure, due to these graphlet types having more orbits than original (static and homogeneous) graphlets.

Note that the original GDV-similarity concept was extended to its directed [43], dynamic [30], heterogeneous [13], or ordered [19] counterpart.

Analogous to GDV-similarity, a measure called *edge-GDV-similarity* was designed to measure topological similarity of two edges [17].

5.2.2.4 GDV-centrality

Chapter 3 discusses the notion of node centrality and a number of existing measures for this purpose, such as degree centrality, betweenness centrality, and subgraph centrality. An additional measure of node centrality exists that is based on graphlets, called *GDV-centrality* [29]. This measure aims to identify as central those nodes that occupy large and dense extended network neighborhoods, i.e., that participate in many graphlets and in large and dense graphlets. If v_i is the ith element of GDV of node v and w_i is the weight of orbit i (as defined above), then *node-GDV-centrality*$(v) = \sum_{i=0}^{72} w_i \times \log(v_i + 1)$ [29].

Analogous to GDV-centrality, *edge-*, *non-edge-*, and *node-pair GDV-centrality* measures were designed to assign higher centralities to edges, non-edges, and node pairs, respectively, that participate in many different graphlets [28].

5.2.3 Graphlet-based Measures of Entire Network Topology

5.2.3.1 Graphlet Frequency Vector (GFV)

As defined originally, GFV of a network counts, for each of the 29 3–5-node graphlets, how many times the given graphlet is present in the network [12]. Formally, GFV of network N, $GFV(N)$, is a 29-dimensional vector, where for each position i ($i = 1, 2, \ldots, 29$), $GFV_i(N)$ is the frequency of appearance of graphlet G_i in the network N.

It was recently recognized that, when comparing GFVs of two networks (discussed later in the chapter), GFVs are network-size dependent. For example, two networks that are actually similar can have different GFVs just because of their size difference, and consequently, GFV-based comparison would mistakenly identify the networks as dissimilar. To avoid this problem, the notion of the relative GFV (RGFV) was introduced [12]. RGFV of network N, $RGFV(N)$, is defined similar to GFV, except that at each position i of the vector $RGFV(N)$, the frequency of graphlet G_i, $GFV_i(N)$, is divided by with the total number of graphlets in the network, $T(N)$ (formally, $T(N) = \sum_{i=1}^{29} GFV_i(N)$).

Note that analogous to GFV and RGFV in the case of original graphlets, similar concepts were also designed for directed [43], dynamic [30], and ordered [37] graphlets.

5.2.3.2 Graphlet Degree Distributions (GDDs)

Recall that the degree distribution of a network measures, for each value of k, the number of nodes of degree k. In other words, the degree distribution counts, for each value of k, the number of nodes that touch k edges, where an edge is the only 2-node graphlet (G_0 in Figure 5.1 (a)). Naturally, degree distribution was generalized from the only 2-node graphlet to all other graphlets, i.e., to all of their orbits, into a notion of GDDs, one GDD for each orbit. Specifically, GDD of orbit i counts, for each value of k, the number of nodes that touch orbit i k times [14]. Given 15 orbits of all 2–4-node graphlets, there are 15 corresponding GDDs, the first one being the degree distribution (Figure 5.7). Similarly, given 73 orbits of 2–5-node graphlets, there are 73 corresponding GDDs. Note that GDD of orbit i can easily be computed from the column of the GDV matrix that corresponds to that orbit (Figure 5.7). Later on in this chapter, we discuss how GDDs of two or more networks can be used to compare and quantify the similarity of the networks.

Note that analogous to GDDs in the case of original graphlets, similar concepts were also designed for dynamic [30], ordered [19], or directed [43] graphlets.

5.2.3.3 Graphlet Correlation Matrix (GCM)

Recall that in the GDV matrix, the value at position (i, j) represents the number of times that node i touches orbit j (Figure 5.7). Also, recall that some of the orbit counts are redundant to each other, and that removing the redundant orbits (i.e., the corresponding columns) from the GDV matrix results in a filtered GDV matrix. Now, consider a network N with n nodes. Given this network's filtered GDV matrix of size $n \times o$ ($o = 11$ for 2–4-node graphlets and $o = 56$ for 2–5-node graphlets), GCM of the network N, GCM_N, is defined as a matrix of size $o \times o$, where the value at position (i, j) in GCM_N represents the Spearman's correlation coefficient between columns i and j

of the filtered GDV matrix (Figure 5.7) [15]. In other words, correlations are computed between all pairs of columns of the filtered GDV matrix. Motivation behind computing GCM is that different networks may have very different orbit dependencies and thus very different GCMs. Later on in the chapter, we discuss how GCMs of two or more networks can be used to compare and quantify the similarity of the networks.

5.3 Computational Approaches Based on the Graphlet Measures

5.3.1 Clustering of Nodes or Edges in a Network

Network clustering is an extremely popular topic in the network science field. Chapter 6 discusses this topic in detail. Here, we focus on graphlet-based network clustering.

Traditionally, network clustering has aimed to divide the network into groups (clusters or communities) of "topologically related" nodes, where the resulting topology-based groups are expected to "correlate" well with node label information, i.e., metadata, such as functions of genes/proteins in BNs. More recently, clustering of a network into groups of edges (rather than nodes) has been explored [17, 57], with the following motivation. Many network clustering methods partition nodes into non-overlapping clusters [60]. While nodes typically belong to multiple functional groups (i.e., have multiple labels), with such methods, clusters of nodes typically cannot capture the node group overlap. But, if these methods are used to cluster edges instead of nodes, the node group overlap can be captured (Figure 5.8). Below, we first discuss node-based network clustering and then we comment on edge-based network clustering.

Going back to the notion of "topological relatedness" of nodes in a network, traditional definition is that nodes should be in the same cluster if they are densely intercon-

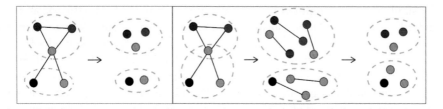

Figure 5.8: An illustration of node (left) versus edge (right) clustering of a toy network containing five nodes. By dividing the network into non-overlapping node clusters (where lack of cluster overlap is an assumption of many real-world clustering methods), each node can by definition end up in at most one node cluster. By dividing the network into non-overlapping edge clusters, each edge can by definition end up in at most one edge cluster, but each node can end up in multiple node clusters (specifically, in up to as many node clusters as the value of its degree is); for example, the blue node is present in each of the resulting two node clusters, because its adjacent edges are present in each of the two corresponding edge clusters. Note that in order to convert an edge cluster to a node cluster, one simply takes the union of all end-nodes of the edges in the cluster, as illustrated on the very right of the figure.

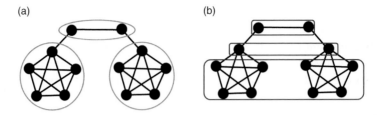

Figure 5.9: An illustration of the difference between denseness-based (a) and topological similarity-based (b) clustering. According to denseness-based clustering, nodes have to be highly interconnected with (and consequently close in the network to) each other in order to be eligible to be placed into the same cluster. For example, each of the 5-node cliques (complete graphs) and the only 2-node clique will likely be in its own cluster. The three denseness-based clusters are shown in blue. In contrast, according to similarity-based clustering, nodes only have to be topologically similar to each other, independent on whether the nodes are densely interconnected or close to each other. The three likely similarity-based clusters are shown in red, where within each cluster, all nodes are topologically identical to each other, i.e., have GDV-similarity of 1, and no nodes across the clusters are topologically identical to each other. Clearly, similarity-based clusters can contain nodes that are not necessarily densely interconnected and close to each other in the network (the bottom two red clusters). At the same time, similarity-based clusters can contain nodes that are densely interconnected or close to each other (the top red cluster, which is the same as the top blue denseness-based cluster).

nected with each other within the network, which is known as structural equivalence of the nodes (Figure 5.9; see Chapter 6 for details). However, it was hypothesized that in some applications, it might be more beneficial to cluster nodes that are topologically similar to each other, which is known as regular equivalence of the nodes (Figure 5.9) [16, 33]. Specifically, graphlets, i.e., nodes' GDV-similarities (Section 5.2.2), were used for the purpose of topological similarity-based clustering. In more detail, GDV-similarity was computed between every pair of nodes in a network, and then standard clustering techiques such as hierarchical or k-means clustering (Chapter 6) were used to group nodes with similar GDVs, i.e., similar extended network neighborhoods, and to put into different groups nodes with dissimilar GDVs [16, 33]. Similar was done by using the edge-GDV-similarity measure (Section 5.2.2) to cluster together topologically similar edges in a network. Whether it was node or edge clustering that was superior, depended on the choice of the clustering method. Later on in this chapter, we discuss biological applications of the GDV-similarity-based clusters of PPI networks.

5.3.2 Dominating Set of a Network

A dominating set (DS) of a network is a set of nodes such that every node in the network is either present in the DS or is adjacent (i.e., directly connected) to at least one of the nodes from the DS [29]. Formally, let $G(V, E)$ be a network, where V is the set of nodes and E is the set of edges of G. A DS of G is a set S, such that $S \subseteq V$ and $\forall v \in V$, either $v \in S$ or $\exists u \in S$ such that u and v are adjacent in G.

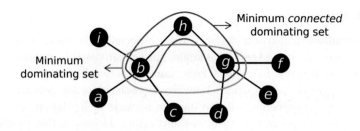

Figure 5.10: An illustration of the concepts of an independent, connected, and minimum dominating set (DS) of a toy network. The set of nodes circled in blue, as well as the set of nodes circled in red, each form a DS of the network, as all nodes of the network are either in the given DS or are directly linked to a node from the given DS. The nodes circled in blue, i.e., $\{b, g\}$, constitute the minimum but independent (i.e., not connected) DS of the network. The nodes circled in red, i.e., $\{b, h, g\}$ form a connected DS of the toy network, and this also happens to be the minimum connected DS of the network.

A DS is said to be *independent* if no two nodes in it are adjacent. A DS is said to be *connected* if for every pair of nodes in the set, there is a path between the two nodes in question. Connected DSs are of interest in BN research, because DSs of PPI networks might correspond to signaling pathways, which are connected [29]. A DS is said to be *minimum* if it contains the smallest possible number of nodes. See Figure 5.10 for an illustration of these concepts. Obviously, the set consisting of all nodes of the network is a DS of the network, but it is not necessarily a minimum DS. Identifying a minimum DS is a challenging task. In fact, finding a minimum DS of a network is an NP-hard problem [61]. Hence, several approximate algorithms are sought for this purpose. One such relatively simple algorithm is as follows, which was generalized to use graphlets [29].

The algorithm in question, as originally designed, finds a DS of a network via the following steps. First, starting with the initial set of all nodes in network G as the network's DS S, select a node u with the minimum degree and remove u from S only if the nodes in the set $S - \{u\}$ remain a connected DS of G. Second, repeat the above step for all nodes in S in the order of their increasing degrees [29]. Intuitively, given that real-world networks have many low-degree nodes and few high-degree nodes, this procedure aims to remove as many lower-degree nodes as possible from the initial DS, in order to decrease the size of the DS as much as possible, ideally resulting in a small high-degree "backbone" of the network that connects the rest of the network. Since the degree of a node is nothing but a measure of node centrality (the higher the degree of a node, the higher the node's centrality), and since GDV-centrality was shown to be superior to degree centrality in the sense that GDV-central genes are more enriched in biologically important (aging-related, cancer-related, HIV-related, or pathogen-interacting) genes than degree-central genes, the above algorithm was modified to use GDV-centrality instead of degree centrality [29].

Proteins that are in a DS of a protein–protein interaction (PPI) network were found to be rich in biologically central proteins (i.e., proteins involved in key biological processes such as aging, cancer, or pathogenicity) [29]. Additionally, the proteins in the DS were found to be significantly enriched in proteins that are targeted by drugs, unlike proteins that are not in the DS.

5.3.3 Link Prediction

Currently available real-world network data, especially BN data, are often incomplete (i.e., have missing edges) or contain false positives (i.e., contain edges that do not exist in reality). In other words, current network data are noisy. In BN data, where edges (i.e., interactions between biomolecules) are generally identified using biological experiments, this can happen because of limitations of the current biotechnologies as well as human biases and errors. However, this problem is not limited to BNs and instead has been studied extensively in other network domains, such as social networks [62, 63, 64]. Hence, many computational approaches, so called link prediction methods (defined below), have been used to help de-noise a network. This is important because the noisiness of network data can hinder the extraction of meaningful knowledge and accuracy of predictions made from network topology.

In simple words, link prediction can be defined as computational de-noising of a network data by identifying missing and spurious links, in order to improve network quality [65, 66, 67]. Intuitively, for every two nodes in a network, a link prediction method uses some measure of network topology of the given node pair, in order to determine whether the two nodes should be connected or not. So, for an input network, some of its connected nodes might remain connected, while others might get disconnected if judged to be false positives by the given link prediction approach. Similarly, some of the network's disconnected nodes might remain disconnected, while others might get connected if judged to be false negatives by the given link prediction approach.

Traditional link prediction methods assume that it is desirable to link nodes with high degrees as measured by preferential attachment, nodes that share many neighbors as measured by Jaccard or Adamic/Adar coefficients, or nodes that share many paths as measured by Katz index ([28] give formal definitions of these and other link prediction measures). See Figure 5.11 for an illustration. Clearly, traditional methods typically link nodes that have large shared neighborhoods, typically only immediate ones. However, a shared neighbor between two nodes is simply a 3-node path whose end-nodes are the two nodes in question, and a 3-node path is simply one of the 3-node graphlets (specifically, graphlet G_1). So, recently, it was argued that in addition to considering only 3-node paths that two nodes share, i.e., only the nodes' immediate shared neighborhoods, larger shared graphlets should also be considered, in order to capture the nodes' extended shared neighborhoods. So, graphlets were used as a basis for a new link prediction approach [28]. In addition to linking nodes that have large shared extended neighborhoods, as captured by the measure of node-pair-GDV-centrality (Section 5.2.2), this new approach also linked nodes that are topologically similar, as captured by the GDV-similarity measure. The graphlet-based approach outperformed other state-of-the-art existing methods in many tests. In the process, it was shown that the graphlet-based approach improved the quality of the currently noisy BN (specifically PPI network) data, in the sense that its predicted edges resulted in higher gene ontology (GO) enrichment than the edges in the original PPI network data. Also, a statistically significantly high portion of its predicted edges could be validated in independent, external PPI data sources [28].

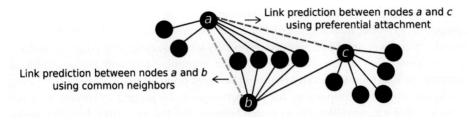

Figure 5.11: An illustration of network de-noising using two popular types of link prediction: shared neighbors and preferential attachment. Because nodes a and b share many neighbors (and also because most of each node's neighbors are shared with the other node), shared neighborhood-based methods will likely predict the existence of a link between a and b (shown in blue). Because nodes a and c share no common neighbors, shared neighborhood-based methods will unlikely predict the existence of a link between a and c. However, because nodes a and c both have high degrees, preferential attachment-based methods will likely predict the existence of a link between a and c (shown in red). Regarding shared neighborhood-based methods, which quantify in some way the amount of shared neighbors between two nodes, note that a shared neighbor between two nodes is nothing but a 3-node path whose end-nodes are the two nodes in question. And a 3-node path is nothing but one of the 3-node graphlets (specifically, graphlet G_1). So, why just count the number of 3-node paths that two nodes share? Why not also count the number of other 3-, 4-, or 5-node graphlets that the two nodes share? In this way, one can more capture the size of the shared neighborhood of two nodes in more detail, and in particular the size of the extended (as opposed to direct, as with 3-node paths only) shared neighborhood. This is exactly what motivated the development of the graphlet-based link prediction approach by Hulovatyy et al. [28] discussed in this chapter.

5.3.4 Network Comparison

Of all computational problems, graphlets have had the most impact on the problem of network comparison. Network comparison is an extremely popular network science problem with a number of practical applications. There are two types of network comparison: Alignment-free and alignment-based. Given two networks, alignment-free network comparison aims to quantify similarity between the compared networks independent of any mapping between their nodes, typically by extracting from each network some network patterns (also called network properties, network features, network fingerprints, or measures of network topology) and comparing the patterns between the networks. For example, if graphlet counts of two networks match, then the networks can be considered to be similar. On the other hand, alignment-based network comparison explicitly aims to map nodes between the compared network in a way that conserves the maximum amount of common substructure. In the process, alignment-based comparison approaches can be used to quantify similarity between the compared networks, but they do so under their resulting node mappings, even though they are not necessarily directly designed for this purpose. Importantly, alignment-free and alignment-based network comparison differ in their computational complexities. Usually, alignment-based comparison is much more computationally complex than alignment-free comparison, because of the need to find a node mapping. Hence, if the purpose is to just quantify the network similarity, without the explicit requirement of knowing an actual node mapping between the compared networks, then alignment-free comparison should be used while alignment-based comparison should be avoided, especially when dealing with large networks.

An illustration of the difference between alignment-free and alignment-based comparison is as follows. Given a mapping between nodes of two networks, according to alignment-based comparison, for the networks to be judged as similar, it is not sufficient that e.g., graphlet counts match between the two networks, but also, nodes that are mapped to each other must have similar graphlet profiles. Let us use a more specific illustration of a network G with nodes $g_1, g_2, g_3, \ldots, g_8$ and a network H with nodes $h_1, h_2, h_3, \ldots, h_8$, where g_1 is mapped to h_1, g_2 is mapped to h_2, g_3 is mapped to h_3, $\ldots$, and g_8 is mapped to h_8. Let us assume that the two networks have the same number of triangles, where we are focusing only on a triangle as a representative graphlet. Also, let us assume that the triangles appear in G only between nodes g_1, g_2, g_3, and g_4, but they appear in H only between nodes h_5, h_6, h_7, and h_8. Then, according to alignment-free comparison that ignores any node mapping, the two networks will still be similar in terms of their triangle counts, but according to alignment-based comparison that accounts for the node mapping, the two networks will be dissimilar, as the triangles appear between different nodes across the two networks. Note, however, that the key goal of alignment-based comparison is to find a good node mapping that identifies similar (conserved) network regions, and a good alignment-based approach would thus recognize the above node mapping from our illustration to be a poor one, as it maps triangles in one network to non-triangles in another network, i.e., it maps dissimilar network regions to each other.

Below, we discuss first alignment-free and then alignment-based network comparison approaches that use graphlets (the latter are also discussed in Chapter 9). Importantly, while in addition to graphlet-based network comparison approaches, other types of approaches exist, the graphlet-based approaches are the state-of-the-art, in both alignment-free and alignment-based network comparison.

5.3.4.1 Alignment-free Network Comparison

Relative graphlet frequency distance (RGFD). Recall from Section 5.2.3.1 that RGFV of network N, $RGFV(N)$, counts, for each graphlet i, its relative frequency as $RGFV_i(N) = GFV_i(N)/T(N)$. Given two networks G and H and their respective RGFVs, RGFD between the networks is $RGFD(G,H) = \sum_{i=1}^{29} |F_i(G) - F_i(H)|$, where $F_i(G) = -\log(RGFV_i(G))$. The logarithms are used because frequencies of different graphlet types can differ by several orders of magnitude, in order to prevent the distance measure to be dominated by the most frequent graphlets. Intuitively, RGFD captures the difference in the occurrence of each graphlet type between the compared networks. Since RGFD is a distance measure, the lower its value, the more similar the compared networks. For more details, see [12].

Graphlet degree distribution agreement (GDDA). Recall from Section 5.2.3.2 that GDD of orbit i measures the distribution of the number of nodes touching orbit i. GDDA compares GDDs between two networks to quantify the networks' similarity. Specifically, first, GDD of orbit i of one network and GDD of the same orbit of the other network are compared, resulting in the ith GDDA between the two networks.

Then, given i^{th} GDDAs for all orbits, total GDDA is computed as the arithmetic or geometric mean of all orbit-specific GDDAs. GDDA is scaled to always have a value between 0 and 1, where the higher the value, the higher the similarity between the compared networks (because GDDA is an agreement measure, unlike RGFD, which is a distance measure). For more details and a formal mathematical definition of GDDA, see [14].

Netdis. Recall from Section 5.2.3 the notion of GFV of a network, which counts the number of occurrences of each graphlet type in the whole network. Similarly, GFV can be defined for each individual node of a network, which counts the number of different graphlet types that a node touches (this is a version of GDV of a node that ignores graphlet orbit information). Netdis uses all nodes' GFVs to summarize the topology of an entire network into its topological feature vector, as follows. Specifically, first, for each node v in a network, Netdis counts v's *restricted*-GFV, where only those graphlets are counted that include nodes which are at most a distance of two from v. Second, given restricted-GFVs for all nodes, Netdis "normalizes" the counts for each graphlet type according to the expected counts in a "gold-standard" null network. Finally, Netdis sums up the updated counts for each graphlet type over all nodes in the network, obtaining a single representative vector for the entire network. Given two networks, Netdis uses their respective representative vectors to compare the networks [68]. Note that it was argued the reliance of Netdis on a null network model is one of its key drawbacks, because different choices of null model typically yield different results [11].

Graphlet correlation distance (GCD). Recall from Section 5.2.3 the notion of GCM of a network, which is an $o \times o$ matrix containing all pairwise correlations between the o orbits over all nodes in the network. Given two networks G and H and their respective GCMs, GCD is defined as the Euclidean distance of the upper triangle matrix values of GCM_G and GCM_H. GCD has been shown to outperform both RGFD, GDDA, and Netdis in the task of alignment-free network comparison, even though all three measures are based on graphlets and in particular on the GDV matrix (Figure 5.7) [15].

Graphlet-based alignment-free network approach (GRAFENE). Unlike the above approaches, GRAFENE can compare both regular (unordered) and ordered networks, by relying on original (unordered) and ordered graphlets, respectively. Also, unlike the above approaches that can compare only two networks at a time, GRAFENE can compare multiple networks simultaneously (i.e., in a single step) to compute their pairwise similarities. GRAFENE uses several different variations of unordered or ordered GFV of a network (see [37] for details). For a given GFV version, given a set of networks to be compared, GRAFENE first computes GFV for each network and combines all GFVs as rows of a GFV matrix. Then, it performs principal component analysis (PCA) on the GFV matrix to compute principal components for each network and picks the first r principal components, where the value of r is at least two and as low as possible so that the r components account for at least 90% of variation in the data. For every pair of networks G and H, GRAFENE computes their cosine similarity, $S_{cosine}(G,H)$, based on networks' first r principal components. If a distance (rather than similarity) measure is preferred, GRAFENE also computes the distance between the networks G and H as $D_{cosine}(G,H) = 1 - S_{cosine}(G,H)$. In the context of PSN comparison, GRAFENE was shown to outperform the other graphlet-based network comparison methods, namely RGFD, GDDA, and GCD. For details, see [37].

Graph kernels. Another category of computational approaches (rather than a specific approach) that has been extensively used in alignment-free network comparison is that of graph kernels [69, 70]. Intuitively, a graph kernel $k(G,H)$ is a measure of similarity between graphs G and H, which is defined as the inner product $< \phi(G), \phi(H) >$ of functions $\phi(G)$ and $\phi(H)$, where the functions $\phi(G)$ and $\phi(G)$ are feature representatives of networks G and H, respectively. For example, $\phi(G)$ and $\phi(H)$ can be the networks' GFVs or their GDV matrices. Indeed, a graphlet-based graph kernel was used to successfully capture graph (dis)similarities and was shown to outperform other graph kernels [71].

5.3.4.2 Alignment-based Network Comparison: Network Alignment (NA)

Chapter 9 discusses the topic of NA in detail. Here, we complement that discussion by focusing on graphlet-based NA approaches.

Genomic sequence alignment has revolutionized our biological understanding. It aims to identify regions of similarities between compared sequences that may be a consequence of functional or evolutionary relationships between the sequences. Analogously, biological NA aims to identify regions of similarities between compared networks (e.g., PPI networks of different species) that may be a consequence of functional or evolutionary relationships between the species. Then, NA can be used to transfer biological knowledge from a network region of a well-studied species to its aligned (similar, conserved) counterpart in a poorly-studied species [72, 73], resulting in this way in learning new biological knowledge about the poorly-studied species.

Chapter 9 shows that analogous to genomic sequence alignment, NA can be local and global, where each NA category has its own (dis)advantages (which is why recently their integration was proposed [74, 75]). All graphlet-based NA methods are global, so the focus of this section is only on global NA, which aims to find a one-to-one mapping between the nodes of the compared networks that identifies large conserved network regions (e.g., that conserves lots of edges between the networks). Chapter 9 also shows that NA can be pairwise and multiple. All but one of the graphlet-based NA methods are pairwise [73, 76], which is why this section focuses on pairwise NA, unless otherwise mentioned.

Optimizing node conservation (NC) via two-stage NA methods. A traditional algorithmic idea behind NA is to: (1) compute similarities between nodes across compared networks with respect to some node cost function (NCF) and (2) quickly identify via an alignment strategy (AS) from all possible alignments a high-scoring alignment with respect to the node similarities, i.e., total NCF over all aligned nodes, which is also known as NC [73, 77]. The resulting NCF-AS NA methods are known as two-stage aligners. While most of such methods allow for using within their NCF biological information external to network topology, such as protein sequence similarities, in this section, the focus is on topological similarity-based NCFs. Graphlet-based two-stage NA methods are GRAAL [23], H-GRAAL [22], and MI-GRAAL [78], all of which use GDV-similarity as their NCF and differ mostly in their ASs. These methods were shown to produce higher-quality alignments compared to other, non-graphlet-based two-stage NA methods, such as IsoRank [77]. For a description of these and additional methods, see Chapter 9.

Most of the two-stage NA methods have their own NCFs and AS. So, when an NA method is found to be superior to another, it is unclear whether this superiority comes from its NCF, its AS, or both. To address this issue, it was proposed that in order to fairly evaluate different two-step NA methods, their different NCFs should be evaluated against each other under the same AS, for each AS, and also, their different ASs should be evaluated against each other under the same NCF, for each NCF [32, 42]. Only by doing so, it can be properly evaluated which NCF or AS is superior. By mixing and matching NCFs and ASs of prominent NA methods, namely MI-GRAAL (which is graphlet-based), IsoRank (which is not graphlet-based), and GHOST (which is not graphlet-based [41]), MI-GRAAL's GDV-similarity-based NCF was shown to be superior, even though IsoRank's and GHOST's NCFs also measure the similarity of nodes' extended network neighborhoods, just like GDV-similarity. However, unlike MI-GRAAL's NCF that is based on graphlets, IsoRank's and GHOST's NCFs are based on PageRank-like and spectral graph theoretic "topological signatures," respectively. The superiority of MI-GRAAL's NCF over the other two NCFs confirms the effectiveness of using graphlets. When comparing the three methods' ASs, no AS was consistently superior, as the performance was dependent on the choice of data or evaluation criteria [32, 42].

Optimizing edge conservation (EC) via search-based NA methods. Traditional NA methods, which are of the two-stage type, identify from possible alignments the high-scoring alignments with respect to total NCF over all aligned nodes (i.e., NC), but they then evaluate the accuracy of the resulting alignments via some other measure that is different than the NCF used to construct the alignments. Typically, the amount of conserved edges is measured, referred to as EC. Thus, the traditional methods align similar nodes between networks hoping to conserve many edges, but only after the alignment is constructed. Instead, MAGNA [25], a search-based (rather than two-stage) NA method, was recently introduced to directly optimize EC while the alignment is constructed. A search-based method can directly optimize EC or any other alignment quality measure by relying on an optimization strategy such as a genetic algorithm (in the case of MAGNA) or simulated annealing (in the case of e.g., SANA [79]). MAGNA improved upon the two-stage methods that existed at the time of its introduction.

Optimizing both NC and EC via two-stage or search-based NA methods. Even more recently, it was recognized that optimizing a combination of NC and EC could further improve alignment quality. So, MAGNA was recently extended into MAGNA++ framework [26], to allow for simultaneously optimizing both NC and EC, which indeed further improved alignment quality; importantly, its NC is graphlet-based, namely it is the GDV-similarity measure. We note that MAGNA++ provides a user-friendly graphical interface for domain (e.g., biological) scientists, plus source code for easy extensibility by computational scientists. Just like MAGNA, MAGNA++ is a search-based aligner. MAGNA++ was recently generalized into additional graphlet-based NA methods: MultiMAGNA++, to allow for multiple (as opposed to pairwise)

NA [80], as well as into DynaMAGNA++, to allow the first time ever for alignment of dynamic (as opposed static) networks [27].

Even two-stage (rather than search-based) NA methods were proposed that can optimize a combination of NC and EC, namely WAVE [81], L-GRAAL [82], and GREAT [83], all of which are graphlet-based. For more details on these methods, see Chapter 9 as well as [73].

Measures of alignment quality, and evaluation and comparison of different NA methods. See Chapter 9 for a detailed description on how to measure quality of an NA method's output, i.e., its alignment, and how to compare NA methods of the given type against each other (local against local, global against global, pairwise against pairwise, or multiple against multiple NA methods). In addition, for further discussion on across-type comparison, e.g., comparison of local against global NA methods, see [74, 75], where it was shown that local NA is complementary to global NA, because the former yields high functional but low topological alignment quality, while the latter yields high topological but low functional alignment quality, and because the two result in different protein function predictions. Similarly, for further discussion on comparison of pairwise against multiple NA methods, see [76], where it was shown that pairwise NA is both more accurate and computationally more efficient (i.e., faster) than multiple NA, which questions the usefulness of the current multiple NA methods.

5.4 Biological Applications of the Graphlet Measures

5.4.1 Protein Function Prediction

Proteins are the macromolecules of the cell that keep us alive by carrying out important cellular functions. Clearly, understanding protein function is important. However, doing so experimentally is expensive and time consuming. Consequently, many proteins remain functionally uncharacterized (or unannotated) [84, 85]. Computational prediction of protein function can help address this, as it is inexpensive and typically fast. Thus, computational protein function prediction [86], and in particular via network analyses [87], has received significant attention. There are several general strategies of how protein function can be predicted from network topology, and graphlets have been used for this purpose, as follows.

Of all BN types, PPI networks have been analyzed the most to explore physical interactions between proteins and predict functions of currently unannotated proteins based on how well their interacting patterns match those of annotated proteins. In particular, proteins with similar graphlet-based topological characteristics in a PPI network have been shown to be functionally related: GDV-similarity was used to group (cluster) topologically similar proteins in a PPI network (Section 5.3.1), and the proteins within the given cluster were shown to belong to the same protein complexes, carry out the same biological functions, be localized in the same subcellular compartments, and have the same tissue expressions [16]. This was confirmed by clustering proteins in the *Saccharomyces cerevesiae* (baker's yeast) PPI network or the human PPI network.

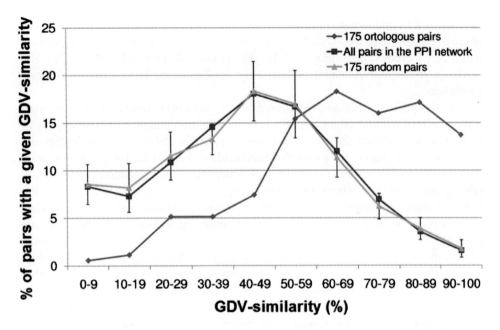

Figure 5.12: GDV-similarity distributions for all 175 orthologous protein pairs (blue), all protein pairs in the PPI network (red), and 175 protein pairs randomly selected from the network (green). The procedure of random pair selection (green) was repeated 30 times, and GDV-similarities were averaged over the 30 runs; green points represent these averages, and bars around the points represent the corresponding standard deviations. The figure is adapted from Memišević et al. [88].

In addition, GDV-similarity was used to measure topological similarity between orthologous proteins (those proteins within a species that have similar sequences), as well as between non-orthologous proteins, in the baker's yeast PPI network, and it was shown that homologous proteins have statistically significantly high GDV-similarities, unlike non-homologous proteins (Figure 5.12) [88]. In other words, sequence similarity correlates well with PPI topological similarity, although each similarity type has some unique aspects, meaning that not all proteins that are topologically similar are also sequence-similar, and vice versa. Given this, and given that traditional computational protein function prediction strategies have heavily relied on sequence-based comparisons (by transferring function from an annotated protein to an unannotated protein that is sequence-similar to it), prediction of protein function based on proteins' topological similarities can complement the sequence-based prediction strategies.

Apart from protein function prediction *within a single PPI network*, i.e., the transfer of function from an annotated protein to a topologically similar unannotated protein from the same network (as above), protein function prediction can be done *across two or more PPI networks*. Specifically, analogous to genomic sequence alignment, biological NA (Section 5.3.4 and Chapter 9) has also been used for protein function prediction. In general, given two or more aligned PPI networks of different species (e.g., *S. cerevesiae* (yeast), *Drosophila melanogaster* (fruit fly), *Caenorhabditis elegans* (worm), and *Homo sapiens* (human)), functional knowledge can be transferred from annotated proteins in one network to unannotated proteins in the other network(s) between the networks' aligned (typically this means GDV-similar) network regions. Indeed, several graphlet-based NA methods, including GRAAL [23], H-GRAAL [22], MI-GRAAL [78], and MAGNA and MAGNA++ [25, 26] have been applied to the task of protein function prediction.

Besides protein function prediction from PPI networks, where in a PPI network nodes are proteins and edges are PPIs between the proteins, protein function can be predicted from protein structures [89], and specifically PSNs, where each PSN is a protein whose nodes are the protein's amino acids and edges model spatial proximity between the amino acids in the protein's 3-dimensional (3D) structure [18, 19, 20]. In other words, each PSN, which is a network itself, is a node of a PPI network. In a recent work, an alignment-based network comparison (i.e., NA) framework called GR-Align was proposed to align and consequently compare topological similarities between pairs of PSNs (i.e., proteins), where GR-Align is based on 2–3-node ordered graphlet GDV-similarities between nodes (i.e., amino acids) of the compared PSNs [19]. In a related work, an alignment-free network comparison framework called GRAFENE was proposed that is based on 3–4-node ordered graphlet RGFVs of the compared PSNs [37]. It was shown that GR-Align and GRAFENE are the two best PSN comparison approaches among a number of graphlet-based network approaches, non-graphlet-based network approaches, non-network 3D-structural approaches, and non-network sequence approaches [37]. Given pairwise PSN similarities given by GR-Align or GRAFENE, analogous to sequence similarity-based protein function prediction [90], function can then be transferred from annotated PSNs (i.e., the corresponding proteins) to unannotated PSNs that are topologically similar to them.

5.4.2 Aging

As an organism ages, its susceptibility to complex diseases increases. Clearly, studying aging is important. However, human aging is difficult to study experimentally in a wet lab due the long human life span and ethical constraints. Hence, computational approaches are necessary for predicting novel human aging-related knowledge. Two traditional ways of computationally predicting aging-related knowledge are: Differential gene expression analyses in human at different ages [91], and transfer of aging-related knowledge from well-studied model species to poorly-studied human between aligned sequence regions [92] (Figure 5.13). However, recently, network-based equivalents of these two analyses types were proposed (Figure 5.13), with focus on graphlet-based analyses [29, 30, 31, 32]. Below, first, we discuss traditional analysis of the static human PPI network, followed by more recent dynamic analysis of the human PPI network at different ages. Second, we discuss the applications of network (rather than sequence) alignment to the transfer of aging-related knowledge from model species to human.

5.4.2.1 Static Analysis of the Human PPI Network in the Context of Aging

Current PPI network data are static. In the static human PPI network, with the hypothesis that topologically central proteins in the network are also biologically central

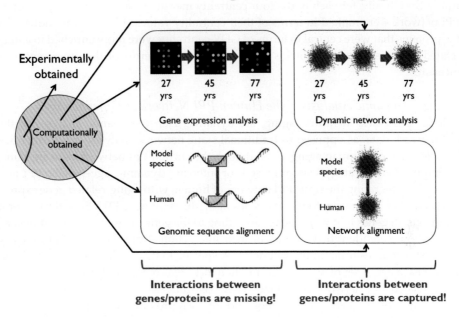

Figure 5.13: Traditional non-network based versus recent network-based analyses of human aging. The current knowledge about human aging has been obtained mostly computationally (rather than experimentally, which is hard to do due to long life span of human as well ethical constraints), by either: (1) identifying human genes that are differentially expressed at different ages, which are then predicted as aging-related candidates (top left), or (2) transferring aging-related knowledge from genes of well-studied model species to genes of poorly-studied human between the genes' aligned sequence regions (bottom left). However, both of these types of analyses ignore valuable information about interactions between genes, even though genes (i.e., their protein products) carry out biological processes by interacting with each other rather than acting alone. So, it was recently proposed to utilize BN data (and specifically PPI network data) to predict aging-related knowledge by: (1) integrating the current static PPI network of human with aging-related gene expression data to infer dynamic, age-specific PPI networks, and then studying changes in network structure (rather than gene expression) with age to predict as aging-related those genes whose PPI network positions (i.e., centralities) significantly change with age (top right), or (2) aligning PPI networks of well-studied model species and poorly studied human via NA and then transferring aging-related knowledge from model species to human between the aligned network (rather than sequence) regions (bottom right). Clearly, both of these types of analyses explicitly account for interactions between proteins. Also, they partly rely on gene expression data (in the context of dynamic network analysis) and sequence data (in the context of NA). Hence, they are likely to further our knowledge about human aging compared to the traditional gene expression- or sequence-based analyses.

(i.e., involved in critical biological processes, such as aging), the measure of GDV-centrality was used to identify topologically central proteins, which were then shown to be significantly more enriched in known aging-related proteins than the non-central ones [29]. In this context, GDV-centrality was shown to outperform other popular node centrality measures, i.e., degree centrality, betweenness centrality, and subgraph centrality. Specifically, for each of the four centrality measures, all proteins in the human PPI network were ranked in terms of their centrality values. Then, it was shown that the proteins that were central in terms of GDV-centrality were more enriched in aging-related proteins than the proteins that were central in terms of the other centrality measures.

5.4.2.2 Dynamic Analysis of the Human PPI Network at Different Ages

Cellular processes, including aging, are dynamic. Hence, dynamic analyses of the PPI network data, as opposed to traditional static network analyses, can further our knowledge about human aging. Even though the current PPI network data are static, a computational strategy was proposed for inferring dynamic, age-specific PPI networks, by integrating the static PPI network of human with aging-related gene expression data that captures activities of genes at 37 different ages [31]. Specifically, for a given age, all genes that were active (according to the gene expression data) at the age in question were identified, and all of their interactions from the static PPI network were extracted, resulting in the PPI network specific to the age in question. By doing this for each age, a set of age-specific PPI networks was obtained, forming a dynamic, aging-related PPI network of human. Then, centralities of proteins were measured in each age-specific network snapshot, and for each protein, it was analyzed how its centralities changed with age. Proteins whose centralities significantly changed with age were predicted as aging-related candidates, and the predictions were validated in several ways, e.g., by contrasting their biological functions against biological functions of known aging-related proteins, or via literature search. In the above analysis, seven different centrality measures were used, including GDV-centrality. Again, GDV-centrality was shown to outperform the other six centrality measures [31]. Specifically, GDV-centrality was able to predict almost as many aging-related candidates as all seven centrality measures combined, meaning that the other six centrality measures rarely contributed any additional aging-related knowledge that GDV-centrality was not able to capture itself.

While the above analysis dealt with dynamic data, each age-specific network snapshot (which itself is static) was analyzed in isolation from the others. Only the time series of the results (i.e., temporal changes of node centrality values) were analyzed in a dynamic fashion, but the results were produced in a static manner, ignoring relationships that exist between the different age-specific network snapshots. To overcome this, the concept of *dynamic*-GDV was used to study aging from the dynamic, age-specific PPI network of human (the same network as above) [30]. Specifically, dynamic-GDV was computed for each protein in the network, where dynamic-GDV spans multiple snapshots at once (because dynamic graphlets capture information on temporal events rather than static edges, and because a given dynamic graphlet can capture multiple temporal events, corresponding to different snapshots). Given the resulting dynamic-GDV matrix, PCA was used to reduce the dimensionality of the matrix. Finally, Euclidean distance was computed between each pair of proteins based on the proteins' PCA-reduced dynamic-GDVs. Intuitively, it was found that known aging-related proteins were highly dynamic-GDV-similar to each other, and this observation was used to predict as novel aging-related candidates those proteins that were not yet implicated in aging but that were dynamic-GDV-similar to known aging-related proteins [30].

5.4.2.3 Transfer of Aging-related Knowledge from Model Species to Human via Network Alignment (NA)

Just as NA was used to transfer protein functional knowledge from a well-studied species to a poorly-studied species between the species' aligned PPI networks (Section 5.4.1), NA was used in a similar manner for across-species transfer of aging-related knowledge [32]. Specifically, PPI networks of yeast, fruit fly, worm, and human were aligned with two popular NA methods at the time, MI-GRAAL [78] and IsoRankN [40], with the goal of transferring aging-related knowledge from known aging-related genes in one species to their aligned partners in another species. In the process, because MI-GRAAL and IsoRankN have both different NCFs and different ASs, their NCFs and ASs were mixed and matched (Section 5.3.4), which led to the finding that MI-GRAAL's NCF, which is based on GDV-similarity, was more accurate under both MI-GRAAL's AS and IsoRankN's AS in the context of predicting aging-related knowledge [32]. Hence, GDV-similarity, i.e., graphlets, were successfully used in the context of NA to study aging.

5.4.3 Disease

5.4.3.1 Cancer

With the hypothesis that cancer proteins would have different PPI network characteristics (i.e., wiring patterns) than non-cancer proteins, it was tested whether cancer proteins are more central than non-cancer proteins. Indeed, it was shown that GDV-central proteins were significantly more enriched in cancer proteins than non-GDV-central proteins were [29]. Additionally, with the same hypothesis, it was tested whether cancer proteins are more topologically similar to each other than to non-cancer proteins. Indeed, clusters of cancer proteins were found to match well with the clusters of GDV-similar proteins [33].

In a follow-up study [34], an RNA-mediated interference (RNAi)-based functional genomics dataset related to melanogenesis, a process of skin pigment production whose misregulation may be linked to skin cancer, was analyzed via a graphlet-based network approach. The goal was to identify novel melanogenesis-related regulators, i.e., targets within the RNAi dataset that may be components of known melanogenesis regulatory pathways. Specifically, the human PPI network was clustered to obtain groups of GDV-similar proteins, and a set of targets was identified that clustered with the known pigment regulator endothelin receptor type B (EDNRB). Importantly, the predicted regulators were validated *phenotypically*, i.e., experimentally [34].

On a related note, it was found that pairs of disease proteins in the human PPI network share significantly more graphlets (specifically, graphlet interactions – see below) than pairs of randomly selected proteins in the network [35]. Intuitively, two nodes in the same graphlet can be considered to be linked with each other even though there might be no direct edge between them. This new definition of node linkage is called graphlet interaction. Based on this observation, new disease proteins were predicted that are related to four common diseases: breast cancer, colorectal cancer, prostate cancer, and diabetes.

5.4.3.2 Pathogenicity

Pathogens generally cause disease by interacting with proteins of host organisms. Identifying these proteins can aid understanding of the disease mechanisms as well as designing new therapeutic strategies (because pathogen-interacting proteins in the host may be likely drug target candidates). Graphlet-based measures of PPI network topology have been used to study pathogen-interacting proteins. First, just as for aging and cancer, it was shown that GDV-central proteins in the human PPI network were significantly more enriched in pathogen-interacting proteins than non-GDV-central proteins [29]. Second, the measures of GDV-similarity and edge-GDV-similarity were used to cluster topologically similar nodes or edges, respectively. Then, new pathogen-interacting proteins were predicted from clusters that were significantly enriched in known pathogen-interacting proteins [17]. Specifically, whenever significantly many nodes in a given node (or edge) cluster were known pathogen-interacting proteins, the remaining nodes in the same cluster were predicted as pathogen-interacting candidate proteins.

5.4.4 Health-related Applications Beyond Computational Biology: Social Networks

Beyond computational biology, network science has applications in many other domains, including social sciences [1, 2]. Social networks capture relations between individuals, where nodes represent individuals and edges represents some type of interactions between individuals. Examples of social networks are a physical contact network (where individuals are linked by an edge if they interact in a face-to-face manner), Facebook network (where individuals are linked if they are friends on Facebook), email network (where individuals are linked if they communicate via email), etc.

A popular research question in social sciences is to understand the relationship between individuals' social networks and their traits such as gender, physical appearance, body mass index (BMI), personality traits, or various types of behaviors. Specifically, the key question is whether people form social links because they are similar, or they become similar after forming social links (or perhaps it is a mix of the two). A particular direction within the above general network–trait question has been to study the relationship between social networks and health. Several studies have been pursued in this direction, which showed, for example, that social network phenomena appear to be relevant to the biological and behavioral traits such as obesity or smoking, as these appear to spread through social interactions [93, 94, 95, 96, 97, 98, 99]. These findings have implications for clinical and public health interventions. A recent such study is graphlet-based [100], which analyzed a dynamic social network resulting from the NetSense study, measured changes in people's social network patterns (captured via the notion of node centrality) over time, clustered nodes whose evolving centrality profiles are similar, and linked the social network-based clusters of people with people's trait-based clusters (Figure 5.14). Five different network (link) types were

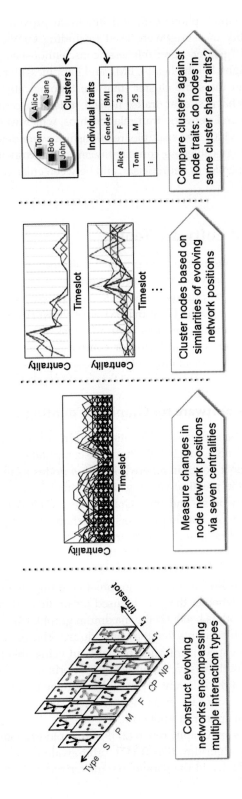

Figure 5.14: A summary of the study by Meng et al. [100] who used dynamic network analysis (including studying changes of nodes' GDV-centralities over time) to link people's social networks to their traits. In the figure, "S," "P," "M," "F," and "CP"/"NP" denote "sms," "phone call," "email," "Facebook," and "close"/"near Bluetooth proximity" communication types, respectively, that exist in the analyzed social network. The figure is adopted from Meng et al. [100].

analyzed (sms, Facebook, email, phone call, and Bluetooth proximity communications), seven node centrality measures were used (including GDV-centrality), and 11 traits were considered (gender, agreeableness, conscientiousness, extraversion, neuroticism, openness, weight, height, health, BMI, and happiness). That is, the study hypothesized that nodes in the same cluster (that are topologically similar in the dynamic network) would have similar traits. Indeed, this was confirmed, which resulted in detection of several network–trait relationships, which were validated in e.g., in the recent literature [100]. For example, relationships were detected between phone call interactions and agreeableness, conscientiousness, or extraversion, between Facebook interactions and gender or neuroticism, and between email interactions and BMI, among several others.

5.5 Graphlet-based Software Tools

Up to this point, we have demonstrated the practical usefulness of graphlets in network science and systems biology. So, how to count graphlets in a network? In this section, first, we focus on general-purpose software tools that "simply" count graphlet measures of nodes or edges (e.g., the GDV matrix) or of networks (e.g., RGFV of a network), whose output can then be used as input into task-specific graphlet-based methods such as those for network clustering or comparison. Second, we provide information about software implementations of the task-specific methods.

5.5.1 General-purpose Software for Graphlet Counting

The original graphlet counting software tool is called GraphCrunch [49]. This tool has its updated version called GraphCrunch 2 [9]. Since the computational complexity of counting a given graphlet on k nodes in a network with n nodes increases exponentially as $\mathcal{O}(n^k)$ – although the practical running time is much lower because of the sparseness of real-world networks – and since GraphCrunch (2) is one of the earliest graphlet counting software tools that implements the exhaustive graphlet counting procedure, several additional more recent software tools and approaches have been proposed that aim at reducing the time complexity of graphlet counting in clever algorithmic ways, as follows.

We discuss and compare several tools for graphlet counting, highlighting their functionalities based on: (1) whether they can be used for undirected graphs [46, 47, 48, 49], directed graphs, or both [44, 45], (2) the maximum graphlet size that they can handle, (3) whether the graphlet counts that they provide are orbit-unaware (i.e., they can "just" count the graphlet-specific information, without being able to distinguish between the different orbits within a given graphlet) [46] or orbit-aware (meaning that they can distinguish between the different orbits within a given graphlet) [44, 45, 47, 48, 49], and (4) whether they count the exact [44, 46, 47, 48, 49] or only approximate [45] number of graphlets that appear in a network.

The tools that we discuss are GraphCrunch (2) [9, 49], rapid graphlet enumerator (RAGE) [48], orbit counting algorithm (Orca) [47], orbit-aware quad census (Oaqc) [44], SAND and SAND-3D [45], and the parallel parameterized graphlet decomposition (PGD) library [46].

Table 5.1: The summary of the functionalities of the considered general-purpose graphlet-counting tools. Under "graph type," "U" stands for "undirected" and "D" stands for "directed," respectively.

Software tool	Graph type	Graphlet size	Orbit aware	Exact count	Link to software
Graph Crunch	U	3–5	Yes	Yes	www0.cs.ucl.ac.uk/staff/natasa/graphcrunch
Graph Crunch 2	U	3–5	Yes	Yes	www0.cs.ucl.ac.uk/staff/natasa/graphcrunch2
RAGE	U	3–4	Yes	Yes	www.eng.tau.ac.il/$\sim$shavitt/RAGE/Rage.htm
Orca	U	4–5	Yes	Yes	http://biolab.si/supp/orca
Oaqc	U/D	3–4	Yes	Yes	https://cran.r-project.org/web/packages/oaqc/index.html
SAND	U	3–4	Yes	No	http://nskeylab.xjtu.edu.cn/dataset/phwang/code
SAND-3D	D	3	Yes	No	http://nskeylab.xjtu.edu.cn/dataset/phwang/code
PGD	U	3–4	No	Yes	http://nesreenahmed.com/graphlets

Of these tools, those that can deal only with undirected graphs (and thus undirected graphlets) are: GraphCrunch (2), RAGE, Orca, SAND, and PGD. SAND-3D can handle only directed networks. Oaqc can handle both directed and undirected networks. Only GraphCrunch (2) and Orca can count graphlets on up to five nodes, while the others can only handle graphlets on up to four nodes, except SAND-3D, which can count only graphlets on up to three nodes. All of these tools are orbit-aware except PGD. Finally, all of the tools but SAND and SAND-3D provide exact graphlet counts. For a visual summary and comparison of the different tools, see Table 5.1.

Next, we comment on some more specific functionalities of each of the above software tools.

GraphCrunch (2). GraphCrunch can compute all of the GDV matrix, GFV, and GDDs of a network (it can do this only from the perspective of nodes but not edges). Unlike any of the other tools discussed in this section, GraphCrunch can compute some of the graphlet-based network similarity measures, such as RGFD and GDDA. Also, unlike the other tools, GraphCrunch allows for comparing a real-world network to a number of random network models that are implemented in it, with respect to several measures of network similarity. GraphCrunch 2, in addition, allows for comparing two real-world networks to each other in both alignment-free and alignment-based fashion, as well as for clustering a network based on the nodes' GDV-similarities. GraphCrunch (2) can be run under Linux, MacOS, and Windows (although we note that for GraphCrunch, its Linux version has been tested the most, while for GraphCrunch 2, its Windows version has been tested the most, and thus, these are the recommended versions).

RAGE. To speed up the exhaustive graphlet counting algorithm of GraphCrunch (2), RAGE was proposed. RAGE counts all up to 4-node induced and non-induced orbit-aware subgraphs in undirected networks; specifically, it can compute the GDV matrix and GFV of a network (it can do this only from the perspective of nodes but not edges). In order to count all 4-node induced subgraphs, RAGE first needs to count all 4-node non-induced subgraphs. Then, it calculates the counts of induced subgraphs given the counts of non-induced subgraphs via a system of equations. The run time complexity of RAGE is $\mathcal{O}(d(G) \cdot m + m^2)$, where $d(G)$ is the maximum degree and m is the total number of edges in the graph G. RAGE is implemented in C++ and is provided as a command-line utility for both Windows- and Unix-based systems. It was demonstrated that RAGE is faster than GraphCrunch (2), as expected due to its clever graphlet counting algorithmic procedures.

Orca. Orca is a newer tool compared to both GraphCrunch (2) and RAGE. It can compute node- and edge-based GDV matrices of a network (but it cannot compute network-based graphlet statistics such as GFV). Orca employs a very clever combinatorial approach for graphlet counting that results in much lower running time, especially when counting 5-node graphlets [47]. Specifically, in order to count all $(k+1)$-node graphlets, Orca needs to compute all k-node graphlet counts and the count for a single $(k+1)$-node graphlet, e.g., the $(k+1)$-node complete graph. Then, it can automatically compute via mathematical equations (rather than by searching for graphlets in the network, like GraphCrunch (2) does) the counts for all other $(k+1)$-node graphlets. For computing counts for 4-node graphlets, the run time complexity of Orca is $\mathcal{O}(d(G) \cdot m + T_4))$, where $d(G)$ is the maximum degree of the graph G, m is the total number of edges in G, and $\mathcal{O}(T_4) = \mathcal{O}(d(G)^2 \cdot m)$ is the time complexity of counting all 4-node complete graphlets. For computing counts for 5-node graphlets, the time complexity is $\mathcal{O}(d(G)^2 \cdot m + T_5))$, where $\mathcal{O}(T_5) = \mathcal{O}(d(G)^4 \cdot n)$ is the time complexity of counting all 5-node complete graphlets and n is the number of nodes in the graph G. Orca is implemented in C++ and is provided as a command-line utility as well as an R package [101]. When compared in the task of counting 4-node graphlets, Orca significantly outperformed RAGE [47], owing to its clever combinatorial approach.

Oaqc. Oaqc is unique in the sense that it is the only one of the considered approaches that can deal with directed networks (although we note that additional implementations exist that can count directed graphlets, such as the approach that originally introduced directed graphlets [43]). Just as Orca, Oaqc can compute node- and edge-based GDV matrices of a network (but it cannot compute network-based graphlet statistics such as GFV). Similar to RAGE, in order to count all 4-node induced subgraphs, Oaqc first needs to count all 4-node non-induced subgraphs. Then, it calculates the counts of induced subgraphs given the counts of non-induced subgraphs via a system of equations. However, Oaqc defines faster algorithmic approaches to count 4-node non-induced subgraphs compared to RAGE, which results in Oaqc's time complexity of $\mathcal{O}(d(G)^2 \cdot m)$. Although theoretically both Oaqc and Orca have same time complexity for computing the 4-node graphlet counts, practically Oaqc was shown to outperform Orca [44], owing to its faster implementation approach.

SAND and SAND-3D. Unlike the above tools, which count exact numbers of subgraphs present in a network, both SAND and SAND-3D use sampling-based methods and can only approximate the different subgraph counts. The working principles of both the methods are similar. Intuitively, for each node v in a network, a given method does the following. First, the method randomly selects a predefined number of nodes in the local neighborhood of v such that the selected nodes, including the node v, are connected. Second, the method then determines the orbit position of node v in the graphlet comprising of the randomly selected nodes and the node v itself. This procedure is performed several times, which essentially gives an estimate of the number of times node v touches an orbit. The predefined number of nodes that needs to be randomly selected depends on the orbit whose count is to be estimated. For example, if the goal is to estimate the number of times node v touches orbits o_1, o_2, or o_3, then this predefined number of nodes equals two. Being sampling-based methods, although the computational time of SAND and SAND-3D is expected to be much lower in comparison to the other tools, the accuracy with which the graphlets are counted can be adversely affected [45]. SAND and SAND-3D can compute the node-based GDV matrix of a network (but they cannot compute its edge-based equivalent nor network-based graphlet statistics such as GFV).

PGD. All of the tools discussed thus far count orbit-aware graphlet statistics. However, if one's goal is to "just" count the orbit-unaware graphlet statistics, then performing orbit-aware counting will unnecessarily increase the running time, if the orbit-unaware statistics can be computed without first computing orbit-aware statistics. One of the tools that performs orbit-unaware graphlet counting, and that does so fast, is PGD [46]. PGD uses a fast parallel algorithm for counting 3–4-node graphlets in undirected networks. Specifically, PGD can compute three types of graphlet-based statistics of a network: its GFV (which by design is orbit-unaware), its RGFV (which by design is orbit-unaware), and the number of times each node or edge touches each graphlet (these are the orbit-unaware versions of the network's node- and edge-GDV matrices, respectively, which are expected to be less informative than their orbit-aware counterparts, as the latter capture more detailed topological information than the former, namely the orbit-specific information). PGD is implemented in C++ and is provided as a command-line utility for Unix-based systems.

5.5.2 Task-specific Graphlet-based Software

With the exception of GraphCrunch (2), all of the above tools output "only" node (or edge) or network-based graphlet statistics, such as the GDV matrix or GFV of a network, but they do not use these statistics for any specific task. Because they produce the graphlet statistics relatively fast, these tools' output can then be used as input to other task-specific software tools that are intended to e.g., compute GDV-similarity between all node pairs in a network, cluster nodes in a network, compute RGFD or GCD between two networks, or align two or more networks. Such task-specific software tools are summarized in Table 5.2.

Table 5.2: The summary of the availability of software implementations of task-specific graphlet-based methods.

Graphlet-based method	Link to software	Section
Directed graphlets	www0.cs.ucl.ac.uk/staff/natasa/DGCD	5.2.1
Dynamic graphlets	https://nd.edu/~cone/DG	5.2.1
Heterogeneous graphlets	https://nd.edu/~cone/colored_graphlets	5.2.1
Ordered graphlets	https://nd.edu/~cone/PSN	5.2.1
Node GDV matrix	http://www0.cs.ucl.ac.uk/staff/natasa/graphcrunch2	5.2.2
Edge, non-edge, and node-pair GDV	http://nd.edu/~cone/downloads/Solava_ECCB2012.zip	5.2.2
Node GDV-similarity	www0.cs.ucl.ac.uk/staff/natasa/	5.2.2
Edge GDV-similarity	http://nd.edu/~cone/downloads/Solava_ECCB2012.zip	5.2.2
Node GDV-centrality	https://nd.edu/ cone/centralities/nodecentrality.html	5.2.2
GFV	http://www0.cs.ucl.ac.uk/staff/natasa/graphcrunch2	5.2.3
GDD	http://http://www0.cs.ucl.ac.uk/staff/natasa/graphcrunch2	5.2.3
GCM, RGFD, GDDA, GCD	http://www0.cs.ucl.ac.uk/staff/N.Przulj/GCD	5.3.4
GRAFENE	https://nd.edu/~cone/PSN	5.3.4

5.6 Exercises

5.1 What are (dis)similarities between graphlets and network motifs? What are (dis)advantages of each?

5.2 Draw all 2–3-node:

(a) Directed graphlets.
(d) Dynamic graphlets with exactly four events.
(b) Heterogeneous node-colored graphlets with two colors.
(c) Ordered graphlets.

5.3 Recall the definition of an automorphism (orbit) of a graph.

(a) Show all automorphisms of the graph in Figure 5.15 (a) and explain how and why they result in a single automorphism orbit for this graph.
(b) Show all automorphism orbits of the graph in Figure 5.15 (b) (there is no need to first identify all of the graph's automorphisms).
(c) Show all automorphism orbits of the graph in Figure 5.15 (c) (there is no need to first identify all of the graph's automorphisms).

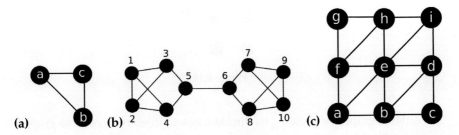

Figure 5.15

5.4 If n is any positive integer, count the number of:

(a) Automorphism orbits of an n-node path, P_n.
(b) Automorphisms of an n-node cycle, C_n, and an n-node clique (complete graph), K_n.

Justify your answers.

5.5 Compute (without using any software tool) the graphlet degree vector (GDV) of node x in Figure 5.16, for 2–4-node graphlets.

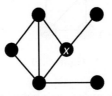

Figure 5.16

5.6 Consider the four graphs in Figure 5.17.

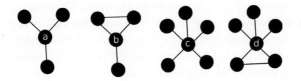

Figure 5.17

(a) What are degree centralities of nodes $a, b, c,$ and d? Order the nodes in terms of their decreasing degree centralities.
(b) What do you think is the order of the same nodes in terms of their decreasing GDV-centralities? Justify your answer. (You do not have to actually compute the nodes' GDV-centralities in order to answer.)

5.7 Compute (without using any software tool) the graphlet frequency vector (GFV) of the graph in Figure 5.18, for 2–4-node graphlets.
5.8 Similar to the definition of the GFV of an undirected, homogeneous, and unordered graph, we can define the GFV of a directed, heterogeneous, or

Figure 5.18

ordered graph, by counting its directed, heterogeneous, or ordered graphlets, respectively.

(a) Compute (without using any software tool) the GFV of the directed graph in Figure 5.19 (a), for 2–3-node directed graphlets.
(b) Compute (without using any software tool) the GFV of the node-colored graph in Figure 5.19 (b), for 2–3-node node-colored graphlets as defined by [13] and discussed in this chapter.
(c) Compute (without using any software tool) the GFV of the ordered graph in Figure 5.19 (c), for 2–3-node ordered graphlets.

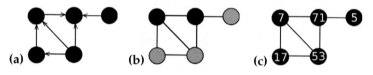

Figure 5.19

5.9 Recall the definition of an isomorphism between two graphs. Consider the two graphs in Figure 5.20.

(a) Are the two graphs isomorphic? Justify your answer.
(b) Compute (without using any software tool) the GFV of each of the two graphs for only 2–3-node graphlets. Do the GFVs of the two graphs match?
(c) Compute (without using any software tool) the GFV of each of the two graphs for 2–4-node graphlets. Now, do the GFVs of the two graphs match?
(d) If any two graphs have mismatching GFVs, can you answer with 100% certainty whether the graphs are isomorphic? Justify your answer.
(e) If any two graphs have matching GFVs, can you answer with 100% certainty whether the graphs are isomorphic? Justify your answer.

5.10 Compute (without using any software tool) the graphlet degree distribution (GDD) of orbit 10 from the GDV matrix shown in Figure 5.7.

5.11 Use the GDV matrix shown in Figure 5.21 to compute (without using any graphlet-related software tool) its corresponding graphlet correlation matrix (GCM), as follows:

(a) Use the entire GDV matrix, i.e., consider all orbits.
(b) Assume that orbits 3, 12, 13, and 14 are redundant to the other orbits (see [15] for details on why). Use the filtered GDV matrix from which the information corresponding to the redundant orbits is removed.

GRAPHLETS IN NETWORK SCIENCE & COMPUTATIONAL BIOLOGY

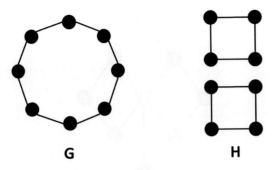

Figure 5.20

Node \ Orbit	o_0	o_1	o_2	o_3	o_4	o_5	o_6	o_7	o_8	o_9	o_{10}	o_{11}	o_{12}	o_{13}	o_{14}
s	3	2	2	1	0	1	0	0	1	0	1	1	0	0	0
t	2	3	1	0	1	1	1	0	1	1	0	0	0	0	1
u	2	3	1	0	1	1	0	1	1	1	0	0	0	0	0
v	3	2	2	1	0	1	0	0	1	0	1	1	0	1	0
w	2	2	0	1	2	0	0	0	0	0	2	0	1	0	0

Figure 5.21

5.12 Is there an advantage of partitioning (as defined mathematically) a graph into edge clusters over partitioning the graph into node clusters, or vice versa? Justify your answer. Ideally, explore relevant literature to answer this.

5.13 Suppose that a link prediction approach predicts a link between two nodes if and only if: (1) the nodes share at least two common neighbors, *and* (2) the nodes' GDV-similarity (for 2–5-node graphlets) is equal to one. Apply this approach to the graph in Figure 5.22 and draw its resulting (de-noised) network. *Hint: You do not need to actually compute GDV-similarity to solve this problem. It is possible to immediately identify topologically identical nodes (i.e., nodes whose GDV-similarity equals one) by looking at the graph – this is related to exercises 5.3 and 5.4 above.*

5.14 Review the methodologies from the Faisal & Milenković 2014 and Meng, Hulovatty, Striegel & Milenković 2016 studies [31, 100], which used dynamic network analysis to study proteins' involvement in aging from BNs and people's network–trait relationships from social networks, respectively. What parts of the two methodologies are (dis)similar? What are (dis)advantages of each?

5.15 Use GraphCrunch to generate an Erdős-Rényi random (ER) network, a geometric random (GEO) network, and a scale-free preferential attachment-based random (SF-BA) network, each with 1,000 nodes and 10,000 edges. Use RAGE and Orca software tools to compute the GDV matrix of each network, for up to 4-node graphlets (i.e., ignore 5-node graphlets in Orca, because RAGE cannot handle 5-node graphlets, in order to make the comparison

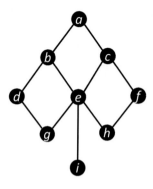

Figure 5.22

between the two tools more fair). Measure the CPU time that each tool takes on each network. Contrast the running times of the different software tools on the same network, for each network. Discuss why one tool might be performing better (faster) than the other (e.g., explore the algorithmic principles behind their graphlet counting procedures by reading their respective publications or looking at their code). Are the given tool's running times consistent across the different networks? If not, explain what could be causing the running times to vary (e.g., explore the differences between the three random graph models that are used to generate the three networks, which are described in the GraphCrunch paper)?

Note: Solutions are available to instructors at www.cambridge.org/bionetworks.

5.7 Acknowledgment

This work was supported by the US National Science Foundation (NSF) CAREER CCF-1452795 grant.

References

[1] Newman, M. (2010):Networks: An Introduction, Oxford University Press.
[2] Barabasi AL. *Network Science*. Cambridge University Press;2016.
[3] Breitkreutz BJ, Stark C, Reguly T, et al. The BioGRID Interaction Database: 2008 update. *Nucleic Acids Research*, 2008;36:D637–D640.
[4] Milenković T, Filippis I, Lappe M, Pržulj N. Optimized null model for protein structure networks. *PLoS ONE*, 2009;4(6):e5967.
[5] Godzik A, Kolinski A, Skolnick J. Lattice representations of globular proteins: How good are they? *Journal of Computational Chemistry*, 1993;14(10):1194–1202.
[6] Lander ES, Linton LM, Birren, et al. Initial sequencing and analysis of the human genome. *Nature*, 2001;409(6822):860–921.
[7] International Human Genome Sequencing Consortium. Finishing the euchromatic sequence of the human genome. *Nature*, 2004;431(7011):931–945.

[8] Koboldt DC, Steinberg KM, Larson DE, Wilson RK, Mardis ER. The next-generation sequencing revolution and its impact on genomics. *Cell*, 2013;155(1):27–38.

[9] Kuchaiev O, Stevanović A, Hayes W, Pržulj N. GraphCrunch 2: Software tool for network modeling, alignment and clustering. *BMC Bioinformatics*, 2011;12:24.

[10] Pržulj N. Protein–protein interactions: Making sense of networks via graph-theoretic modeling. *Bioessays*, 2011;33(2):115–123.

[11] Yaveroglu O, Milenković T, Pržulj N. Proper evaluation of alignment-free network comparison methods. *Bioinformatics*, 2015;31(16):2697–2704.

[12] Pržulj N, Corneil DG, Jurisica I. Modeling interactome: Scale-free or geometric? *Bioinformatics*, 2004;20(18):3508–3515.

[13] Gu S, Johnson J, Faisal FE, Milenković T. From homogeneous networks to heterogeneous networks of networks via colored graphlets. arXiv preprint available at arXiv:1704.01221.2018.

[14] Pržulj N. Biological network comparison using graphlet degree distribution. *Bioinformatics*, 2007;23(2):e177–e183.

[15] Yaveroglu ON, Malod-Dognin N, Davis D, et al. Revealing the hidden language of complex networks. *Scientific Reports*, 2014;4:4547.

[16] Milenković T, Pržulj N. Uncovering biological network function via graphlet degree signatures. *Cancer Informatics*, 2008;6:257–273.

[17] Solava R, Michaels R, Milenković T. Graphlet-based edge clustering reveals pathogen-interacting proteins. *Bioinformatics*, 2012;18(28):i480–i486.

[18] Vacic V, Iakoucheva L, Lonardi, S, Radivojac, P. Graphlet kernels for prediction of functional residues in protein structures. *Journal of Computational Biology*, 2010;17(1):55–72.

[19] Malod-Dognin N, Pržulj N. GR-Align: Fast and flexible alignment of protein 3D structures using graphlet degree similarity. *Bioinformatics*, 2014;30(9):1259–1265.

[20] Lugo-Martinez J, Radivojac P. Generalized graphlet kernels for probabilistic inference in sparse graphs. *Network Science*, 2014;2(2):254–276.

[21] Wong SW, Cercone N, Jurisica I. Comparative network analysis via differential graphlet communities. *Proteomics*, 2015;15(2–3):608–617

[22] Milenković T, Ng W, Hayes W, Pržulj N. Optimal network alignment with graphlet degree vectors. *Cancer Informatics*, 2010;9:121–137.

[23] Kuchaiev O, Milenković T, Memisevic V, Hayes W, Pržulj N. Topological network alignment uncovers biological function and phylogeny. *Journal of the Royal Society Interface*, 2010;7:1341–1354.

[24] Hsieh M, Sze S. Finding alignments of conserved graphlets in protein interaction networks. *Journal of Computational Biology*, 2014;21(3):234–246.

[25] Saraph V, Milenković T. MAGNA: Maximizing accuracy in global network alignment. *Bioinformatics*, 2014;30(20):2931–2940.

[26] Vijayan V, Saraph V, Milenković T. MAGNA++: Maximizing Accuracy in Global Network Alignment via both node and edge conservation. *Bioinformatics*, 2015;31(14):2409–2411.

[27] Vijayan V, Critchlow D, Milenković T. Alignment of dynamic networks. *Bioinformatics*, 2017;33(14):i180–i189.

[28] Hulovatyy Y, Solava RW, Milenković T. Revealing missing parts of the interactome via link prediction. *PloS One*, 2014;9(3):e90073.

[29] Milenković T, Memisevic V, Bonato A, Pržulj N. Dominating biological networks. *PloS One*, 2011;6(8):e23016.

[30] Hulovatyy Y, Chen H, Milenković T. Exploring the structure and function of temporal networks with dynamic graphlets. *Bioinformatics*, 2015;31(12):i171–i180.

[31] Faisal F, Milenković T. Dynamic networks reveal key players in aging. *Bioinformatics*, 2014;30:1721–1729.

[32] Faisal F, Zhao H, Milenković T. Global network alignment in the context of aging. *Transactions on Computational Biology and Bioinformatics*, 2014;12(1):40–52.

[33] Milenković T, Memisević V, Ganesan AK, Pržulj N. Systems-level cancer gene identication from protein interaction network topology applied to melanogenesis-related functional genomics data. *Journal of the Royal Society Interface*, 2010;7(44):423–437.

[34] Ho H, Milenković T, Memisevic V, et al. Protein interaction network topology uncovers melanogenesis regulatory network components within functional genomics datasets. *BMC Systems Biology*, 2010;4(1):84.

[35] Wang XD, Huang JL, Yang L, et al. Identification of human disease genes from interactome network using graphlet interaction. *PloS One*, 2014;9(1):e86142.

[36] Singh O, Sawariya K, Aparoy P. Graphlet signature-based scoring method to estimate protein–ligand binding affinity. *Royal Society Open Science*, 2014;1(4):40306.

[37] Faisal FE, Newaz K, Chaney JL, et al. Grafene: Graphlet-based alignment-free network approach integrates 3d structural and sequence (residue order) data to improve protein structural comparison. *Scientific Reports*, 2017;7(1):14890.

[38] Alon U. Network motifs: Theory and experimental approaches. *Nature Reviews Genetics*, 2007;8(6):450–461.

[39] Estrada E, Rodriguez-Velazquez JA. Subgraph centrality in complex networks. *Physical Review E*, 2005;71(5):056103.

[40] Liao, C, Lu, K, Baym, M, Singh, R, Berger, B. IsoRankN: Spectral methods for global alignment of multiple protein networks. *Bioinformatics*, 2009;25(12):i253–i258.

[41] Patro R, Kingsford C. Global network alignment using multiscale spectral signatures. *Bioinformatics*, 2012;28(23):3105–3114.

[42] Crawford J, Milenković, T, Sun Y. Fair evaluation of global network aligners. *Algorithms for Molecular Biology*, 2015;10(1):19.

[43] Sarajlic A, Malod-Dognin N, Yaveroglu O, Pržulj N. Graphlet-based characterization of directed networks. *Scientific Reports*, 2016;6:35098.

[44] Ortmann M, Brandes U. Efficient orbit-aware triad and quad census in directed and undirected graphs. *Applied Network Science*, 2017;2(1):13.

[45] Wang P, Zhang X, Li Z, et al. A fast sampling method of exploring graphlet degrees of large directed and undirected graphs. arXiv preprint available at https://arxiv.org/pdf/1604.08691v1.pdf. 2016.

[46] Ahmed N, Neville J, Rossi R, Duffield N. Efficient graphlet counting for large networks. In *2015 IEEE International Conference on Data Mining* (ICDM). IEEE; 2015, pp. 1–10.

[47] Hocevar T, Demsar J. A combinatorial approach to graphlet counting. *Bioinformatics*, 2014;30(4):559–565.

[48] Marcus D, Shavitt Y. RAGE: A rapid graphlet enumerator for large networks. *Computer Networks*, 2012;56(2):810–819.

[49] Milenković T, Lai J, Pržulj N. Graphcrunch: A tool for large network analyses. *BMC Bioinformatics*, 2008;9(1):70.

[50] Milo R, Shen-Orr S, Itzkovitz S, et al. Network motifs: Simple building blocks of complex networks. *Science*, 2002;298(5594):824–827.

[51] Shen-Orr SS, Milo R, Mangan S, Alon U. Network motifs in the transcriptional regulation network of *Escherichia coli*. *Nature Genetics*, 2002;31(1):64–68.

[52] Milo R, Itzkovitz S, Kashtan N, et al. Superfamilies of evolved and designed networks. *Science*, 2004;303(5663):1538–1542.

[53] Artzy-Randrup Y, Fleishman SJ, Ben-Tal N, Stone L. Comment on "network motifs: simple building blocks of complex networks" and "superfamilies of evolved and designed networks". *Science*, 2004;305(5687):1107–1107.

[54] Watts DJ, Strogatz SH. Collective dynamics of small-world networks. *Nature*, 1998;393(6684):440–442.

[55] Holme P, Saramaki J. Temporal networks. *Physics Reports*, 2012;519(3):97–125.

[56] Rossi R, Zhou R. Hybrid CPU–GPU framework for network motifs. arXiv preprint available at https://arxiv.org/pdf/1608.05138.pdf. 2016.

[57] Ahn Y, Bagrow J, Lehmann S. Link communities reveal multiscale complexity in networks. *Nature*, 2010;466(7307):761–764.

[58] Evans T, Lambiotte R. Line graphs, link partitions, and overlapping communities. *Physical Review E*, 2009;80(1):016105.

[59] Bishop CM. *Pattern Recognition and Machine Learning*. Springer;2006.

[60] Fortunato S. Community detection in graphs. *Physics Reports*, 2010;486:75–174.

[61] Garey MR, Johnson DS. *Computers and Intractability*, vol. 29, WH Freeman; 2002.

[62] Liben-Nowell D, Kleinberg J. The link prediction problem for social networks, in *Proceedings of the Twelfth International Conference on Information and Knowledge Management*. CIKM 2003, ACM;2003, pp. 556–559.

[63] Liben-Nowell D, Kleinberg J. The link-prediction problem for social networks. *Journal of the American Society for Information Science and Technology*, 2007;58(7):1019–1031.

[64] Bahulkar A, Szymanski BK, Lizardo O, et al. Analysis of link formation, persistence and dissolution in netsense data. In *2016 IEEE/ACM International Conference on Advances in Social Networks Analysis and Mining (ASONAM)*. IEEE;2016, pp. 1197–1204.

[65] Lu L, Zhou T. Link prediction in complex networks: A survey. *Physica A: Statistical Mechanics and its Applications*, 2010;390(6):1150–1170.

[66] Lichtenwalter R, Lussier J, Chawla N. New perspectives and methods in link prediction. In *Proceedings of the 16th ACM SIGKDD International conference on Knowledge Discovery and Data Mining*. ACM;2010, pp. 243–252.

[67] Narayanan A, Shi E, Rubinstein B. Link prediction by de-anonymization: How we won the Kaggle social network challenge. In *Proceedings of the 2011 International Joint Conference on Neural Networks (IJCNN)*. IEEE;2011, pp. 1825–1834.

[68] Ali W, Rito T, Reinert G, Sun F, Deane CM. Alignment-free protein interaction network comparison. *Bioinformatics*, 2014;30(17):i430–i437.

[69] Shervashidze N, Schweitzer P, van Leeuwen EJ, Mehlhorn K, Borgwardt KM. Weisfeiler–Lehman graph kernels. *Journal of Machine Learning Research*, 2011;12(Sep):2539–2561.

[70] Vishwanathan SVN, Schraudolph NN, Kondor R, Borgwardt KM. Graph kernels. *Journal of Machine Learning Research*, 2010;11(Apr):1201–1242.

[71] Shervashidze N, Vishwanathan S, Petri T, Mehlhorn K, Borgwardt K. Efficient graphlet kernels for large graph comparison. In *Proceedings of the Twelfth International Conference on Artificial Intelligence and Statistics (AIStats 2009)*. MIT Press;2009, pp. 488–495.

[72] Sharan R, Ideker T. Modeling cellular machinery through biological network comparison. *Nature Biotechnology*, 200624(4):427–433.

[73] Faisal F, Meng L, Crawford J, Milenković T. The post-genomic era of biological network alignment. *EURASIP Journal on Bioinformatics and Systems Biology*, 2015;2015(1):1.

[74] Meng L, Striegel A, Milenković T. Local versus global biological network alignment. *Bioinformatics*, 2016;32(20):3155–3164.

[75] Guzzi PH, Milenković T. Survey of local and global biological network alignment: The need to reconcile the two sides of the same coin. *Briefings in Bioinformatics*, 2018;19(3):472–481.

[76] Vijayan V, Krebs E, Meng L, Milenković T. Pairwise versus multiple network alignment. arXiv preprint available at https://export.arxiv.org/pdf/1709.04564. 2017.

[77] Singh R, Xu J, Berger B. Pairwise global alignment of protein interaction networks by matching neighborhood topology. In *RECOMB'07 Proceedings of*

the 11th Annual International Conference on Research in Computational Molecular Biology Research in Computational Molecular Biology. Springer;2007, pp. 16–31.

[78] Kuchaiev O, Pržulj N. Integrative network alignment reveals large regions of global network similarity in yeast and human. *Bioinformatics*, 2011;27(10):1390–1396.

[79] Mamano N, Hayes W. SANA: Simulated annealing far outperforms many other search algorithms for biological network alignment. *Bioinformatics*, 2017;33(14):2156–2164.

[80] Vijayan V, Milenković T. Multiple network alignment via multi-MAGNA++. *IEEE/ACM Transactions on Computational Biology and Bioinformatics*, 2017;PP(99). DOI: 10.1109/TCBB.2017.2740381.

[81] Sun Y, Crawford J, Tang J, Milenković T. Simultaneous optimization of both node and edge conservation in network alignment via WAVE. In *International Workshop on Algorithms in Bioinformatics*. Springer;2015, pp. 16–39.

[82] Malod-Dognin N, Pržulj N. L-graal: Lagrangian graphlet-based network aligner. *Bioinformatics*, 2015;31(13):2182–2189.

[83] Crawford J, Milenković T. GREAT: Graphlet edge-based network alignment. In *2015 IEEE International Conference on Bioinformatics and Biomedicine (BIBM)*. IEEE, 2015, pp. 220–227.

[84] Lee D, Redfern O, Orengo C. Predicting protein function from sequence and structure. *Nature Reviews Molecular Cell Biology*, 2007;8(12):995–1005.

[85] Kasabov N. *Springer Handbook of Bio-/Neuro-Informatics*. Springer Science & Business Media;2013.

[86] Radivojac P, Clark WT, Oron TR, et al. A large-scale evaluation of computational protein function prediction. *Nature Methods*, 2013;10(3):221–227.

[87] Sharan R, Ulitsky I, Shamir R. Network-based prediction of protein function. *Molecular Systems Biology*, 2007;3(88):1–13.

[88] Memisevic V, Milenković T, Pržulj N. Complementarity of network and sequence information in homologous proteins. *Journal of Integrative Bioinformatics*, 2010;7(3):275–289.

[89] Cuff A, Redfern O, Dessailly B, Orengo C. Exploiting protein structures to predict protein functions. In Kihara D, ed., *Protein Function Prediction for Omics Era*. Springer;2011, pp. 107–123.

[90] Chitale M, Kihara D. Computational protein function prediction: Framework and challenges. In Kihara D, ed., *Protein Function Prediction for Omics Era*. Springer;2011, pp. 1–17.

[91] Berchtold N, Cribbs D, Coleman et al. Gene expression changes in the course of normal brain aging are sexually dimorphic. *Proceedings of the National Academy of Sciences*, 2008;105(40):15605–15610.

[92] de Magalhaes J, Budovsky A, Lehmann G, et al. The Human Ageing Genomic Resources: Online databases and tools for biogerontologists. *Aging Cell*, 2009;8(1):65–72.

[93] Ueno K. The effects of friendship networks on adolescent depressive symptoms. *Social Science Research*, 2004;34(3):484–510.

[94] Christakis NA, Fowler JH. The spread of obesity in a large social network over 32 years. *New England Journal of Medicine*, 2007;357(4):370–379.

[95] Cohen-Cole E, Fletcher JM. Detecting implausible social network effects in acne, height, and headaches: Longitudinal analysis. *British Medical Journal*, 2008;337:a2533.

[96] Fowler JH, Christakis NA. Dynamic spread of happiness in a large social network: Longitudinal analysis over 20 years in the Framingham heart study. *British Medical Journal*, 2008;337, a2338.

[97] Smith KP, Christakis NA. Social networks and health. *Annual Review of Sociology*, 2008;34:405–429.

[98] O'Malley AJ, Christakis NA. Longitudinal analysis of large social networks: Estimating the effect of health traits on changes in friendship ties. *Statistics in Medicine*, 2011;30(9):950–964.

[99] Christakis N, Fowler F. *Connected: The Surprising Power of Our Social Networks and How They Shape Our Lives–How Your Friends' Friends' Friends Affect Everything You Feel, Think, and Do*. Back Bay Books;2011.

[100] Meng L, Hulovatty Y, Striegel A, Milenković T. On the interplay between individuals' evolving interaction patterns and traits in dynamic multiplex social networks. *IEEE Transactions on Network Science and Engineering*, 2016;3(32–43):679395.

[101] Hocevar T, Demsar J. Computation of graphlet orbits for nodes and edges in sparse graphs. *Journal of Statistical Software*, 2016;71(1):1–24.

6 Unsupervised Learning: Cluster Analysis

Richard Röttger

Clustering in general is long-standing problem in computer science. It seeks to unravel the inherent structure of datasets by "grouping or segmenting a collection of objects into subsets or clusters such that those within each cluster are more closely related to one another than objects assigned to different clusters" [1].

Clustering is a very versatile approach as it can be and in fact is applied in almost every scientific field. Often, the cluster analysis is only the first step in a long chain of downstream analyses. Especially in those cases, it is of utmost importance to ensure that the clustering was performed with great care and correctly since the downstream analysis depends on the quality of the clustering. Even when limited to computational biology, clustering is applied to a vast variety of different problems. Just to name a few examples:

Gene expression analysis One very common example in computational biology is the clustering of gene expression data sets. Here, the goal is to identify groups of, for example, patients which behave similar under a certain condition, e.g., cancer, by means of showing a similar gene expression. This might help to define disease subtypes for which different treatment strategies might be necessary or which show a different survival rate. Having patients with the same condition, one could also ask a different question: Which genes behave similar, i.e., cluster the genes and not the patients. This could give information of genetic modules which then might also be analyzed together with biological networks.

Protein complex detection Large datasets containing information on protein–protein interactions (PPI) in various organisms allow for a different kind of analysis: Can we computationally identify protein complexes? This question is routinely answered by clustering these PPI networks in order to reveal densely connected subgroups within these networks. The main purpose of several clustering algorithms is specifically the detection of those protein complexes.

Protein homology detection Given the protein sequence, can we identify groups of homologous proteins between different species? Here, clustering is also very common and resembles a more advanced approach than only looking at the pairwise sequence similarities. Since the protein sequence is available for a tremendous amount of different species, these datasets might be very large, e.g., when comparing the proteome of all bacteria. Thus, runtime and complexity considerations of the algorithms might play a crucial role in those cases.

This is only a very brief overview of some examples of clustering in computational biology. Clustering in general is also often used for data preprocessing: For example, for *compression* or *summarization* of a large dataset. Normally, a very efficient clustering algorithm is applied to a huge dataset; subsequently, the downstream analysis that would be infeasible on the entire dataset, is performed on a representative subset containing representatives of the different clusters. Depending on the nature of the downstream analysis, the results might be generalized to all objects in the cluster afterwards.

The previous examples should serve as a motivation for a cluster analysis and should also highlight a key problem of clustering: The definition of clustering is very vague (remember from above: Segmenting objects into subsets such that those within each cluster are more closely related to one another than objects assigned to different clusters). There is no unique specification of what closeness between these objects means. In fact, there exists no universally agreed upon and precise definition of the term cluster, partially due to the inherent subjectivity of clustering, which precludes an absolute judgment as to the relative efficacy of all clustering techniques [2]. A definition which is valid in one domain can be meaningless in another domain. Thus, each domain or scientific field has produced their own set of clustering tools and approaches but at the end it remains a highly subjective endeavor. A simple example shall demonstrate this fact: When we seek to cluster the cars parked on a parking lot, we have various ways of relating the cars to each other: By color, size, brand, etc. A grouping based on size might be useful for defining categories of cars but completely useless in assessing popularity of car colors. Here, the key point is to establish a meaningful relationship between the objects. But even when this relationship is well-defined, it is still entirely problem dependent whether the solution requires fine-grained results with many small clusters or rather a coarse separation in a few but large groups. For example, proteins could be grouped in families, folds, classes, etc. All three are meaningful clusterings, but depending on the actual problem, one solution might be more helpful than the other. This should highlight that the reason for the absence of a strict definition of an ideal clustering is simply because their is no one-size-fits-all solution to this problem. This makes clustering inherently complex.

Another point to highlight is that even in the case we have an optimal criteria for the current problem at hand, finding the optimal solution is a problem in itself. The number of possibilities to separate n objects into m groups amounts to

$$\frac{1}{m!}\sum_{k=0}^{m}(-1)^{m-k}\binom{m}{k}k^n \tag{6.1}$$

different options. This is an extremely fast growing number (2.4×10^{15} for only 25 objects and 5 clusters [3]) and an exhausting testing is impossible. This means in turn, that all clustering algorithms are only approximations or heuristics which try to seek a partitioning of the objects in order to optimize an internal fitness function of the tool. Like all heuristics they can be tricked into producing suboptimal results and some heuristics will work better than others on a particular dataset but perform worse on a different problem. This is the reason for the plethora of clustering tools. All those points together lead to the fact that performing a high-quality cluster analysis is tedious, error-prone and complex.

The aim of this chapter is to introduce the reader to clustering in general and highlight its various aspects. We start with basic formal definitions, then introduce the various steps of a cluster analysis, before we discuss the evaluation step by step and highlight various popular standard methods.

This chapter is based on a previous publication on clustering in bioinformatics [4] of the same author and is not cited explicitly throughout the remainder of the chapter.

6.1 Formal Definitions

Before we start discussing specific aspects of a cluster analysis, we have to precisely define the terms and expressions used throughout the rest of this chapter. This includes the various types of clusterings as well as the different forms of data.

6.1.1 Clustering

Let $X = \{x_1, \ldots, x_N\}$ denote a dataset of N objects. This set of objects can be differentiated into groups in various ways which define the different kinds of clusterings [2, 5]:

Partitional clustering (sometimes also called *crisp* or *disjoint* clustering) is the task of seeking a k-partition $C = \{C_1, \ldots, C_k\}$ of X, such that

1. $C_i \neq \emptyset, \quad i = 1, \ldots, k$
2. $\bigcup_{C_i \in C} C_i = X$
3. $C_i \cap C_j = \emptyset \quad i, j = 1, \ldots, k$ and $i \neq j$

Put in words: All objects of X need to be present in the clustering, with each object x_i being in exactly one cluster and every cluster is comprised of at least one object.

Overlapping clustering is a partitional clustering where the rule (3) does not hold. That means, each object might belong to several clusters, thus the name *overlapping* as clusters might share common elements.

Fuzzy clustering can be seen as the generalization of the overlapping clustering. In contrast to the previous, the assignment of objects to a cluster is no longer binary (i.e., object x_i either *is* or *is not* member of the cluster) but each object x_i belongs to every cluster C_j to a certain degree $u_{i,j}$. For a fuzzy k-partitioning $C = \{C_1, \ldots, C_k\}$ the following holds true:

1. $\sum_{j=1}^{k} u_{i,j} = 1 \quad \forall i$
2. $0 < \sum_{i=1}^{k} u_{i,j} \leq N \quad \forall j$

This can be seen as the generalization of the rules above: Each object x_i has to be fully contained in the clustering (Rule 1) and every cluster requires at least one object with a degree of membership greater than zero (Rule 2), i.e., all clusters are non-empty.

Hierarchical clustering greatly differs from the previous definitions. Instead of forming a k-partitioning, hierarchical clustering constructs a nested tree structure $H = \{H_1, \ldots, H_Q\}$ with ($Q \leq N$) of X such that $C_i \in H_m, C_j \in H_l$, and $m > l$ imply $C_i \in C_j$ or $C_i \cap C_j = \emptyset$ for all $i, j \neq i, m, l = 1, \ldots, Q$. The hierarchical clustering yields more information than the partitional clustering as the tree structure additionally reveals the relationship between the clusters. Each hierarchical clustering can be transformed into a partitional clustering by cutting the tree at a given level.

It is important to note that these are only formal definitions of a clustering but do not contain any evaluation of the goodness of a clustering. For example, a clustering consisting of only one cluster containing all objects is a perfectly fine clustering by the definition; from a practical point of view, such a clustering is most likely useless.

6.1.2 Data Formats

As clustering is a data driven analysis method, the preparation and understanding of the data is of utmost importance. The form in which the data is available greatly defines the selection of tools, preprocessing steps, and proximity calculation. The most general classification of a given dataset is the distinction between *one-mode* and *two-mode* datasets.

One-mode data comprises of a $n \times n$ matrix directly describing the relationship between the objects. This can be either a *similarity matrix* or a *distance matrix*, depending on whether the values indicate a measure of similarity or of distance, respectively. As columns and rows in the proximity matrix indexing the same thing (i.e., objects), this form is called "one-mode" [6].

Two-mode data is given in a $n \times d$ data matrix relating n objects to d *features* (also called *variables* or *dimensions*). This data format can also be seen as the raw data as this data type does not yet establish a measure of proximity between the objects. One common challenge in a cluster analysis is the calculation of a meaningful proximity based on the raw data between the objects. Many different domain and data type specific methods exist. At the end, every clustering tool requires a measure of proximity between the objects, either by requiring a one-mode matrix as input or by internally establishing a proximity during the clustering.

In the case of a two-mode dataset, the different features can have different formats. Normally, one distinguishes between the following formats [6, 7, 8]:

Quantitative features which can be subdivided into:

1. continuous values (e.g., fold changes of gene expressions);
2. discrete values (e.g., number of genes);
3. interval values (e.g., timespan of a treatment, 1–2 days, 3–4 days, ...).

Categorical features which can be subdivided into:

1. nominal or unordered (e.g., eye color);
2. ordinal (e.g., qualitative evaluations of pain with "no pain," "some pain," and "strong pain").

Structural data are comprised of repeated measurements of the same variable under different conditions, e.g., time-series data of gene expression.

6.2 Cluster Analysis

After having established the most fundamental definitions, we will first discuss the general workflow of a cluster analysis. It is very important to understand the various steps involved and how they drive the downstream decisions. Generally, a cluster analysis is a highly interconnected process and poor decisions in the beginning might have severe consequences for the entire cluster analysis. A broad structure is given in Figure 6.1 and consists of the following steps:

1. Data preprocessing
2. Proximity calculation
3. Clustering
4. Evaluation and testing

In the following, we will discuss the different steps in detail. Data preprocessing can consist of several steps, including *feature extraction*, *feature selection*, and *normalization*. Here, we will particularly focus on the *principal component analysis* but also briefly mention few other aspects. After the first preprocessing, one of the most crucial steps of a cluster analysis is the establishment of a relationship between the objects (in case the dataset is given as a two-mode dataset). A meaningful proximity calculation between the objects is central to enable the clustering tool to meaningfully separate the dataset into groups. We will discuss a couple of standard measures for the various

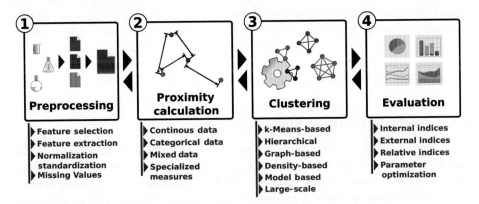

Figure 6.1: Overview of a cluster analysis. The various steps are highly interconnected, meaning poor decisions at the beginning will likely lead to an overall poor result. The organization of the remainder of the chapter is aligned to this figure.

data types mentioned above. Thereafter, the dataset can finally be clustered. Here, we will introduce several general approaches to the clustering problem and discuss a small selection of important algorithms in more detail. The last step of a cluster analysis is the evaluation of the clustering at hand. This is a particularly tricky topic, as normally the ground truth is not known, rendering the objective evaluation very difficult and vague. Further, we will see methods of determining the expected number of clusters like the *GAP statistic*. Generally speaking, the evaluation is the fist point of real feedback and might allow for judging the validity of the decision made before.

6.3 Preprocessing

As described before, the first step of a cluster analysis is the data preprocessing. Here, steps of data cleaning and modifications are performed in such a way that they support the subsequent clustering as good as possible. Even though this is discussed rather in isolation, it is important to acknowledge that decisions made here have huge impact on the efficiency and quality of the downstream steps. Classically, preprocessing might consist of, but is not limited to, the following steps:

Normalization is normally performed when the dataset consists of different kind of features which show a different range of spread and variance of the values.

Feature selection describes the task of selecting the most informative set of features for the clustering.

Feature extraction is the task of constructing new features by the combination of existing features. The goal often is to arrive at fewer but uncorrelated and more informative features helping the clustering algorithm afterwards.

Please note, that this is not an exhaustive overview of possible preprocessing steps. Often, data cleaning is also performed, missing or incomplete values need to be imputed or removed from the sample. Furthermore, one might want to detect outliers, or remove technical biases, e.g., when the samples were using different sampling platforms. In the following, we will discuss some of the most common preprocessing steps and look at principal component analysis in detail.

6.3.1 Normalization and Standardization

Let us begin with the problem of having features of different scales and variances. The similarity function might be highly influenced by the scale of the variables. Let the Euclidean distance serve as an obvious example: A variable which is bound between [0, 1] will be almost entirely disregarded when a second feature ranges in the millions [9]. Therefore, it might be crucial to *normalize/standardize* the variables before usage, i.e., bring them on a comparable scale and variance. The most common methods are [9, 10]:

Min-max normalization is a linear transformation of a feature F to a predefined value range, normally between $[0, 1]$:

$$f' = \frac{f - \min_F}{\max_F - \min_F}, \tag{6.2}$$

where $\min_F$ and $\max_F$ denotes the global minimum/maximum of the feature F in the dataset and f represents some value of F.

Autoscaling or Z-score scaling tackles a problem arising with the normalization: In case the dataset has an extreme outlier, it would result in squeezing all of the other normalized values around zero. In order to be more robust against outliers, autoscaling centers the values of the feature F and scales a given value f with the standard deviation σ_F of the feature:

$$f' = \frac{f - \overline{F}}{\sigma_F}. \tag{6.3}$$

As already mentioned, in the presence of extreme values autoscaling should be favored over the min-max normalization [9]. After autoscaling, the feature will have a mean of 0 (due to the centering) and a standard deviation of 1 [11].

There exists many more methods for standardizing and normalizing your data, e.g., decimal scaling or quantile normalization, just to mention two. Furthermore, for many different types of biomedical datasets, specialized normalization methods exist. For example, Bullard et al. [12] have performed a review on normalization methods for expression data in RNA-seq experiments; similarly for microarray data in Kim et al. [13]. Nevertheless, normalization only for the sake of normalization is also discouraged since it necessarily results in a loss of location and scale of the original values [11].

6.3.2 Feature Selection

The general task of feature selection can be described as the reduction of the feature space to only the most informative features. In a first thought, one could say that the more information, i.e., the more features used, the better the clustering. This is often misleading, as features can be correlated and thus bias the proximity; or if there are features which are not exhibiting any cluster structure, it can also blur otherwise well-separable clusters within each other. At the end, we aim to remove irrelevant, redundant, or noisy features [14] in order to (1) to improve the cluster performance, (2) to reduce the dataset size, (3) to learn about the importance of the features [15].

Feature selection is most widely applied to supervised learning methods like classification tasks [16]; thus, our discussion of feature selection for unsupervised clustering will be quite limited. In supervised learning, we can split our dataset into training and testing data. The general idea now is to create various feature sets (i.e., select a subset of the available features) and evaluate the classification performance on the testing data. The feature set resulting in the best performance is chosen for the final classifier. This approach is a so-called *wrapper approach* because the feature selection is combined "around" the actual machine learning approach. The main challenge of these methods is to efficiently chose the set of features; testing all combinations is normally infeasible due to the combinatorial explosion. In terms of clustering (due to the lack of a training and testing set), this approach needs to be coupled with a method of evaluating the cluster result. Evaluating the performance of a clustering is a problem in itself and will be discussed in Section 6.6.

A different way of selecting features are the so-called *filter* approaches. These methods decide on the feature importance solely on the features themselves and do not couple them with a machine learning approach. This is more suitable for clustering, since there is no need for the tricky evaluation of the clustering. On the other hand, using a filter approach, it is crucial to determine the number of features to use for the final clustering. There should be a trade-off between the number of features and maintaining the inherent structure of the dataset. Thus, many of these methods restrain themselves to removing highly correlated features and thus maintain the overall variance in the dataset which is crucially required in order to separate the clusters later on. Overviews especially for the application of feature selection for clustering can be found in the reviews of Liu et al. [14] and Alelyani et al. [17].

6.3.3 Principal Component Analysis

Feature extraction is a very broad topic and could easily fill books in its own right. In general, feature extraction is used in order to create new, uncorrelated, and more informative features from the given feature set. Please note, that there are also meaningful applications which expand the number of features but in this chapter we focus on feature reduction.

Probably the most well-known and most widely applied feature extraction method is principal component analysis (PCA). The goal is to project the feature space onto the so-called principal components, preserving as much variance of the dataset as possible. As already stated, preserving the variance in the dataset is crucial in order to separate the objects in the dataset. Consider the extreme case of having no variance at all: The dataset would collapse to a single point with no meaningful way to separate the objects into clusters. Thus, when we are able to reduce the dimensionality of the dataset in such a way that we still capture the same (or almost the same) variance, the remaining dimensions do not yield any information for separating the objects. The PCA automatizes this approach by calculating the most informative dimensions, i.e., those accounting for the greatest variance in the dataset first; all following dimensions capture a gradually decreasing amount of the variance.

In order to understand the PCA, we have first to understand the variance of a dataset. Given a dataset $X = \{x_1, \ldots, x_n\}$, the observed (and Bessel corrected, i.e., unbiased) sample variance is defined as

$$\text{Var}(X) = \frac{1}{n-1} \sum_{i=1}^{n} (x_i - \bar{x})^2, \tag{6.4}$$

with $\bar{x}$ being the arithmetic mean of X.

Now let us consider the multi-dimensional case. In this case, the dataset X can be represented as a $n \times d$ two-mode matrix with each row corresponding to one point. Let $\mathbf{x_i} \in X$ be a d dimensional vector, i.e., $\mathbf{x_i} = \{x_{i1}, \ldots, x_{id}\}$, we can calculate the covariance between two dimensions K and L by:

$$\text{Cov}(K,L) = \frac{1}{n-1}\sum_{i=1}^{n}(x_{iK} - \bar{x}_{\bullet K})(x_{iL} - \bar{x}_{\bullet L}) \tag{6.5}$$

with $\bar{x}_{\bullet K}$ being the mean of X in dimension K, $\bar{x}_{\bullet L}$ in dimension L respectively. This allows us to define the covariance matrix:

$$Q = \begin{pmatrix} \text{Cov}(D_1, D_1) & \cdots & \text{Cov}(D_1, D_d) \\ \vdots & \ddots & \vdots \\ \text{Cov}(D_d, D_1) & \cdots & \text{Cov}(D_d, D_d) \end{pmatrix}, \tag{6.6}$$

with the d dimensions $D_1, \ldots, D_d$. The $d \times d$ matrix is symmetric, positive semi-definite, and the diagonals represent the observed sample variance in each dimension. The entire matrix can also be simply calculated by a matrix multiplication of:

$$Q = \frac{1}{n-1}(X - \bar{X})^\top (X - \bar{X}), \tag{6.7}$$

with $\bar{X}$ being the mean vector of the dataset X. Normally, the dataset is standardized (using the autoscaling as described earlier) before applying the PCA; this is particularly recommended when the different features are measured in different units on different scales, e.g., height in meters and weight in kilograms.

This covariance matrix describes the nature and spread of the variance in the dataset. Figure 6.2a depicts a small toy dataset and roughly the spread of the data. In the depicted example, one could quite easily discover the main direction of the spread by eye, namely along the main diagonals of the ellipse (Figure 6.2 (b)). In complex, high-dimensional cases, this is no longer possible; thus we require an automatized way of determining these axes. It turns out to be that the main directions of the spread of a dataset is in the same direction as the eigenvectors of the covariance matrix. We will not discuss this with a mathematical proof, we will just briefly examine an argument which should illustrate this fact. When we turn back to Figure 6.2, we can calculate the covariance matrix of the dataset:

$$Q = \begin{pmatrix} 3 & 1 \\ 1 & 3 \end{pmatrix}. \tag{6.8}$$

In order to understand the type of linear projection this covariance matrix represents, we look at the projection of the unit circle under the matrix Q (Figure 6.2 (c)). We see that the points from the unit circle get exactly projected onto the ellipse describing the spread of the dataset. Further, when we follow the trajectories of the points (Figure 6.2 (d)) onto their projections, we can see that exactly four vectors are mapped onto themselves adjusted with some scaling factor, i.e., fulfilling $Qv = \lambda v$: The vectors which are aligned with the main diagonals describing the ellipse. And those vectors being projected onto themselves are by definition the eigenvectors of the matrix Q. So we can see that the eigenvectors of the covariance matrix Q describes the ellipse which in turn describes the spread of the data. Once again, this is not a proof, only an illustration. A formal proof and further illustraions and descriptions can be found, for example in [18].

Without loss of generality, let $\lambda_1 > \lambda_2 > \cdots > \lambda_d$ the d eigenvalues ordered by magnitude and $v_1, v_2, \ldots, v_d$ the corresponding eigenvectors. Then, v_1 corresponds to

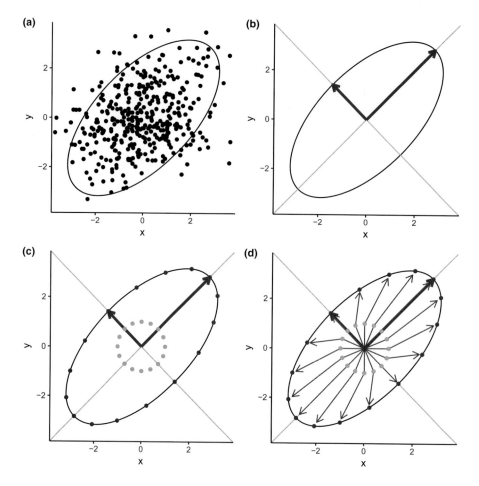

Figure 6.2: (a) Toy dataset and an ellipse roughly describing the spread of the dataset. (b) When we consider the ellipse alone, we can see that an ellipse is described by two perpendicular vectors across the main diagonal. (c) Points of the unit circle (green) are projected onto the eclipse (blue). (d) The projection including the projection trajectory of each point of the unit circle onto the ellipse.

the direction capturing the most variance, v_2 to the dimension of the second most and so forth. As Q is a positive semi-definite matrix, the eigenvectors will form a basis of the vector space and thus we can represent each object in coordinates relative to the eigenvectors. Let $E = (v_1, \ldots, v_d)$ denote the $d \times d$ eigenvector matrix with each column corresponding to the respective eigenvector, then

$$P = E^\top \cdot X^\top \tag{6.9}$$

projects the dataset into the so-called eigenspace. Figure 6.3 (b) depicts such a projection for a two-dimensional example.

This projection is the result of a principal component analysis. Nevertheless, the resulting projection has the same dimensionality as the original dataset. Basically, we have performed a rotation of the dataset (Figure 6.3 (b)). In order to reduce the

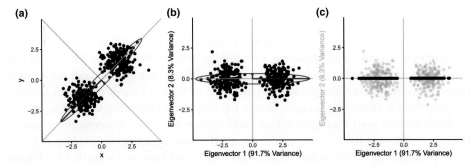

Figure 6.3: (a) Small toy dataset with two distinct clusters. Indicated are the ellipse of the covariance matrix (projection of the unit circle) and the two principal components. (b) The projection of the dataset into the eigenspace. Note, the two axes now represent the principal components and not x and y. (c) The one-dimensional projection of the dataset onto the first principal component. The original points and the second principal component are still indicated in gray.

dimensionality of the dataset, only the first l eigenvectors of the l largest eigenvalues are used to construct the projection matrix E. This projection does lead to an actual dimensionality reduction, as desired (Figure 6.3 (c)). Furthermore, the eigenvalues provide a measure for the amount of variance persevered in the lower dimensional projection. When using the first l eigenvalues, the ratio of preserved variance can be calculated by

$$r = \frac{\sum_{i=1}^{l} \lambda_i}{\sum_{i=1}^{d} \lambda_i} \tag{6.10}$$

In the example of Figure 6.3, you can see that the transformation preserves 91.7% of the original variance when removing one dimension and the dataset still nicely decomposes into two clusters. The amount of preserved variance is a good measure for the quality of the projection but by no means a guarantee that the structure of the dataset is still intact. Furthermore, the eigenvectors themselves provide significant information about the importance of the original dimensions. The eigenvector $v_j = \{\alpha_1, \ldots, \alpha_d\}$ can be read as a linear combination of the original dimensions, i.e.,

$$v_j = \sum_{i=1}^{d} \alpha_i \cdot e_i, \tag{6.11}$$

with $e_1, \ldots, e_d$ being the original features. Here, the magnitude of the different α_i (also called the factor loadings) denote the importance of the ith original dimension to the eigenvector v_j. To make the loadings comparable and thus meaningful, the different features of the dataset should have been standardized before the PCA.

It remains to mention, that PCA is not able to meaningfully reduce the dimension of all datasets. For these purposes, there exists methods like a kernel PCA or other non-linear feature extraction methods which would exceed the purpose of this chapter. For a good overview of dimensionality reduction techniques consider for example the book by Guyon et al. [19].

6.4 Proximity Calculation

The center of each cluster analysis is not necessarily the clustering algorithm itself but often the process of establishing a proximity between the objects. Without those proximities, no clustering tool can assess the structure of the dataset and derive any meaningful clustering. Further, with an unsuited proximity measure it might happen that the cluster attempt fails, regardless of the employed clustering algorithm. So far, we have mainly seen examples which were drawn on a two-dimensional plane in Euclidean space. In these cases, establishing a distance is rather straightforward. But most instances, especially in biomedical context, do not have any so-called Euclidean embedding. How would you place patients, protein structures, or biological networks in such a space? For these cases, different means of establishing similarities need to be found.

6.4.1 Continues Variables

Probably the most prevalent type of features are continues variables. A dataset only consisting of continues variables can also be interpreted as points in a Euclidean space as we have done with the examples before. This allows for two generally different ways of defining the proximity between two objects: One is to assess the distance between the two points (e.g., Minkowski distances, see below), the other is to compare the angle of the trajectory of the points from the null regardless of their distance from the null.

6.4.1.1 Euclidean Distance

This is the most common distance measure when presented with continuous variables and probably also the most natural one since it corresponds to the distance of two points when measured with a ruler. More formally, the Euclidean distance between two d-dimensional objects (or vectors) $u = \{u_1, \ldots, u_d\}, v = \{v_1, \ldots, v_d\} \in X$ is defined as

$$d(u,v) = \sqrt{\sum_{i=1}^{d}(u_i - v_i)^2}. \tag{6.12}$$

Despite its popularity, the Euclidean distance is only one member of a wider group of general distance measures, the so-called Minkowski distances.

6.4.1.2 Minkowski Distance

The family of the Minkowski distances can be regarded as the generalization of the Euclidean distance. When we are considering the same objects u and v as before, the Minkowski distance is defined as:

$$d(u,v) = \left(\sum_{i=1}^{d}|u_i - v_i|^p\right)^{1/p}, \tag{6.13}$$

with the parameter $p \geq 1$. For values with $p < 1$, the Minkowski distance violates the triangle equation and thus is no longer a metric. The Euclidean distance is the

Minkowski distance with $p = 2$. There are further special cases of p which are worth mentioning: For $p = 1$, we receive the so-called city block distance or Manhattan distance [20] which corresponds to the distance between objects as the shortest way only in directions of the main axis of the space; thus "city block" distance since you cannot diagonally cross through a city block. With $p = \infty$, the so-called Chebyshev distance or chessboard distance is defined. Here, the distance between two points is defined as the maximal distance along any of the dimensions. This corresponds to the number of moves required in order to move the king from point u to v in chess. Generally, if we expect to have compact and/or isolated clusters, the Minkowski distances have proven to work very well [11, 21].

Nevertheless, in very high-dimensional cases, the Minkowski distances are less ideal, since for most data distributions the ratio of the distance of the closest points and furthest points reaches one [22]. In other words, there is almost no usable differentiation between very close and very distant objects anymore. In those cases a preceding dimensionality reduction should be performed.

6.4.1.3 Correlation

The Minkowski distances measure the distances in terms of the absolute difference in the values of the features. Nevertheless, sometimes the absolute values in the features are of less interest than the relationship between them. As an example, one can think of gene expression, where the absolute read count is highly dependent on the utilized platform and the applied processing of the raw data and might not perfectly represent the situation in the cell. It might rather be of interest whether a higher value of gene A relates to a low value of gene B in two patients regardless of the actual absolute values. In other words, when considering the objects as two vectors, we do not want to measure the distance between them but rather the similarity of their trajectory from 0. In these cases, correlation based similarity functions are recommended. The best-known correlation coefficient is the so-called Pearson correlation coefficient which is defined for two vectors $u, v \in X$ as:

$$\Phi(u,v) = \frac{\sum_{i=1}^{d}(u_i - \overline{u})(v_i - \overline{v})}{\sqrt{\sum_{i=1}^{d}(u_i - \overline{u})^2 \cdot \sum_{i=1}^{d}(v_i - \overline{v})^2}}. \tag{6.14}$$

The values range between $[-1, 1]$ with 1 indicating perfect linear correlation, -1 perfect linear anti-correlation, and values around 0 indicate the absence of a correlation.

It is important to note, that the Pearson correlation is only perfect when there is a linear relationship between two variables, i.e., if u_i is twice as large as u_j than v_i must also be twice as large as v_j. This is often not a realistic assumption and it is enough if a larger value in u is connected with a larger value in v but not necessarily with the same factor. That means, one seeks to capture a monotonic relationship instead of a linear relationship. In order to achieve that, the Spearman's rank correlation replaces the actual values of the vector with their rank within the vector and than calculates the linear relationship between the ranks of the variables using the Pearson correlation. Figure 6.4 depicts the difference of these two approaches.

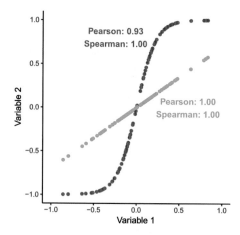

Figure 6.4: A perfect linear relation between some variable 1 and variable 2 (blue) as well as a non-linear but monotonic relationship (red). In case of the non-linear relationship, the value for the Pearson correlation drops to 0.93 while the Spearman rank correlation remains at 1.00.

6.4.2 Categorical Values

Besides continues variables, another important type of variables are categorical variables which can only take values of a finite set of discrete values. These types of variables are less common in technical measurements but are frequent for example in patient records. Most of the time, these variables do not possess a meaningful ordering (e.g., eye color), thus the classical measures for continues variables can not be employed anymore. Further, it is normally distinguished between two different types of categorical variables: Binary and general categorical variables. We will begin the discussion with boolean variables, which are a special case of categorical variables and extend from there to general categorical variables.

6.4.2.1 Boolean Variables

Boolean variables only have two categories which are often used to indicate the presence or absence of a given feature. Let us assume, we have two objects u, v for which we have d boolean features $F_1, \ldots, F_d$. When we seek to compare the two variables on feature F_i, there are only four different outcomes possible (presence of the feature is encoded as 1, absence as 0):

	Object v		
F_i		1	0
Object u	1	a	b
	0	c	d

When we now compare u and v on all d features, we count with the numbers **a,b,c,d** the absolute count of the different outcomes of the d comparisons.[1] Overall, **a** represents

---

[1] Do not mix the dimensionality of the dataset, which is defined as d throughout the entire chapter with the number of negative matches **d** used in this subsection.

the number of positive matches, **d** the number of negative matches, and **b** and **c** the number of mismatches.

The intuition of most measures is to relate the number of positive outcomes with the number of negative outcomes in order to establish a similarity between u and v.

This sounds trivial, but we have to think about the meaning of the different outcomes. We can see that **b** and **c** are equivalent as it does not matter which of the objects has the feature and which does not. In contrast, **a** and **d** are not equivalent. It might be a difference whether two objects agree on having a feature or agree in not having it. This fact becomes obvious when we consider a very rare feature, let's say which has a prevalence in the population of 1 in 1,000. The event that two objects have this feature is so much rarer than both objects not having the feature that the influence of these events should be judged differently. In this case, the presence of the feature has a much higher explanatory power than the absence. Please note, it is only by convention that the common presence of the feature is the event with the higher explanatory power. In case the common absence is the rare event, we count the common absence as **a**.

Nevertheless, there are also cases were the absence or presence have the same explanatory power, often when they are equally prevalent. When we for example encode "heads" and "tail" as our boolean feature for a row of coin flips, the agreement of showing "heads" in the ith throw has no more explanatory power than showing "tails" (considering a fair coin).

This is the reason, why most measures can be classified in two general approaches whether they treat **a** and **d** as equivalent or not [6]:

a and d are equivalent

$$s(u,v) = \frac{\mathbf{a}+\mathbf{d}}{\mathbf{a}+\lambda(\mathbf{b}+\mathbf{c})+\mathbf{d}}. \tag{6.15}$$

The parameter λ basically defines the influence of the disagreements on the measure. Common values for λ are 1 (then called the matching coefficient), 2 (coefficient by Rogers and Tanimoto [23]), and $\lambda = 1/2$ (coefficient by Gower and Legendre [24]).

a and d are different

$$s(u,v) = \frac{\mathbf{a}}{\mathbf{a}+\lambda(\mathbf{b}+\mathbf{c})}. \tag{6.16}$$

Here, the parameter λ has the same function as above. Again, common values are $\lambda = 1$ (Jaccard coefficient [25]), $\lambda = 2$ (coefficient by Sneath and Sokal [26]), and $\lambda = 1/2$ (coefficient by Gower and Legendre [24]).

6.4.2.2 General Categorical Variables

When dealing with categorical variables with more than two levels (the possible values a variable can assume), different measures need to be employed. In a naive approach we could simply dichotomize the categorical levels, i.e., creating for every level a separate variable indicating whether this particular level is present or not. This allows for the usage of the aforementioned coefficients for binary variables, but will consequently lead to a multitude of negative matches [6]. A different approach is deriving a score by counting the categorical features the two objects agree upon:

$$s(u,v) = \sum_{i=1}^{p} w_i \delta_i(u,v), \qquad (6.17)$$

where $\delta_i(u,v)$ is the indicator function whether the two variables agree on category i. Furthermore, w_i allows for weighting of the different categories, normally set to 1. The main purpose of this weighting factor is allowing the comparison of objects with missing value, i.e., objects without any value for some of the categories. In these cases, w_i is simply set to 0. There have been several extensions to this simplistic model. One approach is the normalization with the similarity you would archive by chance with randomly sampled objects [27]. A comprehensive overview can be found in the review of Boriah et al. [28] and the book of Gan et al. [11].

6.4.3 Practical Issues

A pressing issue in practice is the selection of the best suitable distance measure. First, the list presented above is only a very small selection of available proximity functions. Further, there exist several highly specialized measures for particular data types which are not yet included in this list. This is of particular relevance for clustering of biological data, as there exists a plethora of potential objects which are neither categorical nor continues. For example, protein sequences require entire different measures in order to establish a relationship between them. In the case of sequence similarity, there are many different measure available; probably the most used tool is NCBI BLAST [29].

Another example is the similarity between biological networks: Their similarity can be established for example by comparing the key properties of the networks (e.g., number of nodes, edges, shape of the node degree distribution), or more complicated functions like network alignment (detailed in Chapter 9) could be employed. These are just some examples, in practice many specialized measures exist. Discussing all of them would be beyond the scope of this chapter. Thus, as a piece of advise, when confronted with non-standard objects, extensive literature research for potential similarity functions is necessary.

6.5 Clustering Algorithms

This section will discuss various approaches to solve the clustering problem. As already discussed in the introduction, due to the versatility, hardness, and subjectivity of the clustering task, an incredible number of algorithms has been developed. While discussing all of them is unfeasible, we will briefly discuss the most common types of clustering algorithms and then will discuss four tools in more detail afterwards.

6.5.1 Cluster Approaches

In order to give the reader a broad overview of the variety of clustering algorithms, we will briefly discuss the different fundamental concepts most of the tools follow.

There exists no "one-size-fits-all" strategy delivering optimal results in all circumstances. Thus, the classification also discusses some weaknesses of the algorithms which widely apply to most of the tools belonging to that particular category. Broadly, clustering algorithms can be separated into the following categories:

Prototype based Algorithms in this category (sometimes also called squared error-based clustering) seek to identify an optimal k-clustering by means of a cluster prototype. The most well-known algorithm in this category is k-means which will be discussed in Section 6.5.2 and has shown to perform reasonably well on certain problems [30]. The general strategy is to iteratively update cluster centers and object memberships optimizing a given cluster criteria (for instance the sum-of-squared-error criteria).

Hierarchical These methods create an entire nested structure of the clustering. In order to receive a partitional clustering, the retrieved tree structure is cut at a certain level in order to produce k clusters. Normally those methods apply a local criteria (the linking function) in order to join the nearest pair of clusters until all objects are joined. Hierarchical clustering is a very common clustering technique and has been applied in a multitude of different biomedical studies, from multiple sequence alignment [31], protein-protein interaction networks [32], protein evolution [33, 34, 35] to gene expression analysis [36, 37, 38] to name just a few. We will discuss the most common forms of hierarchical clustering in Section 6.5.3.

Graph-based These algorithms represent the input data internally as a graph, with the objects corresponding to the nodes and the (normally weighted) edges to the similarities between the objects. Most algorithms then seek to identify densely connected subgraphs by incorporating the neighborhood of the nodes in the graph or by performing random walks on the given graphs. Most of the algorithms do not require the desired number of clusters k as parameter but are tuned by more indirect means. The probably most prominent examples are affinity propagation [39], clusterONE [40], Markov clustering [41], spectral clustering [42], and transitivity clustering [43]. Graph-based algorithms have been widely applied to biomedical datasets, particularly in the context of biological network and complex analysis [40, 44, 45, 46, 47, 48] but also for protein homology detection [49, 50]. We will discuss transitivity clustering in Section 6.5.5.

Density-based Here, the algorithms seek to identify arbitrarily shaped clusters by separating high-density areas from low-density areas. Normally, density-based clustering algorithms require only one scan of the dataset, are insensitive to noise [11] and can also detect outliers. Density-based clustering methods can be very efficient in terms of memory consumption and computational time and are thus often used for large-scale datasets [51]. The probably most prominent examples are density-based spatial clustering of applications with noise (DBSCAN) [52] and density clustering (DENCLUE) [53]. The primary biomedical application area of these clustering tools is the identification of dense subspaces in interactome data [51]. We will discuss DBSCAN in Section 6.5.4.

Model-based The main assumption of these methods is that the given dataset was generated by an underlying probabilistic model. The aim is to maximize the model in such a way that it best describes the observed data. One advantage of model-based clustering tools is that they allow for the integration of background distributions which is especially in a biomedical context of great importance. The most well-known representatives are hidden-Markov-model-based clustering [54], self-organizing maps [55] or finite mixture models [56]. Typical application areas are gene expression analysis [57, 58, 59] and sequence analysis [60].

Large-scale This is not a specific clustering approach, since tools in this category might belong to any of the above groups. Nevertheless, these methods describe approaches which were developed particularly for coping with large datasets, often running on distributed computer systems. Especially in the era of big data, coping with large-scale datasets becomes a crucial element. Normally, these approaches achieve their performance by either random sampling, data condensation, divide and conquer, incremental learning, or use density-based approaches, grid-based approaches, or a combination thereof [2]. This might result in a less accurate clustering but in cases were other methods fail, those methods might be a beneficial solution. Noteworthy examples of this category are the hierarchical approach BIRCH (Balanced Iterative Reducing and Clustering using Hierarchies) [61], CURE (Clustering Using Representatives) [62], CLARA [63], or the aforementioned tools DBSCAN [52] and DenCLUE [53].

Other strategies There exist several more general strategies which are less common and less relevant for this chapter as the focus is on the most popular clustering approaches applied in the biomedical context. Refer to the aforementioned reviews for an overview of additional approaches, e.g., grid-based clustering, neural network-based clustering, evolutionary algorithms-based clustering, and several more.

6.5.2 *k*-means

One of the most well-known and one of the oldest approaches to clustering is k-means. It is based on an inherently simple and comprehensible idea with a very good runtime complexity (basically linear in the number of objects). The simplicity and efficiency renders k-means to one of the most commonly used clustering approaches. Nevertheless, while useful and meaningful in many approaches, k-means also suffers some well-known drawbacks which will be discussed in this section.

6.5.2.1 *Algorithm*

The central idea of k-means is a so-called prototype based clustering approach, meaning each cluster is represented by a prototype object. There are several approaches on determining such prototypes, but standard k-means assumes objects to be points in a d-dimensional Euclidean space and uses the cluster center as the prototype for each cluster. This is not ideal for clustering networks since k-means requires the d-dimensional embedding. We discuss k-means because of its general importance to the field of clustering and briefly introduce in subsection 6.5.2.3 a variant which does not require the embedding in Euclidean space.

The algorithm itself then iteratively runs a two-step process. After the random initialization of initial cluster centers, each object is assigned to its closest center. Thereafter, the centers are updated based on the points assigned to the cluster. The interplay between updating the cluster centers and re-assigning the objects to

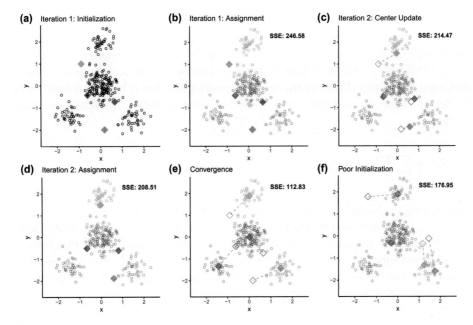

Figure 6.5: This figure represents the general function principle of k-means. Solid diamonds depict the current cluster centers, non-solid diamonds the initial choice of cluster centers. The cluster assignment of the points is indicated by their color. (a) The given dataset with four randomly chosen initial cluster centers. (b) The points get associated to the closest cluster center. (c) The cluster centers get updated by calculating the mean based on the assigned points. (d) All objects get reassigned to the updated cluster centers. (e) After several of those iterations, the algorithm reaches convergence and the final clustering is presented. (f) In case bad initial centers are chosen, a suboptimal clustering can emerge.

their closest cluster centerer is repeated until convergence. The algorithm is formally described in Algorithm 6.1 as pseudocode. Figure 6.5 (a–e) depicts the process on a small example.

Algorithm 6.1 k-means algorithm

Require: $n \times d$ two-mode data matrix, number of desired clusters d
 1: Initialize k cluster centers.
 2: **repeat**
 3:     Assign all points to closest cluster center.
 4:     Update cluster centers based on assigned points.
 5: **until** Centers do not change.
 6: **return** Clusters and cluster centers

Let us begin the analysis of the algorithm with several observations: The choice of the initial cluster centers will have a huge impact on the final result of the clustering. Figure 6.5 (f) depicts the consequences of a poor initial choice of the cluster centers leading to a suboptimal result. We will discuss better initialization methods in the next subsection.

Further, the distance to the closest cluster center has to be calculated. Here, most commonly the Euclidean distance (or other variants of the Minkowski distances) is employed but different measures are possible as well, e.g., correlations or cosine measures. When using the Euclidean distance, it can be shown that k-means seeks to minimize the so-called sum of the squared errors (SSE). Let c_i be the centroid of the ith cluster C_i, the SSE is defined on the overall distances of the objects to their cluster center:

$$\text{SSE} = \sum_{i=1}^{k} \sum_{x \in C_i} d(x, \bar{c}_i)^2, \tag{6.18}$$

with d being the Euclidean distance and $\bar{c}_i$ the center of cluster C_i.

This fact has several consequences in practice. First of all, the SSE is minimal when the clusters have a spherical and compact shape. That means in turn, that k-means seeks to form spherical clusters regardless of whether this is meaningful for the dataset or not. This can be observed in Figure 6.6 where the three elongated clusters are split in a suboptimal manner in order to produce spherical shaped clusters, regardless of the initialization.

Nevertheless, the SSE can be used to judge the relative quality of two clusterings obtained with the same k: In this case we would prefer the clustering with the lower SSE as this solution is closer to the optimal clustering with respect to the SSE. Please keep in mind that when we talk about the optimal clustering with respect to SSE does not mean that the ground truth is discovered optimally. When we compare the clusterings from Figure 6.5 (e) and 6.5 (f), we should clearly prefer the clustering shown in 6.5 (e) which has a smaller SSE. Nevertheless, when we compare with Figure 6.6 (b) and 6.6 (c), there is not really a reason why we should prefer one clustering over the other since both are fundamentally poor clusterings.

To sum-up, k-means is very efficient for detecting clusters of spherical shape of similar size and density. If these conditions do not hold in the presented dataset, one should consider utilizing a different clustering tool. Nevertheless, even in favorable cases for k-means, the optimality is not guaranteed and sophisticated initialization

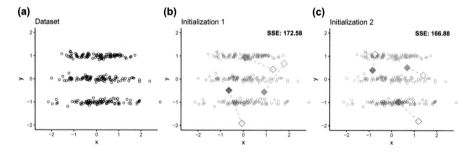

Figure 6.6: (a) Dataset used in this example to demonstrate shortcomings of k-means. Solid diamonds depict the final cluster centers, non-solid diamonds the initial choice of the cluster centers. The cluster assignment of the points is indicated by their color. (b) and (c) Regardless of the initialization, k-means does not find an ideal clustering.

strategies should be employed. Further, outliers in the dataset can distort the clustering quite heavily and should be disregarded. This can be done, for example, by removing points showing an unusual large squared error over several clusterings.

6.5.2.2 Initialization Strategies

As seen in the previous section, the result of k-means highly depends on the initialization of the centroids. As a rule of thumb, natural clusters are wrongfully split when two or more centroids are placed in the same cluster (compare to Figure 6.5 and Figure 6.6). With an increasing number of k this will become more likely the case and suboptimal clusterings are the result. The most straightforward approach is the repeated random initialization leading to a multitude of different clusterings. The "winning" clustering is the one with the least SSE measure. Nevertheless, repeated clustering might be too expensive when using large datasets; further, for large k optimality is hard to archive with random initialization.

A common and more sophisticated initialization method is the so-called *furthest first* strategy (sometimes also called *farthest first*): The first cluster center is chosen at random. All subsequent cluster centers are chosen to be the objects which are furthest away from all previously chosen cluster centers. This strategy ensures that the object space is quickly populated and prevents cluster centers being too close to each other, which would lead to an unnatural split. Nevertheless, this method has a significant disadvantage: Outliers tend to be exactly those points furthest away from the majority of objects, thus this initialization process favors outliers as cluster centers. Naturally, outliers are poor cluster centers as they are by definition the opposite of a cluster center.

In order to tackle the problem of outlier selection, the strategy can be refined to the so-called *subset furthest first* strategy. Here, a subsample of the entire dataset is generated. As outliers are by definition rare, it is likely that no or only a very limited number of actual outliers are in the subsample. When then applying the furthest first strategy, the likelihood of selecting meaningful cluster centers is elevated. Nevertheless, this method requires substantially more objects in the dataset than clusters; it would fail if we expect, for instance, an average cluster size of two or three.

6.5.2.3 Other Variants

The presented prototype-based (i.e., one cluster center acts as the prototype for all elements in the cluster) approach of k-means is a very broad and general approach to the problem (in fact it can be seen as a special case of an expectation maximization algorithm) and many different flavors have been developed over the years. Some are tuned for an increased efficiency by, for example, only clustering a subset of the clusters or by further reducing the amount of required similarity calculation between the objects and cluster centers.

A noteworthy other variant of k-means is the so-called *partitioning around medoids* (PAM). This methods differs from k-means by that it does not calculate cluster centers but uses objects of the dataset as centers. This allows the usage of PAM also for one-mode datasets (e.g., similarity matrix with the pairwise interaction scores of a PPI network). The procedure follows the same concept as k-means by iteratively assigning objects to the closest medoid followed by updating the medoid. It is important to acknowledge that PAM is computationally more expensive than k-means: In order to find the optimal medoid for each cluster, each object of the cluster has to be considered the medoid and the sum of the distances of all other points has to be calculated. This is more expensive compared to averaging over all points of a cluster. Thus, for larger datasets, subsamples of the dataset are employed in order to reduce the number of comparisons.

Among others, k-means was also extended to facilitate fuzzy clustering, meaning each object belongs to each cluster to a certain degree. Probably the most popular approach is the so-called *fuzzy C-means* (FCM) clustering which we briefly discuss here. As in the definition of the fuzzy clustering, let $u_{i,j}$ denote the degree of membership of x_i to cluster C_j. The cluster center c_j for cluster C_j is calculated according to the degree of membership of the objects:

$$c_j = \frac{\sum_{x_i \in X} u_{i,j}^m \cdot x_i}{\sum_{x_i \in X} u_{i,j}^m}, \qquad (6.19)$$

with $m \geq 1$ being the *fuzzifier* which will be discussed below. The degree of memberships of x_i is updated according to the distances to all cluster centers with

$$u_{i,j} = \frac{1}{\sum_{l=1}^{k} \left(\frac{d(x_i, c_j)}{d(x_i, c_l)} \right)^{\frac{2}{m-1}}}, \qquad (6.20)$$

with $d(x_i, c_j)$ being the employed distance measure, e.g., the Euclidean distance. These formulas define the update process and a similar strategy as for k-means can be applied in order to find the optimal clustering with respect to the weighted sum of the squares error:

$$\text{WSSE} = \sum_{i=1}^{n} \sum_{j=1}^{k} u_{i,j} \cdot d(x_i, c_j)^2. \qquad (6.21)$$

The fuzzifier m defines how fuzzy the resulting clustering will be. Note that for $m = 1$ the memberships $u_{i,j}$ converge to either 1 or 0 resulting in a crisp clustering. Without any prior knowledge, m is normally set to 2 and then adjusted depending on the needs of the researcher. Please also note, that FCM generally suffers the same problems as k-means, so it has a tendency to produce same-sized spherical clusters and does not guarantee to discover the optimal partitioning.

6.5.3 Hierarchical Clustering

Similar to k-means, hierarchical clustering is a long standing approach to clustering. Nevertheless, it is still widely used, also for biomedical datasets. Hierarchical clustering has the advantage that it does not only provide a separation of the dataset into groups but also presents the hierarchical relationship between the clusters. This is normally presented in a dendrogram as depicted in Figure 6.7. A crisp clustering can be archived by cutting the dendrogram at a certain level.

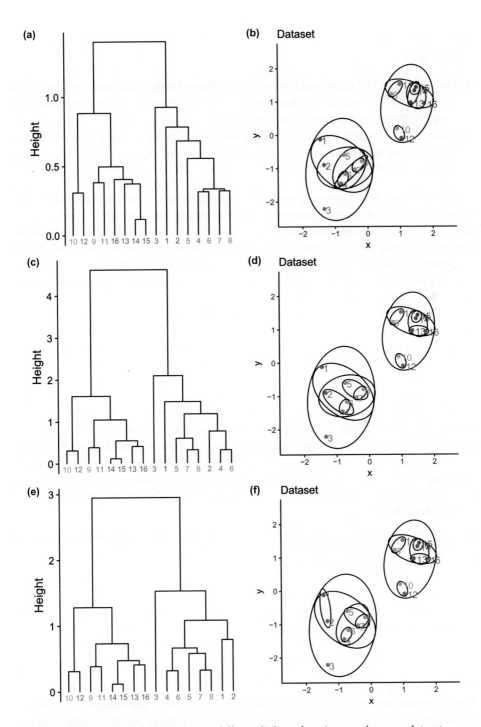

Figure 6.7: Hierarchical clustering using different linkage functions on the same dataset. (a)–(b) Clustering using single linkage. (c)–(d) Clustering using complete linkage. (e)–(f) Clustering using average linkage.

6.5.3.1 Algorithm

When employing hierarchical clustering, two fundamentally different approaches can be distinguished:

Agglomerative The clustering starts with all objects being singletons, i.e., clusters of size one and consecutively joins the closest clusters until all clusters are joined into one big cluster containing all objects.

Divisive The clustering proceeds in the opposite way by starting with all objects being in one big cluster and subsequently divides this cluster into smaller clusters until all objects are singletons.

Even though both clustering methods appear to be equivalent to some degree, the divisive method is in practice computationally more expensive than the agglomerative approach. While the agglomerative clustering uses the minimal distance between the clusters as a cheap heuristic, the divisive approach is required to find the optimal split of a large cluster into smaller clusters. For the latter problem there are no as cheap heuristics as for the agglomerative case; recall the number of possibilities one has to split a dataset into a given number of groups.

On the other hand, the agglomerative methods suffer the problem that they base their decision locally by greedily seeking the current minimal distance while disregarding the overall picture of the entire dataset. Once two clusters are joined they cannot be separated anymore. That can lead to the consequence that an early local suboptimal joining would have lead to a globally better solution; impossible for an agglomerative method to detect. This problem does not occur with the divisive approach as here the decisions for the large clusters are made early on before going into the fine-grained regions. Nevertheless, the computational efficiency has made the agglomerative approach the standard approach when hierarchical clustering is used.

The actual algorithm is again comparably simple as can be seen in Algorithm 6.2. The different variants only employ different measures, the so-called *linkage function*, for establishing the closest pair of clusters. We will discuss the different linkage functions in detail below.

Algorithm 6.2 Agglomerative hierarchical clustering

Require: $n \times n$ one-mode proximity matrix.
1: **repeat**
2:   Merge closest clusters according to linkage function.
3:   Update the distance matrix with the new merged cluster.
4: **until** All objects are in the same cluster.

6.5.3.2 Linkage Functions

The main operation of the hierarchical clustering algorithm is the linkage function, i.e., defining the distance between a pair of clusters. The linkage function highly defines the preferred shape and properties of the clustering; thus it is of utmost importance that the researcher is aware of these facts and whether the dataset is appropriate for the selected linkage function.

Single linkage The single linkage function defines the distance between two clusters as the minimal distance between any two objects of the two clusters. Single linkage clustering can handle clusters of arbitrary shape and sizes. Nevertheless, it has the tendency to form elongated clusters and tends to produce unwanted fading of overlapping clusters, the so-called chaining phenomena. Further, it is very sensitive to noise.

Complete linkage The complete linkage function is the opposite approach to single linkage clustering. Here, the distance between two clusters is defined as the maximal distance between any two points of the two clusters. The name of the linking function originates from a graph theoretical point of view: When repeatedly adding edges into the graph with ascending weights (i.e., distances), two clusters are joined in the moment they form a clique, i.e., they are completely linked. Complete linkage can handle noise and overlapping clusters to a better degree but produces clusters with equal diameter.

Average linkage The average linkage functions seeks to get the "best of both worlds" and can be seen as an intermediate between single and complete linkage. Here, the two clusters with the lowest pairwise average distance between all points of both clusters are joined. This method is less susceptible to outliers than single linkage but tends to produce spherical clusters similar to k-means.

Lance–Williams recurrence formula There exist several additional linkage functions, most notably Ward's method [64] which uses the least increase in the SSE and thus seeks to optimize the same criteria as k-means. True for all methods is that a recalculation of all pairwise similarities after each join would be a waste of resources. Normally, instead of recalculating the new distances, the proximity matrix can easily be updated while using the distances of the previous steps.

Let us assume we are merging the clusters A and B in the new cluster R. Now, all proximities of the cluster R to all other clusters must be established. The calculation of the new proximity $p(R, Q)$ of the cluster R to any existing cluster Q can be done solely based on already known proximities by applying

$$p(R, Q) = \alpha_A p(A, Q) + \alpha_B p(B, Q) + \beta p(A, B) + \gamma |p(A, Q) - p(B, Q)|, \tag{6.22}$$

with $\alpha_A, \alpha_B, \beta, \gamma$ being parameters. The single linkage function can be archived by choosing $\alpha_A = \alpha_B = 1/2$, $\beta = 0$, and $\gamma = -1/2$. All the parameters for all other linkage functions can be seen in Table 6.1.

6.5.3.3 Discussion

As already mentioned, hierarchical clustering is a very popular approach to clustering. The benefits of additionally receiving a dendrogram unraveling the relationship between the clusters can reveal further information which might have remained hidden from the practitioner otherwise. In contrast to k-means, hierarchical clustering only requires a $n \times n$ one-mode proximity function and thus can be applied to all datasets, in particular for the clustering of biological networks.

Table 6.1: List of different parameters for the Lance–Williams recurrence formula for the most commonly used linking functions. $n_{A/B/Q}$ denotes the number of elements in clusters A, B, or Q, respectively. The methods indicated by a * are methods which were not discussed in this chapter and are only mentioned for the sake of the interested reader.

Method	Parameters		
	$\alpha_{A/B}$	β	γ
Single linkage	$1/2$	0	$-1/2$
Complete linkage	$1/2$	0	$1/2$
Average linkage	$\dfrac{n_A}{n_A + n_B}$	0	0
Centroid linkage*	$\dfrac{n_A}{n_A + n_B}$	$-\dfrac{n_A n_B}{(n_A + n_B)^2}$	0
Median linkage*	$1/2$	$1/4$	0
Ward's method	$\dfrac{n_A + n_Q}{n_A + n_B + n_Q}$	$-\dfrac{n_Q}{n_A + n_B + n_Q}$	0

The user has to carefully choose the best suited linkage function as they greatly influence the resulting clustering. Another disadvantage is as previously mentioned, that any decision to merge a cluster is final and cannot be reverted. This might lead to suboptimal results as the fine-grained decisions are made first. There are some attempts which try to ease on that problem at the expense of increased computational complexity. Furthermore, in order to receive a partitional clustering, the tree has to be cut on a certain level. Here, again the user has a wide variety of different approaches to cut the tree as efficient as possible, e.g., not on the same level everywhere but at varying levels.

Additionally, compared to k-means, hierarchical clustering is more expensive in terms of runtime and memory requirements as all pair-wise proximities between the objects need to be stored and managed. There exist methods (e.g., BIRCH [61]) which attempt to tackle this problem by employing a highly efficient data structure which is created on-the-fly and thus enables the efficient clustering of large-scale datasets.

6.5.4 DBSCAN

Another important category of clustering algorithms are the so-called density-based algorithms. All discussed methods so far favor certain shapes or sizes of the clusters. Density-based approaches normally are able to cluster entirely arbitrarily shaped clusters of various sizes. Figure 6.8 depicts such an instance and compares the performance of DBSCAN to k-means and hierarchical clustering.

Generally, density cluster approaches work by dissecting dense areas of the dataset from sparse areas in order to define clusters. Normally these algorithms are very efficient and capable of clustering large datasets. In this section, we will discuss DBSCAN [52], one of the probably most well-known representatives of density-based clustering.

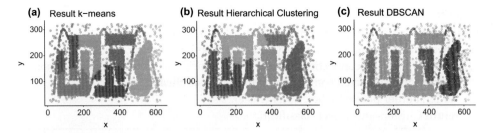

Figure 6.8: Clustering of the chameleon dataset published by Karypis et al. [65] consist of 6 distinct clusters with background noise. The results of k-means (a) with $k = 6$ as well as hierarchical clustering (b) with average linkage function deliver only suboptimal results. Only DBSCAN (c) is able to correctly identify all clusters and further classifies the background as noise (red color) with *MinPts*= 30 and *Eps*= 13.

6.5.4.1 Algorithm

All density-based approaches have in common that they need a definition of the density in order to dissect areas with high density from areas with low density. In case of DBSCAN, the density is defined as the number of other points in the immediate surrounding of the point. DBSCAN now classifies each point based on a user-given radius *Eps* and a minimum number of points within the radius *MinPts* as:

Core point   A point which has at least *MinPts* other points within the radius *Eps*, i.e., points with a lesser distance than *Eps* from the point in question. These points will later form the inner part of the cluster.

Border point   A point which has less than *MinPts* points in its *Eps*-surrounding but has at least one core point in its neighborhood. Clearly, this will most commonly happen in the areas where there is a transition from high-density areas into areas of low density, thus they are border points.

Noise points   Points which do not fulfill the criteria for both aforementioned types of points. These points can then also be seen as noise as they do not have a sufficient number of points in their neighborhood as well as they are not close to dense areas.

It is apparent that the results of DBSCAN highly depends on the choice of *Eps* and *MinPts*. The algorithm itself is straightforward (refer to Algorithm 6.3). The computationally most expensive part is the search of the number of points within the *Eps* neighborhood. Naively implemented, this would require a complexity of $\mathcal{O}(n^2)$ as each point need to be compared to each other point. But in lower dimensional cases, efficient data structures exist which accelerate the neighborhood search significantly, e.g., by the use of *kd*-trees [66].

6.5.4.2 Discussions

The advantages of density-based clustering methods is their ability to recover arbitrarily shaped clusters of arbitrary size. Also, the classification of points as core, border, and noise may lead to meaningful insights into the clustering problem which other clustering approaches normal cannot provide.

Nevertheless, density-based approaches also have shortcomings. Since the choice of parameters highly influences the classification of the various points, these

Algorithm 6.3 DBSCAN algorithm

Require: $n \times d$ two-mode proximity matrix, *MinPts*, and *Eps*
 1: Check surrounding of each object and classify as core, border, noise.
 2: Create a graph by connecting all core points which are in the *Eps* of each other.
 3: Report each connected component of the graph as cluster.
 4: Assign border points to the closest cluster.
 5: **return** Clusters and object classification.

approaches might have difficulties with clusters of various density, especially when the clusters are surrounded by noise. Another shortcoming is that these methods normally do not work very well in high-dimensional datasets. This fact can easily be explained with the curse of dimensionality: When we want a cluster to maintain the same density in a high-dimensional space as in a low-dimensional space, we would require an exponentially growing number of objects with each added dimension. Since the number of objects in clustering problems is fixed, this results in only low-density areas in high-dimensional spaces. In those cases, there won't be a meaningful definition of density anymore.

6.5.5 Transitivity Clustering

The last clustering tool described in this chapter will be transitivity clustering [49, 67, 43, 68], a graph based approach to the clustering problem. Graph based approaches consider the input as a graph in which densely connected areas are identified and reported as clusters. Normally, the graph $G = (V, E)$ is constructed with the objects X forming the nodes V and each edge in E between a pair of nodes corresponds to the similarity between the points. Compare to Figure 6.9 (a) for an example. Naturally, graph-based approaches lend themselves to network clustering problems since they already interpret the input as a network.

6.5.5.1 *Transitive Graph Projection Problem*

Transitivity clustering is based on the so-called weighted transitive graph projection problem (WTGPP). Given an input graph $G = (V, E)$ as described above and a user-given threshold t, the graph is transformed into a graph G' by removing all edges with a similarity below the threshold from the graph G (Figure 6.9 (b)). This graph might fall into several connected components and generally will not be transitive anymore. Transitive means that for all triplets $u, v, w \in V$ holds true: If $(u, v) \in E$ and $(v, w) \in E$ then there also must be $(u, w) \in E$. In other words, each connected component of the graph must be a fully connected clique.

In general, this is not fulfiled by the graph G'. The idea of transitivity clustering is to transform the graph G' into a transitive graph G'' with the least number of modifications. This is done by adding and removing edges from the graph until the transitivity condition holds for all triplets u, v, w. Each of these so-called edit operations is attached with certain costs which are normally calculated by $|t - s(u, v)|$, the difference between the user-given threshold and the actual similarity between the points u and v.

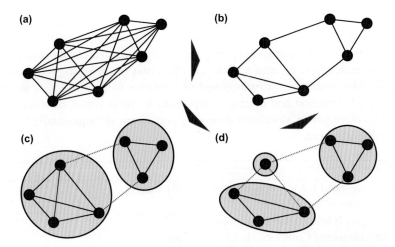

Figure 6.9: The principle of the weighted transitive graph projection problem. (a) The input is considered as a graph G. (b) The graph is transformed into a graph G' were all edges below the user given threshold t are removed. (c) and (d) Two possible version of the modified transitive graph G''. Edge insertions are marked in red, edge deletions with a dashed red line. Depending on the similarities one solution might have less costs than the other solution. The cliques of G'' are reported as clusters (marked in gray).

The solution requiring the least overall costs is the result for the clustering problem (Figure 6.9 (c) and 6.9 (d)). By assigning the costs to the edit operations it is assured that removing edges of highly similar objects is very costly as well as adding edges between very dissimilar objects.

Each connected component of G'' is then reported as a cluster. Solving the weighted transitive graph projection problem is computationally very expensive and thus in most practical instances not feasible to be solved exactly. Thus, a heuristic, described in the next section, is applied.

6.5.5.2 Heuristic Solution

The applied heuristic employs a nature-inspired graph layouting approach. The general procedure is as follows: The objects are projected into an d-dimensional space by arranging them uniformly on a d-dimensional sphere and then forces are applied in such a way that objects with a high similarity attract each other while dissimilar objects repulse each other. Once convergence is reached, a standard clustering tool like k-means or hierarchical clustering can be applied to the projected objects. Algorithm 6.4 describes the general procedure of the algorithm.

Once the objects are arranged on a d dimensional sphere (normally $d = 2$ or 3), the algorithm calculates the force object v inflicts on object u by [67]

$$f_{u \leftarrow v} = \begin{cases} \dfrac{c(uv) \cdot f_{\text{att}} \cdot \log(d(uv) + 1)}{|V|} & \text{for attraction (i.e. } s(uv) > t\text{),} \\ \dfrac{c(uv) \cdot f_{\text{rep}}}{|V| \cdot \log(d(uv) + 1)} & \text{for repulsion (i.e. } s(uv) \leq t\text{).} \end{cases} \quad (6.23)$$

Here, $c(uv)$ refers to the costs of adding or removing an edge which is calculated by $|t - s(uv)|$ with t being the user-given threshold and $s(uv)$ the similarity between the objects u and v. The distance $d(uv)$ corresponds to the Euclidean distance of the objects in the projection space; f_{att} and f_{rep} are the attraction factor or the repulsion factor, receptively. This results in a displacement of u relative to the force $f_{u \leftarrow v}$ in direction of v in case of attraction and in case of repulsion in the opposite direction. In order to improve convergence, simulated annealing is performed by gradually limiting the maximal displacement.

Algorithm 6.4 Transitivity clustering algorithm

Require: Threshold t, $n \times n$ one-mode similarity matrix.
 1: Create graph G' based on t
 2: Dissect graph into connected components CCs
 3: **for all** Connected Components $CC \in CCs$ **do**
 4:     Arrange objects of CC on a d-dimensional sphere
 5:     **repeat**
 6:        Calculate pairwise attraction and repulse forces
 7:        Move objects according to forces
 8:     **until** Convergence reached
 9:     Cluster connected component based on the layout
11: **end for**
10: **return** All clusters of all connected components.

6.5.6 Discussion

Transitivity clustering has proven to work very well on biological as well as artificial datasets [69]. The concept of transitivity suits many clustering problems, since it enforces the intuition that if the objects u and v are similar as well as v and w then, in order to form a meaningful cluster, the objects u and w also should be similar.

Nevertheless, this way defining a cluster prefers rather spherical and compact clusters. Furthermore, since the attraction and repulsion forces need to be calculated for all pairs of objects, the runtime is necessarily quadratical. This might lead to runtime issues for massive datasets. This effect is reduced by the fact that only connected components of the graph G' need to be clustered since two distinct connected components within G' will never be joined in an optimal solution for the graph projection problem. This reduces the runtime to being quadratic in the number of objects of the largest connected components.

6.6 Cluster Evaluation

After the clustering of the dataset, the researcher has arrived at the point of the first full feedback. The results are ready for inspection and are the consequence of the decisions made previously. It is of utmost importance that the results get evaluated in a strategic and unbiased fashion. This makes the evaluation to probably the most critical step in a cluster analysis. The inherent problem of this endeavor is that normally a gold standard, i.e., the ground truth, is missing. Clustering results are normally evaluated by means of so-called cluster validity indices (or measures). Generally, three different types of validity indices are distinguished [70]:

External validity indices These compare a clustering result against a gold standard.
Internal validity indices Normally, these are employed when a gold standard is absent. In such case, the clustering can only be evaluated by means of the clustering result itself and the unlabeled input.
Relative validity indices The last type of indices compare several clusterings from the same algorithm, e.g., when run with different parameters, or different algorithms with each other. This is a common method for parameter training or determining the number of clusters k in a dataset.

6.6.1 External Cluster Evaluation

External validity indices are used when the ground truth, i.e., a gold standard is given. In most practically relevant cases, this won't be the case: When the answer is already known, why cluster in the first place? Nevertheless, there are several reasons for the meaningfulness of these external validity indices. First, it allows for a clustering tool performance evaluation on test data. Here, it makes more sense to use a gold standard in order to objectively judge the performance of a clustering tool. Second, sometimes a gold standard for a similar dataset or for a subset of the dataset is available. In those cases, the external validity indices are a very powerful means of determining the best tool and parameter configuration before turning to the full dataset.

All external indices seek to establish a similarity between the clustering C and the given gold standard K. One approach is considering all pairs of objects u, v and determine whether they are clustered together in C and K. Comparable to the treatment of binary features, let **a** define the number of pairs being clustered together in C and K, **d** the number of pairs which clustered apart in C and in K, **b** the number of pairs which are clustered together in C but not in K, and **c** the number of pairs which are clustered apart in C but are clustered together in K. The most common measure based on this definition are (also note the resemblance to the similarity measures of categorical variables in Section 6.4.2):

Rand index The Rand index is defined as follows [71]:

$$R(C, K) = \frac{a + d}{a + b + c + d}. \tag{6.24}$$

Jaccard index The Jaccard index is very similar to the Rand index but does not count negative matches [25]:

$$J(C, K) = \frac{a}{a + b + c} \tag{6.25}$$

The Jaccard index can be seen as a stricter definition of the Rand index because it only counts positive agreements [72]. The main criticism on these indices, particularly on the Rand index, is their rather small range of actually achieved values and their tendency to generally report higher values with an increasing number of clusters [73, 6]. The most notable extensions to the Rand index is the so-called adjusted Rand index which seeks to counter the aforementioned shortcomings by statistically correcting for chance [74].

Another way of comparing a clustering to a gold standard is by mapping every cluster $C_i \in C$ to the cluster $K_j \in K$ with the greatest overlap. Assuming that the cluster C_i has the largest overlap with K_j, each element $u \in C_i \cup K_j$ can be defined as true positive (TP) if $u \in C_i$ and also $u \in K_i$, as false positive (FP) if $u \in C_i$ but not in K_i and as false negative (FN) if $u \in K_j$ but not in C_i. There is no meaningful definition of true negatives in this scenario. Based on these findings, we can define different validity indices:

Precision basically measures the ratio of TP in the set of all positively classified object pairs (TP and FP) in the clustering:

$$\text{Prec}(C, K) = \frac{TP}{TP + FP}. \quad (6.26)$$

Recall follows a similar approach and measures the ratio of the TP object pairs of all object pairs which should have been positively classified (TP and FN):

$$\text{Rec}(C, K) = \frac{TP}{TP + FN}. \quad (6.27)$$

Precision as well as recall have the problem that they could easily be optimized by producing a trivial solution; a clustering with only singletons would give a perfect precision due to the lack of FPs; a clustering with one single cluster containing all objects would result in an optimal recall due to the lack of FNs.

F-measure is probably the most commonly used external validity index and corrects the obvious shortcomings of precision and recall as it is defined as the harmonic mean of both:

$$F_\beta(C, K) = \frac{(1 + \beta^2) \cdot \text{Prec}(C, K) \cdot \text{Rec}(C, K)}{(\beta^2 \cdot \text{Prec}(C, K)) + \text{Rec}(C, K)}. \quad (6.28)$$

The parameter β is used to influence the weight of the precision or recall. In most cases, $\beta = 1$ (in that case, it is also often called the F1-measure) meaning both precision and recall have the same importance.

This selection only represents a very small portion of all available external measures. More measures are discussed for example in the review of Rendón et al. [75].

6.6.2 Internal Cluster Evaluation

Internal validity indices are more common in real-world applications, since they do not require the existence of a gold standard. The problem of defining a good validity index is analogous to defining a good cluster criteria and thus suffers the same problems of being rather subjective and problem dependent. Generally speaking, most internal measures seek to reward *compactness* and *separation* of the clusters. Due to the similarity to the clustering problem itself, it is crucial that the selected internal measure fits the data. Exactly like the cluster criteria, there are validity indices which prefer compact, spherical clusters, other expect an equal density within the cluster, etc. That means in turn, the validity index must fit the data by means of what shape and type of clusters are expected. Most commonly, the following internal measures are used:

Dunn index relates the maximal cluster diameter $\max_\emptyset$ to the minimal distance between any pair of clusters $\min_d$ [76, 77]:

$$D(C) = \frac{\min_d}{\max_\emptyset}. \quad (6.29)$$

One obvious disadvantage of this index is the sensitivity to outliers as they might tremendously influence the maximal diameter or minimal distance.

Davies–Bouldin index can be seen as a relaxed version of the Dunn index as it relates the average distance of objects in a cluster to the cluster center to the distance between cluster centers [78]. Let $\overline{C_i}$ be the center of cluster C_i and $\sigma_i = \sqrt{1/|C_i| \sum_{u \in C_i} |u - \overline{C_i}|}$ and $d(C_i, C_j)$ the Euclidean distance between two cluster centers, then the Davies–Bouldin index is defined as:

$$DB(C) = \frac{1}{n} \max_{C_i \neq C_j \in C} \left(\frac{\sigma_i \cdot \sigma_j}{d(|\overline{C_i}|, |\overline{C_j}|)} \right). \quad (6.30)$$

Silhouette value is calculated for each object $u \in C_i$ individually. It relates the average distance of the object to all objects within the cluster to the average distance to all objects in the closest foreign cluster. Let $a(u)$ denote the average distance of u within its cluster and $b(u)$ the minimal average distance to all objects of a different cluster, then the silhouette value is defined as

$$S(u) = \frac{b(u) - a(u)}{\max\{a(u), b(u)\}} \quad (6.31)$$

and for the entire dataset as

$$S(C) = \frac{1}{n} \sum_{u \in X} S(u). \quad (6.32)$$

The silhouette value can assume values between $[-1, 1]$ with negative values indicating that a point is on average closer to a foreign cluster than to its own cluster. In a large-scale study evaluating several popular clustering tools it has been shown that the silhouette value corresponds well to external indices for real-world datasets but has problems with intertwined clusterings [69]. Figure 6.10 depicts a typical silhouette plot.

A good review of many more internal cluster validity indices can be found in the work of Liu et al. [79].

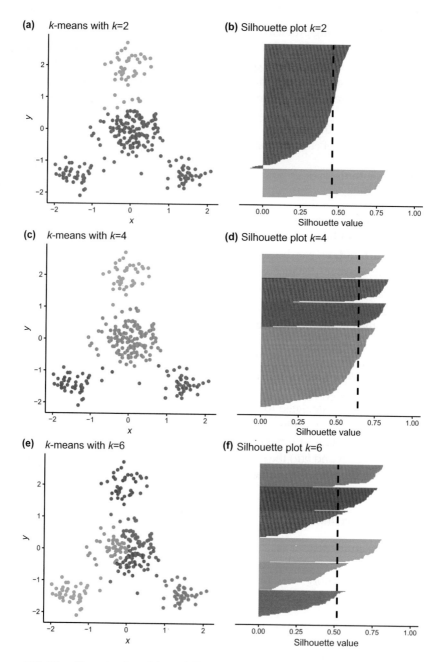

Figure 6.10: The silhouette plot of three different clusterings of the same dataset with varying k. The average silhouette value of the entire clustering is indicated with the black dotted line and shows the best value at the true number of clusters. The reader should be aware that there is no guarantee that the best silhouette value corresponds to the best clustering.

6.6.3 Optimization Strategies

In real world applications, a gold standard is not available. Thus, the researcher has to resort to internal validity indices in order to judge the quality of the clustering as well as to tune the parameters of the clustering tool. Since each clustering tool has at least one parameter, this can be a quite time-consuming and tedious work.

The main problem is that internal indices do not give an absolute measure of the quality of the clustering. That means, a good silhouette value by itself does not guarantee for a good clustering. Nevertheless, the internal indices allow for relative comparisons between several clusterings (compare to Figure 6.10). It is important to note that just testing for a good internal index might be misleading: Several indices show a monotonic behavior with respect to the number of clusters or other parameters. One example is the SSE of k-means: The more clusters, the better this measure, regardless of the actual quality of the clustering (compare to Figure 6.11 (c)).

Thus, more meaningful and strategical approaches are required. Since the number of clusters k is such a crucial parameter, the strategies for optimizing the clustering are actually distinguished whether k is a direct parameter of the tool or not [80, 81].

6.6.3.1 k is not a Parameter

For clustering tools where k is not a direct parameter but the number of clusters is rather influenced indirectly by means of other parameters, a typical optimization process can be as follows: All parameters are varied in a wide range. Depending on the number of parameters, this can be a very time-consuming process. Then, instead of using an internal measure, the various results are compared to each other. The widest range of parameters leading to the exact same (or almost the same) clustering is a good indication for having unraveled the actual structure of the dataset. In other words, when the same result is produced by a wide range of parameters it is likely caused by the inherent structure of the clustering and not by spurious noise. Additionally, internal measures can be used in order to check that the discovered clustering also produces the highest quality by means of internal validity indices.

6.6.3.2 k as a Parameter

If the tool has the desired number of clusters k as a direct parameter the application of an internal validity index is inevitable. In these cases, k is varied in a wide range. Additionally, for each k tested, the remaining parameters of the tool are also tested in an as wide as possible range. For each of the clustering results, an internal validity index is evaluated and the parameter configuration yielding the best result is used for the final clustering. As already stated, there are validity indices which constantly increase (or decrease) with a growing number of clusters regardless of the actual clustering quality. When using these indices, the researcher can look for a so-called knee (or elbow) in the plot of the validity index, i.e., the point where after a rapid decrease the validity index's decrease levels off. Figure 6.11 (c) depicts the elbow on a toy dataset. This concept was generalized in the so-called gap statistic (discussed below). A more exhaustive discussion can be found in the work of Halkidi et al. [81].

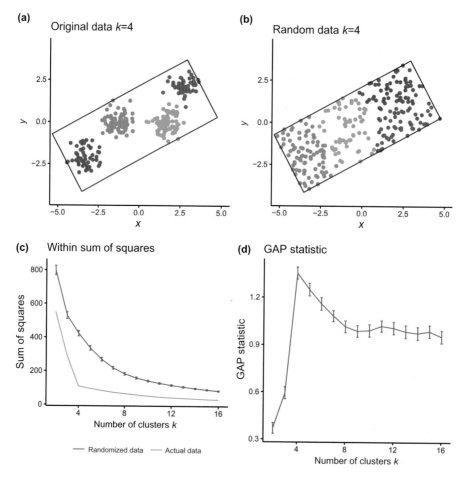

Figure 6.11: (a) Toy dataset consisting of four spherical clusters with 260 objects in total. (b) One sample of the randomly created noise datasets which have the same spread along the principal components as the actual dataset (bounding box is indicated in blue). (c) The sum of squares of the actual clustering (blue) and the average of several (100) runs on randomized data. Note, that both sum of squares monotonically decrease with an increasing k, but a clear elbow can be observed at $k = 4$. (d) Value of the gap statisitc. Here, a clear peak can be observed at the true number of clusters $k = 4$.

6.6.3.3 The Gap Statistic

The gap statistic [82] is a more strategic approach of discovering the "elbow" of a monotonically decreasing validity index. The general idea is that the validity index is statistically corrected by a background model. In order to create a background model, normally a dataset of the same size is sampled uniform at random. The spread of the data is in the same bounding box as the original dataset, aligned to its principal components (See Figure 6.11 (a)–(b)). Several of these random datasets are clustered in the same way as the original datasets (e.g., using k-means with the same k) and the values of the internal validity index is calculated. In Figure 6.11 (c) the within sum of squares error of the actual dataset is plotted together with the value achieved by the random "noise" datasets. The idea is that the difference between the observed statistic and the background clustering is due to the actual structure of the dataset. The gap is calculated as

$$\text{Gap}(k) = E\left[\log(W_k)\right] - \log(W_k) \tag{6.33}$$

where $E\left[\log(W_k)\right]$ denotes the expected SSE based on the random background model and W_k the actual observed SSE. Figure 6.11 (d) depicts the value of the gap for various k. This method is quite general and the idea can be adapted to different distance measures or internal indices [82].

6.7 Final Remarks

This chapter aimed to give the reader a good understanding of the most common techniques and approaches to clustering. It represents only a condensed overview and it is highly advisable when confronted with a clustering problem to seek advice in the extended literature. Especially in bioinformatics, several tailored approaches to specific datasets or problems exist.

It is of importance that each decision is made consciously and with full awareness of the consequence to the downstream methods. For example, the choice of similarity function might influence the performance of the clustering tool. Generally, when confronted with a clustering task, try to gather as much information as possible about the dataset. What shape of clusters do you expect? Will they have an equal density? Equal size? Equal shape? How many dimensions are you dealing with? The more of this information you are able to gather the better, and targeted decisions can be made. When presented with the clustering results, carefully evaluate the results and try to find an objective measure to compare the different clustering tools. Again, the choice of the internal validity index is also highly dependent on the shape of your clustering.

6.8 Exercises

6.1 There exist clustering tools which claim to not require any parameter to be set. Those tools use an internal quality control returning the optimal result. Can this be a realistic claim for all datasets and can you describe a situation where this might not hold?

6.2 Can a principal component analysis be meaningfully applied to all datasets? Can you think of examples where a PCA might fail to reduce dimensions?

6.3 Describe k-means in your own words. Argue what you think are the greatest advantages and greatest disadvantages of k-means and present an example illustrating those.

6.4 Download the dataset from the web-resources of the book (available at www.cambridge.org/bionetworks) containing data on the food consumption amongst different areas in the UK. The data was collected from the 2017 version of the family food datasets from Department for Environment Food and Rural

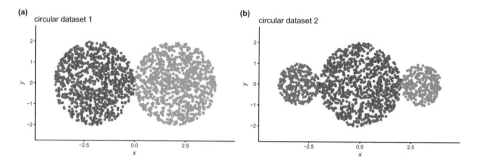

Figure 6.12: Two toy datasets for learning the effects of the different linkage functions of hierarchical clustering. (a) Dataset with uniformly distributed points in two touching equal sized circles. (b) Dataset with uniformly distributed points in three differently sized circles.

Affairs.[2] The data describes the weekly intake of consumables averaged over the last three years (2014–2016) of British families. Download the dataset and perform a PCA on the entire dataset.

(a) Project the dataset onto the first and second principal component.

　i. How much variance is lost during the projection?
　ii. What areas cluster together? Are those areas also geographically close together, or even contained within each other (e.g., England is contained as aggregate of several subregions)?
　iii. What are the most important consumables of the first principal component?

(b) Now let's clean the dataset a little. We will now only look at the four countries of the UK and London (i.e., England, London, Northern Ireland, Scotland, and Wales)

　i. How does the composition of the first principal component change?
　ii. How much variance is lost now?
　iii. What is the most distinct country and which consumable is the most discriminating one?

6.5 Download the two circular datasets which are depicted in Figure 6.12 from the web-resources of the book (available at www.cambridge.org/bionetworks). Obviously, both datasets are artificial datasets but they should demonstrate the influence of the different linkage functions for hierarchical clustering.

(a) Without performing a clustering, what is your intuition which linkage function might work best on these datasets? Justify your thought process.
(b) Now perform a hierarchical clustering on both datasets, using single, average, and complete as linkage functions. Cut the tree such it delivers two

---

[2] www.gov.uk/government/statistical-data-sets/family-food-datasets

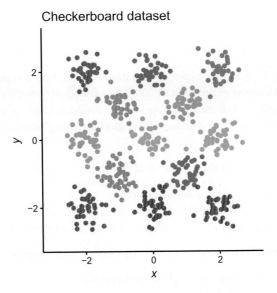

Figure 6.13: Toy dataset consisting of 13 spherical clusters distributed in a checkerboard fashion.

clusters for dataset 1 and three clusters for dataset 2. Does the observation support your thought from the previous question?

(c) What problems of the different linkage functions are highlighted in the example? Do you observe a particular problem with the single linkage clustering?

6.6 The checkerboard dataset, depicted in Figure 6.13 is an artificial dataset with 13 clusters. You can download the dataset from the book's web resources.

(a) Cluster the dataset using k-means with $k = 13$. Repeat the clustering several times with different random initializations. What do you observe? On average, how many restarts do you need in order to achieve the optimal clustering?

(b) Now implement the subset furthest-first initialization. Do you need fewer repeats (on average) in order to achieve the ideal clustering?

(c) What particular problem of k-means is highlighted here?

6.7 Use again the checkerboard dataset from the exercise before and download the gold standard from the book's web resources. In this case, we are in the unique position of knowing the ground truth and we can compare several datasets with each other. Perform the following evaluations:

(a) Use k-means and define the ideal number of clusters using the gap statistic.

(b) Calculate the silhouette value for various k. Plot the average silhouette value in dependency of the number of clusters.

(c) In the following, use at least two different clustering approaches. Optimize the tool's parameters using the F-measure and the Rand index.

i. Is the optimal parameter for the best F-measure also the optimal parameter for the best Rand index?
  ii. What is the optimal F-measure you can achieve?

6.8 Next, we look at a gene expression dataset resembling a typical situation. The dataset contains the expression of 700 genes for 53 patients. One common task is the identification of disease subtypes. You can download the dataset and a gold standard assigning three disease subtypes to the samples from the web-resources of the book.

(a) Perform a PCA of the dataset and create a 2-dimensional projection.
   i. How much of the variance is lost?
   ii. Does the shape of the clusters (using the gold standard) give you a hint of which tool might be suitable and which tool might be inappropriate?

(b) Transform the two-mode matrix into a one-mode distance matrix by
   i. using a version of the Minkowski distance;
   ii. using a correlation coefficient.

(c) Cluster the dataset using hierarchical clustering.
   i. What is the optimal linkage function and tree-cut depth for each of them?
   ii. What is the optimal F-measure achieved?
   iii. What is the optimal silhouette value?

(d) Repeat the previous task, using a clustering tool of your choice.
   i. Justify your choice of clustering tool.
   ii. After parameter optimization, what is the optimal F-measure achieved?

Note: Solutions (for instructors only) and datasets are available at www.cambridge.org/bionetworks.

References

[1] Hastie T, Tibshirani R, Friedman J. Unsupervised learning. In *The Elements of Statistical Learning*, Springer;2009, pp. 485–585.

[2] Xu R, Wunsch D. Clustering algorithms in biomedical research: a review. *IEEE Reviews in Biomedical Engineering*, 2010;3:120–154.

[3] Anderberg MR. *Cluster Analysis for Applications: Probability and Mathematical Statistics: A Series of Monographs and Textbooks*, vol. 19. Academic Press, 2014.

[4] Röttger R. Clustering of biological datasets in the era of big data. *Journal of Integrative Bioinformatics*, 2016;13(1):300.

[5] Xu R, Wunsch D. Survey of clustering algorithms. *IEEE Transactions on Neural Networks*, 2005;16(3):645–678.

[6] Everitt BS, Landau S, Leese M, Stahl D. *Cluster Analysis*, Wiley Series in Probability and Statistics. John Wiley & Sons, Ltd;2011.

[7] Jain AK, Murty MN, Flynn PJ. Data clustering: A review. *ACM Computing Surveys (CSUR)*, 1999;31(3):264–323.

[8] Chidananda Gowda K, Diday E. Symbolic clustering using a new similarity measure. *IEEE Transactions on Systems, Man, and Cybernetics*, 1992;22(2):368–378.

[9] Karthikeyani Visalakshi N, Thangavel K. Impact of normalization in distributed k-means clustering. *International Journal of Soft Computing*, 2009;4(4):168–172.

[10] Al Shalabi L, Shaaban Z, Kasasbeh B. Data mining: A preprocessing engine. *Journal of Computer Science*, 2006;2(9):735–739.

[11] Gan G, Ma C, Wu J. *Data Clustering: Theory, Algorithms, and Applications*, ASA-SIAM Series on Statistics and Applied Probability vol. 20 SIAM;2007.

[12] Bullard JH, Purdom E, Hansen KD, Dudoit S. Evaluation of statistical methods for normalization and differential expression in mrna-seq experiments. *BMC Bioinformatics*, 2010;11(1):94.

[13] Kim SY, Lee JW, Bae JS. Effect of data normalization on fuzzy clustering of DNA microarray data. *BMC Bioinformatics*, 2006;7(1):134.

[14] Liu H, Yu L. Toward integrating feature selection algorithms for classification and clustering. *IEEE Transactions on Knowledge and Data Engineering*, 2005;17(4):491–502.

[15] Guyon I, Elisseeff A. An introduction to variable and feature selection. *Journal of Machine Learning Research*, 2003;3(Mar):1157–1182.

[16] Dash M, Liu H. Feature selection for clustering. In *Pacific-Asia Conference on Knowledge Discovery and Data Mining*, Springer, 2000; pp. 110–121.

[17] Alelyani S, Tang J, Liu H. Feature selection for clustering: A review. *Data Clustering: Algorithms and Applications*, 2013;29.

[18] Shlens J. A tutorial on principal component analysis. arXiv preprint available at https:arxiv.org/pdf/1404.1100pdf, 2014.

[19] Guyon I, Elisseeff A. An introduction to feature extraction. In *Feature Extraction*, Springer, 2006; pp. 1–25.

[20] Pandit S, Gupta S. A comparative study on distance measuring approaches for clustering. *International Journal of Research in Computer Science*, 2011;2(1):29–31.

[21] Mao J, Jain K. A self-organizing network for hyperellipsoidal clustering (hec). *IEEE Transactions on Neural Networks*, 1996;7(1):16–29.

[22] Aggarwal CC, Hinneburg A, Keim DA. On the surprising behavior of distance metrics in high dimensional space. In van den Bussche J, Vianu V, eds., *International Conference on Database Theory*, Springer, 2001; pp. 420–434.

[23] Rogers DJ, Tanimoto TT. A computer program for classifying plants. *Science (New York, NY)*, 1960;132(3434):1115–1118.

[24] Gower JC, Legendre P. Metric and Euclidean properties of dissimilarity coefficients. *Journal of Classification*, 1986;3(1):5–48.

[25] Jaccard P. Nouvelles recherches sur la distribution floral. *Bulletin Societe Vaudoise des Sciences Naturelles*, 1908;44:223–270.

[26] Sneath HA P, Sokal RR. *Numerical Taxonomy: The Principles and Practices of Numerical Classification*. WF Freeman and Co.;1973,573.

[27] Goodall DW. A new similarity index based on probability. *Biometrics*, 1996;22(4):882–907.

[28] Boriah S, Chandola V, Kumar V. Similarity measures for categorical data: A comparative evaluation. In *Proceedings of 2008 Siam International Conference on Data Mining 2008. SIAM*, 2008; pp. 243–254.

[29] Altschul SF, Gish W, Miller W, Myers EW Lipman DJ. Basic local alignment search tool. *Journal of Molecular Biology*, 1990;215(3):403–410.

[30] Emre Celebi M, Kingravi HA, Vela PA. A comparative study of efficient initialization methods for the k-means clustering algorithm. *Expert Systems with Applications*, 2013;40(1):200–210.

[31] Corpet F. Multiple sequence alignment with hierarchical clustering. *Nucleic Acids Research*, 1988;16(22):10881–10890.

[32] Wang J, Li M, Chen J, Pan Y. A fast hierarchical clustering algorithm for functional modules discovery in protein interaction networks. *IEEE/ACM Transactions on Computational Biology and Bioinformatics*, 2011;8(3):607–620.

[33] Loewenstein Y, Portugaly E, Fromer M, Linial M. Efficient algorithms for accurate hierarchical clustering of huge datasets: Tackling the entire protein space. *Bioinformatics*, 2008;24(13):i41–i49.

[34] Jothi R, Zotenko E, Tasneem A, Przytycka TM. Coco-cl: Hierarchical clustering of homology relations based on evolutionary correlations. *Bioinformatics*, 2006;22(7):779–788.

[35] Saunders DGO, Win J, Cano LM, et al. Using hierarchical clustering of secreted protein families to classify and rank candidate effectors of rust fungi. *PLOS One*, 2012;7(1):e29847.

[36] Sturn A, Quackenbush J, Trajanoski Z. Genesis: Cluster analysis of microarray data. *Bioinformatics*, 2002;18(1):207–208.

[37] Brazma A, Vilo J. Gene expression data analysis. *FEBS Letters*, 2000;480(1):17–24.

[38] Eisen M, Spellman PT, Brown PO, Botstein D. Cluster analysis and display of genome-wide expression patterns. *Proceedings of the National Academy of Sciences*, 1998;95(25):14863–14868.

[39] Frey BJ, Dueck D. Clustering by passing messages between data points. *Science*, 2007;315(5814):972–976.

[40] Nepusz T, Yu H, Paccanaro A. Detecting overlapping protein complexes in protein-protein interaction networks. *Nature Methods*, 2012;9(5):471–472.

[41] Dongen SV. A cluster algorithm for graphs. *Report-Information Systems*, 2000;(10):1–40.

[42] Von Luxburg U. A tutorial on spectral clustering. *Statistics and Computing*, 2007;17(4):395–416.

[43] Wittkop T, Emig D, Lange S, et al. Partitioning biological data with transitivity clustering. *Nature Methods*, 2010;7(6):419–420.

[44] Aittokallio T, Schwikowski B. Graph-based methods for analysing networks in cell biology. *Briefings in Bioinformatics*, 2006;7(3):243–255.

[45] Vlasblom J, Wodak SJ. Markov clustering versus affinity propagation for the partitioning of protein interaction graphs. *BMC Bioinformatics*, 2009;10(1):1.

[46] Satuluri V, Parthasarathy S, Ucar D. Markov clustering of protein interaction networks with improved balance and scalability. In *Proceedings of the 1st ACM International Conference on Bioinformatics and Computational Biology*, ACM; 2010,pp. 247–256.

[47] Shih YK, Parthasarathy S. Identifying functional modules in interaction networks through overlapping markov clustering. *Bioinformatics*, 2012;28(18):i473–i479.

[48] Liao CS, Lu K, Baym M, Singh R, Berger B. Isorankn: Spectral methods for global alignment of multiple protein networks. *Bioinformatics*, 2009;25(12):i253–i258.

[49] Wittkop T, Emig D, Truss a, et al. Comprehensive cluster analysis with transitivity clustering. *Nature Protocols*, 2011;6(3):285–295.

[50] Röttger R, Kalaghatgi P, Sun P, et al. Density parameter estimation for finding clusters of homologous proteins-tracing actinobacterial pathogenicity life styles. *Bioinformatics*, 2013;29(2):215–222.

[51] Andreopoulos B, An A, Wang X, Schroeder M. A roadmap of clustering algorithms: Finding a match for a biomedical application. *Briefings in Bioinformatics*, 2009;10(3):297–314.

[52] Ester M, Kriegel HP, Sander J, et al. A density-based algorithm for discovering clusters in large spatial databases with noise. In *Proceedings of the 2nd International Conference on Knowledge Discovery and Data Mining*, AAAI Press;1996, vol. 96, pp. 226–231.

[53] Hinneburg A, Keim DA. An efficient approach to clustering in large multimedia databases with noise. In *Proceedings of the 4th International Conference an Knowledge and Data Mining*. AAAI Press; 1998, vol. 98, pp. 58–65.

[54] Rabiner L, Juang B. An introduction to hidden Markov models. *IEEE ASSP Magazine*, 1986;3(1):4–16.

[55] Van Hulle MM. Self-organizing maps. In Rozenbwg G, Bäck T, Kok JN, eds., *Handbook of Natural Computing*, Springer, 2012;pp. 585–622.

[56] Fraley C, Raftery AE. Model-based clustering, discriminant analysis, and density estimation. *Journal of the American Statistical Association*, 2002;97(458):611–631.

[57] Yeung KY, Fraley C, Murua A, Raftery AE, Ruzzo WL. Model-based clustering and data transformations for gene expression data. *Bioinformatics*, 2001;17(10):977–987.

[58] Medvedovic M, Sivaganesan S. Bayesian infinite mixture model based clustering of gene expression profiles. *Bioinformatics*, 2002;18(9):1194–1206.

[59] McLachlan GJ, Bean RW, Peel D. A mixture model-based approach to the clustering of microarray expression data. *Bioinformatics*, 2002; 18(3):413–422.

[60] Smyth P. Clustering sequences with hidden markov models. In Jordan MI, Kearns MJ, Solla SA, eds., *Advances in Neural Information Processing Systems*, MIT Press;1997, pp. 648–654.

[61] Zhang T, Ramakrishnan R, Livny M. Birch: An efficient data clustering method for very large databases. In *ACM Sigmod Record*, ACM; 1996;vol. 25, pp. 103–114.

[62] Guha S, Rastogi R, Shim K. Cure: an efficient clustering algorithm for large databases. In *ACM SIGMOD Record*, ACM 1998;vol. 27, pp. 73–84.

[63] Massart DL, Kaufman L, Rousseeuw PJ, Leroy A. Least median of squares: a robust method for outlier and model error detection in regression and calibration. *Analytica Chimica Acta*, 1986;187:171–179.

[64] Ward Jr JH. Hierarchical grouping to optimize an objective function. *Journal of the American Statistical Association*, 1963;58(301):236–244.

[65] Karypis G, Han EH, Kumar V. Chameleon: Hierarchical clustering using dynamic modeling. *Computer*, 1999;32(8):68–75.

[66] Patwary MA, Palsetia D, Agrawal A, Fredrik et al. A new scalable parallel dbscan algorithm using the disjoint-set data structure. In *Proceedings of the International Conference on High Performance Computing, Networking, Storage and Analysis*, IEEE Computer Society Press;2012, p. 62.

[67] Wittkop T, Baumbach J, Lobo FP, Rahmann S. Large scale clustering of protein sequences with force: A layout based heuristic for weighted cluster editing. *BMC Bioinformatics*, 2007;8:396.

[68] Wittkop T, Rahmann S, Röttger R, Böcker R, Baumbach J. Extension and robustness of transitivity clustering for protein–protein interaction network analysis. *Internet Mathematics*, 2011;7(4):255–273.

[69] Wiwie C, Baumbach J, Röttger r. Comparing the performance of biomedical clustering methods. *Nature Methods*, 2015;12(11):1033–1038.

[70] Jain AK, Dubes RC. *Algorithms for Clustering Data*. Prentice-Hall, Inc., 1988.

[71] Rand WM. Objective criteria for the evaluation of clustering methods. *Journal of the American Statistical Association*, 1971;66(336):846–850.

[72] Handl M, Knowles J, Kell DB. Computational cluster validation in post-genomic data analysis. *Bioinformatics*, 2005;21(15):3201–3212.

[73] Fowlkes EB, Mallows CL. A method for comparing two hierarchical clusterings. *Journal of the American Statistical Association*, 1983;78(383):553–569.

[74] Hubert L, Arabie P. Comparing partitions. *Journal of Classification*, 1985;2(1):193–218.

[75] Rendón E, Abundez I, Arizmendi A, Quiroz EM. Internal versus external cluster validation indexes. *International Journal of Computers and Communications*, 2011;5(1):27–34.

[76] Dunn JC. A fuzzy relative of the ISODATA process and its use in detecting compact well-separated clusters. *Journal of Cybernetics*, 1973;3(3):32–57.

[77] Dunn JC. Well-separated clusters and optimal fuzzy partitions. *Journal of Cybernetics*, 1974;4(1):95–104.

[78] Davies DL, Bouldin DW. A cluster separation measure. *IEEE Transactions on Pattern Analysis and Machine Intelligence*, 1979;(2):224–227.

[79] Liu Y, Li Z, Xiong H, Gao X, Wu J, Wu S. Understanding and enhancement of internal clustering validation measures. *IEEE Transactions on Cybernetics*, 2013;43(3):982–994.

[80] Halkidi M, Batistakis Y, Vazirgiannis M. Cluster validity methods: Part i. *ACM Sigmod Record*, 2002;31(2):40–45.

[81] Halkidi M, Batistakis Y, Vazirgiannis M. Clustering validity checking methods: Part ii. *ACM Sigmod Record*, 2002;31(3):19–27.

[82] Tibshirani R, Walther G, Hastie T. Estimating the number of clusters in a data set via the gap statistic. *Journal of the Royal Statistical Society: Series B (Statistical Methodology)*, 2001;63(2):411–423.

7 Machine Learning for Data Integration in Cancer Precision Medicine: Matrix Factorization Approaches

Noël Malod-Dognin, Sam F. L. Windels, and Nataša Pržulj

7.1 Introduction

With the advances in capturing technologies, we have access to large scale biomedical data. While these datasets have long been investigated in isolation from each other, the emerging approach is to mine all datasets collectively to benefit from the complementary knowledge that they contain. This is the underlying principle of precision medicine, which aims at individualizing the practice of medicine in the light of all available clinical and molecular data. In this chapter, we present key machine learning based data-integration approaches that have been used in cancer precision medicine. In particular, we focus on two recent data integration approaches: Non-negative matrix factorization (NMF) and non-negative matrix tri-factorization (NMTF).

In Section 7.2, we introduce the basic concepts and key challenges in precision medicine. In Section 7.3, we present the general classes of machine learning based methods for data integration. In Section 7.4, we provide an overview of network-based integration, Bayesian networks, and kernel-based methods. In Sections 7.5 and 7.6, we present non-negative matrix factorization and non-negative matrix tri-factorization in detail.

7.2 Precision Medicine

Systems biology is flooded with large-scale "omics" data, e.g., from genomics, proteomics, interactomics, metabolomics, etc. (we refer the reader to Chapter 3 for an overview of available biological datasets). These datasets, which are usually modeled as networks, have been extensively studied in isolation from each other (e.g., for protein–protein interaction networks, see Chapter 10). However, each dataset only represents one view of the complex system at study, and thus provides only limited information. To benefit from the complementary knowledge hidden in each dataset, the current challenge in network analytics is in designing data integration methods that can mine all data collectively [1].

This is the basis of *precision medicine*, which is an emerging approach for individualizing the practice of medicine in the light of all clinical and molecular data available [2]. Nowadays, biological data fully qualify as big data, and are characterized by the three Vs: *volume*, which is the large size of the data, *velocity*, which is the speed at which new data is generated, and *variety*, which is the heterogeneity of the data coming from different data-sources and capturing technologies [3]. Therefore, there is an increasing gap between our ability to generate biomedical data and our ability to extract relevant biomedical knowledge from them. The key issue in precision medicine is mining these big biological datasets that, due to their average sizes and complexity, cannot be directly interpreted by medical practitioners.

The large heterogeneity of cancers [4, 5] makes oncology very challenging. Cancer precision medicine uses machine learning based data integration methods to collectively mine the big biological data with the following three objectives:

Patient subtyping (also called patient stratification). This is the classification problem of uncovering subgroups within a set of patients. These subgroups can be used to uncover disease mechanisms, to guide the treatments of patients belonging to the subgroup, or to predict disease outcomes. Patients grouped together based on the same underlying disease mechanism are said to be of the same "endotype" [6]. Alternatively, Boland et al. [7] introduced the notion of "verotype," which is the true grouping of patients for treatment definition. In practice, as what constitute verotypes and how to obtain them is not known, endotypes are used as a proxy.

Biomarker and disease mechanism discovery. Communalities between patients from the same subgroup can be used to identify *biomarkers*, which are measurable diagnostic indicators used for assessing the risk, or the presence of a disease [8]. Biomarker discovery is key for improving healthcare and for reducing medical cost [9]. Communalities between the molecular profiles of patients from the same subgroup can also be used to identify the molecular bases of diseases and thus to improve our medical knowledge.

Drug repurposing and personalized treatments. Drug repurposing is the problem of identifying new uses for the existing or experimental pharmacotherapies. It is a prediction problem of predicting new drug-target interactions between known drugs and cancer-affected genes products. Drug repurposing reduces the cost of developing new pharmacotherapies compared with de novo drug discovery and development [10]. Personalized treatment also involves predicting the safety of

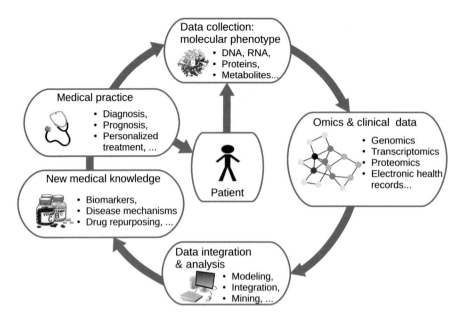

Figure 7.1: Precision medicine. *Molecular phenotypes*, which is the underlying molecular mechanisms of a disease can be revealed through the *collection and analysis* of large scale omics data (e.g., genomic, proteomic and metabolomic), as well as clinical data (e.g., electronic health records). *Data integration and analysis* allows for discovering new *medial knowledge*, such as the identification of new disease genes and biomarkers. This knew knowledge can then be transferred to *medial practice*, leading to personalized medical treatment tailored to an individual.

treatment (e.g., identifying potential side effects, predicting drug response), as well as proposing therapeutic dosage and frequency.

Hence, precision medicine can be seen as a loop of patient data collection, data integration for extraction of new disease relevant knowledge, treatment prediction and ultimately, biological, and clinical validation (see Figure 7.1).

7.3 The Different Types of Data Integration Methods

As described in the Introduction, big systems biology data have increasingly large sizes, diversity, and heterogeneity. Hence, their analysis requires efficient algorithms for extracting the new knowledge hidden in their complexity. Data integration methods have demonstrated the ability to bridge the gap between data production and interpretation in precision medicine, which we elaborate on below. These methods can be classified according to three different criteria.

7.3.1 Homogeneous and Heterogeneous Data Integration

First, data integration methods can be divided into two groups depending on the type of data that they integrate.

On one hand, *homogeneous* approaches integrate data of the same type, but across multiple views (e.g., different experimental conditions). Such data can be represented by a collection of networks that all have the same types of nodes but each network may contain different types of links between the nodes. Despite being very useful, homogeneous data do not fully capture the complexity of a biological system.

On the other hand, *heterogeneous* approaches can integrate networks having varying types of nodes and interactions (e.g., integrating somatic mutation data linking patients to genes, with drug-target interaction data linking genes to drugs). Heterogeneous data integration is computationally more challenging, as it requires developing a data integration framework that can capture the data heterogeneity without much data transformation, that may incur information loss.

7.3.2 Early, Intermediate, and Late Integration

Second, data integration methods can be grouped based on how they collectively mine the networked datasets to build a unified, integrated model.

Early (full) data integration approaches first combine all datasets into a single dataset from which the model is built. Combining the datasets often requires representing all data in a common feature space, which may lead to information loss [11, 12].

Late (decision) data integration approaches first build models for each dataset in isolation from others, and then combine these models into an integrated model. As building models for each dataset in isolation from others disregards their complementary information, late data integration may result in reduced performances of the integrated model [11, 12].

Intermediate (partial) data integration approaches combine datasets through the inference of a joint model. Intermediate data integration does not require data transformation, and thus, does not result in information loss. These methods are also reported to have higher predictive accuracy [11, 12, 13, 14].

7.3.3 Supervised, Unsupervised, and Semi-supervised Data Integration

Finally, data integration methods can be classified according to the training process that they use to learn their integrated models, as illustrated in Figure 7.2.

Supervised methods require training using input data samples with known labels. These methods learn a model via a training process that maximizes the model accuracy on the training data. Once trained, the learned model can be used to assign known labels to new data samples. In precision medicine, input data can represent patients that need to be classified either as control (healthy), or as cases (diseased), or into many disease subtypes (patient groups having similar clinical outcomes). Supervised methods include classification and regression methods, such as logistic regression [15] and kernel-based methods (KBs) [16] (including the well known support vector machines (SVMs) [17]).

Unsupervised methods use unlabeled data set as input. A model is learned by uncovering hidden latent patterns in the data, which are then used to organize the

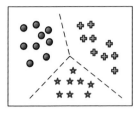 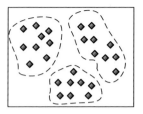 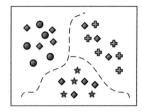

(a) Supervised learning   (b) Unsupervised learning   (c) Semi-supervised learning

Figure 7.2: Supervised, unsupervised, and semi-supervised learning. Labeled data points can be part of 1 of 3 categories, identified by either a blue cross, yellow star or a red circle. Unlabeled data points are represented by gray diamonds. (a) Supervised learning. A model is learned to maximized the accuracy of grouping together data points that have similar labels. (b) Unsupervised learning. The data points are unlabeled, and a model is learned to reveal hidden patterns in the dataset. (c) Semi-supervised learning. Using a mixture of labeled and unlabeled data points, a model is learned to reveal the hidden patterns in the data while maximizing the accuracy of grouping together data points with similar labels.

data into relevant subsets. In precision medicine, these methods are often used for molecular subtyping of patients (groupings of patients according to the similarity of the molecular mechanism of their diseases). Unsupervised methods include clustering approaches, such as hierarchical clustering [18], k-means [18] and its generalizations, including non-negative matrix factorization [19] and non-negative matrix tri-factorization [20]. The latter two are also used for dimensionality reduction.

Semi-supervised methods use as input a combination of labeled and unlabeled data. A model is learned to both uncover the organization of the data and to make new predictions of unlabeled data. For example, when predicting new drug-target interactions for drug repurposing, semi-supervised methods learn known drug-target interactions from labeled data (used as "prior knowledge") to predict new interactions. Semi-supervised learning is particularly suitable for data integration, as it allows for incorporating various data types as prior knowledge. In matrix factorization based methods, addition of prior knowledge can be done via graph regularization [21] (detailed in Section 7.5.3).

7.4 Summary of Data-Integration Methods

While the core of this chapter is about matrix factorization and matrix tri-factorization approaches, here we briefly describe the underlying principles of key machine learning approaches that have been used to integrate biological datasets.

7.4.1 Network-based Data Integration

Network-based methods offer simple ways to integrate different network data. For homogeneous integration of d different networks $N_1 = (V, E_1)$, $N_2 = (V, E_2)$, ..., $N_d = (V, E_d)$, that share the same set of nodes, V, but have different sets of edges ($E_i, 1 \leq i \leq d$) connecting their nodes, the simplest network-based data integration

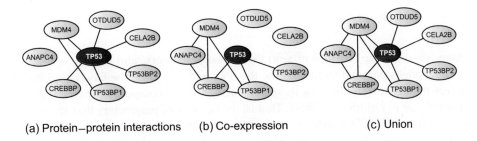

(a) Protein–protein interactions (b) Co-expression (c) Union

Figure 7.3: Illustration of network-based integration. Part of the local interactions arround protein TP53 (a tumor suppressor protein that regulates cell division by preventing cells from growing and dividing in an uncontrolled way), as taken from STRING database [23]. (a) shows the protein-protein interactions between the proteins, while (b) shows the proteins that are co-expressed. (c) shows the union of the two networks.

consist of taking the union of all networks, i.e., $N_{union} = (V, \bigcup_i E_i)$ [22] (illustrated in Figure 7.3). However, this simple approach does not consider varying noisiness and incompleteness of the different datasets, as all input networks contribute equally to the integrated model.

To overcome this limitation, another approach consists of considering the adjacency matrices of each network, $A_i, 1 \leq i \leq d$, and combining them using a linear combination: $A_{union} = \sum_i w_i A_i$, where $w_i \geq 0$ is the weight associated to network N_i so that the quality of the integrated model is optimized [24, 25]. Finding weights w_i requires solving a system of linear equations, which will assign lower weights to "less contributing" networks.

More advanced methods use message passing theory [26] to iteratively update the input networks, making them more similar to each other after each iteration, until they converge to a single, integrated model [27].

Application of network-based approaches for heterogeneous integration (e.g., of networks having different sets of nodes and edges) often requires a preliminary step in which all networks are projected on a set of common nodes [28, 29]. While this preliminary step effectively converts heterogeneous data into homogeneous data, it may result in information loss.

More advanced approaches, called *network propagation* methods, can directly integrate heterogeneous data using diffusion processes that spread information along the edges of the networks [30, 31].

7.4.2 Bayesian Approaches

A Bayesian network is a probabilistic graphical model that represents a set of random variables and their conditional dependencies (also see Chapter 1, e.g. Box 1.16). A Bayesian network is based on a directed acyclic graph (DAG, see Chapter 3) in which nodes represent variables and in which a directed edge from node y to node x represents the conditional probability between x and y. The conditional probability between x and y is denoted by $p(x|y)$ and is the probability of x given the value of y. A joint probability distribution (JPD) of a Bayesian network having n nodes, $x = \{x_1, x_2, \ldots, x_n\}$, is:

$$p(x|\theta) = \prod_{i=1}^{n} p(x_i|Pa(x_i)), \tag{7.1}$$

where $Pa(x_i)$ are the ancestor of x_i in the Bayesian network, and $\theta = \{\theta_1, \theta_2, \ldots, \theta_n\}$ are the model's parameters that define the JPD. Not only does the Bayesian network capture the structure of the data, its sparsity also eases the computation of the JPD over the whole set of random variables. That is, the number of parameters that are needed to characterize the JPD is reduced in the Bayesian network representation [32, 33].

Constructing a Bayesian networks requires (1) learning the network's wiring patterns, which is called *structure learning*, and (2) learning the parameters of its JPD, which is called *parameter learning* [32, 33]. Structure learning consist of identifying the statistical dependencies (represented by edges in the DAG) between the variables. Because the number of possible wirings in a network is super-exponential in its number of nodes, learning the Bayesian network that best represents a dataset is an NP-hard problem for which heuristic algorithms have been proposed [32]. Once the structure and the parameters of a Bayesian network have been learned, it can be used for inference (prediction) about its variables. For example, with a Baysian network representing the regulatory interactions between genes (as illustrated in Figure 7.4), one can ask for the likelihood of a gene to be expressed given the expression status of the other genes. Note that exact inference, which requires the summation of the JPD over all possible values of unknown variables, is also an NP-hard problem [34] for which approximate solutions have been proposed [33].

While Bayesian networks have been successfully used to integrate biological data (e.g., for gene regulatory network inference [35], or for cancer prognosis prediction [36]), they suffer from the following limitations. First, in the structure learning, their sparse representation only captures the most important associations between the variables, discarding all other weaker associations. Second, the directed acyclic graph

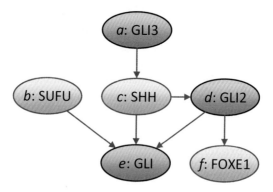

Figure 7.4: Bayesian network: Example of a gene regulatory network. The presented Bayesian network encodes part of the regulatory relations of GLI proteins (in red), which are transcription factors whose mutations are involved in many congenital malformations. In this sparse representation, the expression of a gene depends only on its parents: e.g., the expression of SHH (c) only depends on GLI3 (a), so the conditional probability distribution of c is $p(c|a)$. The joint probability distribution of the Bayesian network is:
$p(a,b,c,d,e,f) = p(a)p(b)p(c|a)p(d|c)p(e|b,c,d)p(f|d)$.

representation does not allows for loops, which are important components of biological networks (e.g., for representing control and feed-back loops). Finally, as already mentioned, their computational complexity limits the usage of Bayesian networks to small datasets.

7.4.3 Kernel-based Methods

Kernel-based methods are machine learning approaches for pattern analysis (also introduced in Chapters 1, 5, and 6). A kernel-based approach works by embedding the original data from its original input space, $\mathcal{X}$, into a higher dimensional space, called the feature space, $\mathcal{F}$, in which the analysis is performed (see illustration in Figure 7.5).

$\mathcal{F}$ is a vector space in which data points are represented by vectors, called *feature vectors*. The embedding of $\mathcal{X}$ in $\mathcal{F}$ is represented by a *kernel matrix* [37], $\mathcal{K}$, which is a symmetric, positive semi-definite matrix whose entries $\mathcal{K}_{i,j} = k(x_i, x_j)$ represent the similarities between any two data points x_i and x_j, which are computed as the inner product between their representations $\phi(x_i)$ and $\phi(x_j)$ in the feature space:

$$k(x_i, x_j) = \langle \phi(x_i), \phi(x_j) \rangle, \tag{7.2}$$

where ϕ maps data points from $\mathcal{X}$ to $\mathcal{F}$ [37, 38]. In practice, only the definition of the *kernel function* $k(x_i, x_j)$ is required.

In kernel-based approaches, a network dataset is frequently represented by using a *diffusion kernel* [39]:

$$\mathcal{K} = e^{-\beta L}, \tag{7.3}$$

where $\beta > 0$ is a parameter controlling the diffusion, and L is the Laplacian matrix of the input network (we refer the reader to Chapter 5 for the definition of a Laplacian matrix). Entries of the diffusion kernel quantify the closeness between any two nodes in the network. Alternatively, in relation to Chapter 5 (graphlets), a network dataset can be represented by a *graphlet kernel* [40], in which a node x_i of the network is

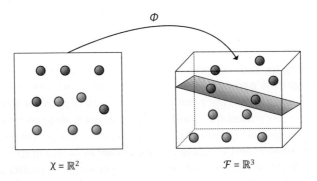

Figure 7.5: Illustration of kernel-based methods. To simplify a classification problem of data points in input space $\mathcal{X} = \mathbb{R}^2$, a kernel function ϕ is used to map $\mathcal{X}$ into a higher dimensional feature space $\mathcal{F} = \mathbb{R}^3$, in which the two clusters of data points are easily separated by a plane.

represented by its graphlet degree vector, $GDV(x_i)$ in the feature space $\mathcal{F}$ (we refer the reader to Chapter 5 for the definition of graphlets and GDV). The kernel matrix is then computed from the following graphlet-based kernel function:

$$k(x_i, x_j) = \langle GDV(x_i), GDV(x_j) \rangle, \tag{7.4}$$

To integrate multiple network datasets, *kernel data-fusion* [41] consists of representing all network data in the same feature space and in linearly combining the corresponding kernel matrices before analysis. Kernel matrices are then mined using traditional statistical and machine learning methods, such as support vector machines (SVMs) [17], principal component analysis (PCA) [42] and canonical correlation analysis (CCA) [43] to produce clusterings, rankings, principal components, and correlations (additional examples are given in Chapters 1, 5, and 6).

While kernel-based approaches have been used to integrate biological data (e.g., for cancer prognosis prediction [44], or for drug repurposing [45]), they suffer from the following limitations. First, there is no guideline for choosing the right kernel function to best represent a given dataset. This shortcoming is partially overcome by *multiple kernel learning* approaches, in which one linearly combines several kernel representations of the same dataset, each capturing different notions of similarity between the nodes of the network [46, 47]. Second, kernel data-fusion implies to transform, or project all network datasets into the same feature space, which may results in information loss.

7.5 Homogeneous Data Integration with Non-Negative Matrix Factorization

7.5.1 Principles and Properties

Non-negative matrix factorization (NMF) is a machine learning method for clustering and dimensionality reduction. In NMF, a network dataset is represented by a non-negative matrix, $A \in \mathbb{R}^{n_1 \times n_2}$, the adjacency matrix of the network. As illustrated in Figure 7.6, this matrix is approximated by the product of two lower-dimensional, non-negative matrix factors, $U \in \mathbb{R}^{n_1 \times k}$ and $V \in \mathbb{R}^{k \times n_2}$, where $k \ll min(n_1, n_2)$ [19]:

$$A \simeq UV. \tag{7.5}$$

In NMF, setting the *rank parameter* $k \ll min(n_1, n_2)$ provides dimensionality reduction [48]. NMF gained particular interest because of its relationship with k-means clustering [49]. From a clustering point of view, the non-negative matrix A represents n_1 data points (e.g., n_1 patients) by their n_2 dimensional feature vectors (e.g., for each patient, a vector representing the expression levels of n_2 genes), this matrix is factorized into two matrices, U and V, where U is the cluster indicator matrix, which assigns the n_1 data points into k clusters, and where V represents the cluster centroids [49]. (From a dimensionality reduction point of view, V is also called a base matrix, as it represents the k-dimensional space defined by the cluster centroids.) Note that the role of U and V can be exchanged, making V the cluster indicator matrix assigning the n_2 data points (e.g., genes) into k clusters, and making U represent the cluster centroids.

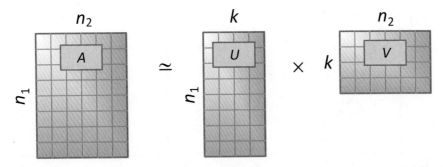

Figure 7.6: Illustration of NMF. In NMF, $n_1 \times n_2$-dimensional matrix A (e.g., the adjacency matrix of a networked dataset) is decomposed into the product of two lower dimensional matrix factors, U and V, where $k \ll min(n_1, n_2)$.

Extracting clusters from V can be done with a procedure such as "hard clustering" [50], in which each data point i is assigned to cluster j, $1 \leq j \leq k$, such that $U_{i,j}$ is the maximum value in row i (e.g., $j = \mathrm{argmax}_{j=1}^{k} U_{i,j}$). Note that to improve clustering interpretation, one can add an orthogonality constraint on a cluster indicator matrix (e.g., for the n_1 data points, $U^T U = I$, where I is the identity matrix), resulting in so-called *orthogonal NMF* [20].

Apart from clustering, another important property of NMF is the *completion property*. Namely, after solving NMF, the reconstructed matrix $\hat{A} = UV$ features new entries, not observed in A, but emerging from the latent factors. As opposed to observable factors, latent factors are factors that are not directly observed, but are inferred. Here, U and V are latent representations for dimension n_1 and n_2 respectively (e.g., patients and genes).

7.5.2 Solving NMF

First, an objective function needs to be defined to quantify the quality of the approximation $A \simeq UV$. Because of the ease of further derivations, a popular objective function is to measure the distance between matrices A and UV as half of the the square of the Euclidean distance between them. The Euclidean distance between two matrices $A, B \in \mathbb{R}^{m \times n}$, also referred to as the Frobenius norm, is defined as:

$$||A - B||_F = \sqrt{\sum_{ij}(A_{ij} - B_{ij})^2}. \tag{7.6}$$

Using half of the square of the Frobenius norm leads to the following objective function for NMF:

$$\min_{U \geq 0, V \geq 0} f(U, V) = \min_{U \geq 0, V \geq 0} \frac{1}{2}||A - UV||_F^2 = \min_{U \geq 0, V \geq 0} \frac{1}{2} \sum_{ij}(A_{ij} - (UV)_{ij})^2, \tag{7.7}$$

This objective function, f, simply measures the sum of squared difference between A and UV. It is lower bounded by 0, and equal to 0 when $A = UV$. Furthermore, f is convex in either U or V (since it is quadratic), but is not jointly convex in both U and V

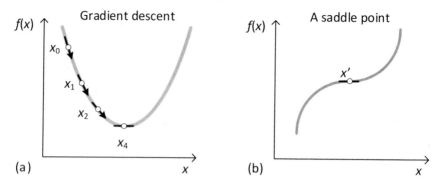

Figure 7.7: Illustration of gradient descent. (a) Starting from a randomly chosen point x_0, following the gradient of f, Δf, represented by the black arrows for each x_i, allows for minimizing f until reaching a stationary point, x_4, in this case the minimum for which the gradient is zero. (b) A saddle point, x', which is also characterized by a zero gradient.

(since it is an order 4 polynomial). It has been shown that solving NMF is an NP-hard non-linear programming problem [51]. Because of this, many heuristic solvers have been proposed to compute locally optimal solutions [8], including the widely used *multiplicative update rules* [52] that we detail below.

A fruitful strategy in continuous optimization is to follow the slope (i.e., the gradient) of the objective function to move in the search space toward a local optimum (as illustrated in Figure 7.7 (a)), resulting in so called *gradient descent* approaches. However, for non-convex problems, such methods can only guarantee to produce stationary points (for which the gradient is equal to zero), but such points may not necessarily be local optima (they can also be saddle points, as illustrated in Figure 7.7(b)).

In NMF, the gradient of $f(U, V)$, $\Delta f(U, V)$, has two components: $\Delta_U f(U, V)$ and $\Delta_V f(U, V)$, which are the partial derivatives of f according to the elements in U and V, respectively:

$$\Delta_U f(U, V) = \frac{\delta \|A - UV\|_F^2}{\delta U} = -(A - UV)V^T = -(AV^T - UVV^T)$$
$$\Delta_V f(U, V) = \frac{\delta \|A - UV\|_F^2}{\delta V} = -U^T(A - UV) = -(U^T A - U^T UV). \quad (7.8)$$

NMF is typically solved using gradient descent. That is, starting from a randomly generated initial solution (U, V), gradient descent is applied such that at each iteration t, V^t is fixed so that U^t can be updated into U^{t+1} by moving according to the gradient of f, and then U^{t+1} is fixed so that V^t can also be updated into V^{t+1}. This process is repeated until convergence, e.g., $U^{t+1} = U^t$ and $V^{t+1} = V^t$, or until an iteration limit (t_{limit}) has been reached. This process is summarized by the following two *additive update rules*:

$$U^{t+1} = U^t - \alpha \Delta_U f(U^t V^t),$$
$$V^{t+1} = V^t - \beta \Delta_V f(U^{t+1} V^t), \quad (7.9)$$

where α and β are the step size parameters controlling the convergence rate of the gradient descent, and $\Delta_U f(U, V)$ and $\Delta_V f(U, V)$ are the partial derivatives of f according to the elements in U and V, respectively. Inserting the partial derivatives presented in Equation 7.8 into the additive update rules presented in Equation 7.9 leads to the following update rules that could be used to minimize the objective function of NMF:

$$U^{t+1} = U^t + \alpha[A(V^t)^T - U^t V^t (V^t)^T]$$
$$V^{t+1} = V^t + \beta[(U^{t+1})^T A - (U^{t+1})^T U^{t+1} V^t].$$
(7.10)

However, applying gradient descent straightforwardly as presented here suffers from two major drawbacks. First, the (fixed) step size parameters α and β need to be tuned. As they control the step size of the gradient descent, higher values of α and β should improve the convergence rate of the gradient descent, but at the risk of divergence when set too high. Second, due to the negative components in $\Delta_U f(U, V)$ and $\Delta_V f(U, V)$ ($-UVV^T$ and $-U^T UV$, respectively), these additive update rules do not ensure that matrix factors U and V will remain non-negative along the gradient descent.

To remedy both issues simultaneously, Lee and Seung rescale the gradient descent by setting α and β to [52]:

$$\alpha = \frac{U^t}{U^t V^t (V^t)^T},$$
$$\beta = \frac{V^t}{(U^{t+1})^T U^{t+1} V^t}.$$
(7.11)

Plugging these scaling factors in the additive update rules presented in Equation 7.10 leads to the following *multiplicative update rules*:

$$U^{t+1} = U^t + \frac{U^t}{U^t V^t (V^t)^T} \times \left[A(V^t)^T - U^t V^t (V^t)^T\right] = U^t \odot \frac{A(V^t)^T}{U^t V^t (V^t)^T},$$
$$V^{t+1} = V^t + \frac{V^t}{(U^{t+1})^T U^{t+1} V^t} \times \left[(U^{t+1})^T A - (U^{t+1})^T U^{t+1} V^t\right] = V^t \odot \frac{(U^{t+1})^T A}{(U^{t+1})^T U^{t+1} V^t},$$
(7.12)

where $\odot$ symbolizes the Hadamard (entry wise) product. These multiplicative update rules both contain the positive component of the gradient in the denominator and the absolute value of the negative component in the numerator of the factor. As these update rules no longer have any negative component, U and V are sure to remain non-negative when initialized as non-negative. Finally, Lee and Seung prove that the objective function, $f(U, V) = \frac{1}{2}\|A - UV\|_F^2$, is non-increasing after each iteration of the multiplicative update rules [52], i.e.,

$$f(U^t, V^t) \geq f(U^{t+1}, V^t) \geq f(U^{t+1}, V^{t+1}).$$
(7.13)

The algorithm for solving NMF by multiplicative update rules has been summarized in Algorithm 7.1.

The computational time complexity of multiplicative update rules is in $O(tn_1 n_2 k)$, where t is the number of iterations of the multiplicative update rules, n_1 and n_2 are the dimensions of the input matrix A and k is the rank parameter (i.e., the number of clusters). Because of the random initialization step of (U^0, V^0), multiplicative update rules are a randomized algorithm for which each different run may result in a different

---
Algorithm 7.1 Multiplicative update rules for U and V in solving NMF, $A \simeq UV$

1: Randomly generate $U^0 > 0$ and $V^0 > 0$
2: **for all** $t = 0, 1, \ldots, t_{\text{limit}-1}$, or until $U^{t+1} = U^t$ and $V^{t+1} = V^t$ **do**
3: $\quad U^{t+1} = U^t \odot \dfrac{A(V^t)^T}{U^t V^t (V^t)^T};$
4: $\quad V^{t+1} = V^t \odot \dfrac{(U^{t+1})^T A}{(U^{t+1})^T U^{t+1} V^t};$
5: Return the last computed (U, V).

---

decomposition. Also, random initialization may result in very poor starting points for the minimization algorithm. Thus, alternative, non-random initialization approaches have been proposed that exploit the relationships between NMF and k-means [53], or with singular value decomposition [54].

7.5.3 Homogeneous Data Integration with NMF

Now, we detail how NMF can be used to integrate multiple datasets. This can be done in two ways, by using *simultaneous decomposition*, or by using *graph regularization* penalties.

7.5.3.1 Simultaneous Decomposition

Given homogeneous networked datasets (with the same set of nodes, but different types of edges) that are represented by the collection of their adjacency matrices, $A_1, A_2, \ldots, A_h$, NMF can be used to integrate all datasets by simultaneously decomposing each matrix while sharing the same cluster indicator matrix factor over all decompositions, i.e., $A_i \simeq UV_i$ for all $i \in \{1, \ldots, h\}$. Sharing a common factor is the key point that allows learning from all datasets.

Simultaneous decomposition requires solving a new objective function,

$$\min_{U \geq 0, V \geq 0} f(U, V) = \min_{U \geq 0, V_i \geq 0 \forall i} \frac{1}{2} \sum_i \|A_i - UV_i\|_F^2. \tag{7.14}$$

Similar to NMF, simultaneous decomposition Equation 7.14 is approximately solved using dedicated multiplicative update rules [55]. Note that the computational complexity of simultaneously decomposing d datasets is d times the cost of a single decomposition.

For example, simultaneous decomposition has been used by iCluster [55] (illustrated in Figure 7.8) in the context of cancer patient stratification. iCluster simultaneously factorizes multiple cancer-relevant relations between patients and their genes, namely DNA copy numbers, DNA methylation, mRNA expression, and micro RNA expression. On breast and lung cancer patients, iCluster identified novel patient subgroups with statistically significantly different clinical outcomes as a result of combining all data in this way.

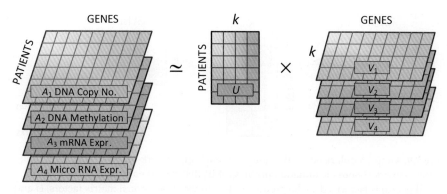

Figure 7.8: Simultaneous decomposition example of iCluster. In iCluster, four patients × genes datasets, namely DNA copy numbers, DNA methylations, mRNA expression, and micro-RNA expression, are simultaneously decomposed while sharing the same cluster indicator matrix, U, across all decompositions.

7.5.3.2 Graph Regularization

Assume that we have two networked datasets. The first one is represented by an $n_1 \times n_2$ matrix, A, describing relationships between nodes of type 1 and type 2 (e.g., patients and genes), and the second one is an $n_2 \times n_2$ adjacency matrix, B, describing relationships between nodes of type 2 (note that a similar reasoning holds if B describes relationships between nodes of type 1). The idea behind graph regularization is to decompose A as the product of two matrix factors [56], while constraining the cluster indicator matrix of nodes of type 2 to group together nodes that are connected in B. This is formally expressed by the following objective function:

$$\min_{U \geq 0, V \geq 0} f(U, V) = \min_{U \geq 0, V \geq 0} \left[\frac{1}{2} \|A - UV\|_F^2 + \frac{1}{2} \alpha \operatorname{tr}(VLV^T) \right], \quad (7.15)$$

where tr denotes the trace of a matrix, L is the Laplacian of matrix B, and $\alpha \geq 0$ is a parameter balancing the influence of B when decomposing A. Intuitively, grouping together elements of type 2 that are connected in B, i.e., for which the Laplacian matrix is negative, will minimize the graph regularization penalty $\operatorname{tr}(VLV^T)$ in the objective function. From a mathematical perspective, graph regularization embeds objects that are similar according to B (i.e., objects i and j that have large weight B_{ij}), as close as possible in the k-dimensional latent space spanned by V [57]:

$$\min_{V \geq 0} \operatorname{tr}(VLV^T) = \min_{V \geq 0} \sum_{ijl} (V_{il} - V_{jl})^2 B_{ij}. \quad (7.16)$$

Similar to simultaneous decomposition, solving graph regularized NMF (Equation 7.15) requires deriving dedicated update rules [56]. The advantage of graph regularized NMF over simulatenous NMF is its lower time complexity. The time complexity of NMF is the same as the time complexity of traditional NMF, presented in Section 7.5.2 . On the other hand, graph regularized NMF does not allow for predicting new entries in B (as matrix completion can only be applied to matrices that are decomposed).

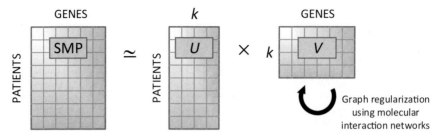

Figure 7.9: Graph regularized NMF example of NBS. In network-based stratification (NBS) [58], the network of somatic mutation profiles (SMP, that relates patients and their cancer mutated genes) is factorized as the product of two lower dimensional matrix factors, U and V. Using graph regularization, the cluster indicator of the genes, V, is constrained to group together genes that interact in a molecular interaction network.

For instance, graph regularized NMF has been used by network-based stratification (NBS) [58] (illustrated in Figure 7.9) in the context of cancer patient stratification. Based on the observation that clinically similar cancer patients may have different sets of mutated genes, but that the mutated genes tend to belong to same biological pathways, NBS factorizes somatic mutation profiles (connecting patients to their mutated genes), while constraining the base matrix of the patients (i.e., the cluster indicator of the genes) to group together genes that interact in a molecular interaction network that capture pathways. Applied on ovarian, uterine, and lung cancer patients, NBS produced patient subtypes with statistically significantly different clinical outcomes, responses to therapies, and tumor histologies.

7.6 Heterogeneous Data Integration with Non-Negative Matrix Tri-Factorization

7.6.1 Principle and Properties

NMF was extended into non-negative matrix tri-factorization (NMTF) [20] (illustrated in Figure 7.10). In NMTF, an $n_1 \times n_2$ matrix, A, which describes the relationships between two types of objects (e.g., the relationships between n_1 patients and their n_2 genes), is decomposed as the products of three non-negative, lower dimensional matrix factors:

$$A \simeq USV^T, \tag{7.17}$$

where $U \in \mathbb{R}^{n_1 \times k_1}$ is the cluster indicator matrix of the objects of type 1 (grouping the n_1 objects of type 1 into k_1 clusters), $V \in \mathbb{R}^{n_2 \times k_2}$ is the cluster indicator matrix of the objects of type two (grouping the n_2 objects of type 2 into k_2 clusters), and where $S \in \mathbb{R}^{k_1 \times k_2}$ is the compressed representation of A that relates the clusters in U to the clusters in V.

NMTF is a *co-clustering* approach, which means that one can extract from U clusters that group together objects of type 1 according to their relationships with objects of type 2, while extracting from V clusters that group together objects of type 2 according

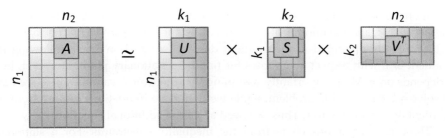

Figure 7.10: Illustration of NMTF. In NMTF, matrix A (e.g., the adjacency matrix of a networked dataset) is decomposed as the product of three lower dimensional matrix factors, U, S, and V, where $k_1, k_2 \ll min(n_1, n_2)$.

to their relationships with the objects of type 1. Such clusters can be extracted from U and V using the same hard-clustering procedure as described in Section 7.5.1.

Similar to NMF, NMTF can be used for matrix completion: after decomposition, the reconstructed matrix $\hat{A} = USV^T$ features new entries, not observed in data matrix A, which can be used for prediction.

7.6.2 Solving NMTF

Similar to solving NMF (presented in Section 7.5.2), NMTF is solved by algorithms based on multiplicative update rules. In Section 7.5.2, we focused on the basic concepts behind gradient descent, additive update rules and multiplicative update rules. Here, we present the advanced concepts from non-linear constrained continuous optimization, as well as the methodology that is used to derive the multiplicative update rules for NMTF (this methodology is also used in practice to obtain the multiplicative update rules of NMF). This section is thus oriented towards advanced readers, while Section 7.5.2 is more accessible to the general audience.

7.6.2.1 Optimizing Non-Linear Constrained Continuous Optimization Problems

In continuous optimization, when searching for a solution x^* that minimizes an objective function, $f(x)$, under m inequality constraints, $g_i(x) < 0, \forall i = 1, \ldots, m$, the common approach is the Karush–Khun–Tucker (KKT) conditions [59].[1]

First, the objective function and the constraints are combined into a single minimization problem, L, in which each inequality is multiplied by a factor η_i called a KKT multiplier:

$$x^* = \text{argmin}_x L(x, \eta) = \text{argmin}_x \left[f(x) - \sum_{i=1}^{m} \eta_i g_i(x) \right]. \tag{7.18}$$

---

[1] As the NMTF optimization problem does not contain any equality constraints, we leave the case of equality constraints out of this discussion.

Notation wise, the minimization of f is called the *primal problem*, L is called the *Lagrangian of f*, and the minimization of L is called the *Lagrange dual function of f*.

As already presented in the gradient descent (see Section 7.5.2), following the gradient of L with respect to x allows for finding a stationary point. However, in L, x depends on η. More importantly, assuming that there are k variables in x, we have to solve a $k + m$ variable problem, while we only have k equations coming from the gradient of L according to x. Thus, we need at least m additional equations.

These new equations come from the inequality constraints: For a stationary point, x^*, if $g_i(x^*) \leq 0$, then constraint i is not involved in changing the solution of the problem, with or without the constraint. In this case, η_i is set to 0. Otherwise, $g_i(x^*) = 0$. Hence, $\eta_i g_i(x^*) = 0$ holds for any stationary point, i.e., the constraint terms are always zero in the set of possible solutions.

This leads to the KKT conditions, which state that a stationary point of a minimization problem under inequality constraints must satisfy:

- **Stationarity condition** (the partial derivative should be equal to 0)

$$\Delta_x f(x) + \sum_{i=1}^{m} \Delta_x \eta_i g_i(x) = 0. \tag{7.19}$$

- **Complementary slackness** (from the inequality constraints)

$$\eta_i g_i(x) = 0, \forall i = 1, \ldots, m. \tag{7.20}$$

These two sets of equations (stationarity condition and complementary slackness) are then used to solve the optimization problem, as illustrated bellow on the specific case of NMTF.

7.6.2.2 Applying KKT Conditions to NMTF

NMTF can be formulated as the following minimization problem:

$$\min_{U \geq 0, S \geq 0, V \geq 0} f(U, S, V) = \min_{U \geq 0, S \geq 0, V \geq 0} \frac{1}{2} \|A - USV^T\|_F^2. \tag{7.21}$$

However, the non-negativity constraint on $S \geq 0$ is often removed (leading to so-called *semi-NMTF* [60]), because the constraint is not needed for a co-clustering interpretation and solving semi-NMTF is more computationally tractable. This leads to solving:

$$\min_{U \geq 0, V \geq 0} f(U, S, V) = \min_{U \geq 0, V \geq 0} \frac{1}{2} \|A - USV^T\|_F^2. \tag{7.22}$$

Following the KKT condition approach, we obtain the Lagrange dual of f, L, by combining the objective function (f) and constraints (the non-negativity of U and V) into a single minimization problem in which the non-negativity constraints on U are associated with KKT multiplier η_1 and the non-negativity constraint on V are associated with KKT multiplier η_2:

$$L = f - \sum_i^{n1} \sum_j^{k1} (\eta_1 \odot U)_{ij} - \sum_i^{n2} \sum_j^{k2} (\eta_2 \odot V)_{ij}, \tag{7.23}$$

where $\odot$ is the Hadamard (element wise) product.

To obtain stationary conditions, we compute the gradient (partial derivatives) of L, ΔL, according to S, U, and V (denoted by $\Delta_S L$, $\Delta_U L$, and $\Delta_V L$, respectively):

$$\Delta L \begin{cases} \Delta_S L = -U^T A V + U U^T S V^T V \\ \Delta_U L = -A V S^T + U S V^T V S^T - \eta_1 \\ \Delta_V L = -A^T U S + V S^T U^T U S - \eta_2. \end{cases} \quad (7.24)$$

This leads to the stationary conditions:

$$\begin{aligned} \Delta_S L &= -U^T A V + U U^T S V^T V = 0 \\ \Delta_U L &= -A V S^T + U S V^T V S^T - \eta_1 = 0 \\ \Delta_V L &= -A^T U S + V S^T U^T U S - \eta_2 = 0. \end{aligned} \quad (7.25)$$

Also, we obtain the following complementarity slackness conditions:

$$\begin{aligned} \eta_1 \odot U &= 0 \\ \eta_2 \odot V &= 0. \end{aligned} \quad (7.26)$$

7.6.2.3 From KKT Conditions to Multiplicative Update Rules

NMTF and semi-NMTF are NP-hard [51], non-convex, continuous optimization problems, for which heuristic algorithms have been proposed, including extension of the multiplicative update rules that are used in NMF [52, 20]. Similar to NMF, NMTF is solved in a fixed point fashion, where starting from initial (randomly generated) matrix factors U, S, and V, an iterative algorithm successively updates each matrix factor whilst keeping the other matrix factors fixed.

As there is no non-negativity constraint on S, by fixing U and V, S can directly be obtained by applying the stationarity condition on S:

$$\begin{aligned} 0 &= \Delta_S L \\ \iff 0 &= -U^T A V + U U^T S V^T V \\ \iff S &= (U^T U)^{-1} (U)^T A V (V^T V)^{-1}. \end{aligned} \quad (7.27)$$

To update U, note that the derivative $\Delta_U L$ contains the KKT multiplier η_1, whose value is obtained by applying the stationarity condition on U:

$$0 = \Delta_U L \iff \eta_1 = -A V S^T + U S V^T V S^T. \quad (7.28)$$

Then, by applying the complementary condition on U and assuming U is non-negative, we get the following update rule for U:

$$\begin{aligned} 0 &= \eta_1 \odot U \\ \iff 0 &= (-A V S^T + U S V^T V S^T) \odot U \\ \iff 0 &= (A V S^T - U S V^T V S^T) \odot U. \end{aligned} \quad (7.29)$$

To obtain an update rule that ensures the non-negativity of U, we make use of the property that any matrix M can be decomposed as the sum $M = M^+ - M^-$, where the $^+$ superscript indicates all positive entries in the matrix, and the $^-$ superscript indicates the absolute value of all negative values in the matrix (i.e., matrices M^+ and M^- are both non-negative). Then, by using this property:

$$0 = (AVS^T - USV^TVS^T) \odot U$$
$$\iff 0 = (AVS^T - USV^TVS^T) \odot U^2$$
$$\iff 0 = ((AVS^T)^+ - (AVS^T)^- - U(SV^TVS^T)^+ + U(SV^TVS^T)^-) \odot U^2$$
$$\iff 0 = \left[(AVS^T)^+ + U(SV^TVS^T)^-\right] \odot U^2 - \left[(AVS^T)^- + U(SV^TVS^T)^+\right] \odot U^2$$
$$\iff \left[(AVS^T)^- + U(SV^TVS^T)^+\right] \odot U^2 = \left[(AVS^T)^+ + (USV^TVS^T)^-\right] \odot U^2$$
$$\iff U^2 = \frac{(AVS^T)^+ + U(SV^TVS^T)^-}{(AVS^T)^- + U(SV^TVS^T)^+} \odot U^2$$
$$\iff U = \sqrt{\frac{(AVS^T)^+ + U(SV^TVS^T)^-}{(AVS^T)^- + U(SV^TVS^T)^+} \odot U}, \quad (7.30)$$

where $\sqrt{\cdot}$ is the element wise square root and $\div$ is the right-array (element wise) division.

The update rule for V is obtained in a similar way: The value of η_2 is obtained by applying the stationarity condition on V, then the multiplicative update rule is obtained by applying the complementary condition on V, whilst keeping U and S fixed. This leads to the following multiplicative update rule for V:

$$0 = \eta_2 \odot V$$
$$\iff 0 = (-A^TUS + VS^TU^TUS) \odot V \qquad (7.31)$$
$$\iff V = V \odot \sqrt{\frac{(A^TUS)^+ + V(S^TU^TUS)^-}{(A^TUS)^- + V(S^TU^TUS)^+}}.$$

In summary, starting from randomly initialized matrix factors (U^0, S^0, V^0) such that $U^0 > 0$, and $V^0 > 0$, Algorithm 7.2 below is used to iteratively update S, U and V either until convergence, or until an iteration limit, t, is reached.

Algorithm 7.2 Multiplicative update rules for S, U and V in solving NMTF, $A \simeq USV^T$

1: Randomly generate $S^0 > 0$, $U^0 > 0$ and $V^0 > 0$
2: **for all** $t = 0, 1, \ldots, t_{limit-1}$, or until $S^{t+1} = S^t$, $U^{t+1} = U^t$ and $V^{t+1} = V^t$ **do**
3: $\quad S^{t+1} = ((U^t)^TU^t)^{-1}((U^t)^TAV^t((V^t)^TV^t)^{-1};$
4: $\quad U^{t+1} = U^t \odot \sqrt{\frac{(AV^t(S^{t+1})^T)^+ + U^t(S^{t+1}(V^t)^TV^t(S^{t+1})^T)^-}{(AV^t(S^{t+1})^T)^- + U^t(S^{t+1}(V^t)^TV^t(S^{t+1})^T)^+}};$
5: $\quad V^{t+1} = \odot V^t \sqrt{\frac{(A^TU^{t+1}S^{t+1})^+ + V((S^{t+1})^T(U^{t+1})^TU^{t+1}S^{t+1})^-}{(A^TU^{t+1}S^{t+1})^- + V^t((S^{t+1})^T(U^{t+1})^TU^{t+1}S^{t+1})^+}}.$
6: Return the last computed (S, U, V).

These update rules have the time complexity in $O(tn_1n_2k_1 + tn_1n_2k_2)$, where t is the number of iterations, which is comparable to the time complexity of NMF. Finally, Wang et al. [21] showed that the objective function, $f(U, S, V) = ||A - USV^T||_F^2$, is nonincreasing after each iteration of the multiplicative update rules, i.e.,

$$f(U^t, S^t, V^t) \geq f(U^t, S^{t+1}, V^t) \geq f(U^{t+1}, S^{t+1}, V^t) \geq f(U^{t+1}, S^{t+1}, V^{t+1}). \tag{7.32}$$

7.6.3 Heterogeneous Data Integration with NMTF

Similar to NMF, NMTF can use both simultaneous decompositions and graph regularization penalties to integrate homogeneous and heterogeneous datasets. Given the NMTF of a single matrix $A_{1,2}$ representing the relationships between objects of types 1 and 2, which is decomposed as the product of three matrix factors: $A_{1,2} \simeq G_1 S_{1,2} G_2^T$. Any new dataset can be simultaneously decomposed with $A_{1,2}$ as long as it relates to at least one of the object types that is already in the decomposition. For example, $A_{2,3}$, which represents the relationships between objects of types 2 and 3, can be simultaneously decomposed with $A_{1,2}$ as $A_{2,3} \simeq G_2 S_{2,3} G_3^T$, while sharing the cluster indicator matrix of the objects of types 2, G_2, across the two decompositions. In this way, NMTF can be used to simultaneously decompose any combination of datasets.

Furthermore, graph regularization penalties can be added so that each cluster indicator matrix, G_i, can benefit from the prior knowledge encoded in a network between nodes of type i (represented by Laplacian matrix L_i).

A generic formulation of this problem is:

$$f = \min_{G_i \geq 0 \forall i, S_{i,j} \geq 0 \forall i,j} \frac{1}{2} \sum_{i \neq j} ||A_{i,j} - G_i S_{i,j} G_j^T||_F^2 + \frac{1}{2} \sum_i \alpha_i \text{tr}(G_i^T L_i G_i), \tag{7.33}$$

where α is a *hyper-parameter* weighing the influence of the regularization. Thus, NMTF provides a principled framework for networked data integration. However, note that each different NMTF-based data integration scheme requires computing the gradient of the corresponding objective function and deriving the corresponding multiplicative update rules.

For example, as illustrated in Figure 7.11, NMTF has been used in the patient specific data-fusion (PSDF) framework [61] in the context of cancer precision medicine. In PSDF, somatic mutation profiles (that represent relationships between patients and their mutated genes) and drug target interactions (that represent relationships between genes and the drugs that target their protein products) are simultaneously decomposed while sharing the cluster indicator matrix of genes (G_2 in Figure 7.11). PSDF uses graph regularization constraints so that the cluster indicator matrix of genes favors grouping together genes that interact in a molecular interaction network and so that the cluster indicator matrix of the drugs favors grouping together drugs that are chemically similar. On serous ovarian cancer patients' data, the patient specific data-fusion framework allows for simultaneously uncovering patient subtypes having statistically significantly different disease outcomes, predicting novel cancer-related genes and predicting drugs for potential repurposings.

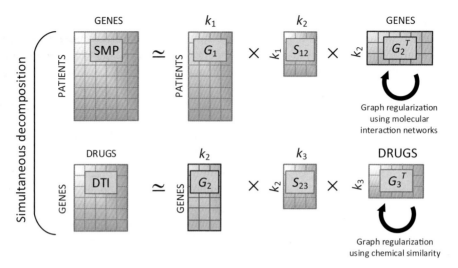

Figure 7.11: Heterogeneous integration with NMTF: Example of the patient specific data-fusion (PSDF) framework. In the PSDF framework, somatic mutation profiles (SMP) and drug target interactions (DTI) are simultaneously decomposed as the product of three matrix factors. Both decompositions share the same cluster indicator matrix of genes, G_2 (in blue), which allows learning from all datasets. The cluster indicator matrix, G_2, is constrained by a graph regularization penalty to favor grouping together genes that interact in a molecular interaction network. Similarly, the cluster indicator matrix G_3 is constrained to favor grouping together drugs that are chemically similar.

7.7 Concluding Remarks

In this chapter we described precision medicine as the study that aims at individualizing the practice of medicine by leveraging the large amounts of clinical and molecular data. We explained how it is enabled by data integration, and provided an overview of key machine learning based data-integration approaches that are currently applied in cancer precision medicine. We focused on describing in detail non-negative matrix factorization and non-negative matrix tri-factorization based data integration methods and illustrated how they can be used for patient subtyping, biomarker discovery, and drug repurposing. We showed how both approaches are NP-hard continuous optimization problems, and how they can be approximately solved computationally.

7.8 Exercises

The number of stars refers to the difficulty level of the exercise, ranging from easy (*) to advanced (***).

7.1 Definitions (*)

  (a) What is precision medicine and what are its goals?
  (b) What are the three different ways of classifying machine learning data integration approaches? Explain them.

(c) Using your answer in (c), classify the patient specific data-fusion framework (described in Section 7.6.3).

7.2 Explain statements (**)

(a) Network based methods, kernel-based methods and Bayesian approaches all suffer from possible information loss. Explain.
(b) Depending on the interpretation of the matrix factors, NMF and NMTF can be used for both dimensionality reduction and clustering. Explain.

7.3 Data integration modeling for gene-disease co-clustering (**)
Gene-disease associations (GDA) can be modeled in the form of a network $G = (N_1 \bigcup N_2, E)$, where N_1 nodes represent genes, N_2 nodes represents diseases and E represents the associations between them. We are interested in co-clustering this network for the following reasons. First, clustering genes based on disease associations, allows for the discovery of distinct disease-specific functional modules, as gene products of genes associated with similar disorders show higher likelihood of physical interactions and higher expression similarity for their transcripts. Second, clustering diseases based on their gene associations can indicate the common genetic origin of diseases.

(a) Both NMF and NMTF could be applied to solve this co-clustering problem. Define both data models and motivate for each why you would pick one over the other.
(b) Assuming you use the NMTF based model from (a) to perform the co-clustering, explain how clustering quality could be improved by integrating additional data, such as PPI data, genetic interaction data and co-expression data on genes and chemical similarity data on drugs. Expanded the current model accordingly.
(c) Additionally, you could also include drug-target interaction data. Expand your model proposed in (b) accordingly.
(d) Explain whether the data integration models presented in (b) and (c) are examples of homogeneous data integration or heterogeneous data integration?

7.4 Data integration modeling for PPI prediction (**)
Given a protein–protein interaction network, matrix factorization methods could be used to predict novel protein–protein interactions (PPI).

(a) Define the data model when solving this problem using NMTF.
(b) What property of matrix factorization modeling is used to make these predictions.
(c) As in the current model, PPIs for a given protein are predicted based on its known interaction pattern, the current model cannot predict PPIs for proteins for which no PPI data is available. Explain how this problem can be overcome.

7.5 Deriving multiplicative update rules (***)

(a) Derive the multiplicative update rules for NMF (as formulated in Section 7.5.2, Equation 7.12) using the derivation procedure presented for the update rules used for NMTF in Section 7.6.2.

(b) In Section 7.6.2, we derived the multiplicative update rules to solve NMTF using the following objective function:

$$\min_{U \geq 0, V \geq 0} f(U, S, V) = \min_{U \geq 0, V \geq 0} \frac{1}{2} \|A - USV^T\|_F^2. \quad (7.34)$$

To account for prior knowledge, NMTF can be expanded to include graph regularization. Assuming we include one graph-regularization term for each of the two object types captured in A, we get the following objective function:

$$\min_{U \geq 0, V \geq 0} f(U, S, V) = \min_{U \geq 0, V \geq 0} \frac{1}{2} \|A - USV^T\|_F^2 + \frac{1}{2} \alpha \operatorname{tr}(U^T \theta_1 U) + \frac{1}{2} \beta \operatorname{tr}(V^T \theta_2 V), \quad (7.35)$$

where θ_1 and θ_2 are regularization terms, and where α and β are hyperparameters weighting the influence of the regularization terms. Derive the multiplicative update rules for this objective function.

Note: Solutions are available to instructors at www.cambridge.org/bionetworks.

7.9 Acknowledgments

This work was supported by the European Research Council (ERC) Starting Independent Researcher Grant 278212, the European Research Council (ERC) Consolidator Grant 770827, the Serbian Ministry of Education and Science Project III44006, the Slovenian Research Agency project J1-8155 and the awards to establish the Farr Institute of Health Informatics Research, London, from the Medical Research Council, Arthritis Research UK, British Heart Foundation, Cancer Research UK, Chief Scientist Office, Economic and Social Research Council, Engineering and Physical Sciences Research Council, National Institute for Health Research, National Institute for Social Care and Health Research, and Wellcome Trust (grant MR/K006584/1).

References

[1] Pržulj N, Malod-Dognin N. Network analytics in the age of big data. *Science*, 2016;353(6295):123–124.

[2] Mirnezami R, Nicholson J, Darzi A. Preparing for precision medicine. *New England Journal of Medicine*, 2012;366(6):489–491.

[3] Beyer M, Laney D. The importance of "big data": A definition. Gartner Inc. Report ID number: G00235055; 2012.

[4] Marusyk A, Polyak K. Tumor heterogeneity: Causes and consequences. *Biochimica et Biophysica Acta (BBA)-Reviews on Cancer*, 2010;1805(1):105–117.

[5] Vogelstein B, Papadopoulos N, Velculescu VE, et al. Cancer genome landscapes. *Science*, 2013;339(6127):1546–1558.

[6] Lötvall J, Akdis CA, Bacharier LB, et al. Asthma endotypes: A new approach to classification of disease entities within the asthma syndrome. *Journal of Allergy and Clinical Immunology*, 2011;127(2):355–360.

[7] Boland MR, Hripcsak G, Shen Y, Chung WK, Weng C. Defining a comprehensive verotype using electronic health records for personalized medicine. *Journal of the American Medical Informatics Association*, 2013; e232–e238.

[8] Gutman S, Kessler LG. The US Food and Drug Administration perspective on cancer biomarker development. *Nature Reviews Cancer*, 2006;6(7):565–572.

[9] Davis JC, Furstenthal L, Desai AA, et al. The microeconomics of personalized medicine: today's challenge and tomorrow's promise. *Nature Reviews Drug Discovery*, 2009;8(4):279–286.

[10] Ashburn TT, Thor KB. Drug repositioning: Identifying and developing new uses for existing drugs. *Nature Reviews Drug Discovery*, 2004;3(8):673–683.

[11] Lanckriet GR, De Bie T, Cristianini N, Jordan MI, Noble WS. A statistical framework for genomic data fusion. *Bioinformatics*, 2004;20(16):2626–2635.

[12] Žitnik M, Zupan B. Data fusion by matrix factorization. *IEEE Transactions on Pattern Analysis and Machine Intelligence*, 2015;37(1):41–53.

[13] Pavlidis P, Weston J, Cai J, Noble WS. Learning gene functional classifications from multiple data types. *Journal of Computational Biology*, 2002;9(2):401–411.

[14] Van Vliet MH, Horlings HM, Van De Vijver MJ, Reinders MJ, Wessels LF. Integration of clinical and gene expression data has a synergetic effect on predicting breast cancer outcome. *PLOS ONE*, 2012;7(7):e40358.

[15] Freedman DA. *Statistical Models: Theory and Practice*. Cambridge University Press; 2009.

[16] Scholkopf B, Smola AJ. *Learning with Kernels: Support Vector Machines, Regularization, Optimization, and Beyond*. MIT Press; 2001.

[17] Vapnik VN, Vapnik V. Statistical Learning Theory, vol. 1. Wiley New York; 1998.

[18] Hartigan JA, Hartigan J. *Clustering Algorithms*, vol. 209. Wiley New York; 1975.

[19] Lee DD, Seung HS. Learning the parts of objects by non-negative matrix factorization. *Nature*, 1999;401(6755):788–791.

[20] Ding C, Li T, Peng W, Park H. Orthogonal nonnegative matrix tri-factorizations for clustering. In *Proceedings of the 12th ACM SIGKDD International Conference on Knowledge Discovery and Data Mining*, ACM SIGKDD; 2006. pp. 126–135.

[21] Wang F, Li T, Zhang C. Semi-supervised clustering via matrix factorization. In *Proceedings of the 2008 SIAM International Conference on Data Mining*. SIAM 2008; pp. 1–12.

[22] Dutkowski J, Kramer M, Surma MA, et al. A gene ontology inferred from molecular networks. *Nature Biotechnology*, 2013;31(1):38–45.

[23] Szklarczyk D, Franceschini A, Wyder S, et al. STRING v10: Protein–protein interaction networks, integrated over the tree of life. *Nucleic Acids Research*, 2014;43(D1):D447–D452.

[24] Mostafavi S, Ray D, Warde-Farley D, Grouios C, Morris Q. GeneMANIA: A real-time multiple association network integration algorithm for predicting gene function. *Genome Biology*, 2008;9(1):S4.

[25] Chen Y, Hao J, Jiang W, et al. Identifying potential cancer driver genes by genomic data integration. *Scientific Reports*, 2013;3:3538.

[26] Pearl J. *Probabilistic reasoning in intelligent systems: Networks of plausible inference.* Morgan Kaufmann;2014.

[27] Wang B, Mezlini AM, Demir F, et al. Similarity network fusion for aggregating data types on a genomic scale. *Nature Methods*, 2014;11(3):333–337.

[28] Davis DA, Chawla NV. Exploring and exploiting disease interactions from multi-relational gene and phenotype networks. *PLOS ONE*, 2011;6(7):e22670.

[29] Sun K, Buchan N, Larminie C, Pržulj N. The integrated disease network. *Integrative Biology*, 2014;6(11):1069–1079.

[30] Guo X, Gao L, Wei C, Yang X, Zhao Y, Dong A. A computational method based on the integration of heterogeneous networks for predicting disease-gene associations. *PLOS ONE*, 2011;6(9):e24171.

[31] Huang YF, Yeh HY, Soo VW. Inferring drug-disease associations from integration of chemical, genomic and phenotype data using network propagation. *BMC Medical Genomics*, 2013;6(3):S4.

[32] Needham CJ, Bradford JR, Bulpitt AJ, Westhead DR. A primer on learning in Bayesian networks for computational biology. *PLOS Computational Biology*, 2007;3(8):e129.

[33] Ben-Gal I. Bayesian networks. In Ruggeri F, Faltin F, Kenett R, eds., *Encyclopedia of Statistics in Quality and Reliability*, Wiley & Sons 2008.

[34] Cooper GF. The computational complexity of probabilistic inference using Bayesian belief networks. *Artificial Intelligence*, 1990;42(2-3):393–405.

[35] Zhu J, Zhang B, Smith EN, et al. Integrating large-scale functional genomic data to dissect the complexity of yeast regulatory networks. *Nature Genetics*, 2008;40(7):854–861.

[36] Gevaert O, Smet FD, Timmerman D, Moreau Y, Moor BD. Predicting the prognosis of breast cancer by integrating clinical and microarray data with Bayesian networks. *Bioinformatics*, 2006;22(14):e184–e190.

[37] Schölkopf B, Tsuda K, Vert JP. *Kernel Methods in Computational Biology.* MIT Press;2004.

[38] Borgwardt KM. Kernel methods in bioinformatics. In Lu HHS, Schölkopf B, Zhao h, eds., *Handbook of Statistical Bioinformatics.* Springer;2011, pp. 317–334.

[39] Kondor RI, Lafferty J. Diffusion kernels on graphs and other discrete input spaces. In *Proceedings of the 19th International Conference on Machine Learning.* Morgan Kaufmann vol. 2; 2002, pp. 315–322.

[40] Vacic V, Iakoucheva LM, Lonardi S, Radivojac P. Graphlet kernels for prediction of functional residues in protein structures. *Journal of Computational Biology*, 2010;17(1):55–72.

[41] Yu S, Tranchevent LC, De Moor B, Moreau Y. *Kernel-Based Data Fusion for Machine Learning.* Springer;2013.

[42] Jolliffe IT. Principal component analysis and factor analysis. In *Principal Component Analysis.* Springer;1986. pp. 115–128.

[43] Hardoon DR, Szedmak S, Shawe-Taylor J. Canonical correlation analysis: An overview with application to learning methods. *Neural Computation,* 2004;16(12):2639–2664.

[44] Daemen A, Gevaert O, De Moor B. Integration of clinical and microarray data with kernel methods. In *Engineering in Medicine and Biology Society, 2007. EMBS 2007. 29th Annual International Conference of the IEEE.* IEEE;2007, pp. 5411–5415.

[45] Napolitano F, Zhao Y, Moreira VM, et al. Drug repositioning: A machine-learning approach through data integration. *Journal of Cheminformatics,* 2013;5(1):30.

[46] Wang X, Xing EP, Schaid DJ. Kernel methods for large-scale genomic data analysis. *Briefings in Bioinformatics,* 2014;16(2):183–192.

[47] Speicher NK, Pfeifer N. Integrating different data types by regularized unsupervised multiple kernel learning with application to cancer subtype discovery. *Bioinformatics,* 2015;31(12)i268–i275.

[48] Cichocki A, Zdunek R, Phan AH, Amari Si. *Nonnegative Matrix and Tensor Factorizations: Applications to Exploratory Multi-way Data Analysis and Blind Source Separation,* John Wiley & Sons;2009.

[49] Ding C, He X, Simon HD. On the equivalence of nonnegative matrix factorization and spectral clustering. In *Proceedings of the 2005 SIAM International Conference on Data Mining.* SIAM;2005. pp. 606–610.

[50] Zass R, Shashua A. A unifying approach to hard and probabilistic clustering. In [ICCV 05'] *Proceedings of the Tenth IEEE International Conference on Computer Vision* vol. 1. IEEE;2005, pp. 294–301.

[51] Vavasis SA. On the complexity of nonnegative matrix factorization. *SIAM Journal on Optimization,* 2009;20(3):1364–1377.

[52] Lee DD, Seung HS. Algorithms for non-negative matrix factorization. In *Advances in Neural Information Processing Systems: Proceedings of NIPS 2000;* 2001, pp. 556–562.

[53] Wild S. Seeding non-negative matrix factorizations with the spherical k-means clustering. PhD thesis, University of Colorado; 2003.

[54] Boutsidis C, Gallopoulos E. SVD based initialization: A head start for nonnegative matrix factorization. *Pattern Recognition,* 2008;41(4):1350–1362.

[55] Shen R, Olshen AB, Ladanyi M. Integrative clustering of multiple genomic data types using a joint latent variable model with application to breast and lung cancer subtype analysis. *Bioinformatics,* 2009;25(22):2906–2912.

[56] Cai D, He X, Han J, Huang TS. Graph regularized nonnegative matrix factorization for data representation. *IEEE Transactions on Pattern Analysis and Machine Intelligence,* 2011;33(8):1548–1560.

[57] Luo D, Ding C, Huang H, Li T. Non-negative Laplacian embedding. In *Ninth IEEE International Conference on Data Mining*. IEEE; 2009. pp. 337–346. Available from: http://ieeexplore.ieee.org/document/5360259/.

[58] Hofree M, Shen JP, Carter H, Gross A, Ideker T. Network-based stratification of tumor mutations. *Nature Methods*, 2013;10(11):1108–1115.

[59] Kuhn H, Tucker A. Nonlinear programming. In *Proceedings of the 2nd Berkeley Symposium on Mathematical Statistics and Probability*. University of California Press;1951, pp. 481–492.

[60] Ding CH, Li T, Jordan MI. Convex and semi-nonnegative matrix factorizations. *IEEE Transactions on Pattern Analysis and Machine Intelligence*, 2010;32(1):45–55.

[61] Gligorijević V, Malod-Dognin N, Pržulj N. Patient-specific data fusion for cancer stratification and personalised treatment. In *Proceedings of the Pacific Symposium on Biocomputing*, World Scientific;2016, pp. 321–332.

8 Machine Learning for Biomarker Discovery: Significant Pattern Mining

Felipe Llinares-López and Karsten Borgwardt

8.1 Introduction

Biomarker discovery, the search for measurable biological indicators of a phenotypic trait of interest, is one of the fundamental data analysis problems in computational biology and healthcare. The challenging nature of datasets in those fields, typically containing many more features than samples, has motivated the development of novel tools for statistical inference in high-dimensional spaces. An example of such datasets may contain gene expression profiles of hundreds of individuals (samples), in which tens of thousands of genes (features) are surveyed for each individual. In this example, the goal would be to identify the biomarkers (e.g., genes) that help us differentiate between individuals of distinct phenotypes. Existing work has predominantly focused on either univariate feature selection methods (e.g., [1]), which consider the effect of each candidate marker in isolation from the others, or multivariate linear models with sparsity-inducing regularisers (e.g., [2, 3, 4, 5, 6, 7]), which jointly model the effect of all candidate markers as a weighted additive combination of individual effects. The application of these techniques in biology and medicine has led to many successes. For instance, they have been extensively used in genome-wide association studies, leading to the discovery of more than 30,000 variant-trait associations [8], many of which led to substantial biological insight and even clinical applications [9]. However, a common limitation of both families of approaches is their inability to discover non-linear signals due to interactions between features. For instance, this "blind spot" has been hypothesized as a factor that could account for at least a fraction of the "missing heritability" in genome-wide association studies [10, 11, 12, 13], that is, the phenomenon that loci discovered by genome-wide association studies only account for a small proportion of the estimated heritability of the phenotypes.

However, assessing the statistical association of *all* high-order feature interactions with a phenotypic trait of interest is a difficult problem. By explicitly considering all high-order feature interactions, the effective number of features in the model explodes combinatorially, further exacerbating the gap between sample size and number of features. To give a sense of scale, in a dataset with $p = 266$ features, which could be considered small by current standards, one could enumerate up to $2^p \approx 10^{80}$ high-order feature interactions; as many as the estimated number of electrons in the observable universe [14, appendix C.4]. The enormous number of feature interactions that would need to be tested for association with the trait of interest creates two fundamental challenges:

1. A *statistical* challenge, caused by the need to perform statistical inference in the extremely high-dimensional space resulting from explicitly considering all high-order feature interactions. This is the so called *multiple hypothesis testing problem*, a phenomenon that makes it considerably challenging to control the probability to report false associations while maintaining enough statistical power to discover the truly significant high-order feature interactions.
2. A *computational* challenge, derived from the need to explore the vast search space consisting of all candidate feature interactions.

Significant pattern mining is a novel branch of machine learning developed specifically to tackle these challenges [15, 16, 17, 18, 19, 20]. The goal of this chapter is to provide readers with the necessary tools to understand state-of-the-art significant pattern mining approaches and to apply these algorithms in data analysis tasks in their respective fields of research. The rest of this chapter is organised as follows:

Section 8.2 is devoted to introduce, in a self-contained manner, the key background concepts the rest of this chapter relies on. Section 8.2.1 defines a general framework for significant pattern mining and details two of its most common instances: (i) *significant itemset mining*, which aims at finding significant high-order interactions between binary features, and ii) *significant subgraph mining*, which looks for significant subgraphs among a database of graph-structured samples. Sections 8.2.2 and 8.2.3 deal with the statistical aspects of significant pattern mining. Section 8.2.2 provides background on statistical association testing for binary random variables, one of the key sub-components of significant pattern mining. Section 8.2.3 discusses the multiple hypothesis testing problem, the main statistical difficulty that significant pattern mining faces, and presents Tarone's improved Bonferroni correction for discrete data. This is the key statistical tool that significant pattern mining algorithms use to solve that difficulty.

Section 8.3 builds upon previous sections and details how to combine Tarone's improved Bonferroni correction with classical data mining techniques, leading to an efficient approach to analyse large databases using Tarone's method.

Sections 8.4 and 8.5 present recent advances in significant pattern mining: Section 8.4 shows how the inherent redundancy between candidate feature interactions can be exploited to gain additional statistical power and Section 8.5 describes how confounding factors such as age, gender, or population structure can be accounted for in significant pattern mining.

Finally, Section 8.6 synthesizes the main concepts of this chapter, gives pointers to existing software and discusses the main open problems in the field.

8.2 The Problem of Significant Pattern Mining

8.2.1 Terminology and Problem Statement

Suppose we are given a dataset $\mathcal{D} = \{(x_i, y_i)\}_{i=1}^{n}$ consisting of n independent and identically distributed (i.i.d.) samples x taking values in an input domain $\mathcal{X}$ and assigned to one of two classes $y \in \{0, 1\}$. Informally, we call any discrete substructure of the input samples $x \in \mathcal{X}$ a *pattern*.

The goal of significant pattern mining is to find, given a search space $\mathcal{M}$ containing candidate patterns of interest, all patterns $\mathcal{S} \in \mathcal{M}$ that occur statistically significantly more often in one class of samples than in the other.

Let the *pattern occurrence indicator* $g_\mathcal{S}(x)$ be defined as the binary random variable that indicates whether pattern $\mathcal{S}$ occurs in an input sample x or not:

$$g_\mathcal{S}(x) = \begin{cases} 1, & \text{if } \mathcal{S} \subseteq x, \\ 0, & \text{otherwise.} \end{cases} \tag{8.1}$$

Then, determining if pattern $\mathcal{S}$ is statistically significantly enriched in input samples x belonging to one of the two classes is equivalent to testing the statistical association of two binary random variables, the class labels y and the indicator $g_\mathcal{S}(x)$, based on n observations $\{(g_\mathcal{S}(x_i), y_i)\}_{i=1}^{n}$ from $\mathcal{D}$.

The precise definition of a pattern $\mathcal{S}$, the corresponding search space of candidate patterns $\mathcal{M}$ and the concept of inclusion $\mathcal{S} \subseteq x$ all depend on the nature on the input domain $\mathcal{X}$. This leads to multiple instances of significant pattern mining, all of which fit in the same abstract framework described above.

8.2.1.1 Significant Itemset Mining

One of the most common instances of significant pattern mining arises when the input samples are p-dimensional binary vectors, i.e., $\mathcal{X} = \{0, 1\}^p$. Conceptually, this means that each input sample x is being represented by p distinct features $x = (u_1, u_2, \ldots, u_p)$, each of which can be either active ($u_j = 1$) or inactive ($u_j = 0$).

This type of representation can be readily applied to a wide variety of applications of interest to computational biology and clinical data analysis. For instance, in a genome-wide association study, each of these p binary features could be the result of a dominant/recessive/over-dominant encoding of single nucleotide polymorphisms (SNPs) or be obtained based on prior knowledge such as functional annotations. In functional genomics, each of the n input samples could correspond to a different genomic region while each of the p binary features could indicate whether a certain property applies to the genomic region or not, such as containing a specific transcription factor binding motif or exhibiting a particular chromatin modification in a given cell type. As another example, in a clinical setting, this type of representation could be used to describe the results of laboratory tests for a set of patients, with each of the p binary features encoding the result of a certain test as either abnormal or normal. Electronic health records (EHRs) can also be partly represented in this form, with each of the p binary features indicating which medical codes from a certain medical ontology (e.g., ICD-9, SNOMED-CT) apply to the record.

In *significant itemset mining*, each pattern S corresponds to a specific high-order feature interaction. Let $S \subseteq \{1, 2, \ldots, p\}$ index an arbitrary subset of the p binary features. Define the *high-order interaction* between the features indexed by S as $z_S(x) = \prod_{j \in S} u_j$. Note that $z_S(x) = 1$ if and only if *all* features in S are simultaneously active in the input sample x and $z_S(x) = 0$ otherwise. We define a pattern S to occur in a sample $x \in \{0, 1\}^p$, i.e., $S \subseteq x$, if the high-order feature interaction induced by S is active ($z_S(x) = 1$). In other words, $g_S(x) = z_S(x)$. This definition can be justified as follows. Suppose each input sample $x \in \{0, 1\}^p$ is alternatively represented as the set containing the indices of active features in x. For example, with $p = 5$, we can represent an input sample $x_1 = (1, 0, 1, 1, 0)$ as $x_1 = \{1, 3, 4\}$ while $x_2 = (0, 1, 0, 1, 0)$ can be represented as $x_2 = \{2, 4\}$. Since this is a one-to-one mapping between the set of p-dimensional binary vectors $\{0, 1\}^p$ and the set of all subsets of $\{1, 2, \ldots, p\}$, denoted the *powerset* $\mathcal{P}(\{1, 2, \ldots, p\})$, both representations are equivalent and the input domain can also be defined as $\mathcal{X} = \mathcal{P}(\{1, 2, \ldots, p\})$. However, we have $S \subseteq x$, with x represented as a set of active features and inclusion defined in the traditional sense, if and only if the high-order feature interaction $z_S(x)$ as defined above takes value 1. Hence, both representations lead to the same pattern occurrence indicator $g_S(x)$. This notation, which represents input samples as a set of active features or *items*, is widespread in the data mining field. This is primarily the reason why this particular instance of significant pattern mining is commonly referred to as significant *itemset* mining.

In applications in which *all* feature interactions are of potential interest, the search space of candidate patterns $\mathcal{M}$ would contain all possible feature subsets, i.e., $\mathcal{M} = \mathcal{P}(\{1, 2, \ldots, p\})$. Thus, the search space would contain $|\mathcal{M}| = 2^p$ patterns.[1] While this is the most common case, one can incorporate certain forms of prior knowledge by explicitly restricting the feature interactions contained in $\mathcal{M}$. For example, in the context of a genome-wide association study, a researcher might wish to focus the analysis on feature interactions occurring between variants belonging to the same gene or between variants belonging to the same biological pathway. This flexibility to encode prior knowledge leads to a reduction in the number of candidate patterns, thus improving the statistical power and reducing runtime. It also opens the door to the development of novel variations of significant itemset mining, targeting specific applications in computational biology and medicine.

Figure 8.1 illustrates a canonical application of significant itemset mining to a toy dataset containing $n = 12$ samples, $n_1 = 6$ of which belong to class $y = 1$ (e.g., cases) and $n_0 = 6$ of which belong to class $y = 0$ (e.g., controls). Each sample is represented by a binary vector with $p = 10$ features, $u_1, u_2, \ldots, u_{10}$. Two patterns, $S_1 = \{1, 3, 5, 6\}$ (yellow) and $S_2 = \{2, 9, 10\}$ (blue), and their respective high-order feature interactions, $z_{S_1}(x)$ and $z_{S_2}(x)$, are highlighted in the figure. In this example, none of the 10 features is individually associated with the class labels. Therefore, a univariate analysis or a multivariate linear model would not capture any association. However, visual inspection reveals that the high-order feature interactions $z_{S_1}(x)$ and $z_{S_2}(x)$ are active considerably more often for samples in class $y = 1$ than for samples in class $y = 0$. The fact that significant high-order interactions can be present

---

[1] We denote by $|\mathcal{M}|$ the cardinality of a set $\mathcal{M}$, that is, the number of elements in $\mathcal{M}$.

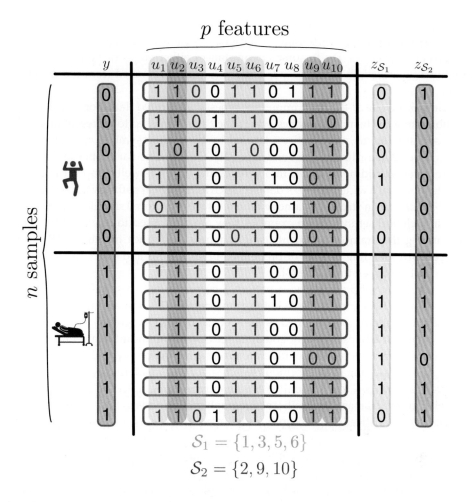

Figure 8.1: Illustration of significant itemset mining. A toy dataset with $n = 12$ samples, divided into $n_1 = 6$ cases and $n_0 = 6$ controls, is shown. Each sample is represented by $p = 10$ binary features, $u_1, u_2, \ldots, u_{10}$. The high-order interactions $z_{S_1}(x)$ and $z_{S_2}(x)$, indexed by patterns $S_1 = \{1, 3, 5, 6\}$ (yellow) and $S_2 = \{2, 9, 10\}$ (blue), can be seen to be active considerably more often in cases than controls. However, none of the binary features in these patterns are individually associated with the case/control status.

even in the absence of univariate associations strongly motivates the development of methods that are able to explore arbitrary feature interactions in high-dimensional datasets. Otherwise, many signals of practical importance might remain undiscovered by standard analyses.

8.2.1.2 Significant Subgraph Mining

Another instance of significant pattern mining of great relevance occurs when each input sample is a graph, i.e., $\mathcal{X} = \{x \mid x = (V, E, l_V, l_E)\}$ where V is a set of vertices, $E \subseteq V \times V$ is a set of edges, $l_V : V \to \Sigma_V$ is a function that labels each vertex in the graph with a categorical attribute from a finite set of categories Σ_V and $l_E : E \to \Sigma_E$ is a function that labels each edge in the graph with a categorical attribute from a finite set of categories Σ_E. In some circumstances either edges or vertices (or both) might be unlabeled, in which case l_E and/or l_V can be removed from the definition of $\mathcal{X}$.

Graphs are one of the most important types of structured data. For instance, they are ubiquitous in chemoinformatics, where they are often used to represent molecules by associating atoms with vertices, labeled by the atomic symbol, and edges with bonds, labeled according to the bond type [21]. Other molecular properties might be incorporated as additional labels into vertices and edges as well. Graphs can also be used to represent a wide variety of datasets such as biological pathways, co-expression networks or protein structures, among others. Graph-structured samples are also common in a clinical setting. As an example, they are often used to represent the result of scanning the brain of a patient using magnetic resonance imaging (MRI), with vertices corresponding to a set of predefined brain regions and edges modelling the connectivity between them [22, 23].

In *significant subgraph mining*, a pattern $\mathcal{S}$ corresponds to a specific induced subgraph of at least one input sample in the graph dataset $\mathcal{D}$.[2] Typically, the search space of candidate patterns $\mathcal{M}$ is defined as the set of *all* distinct subgraphs of input graphs in the dataset of interest $\mathcal{D}$. Analogously to the case of significant itemset mining, the number of subgraphs one can enumerate grows combinatorially with the size of the input graphs, resulting in an enormous search space $\mathcal{M}$. A pattern $\mathcal{S}$ occurs in an input sample $x \in \mathcal{X}$, i.e. $\mathcal{S} \subseteq x$, if and only if $\mathcal{S}$ is an induced subgraph of x. Therefore, given a labeled graph database $\mathcal{D}$, significant subgraph mining aims to find those subgraphs $\mathcal{S}$ that occur statistically significantly more often in one class of graphs than in the other.

Figure 8.2 shows a standard application of significant subgraph mining to a toy dataset with $n = 8$ graphs, $n_1 = 4$ of which belong to class $y = 1$ (successful drugs) and $n_0 = 4$ of which belong to class $y = 0$ (drugs that generate an adverse reaction). Each sample (drug) is represented by a graph with labeled vertices ($|\Sigma_V| = 3$) and labeled edges ($|\Sigma_E| = 2$). Two patterns, or subgraphs, $\mathcal{S}_1$ (yellow) and $\mathcal{S}_2$ (blue), are highlighted in the figure. Their respective occurrence indicators $g_{\mathcal{S}_1}$ and $g_{\mathcal{S}_2}$ are shown on the right of the figure, with entries in the vectors corresponding to graphs in clockwise order for each of the two classes separately. In this example, subgraph $\mathcal{S}_1$ is strongly overrepresented in class $y = 1$ while $\mathcal{S}_2$ is enriched in class $y = 0$. Note that, as in significant itemset mining, univariate associations analyses that only inspect individual nodes and/or edges would miss those associations.

8.2.2 Statistical Association Testing in Significant Pattern Mining

Regardless of the specific type of input domain $\mathcal{X}$ and patterns $\mathcal{S} \in \mathcal{M}$ of interest, the general problem to be solved in significant pattern mining is to efficiently find

---

[2] An induced subgraph $\mathcal{S}$ of a graph x is another graph, formed by selecting a subset of vertices in x and keeping only the edges which connect any pair of the selected vertices.

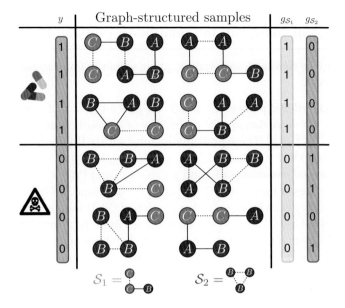

Figure 8.2: Illustration of significant subgraph mining. A toy dataset with $n = 8$ samples, divided into $n_1 = 4$ successful drugs and $n_0 = 4$ drugs that trigger an adverse reaction, is shown. Each sample is represented by a graph with labeled vertices and edges. In this simple example, vertex labels belong to an alphabet with $|\Sigma_V| = 3$ characters (blue/red/green) while edge labels belong to an alphabet with $|\Sigma_E| = 2$ characters (solid/discontinuous). Two patterns, S_1 (yellow) and S_2 (blue), can be seen to occur (be induced subgraphs) significantly more often in one class of drugs than in the other.

all patterns $S \in \mathcal{M}$ such that the random variable $G_S(X)$ is statistically associated with the random variable Y. This naturally leads to the problem of testing the statistical association of two binary random variables, which we discuss in detail in this section.

Two random variables G and Y are said to be *statistically associated* if they are *not* statistically independent, i.e., if their joint probability distribution $\Pr(G = g, Y = y)$ does *not* factorize as $\Pr(G = g, Y = y) = \Pr(G = g)\Pr(Y = y)$. This is equivalent to the conditional probability distributions not being identical to the marginal distributions, i.e., if there exists (g, y) such that $\Pr(G = g \mid Y = y) \neq \Pr(G = g)$ and there exists (g, y) such that $\Pr(Y = y \mid G = g) \neq \Pr(Y = y)$. We denote the fact that random variables $G_S(X)$ and Y are statistically independent as $G_S(X) \perp\!\!\!\perp Y$. See [24] for a general introduction to statistical independence. (See Box 8.1.)

When both random variables G and Y are binary, as is the case in significant pattern mining, their joint distribution $\Pr(G = g, Y = y)$ is fully determined by four quantities:

1. $p_{1,1} = \Pr(G = 1, Y = 1)$.
2. $p_{0,1} = \Pr(G = 0, Y = 1)$.
3. $p_{0,0} = \Pr(G = 0, Y = 0)$.
4. $p_{1,0} = \Pr(G = 1, Y = 0)$.

> **Box 8.1: Joint, marginal, and conditional probabilities**
>
> Let G and Y be two random variables taking values in finite alphabets $\mathcal{G}$ and $\mathcal{Y}$, respectively.
>
> - For each ordered pair $(g, y) \in \mathcal{G} \times \mathcal{Y}$, the *joint probability* of g and y, $\Pr(G = g, Y = y)$, denotes the probability of the joint event $[G = g] \cap [Y = y]$. These probabilities must satisfy $\Pr(G = g, Y = y) \geq 0$ for all $(g, y) \in \mathcal{G} \times \mathcal{Y}$ and $\sum_{g \in \mathcal{G}} \sum_{y \in \mathcal{Y}} \Pr(G = g, Y = y) = 1$.
> - The *marginal probabilities* of g and y, $\Pr(G = g)$ and $\Pr(Y = Y)$, denote the probabilities of the events $[G = g]$ and $[Y = y]$, respectively. Since the event $[G = g]$ can be expressed as a union of $|\mathcal{Y}|$ disjoint events, $[G = g] = \bigcup_{y \in \mathcal{Y}} [G = g] \cap [Y = y]$, it follows from the axioms of probability that the joint and marginal probabilities are related by $\Pr(G = g) = \sum_{y \in \mathcal{Y}} \Pr(G = g, Y = y)$, an operation commonly referred to as *marginalization*. Marginalization leads to valid probabilities, as it can be readily seen that $\Pr(G = g) \geq 0$ for all $g \in \mathcal{G}$ and that $\sum_{g \in \mathcal{G}} \Pr(G = g) = 1$. Analogously, $\Pr(Y = y) = \sum_{g \in \mathcal{G}} \Pr(G = g, Y = y)$, where $\Pr(Y = y) \geq 0$ for all $y \in \mathcal{Y}$ and $\sum_{y \in \mathcal{Y}} \Pr(Y = y) = 1$.
> - Finally, the *conditional probabilities* of g given that $[Y = y]$ and of y given that $[G = g]$, $\Pr(G = g \mid Y = y)$ and $\Pr(Y = y \mid G = g)$, denote the probability of the joint event $[G = g] \cap [Y = y]$ when the sample space has been reduced from the original set of all possible outcomes, $\bigcup_{(g,y) \in \mathcal{G} \times \mathcal{Y}} [G = g] \cap [Y = y]$, to the sets of outcomes $[Y = y] = \bigcup_{g \in \mathcal{G}} [G = g] \cap [Y = y]$ and $[G = g] = \bigcup_{g \in \mathcal{G}} [G = g] \cap [Y = y]$, respectively. Owing to the definition of conditional probability, it follows that $\Pr(G = g \mid Y = y) = \frac{\Pr(G=g, Y=y)}{\Pr(Y=y)}$ and $\Pr(Y = y \mid G = g) = \frac{\Pr(G=g, Y=y)}{\Pr(G=g)}$. Note that these probabilities are undefined if $\Pr(Y = y) = 0$ or $\Pr(G = g) = 0$, respectively.

This can be represented by means of a 2×2 contingency table:

Variables	$G = 1$	$G = 0$	Row totals
$Y = 1$	$p_{1,1}$	$p_{0,1}$	p_Y
$Y = 0$	$p_{1,0}$	$p_{0,0}$	$1 - p_Y$
Col. totals	p_G	$1 - p_G$	1

Since $\Pr(G = g, Y = y)$ must be a valid probability distribution, $p_{1,1} + p_{0,1} + p_{0,0} + p_{1,0} = 1$, thus leaving three degrees of freedom. The marginal distributions of G and Y follow from the joint distribution: $\Pr(G = 1) = p_{1,1} + p_{1,0} = p_G$ and $\Pr(Y = 1) = p_{1,1} + p_{0,1} = p_Y$.

The binary random variables G and Y will be statistically independent and, hence, *not* statistically associated, if the joint distribution factorizes as $\Pr(G = g, Y = y) = \Pr(G = g)\Pr(Y = y)$ for all $(g, y) \in \{0, 1\}^2$. It can be shown that if the distribution factorizes for any of the four possible values of $(g, y) \in \{0, 1\}^2$, it will also factorize for the other three.[3] Therefore, without loss of generality, two binary random variables G and Y are statistically independent if and only if $\Pr(G = 1, Y = 1) = \Pr(G = 1)\Pr(Y = 1)$, that is, if and only if $p_{1,1} = p_Y p_G$. Equivalently, the independence condition can also be expressed in terms of conditional distributions. Two binary random variables G and Y are statistically independent if and only if $\Pr(G = 1 \mid Y = 1) = \Pr(G = 1)$. Again, the condition needs to be verified only for one of the four possible values of $(g, y) \in \{0, 1\}^2$ as all other cases follow automatically.[4]

In practice, however, the joint distribution between the two binary random variables whose association is to be tested is unknown. Therefore, none of those independence conditions can be readily checked. In the context of significant pattern mining, when testing the association between the occurrence of a pattern $G_S(X)$ and the class labels Y, all we have access to are n i.i.d. samples from the joint distribution $\{(g_S(x_i), y_i)\}_{i=1}^n$ obtained from a dataset $\mathcal{D} = \{(x_i, y_i)\}_{i=1}^n$. Using those n samples, the unknown joint distribution $\Pr(G_S(X) = g_S(x), Y = y)$ can be approximated by counting the number of times the events $(g_S(x) = 1, y = 1)$, $(g_S(x) = 0, y = 1)$, $(g_S(x) = 0, y = 0)$, and $(g_S(x) = 1, y = 0)$ are observed. This procedure is also typically represented as a 2×2 contingency table:

Variables	$g_S(x) = 1$	$g_S(x) = 0$	Row totals
$y = 1$	a_S	b_S	n_1
$y = 0$	d_S	c_S	n_0
Col. totals	r_S	q_S	n

where,

1. $a_S = \sum_{i=1}^n g_S(x_i) y_i$ is the number of times the event $(g_S = 1, y = 1)$ occurs.
2. $b_S = \sum_{i=1}^n (1 - g_S(x_i)) y_i$ is the number of times the event $(g_S = 0, y = 1)$ occurs.
3. $c_S = \sum_{i=1}^n (1 - g_S(x_i))(1 - y_i)$ is the number of times the event $(g_S = 0, y = 0)$ occurs.
4. $d_S = \sum_{i=1}^n g_S(x_i)(1 - y_i)$ is the number of times the event $(g_S = 1, y = 0)$ occurs.

These counts can be used to construct a frequentist approximation to the unknown joint probability distribution $\Pr(G_S(X) = g_S(x), Y = y)$: $\hat{p}_{1,1} = a_S/n$, $\hat{p}_{0,1} = b_S/n$, $\hat{p}_{0,0} = c_S/n$ and $\hat{p}_{1,0} = d_S/n$. Again, all four quantities must add up to the total number of observations, $a_S + b_S + c_S + d_S = n$. Thus, the empirical approximation to the joint distribution also has three degrees of freedom. The empirical approximations of the marginal distributions of $G_S(X)$ and Y both follow from the approximation of the joint distribution: $\hat{p}_G = (a_S + d_S)/n = r_S/n$ and $\hat{p}_Y = (a_S + b_S)/n = n_1/n$.

---

[3] See Exercise 8.2.
[4] See Exercise 8.3.

Since the empirical approximation to the joint distribution will always incur some error due to random sampling, it is necessary to derive a procedure that accounts for the uncertainty introduced by the approximation. Roughly, *statistical association testing* involves the following sequence of steps:

1. Define a *test statistic* $T: \{(g_S(x_i), y_i)\}_{i=1}^n \to \mathbb{R}$, i.e., a function that maps the n i.i.d. samples to a scalar value. While any such function is a valid test statistic, not all choices will be equally useful. Qualitatively, a good test statistic must produce output values for data samples $\{(g_S(x_i), y_i)\}_{i=1}^n$ obtained from statistically independent $(G_S(X), Y)$ which are sufficiently different from the output values it produces for data samples $\{(g_S(x_i), y_i)\}_{i=1}^n$ obtained from statistically associated $(G_S(X), Y)$. In short, a useful test statistic can be thought as a scalar measure of association based on the n observed samples.

2. Derive the *null distribution* of the test statistic T, $\Pr(T = t \mid H_0)$. This is the probability distribution of the test statistic T under the *null hypothesis* $H_0 = G_S(X) \perp\!\!\!\perp Y$, i.e., under the assumption that the sample $\{(g_S(x_i), y_i)\}_{i=1}^n$ is obtained from statistically independent $G_S(X)$ and Y.

3. Evaluate the test statistic T on the real data sample $\{(g_S(x_i), y_i)\}_{i=1}^n$, obtaining a scalar value t. In order to take the uncertainty due to random sampling into account, the value of t is related to the null distribution $\Pr(T = t \mid H_0)$ by computing a *P-value*. The P-value is defined as the probability that the test statistic takes a value at least as extreme as t, i.e., a value representing an association at least as strong, under the null hypothesis of independence. Assuming that larger values of t correspond to stronger associations, the P-value is then computed as $p = \Pr(T \geq t \mid H_0)$.

4. If the P-value falls below a predefined significance threshold α, i.e., $p \leq \alpha$, the random variables $G_S(X)$ and Y are deemed *significantly associated*. When testing a single pair of random variables for association, the significance threshold α has a clear interpretation as the *Type I error* of the procedure: it is the probability that random variables $G_S(X)$ and Y are deemed to be significantly associated given that they are actually statistically independent. Therefore, adjusting α allows the practitioner to trade off Type I error and statistical power, that is, to trade off the probability to report an association given that no association exists with the probability to discover an association given that it exists.

One must proceed with care when reporting an association according to the approach described above. From a statistical point of view, a low P-value only means that the null distribution $\Pr(T = t \mid H_0)$ is a poor fit to the observed value t. While this can occur due to $G_S(X)$ and Y being in fact statistically associated and, hence, the null hypothesis H_0 being false, there are other reasons why the model fit might be poor. For example, often the null distribution $\Pr(T = t \mid H_0)$ cannot be derived exactly and one must resort to approximations. Low P-values can therefore arise also as a consequence of the approximation to the null distribution being insufficiently precise. Besides, the statistical model chosen for the null distribution might be a poor fit to the observed value t even when H_0 is true and $\Pr(T = t \mid H_0)$ can be evaluated exactly. This can occur due to mismatches between model assumptions and the ground-truth such as assuming the data to be Gaussian-distributed when the real distribution exhibits heavy tails or if the n samples are not i.i.d. draws. In summary, statistically significant associations should *not* be treated as unquestionable discoveries but rather as potential associations that are worth investigating further.

In the remainder of this section, we will introduce two of the most commonly used test statistics for statistical association testing between two binary random variables: Pearson's χ^2 test [25] and Fisher's exact test [26]. While a mathematically rigorous derivation of either test is outside the scope of this chapter, we will provide informal derivations, with emphasis on building intuition about both test statistics. For a formal derivation of the tests, we refer the reader to the original articles [25, 26].

Both Pearson's χ^2 test and Fisher's exact test are based on 2×2 contingency tables. Without loss of generality, the data sample $\{(g_S(x_i), y_i)\}_{i=1}^n$ can be summarized by the counts $a_S, r_S,$ and n_1 which, together with the sample size n, uniquely determine all other counts in the contingency table. Both Pearson's χ^2 test and Fisher's exact test make the assumption that the counts r_S and n_1 contain little information about the potential existence of an association between the two binary random variables. This is motivated by the observation that r_S is related to the marginal distribution of $G_S(X)$ while n_1 is related to the marginal distribution of Y. As a consequence, both test statistics consider r_S and n_1 as fixed quantities, leaving the count a_S as the only random quantity in the model. Thus, the null distribution for both tests will be of the form $\Pr(T(A_S) = t \mid R_S = r_S, N_1 = n_1, H_0)$, with the specific choice of function $T(A_S)$ being the only aspect differentiating both tests.

In order to derive the null distributions of Pearson's χ^2 test and Fisher's exact test, we will first derive the conditional probability distribution $\Pr(A_S = a_S \mid R_S = r_S, N_1 = n_1, H_0)$ under the null hypothesis of independence $H_0 = G_S(X) \perp\!\!\!\perp Y$. We summarize the result in the following proposition.

Proposition 8.1 *The conditional probability distribution of A_S given that $R_S = r_S$, $N_1 = n_1$ and sample size n under the null hypothesis of independence $H_0 = G_S(X) \perp\!\!\!\perp Y$ is a hypergeometric distribution with parameters n, n_1 and r_S:*

$$\Pr(A_S = a_S \mid R_S = r_S, N_1 = n_1, H_0) = \text{Hypergeom}(a_S \mid n, n_1, r_S)$$

$$= \frac{\binom{n_1}{a_S}\binom{n-n_1}{r_S-a_S}}{\binom{n}{r_S}}. \quad (8.2)$$

Proof By definition of conditional probability distribution we have:

$$\Pr(A_S = a_S \mid R_S = r_S, N_1 = n_1, H_0) = \frac{\Pr(A_S = a_S, R_S = r_S \mid N_1 = n_1, H_0)}{\Pr(R_S = r_S \mid N_1 = n_1, H_0)}. \quad (8.3)$$

Since $R_S = A_S + D_S$, the joint distribution in the numerator can be rewritten as $\Pr(A_S = a_S, R_S = r_S \mid N_1 = n_1, H_0) = \Pr(A_S = a_S, D_S = r_S - a_S \mid N_1 = n_1, H_0)$. Note that A_S depends only on samples for which $y_i = 1$ while D_S depends only on samples for which $y_i = 0$. Therefore, under the assumption that all n samples are i.i.d. draws, the random variables A_S and D_S are statistically independent. This allows the joint distribution in the numerator to be decomposed as $\Pr(A_S = a_S, R_S = r_S \mid N_1 = n_1, H_0) = \Pr(A_S = a_S \mid N_1 = n_1, H_0)\Pr(D_S = r_S - a_S \mid N_1 = n_1, H_0)$.

Let $p_{1|1} = \Pr(G_S = 1 \mid Y = 1)$ and $p_{1|0} = \Pr(G_S = 1 \mid Y = 0)$. If the null hypothesis of independence H_0 holds, then $p_{1|1} = p_{1|0} = p_G$. If the n samples are obtained as i.i.d. draws, A_S can be modeled as the sum of n_1 independent Bernoulli random variables,[5] each with success probability p_G. Hence,[6] $\Pr(A_S = a_S \mid N_1 = n_1, H_0) = \text{Binomial}(a_S \mid n_1, p_G) = \binom{n_1}{a_S} p_G^{a_S}(1-p_G)^{n-a_S}$. Analogously, D_S can be modeled as the sum of $n - n_1$ independent Bernoulli random variables, each with success probability p_G. Hence, $\Pr(D_S = r_S - a_S \mid N_1 = n_1, H_0) = \text{Binomial}(r_S - a_S \mid n - n_1, p_G) = \binom{n-n_1}{r_S - a_S} p_G^{r_S - a_S}(1-p_G)^{(n-n_1)-(r_S-a_S)}$. Finally, since $R_S = A_S + D_S$, with $A_S \perp D_S$, R_S corresponds to a sum of n Bernoulli variables with success probability p_G, leading to $\Pr(R_S = r_S \mid N_1 = n_1, H_0) = \text{Binomial}(r_S \mid n, p_G) = \binom{n}{r_S} p_G^{r_S}(1-p_G)^{n-r_S}$.

Substituting those distributions into Equation (8.3) leads to the final result. In particular, the conditioning on $R_S = r_S$ eliminates the influence of the nuisance parameter p_G, leading to a distributional form that depends only on n_1, r_S, and the sample size n. □

Building on Proposition 8.1, Pearson's χ^2 test and Fisher's exact test can be motivated as shown in the following subsections.

8.2.2.1 Pearson's χ^2 Test

Pearson's χ^2 test can be understood as a Z-score squared:

$$T_{\text{Pearson}}(a_S \mid n, n_1, r_S) = \left(\frac{a_S - \mathbb{E}[a_S \mid R_S = r_S, N_1 = n_1, H_0]}{\text{Std}[a_S \mid R_S = r_S, N_1 = n_1, H_0]} \right)^2, \tag{8.4}$$

where $\mathbb{E}[a_S \mid R_S = r_S, N_1 = n_1, H_0] = r_S \frac{n_1}{n}$ and $\text{Std}[a_S \mid R_S = r_S, N_1 = n_1, H_0] = \sqrt{\frac{r_S}{n} \frac{n-r_S}{n} \frac{n-n_1}{n-1} n_1}$ are the mean and standard deviation of a hypergeometric distribution with parameters n, n_1, and r_S. This leads to the following expression for the test statistic:

$$T_{\text{Pearson}}(a_S \mid n, n_1, r_S) = \frac{\left(a_S - r_S \frac{n_1}{n}\right)^2}{\frac{r_S}{n} \frac{n-r_S}{n} \frac{n-n_1}{n-1} n_1}. \tag{8.5}$$

Intuitively, large values of $T_{\text{Pearson}}(a_S \mid n, n_1, r_S)$ are less likely to occur under the null hypothesis, hence being suggestive of the potential existence of an association.

If the sample size n is sufficiently large, the central limit theorem implies that the underlying Z-score defining Pearson's χ^2 test will be approximately distributed as a standard Gaussian under the null hypothesis H_0.[7] Therefore, for large n, the null

---

[5] The Bernoulli distribution represents the probability distribution of a random variable Y which can only attain two possible outcomes: $Y = 1$, often generically denoted as "success" and $Y = 0$, often referred to as "failure." Typical examples of phenomena which can be modelled with a Bernoulli distribution include a coin toss or the answer to a yes/no question. Given the success probability p, the Bernoulli distribution can be written as $\Pr(Y = y \mid p) = \text{Bernoulli}(y \mid p) = p^y(1-p)^{1-y}$.

[6] The binomial distribution arises as the probability distribution of an integer-valued random variable Z describing the total number of "successes" out of n independent Bernoulli trials, all having the same success probability p. The relationship between the Bernoulli distribution and the binomial distribution is explored in-depth in Exercise 8.7.

[7] Informally, the central limit theorem asserts that the average of n i.i.d. random variables with zero mean and unit variance is approximately distributed as a standard Gaussian provided that n is large enough.

distribution of the test statistic $\Pr(T_{\text{Pearson}}(A_S) = t \mid R_S = r_S, N_1 = n_1, H_0)$ can be approximated as a χ_1^2 distribution. The corresponding two-tailed P-value can then be computed from the survival function of a χ_1^2 distribution, i.e.,

$$p_{\text{Pearson}}(a_S \mid n, n_1, r_S) = 1 - F_{\chi_1^2}(T_{\text{Pearson}}(a_S \mid n, n_1, r_S)), \quad (8.6)$$

where $F_{\chi_1^2}(\bullet)$ is the cumulative density function of a χ_1^2 distribution.

8.2.2.2 Fisher's Exact Test

While there are multiple ways to compute two-tailed P-values for Fisher's exact test, in this chapter we focus on one particular form for the sake of clarity. However, all subsequent derivations can be adapted to work with other definitions of the two-tailed P-value, as well as for one-tailed P-values.

A concise way to motivate Fisher's exact test is to consider the probability $\Pr(A_S = a_S \mid R_S = r_S, N_1 = n_1, H_0)$ as the test statistic itself:

$$T_{\text{Fisher}}(a_S \mid n, n_1, r_S) = \text{Hypergeom}(a_S \mid n, n_1, r_S) = \frac{\binom{n_1}{a_S}\binom{n-n_1}{r_S-a_S}}{\binom{n}{r_S}}. \quad (8.7)$$

This definition is intuitive, as $T_{\text{Fisher}}(a_S \mid n, n_1, r_S)$ will be small for values of a_S which are improbable under the conditional null distribution $\Pr(A_S = a_S \mid R_S = r_S, N_1 = n_1, H_0)$. Thus, in this case, low values of the test statistic $T_{\text{Fisher}}(a_S \mid n, n_1, r_S)$, rather than large values, are indicative of a potential association. As a consequence, to obtain a two-tailed P-value, we will compute $p = \Pr(T \leq t \mid H_0)$ rather than $p = \Pr(T \geq t \mid H_0)$.

Let $A(a_S) = \{a'_S \mid \text{Hypergeom}(a'_S \mid n, n_1, r_S) \leq \text{Hypergeom}(a_S \mid n, n_1, r_S)\}$ be the set of all possible counts a'_S that are at least as improbable as a_S under the conditional null distribution. Then:

$$p_{\text{Fisher}}(a_S \mid n, n_1, r_S) = \sum_{a'_S \in A(a_S)} \text{Hypergeom}(a'_S \mid n, n_1, r_S). \quad (8.8)$$

Unlike Pearson's χ^2 test and as its name indicates, Fisher's exact test does not need to resort to approximations to model the null distribution. Therefore, it is often the preferred choice, specially in situations where the sample size n is small.

8.2.3 Multiple Testing Correction

In the previous section, we introduced Pearson's χ^2 test and Fisher's exact test, two methods to test the statistical association of two binary random variables based on n i.i.d. samples. In the context of significant pattern mining, these can be used to test if the occurrence of a pattern $S \in \mathcal{M}$ in an input sample $x \in \mathcal{X}$ is statistically associated with the sample class y based on a dataset $\mathcal{D} = \{(x_i, y_i)\}_{i=1}^n$. In principle, a P-value p_S could be computed for each pattern $S \in \mathcal{M}$, deeming those patterns $S \in \mathcal{M}$ for which $p_S \leq \alpha$ statistically significant. However, leaving the matter of computational feasibility temporarily aside, such an approach also has an important statistical caveat. A procedure that, independently of the number $|\mathcal{M}|$ of association tests being performed, uses a fixed significance threshold α to assess the significance of each pattern $\mathcal{S} \in \mathcal{M}$, will generate on average $\alpha |\mathcal{M}_{\text{null}}|$ false positives, where $\mathcal{M}_{\text{null}}$ is the set of patterns that are statistically independent of the class labels. In most applications of interest, there is an enormous number $|\mathcal{M}|$ of candidate patterns to be tested, often in the order of trillions, most of which belong to the set of null patterns $\mathcal{M}_{\text{null}}$. Therefore, if the statistical association testing procedure is not modified to account for the vast number of association tests being performed simultaneously, billions of false positives will be reported on average, rendering the results entirely unreliable. This phenomenon, commonly referred to as the *multiple hypothesis testing problem*, has been thoroughly studied by statisticians (e.g., [27, 28, 29]). However, significant pattern mining takes the multiple hypothesis testing problem to a completely new level, dealing with a number $|\mathcal{M}|$ of association tests many orders of magnitude larger than it had ever been considered before in other domains.

A key concept when dealing with the multiple hypothesis testing problem is the *family-wise error rate (FWER)*, defined as the probability of producing any false positives in the analysis. Mathematically, $\text{FWER}(\delta) = \Pr(\text{FP}(\delta) > 0)$, where $\text{FP}(\delta)$ is the number of false positives at adjusted significance threshold δ, i.e., the number of patterns $\mathcal{S} \in \mathcal{M}_{\text{null}}$ for which $p_\mathcal{S} \leq \delta$. A common way to account for the multiple hypothesis testing problem is to modify the statistical association testing procedure in order to *control* the FWER at level α, i.e., to guarantee that the FWER is bounded above by a user-defined level α. Since, by definition, $\text{FWER}(\delta)$ is monotonically increasing in the adjusted significance threshold δ, the optimal significance threshold that controls the FWER at level α is given by:

$$\delta^* = \max \{\delta \mid \text{FWER}(\delta) \leq \alpha\}. \tag{8.9}$$

Any values of the adjusted significance threshold δ that are larger than δ^* will fail the control the FWER at level α, while values that are smaller than δ^* will also control the FWER at level α but yield less statistical power to discover the true associations.

While FWER control offers an elegant framework to account for the multiple hypothesis testing problem, a fundamental limitation remains: $\text{FWER}(\delta)$ is an unknown quantity that depends on the (unknown) joint distribution of $\{p_\mathcal{S} \mid \mathcal{S} \in \mathcal{M}_{\text{null}}\}$. Therefore, neither $\text{FWER}(\delta)$ nor δ^* can be computed exactly. Existing approaches resort to approximating $\text{FWER}(\delta)$, often with a quantity that is provably an upper bound, and use that approximation to propose an adjusted significance threshold $\hat{\delta}^*$. Distinct procedures differ on the quality of their approximation of the FWER, which ultimately determines how close the resulting $\hat{\delta}^*$ is to the optimal adjusted significance threshold δ^*.

8.2.3.1 The Bonferroni Correction

The most popular procedure to control the FWER is the *Bonferroni correction* [30, 31]. The Bonferroni correction approximates the unknown value of $\text{FWER}(\delta)$ with

$\widehat{\text{FWER}}(\delta) = \delta|\mathcal{M}|$, a quantity that can be readily shown to be an upper bound on FWER(δ):

$$\text{FWER}(\delta) = \Pr\left(\bigcup_{S \in \mathcal{M}_{\text{null}}} \{p_S \leq \delta\}\right) \leq \sum_{S \in \mathcal{M}_{\text{null}}} \Pr(p_S \leq \delta)$$
$$\leq \delta|\mathcal{M}_{\text{null}}| \leq \delta|\mathcal{M}|, \tag{8.10}$$

where $\{p_S \leq \delta\}$ is the event that pattern S is deemed significantly associated. Based on the upper bound $\widehat{\text{FWER}}(\delta) = \delta|\mathcal{M}|$, the Bonferroni correction proposes an adjusted significance threshold $\delta_{\text{bonf}} = \max\{\delta \mid \delta|\mathcal{M}| \leq \alpha\} = \alpha/|\mathcal{M}|$. Since $\widehat{\text{FWER}}(\delta) \geq \text{FWER}(\delta)$, it follows that $\delta_{\text{bonf}} \leq \delta^*$ and, therefore, the Bonferroni correction controls the FWER at level α.

In practice, the Bonferroni correction tends to be an over-conservative procedure. Its approximation $\widehat{\text{FWER}}(\delta) = \delta|\mathcal{M}|$ to the FWER often greatly overestimates the real value of FWER(δ), causing δ_{bonf} to be considerably smaller than δ^* and leading to a loss of statistical power. However, the Bonferroni correction has strong points due to its simplicity: It introduces no computational overhead and ensures FWER-control for any kind of data, without the need to satisfy any assumptions. Those aspects have made the Bonferroni correction the most widespread tool for FWER-control, making the loss of statistical power it entails a price happily paid by many practitioners.

Unfortunately, the multiple hypothesis testing problem that arises in significant pattern mining is way beyond what the Bonferroni correction can successfully handle. Since the number $|\mathcal{M}|$ of candidate patterns in a typical significant pattern mining problem is enormous, the resulting adjusted significance threshold δ_{bonf} will be too small and the resulting procedure will have virtually no statistical power. The lack of approaches to control the FWER when performing such a large number of association tests caused traditional pattern mining methods to either (1) ignore the multiple hypothesis testing problem, providing a ranking of patterns by association but no statistical guarantees [32, 33, 34] or (2) artificially limit the size of the search space $\mathcal{M}$ by setting an arbitrary maximum pattern size, in order to be able to apply a Bonferroni correction in the much smaller resulting search space [35, 36, 37].

8.2.3.2 Tarone's Improved Bonferroni Correction for Discrete Data

It was only recently that the first algorithm able to control the FWER in significant pattern mining without imposing *any* limits on the search space $\mathcal{M}$ was proposed [15]. A key component of their approach is the use of Tarone's improved Bonferroni correction for discrete data [38], an alternative way of approximating the FWER that results in a drastically more accurate bound when applying the procedure to pattern mining problems.

The core idea behind Tarone's method is to exploit the nature of test statistics for discrete data. Consider a 2×2 contingency table summarizing n i.i.d. samples of two binary random variables with the margins n_1 and r_S treated as fixed. In order to be consistent with the fixed margins, the cell count a_S must be smaller or equal than $a_{S,\max} = \min(n_1, r_S)$ and larger or equal than $a_{S,\min} = \max(0, r_S - (n - n_1))$. Therefore,

a_S can only take $a_{S,\max} - a_{S,\min} + 1$ distinct values. In turn, this implies that there are at most $a_{S,\max} - a_{S,\min} + 1$ distinct P-values that can be obtained as an outcome of applying a test statistic such as Pearson's χ^2 test or Fisher's exact test to the 2×2 contingency table. Since there is a finite number of P-values that can be observed, there exists a *minimum attainable P-value*, $p_{S,\min} = \min \{p_S(a'_S \mid n, n_1, r_S) \mid a'_S \in [\![a_{S,\min}, a_{S,\max}]\!]\}$,[8] where $p_S(a'_S \mid n, n_1, r_S)$ is the P-value obtained by applying the test statistic of choice to a 2×2 contingency table with cell count a'_S, fixed margins n_1 and r_S and sample size n. The existence of a minimum attainable P-value $p_{S,\min}$ strictly larger than zero is a special property of discrete data. In principle, P-values obtained when testing the association between two continuous random variables can be arbitrarily close to zero. Moreover, the minimum attainable P-value $p_{S,\min}$ depends only on n, n_1 and r_S. In particular, $p_{S,\min}$ does *not* depend on the actual value of the cell count a_S.

The existence of a minimum attainable P-value $p_{S,\min}$ strictly larger than zero has profound implications. Suppose that the adjusted significance threshold δ is smaller than the minimum attainable P-value, $p_{S,\min} > \delta$. Then, by definition, regardless of the value of a_S, the corresponding association will not be deemed statistically significant. Consequently, it can also never cause a false positive at adjusted significance threshold δ. Patterns $S \in \mathcal{M}$ for which this occurs are said to be *untestable* at level δ while the remaining patterns are said to be *testable* at level δ. Let $\mathcal{M}_{\text{test}}(\delta) = \{S \in \mathcal{M} \mid p_{S,\min} \leq \delta\}$ be the set of testable patterns at level δ. Tarone's improved Bonferroni correction for discrete data approximates FWER(δ) as $\widehat{\text{FWER}}(\delta) = \delta |\mathcal{M}_{\text{test}}(\delta)|$, which is also an upper bound on FWER(δ):

$$\text{FWER}(\delta) = \Pr\left(\bigcup_{S \in \mathcal{M}_{\text{null}}} \{p_S \leq \delta\}\right) = \Pr\left(\bigcup_{S \in \mathcal{M}_{\text{test}}(\delta)} \{p_S \leq \delta\}\right)$$
$$\leq \sum_{S \in \mathcal{M}_{\text{test}}(\delta)} \Pr(p_S \leq \delta) \leq \delta |\mathcal{M}_{\text{test}}(\delta)|, \qquad (8.11)$$

where the first step follows from the fact that untestable patterns $S \in \mathcal{M} \setminus \mathcal{M}_{\text{test}}(\delta)$ cannot cause a false positive at adjusted significance threshold δ. Thus, Tarone's method also guarantees FWER control without the need for additional assumptions.

In significant pattern mining, often the number of testable patterns $|\mathcal{M}_{\text{test}}(\delta)|$ is drastically smaller than the total number of candidate patterns $|\mathcal{M}|$. By exploiting the discrete nature of the data, Tarone's method often results in a much tighter upper bound of FWER(δ) than the original Bonferroni correction. This leads to the adjusted significance threshold $\delta_{\text{tar}} = \max \{\delta \mid \delta |\mathcal{M}_{\text{test}}(\delta)| \leq \alpha\}$ being much closer to the optimal δ^* than δ_{bonf}, bringing forth a drastic gain of statistical power over the original Bonferroni correction and making significant pattern mining on the entire search space $\mathcal{M}$ statistically feasible.

---

[8] Given any two integers a and b such that $a < b$, we denote by $[\![a, b]\!]$ the set of consecutive integers $\{a, a+1, \ldots, b\}$.

8.3 A Framework for Significant Pattern Mining Using Tarone's Method

In the previous section, we have identified the multiple hypothesis testing problem as the main statistical challenge that needs to be overcome to make significant pattern mining a reality. We introduced the concept of FWER control at level α as a criterion for statistical association testing that corrects for the multiple hypothesis testing problem and discussed the feasibility of controlling the FWER in the setting of significant pattern mining. Standard techniques, such as the ubiquitous Bonferroni correction, are unable to maintain statistical power when exploring the massive search space of candidate patterns $\mathcal{M}$ present in typical significant pattern mining problems. However, Tarone's method leverages the concept of *testability*, the phenomenon that only a subset $\mathcal{M}_{\text{test}}(\delta) \subseteq \mathcal{M}$ of patterns ($|\mathcal{M}_{\text{test}}(\delta)| \ll |\mathcal{M}|$) can be significant or cause a false positive, to dramatically enhance statistical power. However, evaluating the adjusted significance threshold δ_{tar} naively is computationally an extremely demanding task.

In order to obtain δ_{tar}, the largest value of δ that satisfies $\delta|\mathcal{M}_{\text{test}}(\delta)| \leq \alpha$ needs to be found. Obtaining $\mathcal{M}_{\text{test}}(\delta)$ by brute force would require enumerating all patterns $\mathcal{S} \in \mathcal{M}$, computing all corresponding minimum attainable P-values $p_{\mathcal{S},\min}$ and keeping those patterns $\mathcal{S} \in \mathcal{M}$ for which $p_{\mathcal{S},\min} \leq \delta$. Due to the enormous number $|\mathcal{M}|$ of candidate patterns, this strategy is computationally unfeasible except in very small datasets. This difficulty kept Tarone's method from being used in significant pattern mining until the recent breakthrough presented in [15]. In their work, the authors propose a way to exploit properties of specific test statistics, such as Pearson's χ^2 test and Fisher's exact test, to derive an algorithm that can efficiently compute δ_{tar} by exploring only a small subset of all candidate patterns in the search space $\mathcal{M}$. Since its publication, the original algorithm in [15] has been subsequently enhanced by follow-up work [16, 17], improving its computational efficiency even further. In the remainder of this section, we will present a generic significant pattern mining algorithm that incorporates all these recent developments.

Algorithm 8.1 Significant pattern mining

Input: Dataset $\mathcal{D} = \{(x_i, y_i)\}_{i=1}^{n}$, target FWER α
Output: $\{\mathcal{S} \in \mathcal{M} \mid p_{\mathcal{S}} \leq \delta_{\text{tar}}\}$
1: $\delta_{\text{tar}} \leftarrow \texttt{tarone\_spm}(\mathcal{D}, \alpha)$
2: Return $\{\mathcal{S} \in \mathcal{M}_{\text{test}}(\delta_{\text{tar}}) \mid p_{\mathcal{S}} \leq \delta_{\text{tar}}\}$

Algorithm 8.1 describes a generic significant pattern mining algorithm making use of Tarone's method. Mainly, it proceeds in two key steps:

1. Obtain Tarone's corrected significance threshold δ_{tar}. This step is performed in Line 1, which invokes the routine `tarone_spm`. This routine makes use of certain properties of the minimum attainable P-value to dynamically prune a large proportion of the search space $\mathcal{M}$ as patterns are enumerated. Effectively, this allows to exactly compute δ_{tar} while only enumerating a small subset of patterns comprising the search space, as described in detail in Algorithm 8.2 below.

Algorithm 8.2 `tarone_spm`

Input: Dataset $\mathcal{D} = \{(x_i, y_i)\}_{i=1}^{n}$, target FWER α
Output: Adjusted significance threshold δ_{tar}

1: **function** TARONE_SPM($\mathcal{D},\alpha$)
2:     Initialize global variables $\hat{\delta}_{\text{tar}} \leftarrow 1$ and $\widehat{\mathcal{M}}_{\text{test}}(\hat{\delta}_{\text{tar}}) \leftarrow \emptyset$
3:     NEXT($\emptyset$)   ▷ Start pattern enumeration
4:     $\delta_{\text{tar}} \leftarrow \hat{\delta}_{\text{tar}}$
5:     Return δ_{tar}
6: **procedure** NEXT($\mathcal{S}$)
7:     Compute the minimum attainable P-value $p_{\mathcal{S},\text{min}}$ ▷ see Section 8.3.1
8:     **if** $p_{\mathcal{S},\text{min}} \leq \hat{\delta}_{\text{tar}}$ **then**   ▷ if pattern $\mathcal{S}$ is testable at level $\hat{\delta}_{\text{tar}}$ then
9:         Append $\mathcal{S}$ to $\widehat{\mathcal{M}}_{\text{test}}(\hat{\delta}_{\text{tar}})$
10:         $\widehat{\text{FWER}}(\hat{\delta}_{\text{tar}}) \leftarrow \hat{\delta}_{\text{tar}}|\widehat{\mathcal{M}}_{\text{test}}(\hat{\delta}_{\text{tar}})|$
11:         **while** $\widehat{\text{FWER}}(\hat{\delta}_{\text{tar}}) > \alpha$ **do**
12:             Decrease $\hat{\delta}_{\text{tar}}$
13:             $\widehat{\mathcal{M}}_{\text{test}}(\hat{\delta}_{\text{tar}}) \leftarrow \{\mathcal{S}' \in \widehat{\mathcal{M}}_{\text{test}}(\hat{\delta}_{\text{tar}}) \mid p_{\mathcal{S}',\text{min}} \leq \hat{\delta}_{\text{tar}}\}$
14:             $\widehat{\text{FWER}}(\hat{\delta}_{\text{tar}}) \leftarrow \hat{\delta}_{\text{tar}}|\widehat{\mathcal{M}}_{\text{test}}(\hat{\delta}_{\text{tar}})|$
15:     **if not** pruning_condition($\mathcal{S}, \hat{\delta}_{\text{tar}}$) **then**   ▷ see Section 8.3.2
16:         **for** $\mathcal{S}' \in$ Children($\mathcal{S}$) **do**
17:             NEXT($\mathcal{S}'$)   ▷ Recursively visit nodes in the tree depth-first

2. Compute the P-value $p_{\mathcal{S}}$ for all testable patterns $\mathcal{S} \in \mathcal{M}_{\text{test}}(\delta_{\text{tar}})$, using a test statistic such as Pearson's χ^2 test or Fisher's exact test, and output those that are statistically significant at level δ_{tar}. Note that, by definition, untestable patterns $\mathcal{S} \in \mathcal{M} \setminus \mathcal{M}_{\text{test}}(\delta_{\text{tar}})$ cannot be statistically significant at adjusted significance threshold δ_{tar} and, therefore, do not need to be taken into account in this step. Since the number of testable patterns $|\mathcal{M}_{\text{test}}(\delta_{\text{tar}})|$ tends to be much smaller than the total number of patterns $|\mathcal{M}|$, the computational complexity of this step, performed in Line 2, is less critical than the previous step.

The routine `tarone_spm` is the core of the significant pattern mining algorithm described in Algorithm 8.1. The method explores the search space of candidate patterns $\mathcal{S} \in \mathcal{M}$ recursively, arranging all patterns as nodes of a tree such that the descendants $\mathcal{S}'$ of a pattern $\mathcal{S}$ are super-patterns of $\mathcal{S}$, i.e., $\mathcal{S} \subseteq \mathcal{S}'$. We call such a tree a *pattern enumeration tree*. This particular way of arranging the search space of patterns has been extensively used in pattern mining [39], and is applicable in most instances of significant pattern mining, including both significant itemset mining and significant subgraph mining. For instance, Figure 8.3 illustrates one of the many possible ways to obtain a pattern enumeration tree in a significant itemset mining problem with $p = 5$ features. In this example, each node in the tree corresponds to a high-order feature interaction $\mathcal{S}$ and children $\mathcal{S}'$ of a feature interaction $\mathcal{S}$ are obtained by adding an additional feature to the interaction. Note that, depending on the ordering of features, multiple valid trees can be obtained. Analogously, in significant subgraph mining,

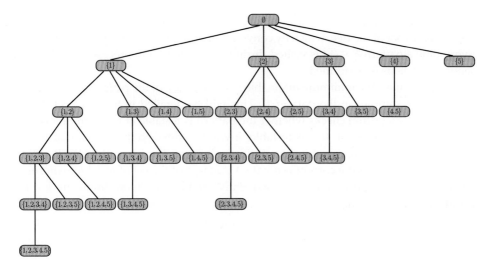

Figure 8.3: Depiction of a valid pattern enumeration tree for a significant itemset mining problem with $p = 5$ features. Each of the $|\mathcal{M}| = 2^5$ high-order interactions $\mathcal{S} \in \mathcal{M}$ is mapped to a node of the tree in such a way that $\mathcal{S}' \in \text{Children}(\mathcal{S})$ implies that $\mathcal{S} \subseteq \mathcal{S}'$.

each node in the tree corresponds to a subgraph $\mathcal{S}$ and children $\mathcal{S}'$ of $\mathcal{S}$ can be obtained by appending additional nodes to subgraph $\mathcal{S}$.

A consequence of this particular way to arrange the search space $\mathcal{M}$, with profound algorithmic implications, is the so called *apriori property* of pattern mining.

Proposition 8.2 (Apriori property)   *Let $\mathcal{S}, \mathcal{S}' \in \mathcal{M}$ be two patterns such that $\mathcal{S}'$ is a descendant of $\mathcal{S}$ in a pattern enumeration tree. Then, $r_{\mathcal{S}'} \leq r_{\mathcal{S}}$, where $r_{\mathcal{S}}$ and $r_{\mathcal{S}'}$ are the marginal counts of patterns $\mathcal{S}$ and $\mathcal{S}'$ in 2×2 contingency tables computed from a dataset $\mathcal{D} = \{(x_i, y_i)\}_{i=1}^{n}$.*

Proof By definition of pattern enumeration tree, if $\mathcal{S}'$ is a descendant of $\mathcal{S}$ then $\mathcal{S} \subseteq \mathcal{S}'$. Therefore, $\mathcal{S}' \subseteq x$ implies $\mathcal{S} \subseteq x$ or, equivalently, $g_{\mathcal{S}'}(x) = 1$ implies $g_{\mathcal{S}}(x) = 1$. Since $r_{\mathcal{S}} = \sum_{i=1}^{n} g_{\mathcal{S}}(x)$ and $r_{\mathcal{S}'} = \sum_{i=1}^{n} g_{\mathcal{S}'}(x)$ the result follows. □

The apriori property is the formalization of an intuitive fact: If patterns are arranged in a tree guaranteeing that descendant nodes correspond to increasingly complex super-patterns, as we descend along the tree, patterns will be less frequently observed in a dataset $\mathcal{D}$. Despite its simplicity, the apriori property will play a central role in the development of a pruning criterion to avoid the need to explicitly enumerate all patterns in $\mathcal{M}$.

As shown in Line 2 of Algorithm 8.2, the routine `tarone_spm` begins by initialising the estimate $\hat{\delta}_{\text{tar}}$ to 1, the largest value δ_{tar} could take, and the set of testable patterns at level $\hat{\delta}_{\text{tar}}$ to the empty set, $\widehat{\mathcal{M}}_{\text{test}}(\hat{\delta}_{\text{tar}}) \leftarrow \emptyset$. In order to compute $\delta_{\text{tar}} = \max\{\delta \mid \delta|\mathcal{M}_{\text{test}}(\delta)| \leq \alpha\}$, candidate patterns $\mathcal{S} \in \mathcal{M}$ will be explored recursively by traversing the corresponding enumeration tree in a depth-first manner. The estimate $\hat{\delta}_{\text{tar}}$ and the corresponding set of testable patterns $\widehat{\mathcal{M}}_{\text{test}}(\hat{\delta}_{\text{tar}})$ will be incrementally adjusted as patterns are enumerated in such a way that, at the end of the process,

$\hat{\delta}_{tar} = \delta_{tar}$ and $\widehat{\mathcal{M}}_{test}(\hat{\delta}_{tar}) = \mathcal{M}_{test}(\delta_{tar})$ holds. This enumeration procedure starts in Line 3 at the root of the tree, which by convention represents the empty pattern $S = \emptyset$.[9] Every time a pattern $S \in \mathcal{M}$ is visited during the enumeration process, the following sequence of steps is performed.

First, in Line 7 the minimum attainable P-value $p_{S,\min}$ of the pattern S is evaluated based on the samples in the dataset $\mathcal{D}$. This will be detailed in Section 8.3.1. In the next line, the algorithm checks whether the pattern is testable at the current level $\hat{\delta}_{tar}$ (i.e., $p_{S,\min} \leq \hat{\delta}_{tar}$) or not. If S is testable, it will be added to the estimate of the set of testable patterns $\widehat{\mathcal{M}}_{test}(\hat{\delta}_{tar})$ in Line 9. Tarone's upper bound on the FWER will be evaluated next, using the current estimate $\widehat{\mathcal{M}}_{test}(\hat{\delta}_{tar})$ of the set of testable patterns at level $\hat{\delta}_{tar}$. In Line 11, this value is subsequently used by the algorithm to check if the FWER condition $\hat{\delta}_{tar}|\widehat{\mathcal{M}}_{test}(\hat{\delta}_{tar})| = \widehat{\text{FWER}}(\hat{\delta}_{tar}) \leq \alpha$ is violated at level $\hat{\delta}_{tar}$. Note that, since the enumeration process is not yet completed, $\widehat{\mathcal{M}}_{test}(\hat{\delta}_{tar}) \subseteq \mathcal{M}_{test}(\hat{\delta}_{tar})$ holds. As a consequence, the FWER approximation $\widehat{\text{FWER}}(\hat{\delta}_{tar})$ evaluated in Line 10 satisfies $\widehat{\text{FWER}}(\hat{\delta}_{tar}) = \hat{\delta}_{tar}|\widehat{\mathcal{M}}_{test}(\hat{\delta}_{tar})| \leq \hat{\delta}_{tar}|\mathcal{M}_{test}(\hat{\delta}_{tar})| = \text{FWER}(\hat{\delta}_{tar})$. If $\widehat{\text{FWER}}(\hat{\delta}_{tar}) \geq \alpha$, then it follows that $\text{FWER}(\hat{\delta}_{tar}) \geq \alpha$ and we can immediately conclude that adjusted significance threshold $\hat{\delta}_{tar}$ is too large and violates the FWER condition for target FWER α. Therefore, in that case, the current estimate of the adjusted significance threshold $\hat{\delta}_{tar}$ is decreased in Line 12. This causes some patterns currently in $\widehat{\mathcal{M}}_{test}(\hat{\delta}_{tar})$ to no longer be testable (Line 13), thereby reducing $\widehat{\text{FWER}}(\hat{\delta}_{tar})$, which is re-evaluated in Line 14. This process is repeated, incrementally reducing $\hat{\delta}_{tar}$, until the FWER condition is satisfied again, i.e., $\widehat{\text{FWER}}(\hat{\delta}_{tar}) \leq \alpha$. Finally, the enumeration process continues recursively by visiting the children of the pattern S currently being processed (Lines 16–17). However, prior to that a pruning condition is evaluated in Line 15. This step, discussed in detail in Section 8.3.2, is essential to the computational feasibility of the algorithm. If the pruning condition applies, no descendant of pattern S will be testable and, hence, they do not need to be enumerated, drastically reducing computational complexity. As the algorithm enumerates patterns, $\hat{\delta}_{tar}$ progressively decreases, making more and more patterns become untestable and the pruning condition in Line 15 to be more stringent. Eventually, the algorithm converges when all patterns $S \in \mathcal{M}$ that have not been pruned from the search space have been visited. At that point, $\hat{\delta}_{tar} = \delta_{tar}$ and $\widehat{\mathcal{M}}_{test}(\hat{\delta}_{tar}) = \mathcal{M}_{test}(\delta_{tar})$, allowing the exact value of δ_{tar} to be returned in Line 5.

The framework described in Algorithms 8.1 and 8.2 can be readily applied to multiple instances of significant pattern mining and choices of test statistic to assess the significance of patterns. Two key steps of the algorithm that remain to be discussed are how to efficiently evaluate the minimum attainable P-value $p_{S,\min}$ and how to propose a valid pruning condition that results in a large reduction of the number of candidate patterns. In the remainder of this section, we discuss each of these issues for the two particular choices of the test statistic that were discussed in Section 8.2.2: Pearson's χ^2 test and Fisher's exact test.

---

[9] We define the empty pattern to occur in every input sample, i.e., $g_\emptyset(x) = 1$ for all $x \in \mathcal{X}$. The empty pattern will thus never be statistically significant; its only purpose is to act as the starting point of the recursive enumeration.

8.3.1 Evaluating Tarone's Minimum Attainable *P*-value

As described in Section 8.2.3, for a test statistic based on a 2×2 contingency table that treats margins n_1 and r_S as fixed quantities, such as Pearson's χ^2 test and Fisher's exact test, the minimum attainable *P*-value $p_{S,\min}$ can be defined as:

$$p_{S,\min} = \min\left\{p_S(a'_S \mid n, n_1, r_S) \mid a'_S \in [\![a_{S,\min}, a_{S,\max}]\!]\right\}, \tag{8.12}$$

where $p_S(a'_S \mid n, n_1, r_S)$ is the *P*-value corresponding to a 2×2 contingency table with cell count a'_S, margins n_1 and r_S, and sample size n.

Evaluating the minimum attainable *P*-value $p_{S,\min}$ of a pattern S based on the n samples in the input dataset $\mathcal{D}$ is a key operation in Algorithm 8.2. In particular, the minimum attainable *P*-value needs to be evaluated for every single pattern $S \in \mathcal{M}$ that is not eliminated by the pruning condition. In a typical execution of Algorithm 8.2, this can amount to billions of evaluations of $p_{S,\min}$, making the computation of $p_{S,\min}$ a critical point to maintain computational efficiency. Obtaining $p_{S,\min}$ by evaluating $p_S(a'_S \mid n, n_1, r_S)$ for each $a'_S \in [\![a_{S,\min}, a_{S,\max}]\!]$, as suggested by a naive application of Equation 8.12, might require $O(n)$ *P*-value computations for each evaluation of $p_{S,\min}$, an unacceptable computational overhead. In this section, we describe how $p_{S,\min}$ can be obtained with $O(1)$ complexity for Pearson's χ^2 test and Fisher's exact test.

First, note that $p_{S,\min}$ is a function of the number r_S of occurrences of pattern S in $\mathcal{D}$, the number n_1 of samples in $\mathcal{D}$ that belong to the positive class and the total sample size n. Since, for a given dataset $\mathcal{D}$, only r_S varies from one pattern to another, we will simply write $p_{S,\min}(r_S)$ to avoid notation clutter, leaving the dependence of $p_{S,\min}$ on n_1 and n implicit.

In order to compute $p_{S,\min}(r_S)$, the minimizer a^*_S of $p_S(a'_S \mid n, n_1, r_S)$ in $[\![a_{S,\min}, a_{S,\max}]\!]$ needs to be obtained. Due to the way the *P*-values are defined for Pearson's χ^2 test (Equation 8.6) and Fisher's exact test (Equation 8.8), the minimizer a^*_S must be in the boundary of $[\![a_{S,\min}, a_{S,\max}]\!]$, i.e., either $a^*_S = a_{S,\min}$ or $a^*_S = a_{S,\max}$. All that remains to be shown is which of the two cases holds for each value of r_S, leading to a closed-form of $p_{S,\min}(r_S)$ for each of the two test statistics.

Proposition 8.3 (Minimum attainable *P*-value for Pearson's χ^2 test) *Define $n_a = \min(n_1, n - n_1)$ and $n_b = \max(n_1, n - n_1)$. Then, the minimum attainable P-value for Pearson's χ^2 test is given by:*

$$p_{S,min}(r_S) = \begin{cases} 1 - F_{\chi^2_1}\left((n-1)\frac{n_b}{n_a}\frac{r_S}{n-r_S}\right), & \text{if } 0 \leq r_S < n_a, \\ 1 - F_{\chi^2_1}\left((n-1)\frac{n_a}{n_b}\frac{n-r_S}{r_S}\right), & \text{if } n_a \leq r_S < \frac{n}{2}, \\ 1 - F_{\chi^2_1}\left((n-1)\frac{n_a}{n_b}\frac{r_S}{n-r_S}\right), & \text{if } \frac{n}{2} \leq r_S < n_b, \\ 1 - F_{\chi^2_1}\left((n-1)\frac{n_b}{n_a}\frac{n-r_S}{r_S}\right), & \text{if } n_b \leq r_S \leq n. \end{cases} \tag{8.13}$$

Proof Let $T_{\max}(r_S)$ be the maximum value of Pearson's χ^2 test statistic for a 2×2 contingency table with sample size n and margins r_S and n_1. As discussed above, $T_{\text{Pearson}}(a_S \mid n, n_1, r_S)$ will be maximized either at $a^*_S = a_{S,\min}$ or at $a^*_S = a_{S,\max}$.

Hence:

$$T_{\max}(r_S) = \frac{\max\left((a_{S,\min} - r_S \frac{n_1}{n})^2, (a_{S,\max} - r_S \frac{n_1}{n})^2\right)}{\frac{r_S}{n} \frac{n-r_S}{n} \frac{n-n_1}{n-1} n_1}. \tag{8.14}$$

Suppose $0 \leq r_S < n_a$. Then, $a_{S,\min} = 0$ and $a_{S,\max} = r_S$, leading to:

$$T_{\max}(r_S) = (n-1) \frac{r_S}{n-r_S} \frac{\max^2(n_1, n-n_1)}{n_1(n-n_1)} = (n-1) \frac{r_S}{n-r_S} \frac{n_b}{n_a}. \tag{8.15}$$

Analogously, if $n_b \leq r_S \leq n$, then $a_{S,\min} = r_S - (n - n_1)$ and $a_{S,\max} = n_1$. Thus:

$$T_{\max}(r_S) = (n-1) \frac{n-r_S}{r_S} \frac{\max^2(n_1, n-n_1)}{n_1(n-n_1)} = (n-1) \frac{n-r_S}{r_S} \frac{n_b}{n_a}. \tag{8.16}$$

Finally, suppose $n_a \leq r_S < n_b$. This case can be studied separately depending on whether $n_1 \leq n - n_1$ or $n_1 > n - n_1$.

Let $n_1 \leq n - n_1$, then $a_{S,\min} = 0$ and $a_{S,\max} = n_1$. This leads to:

$$T_{\max}(r_S) = (n-1) \frac{n_1}{n-n_1} \frac{\max^2(r_S, n-r_S)}{r_S(n-r_S)} = (n-1) \frac{n_a}{n_b} \frac{\max^2(r_S, n-r_S)}{r_S(n-r_S)}$$

$$= \begin{cases} (n-1) \frac{n_a}{n_b} \frac{n-r_S}{r_S}, & \text{if } n_a \leq r_S < \frac{n}{2}, \\ (n-1) \frac{n_a}{n_b} \frac{r_S}{n-r_S}, & \text{if } \frac{n}{2} \leq r_S < n_b. \end{cases} \tag{8.17}$$

If $n_1 > n - n_1$, then $a_{S,\min} = r_S - (n-n_1)$ and $a_{S,\max} = r_S$. Therefore:

$$T_{\max}(r_S) = (n-1) \frac{n-n_1}{n_1} \frac{\max^2(r_S, n-r_S)}{r_S(n-r_S)} = (n-1) \frac{n_a}{n_b} \frac{\max^2(r_S, n-r_S)}{r_S(n-r_S)}$$

$$= \begin{cases} (n-1) \frac{n_a}{n_b} \frac{n-r_S}{r_S}, & \text{if } n_a \leq r_S < \frac{n}{2}, \\ (n-1) \frac{n_a}{n_b} \frac{r_S}{n-r_S}, & \text{if } \frac{n}{2} \leq r_S < n_b. \end{cases} \tag{8.18}$$

Since $p_{\text{Pearson}}(a_S \mid n, n_1, r_S) = 1 - F_{\chi_1^2}(T_{\text{Pearson}}(a_S \mid n, n_1, r_S))$, this concludes the proof. $\square$

Proposition 8.4 (Minimum attainable P-value for Fisher's exact test) *Define $n_a = \min(n_1, n - n_1)$ and $n_b = \max(n_1, n - n_1)$. Then, the minimum attainable P-value for Fisher's exact test is given by:*

$$p_{S,\min}(r_S) = \begin{cases} \binom{n_a}{r_S}/\binom{n}{r_S}, & \text{if } 0 \leq r_S < n_a, \\ \binom{n_b}{n-r_S}/\binom{n}{r_S}, & \text{if } n_a \leq r_S < \frac{n}{2}, \\ \binom{n_b}{r_S}/\binom{n}{r_S}, & \text{if } \frac{n}{2} \leq r_S < n_b, \\ \binom{n_a}{n-r_S}/\binom{n}{r_S}, & \text{if } n_b \leq r_S \leq n. \end{cases} \tag{8.19}$$

Proof Analogously to the proof of Proposition 8.3, $p_{\text{fisher}}(a_S \mid n, n_1, r_S)$ is minimized either at $a_S^* = a_{S,\min}$ or at $a_S^* = a_{S,\max}$. Therefore:

$$p_{S,\min}(r_S) = \min(\text{Hypergeom}(a_{S,\min} \mid n, n_1, r_S), \text{Hypergeom}(a_{S,\max} \mid n, n_1, r_S))$$

$$= \frac{\min\left(\binom{n_1}{a_{S,\min}}\binom{n-n_1}{r_S-a_{S,\min}}, \binom{n_1}{a_{S,\max}}\binom{n-n_1}{r_S-a_{S,\max}}\right)}{\binom{n}{r_S}}. \tag{8.20}$$

The problem will be decomposed in three cases, as done for the derivation of the closed-form expression of the minimum attainable P-value for Pearson's χ^2 test.

Let $0 \leq r_S < n_a$, leading to $a_{S,\min} = 0$ and $a_{S,\max} = r_S$. Then:

$$p_{S,\min}(r_S) = \frac{\min\left(\binom{n-n_1}{r_S}, \binom{n_1}{r_S}\right)}{\binom{n}{r_S}} = \frac{\binom{n_a}{r_S}}{\binom{n}{r_S}}. \qquad (8.21)$$

Similarly, if $n_b \leq r_S \leq n$ then $a_{S,\min} = r_S - (n - n_1)$ and $a_{S,\max} = n_1$. Thus:

$$p_{S,\min}(r_S) = \frac{\min\left(\binom{n_1}{r_S-(n-n_1)}, \binom{n-n_1}{r_S-n_1}\right)}{\binom{n}{r_S}} = \frac{\min\left(\binom{n_1}{n-r_S}, \binom{n-n_1}{n-r_S}\right)}{\binom{n}{r_S}}$$

$$= \frac{\binom{n_a}{n-r_S}}{\binom{n}{r_S}}, \qquad (8.22)$$

where the second step follows from $\binom{n}{k} = \binom{n}{n-k}$.

Let $n_a \leq r_S < n_b$ and $n_1 \leq n - n_1$, so that $a_{S,\min} = 0$ and $a_{S,\max} = n_1$. Hence:

$$p_{S,\min}(r_S) = \frac{\min\left(\binom{n-n_1}{r_S}, \binom{n-n_1}{n-r_S}\right)}{\binom{n}{r_S}}$$

$$= \begin{cases} \frac{\binom{n_b}{n-r_S}}{\binom{n}{r_S}}, & \text{if } n_a \leq r_S < \frac{n}{2}, \\ \frac{\binom{n_b}{r_S}}{\binom{n}{r_S}}, & \text{if } \frac{n}{2} \leq r_S < n_b. \end{cases} \qquad (8.23)$$

Alternatively, if $n_a \leq r_S < n_b$ and $n_1 > n - n_1$, then $a_{S,\min} = r_S - (n - n_1)$ and $a_{S,\max} = r_S$, leading to:

$$p_{S,\min}(r_S) = \frac{\min\left(\binom{n_1}{n-r_S}, \binom{n_1}{r_S}\right)}{\binom{n}{r_S}}$$

$$= \begin{cases} \frac{\binom{n_b}{n-r_S}}{\binom{n}{r_S}}, & \text{if } n_a \leq r_S < \frac{n}{2}, \\ \frac{\binom{n_b}{r_S}}{\binom{n}{r_S}}, & \text{if } \frac{n}{2} \leq r_S < n_b, \end{cases} \qquad (8.24)$$

where again the identity $\binom{n}{k} = \binom{n}{n-k}$ was used. This concludes the proof. □

Figure 8.4 illustrates the behavior of $p_{S,\min}$ for Pearson's χ^2 test (blue) and Fisher's exact test (orange) as the number r_S of occurrences of pattern S in a dataset $\mathcal{D}$ varies. In Figure 8.4 (a), the sample size is $n = 60$ and the number of samples in the positive class is $n_1 = 15$, whereas Figure 8.4 (b) shows an example with a balanced class ratio, i.e., $n = 60$ and $n_1 = 30$. First, $p_{S,\min}$ shows a qualitatively identical behaviour for both test statistics, as expected. For fixed n_1 and n, $p_{S,\min}$ as a function of r_S is symmetric around $r_S = n/2$ and has minima at $r_S = n_a = \min(n_1, n-n_1)$ and $r_S = n_b = \max(n_1, n-n_1)$, as evidenced by the closed-form expressions derived in Propositions 8.3 and 8.4. The key insight to be derived from Figure 8.4 is that $p_{S,\min}$ is large, indicating lower potential to result in a statistically significant association, whenever r_S is small or r_S is large.

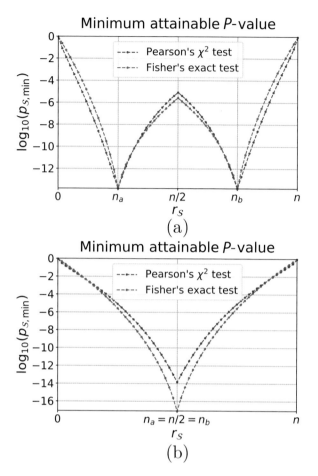

Figure 8.4: Minimum attainable P-value $p_{S,\min}$ as a function of r_S, the number of occurrences of pattern S in a dataset $\mathcal{D}$, for Pearson's χ^2 test (blue) and Fisher's exact test (orange). The number of samples in the positive class n_1 and the sample size n are $n_1 = 15$, $n = 60$ in (a) and $n_1 = 30$, $n = 60$ in (b), respectively.

This is further illustrated in Figure 8.5, which indicates how this relates to the concept of testability at a certain adjusted significance threshold δ. As a direct consequence of the functional form of $p_{S,\min}$, for fixed margin n_1, sample size n and adjusted significance threshold δ, there exists a value $r_{S,\min}(\delta)$ such that patterns S with $r_S < r_{S,\min}(\delta)$ or $r_S > n - r_{S,\min}(\delta)$ are untestable at level δ, while those with $[\![r_{S,\min}(\delta), n - r_{S,\min}(\delta)]\!]$ are testable.[10] This result formalizes the intuition that patterns S with either too small r_S or too large r_S, i.e. patterns S that are either too rare or too common in $\mathcal{D}$, cannot be

---

[10] To be precise, if δ is sufficiently small, the range of values of r_S that lead to pattern S being testable can become the union of two disjoint intervals. However, this poses no practical algorithmic or statistical issues. Moreover, this situation is uncommon in practice, as it corresponds to values of δ that are too small to be of practical relevance in most applications.

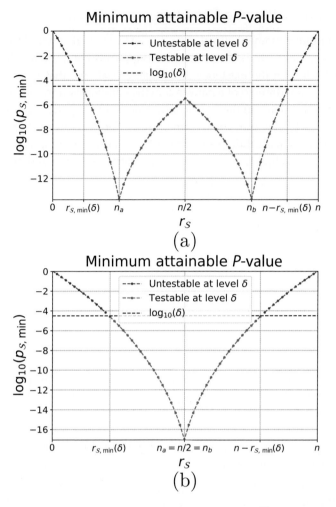

Figure 8.5: Illustration of the concept of testability at level $\delta = 10^{-4.5}$ when Fisher's exact test is the test statistic of choice for two 2×2 contingency tables that differ in the class ratios: (a) unbalanced, with $n_1 = 15$, $n = 60$ and (b) balanced, with $n_1 = 30$, $n = 60$. Values of r_S in the range $[\![r_{S,\min}(\delta), n - r_{S,\min}(\delta)]\!]$ lead to pattern S being testable at level δ (green), while $r_S < r_{S,\min}(\delta)$ or $r_S > n - r_{S,\min}(\delta)$ imply that pattern S is untestable at level δ (red).

statistically significant. Tarone's method can therefore be understood as a statistically principled way to turn this intuition into a filtering criterion to reduce the number of patterns that contribute to the multiple hypothesis testing problem. Most importantly, unlike other alternative approaches, Tarone's method does not use an ad-hoc threshold to filter patterns according to their frequency. It rather learns the threshold in a data-driven manner. This allows Tarone's method to guarantee the filtering of all patterns that have no chance of resulting in a statistically significant association at level δ, regardless of the actual realizations of the class labels, while keeping all patterns that could result in an association.

8.3.2 Designing a Pruning Condition

Even if the minimum attainable P-value $p_{S,\min}$ can be evaluated with $O(1)$ complexity, evaluating $p_{S,\min}$ for all candidate patterns $\mathcal{M}$ is typically computationally unfeasible. Algorithm 8.2 relies on the existence of a *pruning condition*, that is, a way to test if descendants $S' \in \text{Children}(S)$ of a pattern S in the enumeration tree are testable using only information available in the 2×2 contingency table of pattern S. In this section, we show how the specific functional form of $p_{S,\min}$ for Pearson's χ^2 test and Fisher's exact test, combined with the apriori property of pattern mining (Proposition 8.2), leads to the simple yet highly effective pruning criterion summarized in the following proposition:

Proposition 8.5 (Pruning criterion for Pearson's χ^2 test and Fisher's exact test) *Let $S \in \mathcal{M}$ be a pattern satisfying:*

i. $p_{S,\min} > \hat{\delta}_{tar}$, *i.e.*, S *is untestable at level* $\hat{\delta}_{tar}$,
ii. $r_S \leq n_a$, *with* $n_a = \min(n_1, n - n_1)$.

Then, $p_{S',\min} > \hat{\delta}_{tar} \geq \delta_{tar}$ for all $S' \in \text{Children}(S)$. Hence, all descendants of pattern S can be pruned from the enumeration tree.

In conclusion, `pruning_condition`$(S, \hat{\delta}_{tar})$ *in Line 15 of Algorithm 8.2 is true if and only if $r_S \leq n_a$ and $p_{S,\min} > \hat{\delta}_{tar}$.*

Proof First, for fixed sample size n and number n_1 of samples in the positive class, the minimum attainable P-value $p_{S,\min}$ is a monotonically decreasing function of r_S in the range $r_S \in [\![0, n_a]\!]$ for both Pearson's χ^2 test and Fisher's exact test, i.e., $r_{S'} \leq r_S \leq n_a$ implies $p_{S,\min}(r_{S'}) \geq p_{S,\min}(r_S)$. This property, which does not necessarily hold for all test statistics based on contingency tables, can be readily verified from the specific functional form of $p_{S,\min}$ in the range $r_S \in [\![0, n_a]\!]$: $p_{S,\min}(r_S) = 1 - F_{\chi_1^2}\left((n-1)\frac{n_b}{n_a}\frac{r_S}{n-r_S}\right)$ for Pearson's χ^2 test and $p_{S,\min}(r_S) = \binom{n_a}{r_S}/\binom{n}{r_S}$ for Fisher's exact test. Intuitively, this means that in the range $r_S \in [\![0, n_a]\!]$, as the pattern S becomes less rare, the minimum attainable P-value decreases, thereby increasing the potential of pattern S to be statistically significant. Secondly, due to the apriori property, if $S' \in \text{Children}(S)$ then $r_{S'} \leq r_S$.

Combining both facts, if $r_S \leq n_a$ then $p_{S',\min} \geq p_{S,\min}$ for all descendants S' of S in the enumeration tree. Therefore, $p_{S,\min} > \hat{\delta}_{tar}$ implies that $p_{S',\min} > \hat{\delta}_{tar}$ for all $S' \in \text{Children}(S)$. Since $\hat{\delta}_{tar} \geq \delta_{tar}$ at any point during the execution of Algorithm 8.2, this proves the result. $\square$

Search space pruning according to the criterion presented in Proposition 8.5 can only occur for patterns $S \in \mathcal{M}$ satisfying $r_S \leq n_a$. In practice, a large proportion of all candidate patterns in the search space $\mathcal{M}$ are sufficiently rare for this condition to apply. When this happens, Proposition 8.5 simply states that descendants $S' \in \text{Children}(S)$ of an untestable pattern S will also be untestable. As illustrated in Figure 8.6, this can lead to an enormous reduction in computational complexity, allowing one to compute δ_{tar} and (exactly) retrieve all testable patterns $\mathcal{M}_{\text{test}}(\delta_{tar})$ by enumerating only a small subset of all candidate patterns in the search space.

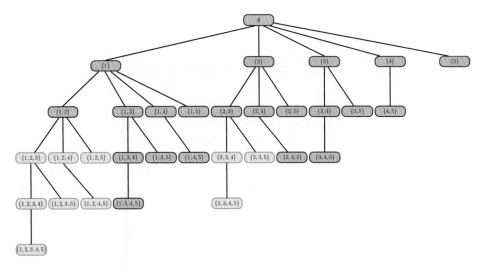

Figure 8.6: Illustration of the effect of search space pruning in a significant itemset mining problem with $p = 5$ features. Suppose that patterns $\mathcal{S}_1$ and $\mathcal{S}_2$ (highlighted in red) satisfy the conditions of Proposition 8.5, i.e., $r_{\mathcal{S}_1} \leq n_a$, $r_{\mathcal{S}_2} \leq n_a$ and $p_{\mathcal{S}_1,\min} > \hat{\delta}_{\text{tar}}$, $p_{\mathcal{S}_2,\min} > \hat{\delta}_{\text{tar}}$. Then, all their descendants (highlighted in orange) can be pruned from the search space, drastically reducing the number of candidate patterns that need to be enumerated.

8.3.3 Implementation Considerations

We will conclude this section by discussing some key design considerations that need to be addressed when implementing a significant pattern mining algorithm.

The first and perhaps most important issue is the choice of algorithm to traverse the enumeration tree containing all patterns in the search space $\mathcal{M}$. Due to the enormous number of candidate patterns in the enumeration tree, the resulting efficiency of the traversal algorithm can greatly depend on the specific strategy and data structures used to perform the enumeration. While this topic is outside the scope of this chapter, the design of efficient algorithms to enumerate patterns has been widely studied by the data mining community, and many highly optimized methods are readily available for itemset mining (e.g., [40, 41, 42], see [43] for a review) and subgraph mining (e.g., [44, 45], see [46] for a review).

In Line 12 of Algorithm 8.2, the estimate $\hat{\delta}_{\text{tar}}$ of the adjusted significance threshold is decreased after the current value has been found to violate the FWER condition. There are multiple ways to implement this step. In theory, using either $\hat{\delta}_{\text{tar}} \leftarrow \hat{\delta}_{\text{tar}} - \Delta$ or $\hat{\delta}_{\text{tar}} \leftarrow 10^{-\Delta}\hat{\delta}_{\text{tar}}$, for sufficiently small Δ, are both valid strategies. A more efficient alternative is to define the sequence of candidate values for $\hat{\delta}_{\text{tar}}$ by sorting, in descending order, the $\lfloor \frac{n}{2} \rfloor + 1$ different values the minimum attainable P-value $p_{\mathcal{S},\min}$ can take.[11] Each time $\hat{\delta}_{\text{tar}}$ needs to be decreased, it is set to the next element of the sequence. This is optimal in the sense that it provides the minimum step size necessary to decrease the FWER approximation $\widehat{\text{FWER}}(\hat{\delta}_{\text{tar}})$.

---

[11] This follows from the fact that, for fixed n_1 and n, $p_{\mathcal{S},\min}(r_{\mathcal{S}})$ as a function of $r_{\mathcal{S}}$ is symmetric around $n/2$.

Another aspect worth considering is how to store in memory the set $\widehat{\mathcal{M}}_{\text{test}}(\hat{\delta}_{\text{tar}})$ of testable patterns at level $\hat{\delta}_{\text{tar}}$ during the execution of Algorithm 8.2. Typically, this set can be rather large, containing hundreds of millions or even billions of patterns. Moreover, this could cause the execution of Line 13 in Algorithm 8.2 to become a computational bottleneck, since in principle the set $\widehat{\mathcal{M}}_{\text{test}}(\hat{\delta}_{\text{tar}})$ of testable patterns needs to be inspected to remove all patterns which become untestable after having decreased $\hat{\delta}_{\text{tar}}$ in the previous line. The key insight to circumvent this problem is that, in order to approximate the FWER using Tarone's method, only the total number of testable patterns $|\widehat{\mathcal{M}}_{\text{test}}(\hat{\delta}_{\text{tar}})|$ is needed, not the patterns themselves. Let $\mathcal{P}_{\min} = \{p_{\mathcal{S},\min}(r_{\mathcal{S}}) \mid r_{\mathcal{S}} \in [\![0, \lfloor \frac{n}{2} \rfloor]\!]\}$ be the set of $\lfloor \frac{n}{2} \rfloor + 1$ distinct values that $p_{\mathcal{S},\min}$ can take for a given sample size n. Rather than explicitly storing $\widehat{\mathcal{M}}_{\text{test}}(\hat{\delta}_{\text{tar}})$, an alternative strategy is to store the number $c(p_{\mathcal{S},\min})$ of patterns enumerated so far that have minimum attainable P-value equal to $p_{\mathcal{S},\min}$, for each of the $\lfloor \frac{n}{2} \rfloor + 1$ different values $p_{\mathcal{S},\min} \in \mathcal{P}_{\min}$. For any given $\hat{\delta}_{\text{tar}}$, $|\widehat{\mathcal{M}}_{\text{test}}(\hat{\delta}_{\text{tar}})|$ can be computed as

$$|\widehat{\mathcal{M}}_{\text{test}}(\hat{\delta}_{\text{tar}})| = \sum_{p_{\mathcal{S},\min} \in \mathcal{P}_{\min} \mid p_{\mathcal{S},\min} \leq \hat{\delta}_{\text{tar}}} c(p_{\mathcal{S},\min}), \tag{8.25}$$

where the summation includes at most $\lfloor \frac{n}{2} \rfloor + 1$ counts. This eliminates the need to store the set $\widehat{\mathcal{M}}_{\text{test}}(\hat{\delta}_{\text{tar}})$ of testable patterns in memory and allows to execute Line 13 of Algorithm 8.2 with $O(1)$ complexity. Once δ_{tar} has been obtained, in Line 2 of Algorithm 8.1, the set of testable patterns $\mathcal{M}_{\text{test}}(\delta_{\text{tar}})$ is needed in order to find those patterns that are statistically significant. To do so, the enumeration process can be repeated, starting again at the root of the enumeration tree, but with fixed $\hat{\delta}_{\text{tar}} = \delta_{\text{tar}}$. As patterns are enumerated, the P-values $p_{\mathcal{S}}$ of patterns testable at level δ_{tar} are computed, and those which are deemed statistically significant are written to an output file. This is summarized in Algorithm 8.3. While this approach requires enumerating patterns twice, therefore approximately doubling runtime, it greatly reduces memory usage. Hence, it is the preferred implementation choice in most situations.

Finally, a strategy that might be useful in some cases is to precompute the n values of $p_{\mathcal{S},\min}(r_{\mathcal{S}})$ and store them as a look-up table. While this requires $O(n)$ additional

---

Algorithm 8.3 find_significant_patterns

Input: Dataset $\mathcal{D} = \{(x_i, y_i)\}_{i=1}^{n}$, Tarone's adjusted significance threshold δ_{tar}
1: **function** FIND_SIGNIFICANT_PATTERNS($\mathcal{D}, \delta_{\text{tar}}$)
2:    NEXT($\emptyset$)   ▷ Start pattern enumeration
3: **procedure** NEXT($\mathcal{S}$)
4:    Compute the minimum attainable P-value $p_{\mathcal{S},\min}$ ▷ see Section 8.3.1
5:    **if** $p_{\mathcal{S},\min} \leq \delta_{\text{tar}}$ **then**   ▷ if pattern $\mathcal{S}$ is testable at level δ_{tar} then
6:       Compute the P-value $p_{\mathcal{S}}$
7:       **if** $p_{\mathcal{S}} \leq \delta_{\text{tar}}$ **then**   ▷ if pattern $\mathcal{S}$ is significant at level δ_{tar} then
8:          Write $\mathcal{S}$ and $p_{\mathcal{S}}$ to an output file
9:    **if not** pruning_condition($\mathcal{S}, \delta_{\text{tar}}$) **then**   ▷ see Section 8.3.2
10:       **for** $\mathcal{S}' \in$ Children($\mathcal{S}$) **do**
11:          NEXT($\mathcal{S}'$)   ▷ Recursively visit nodes in the tree depth-first

memory, it will further speed-up the evaluation of the minimum attainable P-values, which is one of the most repeated operations in Algorithm 8.2. As a consequence, in some situations, it might be a desirable trade-off of slightly increased memory usage for reduced runtime.

8.4 Accounting for the Redundancy Between Patterns

One of the defining characteristics of significant pattern mining is the daunting size of the search space of candidate patterns $\mathcal{M}$. As it has been extensively discussed in this chapter, this leads to two fundamental challenges: (1) the statistical challenge of dealing with an extreme instance of the multiple hypothesis testing problem, and (2) the computational challenge of efficiently exploring this vast search space $\mathcal{M}$. In Section 8.2.3, Tarone's improved Bonferroni correction for discrete data was introduced as a way to achieve FWER control while maintaining sufficient statistical power to discover truly associated patterns. Next, in Section 8.3, it was shown how Tarone's method can be combined with pattern mining techniques. This led to an algorithm that can explore all candidate patterns in $\mathcal{M}$ in a computationally efficient manner and that exhibits a considerable amount of statistical power despite guaranteeing FWER control. However, this approach still has key limitations, providing opportunities to develop new algorithms and to further improve the state-of-the-art in significant pattern mining.

In particular, another defining characteristic of significant pattern mining, which has been overlooked in this chapter so far, is the fact that the search space of candidate patterns $\mathcal{M}$ is not only extremely large, but also harbors non-trivial redundancies among the many patterns $\mathcal{M}$ contains.

One of the main sources of redundancy are subset/superset relationships between patterns. If a pattern $\mathcal{S}'$ contains another pattern $\mathcal{S}$, the random variables $G_\mathcal{S}(X)$ and $G_{\mathcal{S}'}(X)$ that indicate the occurrence of patterns $\mathcal{S}$ and $\mathcal{S}'$ in an input sample X satisfy $G_\mathcal{S}(X) = 0 \Rightarrow G_{\mathcal{S}'}(X) = 0$ or, equivalently, $G_{\mathcal{S}'}(X) = 1 \Rightarrow G_\mathcal{S}(X) = 1$. Therefore, $G_\mathcal{S}(X)$ and $G_{\mathcal{S}'}(X)$ are statistically dependent,[12] being mutually redundant to some degree. The strength of this dependency will depend, in general, on how frequently the "difference pattern" $\mathcal{S}' \setminus \mathcal{S}$ occurs in an input sample X. The more common it is, the more frequently $\mathcal{S}$ and $\mathcal{S}'$ will co-occur, leading to a stronger association between $G_\mathcal{S}(X)$ and $G_{\mathcal{S}'}(X)$ and, therefore, a larger statistical redundancy between $\mathcal{S}$ and $\mathcal{S}'$. This phenomenon is illustrated in Figure 8.7, which represents a toy significant itemset mining problem with $p = 10$ features and $n = 12$ samples. Given a high-order feature interaction $\mathcal{S} = \{2, 9, 10\}$, suppose that another high-order feature interaction $\mathcal{S}' = \mathcal{S} \cup \{7\}$ is formed by adding an additional feature to $\mathcal{S}$. Then, knowing the value of $g_\mathcal{S}(x_i)$ alone is sufficient to know the value of $g_{\mathcal{S}'}(x_i)$ for half of the samples $\{x_i\}_{i=1}^{12}$ in the dataset. In fact, this holds true for any $\mathcal{S}' \subset \mathcal{S}$, regardless of the additional features $\mathcal{S}' \setminus \mathcal{S}$ being added to the high-order feature interaction. Consequently, $G_\mathcal{S}(X)$ is statistically dependent of $G_{\mathcal{S}'}(X)$, for any $\mathcal{S}' \subset \mathcal{S}$.

---

[12] The apriori property of pattern mining, discussed in Proposition 8.2, can also be used to show that $G_\mathcal{S}(X)$ and $G_{\mathcal{S}'}(X)$ are statistically dependent for $\mathcal{S}' \subset \mathcal{S}$.

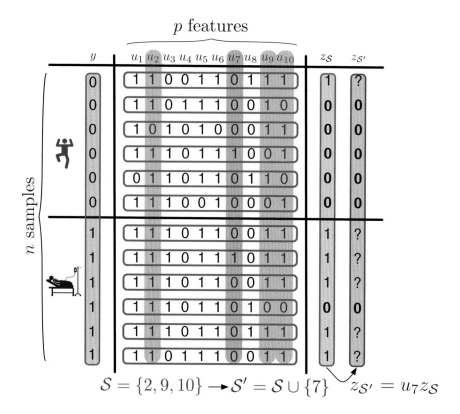

Figure 8.7: Illustration of how redundancy between patterns arises in significant itemset mining as a consequence of inclusion relationships $S' \subset S$ between candidate patterns $S, S' \in \mathcal{M}$.

Subset/superset relationships are not the only potential source of statistical redundancy between any two candidate patterns $S_1, S_2 \in \mathcal{M}$. More generally, as long as the two patterns share some substructures, i.e., $S_1 \cap S_2 \neq \emptyset$, the random variables $G_{S_1}(X)$ and $G_{S_2}(X)$ might be statistically associated. The strength of this association and, therefore, the degree to which patterns S_1 and S_2 are statistically redundant, depends on the probability that the shared substructure $S_1 \cap S_2$ occurs in an input sample X, relative to the probability that the pattern-specific substructures $S_1 \setminus S_2$ and $S_2 \setminus S_1$ occur. For example, if $S_1 \setminus S_2$ and $S_2 \setminus S_1$ are both common substructures that occur in almost every input sample X, the probability that S_1 and S_2 occur in an input sample X will be dominated by the probability that $S_1 \cap S_2$ occurs in X. Therefore, in this situation S_1 and S_2 will be considerably redundant. On the contrary, if $S_1 \cap S_2$ is a common sub-structure, the probability that S_1 occurs in an input sample X will be mostly determined by how frequently $S_1 \setminus S_2$ occurs, while the probability that S_2 occurs in an input sample X will mostly depend on how frequently $S_2 \setminus S_1$ occurs. In this case, $G_{S_1}(X)$ and $G_{S_2}(X)$ will be approximately independent, leading to patterns S_1 and S_2 being barely redundant.

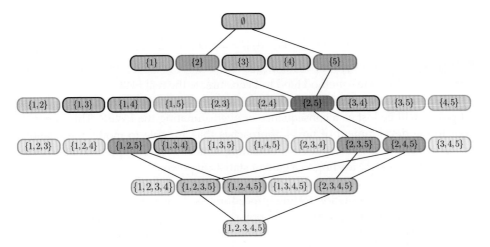

Figure 8.8: Illustration of how the sharing of pattern substructures and subset/superset relationships between patterns leads to any given pattern $S \in \mathcal{M}$ being related to exponentially many other patterns in the search space $\mathcal{M}$. In this significant itemset mining example, a high-order feature interaction $S = \{2,5\}$, highlighted in dark green, is related by subset/superset relationships to 9/32 high-order feature interactions (green) and shares features with other 14/32 high-order feature interactions (orange). The intensity of a node's colour is proportional to the relatedness of the corresponding high-order feature interaction S' with $S = \{2,5\}$.

Due to the combinatorial nature of the search space $\mathcal{M}$, these statistical redundancies that arise due to the sharing of pattern substructures and subset/superset relationships can lead to any given pattern S being statistically associated with exponentially many other patterns S'. This effect is illustrated in Figure 8.8 for a significant itemset mining problem with $p = 5$ features. In this example, a given feature interaction $S = \{2,5\}$ (dark green) is directly related by subset/superset relationships to 9 out of 32 patterns (green) and shares some substructures (in this case, input features) with other 14 out of 32 patterns (orange). In short, in this example, pattern $S = \{2,5\}$ is potentially redundant with more than half of the entire search space $\mathcal{M}$.

The existence of this complex web of inter-dependencies between patterns in the search space $\mathcal{M}$ has profound implications. If the random variables $G_{S_1}(X)$ and $G_{S_2}(X)$ are statistically dependent, the corresponding P-values p_{S_1} and p_{S_2} quantifying the statistical association of $G_{S_1}(X)$ and $G_{S_2}(X)$ with the class labels Y might be statistically dependent as well. Suppose that patterns $S_1, S_2 \in \mathcal{M}$ are *not* associated with the class labels Y, i.e., $S_1, S_2 \in \mathcal{M}_{\text{null}}$ as defined in Section 8.2.3. Then, if p_{S_1} and p_{S_2} were strongly positively correlated, false positives for pattern S_1, i.e., $[p_{S_1} \leq \delta]$, will tend to co-occur with false positives for pattern S_2, i.e., $[p_{S_2} \leq \delta]$. The implications of this observation can be traced back to the derivations of the FWER upper bounds used by the Bonferroni correction and Tarone's method. Both approaches make use of the basic fact that, as a consequence of the axioms of probability, $\Pr([p_{S_1} \leq \delta] \cup [p_{S_2} \leq \delta]) \leq \Pr(p_{S_1} \leq \delta) + \Pr(p_{S_2} \leq \delta)$. However, if p_{S_1} and p_{S_2} had a strong positive correlation, $\Pr([p_{S_1} \leq \delta] \cup [p_{S_2} \leq \delta]) \approx \Pr(p_{S_1} \leq \delta) \approx \Pr(p_{S_2} \leq \delta)$ will hold. More generally, if the set of P-values $\{p_S \mid S \in \mathcal{M}\}$ exhibits exponentially many statistical inter-dependencies, possible extending beyond simple pairwise associations between P-values, then $\Pr\left(\bigcup_{S \in \mathcal{M}_{\text{null}}(\delta)} [p_S \leq \delta]\right) \ll \sum_{S \in \mathcal{M}_{\text{null}}(\delta)} \Pr(p_S \leq \delta)$ and $\Pr\left(\bigcup_{S \in \mathcal{M}_{\text{test}}(\delta)} [p_S \leq \delta]\right) \ll \sum_{S \in \mathcal{M}_{\text{test}}(\delta)} \Pr(p_S \leq \delta)$. Hence, both the Bonferroni correction and Tarone's method tend to overestimate the real FWER.

In summary, by ignoring the existence of redundancies between patterns, statistical power will be lost as a consequence of overestimating the FWER. This opens the door to the development of novel approaches that are able to model these statistical dependencies between patterns, obtaining a more accurate approximation of the FWER and ultimately leading to a gain of statistical power.

The rest of this section is organized as follows. In Section 8.4.1, we present the Westfall–Young permutation-testing procedure as a way to empirically estimate the FWER under the global null hypothesis that no pattern is associated with the class labels. Next, in Section 8.4.2, we introduce an extension of the algorithm presented in Section 8.3 that incorporates permutation-testing to improve statistical power.

8.4.1 Empirically Approximating the FWER Using Random Permutations

Due to the difficulty to exactly evaluate the FWER at a given adjusted significance threshold δ, both the Bonferroni correction and Tarone's method approximate $\text{FWER}(\delta) = \Pr(\text{FP}(\delta) > 0)$ with an upper bound $\widehat{\text{FWER}}(\delta)$. Tarone's method exploits the concept of testability to obtain an upper bound on the true FWER that is considerably more accurate, in significant pattern mining, than the Bonferroni correction. Nonetheless, both methods implicitly assume that all patterns $S \in \mathcal{M}$ are statistically independent. As discussed above, this is rarely the case in significant pattern mining. As a consequence, Tarone's method still tends to overestimate the true FWER, leading to a potential loss of statistical power. An alternative approach to approximate $\text{FWER}(\delta)$ is to use resampling techniques to obtain an empirical estimate $\widehat{\text{FWER}}(\delta)$.

For this purpose, one of the most commonly used resampling schemes consists of applying random permutations to the class labels [47]. Let $\mathcal{D} = \{(x_i, y_i)\}_{i=1}^{n}$ be an input dataset with n i.i.d. samples $x \in \mathcal{X}$ and class labels $y \in \{0, 1\}$. Suppose that $\pi : [\![1, n]\!] \to [\![1, n]\!]$ is a random permutation, i.e. a permutation of the set $[\![1, n]\!]$ selected uniformly at random from the set of all $n!$ permutations of $[\![1, n]\!]$. Define the resampled dataset $\tilde{\mathcal{D}} = \{(x_i, y_{\pi(i)})\}_{i=1}^{n}$ in such a way that the ith input sample x_i is paired with the class label of sample $\pi(i)$, for each $i = 1, \ldots, n$. The effect of obtaining the class labels in the resampled dataset $\tilde{\mathcal{D}}$ by randomly permuting the original labels in $\mathcal{D}$ is to assign a random class label to each input sample while keeping the class ratio n_1/n unchanged. As a consequence, any statistical dependency between patterns and labels which might have existed in the original dataset $\mathcal{D}$ is effectively eliminated by the permutation process. While the ground-truth regarding which patterns $S \in \mathcal{M}$ are statistically associated with the class labels in the original dataset $\mathcal{D}$ is unknown, in the resampled dataset $\tilde{\mathcal{D}}$, no pattern $S \in \mathcal{M}$ can possibly be associated with the class labels.

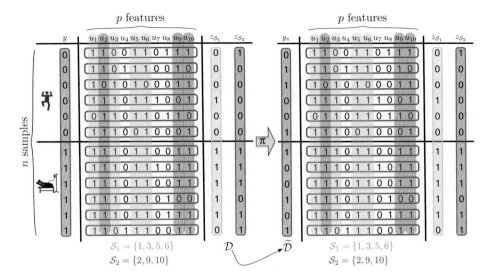

Figure 8.9: A random permutation of the class labels $\pi : [\![1,n]\!] \to [\![1,n]\!]$ can be used to derive a resampled dataset $\widetilde{\mathcal{D}}$ from an input dataset $\mathcal{D}$. In this significant itemset mining example, a dataset $\mathcal{D}$ with $n=12$ samples and $p=10$ features, shown on the left, has been resampled to obtain a new dataset $\widetilde{\mathcal{D}}$, shown on the right. While both datasets share the same samples $\{x_i\}_{i=1}^n$ and class ratio n_1/n, the mapping of class labels and samples is different. A consequence of this is that the occurrence of patterns $\mathcal{S}_1$ and $\mathcal{S}_2$, which is enriched in samples belonging to class $y=1$ in the original dataset $\mathcal{D}$, no longer show any association with the class labels in the resampled dataset $\widetilde{\mathcal{D}}$.

In other words, the *global null hypothesis* $\mathcal{M}_{\text{null}} = \mathcal{M}$ holds for $\widetilde{\mathcal{D}}$. An illustration of the permutation process is shown in Figure 8.9 for a significant itemset mining dataset.

The fact that a resampled dataset $\widetilde{\mathcal{D}}$ obtained in this manner is known to contain no associations can be exploited to obtain an empirical estimate of the FWER under the global null hypothesis that no pattern $\mathcal{S} \in \mathcal{M}$ is associated with the class labels.[13] Suppose that the resampling process described above is repeated a number j_p of times, leading to a set $\{\widetilde{\mathcal{D}}^{(k)}\}_{k=1}^{j_p}$ of j_p resampled datasets. Leaving the matter of computational feasibility temporarily aside, suppose that for each dataset $\widetilde{\mathcal{D}}^{(k)}$, the P-values $p_\mathcal{S}^{(k)}$ for all patterns $\mathcal{S} \in \mathcal{M}$ were obtained in order to find the P-value corresponding to the most significant pattern, i.e. in order to compute $p_{\min}^{(k)} = \min\{p_\mathcal{S}^{(k)} \mid \mathcal{S} \in \mathcal{M}\}$. By construction, $p_{\min}^{(k)} \leq p_\mathcal{S}^{(k)}$ for all patterns $\mathcal{S} \in \mathcal{M}$. Hence, if $p_{\min}^{(k)} > \delta$, no pattern $\mathcal{S} \in \mathcal{M}$ will be deemed significant in the kth resampled dataset $\widetilde{\mathcal{D}}^{(k)}$, leading to no false positives being detected for this resampled dataset ($\text{FP}^{(k)}(\delta) = 0$). On the contrary, if

---

[13] Controlling the FWER under the global null hypothesis is often denoted as *weak control* of the FWER. If the *subset pivotality condition* [47] can be shown to hold, it is possible to prove that permutation-testing also controls the FWER in the strong sense, i.e. when any subset of hypothesis is allowed to be non-null. However, permutation-testing has been extensively applied to problems for which the subset pivotality condition cannot be proven, as is the case of significant pattern mining, nevertheless leading to successful results.

$p_{\min}^{(k)} \leq \delta$, there is at least one pattern $S \in \mathcal{M}$ being deemed significant for the k-th resampled dataset. Since $\widetilde{\mathcal{D}}^{(k)}$ is known to contain no associations, this implies that $\mathrm{FP}^{(k)}(\delta) > 0$. Therefore, an empirical estimator of the FWER can be obtained as:

$$\widehat{\mathrm{FWER}}(\delta) = \frac{1}{j_p} \sum_{k=1}^{j_p} \mathbb{1}\left[p_{\min}^{(k)} \leq \delta\right], \tag{8.26}$$

where $\mathbb{1}[\bullet]$ evaluates to 1 if its input argument is true and to 0 otherwise. Intuitively, the estimator $\widehat{\mathrm{FWER}}(\delta)$ of the FWER at adjusted significance threshold δ is simply given by the proportion of the j_p resampled datasets that contain at least one false positive. If the number of permutations j_p is sufficiently large (e.g., $j_p \approx 10{,}000$), $\widehat{\mathrm{FWER}}(\delta)$ will be a rather accurate estimate of the true value $\mathrm{FWER}(\delta)$. An adjusted significance threshold can then be proposed based on this estimator as:

$$\delta_{\mathrm{perm}} = \max\left\{\delta \mid \frac{1}{j_p} \sum_{k=1}^{j_p} \mathbb{1}\left[p_{\min}^{(k)} \leq \delta\right] \leq \alpha\right\}. \tag{8.27}$$

The empirical estimate of the FWER obtained via permutation testing implicitly accounts for the dependence structure that might exist between patterns. While this can lead to a considerable improvement in statistical power with respect to Tarone's method, a naive application of this procedure to significant pattern mining is entirely unfeasible. First, evaluating $p_{\min}^{(k)}$ for a single resampled dataset $\widetilde{\mathcal{D}}^{(k)}$ is a challenging problem on its own. In principle, evaluating $p_{\min}^{(k)}$ naively would require computing the P-values $p_S^{(k)}$ for all patterns $S \in \mathcal{M}$, which is typically unfeasible due to the size of the search space $\mathcal{M}$. Moreover, in permutation testing, this operation needs to be repeated between $j_p = 1{,}000$ and $j_p = 10{,}000$ times if a sufficiently accurate estimate of the FWER is to be obtained.

In the next section, we will discuss how specific properties of the FWER estimator in Equation 8.26 can be exploited to allow the application of permutation-testing in significant pattern mining.

8.4.2 Permutation Testing in Significant Pattern Mining

The idea of using permutation testing to improve statistical power in significant pattern mining was pioneered by [18]. In their work, the authors explicitly tackle the problem of efficiently computing $p_{\min}^{(k)}$ for a single resampled dataset $\widetilde{\mathcal{D}}^{(k)}$. In order to avoid computing the P-values $p_S^{(k)}$ for all patterns $S \in \mathcal{M}$ in the search space, their method relies again on the concept of minimum attainable P-value. Suppose that $\hat{p}_{\min}^{(k)}$ is an estimate of $p_{\min}^{(k)}$ obtained after having enumerated only a subset of the search space $\mathcal{M}$. If the minimum attainable P-value $p_{S,\min}$ of a pattern S satisfies $p_{S,\min} > \hat{p}_{\min}^{(k)}$, then $\min(\hat{p}_{\min}^{(k)}, p_S^{(k)}) = \hat{p}_{\min}^{(k)}$. In other words, if the minimum attainable P-value $p_{S,\min}$ of a pattern S is larger than the smallest P-value $\hat{p}_{\min}^{(k)}$ among all patterns enumerated so far, pattern S has no chance of having a P-value smaller than $\hat{p}_{\min}^{(k)}$. Hence, $p_S^{(k)}$ does not need to be computed. Most importantly, this idea can be combined

Algorithm 8.4 permutation_testing_spm

Input: Dataset $\mathcal{D} = \{(x_i, y_i)\}_{i=1}^n$, target FWER α, number of permutations j_p
Output: Adjusted significance threshold δ_{perm}

1: **function** PERMUTATION_TESTING_SPM($\mathcal{D},\alpha,j_p$)
2:     Initialize global variable $\hat{\delta}_{\text{perm}} \leftarrow 1$
3:     **for** $k = 1, 2, \ldots, j_p$ **do**
4:         Obtain a random permutation $\pi^{(k)} : [\![1,n]\!] \to [\![1,n]\!]$
5:         Initialize global variable $\hat{p}_{\min}^{(k)} \leftarrow 1$
6:     NEXT($\emptyset$)   ▷ Start pattern enumeration
7:     $\delta_{\text{perm}} \leftarrow \hat{\delta}_{\text{perm}}$
8:     Return δ_{perm}
9: **procedure** NEXT($\mathcal{S}$)
10:     Compute the minimum attainable P-value $p_{\mathcal{S},\min}$ ▷ see Section 8.3.1
11:     **if** $p_{\mathcal{S},\min} \leq \hat{\delta}_{\text{perm}}$ **then**   ▷ if pattern $\mathcal{S}$ is testable at level $\hat{\delta}_{\text{perm}}$ **then**
12:         **for** $k = 1, 2, \ldots, j_p$ **do**
13:             Compute P-value $p_{\mathcal{S}}^{(k)}$ for resampled dataset $\widetilde{D}^{(k)}$
14:             $\hat{p}_{\min}^{(k)} \leftarrow \min(\hat{p}_{\min}^{(k)}, p_{\mathcal{S}}^{(k)})$
15:         $\widetilde{\text{FWER}}(\hat{\delta}_{\text{perm}}) \leftarrow \frac{1}{j_p} \sum_{k=1}^{j_p} \mathbb{1}\left[\hat{p}_{\min}^{(k)} \leq \hat{\delta}_{\text{perm}}\right]$
16:         **while** $\widetilde{\text{FWER}}(\hat{\delta}_{\text{perm}}) > \alpha$ **do**
17:             Decrease $\hat{\delta}_{\text{perm}}$
18:             $\widetilde{\text{FWER}}(\hat{\delta}_{\text{perm}}) \leftarrow \frac{1}{j_p} \sum_{k=1}^{j_p} \mathbb{1}\left[\hat{p}_{\min}^{(k)} \leq \hat{\delta}_{\text{perm}}\right]$
19:         **if not** pruning_condition($\mathcal{S}, \hat{\delta}_{\text{perm}}$) **then**   ▷ see Section 8.3.2
20:             **for** $\mathcal{S}' \in$ Children($\mathcal{S}$) **do**
21:                 NEXT($\mathcal{S}'$)   ▷ Recursively visit nodes in the tree depth-first

with the powerful concept of search space pruning described in Section 8.3: If $p_{\mathcal{S},\min} > \hat{p}_{\min}^{(k)}$ and $r_{\mathcal{S}} \leq \min(n_1, n-n_1)$, no descendant $\mathcal{S}' \in$ Children($\mathcal{S}$) of $\mathcal{S}$ in the enumeration tree can possibly satisfy $p_{\mathcal{S}'}^{(k)} \leq \hat{p}_{\min}^{(k)}$, naturally leading to a valid search space pruning criterion analogous to the one employed by Algorithm 8.2.

While the algorithm proposed in [18] is able to obtain the exact value of $p_{\min}^{(k)}$ while enumerating only a small subset of patterns in the search space $\mathcal{M}$, it is still limited by the fact that this operation needs to be repeated around $j_p \approx 10{,}000$ times. In particular, the whole pattern enumeration process has to be repeated independently for each permutation, leading to a computational overhead that limits the applicability of the method to datasets of small-to-moderate size. Building upon the work in [18], an alternative approach to apply permutation testing in significant pattern mining was proposed in [19]. Unlike the previous method, this approach processes all resampled datasets simultaneously, requiring to enumerate patterns only a single time. In practice, this leads to a drastic reduction in runtime and memory usage that allows scaling-up the method to considerably larger datasets.

Pseudocode describing the approach in [19] is shown in Algorithm 8.4. The skeleton of the method closely parallels Algorithm 8.2. The search space $\mathcal{M}$ of all candidate patterns is explored in the same way: Recursively traversing a pattern enumeration tree that satisfies $\mathcal{S}' \in \text{Children}(\mathcal{S}) \Rightarrow \mathcal{S} \subseteq \mathcal{S}'$ depth-first. The algorithm begins by initializing the estimate $\hat{\delta}_{\text{perm}}$ of the adjusted significance threshold to 1 (Line 2). Next, in Lines 3–5, for each of the j_p resampled datasets, the algorithm precomputes the random permutation of the class labels and initializes the estimate $\hat{p}_{\min}^{(k)}$ of the most significant P-value to 1. After the initialization phase, the algorithm proceeds to start the pattern enumeration procedure at the root of the tree (Line 6). For each pattern $\mathcal{S}$ visited during the traversal of the enumeration tree, Algorithm 8.4 first computes the minimum attainable P-value $p_{\mathcal{S},\min}$ in Line 10. The algorithm then proceeds differently depending on the testability of pattern $\mathcal{S}$.

If pattern $\mathcal{S}$ is testable at level $\hat{\delta}_{\text{perm}}$ then, for each of the j_p resampled datasets $\widetilde{D}^{(k)}$, the algorithm computes the P-value $p_{\mathcal{S}}^{(k)}$ and updates the estimate $\hat{p}_{\min}^{(k)}$ of the most significant P-value for the k-th resampled dataset (Lines 12–14). Next, in Line 15, an estimate $\widehat{\text{FWER}}(\hat{\delta}_{\text{perm}})$ of the FWER at level $\hat{\delta}_{\text{perm}}$ is obtained using the estimates $\hat{p}_{\min}^{(k)}$ of the most significant P-value for each resampled dataset. Since the enumeration process is not yet completed, $\hat{p}_{\min}^{(k)} \geq p_{\min}^{(k)}$ leading to $\frac{1}{j_p}\sum_{k=1}^{j_p} \mathbb{1}\left[\hat{p}_{\min}^{(k)} \leq \hat{\delta}_{\text{perm}}\right] = \widehat{\text{FWER}}(\hat{\delta}_{\text{perm}})$ being a lower bound of $\widehat{\text{FWER}}(\hat{\delta}_{\text{perm}}) = \frac{1}{j_p}\sum_{k=1}^{j_p} \mathbb{1}\left[p_{\min}^{(k)} \leq \hat{\delta}_{\text{perm}}\right]$. Thus, if we have $\widehat{\text{FWER}}(\hat{\delta}_{\text{perm}}) \geq \alpha$, then $\widehat{\text{FWER}}(\hat{\delta}_{\text{perm}}) \geq \alpha$ as well, implying that the FWER condition is violated. In Lines 16–18, Algorithm 8.4 checks this condition and, if found to be violated, decreases he estimate $\hat{\delta}_{\text{perm}}$ of the adjusted significance threshold until the FWER condition is satisfied again.

On the contrary, if pattern $\mathcal{S}$ is untestable at level $\hat{\delta}_{\text{perm}}$, it will not affect the values of the FWER estimator $\widehat{\text{FWER}}(\delta) = \frac{1}{j_p}\sum_{k=1}^{j_p} \mathbb{1}\left[\hat{p}_{\min}^{(k)} \leq \delta\right]$ for any $\delta \leq \hat{\delta}_{\text{perm}}$. Consequently, the computation of the P-values $p_{\mathcal{S}}^{(k)}$ for $k = 1,\ldots,j_p$ can be skipped, as well as the update of the FWER estimator $\widehat{\text{FWER}}(\hat{\delta}_{\text{perm}})$. The fact that untestable patterns cannot modify the value of $\widehat{\text{FWER}}(\delta)$ follows from the definition of testability. If $p_{\mathcal{S},\min} > \hat{\delta}_{\text{perm}}$ and $\hat{p}_{\min}^{(k)} > \hat{\delta}_{\text{perm}}$, then $\min(\hat{p}_{\min}^{(k)}, p_{\mathcal{S}}^{(k)}) > \hat{\delta}_{\text{perm}} \geq \delta_{\text{perm}}$. Hence, an untestable pattern $\mathcal{S}$ is irrelevant as far as the FWER is concerned, also for permutation testing.

Search space pruning is fundamental to the computational feasibility of Algorithm 8.4, as is for other significant pattern mining algorithms. As shown in Section 8.3.2, when Pearson's χ^2 test or Fisher's exact test are used, descendants $\mathcal{S}'$ of an untestable pattern $\mathcal{S}$ will also be untestable provided that $r_{\mathcal{S}} \leq \min(n_1, n - n_1)$. Reiterating the argument above, this implies that those descendants cannot affect the value of $\widehat{\text{FWER}}(\delta)$ and, thus, can be pruned from the search space. In other words, the pruning condition of Algorithm 8.4 is identical to that of Algorithm 8.2. Finally, Lines 20–21 continue the traversal of the tree recursively, visiting the children of patterns for which the pruning condition does not apply.

As was the case for Algorithm 8.2, $\hat{\delta}_{\text{perm}}$ progressively decreases as patterns are enumerated, leading to less patterns being testable and the pruning condition becoming more stringent. Eventually, the algorithm ends after all patterns have either been pruned or visited. Due to the way the algorithm proceeds, once that occurs, $\hat{\delta}_{\text{perm}} = \delta_{\text{perm}}$, allowing the value to be returned in Line 8. Once the adjusted significance threshold δ_{perm} has been obtained, Algorithm 8.3 can be used to retrieve all patterns $\mathcal{S}$ that are statistically significant at level δ_{perm}.

Compared to Algorithm 8.2, the approach proposed by [19] and described in Algorithm 8.4 will exhibit more statistical power, due the use of permutation-testing to obtain a better approximation to the FWER. However, it is a more computationally demanding approach, as j_p P-values need to be computed for each pattern deemed testable during the pattern enumeration procedure. In short, Algorithm 8.2 allows to trade-off computational complexity for statistical power, an option that might be desirable in applications where signals are too weak to be detected by Algorithm 8.2.

8.5 Accounting for a Categorical Covariate

The need to incorporate into the model covariate factors that might have a confounding effect is an ubiquitous problem in computational biology and clinical data analysis. By neglecting to account for such covariates, an algorithm might find many spurious patterns whose association with the class labels is entirely mediated by confounding.

Let $G_\mathcal{S}(X)$ denote the binary random variable indicating the occurrence of pattern $\mathcal{S}$ in an input sample X, Y the binary class label and C be a random variable corresponding to a covariate factor taking values in a domain $\mathcal{C}$. Mathematically, the aforementioned situation occurs when:

1. $G_\mathcal{S}(X)$ and Y are *marginally* statistically associated, i.e., $G_\mathcal{S}(X) \not\perp Y$. As discussed in Section 8.2.2, this is the case if and only if $\exists (x,y) \in \{0,1\}^2$ such that $\Pr(G_\mathcal{S}(X) = g_\mathcal{S}(x), Y = y) \neq \Pr(G_\mathcal{S}(X) = g_\mathcal{S}(x))\Pr(Y = y)$.
2. $G_\mathcal{S}(X)$ and Y are *conditionally independent given C*, i.e. $G_\mathcal{S}(X) \perp Y \mid C$. This occurs if and only if $\Pr(G_\mathcal{S}(X) = g_\mathcal{S}(x), Y = y \mid C = c) = \Pr(G_\mathcal{S}(X) = g_\mathcal{S}(x) \mid C = c)\Pr(Y = y \mid C = c) \ \forall c \in \mathcal{C}$.

Intuitively, condition (1) above implies that $G_\mathcal{S}(X)$ and Y appear to be statistically associated in the absence of information about C, while condition (2) implies that, once the value c taken by the covariate C is known, $G_\mathcal{S}(X)$ carries no further information about Y and, therefore, can be discarded. In most applications, patterns $\mathcal{S} \in \mathcal{M}$ satisfying conditions (1) and (2) are spurious findings that should not be retrieved by a mining algorithm.

These patterns will be of little use to the practitioner, as they provide no additional information about the class membership of an input sample beyond the information that is already contained in the covariate. In many applications, the covariates are quantities that can be measured more easily than the input samples $x \in \mathcal{X}$. For example, while x might represent a set of gene expression profiles or be a representation of a patient's genotype based on a set of single nucleotide polymorphisms, the covariate C often represents simple information such as age, gender, or genetic ancestry. Therefore, from a practical point of view, if a marginally associated pattern $\mathcal{S} \in \mathcal{M}$ is redundant with such a covariate, it might be preferable to simply make use of the covariate when trying to predict the class label Y.

Moreover, not only are patterns satisfying conditions (1) and (2) of little practical use but, in many occasions, they might represent misleading associations. A particularly common example are spurious associations between genotype and phenotype which arise in genome-wide association studies due to population structure [48]. Often, a phenotype might be strongly associated with the genetic ancestry of an individual, which obviously is itself associated with an individual's genotype. Therefore, if population structure is unaccounted for in a genome-wide association study containing individuals with diverse genetic ancestries, a large number of apparently significant patterns might be retrieved. However, a practitioner might later find that, in fact, most of these patterns simply correspond to genotypic motifs that differ among individuals with different genetic ancestries and contain no additional information about the phenotype.

The effect of confounding is illustrated in Figure 8.10. In this example, the same significant itemset mining dataset depicted in Figure 8.1 is shown. Patterns S_1 and S_2 are both marginally associated with the class labels, being clearly enriched among samples of class $y = 1$. However, Figure 8.10 incorporates a factor not present in the original example: a categorical covariate C with $k = 2$ categories, representing the genetic ancestry of a sample. The inclusion of C dramatically changes the interpretation of the associations that patterns S_1 and S_2 represent. While the occurrence of pattern S_1 still carries additional information not present in the covariate, pattern S_2 can be seen to be entirely redundant with the covariate C. Therefore, following our discussion above, pattern S_2 might be considered a spurious association that should ideally not be retrieved.

Most significant pattern mining algorithms, including Algorithm 8.2 in Section 8.3 and Algorithm 8.4 in Section 8.4, are unable to account for covariates. Hence, they are prone to discover many spurious patterns due to confounding, limiting their applicability in computational biology and clinical data analysis. In order to solve this problem, a significant pattern mining approach able to account for the effect of covariates needs to be used. All methods that have been discussed in this chapter so far aim at finding all patterns $S \in \mathcal{M}$ that are (marginally) statistically associated with the class labels, i.e., look for the set of patterns $\{S \in \mathcal{M} \mid G_S(X) \not\perp Y\}$. However, an approach able to correct for the effect of a covariate C would aim to find the set of patterns $\{S \in \mathcal{M} \mid G_S(X) \not\perp Y \mid C\}$ instead. In particular, a pattern $S \in \mathcal{M}$ satisfying conditions (1) and (2) above would *not* be retrieved by the latter formulation yet it would be deemed significant by the former.

Recently, a novel significant pattern mining algorithm able to account for the effect of a categorical covariate C has been proposed [20]. In a follow-up study [49], the techniques proposed in [20] have been used to develop a method able to aggregate weak effects in genome-wide association studies while correcting for confounding effects due to population structure. Unlike existing approaches, such as burden tests [50], the approach in [49] uses significant pattern mining to test *all* genomic regions, regardless of starting position and size. Empirically, this leads to a gain of statistical power over burden tests, whose statistical performance is shown to be sensitive to misspecification of the size of the genomic regions to be tested.

The approach in [20] follows closely Algorithm 8.2, described in Section 8.3. However, in order to incorporate the covariate C into the model, it replaces Pearson's χ^2

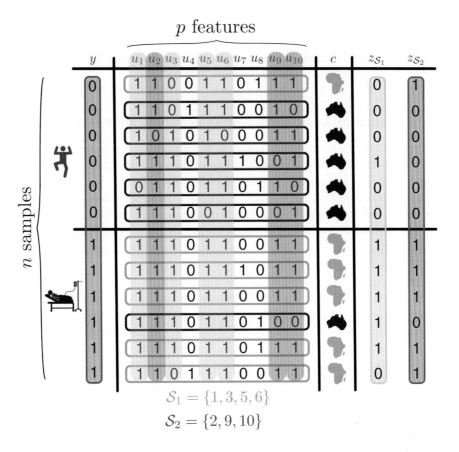

Figure 8.10: A simple illustration of the effect of confounding in a significant itemset mining dataset. A categorical covariate C with $k = 2$ categories (orange and purple) has been included. Two patterns, S_1 and S_2, are marginally associated with the class labels Y. However, only pattern S_1 remains associated with Y given the covariate C. On the contrary, the high-order feature interaction indexed by pattern S_2 is active if and only if the sample is of African ancestry (orange). Thus, pattern S_2 carries no information about Y given that the value of C is known, i.e., S_2 is conditionally independent of Y given C.

test or Fisher's exact test by the Cochran–Mantel–Haenszel (CMH) test [51]. Unlike the former two, the CMH test allows asessing the *conditional* association of two binary random variables $G_S(X)$ and Y given a categorical random variable C with k categories, making it ideal for this task. However, replacing the test statistic has implications for Tarone's method and its integration into the pattern mining algorithm. In particular, it is necessary to:

1. Prove that a minimum attainable P-value exists for the CMH test and derive an efficient expression to compute it.
2. Propose a novel search space pruning criterion that applies to the CMH test.

The remainder of this section is organized as follows. In Section 8.5.1, we introduce the Cochran–Mantel–Haenszel (CMH) test in detail. Next, in Section 8.5.2, we discuss the derivation of a minimum attainable P-value for the CMH test. Finally, Section 8.5.3 introduces a valid pruning condition for the CMH test.

8.5.1 Conditional Association Testing in Significant Pattern Mining

In Section 8.2.2, 2×2 contingency tables were introduced as a way to represent the joint distribution of two binary random variables or its empirical approximation based on counts derived from n i.i.d. samples. Based on these counts, two test statistics to assess the statistical association between two binary random variables were proposed: Pearson's χ^2 test and Fisher's exact test. These were proposed in order to test the (marginal) association between the binary random variable $G_S(X)$ representing the occurrence of a pattern $S \in \mathcal{M}$ in an input sample X and the class labels Y, based on an input dataset $\mathcal{D} = \{(x_i, y_i)\}_{i=1}^{n}$.

In this section, however, we are given a dataset $\mathcal{D} = \{(x_i, y_i, c_i)\}_{i=1}^{n}$ with n i.i.d. samples $x \in \mathcal{X}$ belonging to one of two classes $y \in \{0, 1\}$. Additionally, now each of the n samples is also tagged with a categorical covariate $c \in \{1, 2, \ldots, k\}$, where k is the number of distinct categories that random variable C can take. Unlike in the previous scenario, the new goal is to test the conditional association between $G_S(X)$ and Y given C. This is precisely what the Cochran–Mantel–Haenszel (CMH) test was designed for.

Intuitively, the CMH test can be seen as a way to tackle this problem by reducing it to a set of k instances of Pearson's χ^2 test and then combining the k resulting statistics appropriately. By definition, $G_S(X)$ and Y are conditionally independent given C and, therefore, *not* conditionally associated given C, if $\Pr(G_S(X) = g_S(x), Y = y \mid C = c) = \Pr(G_S(X) = g_S(x) \mid C = c)\Pr(Y = y \mid C = c)$ $\forall c \in \{1, 2, \ldots, k\}$. For each $c = 1, \ldots, k$, define $\mathcal{D}(c) = \{(x_i, y_i) \in \mathcal{D} \mid c_i = c\}$ to be the set of input samples in $\mathcal{D}$ for which the categorical covariate takes value c. The (unknown) joint distribution $\Pr(G_S(X) = g_S(x), Y = y \mid C = c)$ can be empirically approximated using counts derived from $\mathcal{D}(c)$:

Variables	$g_S(x) = 1$	$g_S(x) = 0$	Row totals
$y = 1$	$a_{S,c}$	$b_{S,c}$	$n_{1,c}$
$y = 0$	$d_{S,c}$	$c_{S,c}$	$n_{0,c}$
Col. totals	$r_{S,c}$	$q_{S,c}$	n_c

The interpretation of these counts is similar to the unconditional case described in Section 8.2.2. For instance, $a_{S,c}$ is the number of samples in $\mathcal{D}(c)$ belonging to class $y = 1$ for which pattern S occurs or, equivalently, the number of samples in $\mathcal{D}$ belonging to class $y = 1$ for which pattern S occurs and the covariate takes value c. Consequently, an empirical approximation to $\Pr(G_S(X) = 1, Y = 1 \mid C = c)$ could be obtained as $a_{S,c}/n_c$ for each $c \in \{1, 2, \ldots, k\}$. The remaining counts can be described analogously.

As a consequence of Proposition 8.1 in Section 8.2.2, if $G_S(X)$ is conditionally independent of Y given C, the random variable $A_{S,c}$ given margins $n_{1,c}$ and $r_{S,c}$ and sample size n_c follows a hypergeometric distribution with parameters n_c, $n_{1,c}$ and $r_{S,c}$

for all $c \in \{1, 2, \ldots, k\}$. Besides, under the assumption that all n samples in $\mathcal{D}$ are obtained as i.i.d. draws, it follows that $A_{\mathcal{S},c}$ is statistically independent of $A_{\mathcal{S},c'}$ for any $c \neq c'$, since $\mathcal{D}(c) \cap \mathcal{D}(c') = \emptyset$. Paralleling the derivation of Pearson's χ^2 test described in Section 8.2.2, the following Z-score can be proposed as a way to additively aggregate the individual Z-scores of the k 2×2 contingency tables:

$$Z_{\text{cmh}}(\mathbf{a}_{\mathcal{S}} \mid \mathbf{n}, \mathbf{n}_1, \mathbf{r}_{\mathcal{S}}) = \frac{\sum_{c=1}^{k} a_{\mathcal{S},c} - \mathbb{E}[a_{\mathcal{S},c} \mid R_{\mathcal{S},c} = r_{\mathcal{S},c}, N_{1,c} = n_{1,c}, H_0]}{\sqrt{\sum_{c=1}^{k} \text{Var}[a_{\mathcal{S},c} \mid R_{\mathcal{S},c} = r_{\mathcal{S},c}, N_{1,c} = n_{1,c}, H_0]}}, \quad (8.28)$$

where the term in the denominator follows from the fact that the variance of a sum of independent random variables equals the sum of the variances of each random variable participating in the sum. To simplify the notation, we introduced the vectors $\mathbf{a}_{\mathcal{S}} = (a_{\mathcal{S},1}, \ldots, a_{\mathcal{S},k})$, $\mathbf{n} = (n_1, \ldots, n_k)$, $\mathbf{n}_1 = (n_{1,1}, \ldots, n_{1,k})$ and $\mathbf{r}_{\mathcal{S}} = (r_{\mathcal{S},1}, \ldots, r_{\mathcal{S},k})$, which contain the values of $a_{\mathcal{S},c}$, n_c, $n_{1,c}$ and $r_{\mathcal{S},c}$ for all k 2×2 contingency tables. The final expression for the CMH test can be obtained by squaring this Z-score and plugging in the values of $\mathbb{E}[a_{\mathcal{S},c} \mid R_{\mathcal{S},c} = r_{\mathcal{S},c}, N_{1,c} = n_{1,c}, H_0]$ and $\text{Var}[a_{\mathcal{S},c} \mid R_{\mathcal{S},c} = r_{\mathcal{S},c}, N_{1,c} = n_{1,c}, H_0]$ as the mean and variance of a hypergeometric distribution with parameters n_c, $n_{1,c}$ and $r_{\mathcal{S},c}$:

$$T_{\text{cmh}}(\mathbf{a}_{\mathcal{S}} \mid \mathbf{n}, \mathbf{n}_1, \mathbf{r}_{\mathcal{S}}) = \frac{\left(\sum_{c=1}^{k} a_{\mathcal{S},c} - r_{\mathcal{S},c} \frac{n_{1,c}}{n_c}\right)^2}{\sum_{c=1}^{k} \frac{r_{\mathcal{S},c}}{n_c} \frac{n_c - r_{\mathcal{S},c}}{n_c} \frac{n_c - n_{1,c}}{n_c - 1} n_{1,c}}. \quad (8.29)$$

$T_{\text{cmh}}(\mathbf{a}_{\mathcal{S}} \mid \mathbf{n}, \mathbf{n}_1, \mathbf{r}_{\mathcal{S}})$ therefore aggregates evidence against the null hypothesis H_0 that $G_{\mathcal{S}}(X)$ is conditionally independent of Y given C across all k 2×2 contingency tables. Large values of the test statistic are less likely to occur if the null hypothesis H_0 holds.

An approximation to the null distribution of the CMH test statistic can be obtained using the same arguments employed to derive an approximation to the null distribution of Pearson's χ^2 test statistic. For a sufficiently large sample size n, the null distribution of $Z_{\text{cmh}}(\mathbf{a}_{\mathcal{S}} \mid \mathbf{n}, \mathbf{n}_1, \mathbf{r}_{\mathcal{S}})$ will be approximately a standard Gaussian. Thus, the null distribution of $T_{\text{cmh}}(\mathbf{a}_{\mathcal{S}} \mid \mathbf{n}, \mathbf{n}_1, \mathbf{r}_{\mathcal{S}})$ can be approximated as a χ_1^2 distribution and the corresponding two-tailed P-value can be obtained as:

$$p_{\text{cmh}}(\mathbf{a}_{\mathcal{S}} \mid \mathbf{n}, \mathbf{n}_1, \mathbf{r}_{\mathcal{S}}) = 1 - F_{\chi_1^2}\left(T_{\text{cmh}}(\mathbf{a}_{\mathcal{S}} \mid \mathbf{n}, \mathbf{n}_1, \mathbf{r}_{\mathcal{S}})\right), \quad (8.30)$$

where $F_{\chi_1^2}(\bullet)$ is the cumulative density function of a χ_1^2 distribution.

In Figure 8.11, we illustrate the result of applying the CMH test to assess the statistical significance of pattern $\mathcal{S}_2$ in the dataset previously shown in Figure 8.10. The 2×2 contingency table built using all samples in the dataset $\mathcal{D}$, shown at the top of the figure (light blue), clearly suggests a (marginal) association between $G_{\mathcal{S}_2}(X)$ and Y. Indeed, if Pearson's χ^2 test is used to compute a P-value, one obtains $p_{\text{Pearson}}(a_{\mathcal{S}_2} \mid n, n_1, r_{\mathcal{S}_2}) = 0.021$, a rather significant result taking into account that the sample size is only $n = 12$. However, if this contingency table is split into $k = 2$ distinct tables according to the value of the categorical covariate, leading to the orange and purple contingency tables shown at the bottom of the figure, this association can be seen to disappear. In particular, those tables are so extreme that only one outcome for $a_{\mathcal{S}_2,c}$ is possible in each case, i.e., $a_{\mathcal{S}_2,c,\min} = a_{\mathcal{S}_2,c,\max}$ holds for both. Consequently, the CMH test leads to an entirely non-significant P-value $p_{\text{cmh}}(\mathbf{a}_{\mathcal{S}_2} \mid \mathbf{n}, \mathbf{n}_1, \mathbf{r}_{\mathcal{S}_2}) = 1$, successfully

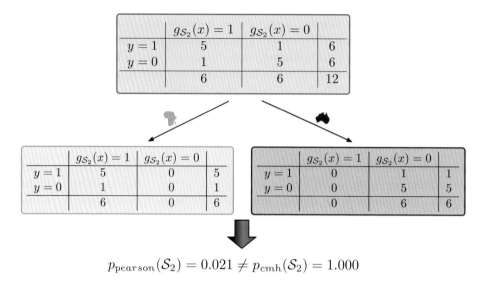

Figure 8.11: Result of using the CMH test to quantify the statistical association between class labels Y and the occurrence of pattern S_2 in the input samples, given the categorical covariate C. The light blue contingency table shown at the top of the figure is built using all samples in the dataset $\mathcal{D}$, while the two contingency tables shown at the bottom of the figure are obtained from the stratified datasets $\mathcal{D}(1)$ and $\mathcal{D}(2)$. In this case, category $c = 1$ has been associated with samples of African ancestry (orange) and category $c = 2$ with samples of Australian ancestry (purple).

correcting the confounding effect of the covariate. In contrast, it can be readily verified that if this analysis is repeated for pattern S_1, which is not affected by confounding in the example of Figure 8.10, the CMH test still returns a rather significant P-value, $p_{cmh}(a_{S_1} \mid \mathbf{n}, \mathbf{n_1}, \mathbf{r}_{S_1}) = 0.029$.

8.5.2 Deriving the Minimum Attainable P-value for the CMH Test

As Pearson's χ^2 test and Fisher's exact test, the CMH test is based on discrete data and, as a consequence, there is a finite number of distinct values it can attain. As discussed in Section 8.2.3, this property is essential for the application of Tarone's method, as it makes it possible to obtain a minimum attainable P-value strictly larger than zero. This leads to the following proposition:

Proposition 8.6 (Minimum attainable P-value for the Cochran–Mantel–Haenszel (CMH) test) *Define* $\mathbf{a}_{S,min} = (a_{S,1,min}, a_{S,2,min}, \ldots, a_{S,k,min})$ *and* $\mathbf{a}_{S,max} = (a_{S,1,max}, a_{S,2,max}, \ldots, a_{S,k,max})$, *where* $a_{S,c,min} = \max(0, r_{S,c} - (n_c - n_{1,c}))$ *and* $a_{S,c,max} = \min(n_{1,c}, r_{S,c})$ *for each* $c = 1, 2, \ldots, k$. *Then, the minimum attainable P-value for the CMH test is given by:*

$$p_{S,min} = 1 - F_{\chi_1^2}\left(\max\left(T_{cmh}(\mathbf{a}_{S,min} \mid \mathbf{n}, \mathbf{n_1}, \mathbf{r}_S), T_{cmh}(\mathbf{a}_{S,max} \mid \mathbf{n}, \mathbf{n_1}, \mathbf{r}_S)\right)\right). \quad (8.31)$$

In particular, this implies that $p_{S,\min}$ can be evaluated in $O(k)$ time, where k is the number of categories for the categorical covariate C.

Proof Equation 8.29 can be rewritten as:

$$T_{\text{cmh}}(\mathbf{a_S} \mid \mathbf{n}, \mathbf{n_1}, \mathbf{r_S}) = \frac{\left(a_{S,\text{tot}} - \sum_{c=1}^{k} r_{S,c} \frac{n_{1,c}}{n_c}\right)^2}{\sum_{c=1}^{k} \frac{r_{S,c}}{n_c} \frac{n_c - r_{S,c}}{n_c} \frac{n_c - n_{1,c}}{n_c - 1} n_{1,c}}, \qquad (8.32)$$

where $a_{S,\text{tot}} = \sum_{c=1}^{k} a_{S,c}$ has been introduced. As described in Section 8.2.3, given fixed margins $n_{1,c}, r_{S,c}$ and sample size n_c, each count $a_{S,c}$ can only take values in the set $a_{S,c} \in [a_{S,c,\min}, a_{S,c,\max}]$, where $a_{S,c,\min} = \max(0, r_{S,c} - (n_c - n_{1,c}))$ and $a_{S,c,\min} = \min(n_{1,c}, r_{S,c})$. Thus, $a_{S,\text{tot}} \in [a_{S,\text{tot},\min}, a_{S,\text{tot},\max}]$, where $a_{S,\text{tot},\min} = \sum_{c=1}^{k} a_{S,c,\min}$ and $a_{S,\text{tot},\max} = \sum_{c=1}^{k} a_{S,c,\max}$. According to Equation 8.32, $T_{\text{cmh}}(\mathbf{a_S} \mid \mathbf{n}, \mathbf{n_1}, \mathbf{r_S})$ will be maximized and, hence, $p_{\text{cmh}}(\mathbf{a_S} \mid \mathbf{n}, \mathbf{n_1}, \mathbf{r_S})$ minimized, either when $a_{S,\text{tot}} = a_{S,\text{tot},\min}$ or $a_{S,\text{tot}} = a_{S,\text{tot},\max}$. Equivalently, this occurs when $\mathbf{a_S} = \mathbf{a_{S,\min}}$ or $\mathbf{a_S} = \mathbf{a_{S,\max}}$, concluding the proof. □

8.5.3 A Search Space Pruning Condition for the CMH Test

In Section 8.3.2, a search space pruning condition valid for Pearson's χ^2 test and Fisher's exact test was derived. The fundamental principle of that pruning criterion is that, for fixed n_1 and n, the minimum attainable P-value $p_{S,\min}$ for these test statistics is a monotonically decreasing function of r_S in the range $r_S \in [0, \min(n_1, n - n_1)]$. This implies that, if a pattern S is untestable at level δ and satisfies $r_S \leq \min(n_1, n - n_1)$, all its descendants $S' \in \text{Children}(S)$ in the pattern enumeration tree will be untestable at level δ as well and can be pruned from the search space. In this section, we will propose an equivalent pruning condition for the CMH test.

The first key observation is that, for fixed $\mathbf{n}$ and $\mathbf{n_1}$, the minimum attainable P-value for the CMH test is a multivariate function of k variables, $r_{S,1}, r_{S,2}, \ldots, r_{S,k}$. We will denote this by $p_{S,\min}(\mathbf{r_S})$, where the dependence of $p_{S,\min}$ on $\mathbf{n}$ and $\mathbf{n_1}$ is kept implicit to avoid cluttering the notation. By applying the apriori property of pattern mining (see Proposition 8.2), it can be shown that $S' \in \text{Children}(S) \Rightarrow r_{S',c} \leq r_{S,c} \forall c = 1, 2, \ldots, k$. Using identical arguments as in Section 8.3.2, a majority of patterns $S \in \mathcal{M}$ in the search space of candidate patterns will be relatively rare, satisfying $r_{S,c} \leq \min(n_{1,c}, n_c - n_{1,c})$ for all $c = 1, 2, \ldots, k$. This naturally leads to the fundamental question of all $c = 1, 2, \ldots, k$. This naturally leads to the fundamental question of whether $\mathbf{r_{S'}} \leq \mathbf{r_S} \Rightarrow p_{S,\min}(\mathbf{r_{S'}}) \geq p_{S,\min}(\mathbf{r_S})$ given that $r_{S,c} \leq \min(n_{1,c}, n_c - n_{1,c}) \forall c = 1, 2, \ldots, k$.[14] If the answer to this question was affirmative, a pruning condition entirely analogous to the one used for Pearson's χ^2 test and Fisher's exact test would also be valid for the CMH test.

Unfortunately, it is easy to come up with non-pathological counterexamples which show that this property does not hold in general. As an example, Figure 8.12 depicts the minimum attainable P-value $p_{S,\min}$ for the CMH test in a problem with $k = 2$

---

[14] The notation $\mathbf{r_{S'}} \leq \mathbf{r_S}$ was introduced as shorthand for $r_{S',c} \leq r_{S,c} \forall c = 1, 2, \ldots, k$.

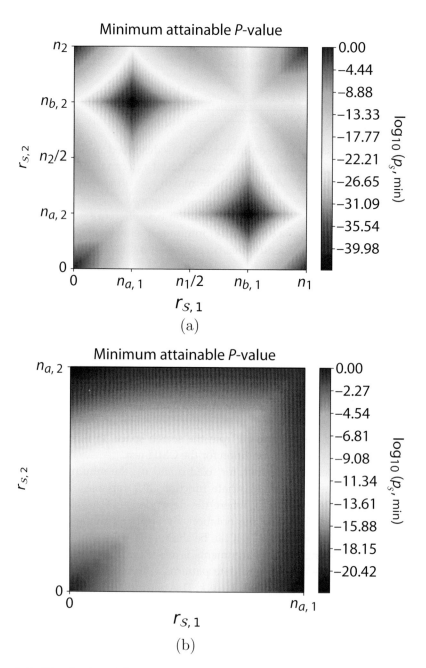

Figure 8.12: Visualization of the minimum attainable P-value $p_{\mathcal{S},\min}$ for the CMH test in a problem with $k = 2$ categories for the covariate. In this example, $n_1 = n_2 = 100$ and $n_{1,1} = 25$, $n_{1,2} = 75$. Thus, category $c = 1$ has a class ratio of $1/4$ while category $c = 2$ has a class ratio of $3/4$. (a) Minimum attainable P-value $p_{\mathcal{S},\min}(\mathbf{r_{\mathcal{S}}})$ over the entire domain $[\![0, n_1]\!] \times [\![0, n_2]\!]$. (b) Minimum attainable P-value $p_{\mathcal{S},\min}(\mathbf{r_{\mathcal{S}}})$ over the region $[\![0, n_{a,1}]\!] \times [\![0, n_{a,2}]\!]$, where $n_{a,1} = \min(n_{1,1}, n_1 - n_{1,1})$, $n_{a,2} = \min(n_{1,2}, n_2 - n_{1,2})$, $n_{b,1} = \max(n_{1,1}, n_1 - n_{1,1})$ and $n_{b,2} = \max(n_{1,2}, n_2 - n_{1,2})$.

categories for the covariate. In particular, Figure 8.12 (b) illustrates the behavior of $p_{S,\min}$ as a function of $r_{S,1}$ and $r_{S,2}$ over the region $[\![0, \min(n_{1,1}, n_1 - n_{1,1})]\!] \times [\![0, \min(n_{1,2}, n_2 - n_{1,2})]\!]$. While the function is approximately monotonic when $r_{S,1}$ and $r_{S,2}$ are both sufficiently far from zero, $p_{S,\min}(\mathbf{r}_S)$ is not monotonically decreasing when one of its arguments is small enough, as can be appreciated from the level curves. This has profound implications for the development of a valid pruning criterion, as in principle there is no simple way to make a statement about the minimum attainable P-value $p_{S',\min}$ of a pattern $S' \in \text{Children}(S)$ based on $p_{S,\min}$ and $\mathbf{r}_S$ alone.

In order to solve this problem, the authors in [20] propose using a monotonically decreasing lower bound on the minimum attainable P-value as a surrogate in the pruning criterion. Borrowing their terminology, we refer to this surrogate as the *lower envelope* of the minimum attainable P-value.

Definition 8.7. Let $S \in \mathcal{M}$ be a pattern satisfying $r_{S,c} \leq \min(n_{1,c}, n_c - n_{1,c})$ $\forall c = 1, 2, \ldots, k$. The lower envelope of the minimum attainable P-value $p_{S,\min}$ is defined as:

$$\tilde{p}_{S,\min} = \min_{S' \supseteq S} p_{S,\min}. \tag{8.33}$$

Note that, as a consequence of the apriori property of pattern mining, $\tilde{p}_{S,\min}$ can be equivalently defined as:

$$\tilde{p}_{S,\min}(\mathbf{r}_S) = \min_{\mathbf{r}_{S'} \leq \mathbf{r}_S} p_{S',\min}(\mathbf{r}_{S'}), \tag{8.34}$$

where the dependence of the minimum attainable P-value $p_{S,\min}$ and its lower envelope $\tilde{p}_{S,\min}$ on $\mathbf{r}_S$ has been made explicit.

Intuitively, $\tilde{p}_{S,\min}(\mathbf{r}_S)$ is defined as the tightest lower bound on the minimum attainable P-value $p_{S,\min}$, hence the term "lower envelope," that satisfies $\mathbf{r}_{S'} \leq \mathbf{r}_S \Rightarrow \tilde{p}_{S,\min}(\mathbf{r}_{S'}) \geq \tilde{p}_{S,\min}(\mathbf{r}_S)$. This notion is illustrated in Figure 8.13 with a conceptual example.

The lower envelope $\tilde{p}_{S,\min}$ is a lower bound of the minimum attainable P-value $p_{S,\min}$ by construction. Also, the fact that $\tilde{p}_{S,\min}(\mathbf{r}_S)$ is monotonically decreasing on $\mathbf{r}_S$, i.e., that $\mathbf{r}_{S'} \leq \mathbf{r}_S \Rightarrow \tilde{p}_{S,\min}(\mathbf{r}_{S'}) \geq \tilde{p}_{S,\min}(\mathbf{r}_{S'}) \geq \tilde{p}_{S,\min}(\mathbf{r}_S)$, is a direct consequence of the way $\tilde{p}_{S,\min}$ is defined. If $\mathbf{r}_{S'} \leq \mathbf{r}_S$, then $\tilde{p}_{S,\min}(\mathbf{r}_{S'}) = \min_{\mathbf{r}_{S''} \leq \mathbf{r}_{S'}} p_{S'',\min}(\mathbf{r}_{S''}) \leq \min_{\mathbf{r}_{S''} \leq \mathbf{r}_S} p_{S'',\min}(\mathbf{r}_{S''}) = p_{S'',\min}(\mathbf{r}_{S''}) = \tilde{p}_{S,\min}(\mathbf{r}_S)$ since $\mathbf{r}_{S''} \leq \mathbf{r}_{S'} \subseteq \mathbf{r}_{S''} \leq \mathbf{r}_S$. These two properties of $\tilde{p}_{S,\min}$ allow proposing the following search space pruning criterion for the CMH test.

Proposition 8.8 (*Pruning criterion for the Cochran–Mantel–Haenszel (CMH) test*) Let $S \in \mathcal{M}$ be a pattern satisfying:

1. $\tilde{p}_{S,\min} > \hat{\delta}_{tar}$, i.e., the lower envelope of the minimum attainable P-value is larger than $\hat{\delta}_{tar}$
2. $r_{S,c} \leq n_{a,c}$ $\forall c \in \{1, \ldots, k\}$, where $n_{a,c} = \min(n_{1,c}, n_c - n_{1,c})$

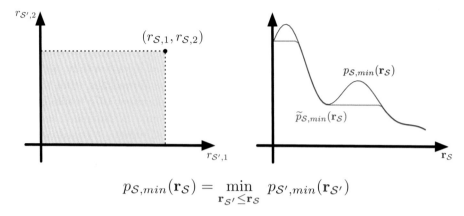

$$p_{S,\min}(\mathbf{r}_S) = \min_{\mathbf{r}_{S'} \leq \mathbf{r}_S} p_{S',\min}(\mathbf{r}_{S'})$$

Figure 8.13: Illustration of the lower envelope $\widetilde{p}_{S,\min}$ of the minimum attainable P-value $p_{S,\min}$. To evaluate $\widetilde{p}_{S,\min}(\mathbf{r}_S)$, the minimum of $p_{S',\min}(\mathbf{r}_{S'})$ over the region $\mathbf{r}_{S'} \leq \mathbf{r}_S$ needs to be computed. In this example, this corresponds to minimizing $p_{S',\min}(\mathbf{r}_{S'})$ over the region shaded in blue on the left of the figure. As a result of this definition, $\widetilde{p}_{S,\min}(\mathbf{r}_S)$ is the tightest lower bound on $p_{S,\min}(\mathbf{r}_S)$ that satisfies $\mathbf{r}_{S'} \leq \mathbf{r}_S \Rightarrow \widetilde{p}_{S,\min}(\mathbf{r}_{S'}) \geq \widetilde{p}_{S,\min}(\mathbf{r}_S)$. This is illustrated, along a one-dimensional slice, on the right of the figure.

Then, $p_{S',\min} > \hat{\delta}_{tar} \geq \delta_{tar}$ for all $S' \in$ Children(S). Hence, all descendants of pattern S can be pruned from the enumeration tree. In conclusion, when using the CMH test, pruning_condition($S, \hat{\delta}_{tar}$) in Line 15 of Algorithm 8.2 evaluates to true if $r_{S,c} \leq n_{a,c}$ $\forall c \in \{1, \ldots, k\}$ and $\widetilde{p}_{S,\min} > \hat{\delta}_{tar}$.

Proof If $S' \in$ Children(S), $\mathbf{r}_{S'} \leq \mathbf{r}_S$ by the apriori property of pattern mining. Hence, as a consequence of the monotonicity of $\widetilde{p}_{S,\min}(\mathbf{r}_S)$, $\widetilde{p}_{S',\min}(\mathbf{r}_{S'}) \geq \widetilde{p}_{S,\min}(\mathbf{r}_S)$. Since the lower envelope $\widetilde{p}_{S,\min}$ is a lower bound on the minimum attainable P-value $p_{S,\min}$ and on the minimum attainable P-value $p_{S,\min}$ and $\widetilde{p}_{S,\min} > \hat{\delta}_{tar}$ by assumption (i), it follows that $p_{S',\min} > \hat{\delta}_{tar}$. Finally, the fact that $\hat{\delta}_{tar} \geq \delta_{tar}$ at any point during the execution of Algorithm 8.2 concludes the proof. □

The resulting pruning condition for the CMH test is mostly analogous to the pruning criterion for Pearson's χ^2 test and Fisher's exact test in Proposition 8.5. However, the minimum attainable P-value $p_{S,\min}$ in condition (i), $p_{S,\min} > \hat{\delta}_{tar}$, is substituted by the lower envelope $\widetilde{p}_{S,\min}$. This allows circumventing the difficulties which arise as a consequence of the minimum attainable P-value $p_{S,\min}$ not being monotonically decreasing for the CMH test. Note that while the concept of lower envelope of the minimum attainable P-value was introduced in [20] in the context of the CMH test, the same idea could be used for other discrete tests statistics with a non-monotonic minimum attainable P-value function. This might help develop new applications of significant pattern mining by using domain-specific test statistics.

Nevertheless, an important aspect that remains to be considered is how to efficiently evaluate $\widetilde{p}_{S,\min}$. A straightforward application of Definition 8.7 implies that a combinatorial optimization problem needs to be solve each time $\widetilde{p}_{S,\min}$ is to be computed. Since the pruning condition is evaluated for every single pattern $S \in \mathcal{M}$ that is enumerated along the execution the algorithm, this would entail an unfeasible computational overhead. In [20] the authors propose a procedure, specific to the CMH test, that allows computing $\widetilde{p}_{\mathcal{S},\min}$ with $O(k \log k)$ operations, where k is the number of categories for the covariate. Consequently, $\widetilde{p}_{\mathcal{S},\min}$ can be evaluated almost as fast as $p_{\mathcal{S},\min}$ for the CMH test, making the entire idea practical. Nonetheless, deriving a more general procedure to efficiently evaluate $\widetilde{p}_{\mathcal{S},\min}$, regardless of the test statistic of choice, is still an open problem.

An alternative approach to account for the effect of a confounding categorical covariate in significant pattern mining was developed by [52]. The main difference with respect to the method described in this section lies in the choice of test statistic. Rather than making use of the CMH test, the approach in [52] uses an exact formulation of logistic regression. By exploiting the discrete nature of the data, this formulation allows to obtain exact P-values for the logistic regression model, avoiding the need for large sample approximations. For additional details, we refer the reader to the original article [52].

8.6 Summary and Outlook

Data analysis in computational biology and healthcare is a challenging task. In order to discover relevant biomarkers from omics and clinical data, the association of a large number of candidate markers with a phenotypic trait of interest needs to be assessed. The small sample size of most available datasets, relative to the number of features, has deterred the search for combinatorial feature interactions and other high-order patterns as potential biomarkers. In this chapter, significant pattern mining has been introduced as a family of machine learning algorithms to systematically explore such high-order structures and assess their statistical association. These techniques are a useful complement for the univariate analyses and linear models typically used for biomarker discovery, which cannot discover such non-linear signals in the data.

At the core of significant pattern mining is the concept of *testability* [38], introduced by R. E. Tarone in 1990. This is the fact that, when testing statistical associations in discrete data, there exists a minimum attainable P-value strictly larger than zero. Consequently, it can be shown that many candidate patterns cannot achieve significance or cause a false positive, leading to an elegant way to solve the statistical and computational challenges that significant pattern mining entails.

Recent advances in the field have focused on solving some of the core limitations of the significant pattern mining algorithm that first advocated the use of Tarone's method. One crucial aspect, not exploited by this algorithm, is the inherent redundancy between candidate patterns. Several novel methods have been proposed to model these dependencies between patterns, allowing to obtain a less conservative significance threshold and gain statistical power. Another fundamental limitation of the original approach is its inability to correct for observed confounding factors such as age, gender, or population structure. These are ubiquitous in biomedical data analysis applications and, if unaccounted for, could give rise to a large number of spurious discoveries. Recently developed significant pattern mining algorithms extend the original approach to incorporate such covariate factors into the model without sacrificing statistical power nor computational efficiency, effectively overcoming this limitation.

Table 8.1: A summary of available significant pattern mining tools.

Algorithm	Reference	Highlights
LAMP, LAMP2.0[1]	[15, 16]	First use of Tarone's method in significant itemset mining
LAMPLINK[2]	[53]	GWAS-oriented user interface for high-order epistasis search
MP-LAMP[3]	[54]	Distributed significant itemset mining algorithm for supercomputers
WY-light[4]	[19]	Permutation-testing based significant itemset and subgraph mining
FACS[4]	[20]	Corrects for a categorical covariate in significant itemset mining
FAIS, FastCMH[4]	[55, 49]	Region-wise GWAS, with optional correction for a categorical covariate

Notes:
[1] Available from http://a-terada.github.io/lamp.
[2] Available from http://a-terada.github.io/lamplink.
[3] Available from http://github.com/tsudalab/mp-lamp.
[4] Available from http://significant-patterns.org.

8.6.1 Software

There exist software tools implementing many of the approaches discussed in this chapter, allowing practitioners to use significant pattern mining to analyze their data (see Table 8.1).

Interested users can obtain both an implementation of LAMP [15], the first pattern mining algorithm using Tarone's method, as well as of LAMP 2.0 [16], a follow-up approach that improves its computational efficiency. Recently, the same authors have also implemented LAMPLINK [53], a toolbox that considerably extends the input/output format of these algorithms to greatly facilitate their use in genome-wide association studies (GWASs). MP-LAMP [54], a parallel implementation of the LAMP 2.0 algorithm, suitable for execution in distributed environments such as computing clusters, is also at the users' disposal. Similarly, users can download implementations of: (1) WY-light and FACS, the significant pattern mining algorithms described in [19] and [20] and (2) FAIS and FastCMH, methods that use significant pattern mining to aggregate weak effects within any genomic region in genome-wide association studies [55, 49]. All these software packages and their main defining characteristics are summarized in Table 8.1. The algorithm names in this table contain hyperlinks to their respective software repositories.

8.6.2 Outlook

The development of methods to correct for the multiple hypothesis testing problem in pattern mining, without resorting to arbitrary limits on pattern size, is a relatively recent field of study. As a consequence, many open problems and future research directions remain to be explored.

In this chapter, family-wise error rate (FWER) control has been the central tool to account and correct for the multiple hypothesis testing problem. However, statisticians have produced a wealth of alternative criteria for this purpose. For instance, false discovery rate (FDR) control [56] has been established as a popular alternative to FWER control in the biomedical field. Unlike FWER control, which aims at keeping the probability of producing *any* false positives low, FDR control aims at upper bounding the expected proportion of false discoveries among all reported findings. An FDR-controlling procedure is effectively allowed to produce false positives with high probability, as long as it also yields sufficiently many true positives. This makes FDR control considerably less stringent than FWER control, leading to a gain of statistical power that is useful in applications for which signals are weak, such as computational biology and healthcare, at the price of increasing the expected number of false positives. Most FDR-controlling procedures require assumptions about the joint distribution of all test statistics corresponding to null hypotheses, often in the form of independence or weak dependence assumptions. Due to the complex inter-dependencies between candidate patterns, these assumptions are hard to justify in the context of significant pattern mining, explaining the fact that most existing methods have focused on FWER control. Extending state-of-the-art significant pattern mining algorithms to perform FDR control is an exciting avenue for future research. Another recently proposed alternative to FWER control is selective inference [57], which aims at developing procedures to properly quantify statistical associations after having performed model selection. Recent work [58] has begun exploring how to apply principles of selective inference to pattern mining, opening yet another interesting research direction.

Discrete data is at the core of all significant pattern mining algorithms discussed in this chapter. While this is sufficient to cover many applications of interest, it would be desirable to develop alternative methods to find high-order feature interactions in datasets with continuous features. In some occasions, a real-valued feature can be binarized without losing too much information, allowing the use of existing significant pattern mining algorithms to analyse the data after a discretization step. However, the need to binarize the data might be undesirable in some settings. Developing significant pattern mining tools that can handle continuous data is a particularly challenging problem as Tarone's concept of testability would, in principle, no longer apply. Solving this limitation remains a key topic for future work.

Finally, significant pattern mining methods are flexible in the way the search space of candidate patterns is defined. While most approaches discussed in this chapter aim at performing an exhaustive search, prior knowledge can be used to drastically reduce the number of candidate high-order structures under study. This would not only lead to improved computational efficiency and statistical power but would also make it easier for practitioners to interpret the results of the analysis. Deriving novel application-specific pattern mining algorithms, combining domain knowledge with the basic principles of general-purpose significant pattern mining methods, is a promising research direction for data analysis in computational biology and medicine.

8.7 Exercises

8.1 Let G and Y be two random variables taking values in the finite alphabets $\mathcal{G}$ and $\mathcal{Y}$, respectively. Further assume that all marginal probabilities of G and Y are nonzero, that is, $\Pr(G = g) > 0$ for all $g \in \mathcal{G}$ and $\Pr(Y = y) > 0$ for all $y \in \mathcal{Y}$. Assuming that one of the three statements below is true, prove that the other two statements must hold as well.

(a) $\Pr(G = g, Y = y) = \Pr(G = g)\Pr(Y = y)$ for all $(g, y) \in \mathcal{G} \times \mathcal{Y}$.
(b) $\Pr(G = g \mid Y = y) = \Pr(G = g)$ for all $(g, y) \in \mathcal{G} \times \mathcal{Y}$.
(c) $\Pr(Y = y \mid G = g) = \Pr(Y = y)$ for all $(g, y) \in \mathcal{G} \times \mathcal{Y}$.

8.2 Let G and Y be two binary random variables. Suppose that there exists $(g, y) \in \{0, 1\}^2$ such that $\Pr(G = g, Y = y) = \Pr(G = g)\Pr(Y = y)$. Prove that this implies that:

(a) $\Pr(G = 1 - g, Y = y) = \Pr(G = 1 - g)\Pr(Y = y)$.
(b) $\Pr(G = g, Y = 1 - y) = \Pr(G = g)\Pr(Y = 1 - y)$.
(c) $\Pr(G = 1 - g, Y = 1 - y) = \Pr(G = 1 - g)\Pr(Y = 1 - y)$.

8.3 Let G and Y be two binary random variables. Suppose that there exists $(g, y) \in \{0, 1\}^2$ such that $\Pr(G = g \mid Y = y) = \Pr(G = g)$. Prove that this implies that G and Y are statistically independent random variables.

8.4 Consider a binary random variable Y that satisfies $\Pr(Y = 1) = p$. We say that Y follows a Bernoulli distribution with success probability p, denoted by $\Pr(Y = y) = \text{Bernoulli}(y \mid p) = p^y(1 - p)^{1-y}$. Suppose that n independent realizations of Y are collected, leading to a sequence of n i.i.d. Bernoulli-distributed random variables $\{Y_i\}_{i=1}^n$, all of which have the same success probability p. Finally, define $Z = \sum_{i=1}^n Y_i$ to be an integer-valued random variable that describes the total number of "successes," i.e., realizations for which $Y_i = 1$, out of all n draws.

The goal of this exercise is to prove that Z follows a binomial distribution with parameters n and p, that is, to show that

$$\Pr(Z = z \mid n, p) = \binom{n}{z} p^z (1 - p)^{n-z},$$

for any integer z between 0 and n. The proof will make use of four intermediate steps:

(a) Show that the joint probability distribution of $\{Y_i\}_{i=1}^n$ can be expressed as

$$\Pr(Y_1 = y_1, \ldots, Y_n = y_n) = p^{\sum_{i=1}^n y_i}(1 - p)^{n - \sum_{i=1}^n y_i}$$

(b) Show that $\Pr(Z = z \mid n, p)$ and $\Pr(Y_1 = y_1, \ldots, Y_n = y_n)$ are related by

$$\Pr(Z = z \mid n, p) = \Pr(Y_1 = y_1(z), \ldots, Y_n = y_n(z)) \, |\mathcal{Y}^n(z)|$$

where $\mathcal{Y}^n(z) = \{\mathbf{y} = (y_1, \ldots, y_n) \mid z = \sum_{i=1}^n y_i\}$ is the set of all realizations $\mathbf{y} = (y_1, \ldots, y_n)$ of $\{Y_i\}_{i=1}^n$ that have exactly z out of n successes and $\mathbf{y}(z) = (y_1(z), \ldots, y_n(z))$ is *any* realization in $\mathcal{Y}^n(z)$.
  (c) Derive a closed-form expression for $\Pr\left(Y_1 = y_1(z), \ldots, Y_n = y_n(z)\right)$.
  (d) Derive a closed-form expression for $|\mathcal{Y}^n(z)|$.

  Finally, show that the result readily follows from (a)–(d).

8.5 A group of researchers wishes to carry out a large-scale experiment to empirically validate a crucial, long-standing hypothesis in their field. The researchers are uncertain of the exact conditions in which the experiment should be performed. However, owing to the impact that a positive finding would have, they decide to exhaustively repeat the experiment $D = 100$ times, using slightly different conditions on each occasion. They compute a P-value for each such repetition of the experiment and consider that a positive finding has occurred if the P-value is smaller than $\alpha = 0.05$. To the researchers' excitement, for one specific set of experimental conditions, the P-value was indeed smaller than $\alpha = 0.05$. Therefore, they conclude that these were the right conditions in which the experiment should be carried out, and prepare a publication in which they describe what they now deem to be the correct experimental setup, as well as the corresponding findings. Unfortunately, in the next years after their article was published, no other research group managed to replicate their results.

  (a) Suppose that the set of $D = 100$ P-values the researchers obtained are jointly statistically independent. Derive a closed-form expression for the exact FWER of their procedure as a function of α, assuming the hypothesis they wish to validate were not true for any of the $D = 100$ experimental conditions they considered.
  (b) What is the corresponding FWER at $\alpha = 0.05$? Should the failure to replicate their results come as a surprise?
  (c) How would the results change if they had only attempted to test $D = 15$ different experimental conditions?
  (d) Based on the expression derived in (a), obtain the exact corrected significance threshold δ which would guarantee that the FWER is equal to α. What is the corresponding value of δ for $\alpha = 0.05$?
  (e) What would be the corrected significance threshold obtained using the Bonferroni correction?
  (f) Suppose now that the $D = 100$ experiments were performed in five different laboratories, rendering the P-values for experiments carried out in the same laboratory no longer statistically independent. Can we guarantee that the corrected significance threshold obtained in (d) will control the FWER at level α? What about the significance threshold obtained in (e)?

8.6 This exercise will explore informally the behavior of Tarone's method in the large sample size regime.

(a) Show that the minimum attainable P-value function for Fisher's exact test $p_{S,\min}(r_S)$ can be approximated by $\left(\frac{n_a}{n}\right)^{r_S}$ provided that n_a is large and that $r_S < n_a$ is small.
The following approximations might be useful:

- $\log n! \approx n \log n - n$ for sufficiently large n (Stirling's approximation).
- $(y - x) \log(y - x) \approx y \log y - (1 + \log y)x$ for $x \ll y$ (first order Taylor series of $(y - x) \log(y - x)$ around $x = 0$).

(b) Assuming that $\Pr(G_S(X) = 1) = p_g$ is constant and that n_a and n grow at the same rate, what does this imply for the behavior of the minimum attainable P-value as the sample size n tends to infinity?

Note: Solutions are available to instructors at www.cambridge.org/bionetworks.

8.8 Acknowledgments and Funding

We thank Damian Roqueiro and Katharina Heinrich for helpful comments, discussions and helping proofread this manuscript. This work was funded in part by the SNSF Starting Grant "Significant Pattern Mining" (KB, FLL) and the Marie Curie ITN MLPM2012, Grant No. 316861 (KB, FLL).

References

[1] Listgarten J, Lippert C, Kadie C. M, et al. Improved linear mixed models for genome-wide association studies. *Nature Methods*, 2012;9(6):525–526.

[2] Tibshirani R. Regression shrinkage and selection via the lasso. *Journal of the Royal Statistical Society. Series B (Statistical Methodology)*, 1996;58(1): 267–288.

[3] H. Zou and T. Hastie, Regularization and variable selection via the elastic net. *Journal of the Royal Statistical Society: Series B (Statistical Methodology)*, 2005;67(2):301–320.

[4] Bondell HD, Reich BJ. Simultaneous regression shrinkage, variable selection, and supervised clustering of predictors with Oscar. *Biometrics*, 2008;64(1):115–123.

[5] Li C, Li H. Network-constrained regularization and variable selection for analysis of genomic data. *Bioinformatics*, 2008;24(9):1175–1182.

[6] Kim S, Sohn KA, Xing EP. A multivariate regression approach to association analysis of a quantitative trait network. *Bioinformatics*, 2009;25(12):i204–i212. DOI: http://dx.doi.org/10.1093/bioinformatics/btp218.

[7] Azencott CA, Grimm D, Sugiyama M, Kawahara Y, Borgwardt K. Efficient network-guided multi-locus association mapping with graph cuts. *Bioinformatics*, 2013;29(13):i171–i179.

[8] Welter D. MacArthur J, Morales J. et al. The NHGRI GWAS catalog: A curated resource of snp-trait associations. *Nucleic Acids Research*, 2014;42(D1):D1001–D1006.

[9] Visscher PM, Brown MA, McCarthy MI, Yang J. Five years of GWAS discovery. *The American Journal of Human Genetics*, 2012;90(1):7–24.

[10] Manolio TA, Collins FS, Cox N J, et al. Finding the missing heritability of complex diseases. *Nature*, 2009;461(7265):747–753.

[11] Zuk O, Hechter E, Sunyaev SR, Lander ES. The mystery of missing heritability: Genetic interactions create phantom heritability. *Proceedings of the National Academy of Sciences*, 2012;109(4):1193–1198.

[12] Hemani G, Shakhbazov K, Westra HJ, et al. Detection and replication of epistasis influencing transcription in humans. Nature, 2014;508(7495):249–253.

[13] Forsberg SK, Bloom JS, Sadhu MJ, Kruglyak L, Carlborg O. Accounting for genetic interactions improves modeling of individual quantitative trait phenotypes in yeast. *Nature Genetics*, 2017;49(4):497–503.

[14] MacKay DJ. *Information Theory, Inference and Learning Algorithms*. Cambridge University Press, 2003.

[15] Terada A, Okada-Hatakeyama M, Tsuda K, Sese J. Statistical significance of combinatorial regulations. *Proceedings of the National Academy of Sciences*, 2013;110(32):12996–13001.

[16] Minato SI, Uno T, Tsuda K, Terada A, Sese J. A fast method of statistical assessment for combinatorial hypotheses based on frequent itemset enumeration. in *Joint European Conference on Machine Learning and Knowledge Discovery in Databases*. Springer;2014; pp. 422–436.

[17] Sugiyama M, Llinares-Lopez F, Kasenburg N, Borgwardt K. Significant subgraph mining with multiple testing correction. In *Proceedings of the 2015 SIAM International Conference on Data Mining 4*, SIAM;2015, pp. 37–45.

[18] Terada A, Tsuda K, Sese J. Fast westfall-young permutation procedure for combinatorial regulation discovery. In *2013 IEEE International Conference on Bioinformatics and Biomedicine*, 12. IEEE BIBM;2013, pp. 153–158.

[19] Llinares-Lopez F, Sugiyama M, Papaxanthos L, Borgwardt K. Fast and memory-efficient significant pattern mining via permutation testing. In *Proceedings of the 21st ACM SIGKDD International Conference on Knowledge Discovery and Data Mining*. ACM;2015, pp. 725–734.

[20] Papaxanthos L, Llinares-Lopez F, Bodenham D, Borgwardt K. Finding significant combinations of features in the presence of categorical covariates. In Jordan MI, LeCun Y, Solla SA, eds., *Advances in Neural Information Processing Systems*, MIT Press;2016, pp. 2271–2279.

[21] Todeschini R, Consonni V. *Handbook of Molecular Descriptors*. John Wiley & Sons;2008.

[22] Bullmore E, Sporns O. Complex brain networks: graph theoretical analysis of structural and functional systems. *Nature Reviews Neuroscience*, 2009;10(3): 186–198.

[23] van den Heuvel MP, Mandl RC, Stam CJ, Kahn RS, Pol, HEH. Aberrant frontal and temporal complex network structure in schizophrenia: A graph theoretical analysis. *Journal of Neuroscience*; 2010;30(47):15915–15926.

[24] Bertsekas DP, Tsitsiklis JN. *Introduction to Probability*. Athena Scientific; 2002.

[25] Pearson K. X. On the criterion that a given system of deviations from the probable in the case of a correlated system of variables is such that it can be reasonably supposed to have arisen from random sampling. *The London, Edinburgh, and Dublin Philosophical Magazine and Journal of* Science, 1900;50(302):157–175.

[26] Fisher RA. On the interpretation of χ^2 from contingency tables, and the calculation of p. *Journal of the Royal Statistical Society*, 1922;85(1):87–94.

[27] Shaffer JP. Multiple hypothesis testing. *Annual Review of Psychology*, 1995;46(1):561–584.

[28] Dmitrienko A, Tamhane AC, Bretz F. Multiple testing problems in pharmaceutical statistics. Chapman and Hall/CRC, 2009.

[29] Noble WS. How does multiple testing correction work? *Nature Biotechnology*, 2009;27(12);1135–1137.

[30] Bonferroni CE. *Teoria statistica delle classi e calcolo delle probabilita*. Libreria Internazionale Seeber; 1936.

[31] Dunn OJ. Estimation of the medians for dependent variables. *The Annals of Mathematical Statistics*, 1959;30(1):192–197.

[32] Bay SD, Pazzani MJ. Detecting group differences: Mining contrast sets. *Data Mining and Knowledge Discovery*, 2001;5(3):213–246.

[33] Hamalainen W. Statapriori: An efficient algorithm for searching statistically significant association rules. *Knowledge and Information Systems*, 2010;23(3):373–399.

[34] Zimmermann A, Nijssen S. *Supervised Pattern Mining and Applications to Classification*. Springer;2014, pp. 425–442. DOI: //doi.org/10.1007/978-3-319-07821-2 17.

[35] Webb GI. Discovering significant rules. In *Proceedings of the 12th ACM SIGKDD International Conference on Knowledge Discovery and Data Mining*. ACM;2006, pp. 434–443.

[36] ———. Discovering significant patterns. *Machine Learning*, 2007;68(1):1–33.

[37] ———. Layered critical values: A powerful direct-adjustment approach to discovering significant patterns. *Machine Learning*, 2008;71(2-3): 307–323.

[38] Tarone RE. A modified Bonferroni method for discrete data. *Biometrics*, 1990;46(2):515–522.

[39] Aggarwal CC, Han J. *Frequent Pattern Mining*. Springer;2014.

[40] Zaki MJ, Parthasarathy S, Ogihara M, Li W. New algorithms for fast discovery of association rules. In *Proceedings of the Third International Conference on*

Knowledge Discovery and Data Mining. ACM;1997, pp. 283–286. Available at www.aaai.org/Library/KDD/1997/kdd97-060.php.

[41] Han J, Pei J, Yin Y. Mining frequent patterns without candidate generation. In *Proceedings of the 2000 ACM SIGMOD International Conference on Management of Data*. ACM;2000, pp. 1–12. DOI://doi.acm.org/10.1145/342009.335372.

[42] Uno T, Kiyomi M, Arimura H. LCM ver. 2: Efficient mining algorithms for frequent/closed/maximal itemsets. In *Proceedings of IEEE ICDM'04 Workshop FIMI'04*. IEEE;2004.

[43] Borgelt C. Frequent item set mining. *Wiley Interdisciplinary Reviews: Data Mining and Knowledge Discovery*, 2012;2(6):437–456.

[44] Yan X, Han J. gSpan: Graph-based substructure pattern mining. In *IEEE International Conference on Data Mining*. IEEE;2002, pp. 721–724.

[45] Nijssen S, Kok JN. The gaston tool for frequent subgraph mining. *Electronic Notes in Theoretical Computer Science*, 2005;127(1):77–87.

[46] Jiang C, Coenen F, Zito M. A survey of frequent subgraph mining algorithms. *The Knowledge Engineering Review*, 2013;28(1):75–105.

[47] Westfall PH, Young SS. *Resampling-Based Multiple Testing: Examples and Methods for p-Value Adjustment*, Wiley Series in Probability and Statistics, vol. 279. John Wiley & Sons;1993.

[48] Vilhjalmsson BJ, Nordborg M. The nature of confounding in genome-wide association studies. *Nature Reviews Genetics*, 2013;14(1):1–2.

[49] Llinares-Lopez F, Papaxanthos L, Bodenham D, et al. Genome-wide genetic heterogeneity discovery with categorical covariates. *Bioinformatics*, 2017;33(12):i1820i1828,.

[50] Dering C, Hemmelmann C, Pugh E, Ziegler A. Statistical analysis of rare sequence variants: An overview of collapsing methods. *Genetic Epidemiology*, 2011;35(S1):S12–S17.

[51] Mantel N, Haenszel W. Statistical aspects of the analysis of data from retrospective studies of disease. *Journal of the National Cancer Institute*, 1959;22(4):719–748.

[52] Terada A, duVerle D, Tsuda K. Significant pattern mining with confounding variables. In Bailey J, Khan L, Washio T, et al., eds., *Pacific-Asia Conference on Knowledge Discovery and Data Mining*. Springer;2016, pp. 277–289.

[53] Terada A, Yamada R, Tsuda K, Sese J. Lamplink: Detection of statistically significant snp combinations from GWAS data. *Bioinformatics*, 2016;32(22):3513–3515.

[54] Yoshizoe K, Terada A, Tsuda K. Redesigning pattern mining algorithms for supercomputers, arXiv preprint available at https://arxiv.org/pdf/1510.07787.pdf. 2015.

[55] Llinares-Lopez F, Grimm D, Bodenham D, et al. Genome-wide detection of intervals of genetic heterogeneity associated with complex traits. *Bioinformatics*, 2015;31(12):i240–i249.

[56] Benjamini Y, Hochberg Y. Controlling the false discovery rate: A practical and powerful approach to multiple testing. *Journal of the Royal Statistical Society. Series B (Statistical Methodology)*, 1995;57(1):289–300.

[57] Taylor J, Tibshirani RJ. Statistical learning and selective inference. *Proceedings of the National Academy of Sciences*, 2015;112(25):7629–7634.

[58] Suzumura S, Nakagawa K, Umezu Y, Tsuda K, Takeuchi I. Selective inference for sparse high-order interaction models. In Precup D, Teh WY, eds., *Proceedings of the 34th International Conference on Machine Learning*, vol. 70. PMLR;2017, pp. 3338–3347.

9 Network Alignment

Noël Malod-Dognin and Nataša Pržulj

9.1 Introduction

In this chapter, we present the problem of network alignment. We start with introducing biological concepts, as well as the biological motivation for aligning networks. We formally define the pairwise network alignment problem, for which we also present standard network alignment algorithms, as well as standard scoring schemes that are used to measure the quality of network alignments. Then, we do the same for the multiple network alignment problem. While the above presented pairwise and multiple network alignment problem focus on undirected, unweighted networks, we conclude this chapter with a brief overview of alignment problems for other types of network data.

9.1.1 Proteins and their Functions

Sequence alignment has revolutionized our understanding of biology. By uncovering relationships between the genomes of species, sequence alignment has given us insights into gene functions and phylogeny. However, the genome is only the blueprint of a cell. Within a genome, which is stored in deoxyribonucleic acid (DNA) molecules, specific regions, called genes, are transcribed into messenger ribonucleic acids (mRNAs), which are the working copies of DNA. The mRNAs are translated into proteins, which are large biomolecules that interact with each other and with other molecules to perform all the biological functions in a cell. We refer the reader to Chapter 10 for an overview of molecular interaction datasets that are available nowadays.

From a chemical point of view, a protein is a linear arrangement of amino acids forming a so-called polypeptide chain. In the aqueous environment in the cell, the polypeptide chain folds and acquires a specific 3-dimensional (3D) shape (sometimes with the help of other proteins called chaperones), which in most cases is globular. This folding process is mainly driven by hydrophobic–hydrophilic effect, as hydrophilic amino acids will tend to move to the surface of the protein while hydrophobic amino

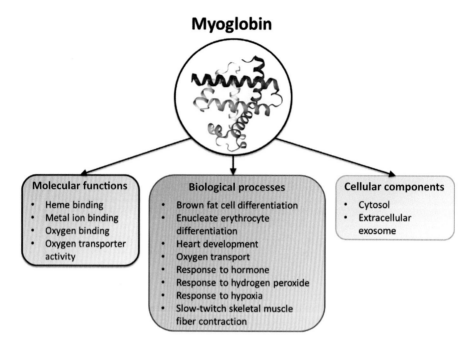

Figure 9.1: An example of protein annotations. Myoglobin, presented here by structure 3RGK from the Protein Data Bank [1], is a protein contributing to intracellular oxygen storage and transcellular facilitated diffusion of oxygen. The three boxes present its molecular function, biological process, and cellular component annotations.

acids will tend to form a hydrophobic core. Other forces include intra-molecular hydrogen bonds and Van der Waals forces. Thus, proteins can be described by their sequences (their linear arrangement of amino acids) and by their 3D structures (the coordinates of their atoms in 3D space).

It has long been shown that proteins with similar sequences, or similar 3D structures may also have similar biological function; e.g., proteins having sequence identity $\geq 70\%$ are expected to share function [2], and this property has widely been used to transfer biological functions across proteins. These protein functions are described with annotations. The most widely used annotation set is the gene ontology (GO) set [3]. In GO, an annotation describes proteins according to the biological processes (BPs) that they participate in, according to the molecular functions (MF) that they perform, and according to the cellular components (CC) where they are located, as illustrated in Figure 9.1. The relationships between annotation terms are encoded in an ontology tree (a directed acyclic graph), from the most generic to the most specific terms (illustrated in Figure 9.2). GO is commonly used as a benchmark for protein functional similarity and for measuring the biological relevance of predicted protein function, as two proteins are considered to be functionally similar if their share sets of GO annotations. Note that some studies additionally consider the pathway specific annotations from KEGG pathways [4] and REACTOME [5] databases.

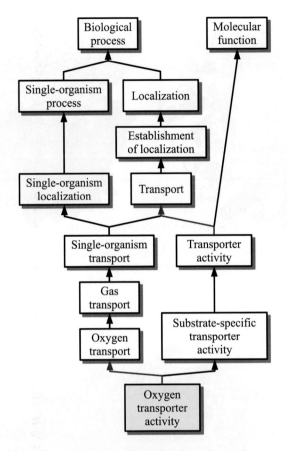

Figure 9.2: An illustration of GO ontology. The directed acyclic graph captures the ancestry of "oxygen transporter activity" annotation in GO ontology, as taken from AmigO web-server [6]. A black arrow from a term A to a term B indicates that term A "is a" B, while a blue arrows indicates that A "is a part of" B.

9.1.2 Protein Interactions and Network Alignment

Because of their complementary shapes and of the physical properties of their amino acids, proteins can bind to each other, forming larger structures called protein complexes (illustrated in Figure 9.3). The pairwise binding between two proteins is called a protein–protein interaction (PPI). PPIs have become increasingly available thanks to the development of high throughput capturing techniques such as yeast two-hybrid [7, 8] and affinity purification coupled with mass spectrometry [9].

The entire set of PPIs of a cell is commonly modeled by a PPI network (illustrated in Figure 9.4), in which nodes represent proteins and edges connect proteins that can interact. PPI networks have modular organization that reveals the molecular machinery of a cell [10] and thus, deciphering the connectivity patterns (also called topology) of these networks is fundamental to understanding the functioning of the cell [11] (detailed also in Chapter 4). A large number of studies focused on determining the commonalities and transferring of annotation between PPI networks of different

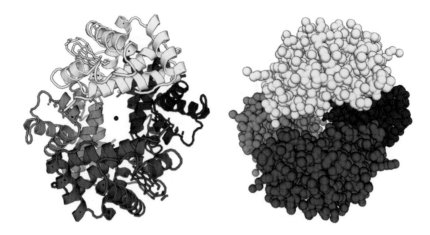

Figure 9.3: An illustration of a protein complex. The protein complex of human deoxy-haemoglobin is composed of two hemoglobin alpha and two hemoglobin beta proteins (each presented in a different color). The drawing on the left is a schematic representation of the polypeptide chains of the involved proteins, while the 3D plot on the right is an all-atom representation. Both drawings are based on structure 4HHB from the Protein Data Bank [1].

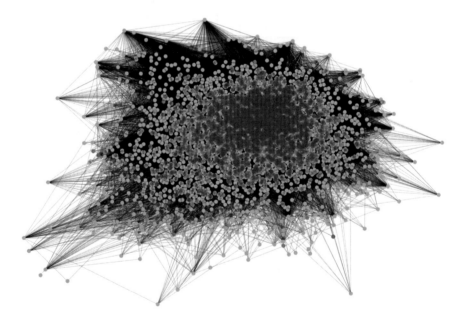

Figure 9.4: An illustration of the PPI network of baker's yeast. This network, constructed from the PPI data from the Integrated Interaction Database [21] (collected in June 2016), contains 5,723 nodes (proteins) and 108,484 edges (interactions between proteins).

species, which is often done by network alignment [12, 13, 14, 15, 16, 17]. In a network alignment, one is interested in finding a mapping between the proteins in PPI networks (called an *alignment*) that maps topologically, functionally, and evolutionary conserved regions of the PPI networks. The mapping algorithm is called an *aligner*.

Network alignments have uncovered valuable information, including evolutionary conserved pathways, protein complexes and functional orthologs [18, 19, 20].

Network alignment has become increasingly studied because of its ability to transfer knowledge across PPI networks of different species. Many proteins have not been functionally annotated [22, 23], especially in human, since biological processes and diseases are hard to study experimentally. Thus, these processes and diseases are experimentally investigated in model organisms, such as yeast *Saccharomyces cerevisiae* (baker's yeast) and mouse *Mus musculus* (a house mouse) and the corresponding knowledge is then transferred from the model organisms to human using network alignments [24, 25].

9.2 Pairwise Network Alignment

9.2.1 Formal Definitions

A *network*, or a graph, is a couple $N = (V, E)$, where V is a set of nodes and E is a set of edges connecting together some of the nodes of V. The exponential number of possible networks with n nodes, 2^{n^2}, makes network classification and comparison problems computationally difficult.

A key comparison problem coming from graph theory is the *subgraph isomorphism problem*, which asks whether network $N_1 = (V_1, E_1)$ exists as an exact subgraph of network $N_2 = (V_2, E_2)$ (for a formal definition, see Chapter 3). In other words, is there a subgraph $N_2' = (V_2', E_2') : V_2' \subseteq V_2, E_2' \subseteq E_2 \bigcap (V_2' \times V_2')$ such that $N_1 \cong N_2'$? This problem is NP-complete in general, which means that no efficient, polynomial-time algorithm can be constructed for solving it [26]. However, we are not interested in exact subgraph isomorphism. Evolutionary relationships between genes are ambiguous and due to limitations in capturing technologies, biological networks are also plagued with noise (i.e., these networks have missing edges and also contain false edges [27]; see Chapter 4 for details).

Thus, *network alignment* [24] is a more general problem of finding the best way to "fit" N_1 into N_2 even if N_1 does not exist as an exact subgraph of N_2. The network alignment problem has two main formulations, as illustrated in Figure 9.5. On one hand, *local* aligners search for small, but highly conserved regions of the networks, called network modules. On the other hand, *global* aligners search for a single comprehensive mapping of the whole set of proteins and protein–protein interactions from different species.

Formally, between two PPI networks, $N_1 = (V_1, E_1)$ and $N_2 = (V_2, E_2)$, for which $|V_1| \leq |V_2|$, an *alignment*, a, is a mapping between the nodes of V_1 and the nodes of V_2:

$$a : V_1^* \to V_2^*, \tag{9.1}$$

with $V_1^* \subseteq V_1$ and $V_2^* \subseteq V_2$. In local alignments, V_1^* and V_2^* are small subsets corresponding to the most similar nodes in the networks (we will define similarity later), while global alignments aim (but do not always succeed) at aligning all nodes from V_1. Furthermore, while local aligners tend to produce confusing *many-to-many* mappings in which a node from V_1^* can be mapped to several nodes in V_2^* and vice-versa, global

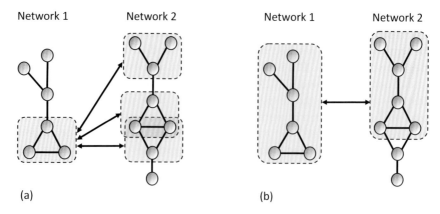

Figure 9.5: Local and global alignments. The alignment presented in panel (a) is local, as it only aligns a small region from the first network. It is also many-to-many, as the small region from Network 1 is aligned to three different regions in Network 2. The alignment presented in panel (b) is global, as it fully aligns Network 1. It is also one-to-one, since Network 1 is aligned to a single region of Network 2.

aligners tend to produce non-ambiguous *one-to-one* mappings in which alignments are bijections between V_1^* and V_2^* (the difference between the two is illustrated in Figure 9.5).

To produce alignments, aligners are guided by objective functions that assign real-valued scores to alignments. While these functions are different across aligners, they can be generically rewritten as:

$$S(a) = \sum_{u \in V_1} n(u, a(u)) + \sum_{(u,v) \in E_1} e(u, a(u), v, a(v)), \tag{9.2}$$

where $n : V_1 \times V_2 \to \mathbb{R}^+$ is the score of mapping a node of V_1 to a node in V_2, and $e : E_1 \times E_2 \to \mathbb{R}^+$ is the score of mapping an edge of E_1 to an edge of E_2.

Because of NP-completeness of the underlying subgraph isomorphism problem, network aligners are *heuristics* (approximate solvers) that use different *objective functions* and *optimization algorithms* to approximately solve the network alignment problem (examples are detailed in the next sections). To uncover alignments that maximize their objective functions, most aligners either rely on alignment graphs or bipartite matchings:

An alignment graph is a graph whose nodes represents the possible mappings between the proteins of the two PPI networks and whose edges represent the possible mapping between their interactions. In the alignment graph, nodes and edges can be weighted according to the contributions of the corresponding protein and interaction mappings to the aligner's objective function. Intuitively, searching for a high scoring alignment corresponds to searching in the alignment graph for a subgraph having a large sum of node and edge weights.

A bipartite graph is graph $B = (V_1 \cup V_2, E)$ in which the node set is partitioned into two sets, V_1 and V_2, such that edges (in E) only connect the nodes of V_1 and the nodes of V_2. Weights can be associated to edges representing the costs of the node

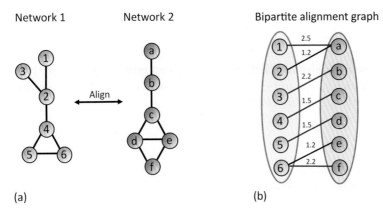

Figure 9.6: Bipartite matching. To align the two networks from panel (a), many global aligners rely on maximum weight bipartite matching. To this aim, they create a bipartite alignment graph as illustrated in panel (b), in which the nodes from network 1 are connected to nodes from network 2 by edges having weights that capture the score of aligning the two nodes (the higher the better), represented here by the black and the red edges. The maximum weight bipartite matching algorithm searches in this bipartite graph for a one-to-one matching between the nodes having the maximum sum of edge weights (illustrated by the red edges).

mappings. The maximum weight bipartite matching is a problem in graph theory that seeks to find in B a one-to-one node mapping V_1 and V_2 having the maximum cost (i.e., the maximum sum of edge weights over all possible sums). The maximum weight bipartite matching can be optimally solved in polynomial time with methods such as the Hungarian algorithm [28] (see Chapter 3). Thus, many global aligners use two-step approaches in which they first compute mapping scores between the nodes of the networks to be aligned and then model their alignment as the maximum weight bipartite matching (see illustration in Figure 9.6).

9.2.2 Scoring Alignments

With the multitude of network alignment methods that are available, a critical issue is to evaluate the quality of the produced alignments. This section reviews *standard alignment scoring schemes* that are used in network alignment studies.

9.2.2.1 Scoring Local Network Alignments

Between two networks, local aligners aim to uncover sets of aligned modules that correspond to biological pathways or protein complexes, which is why local aligners are mainly evaluated according to the biological coherence of their alignments. Local alignments are not assessed in terms of topological coherence (e.g., amount of conserved protein interactions), as edge conservation is only properly defined in the case of the non-ambiguous one-to-one mappings produced by global aligners. The following are commonly used scoring schemes for local alignments:

k-**correctness.** Two proteins, p_1 and p_2, having functional annotation sets s_1 and s_2, are said to be correctly aligned if they share at least k annotations; i.e., if $|s_1 \cap s_2| \geq k$.

k-correctness of two aligned modules is the percentage of their aligned proteins that are correctly aligned. Then, k-correctness is extended to a whole alignment by averaging the k-correctness of all aligned modules.

Functional consistency. Between two aligned proteins, p_1 and p_2, having functional annotation sets s_1 and s_2, functional consistency (FC) is the percentage of annotations that are shared between the two proteins:

$$FC(p_1, p_2) = \frac{|s_1 \cap s_2|}{|s_1 \cup s_2|}. \tag{9.3}$$

FC is extended to two aligned modules by averaging the FC of their aligned proteins. Finally, FC can be extended to a whole alignment by averaging the FC of all the aligned modules.

Semantic similarity. Two different GO terms may describe related contexts, as evidenced by the common terms in their ancestries in the ontology tree. Semantic similarity takes these ancestries into account to measure the similarity between two terms. There exist two types of semantic similarities. *Node-based semantic similarity* defines the information content of a term as a function of its frequency of appearance in the annotated dataset and measures the similarity between two terms according to their most informative ancestor in the ontology [29, 30, 31]. *Edge-based semantic similarity* only uses the ontology tree and measures the similarity between two terms based on the shortest path between them, or based on the depth in the ontology of their common ancestors [32]. Semantic similarity is extended to two aligned proteins p_1 and p_2 having annotation sets s_1 and s_2, by averaging, for each annotation in s_1 and in s_2, the highest semantic similarity that is obtained with a term from the other set (so-called best-match average). The semantic similarity between aligned proteins can then be extended to aligned modules and whole alignments in the same way that is for functional consistency.

Agreement with reference modules. Many scores have been proposed to assess the agreement between an aligned module and a reference cluster of genes (e.g., a known protein complex or a known pathway). These scores are usually based on precision and recall [33]. Precision (Pr) is the percentage of proteins in the aligned module that belong to the reference and recall (Re) is the percentage of proteins from the reference that appear in the aligned module. Precision and recall are opposing measures that can be unified into single scores, such as the F-score:

$$F = 2 \times \frac{P_r \times R_e}{P_r + R_e}. \tag{9.4}$$

Other similar measures have been defined in the field of information retrieval, such as the *Matthews correlation coefficient* and *informedness*. All these measures can then be extended over all aligned modules to assess the quality of a whole alignment.

9.2.2.2 Scoring Global Network Alignments

Let us first introduce some notations. The PPI networks of two species are represented by two networks, $N_1 = (V_1, E_1)$ and $N_2 = (V_2, E_2)$, such that N_1 is smaller than or equal to N_2, i.e., $|V_1| \leq |V_2|$. A global alignment, a, between the two PPI networks defines a common subnetwork, $A = (V_a, E_a)$, whose nodes in V_a are the aligned nodes

$u \leftrightarrow w$, $u \in V_1$, $w \in V_2$, and whose edges in E_a are the aligned edges $(u,v) \leftrightarrow (w,x)$, $(u,v) \in E_1$, $(w,x) \in E_2$. Finally, the aligned region on the larger graph (i.e., the subnetwork of N_2 that is induced by its aligned nodes in V_a) is denoted by $N_2[V_a]$, and the edge set of this subgraph is denoted by $E_{N_2[V_a]}$.

Similar to local alignments, global alignments are assessed according to their biological coherence. Biological coherence is measured with the same scores that are used for local alignments, namely k-correctness, functional consistency, semantic similarity and agreement with reference modules, which are computed for a whole global alignment as the average of the similarity scores between aligned proteins.

However, unlike local alignments, the quality of global alignments is also assessed in terms of *topological coherence*, with the underlying idea being that the aligned regions of the networks should have similar wiring patterns. We briefly present the most widely used topological scores.

Node correctness. When a reference alignment between the nodes of the two networks is known, e.g., for synthetic alignment instances, the quality of an alignment produced by a network aligner can be measured by its node correctness, which is the percentage of the aligned nodes that are correctly mapped according to the reference alignment. However, for most biological network alignment networks, reference alignments between the nodes of the networks are generally not available.

Node coverage. A frequently overlooked measure of the quality of an alignment is how many nodes/proteins it maps between the networks. While global network aligners aim at aligning all nodes of the smaller network to the nodes of the larger one, they often fail to do so. This is measured by *node coverage* (NC), where the number of mapped nodes is normalized in [0,1] according to the number of nodes in the smaller network:

$$NC(a) = \frac{|V_a|}{|V_1|} \times 100\%. \tag{9.5}$$

Edge correctness [19]. A popular measure of an alignment quality is *edge correctness* (EC), the percentage of the interactions (edges) of the smaller network that are aligned to some edges of the larger network:

$$EC(a) = \frac{|E_a|}{|E_1|} \times 100\%. \tag{9.6}$$

Induced conserved sub-structure score [34]. Although EC is an intuitive measure of an alignment quality, it only considers the smaller network. An alignment with large EC may map a sparse small network onto a dense region of the large network. Thus, the *induced conserved substructure score* (ICS) considers the alignment from the larger network's point of view, by measuring the percentage of the interactions from the aligned region of the larger network that are aligned to some interaction from the smaller network. It is defined as:

$$ICS(a) = \frac{|E_a|}{|E_{N_2[V_a]}|} \times 100\%. \tag{9.7}$$

Symmetric sub-structure score [35]. While a large EC allows a sparse small network to be mapped onto a dense region of the larger network, a large ICS allows a sparse region of the larger network to be mapped onto a dense region of the smaller network. The *symmetric substructure score* (S^3) considers both networks by comparing the number of aligned edges to the number of edges of the smaller network and to the number of edges in the aligned region of the larger network. It is defined as:

$$S^3(a) = \frac{|E_a|}{|E_1| + |E_{N_2[V_a]}| - |E_a|} \times 100\%. \tag{9.8}$$

Size of the largest common connected component [19]. Another popular measure of an alignment quality is the size of the largest connected component (LCC) shared by the two networks (i.e., the largest connected component of that is found in A). A larger LCC implies that the alignment contains a larger amount of shared continuous structure between the two networks.

9.2.2.3 Agreement and Trade-off Between Scores

Recall that the goal of network alignment is to produce alignments that simultaneously exhibit topological and biological coherence. As evidenced by network alignment studies, topological scores tend to agree with each other, biological scores tend to agree with each other, but topological and biological scores largely disagree on which alignment is the best (illustrated in Figure 9.7). However, topological scores can be

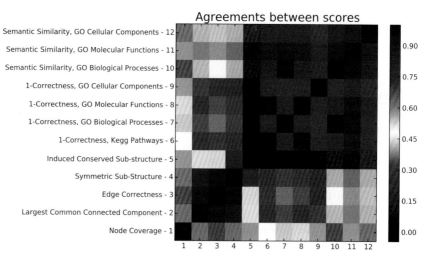

Figure 9.7: Agreements between alignment scores. The heat-map presents the agreements between topological alignments scores (numbered from 1 to 5) and biological alignment scores (numbered from 6 to 12), as measured by their Pearson's correlation coefficients (PCCs) over 2,770 alignments produced by eight different aligners (NATALIE 2.0 [36], SPINAL [37], PISWAP [38], MAGNA++ [39], HUBALIGN [40], L-GRAAL [41], OPTNETALIGN [42] and MODULEALIGN [43]) between the protein–protein interaction networks of eight species (*Homo sapiens, Saccharomyces cerevisiae, Drosophila melanogaster, Arabidopsis thaliana, Mus musculus, Caenorhabditis elegans, Schizosaccharomyces pombe,* and *Rattus norvegicus*). High PCC values (red) highlight the scores that are in good agreements, while low PCC values (blue) highlight the scores that have no agreements. The figure is generated from the data of Malod-Dognin et al. [44].

well subsumed by S^3 score, which strongly correlates with the other topological scores (node coverage, largest common connected component, edge correctness and induced conserved substructure). Also, all biological scores can be subsumed by 1-correctness of KEGG pathway annotations, which strongly correlates with the other biological scores (1-correctnesses and semantic similarities of GO biological process, molecular function and cellular component annotations).

In their study, Meng et al. [45] measure the trade-off between various topological scores $S_1, S_2, \ldots, S_k$, as their geometric mean:

$$\text{TradeOff}(S_1, S_2, \ldots, S_k) = \sqrt[k]{S_1 \times S_2 \times \cdots \times S_k}. \tag{9.9}$$

Because topological scores can be well subsumed by S^3 score and biological scores can be well subsumed by 1-correctness of KEGG pathway annotations, Malod-Dognin et al. [44] proposed to measure the trade-off between biological and topological quality by using Meng et al. trade-off between S^3 and 1-correctness of KEGG pathway annotations.

9.2.3 Example Pairwise Local Alignment Method: PathBlast

PathBlast [18] was the first network aligner that was introduced. It is a local, many-to-many aligner that searches between two networks for highly conserved pathways.

To perform alignment, PathBlast combines the two networks into an alignment graph, in which nodes represent pairs of proteins (one from each network) that can be aligned because of their high sequence homology (BLAST sequence similarity E-value $\leq 10^{-2}$), and edges represent conserved interactions (where interactions exist in both PPI networks between the aligned proteins), gaps or mismatches. Gaps occur when two proteins that are interacting in one network are aligned to proteins that do not interact in the other network, and mismatches occur when two proteins that do not interact in one network are aligned to proteins that do not interact in the other network (see illustration in Figure 9.8 (a)). In the alignment graph, a path P represents a pathway alignment between the two networks and is associated with a log-probability score, $S(P)$:

$$S(P) = \sum_{v \in P} \log_{10} \frac{p(v)}{p_{random}} + \sum_{e \in P} \log_{10} \frac{q(e)}{q_{random}}, \tag{9.10}$$

where $p(v)$ is the probability of true homology between the two proteins represented by node v, given their pairwise BLAST E-value, E_v, as computed by using Bayes' rule:

$$p(v) = p(H|E_v) = \frac{p(E_v|H)p(H)}{p(E_v)}, \tag{9.11}$$

where H represents the case of true homology between the two proteins represented by node v. The probability distribution for $p(E_v)$ is the frequency of each E-value over all nodes in the alignment graph. The probability distribution $p(E_v|H)$ is based on the E-values within the subset of nodes for which both proteins are in the same orthology group according to COG database [46]. The prior probability $p(H)$ is the frequency of nodes representing proteins that are in the same orthology group over all nodes in the alignment graph. The probability of an edge, $q(e)$, is based on the individual probabilities of the protein interactions it represents. In case of a direct match, an edge represents only two aligned protein interactions, but in cases of gaps and mismatches it can represent three or four protein interactions, respectively. The empirical probability $Pr(i)$ of a protein interaction i is based on the number of experiments that validate it: $Pr(i) = 0.1$ if interaction i is supported by one experiment, 0.3 for two experiments and 0.9 for three or more experiments. The edge probability $q(e)$ is the product of these probabilities:

$$q(e) = \prod_{i \in e} Pr(i). \tag{9.12}$$

The background probabilities p_{random} and q_{random} are the expected values of $p(v)$ and $q(e)$ over all nodes and edges in the alignment graph.

To uncover a high scoring path, the alignment graph is first converted into a directed acyclic graph (DAG) by randomly ordering the nodes of the alignment graph and by orienting edges from lower ordered nodes towards high ordered nodes (see illustration in Figure 9.8 (b)). In the DAG, the highest scoring path of length

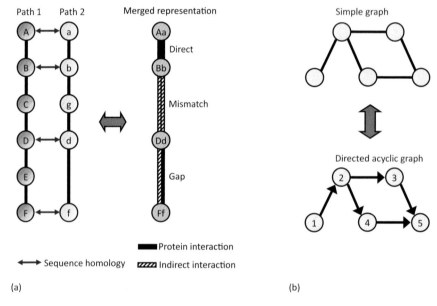

Figure 9.8: An illustration of PathBlast's merged graph representation. (a) In each of the two pathways, solid lines represent direct protein-protein interactions. Between the two pathways, red lines connect proteins that have statistically significant sequence similarity (BLAST E-value ≤ 0.05). The two pathways are combined into a merged representation (also called a global alignment graph) in which nodes represent pairs of homologous proteins and edges represent protein interaction relationships of three types: Direct interaction (black), gap (one interaction is indirect, black and dashed), and mismatch (both interactions are indirect, dashed). (b) A global alignment graph, which is originally a simple graph, is converted into a directed acyclic graph by randomly ordering its nodes and by orienting edges from lower ordered nodes towards higher ordered nodes.

$l = 2\ldots L$ ending at node v is found using the following dynamic programming recurrence, DP:

$$\begin{cases} DP(v,l) = \underset{u \in Pred(v)}{\operatorname{argmax}} \left[S(u, l-1) + \log_{10} \frac{p(v)}{p_{\text{random}}} + \log_{10} \frac{q(e_{u \to v})}{q_{\text{random}}} \right] \\ DP(v,1) = \log_{10} \frac{p(v)}{p_{\text{random}}} \end{cases} \quad (9.13)$$

In practice, PathBlast randomly generates $5L!$ DAGs and searches in them for paths of length L ending in each of their n nodes. This approach is both computationally intensive, with a worst case time complexity of $O(L! n)$, and is restricted to very simple topologies (only paths).

9.2.4 Overview of Other Pairwise Local Alignment Methods

Due to the above mentioned limitations of PathBLAST, several other methods for pairwise local network alignment have been proposed:

NetworkBLAST [18] is an extension of PathBlast that searches between two networks for conserved modules, which are densely connected, clique-like subnetworks that better correspond to conserved protein complexes than the linear paths of PathBlast. To this aim, NetworkBLAST first identifies high-scoring seeds in an alignment graphs similar to the one of PathBlast, and then heuristically extends the seed alignments using a greedy approach.

MaWish [47] is inspired by duplication/divergence models that focus on the principles of evolution of protein interactions. It is based on a mathematical model that extends the concepts of match, mismatch, and gap in sequence alignment to that of match, mismatch, and duplication in network alignment. MaWish encodes these evolutionary principles into the edge weights of its alignment graph, which is mined using a greedy seed-and-extend algorithm.

NetAligner [48] is a network aligner that directly addresses the issue of false negatives (missing interactions) in the current interactomes by predicting likely conserved interactions. For all conserved or likely conserved interactions (when orthologous pairs of proteins are interacting in one network, but not in the other), NetAligner computes the probability (p-value) that the interaction is conserved (the higher the p-value, the more likely it is that the interaction is conserved). Then, it constructs an alignment graph in which nodes represent possible mappings between orthologous proteins, and in which edges represent possible mapping between their interactions and are weighted by the corresponding interaction conservation p-value. The alignment graph is thresholded to filter out unreliable interactions (with low p-value). To uncover alignments, NetAligner searches in the alignment graph for seed connected components, which are greedily extended by connecting vertices of the connected component through gap or mismatch edges.

AlignNemo [49] extends the notion of indirect interactions of PathBLAST, which originally accounted for only one indirect edge. Between two networks, AlignNemo creates a weighted alignment graph in which nodes represent the possible mappings between orthologous proteins from the two networks, and whose weights are the proteins' homology scores from Inparanoid [50]. To account for the interactions between the proteins, edges are weighted according to the number of paths that connect the proteins in their respective PPI networks, so that the more paths connect the proteins, the greater the score in the alignment graph. Then, AlignNemo uses a seed-and-extend strategy to uncover densely connected aligned modules. First, it extracts all the high-scoring 4-nodes subgraphs from its alignment graph, which are then heuristically extended using a greedy approach.

AlignMCL [51] is based on the same philosophy as AlignNemo, but it replaces the computationally intensive subgraph search with a more efficient *Markov clustering* (MCL) on the alignment graph to uncover the aligned modules. In a nutshell, MCL is a graph clustering approach that simulates random walkers on a graph and that partitions the nodes of the graphs into different clusters in order to minimize the probability that random walkers will travel from one cluster to another.

Recently, Meng et al. [52] compared the alignments produced by local and global aligners on the PPI networks of four species, *Homo sapiens, Saccharomyces cerevisiae, Drosophila melanogaster*, and *Caenorhabditis elegans*. The comparison confirmed that local aligners tend to produce smaller alignments in which many regions of the networks are left unaligned, while global network aligners align more regions of the networks. The comparison showed that when using only topological information during the alignment process, global network alignments outperform local network alignments both topologically and biologically. When sequence information is also used, global network alignments again outperform local network alignments in terms of topological quality, but local network alignments are superior in terms of biological quality. All together, local alignments had high biological relevance, but left many regions of the networks unaligned, while global aligners align more regions of the networks, but at the cost of lower biological relevance. To produce alignments having high biological relevance and that also cover the regions of the networks well, Meng et al. proposed IGLOO [45]. It is a novel alignment method that starts from "seed" local alignments (either produced by AlignMCL, or by AlignNemo), which are extended to best cover the two networks to be aligned.

9.2.5 Global Pairwise Network Alignment Methods

In this section, we detail some of the global pairwise network alignment methods. As there are many, we detail three of them and then briefly overview several others that are commonly used.

9.2.5.1 IsoRank

IsoRank [53] was proposed as the first global network aligner. It is a two stage approach which first computes functional similarity scores R_{ij} between any protein i from the first network and any protein j from the second network, and then uses these scores to guide its two different alignment strategies.

Intuitively, in IsoRank's objective function, two proteins from different networks are similar if their neighborhoods are also similar. Inspired by Google's PageRank algorithm [54], IsoRank requires that for any aligned protein pair, $i \leftrightarrow j$, the functional

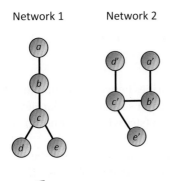

$$R \begin{cases} \bullet\ R_{aa'} = 1/4\ R_{bb'} \\ \bullet\ R_{bb'} = 1/3\ R_{ac'} + 1/3\ R_{a'c} + R_{aa'} + 1/9\ R_{cc'} \\ \bullet\ R_{cc'} = 1/4\ R_{bb'} + 1/2\ R_{be'} + 1/2\ R_{bd'} + 1/2\ R_{eb'} + 1/2\ R_{db'} + R_{ee'} + R_{ed'} + R_{de'} + R_{dd'} \\ \bullet\ R_{dd'} = 1/9\ R_{cc'} \\ \ldots \end{cases}$$

Figure 9.9: An illustration of IsoRank's functional similarity score. Applying Equation 9.14 between the nodes of networks 1 and 2 defines a system of equations that constrains R (only the four first equations are presented in the figure). For example, nodes a and a' have only one neighbor each (b for node a, and b' for node a'), and thus, $R_{aa'}$ only depends on $R_{bb'}$. However, b and b' have two neighbors each, and thus $R_{bb'}$ has to be equally shared across the four corresponding neighbor matching scores ($R_{aa'}$, $R_{ac'}$, $R_{a'c}$, and $R_{cc'}$). This leads to $R_{aa'} = \frac{1}{4} R_{bb'}$. The presented values of R in the table are obtained after solving the system of equations.

similarity score, R_{ij}, should be equal to the total support provided by each of the $|N(i)| \times |N(j)|$ possible matches between the neighbors of i and j. In turn, each aligned protein pair from the neighborhoods of i and j, $u \leftrightarrow v$, must distribute back its score R_{uv} equally among the possible matches between their neighbors, i.e.:

$$R_{ij} = \sum_{u \in N(i)} \sum_{v \in N(j)} \frac{1}{|N(u)| \times |N(v)|} R_{uv}. \tag{9.14}$$

The intuition behind this functional similarity score is illustrated in Figure 9.9.

Finding R that satisfies Equation 9.14 can be done by solving the eigenvalue problem:

$$R = AR, \text{ s.t. } A[i,j][u,v] = \begin{cases} \frac{1}{|N(u)| \times |N(v)|} & \text{if } (i,u) \in V_1, (j,v) \in V_2, \\ 0 & \text{otherwise.} \end{cases} \tag{9.15}$$

This eigenvalue problem is solved using the iterative power method, which repeatedly updates R according to the update rule:

$$R \leftarrow \frac{AR}{|AR|}. \tag{9.16}$$

The functional similarity score can be extended to take into account not only the numbers of shared neighbors, but also the sequence similarity of the aligned proteins:

$$R = \alpha A R + (1 - \alpha) S, 0 \leq \alpha \leq 1, \tag{9.17}$$

where S is a matrix containing sequence-based similarities between the proteins of N_1 and the proteins of N_2, normalized so that the sum of all sequence similarities equal 1. Parameter α balances the weight of the network data (shared neighborhood) versus sequence data.

Then, IsoRank searches for a one-to-one alignment that maximizes the sum of R scores between the aligned proteins. IsoRank can optimally solve its alignment problem by using the maximum weight bipartite matching algorithm on the bipartite network in which the nodes of V_1 are connected to nodes in V_2 by edges having weights from R. Because solving maximum weight bipartite matching problems can be computationally intensive, with a worst time complexity of $O(n^3)$ when aligning networks with n nodes, IsoRank can also produce an alignment with the heuristic iterative procedure described in Algorithm 9.1.

Algorithm 9.1 Isorank'S approximate maximum weight bipartite matching.

1: **Input** R
2: $a = \emptyset$
3: **while** $R \neq \emptyset$ **do**
4:     Identify the highest score R_{pq}
5:     Add $p \leftrightarrow q$ to a
6:     Remove all scores involving p or q from R
7: **Output** alignment a

9.2.5.2 Global Alignment Methods: GRAAL and H-GRAAL

GRAAL [19] is the earliest member of the GRAAL family of network aligners, which includes GRAAL, H-GRAAL [55], MI-GRAAL [56], C-GRAAL [57] and L-GRAAL [41]. GRAAL is the earliest network aligner that aligns a pair of nodes originating from different networks based solely on the similarity of their wiring patterns (topology) in their respective networks.

In GRAALs' approach, the topology around a node in a network is captured by *graphlets*, which are small, connected, non-isomorphic, induced subgraphs of a larger graph (denoted by $G_0, \ldots, G_{29}$ in Figure 9.10) [58] (detailed in Chapter 5). Within each graphlet, because of symmetries, some nodes are topologically identical to each other: such identical nodes are said to belong to the same *automorphism orbit* (denoted by $0, \ldots, 72$ in Figure 9.10) [59].

Graphlets can be used to generalize the notion of the node degree: the ith *graphlet degree* of node v, denoted by v_i, is the number of times node v touches a graphlet at orbit i. The *graphlet degree vector* (GDV), or signature of a node is the 73 dimensional vector containing the graphlet degrees of the node in the network (see illustration in Figure 9.11). The GDV of a node provides a highly constraining measure of the local topology around the node in the network and comparing the GDVs of two nodes provides a measure of local topological similarity between them. The *GDV similarity*

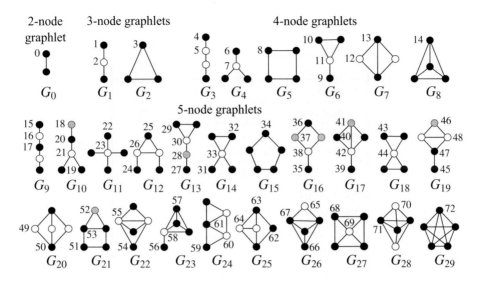

Figure 9.10: An illustration of the 30 2- to 5-node graphlets and their 73 orbits, as defined by Przulj [59]. Within each graphlet (denoted by $G_0, \ldots, G_{29}$), nodes having the same color belong to the same automorphism orbit (denoted by $0, \ldots, 72$).

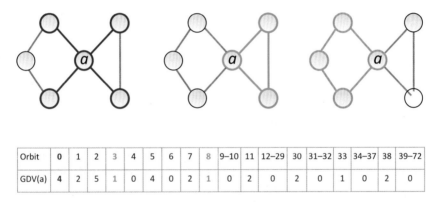

Orbit	0	1	2	3	4	5	6	7	8	9–10	11	12–29	30	31–32	33	34–37	38	39–72
GDV(a)	4	2	5	1	0	4	0	2	1	0	2	0	2	0	1	0	2	0

Figure 9.11: An illustration of the GDV of a node; e.g., node a is touched by four different edges (orbit 0, in red on the left panel), by one triangle (orbit 3, in green in the middle panel) and by one four-cycle (orbit 8, in blue in the right panel), among many other graphlets. These different orbit degrees for node a are encoded in its GDV vector, which quantifies the wiring patterns around the node in the network.

[60] between two nodes is computed as follows. The distance $D_i(u,v)$ between the ith orbits of nodes u and v is defined as:

$$D_i(u,v) = w_i \times \frac{|\log(u_i+1) - \log(v_i+1)|}{\log(\max\{u_i, v_i\} + 2)}, \tag{9.18}$$

where w_i is the weight of orbit i that accounts for dependencies between orbits. Weight w_i is computed as $w_i = 1 - \frac{\log(o_i)}{\log(73)}$, where o_i is the number of orbits that orbit i depends on, including itself. For example, the count of orbit 2 (i.e. being in the middle of a three node path) of a node depends on its count of orbit 0 (i.e., its node degree) and on itself,

Table 9.1: Orbit weights.

Orbits	Weights
o_0	1
o_1, o_2, o_3	2
$o_4, o_6, o_7, o_9, o_{14}$	3
$o_5, o_8, o_{10}, o_{11}, o_{12}, o_{13}, o_{15}, o_{18}$, $o_{22}, o_{23}, o_{24}, o_{27}, o_{31}, o_{56}, o_{72}$	4
$o_{17}, o_{19}, o_{25}, o_{35}, o_{39}, o_{44}, o_{45}, o_{54}, o_{69}$	5
$o_{16}, o_{20}, o_{21}, o_{28}, o_{29}, o_{32}, o_{33}, o_{34}$, $o_{36}, o_{41}, o_{43}, o_{46}, o_{49}, o_{50}, o_{52}, o_{55}$, $o_{57}, o_{58}, o_{61}, o_{62}, o_{65}, o_{70}, o_{71}$	6
$o_{26}, o_{30}, o_{37}, o_{38}, o_{40}, o_{42}, o_{48}, o_{59}$, o_{64}, o_{66}, o_{67}	7
$o_{47}, o_{51}, o_{60}, o_{63}, o_{68}$	8
o_{53}	9

so $o_2 = 2$. For orbit 15, $o_{15} = 4$, since it is affected by orbits 0, 1, 4, and itself. The values of o_i for all 2- to 5-nodes graphblet orbits are listed in Table 9.1.

The similarity $S(u, v)$ between nodes u and v is defined as:

$$S(u,v) = 1 - \frac{\sum_{i=0}^{72} D_i(u,v)}{\sum_{i=0}^{72} w_i}. \qquad (9.19)$$

$S(u, v)$ is in (0, 1], where similarity 1 means that the GDVs of nodes u and v are identical.

In GRAAL, the cost $C(u, v)$ of aligning two proteins, u and v, between two PPI networks takes into account both their GDV similarities and their node degrees, $deg(u)$ and $deg(v)$, so that densely connected regions of the networks are aligned first:

$$C(u,v) = 2 - \left((1-\alpha) \times \frac{deg(u) + deg(v)}{max\_deg(N_1) + max\_deg(V_2)} + \alpha \times S(u,v)\right), \qquad (9.20)$$

where $max\_deg(N)$ is the maximum degree of nodes in network N, $S(u, v)$ is the graphlet degree similarity between the two nodes, and α is a parameter that controls the importance of node degrees and GDV similarity. Between two nodes, the cost of 0 corresponds to a pair of topologically similar nodes, while the cost close to 2 corresponds to a pair of topologically very different nodes.

Between two networks, $N_1 = (V_1, E_1)$ and $N_2 = (V_2, E_2)$, GRAAL heuristically searches for one-to-one alignments having minimum sum of pairwise costs by using a greedy seed-and-extend strategy. Starting from an empty alignment, GRAAL first considers a single "seed" pair of nodes (one node from each network) with low pairwise cost and adds it to the alignment. Then, GRAAL greedily extends the alignment by aligning the neighbors of the already aligned nodes having minimum pairwise costs. The extension process is repeated until no pair of nodes can be added. After the extension steps, some nodes of the smaller network may remain unaligned. For this reason, GRAAL repeats the same seed-and-extend algorithm on the pair of pth power of networks N_1 and N_2, networks N_1^p and N_2^p for $p = 1, 2, 3, \ldots$. Network $N_1^p = (V_1, E_1^p)$ has the same set of nodes as network N_1, but edge (v, x) is in E_1^p if and only if the shortest path distance between nodes v and x in N_1 is less than or equal to p. Network N_2^p is constructed in the same way from network N_2. Aligning N_1^p and $N_2^p, p \geq 1$, allows for aligning a path of length p in one network to a single edge in another network, which is analogous to allowing "gaps" in PathBLAST. GRAAL stops when all nodes from the smaller network are aligned.

On the other hand, H-GRAAL [55] finds an optimal alignment between networks N_1 and N_2 by solving the maximum weight bipartite matching on a bipartite graph (see Chapter 3) in which node $u \in V_1$ is connected to node $v \in V_2$ by edge (u, v) having the weight $2 - C(u, v)$, where $C(u, v)$ is the cost of aligning proteins u and v in GRAAL. Hence, larger weights relate more functionally and topologically similar nodes.

9.2.5.3 Overview of Other Pairwise Global Alignment Methods

MI-GRAAL [56] is the third member of the GRAAL family of network aligners. It is a one-to-one global aligner that integrates together many node-to-node similarities (namely, GDV similarity, relative degree difference, relative clustering coefficient difference, relative eccentricity difference and sequence similarity) using a linear combination. Then, MI-GRAAL searches for alignments that maximize the sum of the integrated similarity scores between the proteins using a seed-and extend strategy.

NATALIE 2.0 [61, 36] is a global, one-to-one network aligner that formalizes network alignment as a quadratic assignment problem (QAP). In operation research, given n facilities, n locations, the distances between any two locations and the possible flows (amount of supplies that can be transported) between any two facilities, QAP is the problem of assigning all facilities to different locations while minimizing the sum of distances multiplied by the corresponding flows. In NATALIE, QAP is solved using an exact integer programming approach. However, to escape from the NP-hardness of solving the integer program, NATALIE 2.0 only considers aligning proteins that are sequence similar.

C-GRAAL [57] is the fourth member of the GRAAL family of network aligner. It is a global, one-to-one network aligner that uses a "common neighborhood" seed-and-extend strategy to align networks according to MI-GRAAL's integrated similarity scores. It first finds "seed" alignments (high scoring pairs of nodes). Then it iteratively extends the seed alignment by greedily aligning the neighborhoods of already aligned nodes. When the alignment cannot be extended in this fashion anymore, C-GRAAL greedily aligns the remaining nodes.

GHOST [34] is a global, one-to-one network aligner that encodes the wiring pattern around a node in a network with a "spectral signature" derived from the eigendecomposition of a matrix encoding the connectivity in the node's local neighborhood. GHOST searches for alignments that maximize the combination of sequence similarity and spectral signature similarity between the aligned nodes by heuristically solving a quadratic assignment problem similar to the one of NATALIE 2.0.

PINALOG [62] is a global, one-to-one network aligner in which the score of aligning two proteins is a linear combination of their sequence similarity and of the similarity between their GO annotations. PINALOG first identifies and aligns communities (regions in the networks in which nodes are more densely connected to each other than with the rest of the networks), and then iteratively extends the alignments by aligning the neighborhoods of already aligned nodes using bipartite matching.

PISWAP [38] is a global, one-to-one network aligner that first relies purely on the sequence similarities between the proteins of the two networks to generate an initial alignment using a maximum weight bipartite matching algorithm. Then, it improves its initial alignment by using the intuition that biologically conserved interactions can compensate for mapping proteins whose sequences are not particularly similar. In this way, the topology of the networks is taken into account and information is propagated from each vertex to its neighbors. The alignment itself is iteratively improved using a local "3-opt" heuristic, which is originally used for solving the traveling salesman problem [63], and which consists of randomly swapping three edges in the bipartite alignment graph when trying to improve the alignment score.

SPINAL [37] is a global, one-to-one network aligner that uses a two-pass matching algorithm. In the first pass, SPINAL iteratively improves the match confidence for each pair of nodes by taking into account the confidence of matching their neighbors that was computed in the previous iteration. This process is repeated until convergence is achieved. Then, in the second pass, SPINAL builds its alignment either by using a seed-and-extend algorithm, or by performing the maximum-weight bipartite matching.

NETAL [64] is a global, one-to-one aligner that instead of computing the similarity between proteins once-for-all before aligning them, iteratively recomputes the similarities between currently non-aligned proteins during the alignment process. At each iteration, the scores of aligning non-aligned nodes depends on the expected number of interactions incident to them that would be added to the alignment if the two proteins were to be matched.

HUBALIGN [40] is a global, one-to-one network aligner that is based on the observation that proteins acting as hubs in the PPI networks are functionally and topologically more important, as their removal may disconnect functional parts of the interactomes [65] (detailed in Chapter 5). HUBALIGN heuristically estimates the "hubbiness" of a protein, which they call its "importance" score, by iteratively peeling-off the lowest degree nodes. Then, HUBALIGN aligns proteins based on a combination of their sequence similarities and their importance scores, using a greedy seed-and-extend algorithm.

MAGNA [35] is a global, one-to-one network aligner that directly aims at maximizing the edge conservation between the aligned networks. MAGNA either optimizes edge correctness, induced conserved substructure, or symmetric substructures scores by using a dedicated genetic algorithm. Genetic algorithms are evolutionary based heuristics relying on three operations, mutation, crossover, and selection, and on a function to measure the quality of a solution, called the fitness function. Starting from a pool of initial solutions (called the first generation), a genetic algorithm iteratively generates a new generation of solutions by using the following steps: (1) individuals in the current generation are randomly selected and undergo slight changes using the mutation operator, (2) the fitness of all individuals is measured using the fitness function and pairs of solutions having the highest fitness are combined to produce new offspring solutions using the crossover operator, and (3) solutions are selected to form the new generation of solutions according to the selection operator to both optimize fitness and to preserve diversity in the solution pool. A genetic algorithm iterates until a maximum number of generations has been produced, or a satisfactory fitness has been reached. MAGNA++ [39] is an updated version of MAGNA that allows for parallel computations, as well as the ability to use node conservation in addition to edge conservation.

L-GRAAL [41] is the last member of the GRAAL family of network aligners. Unlike previous aligners that either do not take into account the mapped interactions, or use naive interaction mapping scoring schemes, L-GRAAL directly optimizes an objective function that takes into account both sequence-based protein conservation and topological, graphlet-based interaction conservation. L-GRAAL formalizes its alignment problem as an integer program, which L-GRAAL solves using an iterative double dynamic programming heuristic based on Lagrangian relaxation [66].

OPTNETALIGN [42] uses a multi-objective memetic (genetic) algorithm, coupling swap-based local search, mutation, and crossover operations to create a population of alignments that optimize the conflicting goals of topological and sequence similarity. OPTNETALIGN uses the concept of Pareto dominance to explore the trade-off between the two objectives as it runs [67]. In multi-objective optimization, Pareto dominance is a binary relationship between two solutions: one solution is Pareto dominant with respect to another solution if for all the objective functions (e.g., for OPTNETALIGN, topological similarity and sequence similarity), it improves on the other solution.

MODULEALIGN [43] uses a hierarchical clustering (detailed in Chapter 6) of functionally related proteins to define its module-based homology scores between proteins. Then, it uses an iterative algorithm to find an alignment that maximizes a linear combination of its homology scores and of HUBALIGN's importance scores.

SANA [68] is a global, one-to-one aligner that directly maximizes the S^3 score between networks. SANA generates its alignments using an heuristic based on simulated annealing (SA). SA is an iterative heuristic that, at each iteration, considers some neighboring solution s' of the current solution s and accepts moving to solution s' or staying in state s according to some probability. Similar to the slow cooling process in annealing, the probability of accepting a bad move decreases over the iterations and the process ends when it converges to a local optimum.

As we can see, a multitude of methods based on different objective functions and optimization techniques has been proposed. Global network aligners have been extensively reviewed and compared (e.g., [69, 70]), but no method emerged as a gold standard. In particular, the following three issues remained unanswered. First, because of the variety of network aligners that are available, choosing an appropriate method is a difficult task. This is made harder by the large number of alignment quality measures. Second, aligners combine sequence similarity with connectivity similarity to guide their alignment processes in order to produce more biologically relevant alignments than by using only sequences. Such aligners use parameters that balance the amount of topological and sequence information that is used to guide their alignment processes. However, there is no clear guidance on how to set-up these parameters to produce the most biologically relevant alignments. Finally, the main limitation of global aligners is the coverage of their alignments. When aligning a small network to a larger one, many proteins of the larger network are left unaligned and no information can be gained for the unaligned proteins.

To investigate these issues, Malod-Dognin et al. [44] comprehensively evaluated eight state of the art network alignment methods (NATALIE 2.0 [36], SPINAL [37], PISWAP [38], MAGNA++ [39], HUBALIGN [40], L-GRAAL [41], OPTNETALIGN [42], and MODULEALIGN [43]) when aligning the protein–protein interaction networks of eight species (*Homo sapiens, Saccharomyces cerevisiae, Drosophila melanogaster, Arabidopsis thaliana, Mus musculus, Caenorhabditis elegans, Schizosaccharomyces pombe,* and *Rattus norvegicus*) using various scoring schemes.

First, on this large scale PPI data, they observed that three methods, HUBALIGN, L-GRAAL and NATALIE 2.0, regularly produce the most topologically and biologically coherent alignments.

Second, when investigating the parameters that balance the amount of topological and sequence information used to guide the alignment processes, they observed that using topological information only results in alignments having the highest topological coherence and the lowest biological coherence. In contrast, using sequence information only results in alignments having the highest biological coherence and the lowest topological coherence. Importantly, their experiments show that global network aligners do not succeed in using topological information (even in combination with sequence information) to produce alignments that are biologically more relevant than the alignments based solely on sequence information.

Finally, when studying the collective behavior of network aligners, they observed that PPI networks were almost entirely aligned with a handful of aligners, which they unified into a new tool, **ULIGN**. This new tool enables complete alignment of two networks, which traditional global and local aligners fail to do. Also, the multiple mappings of ULIGN define biologically relevant soft clusterings of proteins in PPI networks, which can be used for refining the transfer of annotations across networks.

9.3 Multiple Network Alignment

9.3.1 Context and Formal Definitions

Network alignment originated as a pairwise problem. However, with the availability of PPI networks of multiple species came the need for multiple network alignments. In multiple network alignment, given k networks, aligning them means to group together the proteins that are evolutionary, or functionally conserved in all networks. Multiple network alignment allows uncovering network regions that are conserved across all networks, which may correspond to biological processes shared by all members of a family of species.

As the complexity of network alignment grows exponentially with the number of networks to be aligned, the proposed multiple network alignment algorithms use simple and scalable alignment schemes. Multiple network alignment is better seen as a clustering problem in k-partite graphs, in which clusters (also called equivalence classes) should group together evolutionary and functionally related proteins across

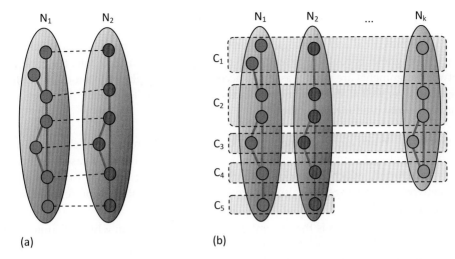

Figure 9.12: Pairwise and multiple alignments. (a) A pairwise alignment, in which some nodes of network N_1 are aligned to nodes of network N_2, as represented by the dashed lines. (b) A multiple network alignment, in which clusters (dashed rectangles) align together nodes coming from different networks.

the input networks (see Figure 9.12) [71]. Let $N = \{N_1, N_2, \ldots, N_k \mid N_i = (V_i, E_i)\}$ be the set of k input PPI networks. The possible matchings between the proteins of the k input networks are represented in a k-partite graph $G = (V_1 \cup V_2 \cup \cdots \cup V_k, E)$ in which edges in E connect nodes coming from different networks and are associated with a weight representing the cost of matching the two nodes. Multiple network alignment is the problem of finding in G a set of clusters $C = \{c_1, c_2, \ldots, c_n\}$ that maximizes both the functional similarity between the aligned proteins (intra-cluster similarity, ICS, which is the sum of edge-weights between the proteins of the same cluster) and the conservation of the interaction (inter-cluster conservation, ICC, which is the number of conserved interactions between the proteins of any two clusters). Thus, multiple network alignment searches in G for a clustering C that maximizes:

$$S(C) = \alpha \sum_i ICS(c_i) + (1-\alpha) \sum_{i \neq j} ICC(c_i, c_j), \qquad (9.21)$$

where α is a parameter that balances the influence of intra-cluster and inter-cluster similarity on the node alignment process.

Such alignment is global if the clusters are non-overlapping and if the set of clusters is maximal (i.e., if no additional cluster can be added to C); otherwise, it is local. Also, it is one-to-one if clusters in C contain at most one protein from each network. Otherwise, it is many-to-many.

Note that even when reduced to a k-partite matching problem (i.e., when $\alpha = 1$), multiple network alignment cannot be optimally solved, as k-partite matching is NP-hard for $k \geq 3$ [72, 73].

9.3.2 Scoring Multiple Network Alignments

While the biological and topological scores that are used to measure the quality of pairwise alignments could be extended to multiple network alignments, the multiple network alignments are mainly assessed according to their ability to produce clusterings that cover well all the input networks and that group proteins of similar functions. The commonly used measures are:

k-coverage: When aligning n networks, k-coverage is the number of clusters in the alignment that contain proteins from $k \leq n$ networks. In many cases, only $k = n$ is reported, which corresponds to clusters containing proteins from all input networks. k-coverage can also be presented as the number of proteins in such clusters.

Exact cluster ratio: This is the percentage of aligned clusters in which all annotated proteins share at least one GO term. Similar to k-coverage, it can also be presented as the percentage of all proteins that are in the exact clusters.

Mean normalized entropy (MNE): This is another measure of the consistency of the annotations of genes in the aligned clusters. The normalized entropy, $NE(c)$, of cluster c is defined as:

$$NE(c) = -\frac{1}{\log d} \sum_{i=1}^{d} p_i \times \log p_i, \tag{9.22}$$

where p_i is the fraction of proteins in c annotated with term t_i and d is the number of different terms that annotate proteins in c. MNE is the average of the normalized entropy of all the clusters containing annotated proteins.

Next, we present examples of multiple network alignment methods. We present in detail the reference one, SMETANA [74], as well as the recent FUSE [75], which outperforms SMETANA in terms of quality of its produced multiple network alignments. Note that SMETANA relies on concepts from probability and its description is more suited for advanced readers. The description of FUSE is more accessible to the general audience.

9.3.3 Multiple Network Alignment Method: SMETANA

SMETANA (semi-Markov random walk scores Enhanced by consistency Transformation for Accurate Network Alignment) is a global, many-to-many multiple network aligner that searches between networks for alignments having maximum expected accuracy (MEA) [74]. It relies on pairwise node scores based on a semi-Markov random walk mode, which provides a probabilistic similarity measure between nodes that belong to different networks.

Let $N_{1,2} = (V_{1,2}, E_{1,2})$ be the product graph of networks $N_1 = (V_1, E_1)$ and $N_2 = (V_2, E_2)$, where $V_{1,2}$ is the Cartesian product $V_1 \times V_2$ and two nodes in $V_{1,2}$ are connected by edges in $E_{1,2}$ if the corresponding nodes in N_1 are connected by an edge in E_1 and if the corresponding nodes in N_2 are connected by an edge in E_2. Between N_1

and N_2, SMETANA estimates $s(u,v)$, the correspondence scores between two proteins $u \in V_1$ and $v \in V_2$, using random walks on $N_{1,2}$ to account for their wiring patterns. Additionally, the semi-Markov random model of SMETANA assumes that the expected amount of time that a random walker spends at node uv is proportional to the sequence-based similarity $h(u,v)$, so that both higher interaction pattern similarity and sequence similarity between nodes would lead to higher correspondence scores. The correspondence scores are computed as:

$$S = \frac{Q \circ H}{\mathrm{tr}(QH^T)}, \tag{9.23}$$

where S, H and Q are $|V_1| \times |V_2|$ dimensional matrices such that $S[u][v] = s(u,v)$, $H[u][v] = h(u,v)$, $Q[u][v] = \pi_1(u)\pi_2(v)$ (π_i being the steady state distribution of random walks on network N_i), tr is the trace of a matrix (i.e., the sum of its main diagonal elements) and $\circ$ denotes the Hadamard (element-wise) product. These scores are computed for all pairs of networks in $N = \{N_1, N_2, \ldots, N_k\}$.

Based on their correspondence scores, the probability of the two nodes being aligned is:

$$p(u \leftrightarrow v | N_1, N_2) = \frac{1}{2}\left[\frac{s(u,v)}{\sum_{v' \in V_2} s(u,v')} + \frac{s(u,v)}{\sum_{u' \in V_1} s(u',v)}\right], \tag{9.24}$$

which is better expressed in terms of matrices:

$$P = \frac{1}{2}[J_1 S + S J_2], \tag{9.25}$$

where P is a $|V_1| \times |V_2|$ matrix such that $P[u][v] = p(u \leftrightarrow v|N_1, N_2)$, J_1 is a $|V_1| \times |V_1|$ diagonal matrix such that $J_1[u][u] = 1/\sum_{j \in V_2} s(u,j)$, and J_2 is a $|V_2| \times |V_2|$ diagonal matrix such that $J_2[v][v] = 1/\sum_{j \in V_1} s(j,v)$.

These probabilities are enhanced through intra- and inter-network consistency transformations. The intra-network consistency transformation is based on the idea that the more likely the neighbors of u are to be aligned to the neighbors of v, the more likely it is that u will be aligned to v. Thus, SMETANA uses the alignment probabilities of the neighbors of u and v to refine the alignment probability of $u \leftrightarrow v$:

$$P' = \alpha P + (1-\alpha) A_1 P A_2^T, \tag{9.26}$$

where P' is the matrix containing the transformed probabilities of matching nodes of the two networks, and where A_1 and A_2 are the transition probability matrices of networks N_1 and N_2, respectively.

In the second consistency transformation, SMETANA incorporates the information from all other networks in the alignment to improve the estimation of pairwise node alignment probabilities. To this aim, SMETANA estimates the probability that two networks N_i and N_j are homologous to each other as:

$$p(N_i \diamond N_j) \approx \frac{1}{|M|} \sum_{u \leftrightarrow v \in M} p(u \leftrightarrow v | N_i, N_j) \tag{9.27}$$

where $N_i \diamond N_j$ denotes that N_i and N_j are homologous and M is a maximum weighted bipartite matching between N_i and N_j. Then, the posterior probabilities of two nodes being aligned given all networks in N is $p(u \leftrightarrow v|N)$, which is computed as:

$$P'' = \frac{\sum_{N_i \in N} P'_{1,i} P'_{2,i} p(N_1 \diamond N_i) p(N_2 \diamond N_i)}{\sum_{N_i \in N} p(N_1 \diamond N_i) p(N_2 \diamond N_i)}, \qquad (9.28)$$

where P'' is a $|V_1| \times |V_2|$ dimensional matrix such that $P''[u][v] = p(u \leftrightarrow v|N)$, and $P'_{i,j}$ are the transformed alignment probabilities between networks N_i and N_j, as computed by Equation 9.26.

Given a set of networks $N = \{N_1, N_2, \ldots, N_k\}$, SMETANA searches for a multiple network alignment A that maximizes the expected accuracy over all networks in G. Let A^* be the true multiple network alignment between networks in G. The accuracy of alignment A is:

$$\text{accuracy}(A, A^*) = \frac{1}{|A|} \sum_{u \leftrightarrow v \in A} q(u, v), \qquad (9.29)$$

where $q(u, v) = 1$ if the two nodes u and v are aligned in the reference alignment A^*, or 0 otherwise. Since A^* is generally not known, A is chosen to maximize the expected accuracy:

$$E_{A^*}[\text{accuracy}(A, A^*)] = \frac{1}{|A|} \sum_{u \leftrightarrow v \in A} P(u \leftrightarrow v|G), \qquad (9.30)$$

where $P(u \leftrightarrow v|G)$ is the posterior probability of aligning u and v. Using the transformed probabilities defined in Equation 9.28, the MEA problem reduces to a maximum-weighted k-partite matching.

SMETANA finds a suboptimal solution using the following greedy approach. Starting from an empty alignment, SMETANA greedily constructs the alignment through successive insertion of the node pair with the largest posterior alignment probability.

When SMETANA was introduced, it clearly outperformed the earlier multiple network alignment methods with its ability to produce alignments that both cover the networks well and that group together proteins that are functionally related, which is why it became a reference aligner to which all recent approaches are compare with. However, as highlighted by Vijayan and Milenković [76], SMETANA suffers from running time and memory issues when aligning the large scale PPI data that are now available.

9.3.4 Multiple Network Alignment Method: FUSE

FUSE [75] is a global, one-to-one multiple network aligner that, similar to SMETANA, defines new pairwise scores between proteins that capture their homological and topological relationships from all networks. However, FUSE does so by integrating all protein sequence similarities and all PPI network topologies together using a non-negative matrix tri-factorization (NMTF) approach, as illustrated in Figure 9.13.

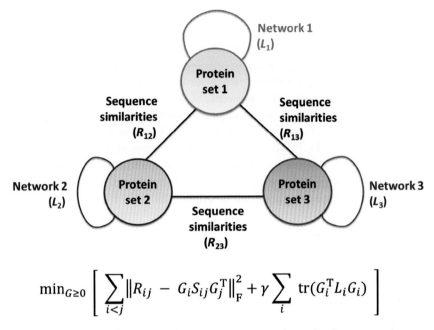

Figure 9.13 An illustration of FUSE's integration scheme for three networks.

NMTF is the problem of approximating a high-dimensional matrix, $\mathbf{R}_{ij} \in \mathbb{R}_+^{n_i \times n_j}$, whose entries relate n_i object of type i with n_j objects of type j, with the product of three low-dimensional non-negative matrix factors: $\mathbf{R}_{ij} \approx \mathbf{G}_i \mathbf{S}_{ij} \mathbf{G}_j^T$, where, $\mathbf{G}_i \in \mathbb{R}_+^{n_i \times k_i}$ is the cluster indicator matrix that groups together the n_i objects of type i into k_i clusters, $\mathbf{G}_j \in \mathbb{R}_+^{n_j \times k_j}$ is the cluster indicator matrix that groups together the n_j objects of type j into k_j clusters, and $\mathbf{S}_{ij} \in \mathbb{R}^{k_i \times k_j}$ is a low-dimensional, compressed version of $\mathbf{R}_{ij}$ that relates the two clusterings together. To obtain the low-dimensional matrix factors, $\mathbf{G}_i, \mathbf{S}_{ij}, \mathbf{G}_j$, one needs to solve the following optimization problem:

$$\min_{\mathbf{G}_i \geq 0, \mathbf{G}_j \geq 0} J = \min_{\mathbf{G}_i \geq 0, \mathbf{G}_j \geq 0} \| \mathbf{R}_{ij} - \mathbf{G}_i \mathbf{S}_{ij} \mathbf{G}_j^T \|_F^2 . \tag{9.31}$$

NMTF has three main applications:

Co-clustering: There is a close relationship between NMTF and k-means clustering; cluster indicator matrix $\mathbf{G}_i$ can be used to cluster objects of type i according to their relationships with objects of type j, and similarly, cluster indicator matrix $\mathbf{G}_j$ can be used to cluster objects of type j according to their relationships with objects of type i.

Dimentionality reduction: This is achieved by choosing rank parameters, $k_i, k_j \ll \min\{n_i, n_j\}$.

Matrix completion: Some entries in the initial relation matrix $\mathbf{R}_{ij}$ are zero (e.g., because of data capturing limitations, or experimental errors) and they can be recovered from the obtained low-dimensional matrix factors using the *reconstructed relation matrix*: $\hat{\mathbf{R}}_{ij} = \mathbf{G}_i \mathbf{S}_{ij} \mathbf{G}_j$.

FUSE uses the matrix completion property to predict new associations between proteins. Between two PPI networks, N_i and N_j, the sequence similarity scores between their proteins are recorded in the high-dimensional relation matrix, $\mathbf{R}_{ij} \in \mathbb{R}^{n_i \times n_j}$, where, n_i is the number of proteins in species i and n_j is the number of proteins in species j. Entries in R_{ij} are the sequence similarities based on BLAST's e-values. Interactions between proteins in PPI network, N_i, are represented by a graph Laplacian matrix, $\mathbf{L}_i = \mathbf{D}_i - \mathbf{A}_i$, where $\mathbf{A}_i$ is the adjacency matrix of network N_i and $\mathbf{D}_i$ is the diagonal degree matrix of i (i.e., diagonal entries in $\mathbf{D}_i$ are the degrees of the nodes).

To align five networks, FUSE simultaneously factorizes all relation matrices, $\mathbf{R}_{ij} \approx \mathbf{G}_i \mathbf{S}_{ij} \mathbf{G}_j^T$, $0 \leq i,j \leq 5$, under the constraints that the cluster indicator matrix of each network should favor grouping together proteins that interact in the network. Hence, FUSE minimizes the following objective function:

$$\min_{G \geq 0} J = \min_{G \geq 0} \left[\sum_{i,j} \| \mathbf{R}_{ij} - \mathbf{G}_i \mathbf{S}_{ij} \mathbf{G}_j^T \|_F^2 + \gamma \sum_i \mathrm{tr}(\mathbf{G}_i^T \mathbf{L}_i \mathbf{G}_i) \right], \quad (9.32)$$

where tr denotes the trace of a matrix (i.e., the sum of its main diagonal elements) and γ is a regularization parameter, which balances the influence of network topologies in the reconstruction of the relation matrices. The second term of Equation 9.32 is the graph regularization term, which works as follows: Connected pairs of proteins are represented with negative entries in the Laplacian matrix of the corresponding PPI network, and these entries act as rewards that reduce the value of the objective function, J, forcing the proteins to belong to the same cluster. The optimization problem (Equation 9.32) is solved by applying an algorithm that follows *multiplicative update rules* (see Chapter 7 for details).

After the convergence of NMTF, FUSE computes the reconstructed relation matrices over all pairs of networks, i and j; $\hat{\mathbf{R}}_{ij} = \mathbf{G}_i \mathbf{S}_{ij} \mathbf{G}_j$, which FUSE thresholds to keep only the top 5% entries in each reconstructed matrix. Because a large number of the initial sequence similarities is not recovered in the thresholded reconstructed matrices, FUSE defines the functional score $w_{u,v}$ between proteins $u \in N_i$ and $v \in N_j$ in the k-partite network as the linear combination of their sequence similarity, $\mathbf{R}_{ij}[u][v]$, and their NMTF-predicted score $\hat{\mathbf{R}}_{ij}[u][v]$:

$$w_{u,v} = \alpha \times \mathbf{R}_{ij}[u][v] + (1 - \alpha) \times \hat{\mathbf{R}}_{ij}[u][v], \quad (9.33)$$

where α is a balancing parameter in $[0,1]$ to either favor the sequence similarities (when $\alpha = 1$, only sequence similarities are used) or the novel predicted associations (when $\alpha = 0$, only NMTF scores are used).

FUSE models the global multiple network alignment between networks N_1, N_2, ..., N_k as the maximum weight k-partite matching problem in a graph G, whose node set is the union of the nodes sets of all input networks and whose edges connect nodes u and v having functional score $w_{u,v} > 0$. Edges in G are associated with the corresponding functional scores. Because the maximum weight k-partite matching problem is known to be NP-hard for $k \geq 3$ [72, 73], FUSE uses a simple heuristic strategy that can be seen as a progressive aligner that first maps and merges the first two networks, and then successively adds into the "merge graph" the remaining networks.

It is based on a so-called *graph merge operation*. Given FUSE's edge-weighted k-partite graph $G = (\bigcup_{i=1}^{k} V_i, E, W)$, $G[V_i, V_j]$ is the edge-weighted bipartite subgraph of G that is induced by the two subsets of nodes V_i and V_j. FUSE uses the maximum weight bipartite matching $F_{i,j} = \{u_1 \leftrightarrow v_1, u_2 \leftrightarrow v_2, \ldots, u_l \leftrightarrow v_l\}$ of $G[V_i, V_j]$, in which $u_k \leftrightarrow v_k$ means that node $u_k \in V_i$ is matched with node $v_k \in V_j$, to merge V_i with V_j into V_{ij}. This is done by identifying the mapped nodes $u_k \leftrightarrow v_k$ and by creating the corresponding *merged node* $u_k v_k \in V_{ij}$. The nodes of V_i and V_j that are not matched are also added into V_{ij} so that they can still be aligned to other networks (forming clusters that will not cover all networks). The created merged nodes inherit the edges from their parent nodes, and multiple edges are replaced by a single edge with the sum of weights of the multiple edges as the new weight of the edge. FUSE iteratively applies this graph merge operation until all graphs have been aligned, as summarized in Algorithm 9.2.

Algorithm 9.2 FUSE's approximate maximum weight k-partite matching.

1: **Input** $G = (\bigcup_{i=1}^{k} V_i, E, W)$
2: **for** $i = \{2, \ldots, k\}$ **do**
3: Find maximum weight bipartite matching $F_{1,i}$ of $G[V_1, V_i]$
4: Construct G_{1i}, the merge of V_1 and V_i from G along $F_{1,i}$
5: Set $G = G_{1i}$, and relabel V_{1i} as V_1
6: $C = \{\emptyset\}$
7: **for** each merged node u in V_1 **do**
8: Cluster C_u is the set of nodes that are merged into u
9: Add C_u to C
10: **Output** C.

FUSE has been compared with the state of the art multiple network aligners (including Isorank-N [77], Beams [71], SMETANA [74] and CSRW[78]) when aligning the five largest and most complete PPI networks from BioGRID (*Homo sapiens*, *Saccharomyces cerevisiae*, *Drosophila melanogaster*, *Mus musculus*, and *Caenorhabditis elegans*). The obtained results showed that FUSE produces the best multiple network alignments thus far, producing clusters that are more biologically relevant and that cover the networks the best. Moreover, FUSE is computationally efficient and thus applicable to modern large scale PPI data.

9.3.5 Overview of Other Multiple Network Alignment Methods

Isorank-N [77] is the direct extension of ISORANK (see Section 9.2.5) to global multiple network alignment. It uses ISORANK's strategy to derive pairwise alignment scores between every pair of networks. Then, it uses a spectral partitioning method to compute global, many-to-many multiple network alignments.

Graemlin [79] produces local multiple network alignments using a progressive alignment scheme, by successively performing pairwise alignments of the closest network pairs.

NetworkBlast-M [80] is a local multiple network aligner that greedily searches for highly conserved local regions in the alignment graph constructed from the pairwise protein sequence similarities.

NetCoffee [81] is a global, many-to-many multiple network aligner. It is the earliest multiple network aligner in which the score of mapping two nodes does not only depend on the scores in pairs of networks, but also on their conservation across all PPI networks being aligned. NetCoffee uses a triplet approach in which the score of mapping node u from network N_i to node v from network N_j, $i \neq j$, is equal to the number of nodes w from networks N_k, $k \neq i$ and $k \neq j$, such that u, v and w are all sequence homologs.

Beams [71] is a fast heuristic that constructs global, many-to-many, multiple network alignments. Beams first constructs a k-partite networks in which proteins from the input networks are connected to each other if they are sequence similar. Beams searches in this k-partite networks for all "seed" alignments that are cliques (sets of nodes such that any two nodes within a set are connected by an edge). Beams constructs a multiple alignment by greedily merging overlapping cliques into the same clusters and by selecting non-overlapping clusters that maximizes both the sequence similarity of the aligned proteins and the number of interactions that are conserved across the selected clusters.

GEDEVO-M [82] is a global, one-to-one multiple network aligner based on graph edit distance (GED). GED is based on graph edit operations, namely node insertion, node deletion, edge insertion, and edge deletion, which are all assigned specific costs. The GED between two networks is the smallest sum of costs required for transforming one network into the other. Finding the sequence of editing operations having the minimum sum of costs is NP-hard [83]. Between multiple networks, GEDEVO-M search for an alignment that minimizes the sum of GEDs between all pairs of aligned networks using a genetic algorithm.

CSRW [78] is a global, many-to-many multiple network aligner that follows the SMETANA methodology (see Section 9.3.3) to produce alignments having maximum expected accuracy (MEA). The main difference with SMETANA is in how the random walkers explore the product graph $N_{1,2}$ to derive the pairwise scores between the nodes of networks N_1 and N_2. In both approaches, random walkers move along the edges of the product graph, which represent conserved edges between the two input networks. In SMETANA, random walkers move along the edges uniformly at random, while CSRW uses a context sensitive approach, in which the probability of using an edge to move towards a node uv depend on the similarity of the degrees of node u in N_1 and of node v in N_2.

MultiMAGNA++ [76] is the extension of MAGNA++ to global multiple network alignment. Using a genetic algorithm, MultiMAGNA++ produce global, one-to-one multiple network alignments that directly maximize edge conservation between the aligned networks.

While pairwise network aligners produce aligned node pairs between two networks, multiple network aligners produce aligned node clusters between more than two networks. Because multiple network alignments uncover conserved regions between more networks than pairwise alignments, multiple network alignments are considered to be more robust and insightful than pairwise alignments. As a consequence, the recent focus of research has been on developing new multiple network alignment methods, despite the higher computational complexity of the multiple network alignment problem. However, due to the differences between pairwise and multiple network alignment outputs, as well as between the corresponding topological and biological quality measures that are used to evaluate them, pairwise and multiple network alignments are rarely compared to each other. Thus, it is not clear whether multiple network alignments are superior to pairwise alignments, or not.

To answer this question, Vijayan et al. [84] introduced a framework to compare pairwise against multiple network aligners in both a pairwise and multiple manner. They observed that pairwise alignment methods can be used to produce multiple alignments that are biologically and topologically more relevant than the multiple alignments produced by multiple network aligners. Also, they observed that using pairwise alignments allows for higher protein function prediction accuracy than using multiple alignments. Finally, pairwise network aligners are shown to be computationally more efficient than multiple network aligners. (See Table 9.2 for a list of available network alignment methods.)

9.4 Aligning Other Types of Networks

Network alignment originated from the problem of finding similarities between protein–protein interaction networks of different species. These are undirected, unweighted graphs. However, these graphs may not be suited to model all complex biological systems. We briefly illustrate alternative network formalisms that have been used to model interaction data, as well as some methods that have been proposed to align them.

9.4.1 Probabilistic Networks

Many interactions in biology have associated probabilities. For instance, an interaction between two proteins may depend on factors such as protein abundance and proximity. Similarly, the reliability of the observed interactions may depend on many methodological and experimental factors. For instance, the STRING database [87] assigns a protein–protein interaction a probability depending on the type and the number of different experiments in which the proteins have been reported to interact. Such probabilistic interactions are better models as probabilistic networks $N = (V, E, P)$, where nodes in V may interact according to edges in E with probabilities (e.g., weights in $[0, 1]$) in P (illustrated in Figure 9.14). To align probabilistic networks, Todor et al. [88] proposed a variant of ISORANK (described in Section 9.2.5) in which the score, R, of aligning proteins is modified to account for the probabilistic nature of the interactions. In ISORANK, the scores of aligning proteins, R, is iteratively computed as $R \to \alpha AR + (1-\alpha)S$, where S is the matrix containing the sequence similarities between the nodes and A represents the local similarity between the edges. A is computed as

$$A[i,j][u,v] = \begin{cases} \frac{1}{|N(u)| \times |N(v)|}, & \text{if } (i,u) \in V_1, (j,v) \in V_2 \\ 0, & \text{otherwise} \end{cases}, \qquad (9.34)$$

Table 9.2: The list of available network alignment methods. Methods in bold are presented in detail in this chapter.

Method	Local or global	Pairwise or multiple	Many-to-many or one-to-one	Availability
PathBLAST [18]	Local	Pairwise	Many-to-many	http://www.pathblast.org
NetworkBLAST [85]	Local	Pairwise	Many-to-many	http://www.cs.tau.ac.il/~bnet/networkblast.htm
MaWISh [47]	Local	Pairwise	Many-to-many	http://compbio.case.edu/koyuturk/software/mawish
NetAligner [48]	Local	Pairwise	Many-to-many	http://netaligner.irbbarcelona.org
AlignNemo [49]	Local	Pairwise	Many-to-many	http://www.sourceforge.net/p/alignnemo/home/Home
AlignMCL [51]	Local	Pairwise	Many-to-many	http://sites.google.com/site/alignmcl
Graemlin [79]	Local	Multiple	Many-to-many	http://graemlin.stanford.edu
NetworkBLAST-M [80]	Local	Multiple	Many-to-many	http://www.cs.tau.ac.il/~bnet/License-nbm.htm
LocalAli [86]	Local	Multiple	Many-to-many	https://code.google.com/p/localali
IsoRank [53]	Global	Pairwise	One-to-one	http://groups.csail.mit.edu/cb/mna
GRAAL [19]	Global	Pairwise	One-to-one	http://www0.cs.ucl.ac.uk/staff/natasa/GRAAL
H-GRAAL [55]	Global	Pairwise	One-to-one	http://www.nd.edu/~cone/software_data.html
MI-GRAAL [56]	Global	Pairwise	One-to-one	http://www0.cs.ucl.ac.uk/staff/natasa/MI-GRAAL
NATALIE 2.0 [36]	Global	Pairwise	One-to-one	http://www.mi.fu-berlin.de/w/LiSA/Natalie
C-GRAAL [57]	Global	Pairwise	One-to-one	http://www0.cs.ucl.ac.uk/staff/natasa/C-GRAAL
GHOST [34]	Global	Pairwise	One-to-one	http://www.cs.cmu.edu/~ckingsf/software/ghost
PINALOG [62]	Global	Pairwise	One-to-one	http://www.sbg.bio.ic.ac.uk/~pinalog
PISwap [38]	Global	Pairwise	One-to-one	http://cb.csail.mit.edu/cb/piswap/webserver
SPINAL [37]	Global	Pairwise	One-to-one	http://code.google.com/p/spinal
NETAL [64]	Global	Pairwise	One-to-one	http://bioinf.modares.ac.ir/software/netal
HUBALIGN [40]	Global	Pairwise	One-to-one	http://ttic.uchicago.edu/~hashemifar/software/HubAlign.zip
MAGNA [35]	Global	Pairwise	One-to-one	http://www.nd.edu/~cone/MAGNA
MAGNA++ [39]	Global	Pairwise	One-to-one	http://www.nd.edu/~cone/MAGNA++
L-GRAAL [41]	Global	Pairwise	One-to-one	http://www0.cs.ucl.ac.uk/staff/natasa/L-GRAAL
OPTNETALIGN [42]	Global	Pairwise	One-to-one	https://github.com/crclark/optnetaligncpp
MODULEALIGN [43]	Global	Pairwise	One-to-one	http://ttic.uchicago.edu/~hashemifar/ModuleAlign.html
SANA [68]	Global	Pairwise	One-to-one	http://sana.ics.uci.edu
IsoRank-N [77]	Global	Multiple	Many-to-many	http://groups.csail.mit.edu/cb/mna
SMETANA [74]	Global	Multiple	Many-to-many	http://www.ece.tamu.edu/~bjyoon/SMETANA
BEAMS [71]	Global	Multiple	Many-to-many	http://webprs.khas.edu.tr/~cesim/BEAMS.tar.gz
NetCoffee [81]	Global	Multiple	Many-to-many	http://code.google.com/p/netcoffee
GEDEVO-M [82]	Global	Multiple	One-to-one	http://gedevo.mpi-inf.mpg.de/multiple-network-alignment
CSRW [78]	Global	Multiple	Many-to-many	Upon request
FUSE [75]	Global	Multiple	One-to-one	http://www0.cs.ucl.ac.uk/staff/natasa/FUSE
MultiMAGNA++ [76]	Global	Multiple	One-to-one	http://nd.edu/~cone/multiMAGNA++

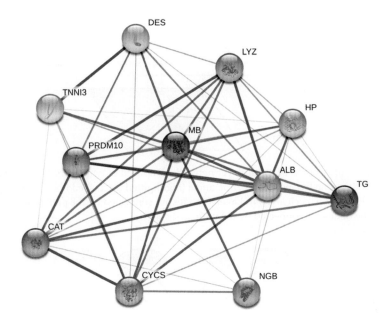

Figure 9.14: An illustration of a probabilistic network representing the protein-protein interactions of human myoglobin (MB, in red). The thickness of the edges is proportional to their probabilities. The figure was generated from the STRING database [87].

where $N(u)$ and $N(v)$ are the neighborhoods of nodes u and v, respectively. In the approach of Todor et al, A is replaced with its expected values, $E(A)$, which are computed as $E(A[i,j][u,v]) = \sum_k kP(A[i,j][u,v] = k)$, where the summation is taken over the possible values k of $A[i,j][u,v]$.

9.4.2 Multilayer Networks

Dynamic data can be represented by consecutive events. Such a dataset is better modeled as a multilayer network, $N = (V, E^1, E^2, \ldots, E^k)$, in which interactions between nodes in V are represented by a series of edge sets $E_1, E_2, \ldots, E_k$, one for each event. In a multilayer network, each event, i, is fully described by a *layer*, which is the network composed of the node set V and of edge set E_i (illustrated in Figure 9.15). For example, in ecology, multilayer networks are used to model the daily dynamics of social interactions between individuals (e.g., between 27 zebra in Kenya over 58 days [89]). To align such networks, Vijayan et al. [90] introduced the notion of the conserved event time (CET): given two multilayer networks $N_1 = (V_1, E_1^1, E_1^2, \ldots, E_1^k)$ and $N_2 = (V_2, E_2^1, E_2^2, \ldots, E_2^k)$ for which $u \in V_1$ is aligned to $u' \in V_2$ and $v \in V_1$ is aligned to $v' \in V_2$, CET is the number of events $1 \leq i \leq k$ during which both nodes (u,v) and (u',v') are interacting in E_1^i and in E_2^i, respectively. In the same way, the non-conserved event time (NCET) is defined as the number of events during which one pair of nodes is interacting, while the other one is not interacting. DynaMAGNA++ [90], an extension of MAGNA++ [39] described in Section 9.2.5.3, uses a dedicated genetic algorithm to uncover alignments that maximize the sum of CETs while minimizing the sum

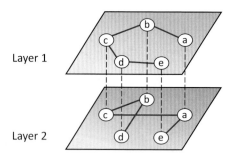

Figure 9.15: An illustration of a multilayer network composed of two layers. The two layers are composed of the same five nodes, a, b, c, d, and e (highlighted by dashed lines), but these nodes are differently wired in each layer. In Layer 1, nodes are connected by the path (a,b,c,d,e), while in Layer 2, the nodes are connected by the path (d,b,c,a,e).

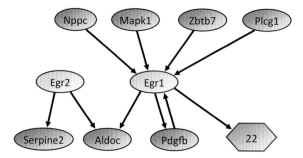

Figure 9.16: An illustration of a directed network, representing the gene regulatory interactions of rat's early growth response (EGR) proteins (in yellow) that regulate genes during cell differentiation and mitogenesis. The hexagon represents a cluster of 22 regulated genes: Dlk1, Eno2, Me1, Serpine1, Abcb1, Pnmt, Slc9a3, Adrb1, Th, Id1, Pdgfa, Chrna7, Lhb, Lhcgr, F3, Ptp4a1, Slc6a6, Myh6, Atp2a2, TSP-1, Fgf2, and Mmp14. The figure was generated from the transcriptional Regulatory Element Database [93].

of NCETs over all aligned pairs of nodes. DynaMAGNA++ can also use dynamic graphlets (an extension of graphlets to multilayer networks, see Chapter 5) to guide its alignment process.

9.4.3 Directed Networks

Unlike PPIs, some biological phenomena are directed. Examples include gene regulatory interactions, in which genes called transcription factors (TFs) regulate the expression of target genes (TGs). Directed interactions are better modeled as directed networks, $N = (V, A)$, where V is the set of nodes and in which two nodes, a and b, can be connected by a directed edge (also called a directed arc) from node a (also called the tail node) to node b (also called the head node) in A (illustrated in Figure 9.16). While traditional topological descriptors have been extended to directed graphs (e.g., directed graphlets [91, 92]), there does not exist a network alignment method dedicated to directed networks.

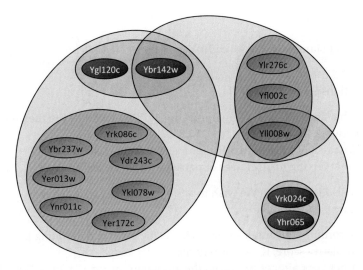

Figure 9.17: An illustration of a hyper-network, which represents protein complexes in yeast. It connects spliceosome related proteins (in pink) and ribosomal proteins (in blue). The figure is based on proteomics data from Gavin et al. [97].

9.4.4 Hyper-Networks

Traditional networks can only model binary (pairwise) relationships between the nodes (e.g, pairwise bindings between proteins). Such a model does not fully capture the higher order organization of the cellular machinery, such as protein complexes and biological pathways that can only function if they involve all of their constituent proteins. To model this multi-scale molecular organization of the cell, Estrada and Rodriguez-Velazquez [94] suggested to use an old hyper-network (hypergraph [95]) model, $N = (V, H)$, in which hyper-edges in H can connect any number of nodes in V (illustrated in Figure 9.17). Similar to directed networks, while topological descriptors have been extended to hyper-networks (e.g., hypergraphlets [96]), no hyper-network aligner has been proposed yet.

9.5 Concluding Remarks

In this chapter, we introduced in Section 9.1 the biological concepts, as well as the biological motivation for aligning networks. Then, in Section 9.2, we defined the pairwise network alignment problem. We also presented the standard network alignment scoring schemes that are used to measure the quality of pairwise network alignment, as well as key network alignment algorithms. In Section 9.3, we similarly define the multiple network alignment problem, the corresponding multiple network alignment scoring schemes and key multiple network alignment methods. While the above presented pairwise and multiple network alignment problem focus on undirected, unweighted networks, we presented in Section 9.4 a brief overview of alignment problems for other types of network data.

9.6 Exercises

The number of stars refers to the difficulty of the exercise, from easy (*) to advanced (***)

9.1 Definitions (*)

   (a) What are protein–protein interactions (PPIs) and how are they modeled as networks?
   (b) What is pairwise network alignment and what is it used for?
   (c) What are the main classes of pairwise network alignments?

9.2 Pairwise alignment (*) We wish to align the two networks shown in Figure 9.18.

   (a) According to Isorank's rules, what is the constraint defining the functional similarity score, R, between nodes $a \in V_1$ and $a' \in V_2$?
   (b) A network aligner produced the alignment presented with dashed lines.
      i. Is the alignment local, or global? Is it one-to-one, or many-to-many?
      ii. To measure if the alignment covers the two networks well, compute the node coverage of the alignment.
      iii. Also, compute the topological coherence of the alignment using edge-correctness.

9.3 Multiple network alignment (*) A multiple network aligner produced the alignment presented in Figure 9.19.

   (a) Is the alignment local, or global? Is it one-to-one, or many-to-many?

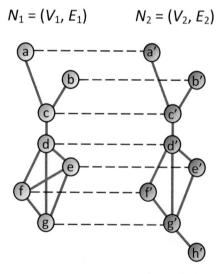

Figure 9.18: Exercise 9.6. The two networks, N_1 and N_2, to be aligned. The dashed lines show an alignment between the two networks.

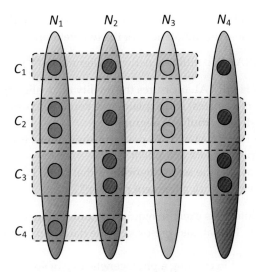

Figure 9.19: Exercise 9.6. A multiple network alignment between four networks, N_1 to N_4. The clusters C_1 to C_4 of nodes within dashed quadrangles are the result of the alignment.

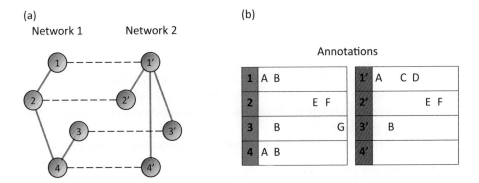

Figure 9.20: Exercise 9.6. (a) an alignment between two networks. (b) The annotations of the nodes.

  (b) To measure if the alignment covers well the four networks, which value of k should be used to compute k-coverage?
  (c) Compute the corresponding k-coverage score, both in terms of clusters and in terms of proteins in the clusters.

9.4 Topological and biological coherence (**)
   A network aligner produced the alignment presented in the left panel of Figure 9.20.

  (a) Is the alignment more topologically coherent (using edge-correctness), or more biologically coherent (using 1-correctness between the node annotations presented in the right panel of Figure 9.20)?
  (b) What could explain the observed topological and biological coherences?

(c) Assuming that biological annotations can be transferred between aligned nodes, complete the annotations of the two networks.

9.5 Extending H-GRAAL to multiple network alignment (***)

(a) Recall how H-GRAAL measures the quality of an alignment (e.g., what H-GRAAL maximizes) and explain how you would extend it to the case of global, one-to-one, multiple alignment.
(b) Recall how H-GRAAL models pairwise alignment. How you would extend it for multiple network alignment?
(c) What are possible the search algorithms that could be applied to solve this new multiple network alignment problem?

9.6 Practical (***) Note: This exercise requires downloading data from databases, making scripts to format, and pre-process the data, and running various software.

We are interested in creating a phylogenetic tree of species based on their protein–protein interaction network similarities rather than the usually used genome similarities. We focus on five species, namely human (*Homo sapiens*), baker's yeast (*Saccaromyces cerevisiae*), round worm (*Caenorhabditis elegans*), fruit fly (*Drosophila melanogaster*), and thale cress (*Arabidopsis thaliana*), which will be aligned using Magna++ (https://www3.nd.edu/~cone/MAGNA++/).

(a) Data collection:

   i. From STRING database (https://string-db.org/cgi/download.pl) collect the protein–protein interaction networks of each species separately (choose the organism first, click update and then download the protein.link file)
   ii. Follow the same procedure to collect the protein sequences files.

(b) Preliminary processing:

   i. For each species, write a script that generates networks from the protein interaction files. In order to focus on the most reliable interations, consider only interactions having combined scores greater than or equal to 0.7 (700 in protein link files). Filter out disconnected proteins and remove duplicated edges. The remaining interactions should be encoded in an edge-list file, in which each a line describes an edge (in the format "node_1 node_2").
   ii. Write a script to compute the sequence similarities between proteins of any two species using NCBI's BLASTP (ftp://ftp.ncbi.nlm.nih.gov/blast/executables/blast+/LATEST/). The sequence similarity that we are using is defined a (1 - [BLASTP's E-value]). Then create a sequence similarity file in which each line contains the sequence similarity between a node from one network with a node from the other network, in the format "node_1 node_2 similarity," but only for pairs of nodes that have sequence similarity greater than zero.

(c) Aligning networks:

   i. Magna++ has a parameter that balances topological (edge) coherence and biological (node) coherence. According to the documentation of Magna++, what should be the value of this parameter if we want to only consider topological similarity?
   ii. Align the networks when maximizing edge conservation only (using S^3 score).

(d) Building the phenetic tree:

   i Using S^3 score as a similarity (or, equivalently, $[1 - S^3]$ as a dissimilarity) between two networks, create the phylogenetic tree of the five species using a hierarchical clustering method such as UPGMA (see Chapter 6).
   ii How different is the obtained classification from the traditional, sequence-based phylogenetic tree of these species?

Note: Solutions are available to instructors at www.cambridge.org/bionetworks.

9.7 Acknowledgments

This work was supported by the European Research Council (ERC) Starting Independent Researcher Grant 278212, the European Research Council (ERC) Consolidator Grant 770827, the Serbian Ministry of Education and Science Project III44006, the Slovenian Research Agency project J1-8155 and the awards to establish the Farr Institute of Health Informatics Research, London, from the Medical Research Council, Arthritis Research UK, British Heart Foundation, Cancer Research UK, Chief Scientist Office, Economic and Social Research Council, Engineering and Physical Sciences Research Council, National Institute for Health Research, National Institute for Social Care and Health Research, and Wellcome Trust (grant MR/K006584/1).

References

[1] Rose PW, Prlić A, Altunkaya A, et al. The RCSB Protein Data Bank: Integrative view of protein, gene and 3D structural information. *Nucleic Acids Research*, 2017;45(D1):D271–D281.

[2] Rost B. Enzyme function less conserved than anticipated. *Journal of Molecular Biology*, 2002;318(2):595–608.

[3] Ashburner M, Ball CA, Blake JA, et al. Gene ontology: Tool for the unification of biology. *Nature Genetics*, 2000;25(1):25.

[4] Kanehisa M, Goto S. KEGG: Kyoto encyclopedia of genes and genomes. *Nucleic Acids Research*, 2000;28(1):27–30.

[5] Joshi-Tope G, Gillespie M, Vastrik I, et al. Reactome: A knowledgebase of biological pathways. *Nucleic Acids Research*, 2005;33(suppl_1):D428–D432.

[6] Carbon S, Ireland A, Mungall CJ, et al. AmiGO: Online access to ontology and annotation data. *Bioinformatics*, 2008;25(2):288–289.

[7] Fields S, Song OK. A novel genetic system to detect protein–protein interactions. *Nature*, 1989;340:245–246.

[8] Petschnigg J, Groisman B, Kotlyar M, et al. The mammalian-membrane two-hybrid assay (MaMTH) for probing membrane-protein interactions in human cells. *Nature Methods*, 2014;11(5):585–592.

[9] Ho Y, Gruhler A, Heilbut A, et al. Systematic identification of protein complexes in *Saccharomyces cerevisiae* by mass spectrometry. *Nature*, 2002;415(6868): 180–183.

[10] Gavin AC, Aloy P, Grandi P, et al. Proteome survey reveals modularity of the yeast cell machinery. *Nature*, 2006;440(7084):631.

[11] Ryan CJ, Cimermančič P, Szpiech ZA, Sali A, Hernandez RD, Krogan NJ. High-resolution network biology: Connecting sequence with function. *Nature Reviews Genetics*, 2013;14(12):865–879.

[12] Alon U. Network motifs: Theory and experimental approaches. *Nature Reviews Genetics*, 2007;8:450–461.

[13] Pržulj N. Protein–protein interactions: Making sense of networks via graph-theoretic modeling. *Bioessays*, 2011;33(2):115–123.

[14] Koh GCKW, Porras P, Aranda B, Hermjakob H, Orchard SE. Analyzing protein–protein interaction networks. *Journal of Proteome Research*, 2012;11(4):2014–2031.

[15] Ji J, Zhang A, Liu C, Quan X, Liu Z. Survey: Functional module detection from protein–protein interaction networks. *IEEE Transactions on Knowledge and Data Engineering*, 2014;26(92):261–277.

[16] Pritykin Y, Singh M. Simple topological features reflect dynamics and modularity in protein interaction networks. *PLOS Computational Biology*, 2013 10;9(10):e1003243.

[17] Nepusz T, Paccanaro A. Structural pattern discovery in protein–protein interaction networks. In Kasabov N, ed. *Springer Handbook of Bio-/Neuroinformatics*. Springer;2014, pp. 375–398.

[18] Kelley BP, Sharan R, Karp RM, et al. Conserved pathways within bacteria and yeast as revealed by global protein network alignment. *Proceedings of the National Academy of Sciences*, 2003;100(20):11394–11399.

[19] Kuchaiev O, Milenković T, Memišević V, Hayes W, Pržulj N. Topological network alignment uncovers biological function and phylogeny. *Journal of the Royal Society Interface*, 2010;7(50):1341–1354.

[20] Bandyopadhyay S, Sharan R, Ideker T. Systematic identification of functional orthologs based on protein network comparison. *Genome Research*, 2006;16(3):428–435.

[21] Kotlyar M, Pastrello C, Sheahan N, Jurisica I. Integrated interactions database: Tissue-specific view of the human and model organism interactomes. *Nucleic Acids Research*, 2016;44(D1):D536–D541.

[22] Sharan R, Ulitsky I, Shamir R. Network-based prediction of protein function. *Molecular Systems Biology*, 2007;3(1):88.

[23] Dwight SS, Harris MA, Dolinski K, et al. Saccharomyces Genome Database (SGD) provides secondary gene annotation using the Gene Ontology (GO). *Nucleic Acids Research*, 2002;30(1):69.

[24] Sharan R, Ideker T. Modeling cellular machinery through biological network comparison. *Nature Biotechnology*, 2006;24(4):427–433.

[25] Malod-Dognin N, Prulj N. GR-Align: Fast and flexible alignment of protein 3D structures using graphlet degree similarity. *Bioinformatics*, 2014;30(9):1259–1265.

[26] Cook SA. The complexity of theorem-proving procedures. In *Proceedings of the Third Annual ACM Symposium on Theory of Computing STOC '71*. ACM;1971, pp. 151–158.

[27] Venkatesan K, Rual JF, Vazquez A, et al. An empirical framework for binary interactome mapping. *Nature Methods*, 2009;6(1):83–90.

[28] Kuhn HW. The Hungarian method for the assignment problem. *Naval Research Logistics (NRL)*, 1955;2(1-2):83–97.

[29] Lin D. An information-theoretic definition of similarity. In *Proceedings of the 15th International Conference on Machine Learning*. Morgan Kaufmann;1998, pp. 296–304.

[30] Resnik, P. Using information content to evaluate semantic similarity in a taxonomy. IJCAI'95 *Proceedings of the 14th International Joint Conference on Artificial Intelligence*, vol. 1. Morgan Kaufmann;1995, pp. 448–453.

[31] Resnik P. Semantic similarity in a taxonomy: An information-based measure and its application to problems of ambiguity in natural language. *Journal of Artificial Intelligence Research*, 1999;11:95–130.

[32] Cheng J, Cline M, Martin J, et al. A knowledge-based clustering algorithm driven by gene ontology. *Journal of Biopharmaceutical Statistics*, 2004;14(3):687–700.

[33] Fawcett T. An introduction to ROC analysis. *Pattern Recognition Letters*, 2006;27(8):861–874.

[34] Patro R, Kingsford C. Global network alignment using multiscale spectral signatures. *Bioinformatics*, 2012;28(23):3105–3114.

[35] Saraph V, Milenković T. MAGNA: Maximizing accuracy in global network alignment. *Bioinformatics*, 2014;30(20):2931–2940.

[36] El-Kebir M, Heringa J, Klau GW. Natalie 2.0: Sparse global network alignment as a special case of quadratic assignment. *Algorithms*, 2015;8(4):1035–1051.

[37] Aladağ AE, Erten C. SPINAL: Scalable protein interaction network alignment. *Bioinformatics*, 2013;29(7):917–924.

[38] Chindelevitch L, Ma CY, Liao CS, Berger B. Optimizing a global alignment of protein interaction networks. *Bioinformatics*, 2013;29(21):2765–2773.

[39] Vijayan V, Saraph V, Milenković T. MAGNA++: Maximizing Accuracy in Global Network Alignment via both node and edge conservation. *Bioinformatics*, 2015;31(14):2409–2411.

[40] Hashemifar S, Xu J. HubAlign: An accurate and efficient method for global alignment of protein–protein interaction networks. *Bioinformatics*, 2014;30(17):i438–i444.

[41] Malod-Dognin N, Pržulj N. L-GRAAL: Lagrangian graphlet-based network aligner. *Bioinformatics*, 2015;31(13):2182–2189.

[42] Clark C, Kalita J. A multiobjective memetic algorithm for PPI network alignment. *Bioinformatics*, 2015;31(12):1988–1998.

[43] Hashemifar S, Ma J, Naveed H, Canzar S, Xu J. ModuleAlign: Module-based global alignment of protein–protein interaction networks. *Bioinformatics*, 2016;32(17):i658–i664.

[44] Malod-Dognin N, Ban K, Pržulj N. Unified alignment of protein–protein Interaction networks. *Scientific Reports*, 2017;7(1):953.

[45] Meng L, Crawford J, Striegel A, Milenkovic T. IGLOO: Integrating global and local biological network alignment. arXiv preprint available at https://arxiv.org/pdf/1604.06111.pdf. 2016.

[46] Tatusov RL, Galperin MY, Natale DA, Koonin EV. The COG database: A tool for genome-scale analysis of protein functions and evolution. *Nucleic Acids Research*, 2000;28(1):33–36.

[47] Koyutürk M, Kim Y, Topkara U, et al. Pairwise alignment of protein interaction networks. *Journal of Computational Biology*, 2006;13(2):182–199.

[48] Pache RA, Aloy P. A novel framework for the comparative analysis of biological networks. *PLoS ONE*, 2012;7(2):e31220.

[49] Ciriello G, Mina M, Guzzi PH, Cannataro M, Guerra C. AlignNemo: A local network alignment method to integrate homology and topology. *PLoS ONE*, 2012;7(6):e38107.

[50] Sonnhammer EL, Östlund G. InParanoid 8: Orthology analysis between 273 proteomes, mostly eukaryotic. *Nucleic Acids Research*, 2014;43(D1):D234–D239.

[51] Mina M, Guzzi PH. AlignMCL: Comparative analysis of protein interaction networks through Markov clustering. In *2012 IEEE International Conference on Bioinformatics and Biomedicine Workshops (BIBMW)*. IEEE;2012. pp. 174–181.

[52] Meng L, Striegel A, Milenković T. Local versus global biological network alignment. *Bioinformatics*, 2016;32(20):3155–3164.

[53] Singh R, Xu J, Berger B. Pairwise global alignment of protein interaction networks by matching neighborhood topology. In Speed T, Huang H, eds., *Research in Computational Molecular Biology*, vol. 4453 of *Lecture Notes in Computer Science*. Springer Berlin Heidelberg;2007. pp. 16–31.

[54] Page L, Brin S, Motwani R, Winograd T. The PageRank citation ranking: Bringing order to the web. Technical Report. Stanford InfoLab;1999.

[55] Milenković T, Leong W, Hayes W, Pržulj N. Optimal network alignment with graphlet degree vectors. *Cancer Informatics*, 2010;9:121–137.

[56] Kuchaiev O, Pržulj N. Integrative network alignment reveals large regions of global network similarity in yeast and human. *Bioinformatics*, 2011;27(10):1390–1396.

[57] Memišević V, Pržulj N. C-GRAAL: Common-neighbors-based global GRAph ALignment of biological networks. *Integrative Biology*, 2012;4(7):734–743.

[58] Pržulj N, Corneil DG, Jurisica I. Modeling interactome: Scale-free or geometric? *Bioinformatics*, 2004;20:3508–3515.

[59] Pržulj N. Biological network comparison using graphlet degree distribution. *Bioinformatics*, 2007;23(2):177–183.

[60] Milenković T, Pržulj N. Uncovering biological network function via graphlet degree signatures. *Cancer Informatics*, 2008;6:257.

[61] El-Kebir M, Heringa J, Klau G. Lagrangian relaxation applied to sparse global network alignment. *Pattern Recognition in Bioinformatics*, 2011;225–236.

[62] Phan HT, Sternberg MJ. PINALOG: A novel approach to align protein interaction networks implications for complex detection and function prediction. *Bioinformatics*, 2012;28(9):1239–1245.

[63] Chandra B, Karloff HJ, Tovey CA. New results on the old k-Opt algorithm for the TSP. In *SODA'94 Proceedings of the 5th Annual ACM-SIAM Symposium on Discrete Algorithms*; 1994. pp. 150–159.

[64] Neyshabur B, Khadem A, Hashemifar S, Arab SS. NETAL: A new graph-based method for global alignment of protein–protein interaction networks. *Bioinformatics*, 2013;29(13):1654–1662.

[65] Dunn R, Dudbridge F, Sanderson CM. The use of edge-betweenness clustering to investigate biological function in protein interaction networks. *BMC Bioinformatics*, 2005;6(1):39.

[66] Guignard M. Lagrangean relaxation. *TOP*, 2003;11(2):151–200.

[67] Deb K. *Multi-Objective Optimization Using Evolutionary Algorithms*, vol. 16. John Wiley & Sons;2001.

[68] Mamano N, Hayes WB. SANA: Simulated annealing far outperforms many other search algorithms for biological network alignment. *Bioinformatics*, 2017;33(14):2156–2164.

[69] Clark C, Kalita J. A comparison of algorithms for the pairwise alignment of biological networks. *Bioinformatics*, 2014;30(16):2351–2359.

[70] Emmert-Streib F, Dehmer M, Shi Y. Fifty years of graph matching, network alignment and network comparison. *Information Sciences*, 2016;346: 180–197.

[71] Alkan F, Erten C. BEAMS: Backbone extraction and merge strategy for the global many-to-many alignment of multiple PPI networks. *Bioinformatics*. 2013;30(4):531–539.

[72] Karp RM. Reducibility among combinatorial problems. In Miller RE, Thatcher JW, Bohlinger JD eds., *Complexity of Computer Computations*. Springer; 1972, pp. 85–103.

[73] Papadimitriou CH. *Computational Complexity*. Addison Wesley Pub. Co.;1994.

[74] Sahraeian SME, Yoon BJ. SMETANA: Accurate and scalable algorithm for probabilistic alignment of large-scale biological networks. *PLoS ONE*, 2013;8(7):e67995.

[75] Gligorijević V, Malod-Dognin N, Pržulj N. Fuse: Multiple network alignment via data fusion. *Bioinformatics*, 2016;32(8):1195–1203.

[76] Vijayan V, Milenkovic T. Multiple network alignment via multiMAGNA++. In *Proceedings of the 15th International Workshop on Data Mining in Bioinformatics (BIOKDD) at the 22nd ACM SIGKDD 2016 Conference on Knowledge Discovery & Data Mining (KDD)*; ACM SIGKDD 2016.

[77] Singh R, Xu J, Berger B. Global alignment of multiple protein interaction networks with application to functional orthology detection. *Proceedings of the National Academy of Sciences*, 2008;105(35):12763–12768.

[78] Jeong H, Yoon BJ. Accurate multiple network alignment through context-sensitive random walk. *BMC Systems Biology*, 2015;19(suppl1): S7.

[79] Flannick J, Novak A, Srinivasan BS, McAdams HH, Batzoglou S. Graemlin: General and robust alignment of multiple large interaction networks. *Genome Research*, 2006;16(9):1169–1181.

[80] Kalaev M, Bafna V, Sharan R. Fast and accurate alignment of multiple protein networks. In Vingron M, Wong L, eds., *Research in Computational Molecular Biology*. Springer;2008. pp. 246–256.

[81] Hu J, Kehr B, Reinert K. NetCoffee: A fast and accurate global alignment approach to identify functionally conserved proteins in multiple networks. *Bioinformatics*, 2013;30(4):540–548.

[82] Ibragimov R, Malek M, Baumbach J, Guo J. Multiple graph edit distance: Simultaneous topological alignment of multiple protein-protein interaction networks with an evolutionary algorithm. In *Proceedings of the 2014 Annual Conference on Genetic and Evolutionary Computation*. ACM;2014. pp. 277–284.

[83] Garey MR, Johnson DS. Computers and Intractability: A Guide to the Theory of NP-Completeness. WH Freeman;1979;pp. 90–91.

[84] Vijayan V, Krebs E, Meng L, Milenkovic T. Pairwise versus multiple network alignment. arXiv preprint available at https://arxiv.org/pdf/1709.04564v1.pdf. 2017.

[85] Sharan R, Suthram S, Kelley RM, Kuhn T, McCuine S, Uetz P, et al. Conserved patterns of protein interaction in multiple species. *Proceedings of the National Academy of Sciences*, 2005;102(6):1974–1979.

[86] Hu J, Reinert K. LocalAli: An evolutionary-based local alignment approach to identify functionally conserved modules in multiple networks. *Bioinformatics*, 2014;31(3):363–372.

[87] Szklarczyk D, Morris JH, Cook H, et al. The STRING database in 2017: Quality-controlled protein–protein association networks, made broadly accessible. *Nucleic Acids Research*, 2017;45(D1):D362–D368.

[88] Todor A, Dobra A, Kahveci T. Probabilistic biological network alignment. *IEEE/ACM Transactions on Computational Biology and Bioinformatics*, 2013;10(1):109–121.

[89] Rubenstein DI, Sundaresan SR, Fischhoff IR, Tantipathananandh C, Berger-Wolf TY. Similar but different: Dynamic social network analysis highlights

fundamental differences between the fission-fusion societies of two equid species, the onager and Grevys zebra. *PLoS ONE*, 2015;10(10):e0138645.

[90] Vijayan V, Critchlow D, Milenkovic T. Alignment of dynamic networks. *Bioinformatics*, 2017;33(14):i180–i189.

[91] Sarajlić A, Malod-Dognin N, Yaveroğlu ÖN, Pržulj N. Graphlet-based characterization of directed networks. *Scientific Reports*, 2016;6:35098.

[92] Aparicio D, Ribeiro P, Silva F. Extending the applicability of graphlets to directed networks. *IEEE/ACM Transactions on Computational Biology and Bioinformatics*, 2016;14(6):1302–1315.

[93] Jiang C, Xuan Z, Zhao F, Zhang MQ. TRED: A transcriptional regulatory element database, new entries and other development. *Nucleic Acids Research*, 2007;35(1):D137–D140.

[94] Estrada E, Rodriguez-Velazquez JA. Complex networks as hypergraphs. arXiv preprint available at https://arxiv.org/pdf/physics/0505137.pdf 2005.

[95] Berge C. *Hypergraphs: Combinatorics of Finite Sets*, vol. 45. Elsevier;1984.

[96] Lugo-Martinez J, Radivojac P. Classification in biological networks with hypergraphlet kernels. arXiv preprint available at https://arxiv.org/pdf/1703.04823.pdf 2017.

[97] Gavin AC, Aloy P, Grandi P, et al. Proteome survey reveals modularity of the yeast cell machinery. *Nature*, 2006;440(7084):631.

10 Network Medicine

Pisanu Buphamalai*, Michael Caldera*, Felix Müller, and Jörg Menche

10.1 Introduction

Since the publication of the first draft of the human genome less than two decades ago [1, 2], rapid technological progress has revolutionized biomedical research. Thanks to a diverse array of "omics" technologies (e.g., genome sequencing, transcriptome mapping, proteomics, metabolomics, and others), we can now quantify both healthy and disease states at molecular resolution. At the same time, it has become clear that the detailed characterization of the individual molecular components alone (genes, proteins, metabolites, etc.) does not suffice to truly understand the nature of (patho-) physiological states and how to modulate them. Indeed, biomolecules do not act in isolation, but within an intricate and tightly coordinated machinery of complex interactions, such as protein–protein, gene regulatory, or signaling interactions. Network medicine is an emerging field that aims to apply tools and concepts from network theory to elucidate this machinery. Network approaches have helped unravel the molecular mechanisms of a broad range of diseases, from rare Mendelian disorders [3], cancer [4] or metabolic diseases [5], to basic attack strategies of viruses [6], to name but a few examples. While the molecular networks that underly biological processes may be the most natural candidate for applying network concepts in biomedical research, they are certainly not the only one. Networks are used across the full spectrum of medicine, from biomarker [7] to drug discovery [8], from the spread of obesity [9] to global outbreaks of infectious diseases [10], and from characterizing the relationships among diseases [11] to those among physicians within the health care system [12].

This chapter aims to give a general introduction to the dynamic field of network medicine. We start with a broad overview of major network types that are relevant to medicine. We then discuss with more detail the cellular network of molecular interactions among proteins and other biomolecules, the perhaps most widely used network in biomedical research. In the last section, we introduce disease module analysis, an important application of network tools to elucidate the molecular mechanisms of a particular disease.

*equal contribution

10.2 Networks in Medicine

10.2.1 Overview

One can distinguish three basic network types that cover different disease-relevant relationships: (*i*) Molecular networks describing the relationships between the molecular constituents of living organisms, for example, maps of all protein–protein interactions or metabolic reactions in a cell. The observation that such molecular maps share certain universal topological features with vastly different systems, e.g., the World Wide Web, collaboration networks, power grids, and many others, was instrumental for the development of network science. Today, it seems almost trivial that networks provide the most natural way of describing and analyzing the large-scale organization of biomolecules and their interactions. (*ii*) Disease networks are a powerful tool to investigate the diverse relationships between diseases. For example, two diseases can be linked if they share genetic associations or if they have similar clinical manifestations. In contrast to molecular networks, in which links often represent direct physical interactions, disease–disease networks represent more abstract relationships. They therefore serve as beautiful examples for the power of networks as a general tool for the analysis, integration, and intuitive visualization of large and complex data. (*iii*) Population-scale networks, i.e., networks describing the complex interactions among humans have been very successful in modeling and predicting the spread of contagious diseases, for example, global swine flu or ebola pandemics. These studies show the enormous potential of networks to serve as a platform for translating exact analytical results from physics and mathematics and translating them to concrete applications in medicine. (See Box 10.1.)

10.2.2 Molecular Networks

There are a plethora of molecular networks describing different aspects of the molecular and cellular organization of living organisms. A broad distinction can be made between physical and functional interaction networks. Physical interactions involve actual physical contact between the participating biomolecules, for example, proteins that assemble in a complex or receptor–ligand binding. Functional interaction, on the other hand, can refer to any kind of biologically relevant relationship. In co-expression networks, for example, genes are connected if their expression patterns are strongly correlated [13]. In the following we introduce the main types of molecular networks that are used to elucidate diverse disease mechanisms. Some of them were introduced in previous chapters, but we also summarize them here for completeness.

10.2.2.1 Protein–Protein Interaction Networks

Many molecular processes within a cell are performed by molecular machines consisting of a large number of protein components organized by their protein–protein interactions (PPIs). PPIs result from biochemical events steered by electrostatic forces leading to physical contacts of high specificity between two or more proteins [14]. Perturbed PPIs are involved in the pathobiology of many diseases, ranging from diabetes and obesity to Crohn's disease or cancer [15]. In analogy to the "genome" representing

Box 10.1: Networks in medicine

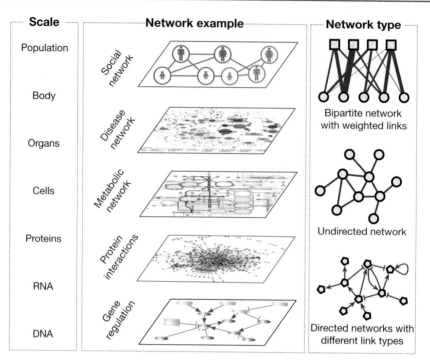

The diverse networks that are studied in network medicine reflect the different levels of organization that are relevant to human disease. From the molecular level, e.g., networks of interacting biomolecules that form the basis of all cellular processes, to the level of social interactions that are involved in the transmission of infectious diseases. Depending on the particular system, different network types are used for their description. Undirected and unweighted networks represent the most basic network type. More complex types may include a link directionality, link weights or use different types of nodes, for example in bipartite networks.

the collection of all genes in an organism, the collection of all molecular interactions is often referred to as the "interactome." The interactome can be represented by a network in which the nodes are proteins and the edges correspond to physical interaction between them. Over the last decade, significant experimental efforts have been made to map out the complete human interactome. High-throughput techniques such as yeast two-hybrid (Y2H) and immunoprecipitation linked to mass spectrometry are capable of mapping thousands of interactions in parallel (see Box 10.2). There has also been substantial work in curating interactions that were identified in small-scale experiments, as well as using computational tools to predict interactions [15].

Box 10.2: Mapping the human interactome

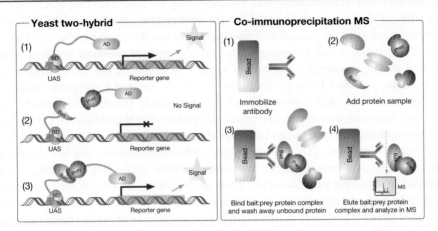

There are two major high-throughput techniques for the identification of protein interactions:

Yeast two-hybrid: (1) the system uses a protein consisting of a DNA binding domain (BD) and an activation domain (AD) that is responsible for activating transcription of DNA. (2) In Y2H, the two domains are separated and fused to proteins whose interaction is investigated. The BD is fused to the so-called bait, the AD to the prey. (3) Upon interaction between the two proteins of interest, the AD comes in close proximity to the reporter gene and the transcription leads to a signal.

Co-immunoprecipitation coupled to mass spectrometry: (1) In a first step, a target (bait) protein-specific antibody is immobilized on beads (e.g., agarose). (2) When the cell lysate is added, the antibody will specifically bind the target protein and indirectly capture proteins (prey) that are capable of binding to it. (3) After washing away unbound proteins, (4) the proteins of interest are eluted and analyzed using mass spectrometry. In short, the sample (the proteins) is first ionized and fragmented into smaller molecules, e.g., amino acids and peptides. Their mass-to-charge ratios can then be determined by accelerating the ions and subjecting them to an electric and/or magnetic field. Finally, the proteins in the sample can be identified by comparing with databases of known masses and characteristic fragmentation patterns.

Despite these promising first steps, our knowledge of the human interactome map remains far from complete, estimates indicate that only 10–30% of the full interactome has been revealed currently [16]. Nevertheless, interactome-based studies have contributed substantially to our understanding of biological processes both in homeostasis and in disease states, see Section 10.3.

10.2.2.2 Metabolic Networks

Metabolism (from Greek μεταβολή for "change") refers to the sum of all processes that are involved in assembling and disassembling the basic building blocks of cells, in particular the biochemical reactions for energy conversion. Traditionally, these reactions have been organized into specific pathways, for example the tricarboxylic acid (TCA) cycle, which corresponds to the sequence of chemical reactions in the cell that produces energy (also known as citric acid – or Krebs cycle, named after Hans Krebs, a Nobel Laureate in 1953). Metabolic networks represent collections of such pathways that connect chemical compounds (metabolites), biochemical reactions, enzymes, and genes. The relationships between the individual components of a given metabolic system can be inferred using comparative genomics combined with metabolomic data [17]. Metabolic networks are the most complete among the different biological networks, i.e., they reflect a near exhaustive knowledge of the involved biochemical processes [18]. They are available for a wide range of species and can be accessed through databases such as the Kyoto Encyclopedia of Genes and Genomes (KEGG) [19] or Reactome [20]. The currently most comprehensive human metabolic network, Recon 2.2 [21], includes 5,324 metabolites, 7,785 reactions, and 1,675 associated genes. Such metabolic networks do not only offer deep insights into the basic machinery of cells, but can also be used for *in silico* simulations to study how different parameters (e.g., metabolite concentrations) affect local and global properties of the biochemical network. The two most commonly used methods employ either (1) deterministic approaches (e.g., systems of ordinary differential equations) or (2) stochastic models (e.g., effect probabilities upon network perturbation) [22]. Metabolic network analyses can yield profound insights into the evolutionary emergence of complex life forms [23, 24], help understand the molecular mechanisms that drive the response to vaccination [25], or elucidate the interplay between metabolism and gene regulation [26]. (See Box 10.3.)

10.2.2.3 Regulatory Networks

Regulatory networks describe the complex machinery of genes and their corresponding proteins and RNAs, as well as the interactions between them that control the level of gene expression across the genome under specific conditions. Of particular importance for expression regulation are transcription factors (TFs), i.e., DNA-binding proteins that modulate the first step in gene expression [27]. In the most common representation of regulatory networks, nodes correspond to genes and links to the regulation of the expression of one gene by the product of the other. The links are typically directed and have either an activating (i.e., an increase in the concentration of one leads to an increase in the expression of the other) or inhibitory effect (increase in the concentration of one leads to decrease in the other) [28, 29]. Several experimental techniques exist to create large-scale data for building genome-wide regulatory networks, such as Chromatin-Immunoprecipitation Chip (ChIP-on-chip) [30] and ChIP-Sequencing [31]. Comprehensive databases include the Universal Protein Binding Microarray Resource for Oligonucleotide Binding Evaluation (UniPROBE) [32] or JASPAR [33].

Gene regulatory networks provide powerful tools to identify key transcription factors that control cell fate, for example in early blood development [34, 35].

Box 10.3: Metabolic and regulatory networks

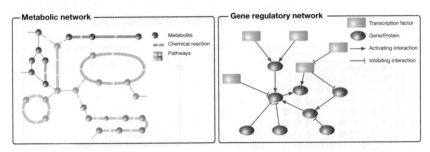

Metabolic networks describe the conversion/transformation of chemicals (metabolites) within a cell, organ, or whole organism. The nodes represent specific molecules while the edges describe the chemical reactions that take place between the nodes. Often these reactions are catalyzed by enzymes. Specific routes/compartments that are known to perform a particular function are called pathways.

Gene regulatory networks consist of genes that regulate each other. Often these genes are transcription factors that are capable of binding to DNA. The type of interactions can be either positive leading to an increase of protein concentration of the regulated gene, or negative, which leads to a decrease in protein concentration.

They can also be used to interpret variants identified in genome-wide association studies (GWAS), as they often perturb regulatory modules that are highly specific to disease-relevant cell types or tissues [36]. Lastly, gene regulatory networks also shed light on evolutionary conditions and pathways by which new regulatory functions emerge [37]. (See Box 10.3.)

10.2.2.4 Co-Expression Networks

In co-expression networks, genes are linked if their expression levels are significantly correlated under different experimental conditions, for example over time, across different tissues or cell types, or across a patient population (see Box 10.4 for an overview of the construction process) [13, 39]. In contrast to regulatory networks, co-expression networks do not offer an immediate causal relationship between genes. They can be used, however, to identify groups of genes that are more broadly functionally related, for example, controlled by the same transcriptional regulatory program, or members of the same pathway or protein complex [40]. Network analyses have been used to identify commonly affected pathways in heterogeneous diseases like autism spectrum disorder [41] or inflammatory bowel disease [42], predict causal GWAS genes associated with bone mineral density [43], or help explain the mechanism of breast cancer development [44].

> **Box 10.4: Co-expression networks**
>
>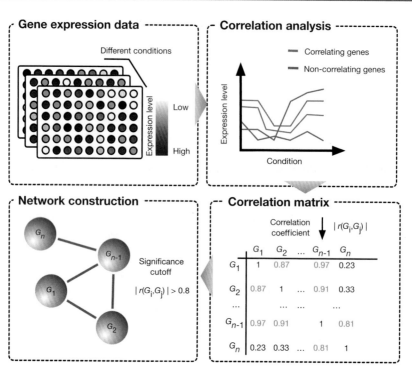
>
> **Construction of a co-expression network:** Creating a co-expression network requires gene expression data over several conditions, for example different treatments, across several tissues or patients. For each gene pair one can then calculate a correlation coefficient for their respective expression values across the different conditions, resulting in a correlation matrix. Extracting biologically meaningful correlations can be quite challenging, as true signals are often masked by noise that can arise, for example, from experimental confounding factors, batch effects, or sample heterogeneity. A widely used alternative to somewhat arbitrary global thresholds preserves the continuous nature of correlation scores and instead applies soft thresholding to identify network subclusters [13]. With recent large-scale resources, such as GTEx [38], noise from sample heterogeneity can be reduced and co-expression networks can be constructed in a tissue-specific manner, thus providing deeper and more robust insights onto the regulatory system in diseases.

10.2.2.5 *Genetic Interactions*

Two genes are linked by a genetic interaction if the effect of a simultaneous alteration (e.g., a mutation or the complete knock-down) of both genes differs from the

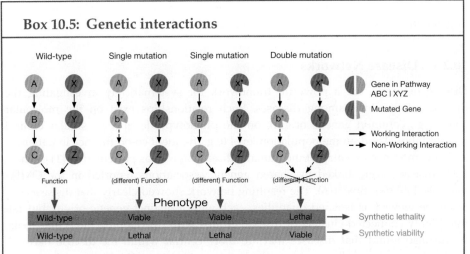

Box 10.5: Genetic interactions

Genetic interactions occur when the phenotype of two combined mutations differs significantly from the expectation based on the individual mutations. These interactions can be either positive (combined effect stronger than expected) or negative (combined effect weaker). The two most extreme outcomes are called "synthetic lethality" and "synthetic viability." In **synthetic lethality** the two individual mutations often occur in two independent, yet redundant pathways, so that the loss of one can be compensated for by the second. Only when targeting both pathways the systems fails. In **synthetic viability** the mutation in one pathway often leads to a toxic gene product. Only by also affecting another pathway the production of this toxic product is stopped and the resulting phenotype is again viable.

expectation based on the individual alterations [45] (see Box 10.5). The most extreme negative genetic interaction, often called "synthetic lethality," occurs when the simultaneous mutation of two genes is lethal, while individually both mutations are viable. Conversely, the most extreme positive genetic interaction ("synthetic viability") occurs, when a combination of two mutations is viable, while both individual mutations are lethal. Genetic interactions imply a functional relationship between the two genes, for example involvement in a common biological process or pathway, or conversely involvement in compensatory pathways with unrelated apparent function [46]. Hence, genetic interactions are an effective tool for biological discovery, e.g., for dissecting signaling pathways. They may also explain a considerable component of undiscovered genetic associations with human diseases and might help identify potential therapeutic targets. Over the last decade, genetic interactions have been investigated using mainly synthetic genetic array technology and RNA interference in yeast and *Caenorhabditis elegans*. A recent yeast based high-throughput screen [47], for example, tested all pairwise combinations of 6,000 genes resulting in almost 1 million interactions. Such maps can be used to study the large-scale organization of functions in a cell [47], identify the hierarchical organization of specific biological processes [48], or generate hypotheses on the function of uncharacterized genes [49].

10.2.3 Disease Networks

Disease networks are a powerful framework for systematically investigating the diverse relationships among diseases. Such relationships exist on the molecular level (e.g., common genetic origin), on the phenotypic level (e.g., similar clinical manifestations) and on the population level (e.g., frequent co-occurrence in patients). A first comprehensive map of the human "diseaseome" was presented in [11], where 1,377 diseases were linked by shared genetic associations reported in the OMIM database [50] (see Box 10.6). The resulting network showed clearly that diseases can rarely be viewed as isolated quantities, each with a distinct genetic origin, but fall into highly connected clusters of disease groups with overlapping molecular roots. It was also found that diseases that are more central within the disease network tend to be more prevalent and have higher mortality rates [51]. The genetic overlap

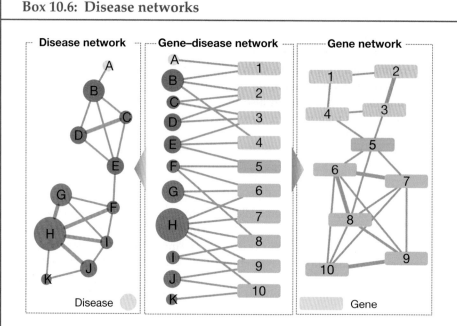

Box 10.6: Disease networks

Disease networks in which diseases are linked if they share a genetic association are based on gene–disease association data that can be represented as a bipartite network (middle panel). This bipartite network can then be projected either onto the diseases, resulting in a disease–disease network (left) or onto a gene–gene network, in which links represent a common disease association.

among diseases also extends towards physical interactions among the respective gene products, as well as similar gene expression profiles.

Similar results were obtained in a disease network in which diseases were linked by the similarity of their clinical manifestations [52] that were extracted from a large-scale screen of the biomedical literature and the annotated Medical Subject Headings (MeSH) metadata [53]. Confirming the strong correlation between the similarity of the symptoms of two diseases, the number of shared genetic associations and the extent to which their corresponding proteins interact, the study further revealed that the diversity of the clinical manifestations of a disease can be related to the degree of localization of the associated genes on the underlying protein interaction network. More detailed analyses that compared disease networks of different disease classes (e.g., complex diseases, Mendelian diseases, or cancer) and protein interaction networks identified interesting differences between diseases with different inheritance modes [54, 55, 56].

Networks can also be used to study comorbidity, i.e., the tendency of certain diseases to co-occur in the same patient. A disease network extracted from over 30 million patient records revealed that disease progression patterns of individual patients can be related to topological properties of the respective diseases within the co-morbidity network, for example, peripheral diseases tend to precede more central diseases [57]. These central, highly connected diseases are in turn associated with a higher mortality rate. More recently, differences in disease progression patterns that are related to age and sex have been characterized [58]. Co-morbidity networks have been used to address a wide range of further biomedical challenges, from drug repurposing [59] to the identification of potential drug side-effects [60], from biomarker identification [61] to approaches how to disentangle genetic and environmental factors of diseases [62].

10.2.4 Social Networks

A third important application of networks in medicine addresses the spread of contagious diseases, such as viral or bacterial infections (Box 10.7). Mathematical models of disease spreading go back as far as the year 1760, when Daniel Bernoulli formulated the first analytical method for quantifying the effectivity of inoculation against smallpox [64] (see Box 10.8 for an overview of important epidemiological models). Some 240 years later, the rise of complex networks made it possible to add a key ingredient to such models, namely realistic topologies of the networks on which diseases propagate, in particular global transportation maps and networks of social interactions [63] (see Box 10.7). Detailed information on interactions between humans on a local scale and on worldwide travel patterns is crucial for accurate predictions of the spatio-temporal spread of infectious diseases. Historically, the mobility of humans was largely confined by geography, such as rivers or mountains that could not be crossed easily. Such geographical borders naturally confined the propagation of epidemics. In present day, however, where both humans and goods can easily and quickly travel worldwide via air traffic, not even oceans can limit contagions [10]. As a consequence, an infection that started in a remote rural region may quickly propagate all over the world once

> **Box 10.7: Networks of disease spread**
>
> Global transportation map
>
>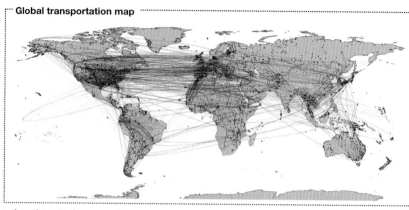
>
> Local contagion map
>
>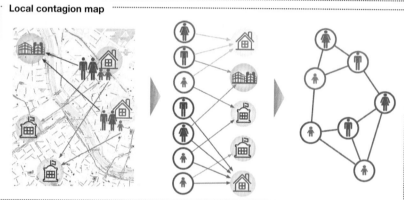
>
> Global air traffic plays a major role in the spread of epidemic disease across the world. Locally, infectious diseases, but also personal traits like happiness or habits like smoking, are transmitted through social interactions. These interactions can occur, for example at home or at work, which can be represented as a bipartite network that can be mapped to a person-to-person network (illustration adapted from [63]).

it has reached an airport, leading to much faster, much wider, and seemingly more erratic patterns of global epidemics.

10.2.4.1 Transportation Networks

Network-based epidemiological models that incorporate the structure of worldwide transportation networks can shed light on the complicated propagation patterns observed in recent pandemic outbreaks, help identify the source of an outbreak, predict future highly affected areas, or design most effective immunization or prevention strategies [67, 68]. Examples for recent outbreaks of infectious diseases that were studied with the help of network models include the SARS pandemic in 2003 [69], the H1N1 outbreak in 2009 [70], the ebola crisis of 2015, or the spread of HIV in the Philippines [71].

Like many other real world networks, air-traffic networks have been found to be approximately scale free [72]. Scale-free networks are therefore the prime model for analytical studies of epidemic outbreaks and for the analysis of real data from past and current epidemics [73]. Important global properties of a pandemic are directly linked to the structure of the underlying networks. For example, the characteristic (super-) hubs of scale-free networks can often be identified with large airports that play an important role in the spread of a disease, both through the large number of people gathering at such airports and through the large number of destinations that they serve. Indeed, scale-free networks are generally more prone to global infections than more regular network structures that do not exhibit the "small word effect." The critical spreading rate at which an infection is likely to propagate through the entire network is given by the ratio between the average degree and its variance. In large scale-free networks with degree distribution $P(k) \sim k^{-\gamma}$, the variance goes to infinity for power coefficients $\gamma < 3$. The critical spreading drops to zero in this case, meaning that a local infection is likely to become global, even for small infection rates [74].

10.2.4.2 Social Contagion

Approaches used to elucidate large-scale properties of infectious disease outbreaks can also be used to study the dynamics of social interactions, such as the spread of ideas, attitudes, and behaviors [75]. Reflecting the complexity of social relationships, links in social networks may represent, for example, friendship, family relationships, common work-place, shared political preferences, and many more. Collectively, these relationships not only define and shape our social relationships, but may also have concrete medical impact as shown in a seminal work on the spread of obesity [9]: The authors quantified how changes in body-mass index correlated among members of a social network of friends and family. Surprisingly, they found that obesity preferentially spreads through close social relationships. This effect is strong between men and between women, but almost negligible between man and woman. Similar studies were carried out to dissect the social component of starting to smoke [76] or of general happiness in life [77]. The results suggest that people surrounded by many happy people and those who are central in the network are more likely to become happy in the future. This effect was not observed among co-workers [77].

Recently there have also been efforts to combine global disease dynamics of transportation networks with contagion occurring on social networks. Multiplex or multilayer networks provide the analytical platform for combining several networks [78, 79, 80]. In such multilayer networks, different types of contact (at work, in the supermarket, at the airport) can be represented by distinct layers. It has been shown that the epidemic threshold is determined by the largest eigenvalue of the contact probability matrices of the different layers [78]. A powerful tool to study the full dynamics of spreading phenomena on networks, both simple or multilayered, are reaction diffusion processes [81].

Box 10.8: Basic mathematical models of disease spread

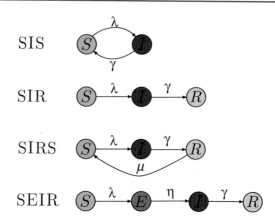

Classical epidemic models aim to determine the fraction of a population affected by a contagious disease over time. Most models represent the disease-status of an individual by one of three basic states [65]:

The **susceptible** (S) state, in which an individual can contract a disease. The **infected** (I) state, in which the individual carries the disease and can transmit it. The **recovered** (R) state, in which an individual is immune to repeated infections. More advanced models may also include further states, such as the **exposed** (E) state, in which an individual is already infected, but cannot yet transmit the disease. The microscopic dynamics of epidemiological models is given by transitions between the different states, macroscopic properties emerge from the interaction of many individuals. The most widely studied models are the following:

The **SIS model**, in which the recovery of a disease does not convey immunization, but renders an individual susceptible again, for example the common cold. The dynamics of the system are completely determined by the two rates of infection λ and recovery γ, respectively.

In the **SIR model** [66] susceptible individuals become infected with rate λ and recover with rate γ. This system exhibits an epidemic threshold $\alpha = \frac{\lambda}{\gamma}$, such that for $\alpha \leq 1$ a disease will die out in the long run, whereas for $\alpha > 1$ it will persist in the population.

The **SIRS model** contains an additional temporary immunity state, so that recovered individuals become susceptible again with rate μ. The impact of the incubation periods can be modeled by adding an exposed state (E), in which an individual has been infected, but is not yet infectious.

In network-based generalizations of these models, the individuals are identified with nodes and diseases spread along the connections of the network. In the simplest case this can be done by substituting the infection rate λ with a degree-dependent rate $\lambda = \lambda(k)$, so that the likelihood of becoming infected grows with the number of infected neighbors.

10.3 Interactome Analysis

As we have seen above, there exists a great variety of molecular interaction networks that can yield important insights into disease mechanisms. In the following, we will focus on "interactome networks" containing only physical interactions. The basic tools and concepts apply readily to other types of networks, however.

10.3.1 Interactome Construction

A large number of publicly available databases provide comprehensive collections of interactions between proteins and other relevant biomolecules (e.g. protein–DNA, protein–RNA, enzyme–metabolite interactions) in human, but also in other species, see [82] for a compendium of available resources. Among the most comprehensive, actively maintained and widely used databases are STRING [83], BioGRID [84], and MIntACT [85]. Note that they may also contain interactions that are not strictly physical, for example co-expression or other types of functional relationships among genes and their products. A well curated collection of only physical interactions has recently been published in the HIPPIE database [86]. Each interaction in HIPPIE is annotated with the original publication(s), details on the experimental protocol and an aggregated confidentiality score, thus allowing the user to adapt the final interactome network to specific requirements and preferences.

Generally, one can distinguish between three main sources of PPIs: (1) **interactions curated from the scientific literature** and typically derived from small-scale experiments, for example using co-immunoprecipitation, X-ray crystallography, or nuclear magnetic resonance. (2) **Interactions from systematic, proteome-scale mapping efforts**. The two main techniques are yeast two-hybrid (Y2H) assays [87] and binding affinity purifications coupled to mass spectrometry (MS) [88, 89], which produce rather different, yet complementary results (see Box 10.2). Y2H can map out precise, binary protein interactions, yet without biological context. It is not guaranteed, for example, that an experimentally observed interaction is biologically relevant, or whether the two respective proteins are in fact never expressed at the same time in the same cell. Co-complexes observed in MS experiments, on the other hand, are derived from a specific biological sample, yet are more difficult to translate into precise pairwise interactions [14]. (3) **Interactions from computational predictions**, for example based on protein structure [90] or other genomic data [91]. All three sources of PPIs have strengths and limitations in terms of comprehensiveness, noise and biases [92], such as biases in the selection of protein pairs [93] or experimental biases, for example towards highly expressed genes [87].

10.3.2 Basic Interactome Properties

Figure 10.1 gives a visual impression of a manually curated interactome from [16] and summarizes its global topological properties. In total, it contains 13,460 proteins connected via 141,296 physical interactions, so on average each protein has about 21 interaction partners. Characteristic not only to this, but also to many other complex

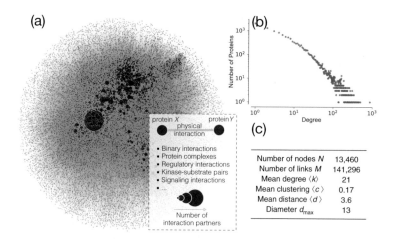

Figure 10.1: (a) A global picture of the interactome (original data curated by [16], figure adapted from [94]). The network consists of 13,460 proteins and 141,296 interactions that have been collected from different sources with various kinds of physical interactions, including binary interactions from systematic yeast two-hybrid screens, protein complexes, kinase-substrate pairs and others. (b) The overall topology is characterized by a highly heterogeneous degree distribution that follows approximately a power-law. (c) Other important structural properties of the interactome.

networks, is the high heterogenity among the degrees of the nodes, i.e., in the number of connections they have to other nodes differs widely (see Box 10.9 for an overview of important terms in network science). While the vast majority of proteins have only few neighbors (more than 2,000 have only a single link), there is also a considerable number of nodes with hundreds of connections, such as *GRB2* (degree $k = 872$), *YWHAZ* ($k = 502$) and *TP53* ($k = 450$), so-called "hubs." The histogram of all nodes' degrees shows "scale-free" properties,[1] i.e., $P(k)$ follows approximately a power-law $P(k) \sim k^{-\gamma}$. As laid out in more detail in Chapter 3, the broad degree distribution and, as a consequence, the presence of hubs have a profound impact on many network properties. Hubs serve as shortcuts that connect distinct parts of the network, resulting in a network property often referred to as the "small word effect" [96] (in some cases of scale-free networks even "ultra-small" [97]). In the interactome, for example, it takes on average less than four steps ($\langle d \rangle = 3.6$) to reach any other protein from any given starting point. This high degree of connectedness is also associated with a remarkable resilience of the overall network structure against random failure of individual nodes and/or edges. Scale-free networks can maintain global connectedness even upon removal of a considerable fraction of nodes and edges [98, 99, 100, 101]. The flipside of this robustness towards random failure, however, is a particular vulnerability towards targeted attack against the hubs [102]. For the interactome, for example,

---

[1] How accurately this and other networks can be described by a power-law is subject to some debate, see [95] for a thorough discussion. For our purposes, however, the precise mathematical nature of the degree distribution plays only a secondary role.

Box 10.9: Basic topological characteristics of networks

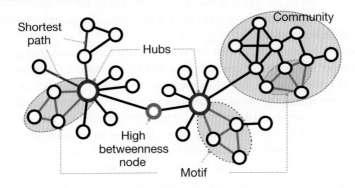

- The degree of a node is the number its direct neighbors. The degree distribution across all nodes is an important global network characteristic.
- Scale free networks are characterized by a degree distribution that follows a power law: While most nodes have few neighbors, there are also a few highly connected hubs with a large number of neighbors.
- A path between two nodes is a sequence of links connecting the two. The minimum number of links needed to connect the two is called shortest path length and represents their network distance.
- Centrality measures quantify the topological importance of a node within the network. There are different types of centrality measures, the betweenness centrality, for example, quantifies how many shortest paths of the full network cross through a certain node.
- Clustering describes a tendency observed in many biological (and other) networks that two neighbors of a node are often also connected to each other, thus forming a triangle.
- Motifs are small recurrent subgraphs in a network that occur particularly frequently.
- Network communities are groups of tightly interconnected nodes that have more connections among themselves than to the rest of the network.

the removal of ~ 30% of the most highly connected nodes is sufficient to completely destroy the network, leaving only disconnected fragments.

10.3.3 Interactome Topology and Biological Function

The degree of connectedness of a protein is directly related to its biological importance: As first shown for the yeast *Saccharomyces cerevisiae* [103], and later confirmed also in human cell lines [49], the products of essential genes, i.e., genes that are critical for the survival of an organism, tend to have a high number of interaction partners and take on central positions in the interactome. In contrast, genes whose loss of function can be more easily compensated for tend to have fewer interactions and are situated at the periphery of the interactome.

Interactome networks have also important structural features that go beyond the degree (or other measures of centrality) of individual nodes: "Network modules," i.e. groups of nodes that are densely interconnected among themselves, but sparsely connected to the rest of the network, can often be identified with proteins that jointly perform a certain function [104, 105, 106]. This relation between functional similarity of genes (see ahead to Box 10.14) and their closeness in interactome networks has also been found for shared pathway membership, co-localization in the same cellular component or co-expression [87, 89]. The local aggregation of cellular function within interactome networks represents a fundamental biological organization principle that forms the basis for many important applications, ranging from the prediction of protein function to disease gene identification and drug target prioritization.

10.3.4 Diseases in the Interactome

The observation that functionally similar proteins are often densely interconnected can be generalized also to other relationships among genes, in particular to shared disease associations. Genes that are implicated in the same disease tend to have more interactions among each other than expected for completely randomly distributed genes [107]. Note, however, that this does not necessarily imply particularly densely interconnected network patterns as those observed for genes involved in the same function. Indeed, *dys*function is typically distributed among several, often only loosely connected functional modules within the interactome [108]. A systematic study on ~ 300 complex diseases showed that currently available interactome networks offer sufficient coverage to identify these "disease modules," thereby confirming a fundamental hypothesis of interactome-based approaches to human disease [16]. The specific topological properties of disease modules differ between classes of diseases (e.g., complex diseases, Mendelian diseases, or cancer) and inheritance modes (autosomal dominant or recessive). Cancer driver genes are often highly central, while recessive disease genes tend to be more isolated at the periphery of the interactome [56].

10.3.5 Localization in Networks

As shown above, network-based localization of (dys)function is a central part of many interactome-based studies. In network science, the identification of densely connected groups of nodes is known as "community detection" [109]. While numerous algorithms exist for this task, they are usually not well suited for the identification of only weakly connected local network neighborhoods such as disease modules [108]. In order to quantify the tendency of a given set of disease genes to be localized in a certain neighborhood, we first need to inspect different possibilities for **measuring distances among a set of nodes in a network**. The simplest way to summarize the localization of a set **S** consisting of s nodes into a single quantity is to compute the network distance d_{ij} for all $\binom{S}{2} = \frac{s(s-1)}{2}$ pairs of nodes i and j and take the average:

$$d_{\text{av}}(\mathbf{S}) = \frac{2}{s(s-1)} \sum_{ij} d_{ij}, \tag{10.1}$$

which can be interpreted as a diameter of the set **S**. As a consequence of the "small-world" nature of many relevant networks, differences in the absolute values of d_{av} for different gene sets are often relatively small. Several variations and extensions of Equation 10.1 have therefore been proposed [110]. For example, instead of taking the average over all possible node pairs, one can consider only the distance to the next closest node, respectively:

$$d_{\text{close}}(\mathbf{S}) = \frac{1}{s} \sum_{i} \min_{j \in \{S \setminus i\}} (d_{ij}). \tag{10.2}$$

This gives different results as d_{av} in situations where a module is split into several "islands," for example due to network incompleteness. Whereas d_{close} correctly reflects the high degree of localization within the individual islands, it is diluted when the distances of all pairs are averaged. Other variations include adding weights to different path lengths d_{ij}, see Box 10.10 for more examples. Complementary to such distance-based measures, one can also use **connectivity-based measures** to determine the degree of connectedness among a set of nodes. The simplest way is to consider the number of links between them. A perhaps more intuitive measure is given by the size of the largest connected component, i.e., the highest number of nodes that are directly connected to one another. We can apply tools from statistical physics to understand many of its properties analytically [111]. It is, however, relatively sensitive to data incompleteness. In extreme cases, a single missing link in the network or a missing node from the set **S**, e.g., a protein, whose disease association is yet unknown, can fragment the connected component into isolated nodes.

The concepts introduced above can be readily extended to measure distances between two node sets **S** and **T**, for example, for quantifying the interactome-based similarity between two diseases [16]. The equivalent of Equation 10.1, i.e., the average over all possible pairs of nodes between two node sets is given by

$$d_{\text{av}}(\mathbf{S}, \mathbf{T}) = \frac{1}{s} \sum_{i \in S} \frac{1}{t} \sum_{j \in T} d_{ij}. \tag{10.3}$$

Similarly to different linkage methods in hierarchical clustering algorithms, there are different ways to compute the distance between two sets of nodes, see Box 10.10 for a number of frequently used options.

10.3.6 Randomization of Network Properties

By themselves, the absolute values of localization or distance as introduced above bring few insights. To judge whether an observed clustering of a particular node set is significant, we need to compare it to suitable random models. Many quantities that

Box 10.10: Distance measures in networks

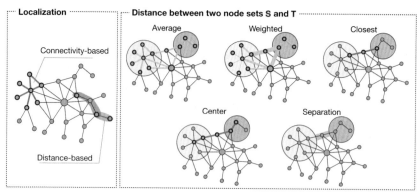

There are different ways to quantify the degree of "localization" of a given set of nodes **S**, i.e., whether or not they aggregate in a certain network neighborhood. Distance-based localization measures are based on different averages over pairwise distances d_{ij} between all nodes in the set, e.g.:

$$d_{\text{av}}(\mathbf{S}) = \frac{2}{s(s-1)} \sum_{ij} d_{ij} \tag{10.4}$$

$$d_{\text{close}}(\mathbf{S}) = \frac{1}{s} \sum_{i} \min_{j \in \{S \setminus i\}} (d_{ij}) \tag{10.5}$$

$$d_{\text{exp}}(\mathbf{S}) = -\frac{2}{s(s-1)} \ln \sum_{ij} \exp(-d_{ij}) \tag{10.6}$$

These measures can be generalized to two node sets **S** and **T**:

$$d_{\text{av}}(\mathbf{S},\mathbf{T}) = \frac{1}{s} \sum_{i \in S} \frac{1}{t} \sum_{j \in T} d_{ij} \tag{10.7}$$

$$d_{\text{close}}(\mathbf{S},\mathbf{T}) = \frac{1}{s+t} \left[\sum_{i \in S} \min_{j \in T}(d_{ij}) + \sum_{i \in T} \min_{j \in S}(d_{ij}) \right] \tag{10.8}$$

$$d_{\text{exp}}(\mathbf{S},\mathbf{T}) = -\frac{1}{s} \sum_{i \in S} \frac{1}{t} \ln \sum_{j \in T} \exp(-d_{ij}) \tag{10.9}$$

Nodes that are common to both sets **S** and **T** are usually taken to contribute with $d_{ij} = 0$ in the above formula. Instead of averaging over all pairs of nodes between **S** and **T** one can also define a center for each and use the distance between them:

$$d_{\text{center}}(\mathbf{S},\mathbf{T}) = d(\text{center}(\mathbf{S}), \text{center}(\mathbf{T})) \tag{10.10}$$

Another option is the separation parameter introduced in [16]:

$$\text{sep}(\mathbf{S},\mathbf{T}) = d_{\text{close}}(\mathbf{S},\mathbf{T}) - \frac{1}{2}(d_{\text{close}}(\mathbf{S}) + d_{\text{close}}(\mathbf{T})) \tag{10.11}$$

Negative values $\text{sep}(\mathbf{S},\mathbf{T}) < 0$ suggest overlapping network modules, while $\text{sep}(\mathbf{S},\mathbf{T}) > 0$ indicates separated modules. Note, however, that the separation parameter is not an intensive quantity, i.e., its magnitude depends on the number of nodes in the respective sets.

occur in the context of network analyses do not follow normal (Gaussian) distributions, such as the scale-free degree distribution, and therefore require particular care when choosing statistical tests. Comparisons with ensembles of randomized networks obtained from simulations are often the best choice. In general, we can distinguish two types of randomizations: (1) **Randomizing the network topology**, for example the interaction partners of a particular protein, and (2) **randomizing node attributes**, such as the disease associations of a group of genes.

10.3.6.1 Randomizing the Network Topology

To exclude that a seemingly interesting observation, for example, the local aggregation of disease genes in the interactome, could be a generic consequence of the overall topology of the underlying network, we need to compare our results from the original network with those obtained from networks with randomized topology. There are numerous randomization procedures. Which one is most suited, depends on the particular reference that is needed for a specific observation. The simplest method is to fix only the number of nodes N and the number of links L of the original network and to redistribute the links completely at random among the nodes. As shown in Chapter 3, this procedure results in an Erdős-Rényi network. Many properties of Erdős-Rényi networks can be calculated analytically and without extensive computer simulations, for example the expected clustering or the size of the largest connected component. However, the topology of most real world networks differs substantially from the one of a corresponding complete random graph, for instance hubs are completely absent in the latter. Hence, comparisons between the two are rarely meaningful and can in fact be rather misleading.

A more adequate reference that is suitable for most applications is given by networks in which the number of neighbors of every node are kept constant, but the specific interaction partners are completely randomized. This ensures that important structural features, in particular the degree distribution and presence of hubs, are preserved in the ensemble of randomized networks. Box 10.11 introduces the two main algorithms that are used to generate such randomized networks: The "switching algorithm" [112], is an iterative method, where at each step two links are selected at random and their endpoints are swapped. For example, the links connecting the nodes $n_1 \leftrightarrow n_2$ and $n_3 \leftrightarrow n_4$, respectively, can be reconnected to $n_1 \leftrightarrow n_3$ and $n_2 \leftrightarrow n_4$. Note that this may result in multiple links between two nodes or self-loops. In an application where such links are not meaningful, the original link pairs should be restored. As we repeatedly apply this procedure, the interactions of the network become more and more randomized, without altering the degree of each node. A drawback of this simple method is that no precise criteria exist as to how many switches should be performed to ensure a good mixing. Empirical results suggest $100\,L$ switching attempts, which can be computationally rather expensive for large networks [113].

A more efficient method for generating random networks with a prescribed degree sequence is to apply a variation of the "configuration model" [114, 115]. The second algorithm introduced in Box 10.11 is the "matching algorithm," in which all links of a given network are broken at once and then randomly reassembled one by one. As in the switching algorithm, the potential creation of self-loops and multiple links may

Box 10.11: Network randomization

Randomizing the network topology

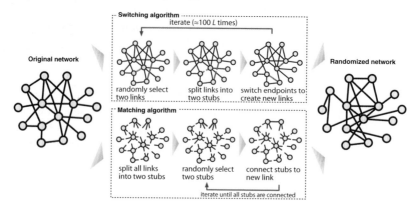

There are two frequently used algorithms to generate an ensemble of randomized networks with fixed degree distribution. In the *switching algorithm*, two links are chosen at random and their endpoints switched. Repeating this procedure will eventually lead to a fully randomized version of the original network. In the **matching algorithm**, all links of the given network are broken at once and then one by one reconnected at random.

Randomizing node attributes

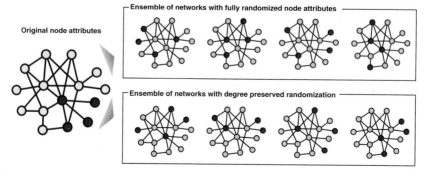

The most basic procedure to randomize node attribues (e.g., disease associations of genes) is to redistribute them completely at random on the network. For more restricted random controls, one can also keep specific topological properties of a node attribute constant, in particular the degree of the annotated node. In this case, only nodes with the same (or at least similar) properties are allowed choices.

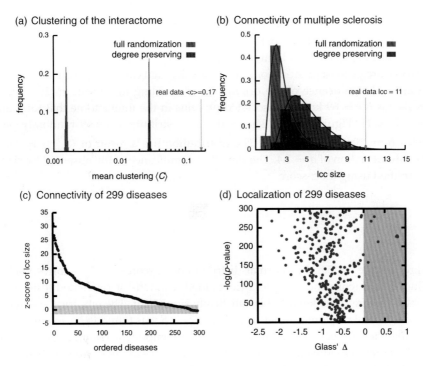

Figure 10.2: Network randomization. (a) Comparison of the clustering coefficient of the interactome (see Figure 10.1) with distributions obtained from complete randomization and degree-preserving randomization. (b) Comparison of the size of the largest connected component (lcc) of proteins associated with multiple sclerosis in the interactome with two distributions obtained from full and degree preserving randomization, respectively. (c) Sorted z-scores of the lcc size of 299 diseases in the interactome. (d) Significance and effect size of the observed localization $d_{av}(S)$ of 299 diseases compared to randomized gene sets. (Data from [16].)

need to be prevented in certain applications. Note that in this case the ensemble of the generated networks is no longer completely unbiased, but the effects are usually small and can often be neglected for large networks [113].

Figure 10.2a shows an application of the two randomization strategies to evaluate the observed mean clustering coefficient $\langle C \rangle = 0.17$ of the interactome. As expected, we find excellent agreement between the values observed in 10,000 simulations of a full random model corresponding to an Erdős-Rényi network and the respective analytical value $\langle C \rangle = p = \frac{2L}{N(N-1)} = 0.0016$. Simulations of the degree preserving matching algorithm yield the considerably higher mean value $\langle C \rangle = 0.03$, which is still significantly smaller than the originally observed clustering, indicating that the clustering of the interactome could not have emerged by chance.

10.3.6.2 Randomizing Node Properties

Instead of rewiring the structure of the network itself, it is often useful to consider randomizing certain node attributes, for example disease associations of individual genes in the interactome. In the simplest case of **random label permutation**, we detach the attribute of interest from their original nodes and redistribute them completely at random among all nodes of the network. For example, to investigate the connectivity of N_d disease proteins in terms of their largest connected component (lcc), we select the same number of proteins randomly from the network and measure their lcc. Repeating this procedure yields a random control distribution that can then be used to determine the statistical significance of the original lcc. According to data from [16], multiple sclerosis has $N_d = 69$ known associated proteins in the interactome that form an lcc of size $S = 11$. Figure 10.2 (b) shows the lcc distribution for 69 randomly picked proteins from 10,000 simulations. The distribution has a mean of $\langle S_{rand}^{full} \rangle = 2.9$ and a standard deviation of $\sigma = 1.4$. The statistical significance of the observed lcc size can be quantified using the z-score

$$z\text{-score} = \frac{S - \langle S_{rand}^{full} \rangle}{\sigma}, \tag{10.12}$$

yielding z-score $= 5.8$. For normal distributions, z-scores > 1.65 correspond to a p-value < 0.05 (corresponding to a right-sided test, left- or two-sided tests are also possible) and are considered to be statistically significant. The empirical p-value, i.e., the fraction of all random simulations with $S_{rand}^{full} \geq S$ was found to be p-value $= 0.003$. Taken together, we conclude that the connected component for multiple sclerosis is unlikely to have emerged by chance or as a trivial consequence of the network topology, indicating the potential presence of a disease module.

10.3.6.3 Degree Preserving Label Permutation

There are also stricter attribute randomization procedures that impose certain constraints on the allowed set of nodes among which an attribute can be distributed. Prominent cancer genes, for example, tend to have a large number of interactions in literature-curated interactome networks, simply because they have been investigated more intensively than other genes. To test whether the high connectivity among such genes can be explained by their high degree alone, we need to generate random distributions of node attributes that maintain the degree of the individual nodes carrying the original annotation. Note that swapping only between nodes of exactly the same degree will be problematic for high-degree nodes, as there may be only few, or even a single node in the entire network that have a certain degree. It is therefore useful to relax the requirement of having exactly the same degree and work with bins of nodes with comparable degree instead. Figure 10.2 (b) shows the distribution S_{rand}^{degree} obtained using such an approach. The mean value $\langle S_{rand}^{degree} \rangle = 5.1$ is larger than the one obtained from the full randomization, but still significantly smaller than the value $S = 11$ from the original data (z-score $= 3.1$, empirical p-value $= 0.009$), indicating that the high degree of the disease proteins alone does not explain their observed high connectivity.

These randomization procedures can also be applied to evaluate the distance-based localization measures introduced above, for example $d_{av}(\mathbf{S})$. From each random simulation we can extract d_{av}^{rand} and then compute the mean $\langle d_{av}^{rand} \rangle$ and corresponding standard deviation $\sigma \left(d_{av}^{rand} \right)$. In analogy to the z-score introduced above, we can use Glass' Δ to quantify the effect size of any difference observed between the true value $d_{av}(S)$ and the values obtained in the respective randomization simulations:

$$\Delta = \frac{d_{av}(S) - \langle d^{rand} \rangle}{\sigma\left(d^{rand}\right)}. \tag{10.13}$$

The statistical significance of an observed difference in the respective means $d_{av}(S)$ and $\langle d^{rand} \rangle$ can be obtained from a Mann–Whitney U test, for example. Figure 10.2 (c–d) shows the results for the randomization valuation of the localization observed among 299 diseases on the interactome.

Numerous more advanced randomization procedures exist that can preserve topological features beyond the degree distribution. For example, there are algorithms to generate randomized networks that maintain the mean clustering coefficient of the original network [116] or the correlation structure between the degrees of adjacent nodes [117, 118]. Another level of sophistication needs to be applied when randomizing metabolic networks, where simple link rewiring would likely generate reactions that are biochemically impossible [119, 120].

10.4 Disease Module Analysis

10.4.1 Overview

Sequencing technology has accelerated the discovery of disease associated genetic variations significantly. For most diseases, however, we are still far from a complete understanding of the underlying molecular mechanisms. Most complex diseases, such as cardiovascular diseases, cancer, or diabetes mellitus (the three most frequent causes of death worldwide), involve hundreds of genes and their complex interactions. It has been estimated, for example, that more than 2,000 genes are involved in intellectual disabilities, yet our current knowledge includes only around 800 genes [121]. The situation is similar for rare Mendelian disorders. Estimates for the total number of rare genetic disorders range from 6,000 to 8,000, a majority of which likely to be caused by a single genetic aberration. Despite this simple genetic architecture, less than half of all suspected diseases and corresponding disease genes are currently known.

Network-based **disease modules** offer a general framework for investigating how the pathobiology of a particular disease may arise from a combination of many genetic (but also epigenetic, environmental, behavioral etc.) variations. Succesful applications range from rare Mendelian disorders [3], to cancer [4] and other complex disorders, like metabolic [5], inflammatory [42], or developmental diseases [122]. A disease module is loosely defined as the comprehensive set of cellular components associated with a certain disease and their interactions. More specifically, the term refers to a connected subgraph of the interactome, whose perturbation causes the disease [18]. Figure 10.3 gives an overview of the disease module analysis process. The first step is to construct an interaction network and collect genes known to be associated with the particular disease of interest. These "seed genes" will serve as starting point for network-based gene prioritization algorithms. The resulting network module can then be validated and enriched with various additional datasets that will also be used in the biological interpretation of the final disease module.

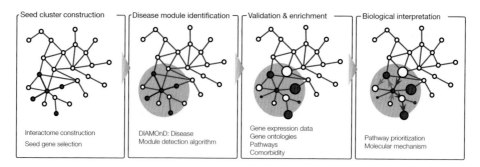

Figure 10.3: The basic steps of a disease module analysis process: First, interactome and seed gene data are collected. Next, a network-based disease gene prioritization method is employed. The performance of the predictions is then validated through comparison and enrichment with independent external data. In the last step, the module is explored for important biological pathways, overlap with other disease modules etc. (Figure adapted from [123].)

10.4.2 Seed Cluster Construction

The first step of the disease module analysis is the construction of a seed cluster, i.e., the curation of a suitable molecular interaction network and a set of genes known to be associated with the particular disease of interest. Box 10.12 lists a number of resources that may serve as a starting point.

10.4.2.1 Interactome Construction

As introduced above, one can make a broad distinction between physical interactions, e.g., protein co-complexes or binary protein–protein interactions, and functional interactions, e.g., genetic interactions or co-expression. By definition, physical interactions represent a direct molecular relationship, thus facilitating the identification of causal molecular mechanisms. Functional interactions, on the other hand, offer a much broader spectrum of potentially relevant associations between genes and gene products and can often be more easily adapted to a particular diseases, for example by incorporating tissue-specific expression data. Incorporating such information can considerably improve disease gene prioritization [124, 125, 126], see also Chapter 11. The choice of interaction type and used data sources will affect coverage (number of contained genes/proteins and their interactions), biases (for example, towards well-studied genes) and signal to noise ratio (number of false positive interactions) of the final interactome. Physical interactions offer more control over biases and signal to noise ratio, but often at the cost of lower coverage. Biases can be reduced by relying only on data obtained from systematic high-throughput studies, e.g., from [87, 89]. False positive interactions can be reduced by filtering for interactions that have been reported by several studies and by different experimental techniques. Several databases, such as HIPPIE [86] or STRING [83] offer integrated interaction scores for this purpose.

Box 10.12: Resources for disease module analyses

Interactome databases:

BIOGRID	thebiogrid.org
BioPlex	bioplex.hms.harvard.edu
HIPPIE	cbdm-01.zdv.uni-mainz.de/~mschaefer/hippie/
IntAct	www.ebi.ac.uk/intact
MatrixDB	matrixdb.univ-lyon1.fr
MINT	mint.bio.uniroma2.it
STRING	string-db.org

A more comprehensive list can be found on EBI's PSICQUIC view that also offers programmatic acces, see www.ebi.ac.uk/Tools/webservices/psicquic/view/

Disease genes:

DGA	dga.nubic.northwestern.edu
GWAS Catalog	www.ebi.ac.uk/gwas
Gene2Mesh	gene2mesh.ncibi.org
HGMD	hgmd.cf.ac.uk
OMIM	omim.org
OrphaNet	www.orpha.net

Integrated and functional web-based services:

DisGeNet	disgenet.org
GeneMANIA	genemania.org
HumanBase	hb.flatironinstitute.org

Ontologies:

Disease ontology (DO)	disease-ontology.org
Gene ontology (GO)	www.geneontology.org
Human phenotype ontology (HPO)	human-phenotype-ontology.github.io
Mammalian phenotype ontology (MPO)	www.informatics.jax.org/vocab/mp_ontology

A comprehensive list of biological ontologies can be accessed from EBI's Ontology Lookup Service under https://www.ebi.ac.uk/ols/ontologies

10.4.2.2 Seed Gene Selection

There are numerous resources that collect genes associated with diseases (see Box 10.12). Note that the term "disease associated gene" itself is only loosely defined and covers a wide spectrum from high penetrance dominant mutations to GWAS variants of rather small effect size or genes observed to be differentially regulated in patient subgroups. Similarly, the level of evidence for reported disease associations may differ greatly, from rare gene variants with a known and experimentally validated functional mechanism, to genes with unknown mechanism, yet repeatedly confirmed in multiple patient cohorts, to rather speculative associations inferred solely from text mining.

10.4.2.3 Evaluation of the Seed Cluster

Both the interactome construction and the seed gene selection involve a certain trade-off between using only highest-confidence data and achieving the highest possible coverage. There is no simple and universally applicable solution to this challenging problem that requires a certain amount of experimentation, ideally guided by a domain expert for the specific disease under study. From a network perspective, however, localization measures introduced above can be used as a rough indicator whether a particular combination of interactome and seed gene data meets the minimal criteria for a meaningful disease module analysis. Figure 10.4 shows the seed cluster for an asthma disease module from [123]. From a total of 129 seed genes that could be mapped to the interactome, 37 form the largest connected component, indicating a highly significant (z-score = 10.7) network localization. This suggests that the seed cluster has sufficient "signal" pinpointing the network neighborhood of the complete asthma module that can then be identified through a network-based expansion algorithm.

10.4.3 Network-Based Disease Gene Prioritization

Network-based disease gene prioritization methods build on the observation that genes associated with the same disease tend to be localized in the same interactome neighborhood. We can therefore use the network topology to extrapolate from a given set of seed genes to identify other genes that are likely to be also involved in the disease or at least strongly affected by the local interactome perturbation. Over the last years, numerous algorithms have been developed for this purpose. They can be broadly classified into three major categories: (1) connectivity based methods (2) path-based methods and (3) diffusion-based method (see Box 10.13).

10.4.3.1 Connectivity-Based Methods

Connectivity-based methods exploit the observed propensity among disease genes to interact with each other. Early pioneering approaches considered all direct neighbors of seed genes as potential candidate genes [127]. As more and more interactome and seed gene data become available, such approaches tend to generate an increasing number of false positives. More recent algorithms therefore utilize more advanced connectivity patterns, such as graphlets [128], or take the degree heterogeneity of the interactome explicitly into account [129]. Indeed, hubs in the network are expected to

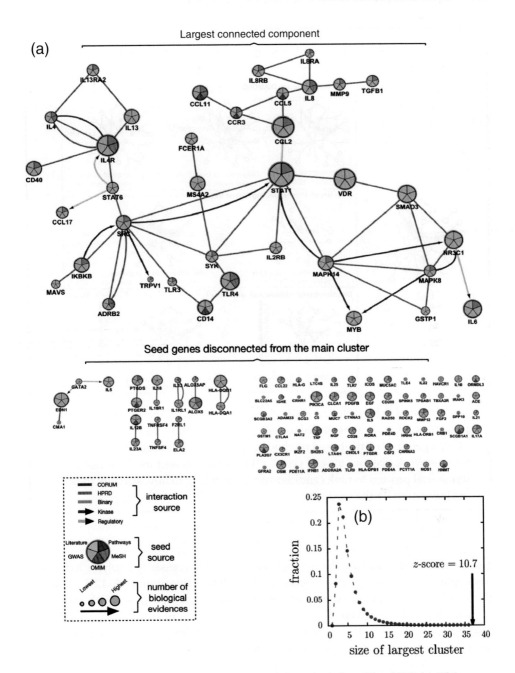

Figure 10.4: Seed cluster of an asthma disease module analysis from from [123]. (a) Of the 129 expert curated seed gene, 37 form the largest connected component, the rest are scattered throughout the interactome. (b) The size of the largest connected component is highly significant (z-score = 10.7) compared to random expectation.

Box 10.13: Network-based disease gene prioritization

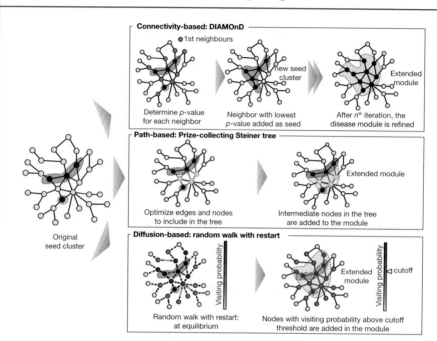

Illustration of three different methodologies for network-based disease gene prioritization: (1) **Connectivity-based** methods evaluate the direct neighbors of seed genes. (2) **Path-based** methods evaluate candidate genes based on their network distance to seed genes. (3) **Diffusion-based** methods use a dynamical process to rank candidate gene according to how strongly they are influenced by the seed genes.

also interact with a large number of seed genes without necessarily implying a disease-association. To correct for these effects, the DIAMOnD algorithm [108, 123] evaluates the *significance* of a given number of connections k_s to s seed genes with respect to the total degree k of a given candidate gene. In a network of size N, with s randomly distributed seed genes, the probability that a gene with degree k connects to exactly k_s seed genes is given by the hypergeometric distribution

$$P(X = k_s) = \frac{\binom{s}{k_s}\binom{N-s}{k-k_s}}{\binom{N}{k}}. \tag{10.14}$$

The significance of a given number of connections is therefore given by the *p*-value

$$p\text{-value} = \sum_{n=k_s}^{k} P(X = n), \tag{10.15}$$

which can then be used to iteratively rank all genes in the network. Note that the resulting disease module may consist of genes without direct connectivity to the initial seed genes.

10.4.3.2 Path-Based Methods

Instead of using the direct connectivity to seed genes, candidate genes can also be ranked according to their network distance to the set of seed genes (compare also with Box 10.10). A versatile set of algorithms that combines different distance measures for prioritizing candidate genes has been proposed in [130]. Instead of ranking the genes iteratively, it is also possible to search for an optimal set of candidate genes that collectively minimize the path lengths between the seed genes. Such approaches often implement variations of minimum spanning tree (or "Steiner tree") search algorithms [131, 132, 133]. Basically, the algorithm will construct a tree consisting of a minimum amount of edges while connecting all the seeds into a single cluster.

10.4.3.3 Diffusion-Based Methods

The methods described above rely only on the static topology of the network. It is also possible to use dynamical models to explore the network neighborhood around the seed genes for gene prioritization [3, 4, 134, 135, 136, 137]. Among the most widely used dynamical models are diffusion processes, such as the random walk with restart (RWR) [138]: Here, the seed genes serve as starting points for a random walk process along the links of the network. At every time step, the walker either proceeds to a randomly picked neighboring gene, or returns with restart probability r to one of the seed genes. The restart ensures that the local neighborhood around the seed genes is emphasized by the walker, otherwise all seed gene information would be lost in the long run of the process. The frequencies with which the individual nodes in the network are visited will eventually converge to a steady state and can then be used to rank all genes in the network according to their "dynamical closeness" to the seed genes. The process can be formalized as follows: Consider the vector $\mathbf{p}_t$ whose elements $p_i \ldots p_N$ represent the probability of the walker visiting node i at time t. The visiting probability at time t can be derived from the visiting probability at time $t-1$ via

$$\mathbf{p}_t = \mathbf{W}\mathbf{p}_{t-1}, \tag{10.16}$$

where $\mathbf{W}$ is the so-called transition matrix and defined as the column normalized adjacency matrix $\mathbf{A}$ with $W_{i,j} = \dfrac{A_{i,j}}{\sum_i k_i}$. At time t_0, only seed genes have (uniform) non-zero probability p, as well after each restart, which happens at a rate r. Equation 10.16 then becomes

$$\mathbf{p}_t = (1-r)\mathbf{W}\mathbf{p}_{t-1} + r\mathbf{p}_0. \tag{10.17}$$

The steady-state solution for Equation 10.17 is given by

$$\mathbf{p}_\infty = r(\mathbf{I} - (1-r)\mathbf{W})^{-1}\mathbf{p}_0. \tag{10.18}$$

The genes in the network can then be ranked according to the visiting probability p_∞. The restarting probability r can be used to adjust the influence of the seed genes on the diffusive process, from free diffusion (walker is not restricted by seed genes, $r = 0$) to no diffusion at all (walker remains at seeds, $r = 1$).

10.4.4 Validation and Enrichment

After completion of the preferred candidate gene ranking procedure, we first need to evaluate its performance. A second, closely related task is to determine a sensible cutoff, i.e. how many ranked genes should be considered for the final disease module, as most prioritization methods rank all genes in the network without offering an intrinsic stopping criterion. There are two complementary approaches: (1) Estimating the predictive power of the disease gene predictions using cross-validation methods. (2) Comparison with independent biological data.

10.4.4.1 Cross-validation of Prediction Performance

In principle, cross-validation of disease gene prioritization algorithms works in the same way as with other classification tasks (compare also with Chapters 6–8): For a basic k-fold cross-validation, the set of seed genes is first randomly divided into k groups (the special case where k equals the number of seed genes is often referred to as "leave-one-out" cross-validation). One of the groups can then serve as the "test-set" of true positives, while the remaining $k-1$ groups are used as modified seed gene pool. The gene prioritization algorithm is then run on this modified pool to test how well the method is able to retrieve the left out genes in the test set. Repeating this procedure k times with each of the k groups serving as test set yields a statistic on the expected average performance of the method. The choice of k determines the trade-off between high bias (large k) and high variance (small k). An important difference to many other classification tasks is the lack of clear true negatives, i.e., genes that we know not to be involved in the disease. Several proxies have been proposed, for example essential genes, genes of high genetic variability or manually curated genes that are unlikely to be involved in a particular disease according to their expression patterns. These gene sets can only offer approximations and remain necessarily incomplete, making the interpretation of standard performance measures difficult, such as receiver operating characteristic curves.

10.4.4.2 Enrichment with Independent Biological Data

A complementary approach for estimating the performance is to test for enrichment of the ranked genes with independent biological data (see Box 10.12). Figure 10.5 shows the biological enrichment of the top 400 ranked genes from an asthma disease module analysis [123]. To compare the biological signal of the ranked genes with the one of the manually curated seed genes, the authors chose a sliding window of ranked genes with the same size of the seed genes and within each window computed the enrichment with five different datasets: (1) Genes differentially expressed in a relevant case/control study, (2) genes participating in expert curated relevant biological pathways, (3) genes contained in general pathways that were found enriched in the seed genes, (4) genes annotated to similar biological processes as the seed genes according the gene ontology (GO, see Box 10.14) and (5) genes that are known to be implicated in

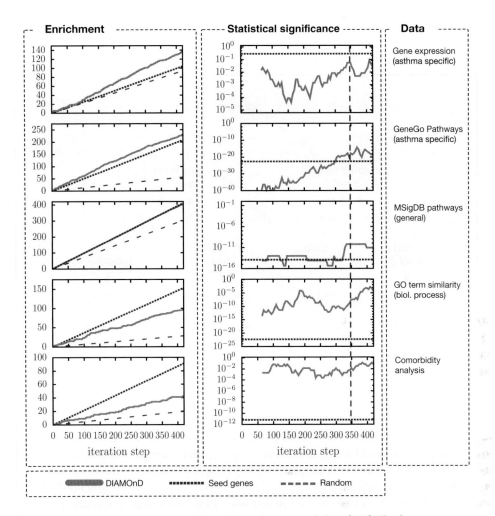

Figure 10.5: Biological enrichment of the asthma disease module in [123]. The first two columns show the number (and the corresponding statistical significance, respectively) of the identified candidate genes that were found in the different validation datasets indicated in the third column. The values for the candidate genes are show in orange, the values for seed genes and random expectation in red and green, respectively.

diseases that show high co-morbidity with asthma. A comparison of the enrichments across different datasets allows for an evaluation of the general plausibility of the ranked genes, but also for an estimation of the border of the disease module.

10.4.5 Biological Interpretation

The data collected for the performance evaluation can further be used for an integrated analysis of the biological mechanisms represented in the disease module. Figure 10.6

Box 10.14: Ontologies

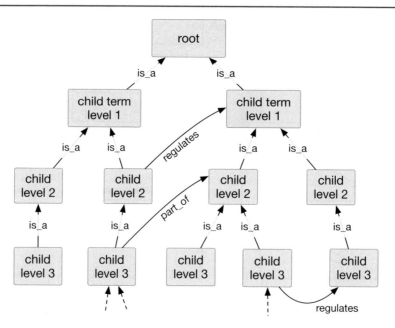

Ontologies are controlled vocabularies to organize the knowledge of a specific field, for example biological pathways, diseases or phenotypes (see Box 10.12 for a list of biomedical ontologies). These vocabularies are usually manually curated by an authoritative consortium of domain experts. An important vocabulary is the gene ontology (GO). It consists of three separate branches: (1) "cellular component" (4,195 terms), (2) "molecular function" (11,120 terms), and (3) "biological process" (29,682 terms), each forming a hierarchical, acyclic tree. The root term at the top is the most general, increasingly specific terms are connected by either `is_a`, `part_of` or `regulates` links that describe the particular relationship between the respectively linked terms.

Ontologies are not only useful for systematic annotation and collection of knowledge, but can also be used to assess the "semantic similarity" among different terms according to their relative position in the tree [139]. A common approach relates the specificity (tree depth) of a term to its information content (IC). The similarity between two terms can then be calculated from the IC of their most informative (i.e., highest IC) common ancestor. Note that most biological entities, such as gene products, are usually annotated with several terms and different strategies can be used to aggregate the similarity among several terms, see [139] for a detailed discussion.

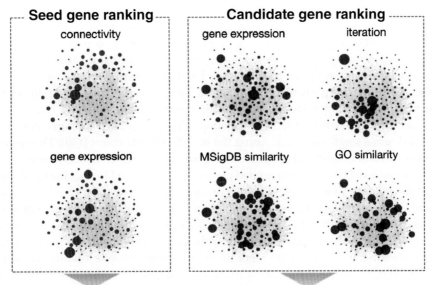
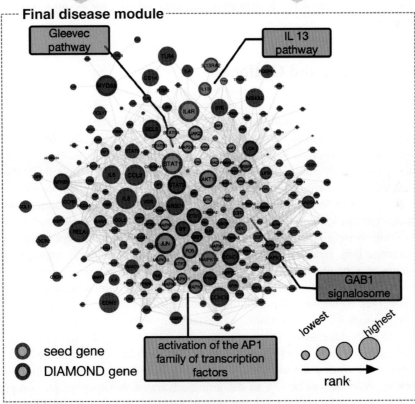

Figure 10.6: Illustration of the ranking procedure (top) and the final asthma disease module (bottom) from [123]. Seed genes and candidate genes are first ranked separately according to their enrichment with different biological datasets. The individual rankings are then combined into a final score for each gene in the disease module, which can then be used to prioritize pathways within the module.

illustrates how the different data are combined into a final score for each gene in the asthma disease module, which in turn can be used to prioritize pathways within the module. The first step is to create a ranking of all genes for each individual data source. Seed genes and candidate genes are often examined separately, which has the advantage that they can be given different weights when they are combined later on. Depending on the particular data type, the ranking can be based on fold-change for differential expression data, GWAS *p*-value or functional similarity with known processes (compare with Box 10.14), for example. The individual rankings can then be combined into a single score, e.g., using the so-called Borda-count [140]: The score of a gene is taken to correspond to its inverted rank and the scores of different rankings are simply added. Finally, the integrated gene score can be used to prioritize pathways within the module, thus complementing commonly used measures, such as coverage of genes in the pathway. The integrated biological relevance of a pathway within the module can be quantified by the average score of its genes. Additional potentially interesting network-based analyses that can be performed with the disease module include identifying overlaps with other diseases or with network modules known to be modulated by drugs, for example using the distance measures above, or applying community detection to identify potential submodules, for example for patient stratification.

10.5 Summary and Outlook

Network medicine is a highly dynamic and rapidly expanding field covering virtually all areas of biomedical research. This brief introduction can therefore only provide a necessarily incomplete and highly subjective selection. We hope that the references we provide may serve as a starting point for further reading and also recommend a recently published textbook focusing exclusively on this subject [141].

An important challenge in current biomedical research is to integrate the ever growing amount of "omics" data (e.g., genomics, epigenomics, proteomics, metabolomics, lipidomics). Network approaches are inherently holistic and integrative, and particularly multilayer networks are very promising candidates for addressing this challenge [79]. First analytical analyses of multilayer networks highlight the importance of a detailed, context-aware mapping of different types of interactions to fully understand the interplay between structure and dynamics of such complex networks [142]. So far, most studies on biomolecular networks focus on structural network properties and a thorough understanding of their dynamical properties remains an important issue. The concept of dynamic controllability, for example, is well established in network theory [143, 144] and could in principle be applied to driving a cell from a disease state to a healthy state [143]. We expect that such network approaches will be key to designing advanced therapeutics for complex diseases that cannot be understood, nor treated, by a simple mono-causal molecular mechanism. The ultimate goal of network medicine is of course to contribute not only to basic research, but to the translation to benefit patients. Based on the pace at which network medicine is progressing, we are confident that this exciting and challenging goal will be reached rather sooner than later.

10.6 Exercises

To familiarize yourself with some basic network-based approaches to human diseases we will perform a rudimentary disease module analysis. The exemplary solution we provide is based on the programming language python and utilizes heavily the excellent `networkx` module, but of course other programming languages offer similar functionalities.

10.1 Constructing the interactome

   (a) Use one of the databases listed in Box 10.12 to construct an interactome network. We suggest using HIPPIE, as it allows for both programmatic access via an API or download of the entire dataset in an easy to parse text format.
   (b) Construct different networks with different parameters, such as different confidence scores or different experimental sources.
   (c) Perform a basic characterization of the overall topology of each network, e.g., overall coverage, degree distribution, number of isolated components, distribution of shortest pathlengths, clustering coefficient, etc.

10.2 Constructing a seed cluster for a particular disease

   (a) Use one of the databases listed in Box 10.12 to assemble a set of seed genes for a specific disease.
   (b) Place the seed genes on the interactome and determine the degree of localization using different measures from Box. 10.10.
   (c) Assess the statistical significance of the measured localization using different randomization schemes, both for the network topology and the seed genes (see Box. 10.11).

10.3 Constructing a disease module

   (a) Implement two different network-based gene prioritization algorithms introduced in Box 10.13.
   (b) Rank all genes in the interactome using both methods and with varying parameters of the respective algorithms.
   (c) Evaluate how the results change when removing various fractions of the seed genes.

10.4 Perform an enrichment analysis of the disease module

   (a) Use the databases listed in Box 10.12 to assemble an independent set of genes with potential relevance to the the disease, e.g., genes found to be differentially expressed in a patient cohort.
   (b) Test whether the ranked candidate genes are enriched for the genes of the independent validation set.
   (c) Perform a gene set enrichment analysis of the disease module using gene ontology to identify prominent biological processes within the module.

Note: Solutions are available to instructors at www.cambridge.org/bionetworks.

References

[1] Craig Venter J, Adams MD, Myers EW, et al. The sequence of the human genome. *Science*, 2001;291(5507):1304–1351.

[2] Lander ES, Linton LM, Birren B, et al. Initial sequencing and analysis of the human genome. *Nature*, 2001;409(6822):860–921.

[3] Smedley D, Köhler S, Czeschik JC, et al. Walking the interactome for candidate prioritization in exome sequencing studies of mendelian diseases. *Bioinformatics*, 2014;30(22):3215–3222.

[4] Leiserson MDM, Vandin F, Wu HT, et al. Pan-cancer network analysis identifies combinations of rare somatic mutations across pathways and protein complexes. *Nature Genetics*, 2015;47(2):106–114.

[5] Chen Y, Zhu J, Lum PK, et al. Variations in DNA elucidate molecular networks that cause disease. *Nature*, 2008;452(7186):429–435.

[6] Pichlmair A, Kandasamy K, Alvisi G, et al. *Nature*, 2012;487:486–490.

[7] Chuang HY, Lee E, Liu YT, Lee D, Ideker T. Network-based classification of breast cancer metastasis. *Molecular Systems Biology*, 2007;3:140.

[8] Csermely P, Korcsmfiaros T, Kiss HJM, London G, Nussinov R. Structure and dynamics of molecular networks: A novel paradigm of drug discovery: A comprehensive review. *Pharmacology & Therapeutics*, 2013;138:333–408.

[9] Christakis NA, Fowler JH. The spread of obesity in a large social network over 32 years. *New England Journal of Medicine*, 2007;357:370–379.

[10] Colizza V, Barrat A, Barthfielemy M, Vespignani A. The role of the airline transportation network in the prediction and predictability of global epidemics. *Proceedings of the National Academy of Sciences USA*, 2006;103:2015–2020.

[11] Goh KI, Cusick ME, Valle D, Childs B, Vidal M, Barabási AL. The human disease network. *Proceedings of the National Academy of Sciences USA*, 2007;104(21):8685–8690.

[12] Landon BE, Keating NL, Barnett ML, et al. Variation in patient-sharing networks of physicians across the United States. *Journal of the American Medical Association*, 2012;308(3):265–273.

[13] Zhang B, Horvath S. A general framework for weighted gene co-expression network analysis. *Statistical Applications in Genetics and Molecular Biology*, 2005;4:Article 17.

[14] De Las Rivas J, Fontanillo C. Protein-protein interactions essentials: Key concepts to building and analyzing interactome networks. *PLOS Computational Biology*, 2010;6:e1000807.

[15] Vidal M, Cusick ME, Barabasi AL. Interactome networks and human disease. *Cell*, 2011;144:986–998.

[16] Menche J, Sharma A, Kitsak M, et al. Disease networks. Uncovering disease–disease relationships through the incomplete interactome. *Science*, 2015;347(6224):1257601.

[17] Thiele I, Palsson, BØ. A protocol for generating a high-quality genome-scale metabolic reconstruction. *Nature Protocols*, 2010;5(1):93–121.

[18] Barabasi AL, Gulbahce N, Loscalzo J. Network medicine: A network-based approach to human disease. *Nature Reviews Genetics*, 2011;12:56–68.

[19] Kanehisa M, Furumichi M, Tanabe M, Sato Y, Morishima K. KEGG: New perspectives on genomes, pathways, diseases and drugs. *Nucleic Acids Research*, 2017;45:D353–D361.

[20] Fabregat A, Jupe S, Matthews L, et al. The reactome pathway knowledgebase. *Nucleic Acids Research*, 2016;44:D481–D487.

[21] Swainston N, Smallbone K, Hefzi H, et al. Recon 2.2: From reconstruction to model of human metabolism. *Metabolomics*, 2016;12:109.

[22] Chan SY, Loscalzo J. The emerging paradigm of network medicine in the study of human disease. *Circulation Research*, 2012;111:359–374.

[23] Goldford JE, Hartman H, Smith TF, Segre D. Remnants of an ancient metabolism without phosphate. *Cell*, 2017;168:1126–1134.e9.

[24] Josephides C, Swain PS. Predicting metabolic adaptation from networks of mutational paths. *Nature Communications*, 2017;8:685.

[25] Li S, Sullivan NL, Rouphael N, et al. Metabolic phenotypes of response to vaccination in humans. *Cell*, 2017;169:862–877.e17.

[26] Klosik DF, Grimbs A, Bornholdt S, Hutt MT. The interdependent network of gene regulation and metabolism is robust where it needs to be. *Nature Communications*, 2017;8:534.

[27] Carninci P, Kasukawa T, Katayama S, et al. The transcriptional landscape of the mammalian genome. *Science* 2005;309(5740):1559–1563.

[28] Zhang Y. Gene regulatory networks: Real data sources and their analysis. In Iba H, Noman N, eds., *Evolutionary Computation in Gene Regulatory Network Research*. John Wiley & Sons, Inc.;2016, pp. 49–65.

[29] Karlebach G, Shamir R. Modelling and analysis of gene regulatory networks. *Nature Reviews Molecular Cell Biology*, 2008;9:770–780.

[30] Blat Y, Kleckner N. Cohesins bind to preferential sites along yeast chromosome III, with differential regulation along arms versus the centric region. *Cell*, 1999;98:249–259.

[31] Furey TS. ChIP-seq and beyond: New and improved methodologies to detect and characterize protein-DNA interactions. *Nature Reviews Genetics*, 2012;13:840–852.

[32] Hume MA, Barrera LA, Gisselbrecht SS, Bulyk ML. UniPROBE, update 2015: New tools and content for the online database of protein-binding microarray data on protein-DNA interactions. *Nucleic Acids Research*, 2015;43: D117–D122.

[33] Mathelier A, Fornes O, Arenillas DJ, et al. JASPAR 2016: A major expansion and update of the open access database of transcription factor binding profiles. *Nucleic Acids Research*, 2016;44:D110–D115.

[34] Moignard V, Woodhouse S, Haghverdi L, et al. Decoding the regulatory network of early blood development from single-cell gene expression measurements. *Nature Biotechnology*, 2015;33:269–276.

[35] Goode DK, Obier N, Vijayabaskar MS, et al. Dynamic gene regulatory networks drive hematopoietic specification and differentiation. *Developmental Cell*, 2016;36:572–587.

[36] Marbach D, Lamparter D, Quon G, et al. Tissue-specific regulatory circuits reveal variable modular perturbations across complex diseases. *Nature Methods*, 201613:366–370.

[37] Friedlander T, Prizak R, Barton NH, Tkačik G. Evolution of new regulatory functions on biophysically realistic fitness landscapes. *Nature Communications* 2017;8:216.

[38] GTEx Consortium. Human genomics: The Genotype-Tissue expression (GTEx) pilot analysis: Multitissue gene regulation in humans. *Science*, 2015;348:648–660.

[39] De Smet R, Marchal K. Advantages and limitations of current network inference methods. *Nature Reviews Microbiology* 2010;8:717–729.

[40] Weirauch MT. Gene coexpression networks for the analysis of DNA microarray data. In Dehmer M, Emmert-Streib F, Graber A, Salvador A, eds., *Applied Statistics for Network Biology*. Wiley-VCH Verlag GmbH & Co. KGaA;2011, pp. 215–250.

[41] Parikshak NN, Swarup V, Belgard TG, et al. Genome-wide changes in lncRNA, splicing, and regional gene expression patterns in autism. *Nature* 2016;540:423–427.

[42] Peters LA, Perriogue J, Mortha A, *et al*. A functional genomics predictive network model identifies regulators of inflammatory bowel disease. *Nature Genetics*, 2017;49:1437–1449.

[43] Calabrese GM, Mesner LD, Stains JP, et al. Integrating GWAS and co-expression network data identifies bone mineral density genes SPTBN1 and MARK3 and an osteoblast functional module. *Cell Systems*, 2017;4:46–59.e4.

[44] Guo X, Xiao H, Guo S, Dong L, Chen J. Identification of breast cancer mechanism based on weighted gene coexpression network analysis. *Cancer Gene Therapy*, 2017;24:333–341.

[45] Boucher B, Jenna S. Genetic interaction networks: Better understand to better predict. *Frontiers in Genetics*, 2013;4:290.

[46] Srivas R, Shen JP, Yang CC, et al. A network of conserved synthetic lethal interactions for exploration of precision cancer therapy. *Molecular Cell*, 2016;63:514–525.

[47] Costanzo M, VanderSluis B, Koch EN, et al. A global genetic interaction network maps a wiring diagram of cellular function. *Science*, 2016;353:aaf1420.

[48] Kramer MH, Farré JC, Mitra K, et al. Active interaction mapping reveals the hierarchical organization of autophagy. *Molecular Cell*, 2017;65:761–774.e5.

[49] Blomen VA, Májek P, Jae LT, et al. Gene essentiality and synthetic lethality in haploid human cells. *Science*, 2015;350:1092–1096.

[50] Amberger JS, Bocchini CA, Schiettecatte F, Scott AF, Hamosh A. OMIM.org: Online mendelian inheritance in man (OMIM R), an online catalog of human genes and genetic disorders. *Nucleic Acids Research*, 2015;43:D789–798.

[51] Lee DS, Park J, Kay KA, et al. The implications of human metabolic network topology for disease comorbidity. *Proceedings of the National Academy of Sciences USA*, 2008;105:9880–9885.

[52] Zhou X, Menche J, Barabasi, AL, Sharma A. Human symptoms–disease network. *Nature Communications*, 2014;5:4212.

[53] NIH: US National Library of Medicine. Medical subject headings. Available online at www.nlm.nih.gov/mesh/.

[54] Barrenas F, Chavali S, Holme P, Mobini R, Benson M. Network properties of complex human disease genes identified through genome-wide association studies. *PLOS ONE*, 2009;4:e8090.

[55] Zhang M, Zhu C, Jacomy A, Lu LJ, Jegga AG. The orphan disease networks. *American Journal of Human Genetics*, 2011;88:755–766.

[56] Pinero J, Berenstein A, Gonzalez-Perez A, Chernomoretz A, Furlong LI. Uncovering disease mechanisms through network biology in the era of next generation sequencing. *Science Reports*, 2016;6:24570.

[57] Hidalgo, CA, Blumm, N, Barabasi, AL, Christakis, NA. A dynamic network approach for the study of human phenotypes. *PLOS Computational Biology*, 2009;5:e1000353.

[58] Chmiel A, Klimek P, Thurner S. Spreading of diseases through comorbidity networks across life and gender. *New Journal of Physics*, 2014;16:115013.

[59] Hu JX, Thomas CE, Brunak S. Network biology concepts in complex disease comorbidities. *Nature Reviews Genetics*, 2016;17: 615–629.

[60] Duran-Frigola, M, Rossell, D, Aloy, P. A chemo-centric view of human health and disease. *Nature Communications*, 2014;5:5676.

[61] Gomez-Cabrero, D. Menche J, Vargas C, et al. From comorbidities of chronic obstructive pulmonary disease to identification of shared molecular mechanisms by data integration. *BMC Bioinformatics*, 2016;17:441.

[62] Klimek, P, Aichberger, S, Thurner, S. Disentangling genetic and environmental risk factors for individual diseases from multiplex comorbidity networks. *Science Reports*, 2016;6:39658.

[63] Pastor-Satorras, R, Castellano, C, Van Mieghem, P, Vespignani, A. Epidemic processes in complex networks. *Reviews of Modern Physics*, 2015;87: 925–979.

[64] Bernoulli, D. Essai d'une nouvelle analyse de la mortalité causée par la petite verole et des avantages de l'inoculation pour la prevenir. *Histoire de l'Academie Royale des Sciences (Paris) Avec les Mémoires de Mathematique & de Physique*, 1760;1:1–45.

[65] Hethcote HW. Three basic epidemiological models. *Applied Mathematical Ecology*, 1989;18:119–144.

[66] Kermack WO, McKendrick AG. A contribution to the mathematical theory of epidemics. *Proceedings of the Royal Society of London A: Mathematical, Physical and Engineering Sciences*, 1927;115:700–721.

[67] Longini Jr IM, Nizam A, Xu S, et al. Containing pandemic influenza at the source. *Science*, 2005;309:1083–1087.

[68] Granell C, Gomez S, Arenas A. Dynamical interplay between awareness and epidemic spreading in multiplex networks. *Physical Review Letters*, 2013;111:128701.

[69] Hufnagel L, Brockmann D, Geisel T. Forecast and control of epidemics in a globalized world. *Proceedings of the National Academy of Sciences USA*, 2004;101:15124–15129.

[70] Brockmann D, Helbing D. The hidden geometry of complex, network-driven contagion phenomena. *Science*, 2013;342:1337–1342.

[71] Verdery AM, Siripong N, Pence BW. Social network clustering and the spread of HIV/AIDS among persons who inject drugs in two cities in the Philippines. *Journal of Acquired Immune Deficiency Syndromes*, 2017;76:26–32.

[72] Barabasi AL, Albert R. Emergence of scaling in random networks. *Science*, 1999;286:509–512.

[73] Pastor-Satorras R, Vespignani A. Epidemic spreading in scale-free networks. *Physical Review Letters*, 2001;86,:3200–3203.

[74] Pastor-Satorras R, Vespignani A. Immunization of complex networks. *Physical Review E: Statistical, Nonlinear, Biological, and Soft Matter Physics*, 2002;65: 036104.

[75] Castellano C, Fortunato S, Loreto V. Statistical physics of social dynamics. *Reviews of Modern Physics*, 2009;81:591–646.

[76] Christakis NA, Fowler JH. The collective dynamics of smoking in a large social network. *New England Journal of Medicine*, 2008;358:2249–2258.

[77] Fowler JH, Christakis NA. Dynamic spread of happiness in a large social network: longitudinal analysis over 20 years in the Framingham heart study. *BMJ*, 2008;337: a2338.

[78] Cozzo E, Banos RA, Meloni S, Moreno Y. Contact-based social contagion in multiplex networks. *Physical Review E: Statistical, Nonlinear, Biological, and Soft Matter Physics*, 2013;88:050801.

[79] Boccaletti S, Bianconi G, Criado R, et al. The structure and dynamics of multilayer networks. *Physics Reports*, 2014;544:1–122.

[80] Kivelä M, Arenas A, Barthelemy M, et al. Multilayer networks. *Journal of Complex Networks*, 2014;2:203–271.

[81] Noh JD, Rieger H. Random walks on complex networks. *Physical Review Letters*, 2004;92:118701.

[82] Bader GD, Cary MP, Sander C. Pathguide: A pathway resource list. *Nucleic Acids Research*, 2006;34:D504–D506.

[83] Szklarczyk, D. et al. The STRING database in 2017: Quality-controlled protein–protein association networks, made broadly accessible. *Nucleic Acids Research*, 2017;45:D362–D368.

[84] Chatr-Aryamontri A, Oughtred R, Boucher L. et al. The BioGRID interaction database: 2017 update. *Nucleic Acids Research*, 2017;45:D369–D379.

[85] Orchard S, Ammari M, Aranda B, et al. The MIntAct project: IntAct as a common curation platform for 11 molecular interaction databases. *Nucleic Acids Research*, 2014;42:D358–63.

[86] Alanis-Lobato G, Andrade-Navarro MA, Schaefer MH. HIPPIE v2.0: Enhancing meaningfulness and reliability of protein–protein interaction networks. *Nucleic Acids Research*, 2017;45:D408–D414.

[87] Rolland T, Taşan M, Charloteaux B, et al. A proteome-scale map of the human interactome network. *Cell*, 2014;159:1212–1226.

[88] Huttlin EL, Ting L, Bruckner RJ, et al. The BioPlex network: A systematic exploration of the human interactome. *Cell*, 2015;162:425–440.

[89] Huttlin EL, Bruckner RJ, Paulo JA, et al. Architecture of the human interactome defines protein communities and disease networks. *Nature*, 2017;545:505–509.

[90] Zhang QC, Petrey D, Deng L, et al. Structure-based prediction of protein–protein interactions on a genome-wide scale. *Nature* 2012;490:556–560.

[91] Jansen, R. Yu H, Greenbaum D, et al. A Bayesian networks approach for predicting protein–protein interactions from genomic data. *Science*, 2003;302:449–453.

[92] Hakes L, Pinney JW, Robertson DL, Lovell SC. Protein–protein interaction networks and biology: What's the connection? *Nature Biotechnology*, 2008;26:69–72.

[93] Gillis J, Ballouz S, Pavlidis P. Bias tradeoffs in the creation and analysis of protein–protein interaction networks. *Journal of Proteomics*, 2014;100:44–54.

[94] Caldera M, Buphamalai P, Muller F, Menche J. Interactome-based approaches to human disease. *Current Opinion in Systems Biology*, 2017;3:88–94.

[95] Clauset A, Shalizi C, Newman M. Power-law distributions in empirical data. *SIAM Review*, 2009;51:661–703.

[96] Watts DJ, Strogatz SH. Collective dynamics of "small-world" networks. *Nature*, 1998;393:440–442.

[97] Cohen R, Havlin S. Scale-free networks are ultrasmall. *Physical Review Letters*, 2003;90:058701.

[98] Callaway DS, Newman ME, Strogatz SH, Watts DJ. Network robustness and fragility: Percolation on random graphs. *Physical Review Letters*, 2000;85:5468.

[99] Newman ME, Strogatz SH, Watts, DJ. Random graphs with arbitrary degree distributions and their applications. *Physical Review E: Statistical, Nonlinear, Biological, and Soft Matter Physics*, 2001;64:026118.

[100] Cohen R, Erez K, Ben-Avraham D, Havlin S. Resilience of the internet to random breakdowns. *Physical Review Letters*, 2000;85:4626.

[101] Dorogovtsev SN, Mendes JF. *Evolution of Networks: From Biological Nets to the Internet and WWW*. Oxford University Press;2003.

[102] Albert R, Jeong H, Barabasi AL. Error and attack tolerance of complex networks. *Nature*, 2000;406:378–382.

[103] Jeong H, Mason SP, Barabasi AL, Oltvai ZN. Lethality and centrality in protein networks. *Nature*, 2001;411:41–42.

[104] Hartwell LH, Hopfield JJ, Leibler S, Murray AW. From molecular to modular cell biology. *Nature*, 1999;402:C47–52.

[105] Spirin V, Mirny LA. Protein complexes and functional modules in molecular networks. *Proceedings of the National Academy of Sciences USA*, 2003;100:12123–12128.

[106] Barabasi AL, Oltvai ZN. Network biology: Understanding the cell's functional organization. *Nature Reviews Genetics*, 2004;5:101–113.

[107] Feldman I, Rzhetsky A, Vitkup D. Network properties of genes harboring inherited disease mutations. *Proceedings of the National Academy of Sciences USA*, 2008;105:4323–4328.

[108] Ghiassian SD, Menche J, Barabasi AL. A DIseAse MOdule detection (DIAMOnD) algorithm derived from a systematic analysis of connectivity patterns of disease proteins in the human interactome. *PLOS Computational Biology*, 2015;11:e1004120.

[109] Fortunato S. Community detection in graphs. *Physics Reports*, 2010;486:75–174.

[110] Guney E, Menche J, Vidal M, Barabasi AL. Network-based in silico drug efficacy screening. *Nature Communications*, 2016;7:10331.

[111] Newman ME. The structure and function of complex networks. *SIAM Review*, 2003;45:167–256.

[112] Maslov S, Sneppen K. Specificity and stability in topology of protein networks. *Science*, 2002;296, 910–913.

[113] Milo R, Kashtan N, Itzkovitz S, Newman MEJ, Alon U. On the uniform generation of random graphs with prescribed degree sequences. arXiv preprint available at https://arxiv.org/pdf/cond-mat/0312028.pdf. 2004.

[114] Bender EA, Canfield ER. The asymptotic number of labeled graphs with given degree sequences. *Journal of Combinatorial Theory, Series A*, 1978;24:296–307.

[115] Bollobás B. Random graphs. In *Graph Theory*, 123–145 (Springer, 1979).

[116] Serrano MA, Boguna M. Tuning clustering in random networks with arbitrary degree distributions. *Phys. Rev. E* 72, 036133 (2005).

[117] Boguna, M, Pastor-Satorras, R. Class of correlated random networks with hidden variables. *Physical Review E: Statistical, Nonlinear, Biological, and Soft Matter Physics*, 2003;68:036112.

[118] Weber S, Porto M. Generation of arbitrarily two-point-correlated random networks. *Physical Review E: Statistical, Nonlinear, Biological, and Soft Matter Physics*, 2007;76:046111.

[119] Samal A, Martin OC. Randomizing genome-scale metabolic networks. *PLOS ONE*, 2011;6:e22295.

[120] Basler G, Ebenhoh O, Selbig J, Nikoloski Z. Mass-balanced randomization of metabolic networks. *Bioinformatics*, 2011;27:1397–1403.

[121] Vissers LELM, Gilissen C, Veltman JA. Genetic studies in intellectual disability and related disorders. *Nature Reviews Genetics*, 2016;17:9–18.

[122] Krishnan A, Zhang R, Yao V, et al. Genome-wide prediction and functional characterization of the genetic basis of autism spectrum disorder. *Nature Neuroscience*, 2016;19: 1454–1462.

[123] Sharma A, Menche J, Chris Huang C, et al. A disease module in the interactome explains disease heterogeneity, drug response and captures novel pathways and genes in asthma. *Human Molecular Genetics*, 2015;24: 3005–3020.

[124] Barshir R, Shwartz O, Smoly IY, Yeger-Lotem E. Comparative analysis of human tissue interactomes reveals factors leading to tissue-specific manifestation of hereditary diseases. *PLOS Computational Biology*, 2014;10:e1003632.

[125] Magger O, Waldman YY, Ruppin E, Sharan R. Enhancing the prioritization of disease-causing genes through tissue specific protein interaction networks. *PLOS Computational Biology*, 2012;8:e1002690.

[126] Li M, Zhang J, Liu Q, Wang J, Wu FX. Prediction of disease-related genes based on weighted tissue-specific networks by using DNA methylation. *BMC Medical Genomics*, 2014;7(Suppl 2):S4.

[127] Oti M, Snel B, Huynen MA, Brunner HG. Predicting disease genes using protein–protein interactions. *Journal of Medical Genetics*, 2006;43:691–698.

[128] Wang XD, Huang JL, Yang L, et al. Identification of human disease genes from interactome network using graphlet interaction. *PLOS ONE* 2014;9:e86142.

[129] Erten S, Bebek G, Ewing RM, Koyuturk M, et al. DADA: Degree-aware algorithms for network-based disease gene prioritization. *BioData Mining*, 2011;4:19.

[130] Guney E, Oliva B. Exploiting protein-protein interaction networks for genome-wide disease-gene prioritization. *PLOS ONE*, 2012;7:e43557.

[131] Bailly-Bechet M, Borgs C, Braunstein A, et al. Finding undetected protein associations in cell signaling by belief propagation. *Proceedings of the National Academy of Sciences USA*, 2011;108:882–887.

[132] Tuncbag N, McCallum S, Huang SSC, Fraenkel E. SteinerNet: A web server for integrating 'omic' data to discover hidden components of response pathways. *Nucleic Acids Research*, 2012;40:W505–W509.

[133] Tuncbag N, Gosline SJC, Kedaigle A, et al. Network-based interpretation of diverse high-throughput datasets through the omics integrator software package. *PLOS Computational Biology*, 2016;12:e1004879.

[134] Krauthammer M, Kaufmann CA, Gilliam TC, Rzhetsky A. Molecular triangulation: Bridging linkage and molecular-network information for

identifying candidate genes in Alzheimer's disease. *Proceedings of the National Academy of Sciences USA*, 2004;101:15148–15153.

[135] Vanunu O, Magger O, Ruppin E, Shlomi T, Sharan R. Associating genes and protein complexes with disease via network propagation. *PLOS Computational Biology*, 2010;6:e1000641.

[136] Vandin F, Upfal E, Raphael BJ. Algorithms for detecting significantly mutated pathways in cancer. *Journal of Compututational Bio*logy, 2011;18:507–522.

[137] Cowen L, Ideker T, Raphael BJ, Sharan R. Network propagation: A universal amplifier of genetic associations. *Nature Reviews Genetics*, 2017;18:551–562.

[138] Kohler S, Bauer S, Horn D, Robinson PN. Walking the interactome for prioritization of candidate disease genes. *The American Journal of HumanGenetics* 82, 949–958 (2008).

[139] Pesquita C, Faria D, Falcao AO, Lord P, Couto FM. Semantic similarity in biomedical ontologies. *PLOS Computational Biology,* 2009;5:e1000443.

[140] Van Erp M, Schomaker, L. Variants of the borda count method for combining ranked classifier hypotheses. In Schomaker L, Vuurpijl L, eds., *Proceedings 7th International Workshop on Frontiers in Handwriting Recognition*. International Unipen Foundation;2000, pp. 443–452.

[141] Loscalzo J, Barabasi AL, Silverman EK, eds. *Network Medicine: Complex Systems in Human Disease and Therapeutics*. Harvard University Press;2017.

[142] De Domenico M, Granell C, Porter MA, Arenas A. The physics of spreading processes in multilayer networks. *Nature Physics*, 2016;12:901–906.

[143] Liu YY, Slotine JJ, Barabasi AL. Controllability of complex networks. *Nature*, 2011;473:167–173.

[144] Liu YY, Slotine JJ, Barabasi AL. Observability of complex systems. *Proceedings of the National Academy of Sciences USA*, 2013;110:2460–2465.

11 Elucidating Genotype-to-Phenotype Relationships via Analyses of Human Tissue Interactomes

Idan Hekselman, Moran Sharon, Omer Basha, and Esti Yeger-Lotem

11.1 Introducing Genotypes, Phenotype, and Molecular Interaction Networks

One of the main enigmas in biology is the genotype to phenotype relationship: How does a certain change in a DNA sequence of an organism lead to a specific phenotype? What are the molecular mechanisms that allow cells harboring certain mutations to proliferate and lead to cancer, while cells harboring other mutations die and lead to neuro-degeneration? All cellular processes, from cell growth and proliferation to metabolism and cell death, involve biochemical interactions among genes, proteins, RNA, metabolites, and other molecules. In the face of specific mutations, some of these interactions become altered and other interactions arise, resulting in distinct phenotypes. Network concepts, algorithms, and tools allow for defining, modeling, and predicting how interactions combine into cellular processes. Consequently, several network-based approaches have been developed and applied to elucidate genotype-to-phenotype relationships. Notably, many previous approaches employed generic molecular interaction networks. Phenotypes, however, and in particular human diseases including cancer and neurodegeneration, tend to occur in specific tissues and cell types. In recent years, the mapping of tissues and cell types has gone through unprecedented expansion, from small numbers of tissue profiles to massive resources containing hundreds of tissue profiles. In this chapter, we present current network-based approaches that harness tissue contexts for elucidating the molecular basis that translates genotypes into phenotypes.

11.1.1 What are Genotypes and Phenotypes?

Each of us is composed of trillions of cells that make up our body. Almost all these cells contain a copy of our genome, the DNA sequence that determines the features that make us who we are. For the large part, we inherited this DNA sequence from our parents. Yet our genome also contains some unique changes to the inherited sequence, including insertions or deletions of DNA fragments, transpositions, and point mutations. When these changes occur in the first stages of our development in the germline, they are shared by cells across our body and become part of our heritable genome. When these changes occur later in life, they are limited to certain cells and are considered as somatic changes. Most of this chapter is focused on understanding the consequences of heritable, germline changes, and in particular the changes that lead to disease.

A *genotype* refers to sub-sequences in the genome that determine a certain characteristic by which we differ from other individuals. A genotype may consist of variants (alleles) of a single gene or variants in multiple genes. One example is the genotype that determines our blood type: Two alleles determine whether we have blood type *A*, *B*, *AB* or *O*. If we have both the *A* and *B* alleles our blood type will be *AB*. However, not in all cases both alleles determine the characteristic equally. An allele that determines a characteristic is called *dominant* and is denoted by a capital letter, as *A* and *B* in the example above. An allele that does not determine the characteristic in the presence of a dominant allele is called *recessive*, and is denoted by a lower-case letter. Individuals with alleles *A* and *o* will have blood type *A*, since *o* represents a recessive allele. Only individuals with two copies of the recessive allele, denoted homozygous recessive individuals, will manifest the characteristic determined by that allele. For example, the only individuals with blood type *O* are those with genotype *oo*.

A *phenotype* is the observable characteristic that is determined by the genotype. In the example above, the phenotype is the individual's blood type. In general, a phenotype refers to any measurable or observable characteristic: It can be a physiological property of the individual, such as blood type or eye color; it can be a certain behavioral feature of the individual; and it can even be a molecular characteristic, such as the expression of a certain protein.

The relationships between genotypes and phenotypes are not straightforward [1]. First, several genotypes may lead to the same phenotype. For example, individuals with blood type genotypes *AA* and *Ao* will both have blood type *A*. Second, a certain genotype may lead to multiple phenotypes. For example, identical twins have the same genotype, yet often have minor physiological differences owing to environmental factors, epigenetic factors, or random effects. In addition, a phenotype may become observable only under certain conditions. For example, certain variants of the BRCA1 gene increase the risk for breast cancer, yet not all carriers of these variants will develop this cancer. A certain phenotype may result from a genotype that involves several genes, as in the case of height or eye color. Consequently, the link between a certain genotype and a certain phenotype can be elusive and hard to pinpoint [2, 3].

Connecting between a genotype and the phenotype that it determines is particularly important from a clinical perspective. Many genetic diseases tend to "run in families," yet until the previous century the genetic basis for this phenomenon has

> **Box 11.1: Genotype and phenotype**
>
> **Genotype:** Part of the genome of an individual that differs in sequence from other individuals and determines a certain characteristic of that individual (phenotype). A genotype is typically associated with variants of a specific gene, or with a combination of genes. Mutation: A change in the nucleotide sequence of a gene that occurs in less than 1% of the population. A change that occurs in more than 1% of the population is called a genetic variation.
>
> **Phenotype:** A specific characteristic of an individual that is observable or measurable, such as physiological properties, behavior, or a molecular characteristic. Phenotypes result from changes in the genome (genotypes), potentially in combination with epigenetic and environmental factors.

been unclear. Today, we know that some of the diseases are monogenic, namely are caused by variants of a single gene, as in sickle cell anemia [4]. Other diseases are complex and involve variants of multiple genes, as in Crohn's disease or asthma [5, 6]. Owing to the huge progress in genetics and in genome sequencing, the genotypes underlying many monogenic diseases have been elucidated. To date, we know the causal alleles for hundreds of diseases, and numbers keep rising owing to the intensive usage of sequencing techniques in the medical arena [7]. Nevertheless, each of us harbors many genomic alterations that for the large part have no phenotypic consequences [8]. Therefore, pinpointing causal alleles out of the many altered alleles we carry remains challenging, particularly for complex and rare diseases [9]. One factor that helps shed a light on the complex relationships between genotypes and phenotypes is the molecular interactions that occur in the human body [10]. (See Box 11.1.)

11.1.2 Molecular Interactions Play a Role in Genotype–Phenotype Relationships

The association between a genotype and a phenotype is the first step in understanding how a specific phenotype is determined. With respect to diseases, it allows for identification of diseased patients and carriers through genotyping assays. However, it often remains to be understood how exactly a phenotype emerges: What is the molecular basis by which the specific genotype causes disease phenotype? What is the chain of events that leads from a causal variant to the observed characteristic? To illustrate this challenge, we describe in Box 11.2 the case of sickle cell anemia, the first disease with a resolved genetic and molecular basis.

However, sickle cell disease is an exception. For many other diseases with known causal genotypes, the molecular mechanisms by which the causal variant leads to pathological phenotypes remain to be elucidated. The elucidation of the mechanism

> **Box 11.2: The molecular basis of sickle cell disease**
>
> Sickle cell disease is a blood disorder caused by a mutation in the oxygen-carrying protein hemoglobin that is found in red blood cells [4]. Owing to this mutation, red blood cells become rigid and sickle-shaped and lack efficient oxygen transfer. The causal mutation is a single change in the β-globin gene, where the nucleotide adenine (A) is replaced by thymine (T). In the resulting protein, this leads to a substitution of the amino acid glutamate by valine at position 6 of the protein sequence. Notably, sickle cell disease was the first human disease shown to be caused by an abnormal protein.
>
> *How does this mutation lead to anemia?* In contrast to the charged glutamate that is present in the wildtype protein, valine is hydrophobic and creates a hydrophobic patch on the surface of the mutated hemoglobin. This patch tends to interact with a hydrophobic patch on the surface of another hemoglobin, thus leading to a hydrophobic bridge between them, and consequently to an occasional polymerization of several hemoglobin molecules. When several hemoglobin molecules polymerize in the red blood cell, the cell becomes sickle-shaped and prone to destruction, leading to anemia.

is not only important for better understanding of the pathogenesis, but also for the development of preventive therapies and cure. A way forward in this direction is to focus on the cellular environment that surrounds causal variants, and specifically on their molecular interactions [11].

Molecular interactions are key in defining, modeling, and understanding a large variety of cellular processes. Metabolic processes, such as the degradation of carbohydrates to create energy, are defined by the interactions between enzymes and their substrates. Transcription regulation processes, which are responsible for the activation or repression of gene expression, are composed of interactions among components of the transcriptional machinery and chromosomal DNA. Signal transduction pathways, by which cells respond to hormones and other stimuli, are composed of interactions between hormones, receptors, and signal transducing molecules.

Molecular interactions are also key in elucidating genotype–phenotype relationships. In general, we can view certain genotypes as causing a change, or perturbation, in the regular pattern of molecular interactions, which subsequently elicits the phenotypes observed. The aberrant polymerization of hemoglobin molecules in red blood cells of sickle cell disease patients is one example of this principle (Box 11.2). Another example is presented by mutations in RAS genes, which are recurrent in different malignancies [12]. RAS genes are involved in the signals passed between cells to regulate their growth. Cancer-causing mutations in RAS genes lead to aberrant forms of RAS proteins, causing them to lose their ability to interact and thus deactivate, and therefore result in uncontrolled proliferation of cells.

Table 11.1: Major molecular interaction databases.

Type of molecular interaction	Representative databases	Number of relationships
Protein–protein interactions (PPIs)	BioGRID	$\sim 3 \times 10^5$
	IntAct	
Transcription regulation		
(transcription factor – DNA interactions)	TRANSFAC	$\sim 10^4$
	ENCODE	$\sim 10^4$
	ENCODE	$\sim 10^4$
microRNA-target transcript	miRecords	$\sim 10^3$
	TarBase	$\sim 10^5$
Transcription factors – microRNAs	TransMIR	10^3
Genetic interactions	BioGRID	$\sim 2 \times 10^3$
Functional interactions	STRING	
Metabolic interactions	BiGGRecon 2	$\sim 8 \times 10^3$

11.1.3 Molecular Interaction Networks

The recognized importance of molecular interactions, as cellular process decipherers, has led to extensive efforts to map them. To date, thousands of interactions between human molecules have been detected experimentally (e.g., [13, 14, 15, 16, 17]). These interactions are typically recorded in public databases, some of which are listed in Table 11.1. These databases typically hold in the order of 10^4 or 10^5 known relationships. The next question is: How to make use of these valuable data?

In a seminal study, Schwikowski, Uetz et al., used a network to portray the few thousands of protein–protein interactions (PPIs) that they had discovered earlier between yeast proteins [18, 19]. They represented each protein by a node in a network, and each interaction by an edge. By viewing these PPIs as one network, they were able to show that yeast proteins are connected to each other through relatively short paths, that proteins involved in the same process tend to interact with each other, and that the role of a protein can be inferred from the roles of its interaction partners [19]. The insights that they gained, along with other seminal studies [20, 21], opened the field of network biology, where graph and network algorithms are developed and applied to answer biological questions [11].

Since these seminal studies, many types of biological network models have been created. In this chapter, we focus mainly on molecular interaction networks (reviewed in [22]; see Box 11.3). These network models follow a similar representation of interactions as the one used by Schwikowski et al., described above [18]. Network nodes represent certain molecules, such as proteins, genes and metabolites, and network edges represent their relationships, such as PPIs, regulatory interactions, genetic interactions or enzyme-metabolite relationships. Network models can include a single type of nodes or relationships, as in PPI networks, or integrate several types of nodes and relationships. They can reflect physical molecular interactions, functional relationships such as co-expression correlations, or even co-citation in the literature.

> **Box 11.3: Molecular interaction networks and interactomes**
>
> **Node:** A certain molecule, such as a protein or a gene.
> **Edge:** A relationship between two nodes, such as a physical interaction between proteins or correlated expression of two genes across samples or conditions.
> **A molecular interaction network:** A network where nodes represent molecules and edges represent their relationships.
> **Interactome:** A molecular interaction network that represents all genes and/or proteins of an organism and their physical interactions.

Network models may include weights over nodes or edges that reflect, for instance, the importance of certain nodes or the reliability of their relationships. Network models may also include additional entities, such as drugs and diseases [22]. Common types of network models are summarized in Table 11.2.

Networks models that describe relationships between human genes and proteins are frequently derived from large-scale experimental assays, such as gene expression profiling, and from databases that gather data from various sources, such as protein interaction databases (Table 11.1). Network sizes tend to be in the order of 10^5 nodes and 10^6 edges. While they constitute detailed maps, networks are also incomplete and noisy. Their usage is therefore worthwhile but also challenging, calling for sophisticated approaches for meaningful analyses and investigation. These approaches are described throughout this book (see Chapters 3–9). In this chapter, we focus specifically on concepts and approaches for illuminating the molecular basis of genotype–phenotype relationships in context of the human tissues.

11.2 Network Approaches for Elucidating Disease-Related Genotype–Phenotype Relationships

In general, causal genotypes lead to disease by altering processes that normally occur within cells or tissues. In the previous section, we mentioned two examples: The aberrant polymerization of mutant hemoglobin molecules that leads to rigid and fragile red blood cells (Box 11.2), and the aberrant activation of mutant RAS proteins that results in uncontrolled cellular proliferation (Section 11.1.2). Another example of diseases that stem from dysfunctional metabolic processes is presented in Box 11.4. The examples show that by studying how causal variants alter their molecular environment, we can gain better understanding of the molecular basis of the respective diseases. In this chapter, we describe efforts that assessed the effect of causal genotypes on their molecular environments. We then describe approaches that, given this effect, use molecular interaction networks to decipher the molecular basis of genotype–phenotype relationships.

Table 11.2: Common types of network models.

Network type	Nodes	Edges	Example
Protein–protein interaction (PPI) network	Proteins	Edges connect interacting proteins. Edge weight can represent confidence of interaction.	In VarWalker, a PPI network is used with genomic and transcriptomic data to identify mutation affected sub-networks in cancer [23].
Transcription regulation network	Transcription factors (TFs) and genes	Edges connect TFs and their target genes	
Metabolic network	Enzymes, substrates and products	Edges connect substrates and products of an enzymatic reaction, or consecutive enzymes in a metabolic pathway	KEGG pathway database contains metabolic pathways in human and other organisms.
Genetic interaction network	Genes	Edges connect genes with positive or negative genetic interactions. Edge weight represents the strength of the epistatic relationship.	In DriverNet, a genetic association network is used to predict functional important driver genes in cancer [24].
Co-expression network	Genes	Edges are weighted by the expression correlation between the interacting genes.	To elucidate gene function on a global scale, pairs of genes that were co-expressed in humans, flies, worms, and yeast were identified [25]
Functional network	Genes	Edges represent the presence of any functional relationship between interacting genes, such as PPI, co-expression, co-mentioning in the literature, etc.	STRING database contains functional relationships among proteins of human and other organisms.
Integrated network	Proteins, genes, microRNAs, metabolites	Each edge represents a specific relationship, such as PPI or transcription regulation. Thus, two interacting molecules can be connected by more than one edge.	ResponseNet identifies regulatory pathways in an integrated network composed of PPIs and transcriptional regulation interactions, and is used to infer how mutations lead to altered gene expression [26].

> **Box 11.4: Causal genotypes uncover the importance of metabolic processes**
>
> Inborn error of metabolism (IEM) diseases often result from inherited recessive loss-of-function mutations in enzymes and transporters. These loss-of-function mutations diminish the activity of the encoded protein, the activity of the metabolic pathways it participates in, thus leading to metabolic catastrophes [27]. Interestingly, the study of IEMs unraveled the importance of certain metabolic pathways to human health. For example, by studying hypoketotic hypoglycemia patients, the role of mitochondrial fatty acid oxidation in human energy homeostasis was deduced [28].

11.2.1 Causal Variants Tend to Perturb Their Surrounding Molecular Network

Before delving into approaches that use interactomes to decipher genotype–phenotype relationships, we first need to understand whether and in what way causal alleles or causal variants alter their molecular interaction network. Theoretically, causal variants can affect their surrounding network in several ways that we illustrate in Figure 11.1. In the most extreme scenario, causal genes are totally eliminated from the network together with their direct interactions. This node removal, or "null" scenario, fits causal alleles that lead to a complete loss of the gene product or its function (Figure 11.1 (a)). In a more moderate scenario, only a subset of the interactions of the causal variant are removed from the network, while other interactions are left intact (Figure 11.1 (b)). This scenario fits partial loss of the gene product or its function. It can result from exon skipping, from a mutation in one of the interaction interfaces of the gene product, and more, and is commonly referred to as an "edgetic perturbation" [29]. A different scenario includes the incorporation of new interactions into the network, also called "gain-of-interaction" scenario (Figure 11.1 (c)). This scenario could result from genetic changes such as point mutations that create a new interaction interface on the gene product [30], as in sickle cell disease (Box 11.2). Notably, these theoretical scenarios are simplistic. The same causal variant can be involved in loss- and gain-of-interactions concurrently. Additionally, its interactions might change in a subtler manner, due to altered binding affinity between the gene product and its interactors, or due to context-sensitive changes, such as the presence of a co-factors [31, 32]. The altered interaction profile of a certain variant is sometime referred to as its *edgotype* [30].

Several studies analyzed edgetic perturbations from a structural perspective [31, 33, 34]. Yu and colleagues analyzed interactions between proteins with resolved interaction interfaces. Upon mapping causal mutations onto the structures of these proteins, they found that in-frame mutations, including missense point mutations, insertions, and deletions, were enriched within interaction interfaces. This suggests that edgetic perturbations indeed alter phenotypes. They also showed that causal

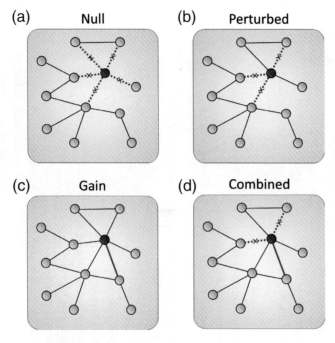

Figure 11.1: Edgotype scenarios. Nodes represent genes or proteins, with red nodes representing causal variants. Edges represent interactions, where interactions that are common to both the wildtype protein and the causal variant appear in black, interactions that do not occur in the presence of the causal variant appear in red, and interactions that occur only in the presence of the causal variant appear in blue. (a) The null scenario, where the causal variant lost all its interactions. (b) The perturbed scenario, where the causal variant lost some of its interactions while other are left intact. (c) The gain scenario, where the variant gained a new interaction compared to wildtype. (d) The combined scenario, where loss and gain of interactions co-occur.

mutations in the same gene that impact distinct interfaces, are twice more likely to cause a different disease. Table 11.3 lists tools for elucidating edgotypes of genetic variants.

Several studies systematically assessed the frequency of the edgotypic scenarios described above (Figure 11.1). Zhong et al. analyzed the PPIs of 29 causal variants belonging to five Mendelian disorders [29]. They found five variants that behaved as null, 16 variants with edgetic perturbations, and eight variants that retained their wildtype PPIs. Additional studies compared the edgotypic profile of causal variants to the edgotypic profile of common variants, namely variants that deviate in sequence from wildtype proteins but have no phenotypic consequences. Rolland et al. showed that disease variants tend to cause more edgetic perturbations than their respective common variants [35]. Sahni et al. expanded this analysis to include 67 common variants and 460 causal variants of 220 genes associated with Mendelian disorders [36]. They found that two-thirds of the causal variants exhibited perturbed PPIs, with half corresponding to edgetic perturbations, and only few causal variants exhibited gain-of-interactions. Common variants, in contrast, rarely affected PPIs. In a related study,

Table 11.3: Available edgotype prediction tools.

Tool name	Interaction identified	Input	Output	Comments
BayesPI-BAR	TF-DNA	Mutant regulatory sequence	Binding affinity of TF	Bayesian method
BeAtMuSiC	Physical PPI	Protein complex structure from Protein Data Bank and protein-coding SNPs	Change in interaction affinity	Coarse-grained modeling
ComPPI	Physical PPI	Protein	Cellular localization and compartment-specific edgotype	Probabilistic disjunction
ConCavity	Ligand-protein	Protein structure from Protein Data Bank (PDB)	Ligand-binding pockets	3D structure based
dSysMap	Physical PPI	Protein/mutant or protein/disease	Detailed interactome with disease and structural annotations of the nodes and edges	3D structure based
HaploReg	TF-DNA	Mutant regulatory sequence	Chromatin state and binding annotation, sequence conservation across mammals, regulatory motifs and expression patterns	
MutaBind	Physical PPI	Protein-protein complex structure from Protein Data Bank (PDB) and protein-coding SNPs	Change in interaction affinity	Molecular mechanics force fields, statistical potentials, and fast side-chain optimization algorithms

Table 11.3: Cont.

Tool name	Interaction identified	Input	Output	Comments
OncoCis	TF-DNA	Mutant regulatory sequence	Chromatin accessibility and histone modifications, conservation, consensus binding motifs and their change	
PolymiRTS	miR-RNA	Query protein-coding gene/GO annotation/phenotypic trait	Mutations that alter miR-RNA interactions and consequent diseases, traits, and pathways	Diseases, traits, and pathways
Predicting DNA recognition by C2H2 zinc finger proteins	TF-DNA	C2H2 protein sequence and target DNA sequence	Zinc-finger protein–DNA scoring	Support vector machines
SomamiR	miR-RNA	Protein-coding gene	Somatic mutations that alter miR-RNA interactions	Cancer (somatic mutations)
Structure-PPi	Physical PPI	Protein	Domains, important sites, structural properties, binding attributes, and mutations	3D structure based and prediction of change in biochemical activity
TissueNet v2.0	Physical PPI	Protein (and optional – tissue)	Tissue-specific edgotype	Tissue-specific pathways

Fuxman Bass et al identified both loss and gain of protein–DNA interactions between 1,086 transcription factors and 109 target genes containing non-coding disease mutations [37]. Interestingly, causal variants that encoded transcription factors and had intact PPIs, frequently had perturbed protein–DNA interactions [36]. Lastly, distinct causal variants of the same gene that were found to have distinct interaction profiles, were often associated with distinct disease phenotypes [36]. To conclude, edgetic perturbations involving PPIs or protein–DNA interactions appear to be widespread and more frequent among causal alleles relative to common alleles. Thus, they are likely to mediate genotype–phenotype relationships.

11.2.2 Network Approaches to Identify the Molecular Basis of Diseases

In the early stages of network biology, the intuitive approach to elucidate how a causal allele leads to disease was to infer from the known functions and cellular roles of the aberrant gene what might be the impact of its aberration. This was achieved by mining the literature and online repositories, such as gene-based databases (e.g., [38]) and gene ontologies (GO) [39]. The merit of this approach remains clear; however, it is also limited by our incomplete knowledge.

As we have shown in the previous section, causal alleles often perturb their surrounding interactions, and distinct edgetic perturbations can lead to different diseases. This suggests that inspecting the interactions of aberrant genes is a good starting point for elucidating the molecular basis of genetic diseases. In addition, since many molecular interactions were detected or inferred from large-scale screens, they tend to be less biased toward known pathways. Thus, network-based approaches can open new and potentially meaningful avenues for study.

Below we describe several network-based approaches to elucidate genotype–phenotype relationships. We focus on conceptual descriptions of the distinct approaches, and not on their algorithmic and technical details. The latter are described in other chapters in this book. Available network-based tools for elucidating genotype–phenotype relationships are listed in Table 11.4.

11.2.2.1 Network Approaches to Monogenic Diseases

Monogenic diseases are caused by mutations in a single gene. Thus, network-based approaches typically focus on the interactions surrounding this gene, and use them to formulate testable hypotheses (Figure 11.2 (a)). A nice illustration of this approach is given by studies of the molecular basis of Huntington's disease. Huntington's disease is a fatal neurodegenerative disorder caused by poly-glutamine repeats in the huntingtin (*Htt*) protein. To illuminate the molecular basis of *Htt* toxicity, Kaltenbach et al. mapped the interactors of *Htt* and identified 234 proteins at high-confidence [40]. Upon testing an arbitrary subset, they found that 27 of the 60 interactors of *Htt* were genetic modifiers of neurodegeneration (45%), an order of magnitude higher than the 1%–4% typically observed in unbiased genetic screens. These interactors were involved in synaptic transmission, cytoskeletal organization, signal transduction, and transcription. A different study examined the interactome surrounding *Htt*,

Table 11.4: Available network tools for elucidating genotype-phenotype relationships.

Tool name	Edgotype	Input	Output	Comments
AtgO	Integrated	Gene	Interactome with ontology of autophagy functions	Progressive procedure
DriverNet	Genetic	Genome data (mutation and CNVs) and expression data	Predicted cancer driver genes	Greedy algorithm
HotNet2	Genetic	Genomic and transcriptomic data	Mutated, cancer-related subnetworks	Diffusion-oriented
Interactome INSIDER	Physical PPI	Protein/disease/set of mutations	Interactome of common and disease variants and edges structure	3D structure based
MATISSE	Any	Edges dataset and expression data	Modules	
MetaReg	Any	Protein	Interactome browser	Probabilistic modeling
ModMap	Any	Two interactomes	Modules	
MyProteinNet	Physical PPI	Edges dataset and tissue expression data	Tissue-specific interactome	Bayesian scoring
NBS	Genetic	Genomic and transcriptomic data from patients	Patients clustering	Random-walk / Network-propagation model and consensus clustering framework

Table 11.4: Cont.

Tool name	Edgotype	Input	Output	Comments
OmicsIntegrator	Any	Proteins, fold change and algorithm parameters	Significant subnetwork	Prize-Collecting Steiner Forest algorithm
OncoIMPACT	Genetic	Genomic and transcriptomic data	Patient-specific cancer driver genes predictions	Model-driven approach
PathBLAST	Physical PPI	Complex of proteins	Same complex in another species	Sequence similarity
PIVOT	Any	Edges data	Interactome visualization	
ResponseNet	Integrated	Two weighted lists of condition-related proteins and genes	Interactome subnetwork connecting them through signaling pathways	Minimum-cost flow optimization problem
SDREM	Integrated	Time series gene expression data and source proteins	Signaling pathways and transcription factors that control stimulus responses	End-to-end response models signaling pathways cross-talk
SPIKE	Any	Protein	Interactome browser	
TieDIE	Genetic	Signaling pathway and	Cancer driver mutation's effect on gene expression	Network kernel diffusion approach
Torque	Physical PPI	Complex of proteins	Same complex in another species	Sequence similarity
VarWalker	Physical PPI	Genomic and transcriptomic data	Mutated, cancer-related subnetworks	Random Walk with Restart algorithm

> **Box 11.5: Analyzing direct interactions of causal genes and variants**
>
> *Spinocerebellar ataxia type 1* (SCA1) is an inherited neurodegenerative disease that is caused by a poly-glutamine repeat in ataxin 1 (*Atxn1*). Lim et al. analyzed the interactions of *Atxn1* [42]. They found that the inclusion of poly-glutamine repeats favors the formation of a specific protein complex and attenuates the formation and function of another protein complex involving *Atxn1*, thus providing a mechanistic insight into the molecular pathogenesis of SCA1.
> *Cystic fibrosis* (CF) is an inherited disease characterized by the buildup of thickened mucus that damages lungs and many other organs. A main causal allele has a deletion of phenylalanine 508 in the CF transmembrane conductance regulator (ΔF508 *Cftr*). By mapping and comparing between the interactions of *Cftr* and ΔF508 *Cftr* in various conditions, Pankow et al. identified mutant-specific interactors whose reduction could rescue the mutant's defect [43].

and revealed that the Git1 protein enhances *Htt* aggregation by recruitment of the protein into membranous vesicles [41]. In both studies, the network surrounding *Htt* provided the first step toward illuminating the molecular basis of *Htt* toxicity. Two other examples are presented in Box 11.5.

11.2.2.2 Network Approaches to Complex Diseases

Complex diseases arise due to variations in multiple genes, potentially in combination with non-genetic factors (Section 11.1.1). Parkinson's disease, for example, has been associated with individual mutations in many genes, such as α-syn, PINK1, and parkin. In complex diseases, each gene can be analyzed independently as described above (Section 11.2.2). However, by focusing on shared interactors or sub-networks, the search space for testable hypotheses can be reduced effectively (Figure 11.2 (b)). An example of this approach is provided by a study of ataxias, a class of neurodegenerative disorders characterized by loss of balance due to cerebellar Purkinje cells (PC) degeneration. To shed light on ataxias and the normal functions of the genes involved, Lim et al. created an interaction network for 54 genes that were shown to be causal for 23 inherited ataxias [42]. Many of the causal genes shared interacting partners. By investigating the interactors of *Atxn1* that is causal for ataxia, they found two interactors that were able to modify neurodegeneration in animal models, thus providing a molecular basis for these diseases.

In many complex diseases, with cancer being a prominent example, causal genes are still emerging owing to genome-wide association studies (GWAS) and whole genome sequencing [7]. In such contexts, network approaches are also used to infer previously hidden causal genes [44]. For example, Pujana et al. used functional networks to identify genes potentially associated with higher risk of breast cancer [45]. Starting with four known genes encoding tumor suppressors of breast cancer, they

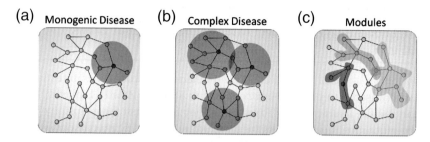

Figure 11.2: Network approaches to identify the molecular basis of genetic diseases. Nodes and edges represent proteins/genes and their interactions, respectively, with red nodes representing disease-related genes. (a) In monogenic diseases, the interaction neighborhood surrounding the causal gene is analyzed. (b) In complex diseases, the interaction neighborhood surrounding each causal gene is analyzed, with particular emphasis given to overlapping neighborhoods. (c) Modules represent distinct subnetworks, each involving a subset of disease-related genes.

combined data of gene expression profiling with functional genomic and proteomic data from model organisms including yeast, fly, and nematode, to generate a network containing 118 genes linked by 866 potential functional associations. This network was more highly connected than expected by chance, suggesting that its components function in biologically related pathways. One of the components of the network was HMMR, a gene encoding a centrosome subunit. Two case-control studies of incident breast cancer indicated that the HMMR locus is indeed associated with higher risk of breast cancer in humans.

11.2.2.3 Disease Modules

Shared interactions are a hallmark of genes that participate in common processes [19], and was shown particularly among disease genes [46, 47]. Thus, several network approaches were developed to identify relatively dense subnetworks enriched for disease genes, denoted as disease modules (Figure 11.2 (c)). One example for this approach is presented by a study of the heterogeneity and the molecular basis of asthma [48]. To create an asthma module, genes previously associated with asthma were mapped onto an interactome, together with other genes that were significantly interconnected with them. The resulting module was shown to be related to asthma by several measures, including gene expression data, gene ontology, pathway information, and comorbidity analysis. Further experimental analysis unraveled the role of the GAB1 signaling pathway in modulating asthma.

11.2.2.4 Inter-Organismal Networks

The study of breast cancer by Pujana et al. described above (Section 11.2.2.2) showed that network information from other species can be valuable [45]. Another example for this approach is presented by a recent study of Parkinson's disease [49]. There, Khurana et al. used various genome-scale networks to identify functional relationships between genes causal for Parkinson's disease (PD). They started with yeast genes that affect the toxicity of α-syn and their connecting subnetwork, and then created a "human" version of this network by relying on homology assignment through sequence, structure, and interactions topology. The resulting network linked α-syn to other Parkinson's disease genes via pathways involved in protein trafficking and ER quality control, as well as mRNA metabolism and translation.

11.3 Tissue-Sensitive Molecular Interaction Networks

The various network approaches that we discussed in Section 11.2 treated networks as static, contextless maps. However, in each organism the cellular composition of gene products, their cellular localization, their post-transcriptional states and their interactions are dynamic and differ between conditions and from one cell to another. This is particularly evident in the human body, which consists of various organs, tissues, and cell types. Within an individual, cells share the same genotype. What gives each cell type, tissue, and organ their unique characteristics and function, is the composition and combination of expressed gene products, their interactors, and other factors. Therefore, including this information within the static network is necessary for obtaining a physiologically relevant understanding and view of the functions of different tissues and organs.

Many hereditary diseases tend to manifest clinically in few tissues and cell types, while they share the same genotype with all other tissues [50]. Moreover, the causal genes for hereditary disease tend to express throughout the body and yet many tissues remain unaffected [51]. This suggests that disease processes that lead to phenotype do not depend on causal genes expression alone, but also depend on the specific molecular environment within each tissue, and on the specific interactions created (Box 11.6). Therefore, tissue and cell type information is particularly important for understanding disease processes. The inclusion of context into networks is a well-known challenge in networks research [53, 54]. Below we describe efforts to characterize the composition of human tissues, and to create tissue and cell-type specific networks. For brevity, tissues will be used to denote cell types as well.

11.3.1 Characterizing the Composition of Human Tissues

As described in Section 11.1, the recognition of the importance of molecular interactions led to intensive experimental efforts for their mapping. These efforts led to the mapping of thousands of molecular interactions between human proteins (Table 11.1). Yet the detection of these interactions typically lacked physiological context. For example, many PPIs between human proteins were actually mapped in yeast, via yeast-two-hybrid experiments [35]. Given the huge effort invested in large-scale PPI detection assays, they are not repeated in many different tissues and cell types. Other screens, such as ChIP-seq were carried in human cell lines [17], which are known to differ from primary tissues [55]. Since interactions appear to be contextless, other strategies to incorporate physiological context into networks are needed.

Molecular interactions cannot occur in a certain tissue if one of the interacting partners is not expressed. This simple observation opened the way for adding tissue contexts into the static, contextless interaction networks [53]. However, this could not

Box 11.6: Examples of tissue-specific interactions that contribute to disease

BRCA1 is a **globally expressed** tumor-suppressor protein with many known PPIs. Familial mutations in BRCA1 increase the risk for breast and ovarian cancers, but do not elicit cancer in other parts of the human body. ESR1 is an estrogen receptor protein that activates cellular proliferation, and is expressed in few tissues throughout the body. The interaction between BRCA1 and ESR1 is therefore limited to few tissues, including breast tissue, and provides a potential basis for the **breast-specific effects** of BRCA1 germline mutations [51].

DAG1 is a **globally expressed** transmembrane cell adhesion receptor dystroglycan 1. It interacts with dystrophin (DMD) and caveolin 3 (CAV3). DMD is over-expressed in muscle, while CAV3 is almost specific to muscle, making the interactions between these three genes muscle-specific. Consequently, mutations in each of these genes, including mutations in the globally expressed DAG1, give rise to various forms of muscular dystrophies [52].

Networks show the PPIs involving BRCA1 (a), and DAG1 (b). Blue nodes: globally-expressed proteins; orange nodes: tissue-specific proteins; gray nodes: intermediately expressed proteins. Diamond-shaped node represents the disease-causing gene.

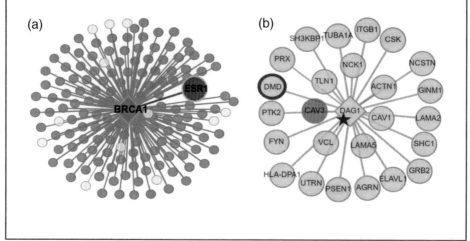

have been achieved without molecular profiling of human tissues, as we describe below.

One of the largest tissue profiling effort to date was published more than a decade ago [56]. Su et al., profiled the transcriptomes of 79 human tissues via DNA microarrays. They measured the expression of thousands of genes per tissue, and provided the most comprehensive tissue expression profiles for several years to come. Additional studies measured the transcriptomes of human tissues by other methods, such as massively parallel signature sequencing [57] and expressed sequence tags (EST) [58]. A turn point in mapping the expression of human tissues occurred in recent years owing to the development of next generation sequencing. The application of RNA sequencing to human tissues revealed that many more genes were expressed per tissue [59]. This led to several mapping efforts [60, 61, 62]. The largest dataset of normal tissues to date was created by the Genotype Tissue Expression (GTEx) consortium [63]. GTEx dataset, which is publicly available through a portal, includes over 8,500 profiles covering about 50 tissues that were sampled from roughly 900 donors [64].

For years, proteomic profiling of tissues was considered a much harder task owing to lack of large-scale experimental proteomic techniques. These techniques have advanced in recent years. Measurements based on immunohistochemistry, where protein-specific antibodies were applied to slices of human tissues, enabled to estimate the expression level of the target protein and its cellular localization in over 40 tissues [60]. Other studies used mass-spectrometry to analyze the proteomic composition of 17 adult tissues, 7 fetal tissues and 6 primary hematopoietic tissues [65, 66]. These efforts resulted in tissue-wide mapping of the expression of about 20,000 proteins. These, in turn, are overall concurrent though weakly correlated with RNA sequencing results [60]. Various tissue profiling resources appear in Table 11.5.

Table 11.5: Resources and datasets of human tissue profiles.

Dataset name	Profiled molecules	Technique	Number of profiles, tissues, molecules
BodyMap 2.0 [61]	RNA	RNA sequencing	16 human tissues
Human Protein Atlas (HPA) [60]	Protein	Immunohisto-chemistry	44 human tissues
Human Protein Atlas (HPA)	RNA	RNA sequencing	32 human tissues
Fantom [62]	Genes Short RNA	Cap analysis gene expression (CAGE)	Total of 1,829 human and 1,029 mouse CAGE libraries [67]
Genotype Tissue Expression (GTEx) [68]	RNA	RNA sequencing	53 tissues More than 8,000 samples
Tissue-specific Gene Expression and Regulation (TiGER) [69]	cDNA/RNA	EST	30 tissues, tissue-specific gene regulation
Cancer Genome Anatomy Project (CGAP)		cDNA libraries, EST	Cancer samples

11.3.2 Constructing Tissue-Sensitive Interactomes

The availability of tissue profiles enabled the identification and characterization of tissue-specific and globally expressed genes and proteins [53, 70, 71]. Moreover, it opened the door for efforts to computationally construct tissue interactomes from existing, contextless network models. The main assumption driving the construction of tissue interactomes is that the in vivo occurrence of a physical interaction depends on multiple factors: co-localization of the interacting molecules, post-transcriptional or post-translational modifications, presence of other interacting partners, etc. Yet, a basic requirement is the co-expression of the interacting molecules. Accordingly, an interaction is considered feasible within a certain tissue, only if the interacting molecules are co-expressed in that tissue (Figure 11.3). This basic requirement, in combination with the available expression profiles of human tissues, underlies the construction of tissue interactomes (e.g., [70]). In particular, the filtering of infeasible interactions precluded about 30% of the total number of known PPIs per tissue [72].

The construction of tissue interactomes by the filtering of infeasible interactions has become a common approach (e.g., [70, 72, 73, 74], and reviewed in [53]). Of note is a study by Emig and Albrecht who were among the first to construct tissue interactomes from RNA-sequenced tissue profiles [71]. Interestingly, they showed that tissue-specific PPIs were less common than previously assumed based on microarray profiles. A study by Liu et al [75] used proteomic data [65] to analyze tissue interactomes, and showed that they are sparser than the generic interactome.

Other studies relaxed the filtering. Instead of completely removing interactions from the network of a certain tissue, they used interaction weights to reflect the likelihood of an interaction (Figure 11.3). Magger et al. penalized PPIs involving non-expressed proteins by multiplying the original confidence score of the interaction by a fixed penalty factor for each pair-mate that was not expressed in the tissue [76]. Greene et al. used a Bayesian weighting scheme to rank functional relationships within a tissue [77]. The inclusion of co-expression as one of the weighting parameters resulted in the assignment of lower weights to infeasible interactions. Table 11.6 presents various tools for obtaining and visualizing tissue interactomes.

Table 11.6: Tools for obtaining tissue interactomes.

Tool name	Type of network	Comment
TissueNet [52]	Protein–protein interaction	Based on data from HPA and GTEx
MyProteinNet [78]	Protein–protein interaction	
GIANT [79]	Functional network	
SPECTRA [80]	Protein–protein interaction	Compares tissue and tumor-specific PPI networks in human
HIPPIE [81]	Protein–protein interaction	
IID [82]	Protein–protein interaction	Tissue-specific views of human and model organism interactomes

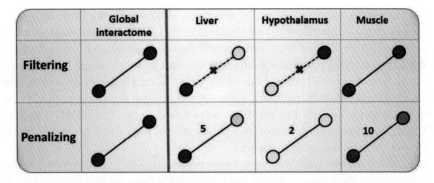

Figure 11.3: Protein-protein interactions in different tissues according to filtering and penalizing methods of tissue network constructions. A node and an edge denote a protein and an interaction between proteins, respectively. Node colors represent the expression level of the protein, with light colors matching lowly expressed proteins.

The benefit of tissue interactomes over the global/generic interactome was demonstrated in several studies. Magger et al. showed that tissue PPI networks, created by complete PPI removal or by penalizing PPIs that involve non-expressed proteins, considerably improved the prioritization of disease genes relative to the generic interactome [76]. Guan et al created functional networks for mouse tissues and used them to predict genes associated with different phenotypes [83]. They too showed that prediction performance improved significantly by using tissue-specific networks as compared to the global functional network. Enhancement in the prediction of disease genes was also obtained in a study that assessed tissue interactomes weighted by DNA methylation data [84]. Greene et al. created functional networks for human tissues, and showed that they can be used to prioritize GWAS hits for hypertension better than GWAS alone or GWAS combined with a generic network [77].

The filtering implies that if a protein is expressed in a tissue below a certain threshold, its interactions are not considered to occur in that tissue (marked by a dashed edge in liver and hypothalamus). The interaction is only considered to occur in muscle. Penalizing implies that interactions are not Boolean, but are scored according to the expression level of the interacting proteins. The numbers presented are arbitrary and used to demonstrate different interaction scores.

11.3.3 Constructing Other Types of Context-Sensitive Interactomes

The same principle that governs the construction of tissue-sensitive interactomes can be applied to obtain other context-sensitive networks. A particularly useful context for studying genotype–phenotype relationships is a disease-sensitive interactome. Accordingly, interactions between genes or proteins that are not co-expressed in the mapped disease state will be down-weighted or removed from the network (e.g., [85]). One of the prominent disease-based mapping initiatives is led by The Cancer Genome Atlas (TCGA) [86]. This initiative provided genomic, transcriptomic, proteomic, and epigenomic profiles of over 30 types of cancer. An example for the usage of this resource is provided by the study of Drake et al [87]. There, differentially expressed master transcriptional regulators, functionally mutated genes, and differentially activated kinases in lethal metastatic castration-resistant prostate cancer (CRPC) tissues, were integrated to create a robust signaling network consisting of druggable kinase pathways. Using MSigDB hallmark gene sets, six major signaling pathways, with phosphorylation of several key residues, were found to be significantly enriched in CRPC tumors after incorporation of phosphoproteomic data. These were subsequently used to derive patient-specific networks.

Most of this section was devoted to tissue profiles due to their increasing availability. Nevertheless, the heterogeneity of the human body does not end with tissues. Each tissue consists of a variety of cells and cell types that differ from each other physiologically, functionally, and molecularly. Several large-scale efforts were carried to profile cell types [62]. Cell-type information contributes to a higher resolution of analysis, and to deeper understanding of phenotypes. Yet, even cell types profiles are an average of multiple cells, and thus may hide important features that are observable only at the single cell level, such as drug resistance in cancer cells [88, 89]. Important technological breakthroughs allow for RNA-sequencing of single cells. While some technical and computational challenges remain to be solved, single cell analyses will undoubtedly provide new views that will help unravel the complexity of human tissues [90, 91].

11.4 Using Tissue Interactomes to Illuminate Genotype–Phenotype Relationship

While genotypes are shared by cells across the human body, phenotypes and in particularly disease phenotypes, are frequently tissue-sensitive [51]. Some examples include germline mutations in the CFTR gene that are causal for cystic fibrosis, which manifests mainly in the lungs; familial mutations in BRCA1 gene that increase the risk for breast and ovarian cancers; and familial mutations in α-syn that lead to Parkinson's disease, which manifests primarily in the brain and particularly in the substantia nigra par compacta. This tissue-selective manifestation of causal variants is intriguing given the pattern of expression of the causal variants. Like most protein-coding genes, these variants tend to be expressed at multiple tissues across the body. However, most of these tissues remain unaffected [51]. Thus, tissue context appears to determine the phenotypic consequence of these aberrant genotypes.

In Section 11.3.2, we highlighted several studies showing that tissue networks provide more informative views than generic interactomes [76, 77, 92]. The usage of tissue interactomes for elucidating genotype-phenotype relationships is indeed becoming more frequent. Thus, instead of applying the approaches described in Section 11.2 to a generic interactome, these approaches are increasingly applied to the interactome of the relevant tissue or cell type. A study that demonstrates nicely the usage of tissue network was published by Jerby et al. [93]. There, a liver metabolic network was constructed by integrating a variety of liver-specific molecular data sources, including literature-based knowledge, transcriptomic, proteomic, metabolomic, and phenotypic data. Predictions based on the resulting model correlated measurements across hormonal and dietary conditions, and performed better than upon using the original, generic human model. Below we describe two types of approaches that extend the concept of tissue networks by applying comparative analysis to these networks in order to pinpoint genotype–phenotype relationships.

11.4.1 Differential Network Analysis

To get a better understanding of the processes that separate a specific tissue from other tissues, we can consider the quantitative changes in gene expression and how they relate to PPIs. Rather than ask "which interactions are most expressed in a certain tissue?" we can ask "which interactions are most altered between tissues?" [94]. This differential network approach aims to identify PPIs that are differentially relevant in the tissue or condition of interest. The power of this approach was nicely illustrated by the usage of differential epistasis mapping for understanding DNA-damage response in yeast [95]. The genetic interaction network that was measured in yeast cells, that were perturbed with a DNA-damaging agent, did not show a clear DNA-damage signature. Only upon focusing on interactions that were altered significantly between this network and the network measured in unperturbed cells, the DNA-damage signature emerged.

Several tools offer schemes for differential network analysis. For example, the Differential Network Analysis (DNA) package analyzes differential co-expression based on microarray data. It calculates expression correlations between genes, and highlights gene pairs with altered co-regulation between conditions [96]. The DINA framework identifies gene pairs whose co-expression is condition-specific based on microarray data, and predicts transcriptional regulators that may be responsible for their co-regulation [97]. DyNet is a Cytoscape application that computes the variance of nodes and interactions in different networks, and highlights those that are highly variable [98]. DINGO is a framework that estimates topological differences between networks and highlights sub-networks that are altered [99]. This was used to elucidate cancer development pathways. DifferentialNet is a novel web-based database for differential network analysis of human tissue interactomes. DifferentialNet associates each PPI with a tissue-specific score that reflects how its likelihood changed in the selected tissue relative to other tissues, based on the expression levels of the interacting proteins across tissues (available at http://netbio.bgu.ac.il/diffnet/).

11.4.2 Meta-Analysis of Tissue Interactomes for Elucidating Genotype–Phenotype Relationship

The availability of multiple tissue interactomes allows for the development of comparative approaches for elucidating genotype–phenotype relationships. Lage et al. analyzed concurrently over 1,000 heritable diseases [10]. They used text mining and expert curation to associate diseases with their affected tissues. Then, they integrated data of gene expression in human tissues [56] with data of protein complexes. They showed that causal genes and their related complexes tend to be over-expressed in their affected tissues, except for cancer-related genes and complexes. Another example is presented by the study of Kitsak et al., which analyzed disease modules in tissues contexts. They created tissue networks which contained only genes that were up-regulated in that tissue relative to other tissues [100]. Then, they analyzed the connectivity between genes associating with the same complex disease, for complex disease associated with 20 genes or more. They compared between different tissues by measuring the mean shortest distance between disease genes in each subnetwork, and by counting the fraction of disease genes that form a single connected component. Both measures were found to be smaller for disease genes relative to random genes. Their results suggested that disease genes tend to cluster in a well-defined network neighborhood, and that the completeness of disease modules determines disease manifestation in selected tissues. Barshir et al. analyzed the tissue-specificity of hereditary diseases. They created tissue interactomes by integrating data of PPIs with tissue RNA-sequencing and protein expression profiles [51]. They showed that causal genes have a significant tendency for PPIs that occur exclusively in the affected tissue. They demonstrated that these tissue-exclusive PPIs can highlight disease mechanisms, and, owing to the small number of these interactions, constitute an efficient filter for identifying genotype–phenotype relationships (Box 11.6). The uniqueness of their approach is in asking not only why a disease manifests in its affected tissue, but also in asking why the disease is not manifesting in other tissues that express the causal gene. By this, they can filter the putative mechanisms and reduce their number considerably.

11.5 Conclusion

The molecular basis of genotype–phenotype relationships remains elusive for hundreds of diseases. Several studies have demonstrated the value of network-based approaches, where the unbiased, large-scale view into molecular relationships exposed perturbed interactions and processes that contribute to disease manifestation and progression. Since diseases are typically specific to cell types, tissues, and organs, it is important to study them in the correct context. The massive profiling of human tissues, disease patients, cell types, and recently single cells via state-of-the-art techniques enables the construction of refined networks that are not only detailed but also highly context-sensitive. This opened the door both for analysis of genotype-phenotype relationships in the correct physiological context, as well as via comparative analyses of different tissue and cell type contexts. The massive profiling has mainly been carried at the transcript level, yet advances in proteomics, phospho-proteomics, metabolomics, and lipidomics suggest that in the coming years additional types of molecules will be profiled in physiological contexts. Future network-based tools will need to meet the challenges raised by these multi-omics datasets, and to harness them effectively toward the identification of causal variants and the molecular basis of their pathological consequences.

11.6 Exercises

11.1 Metabolic interactions can be represented in two ways, define each way and give an example of these representations.

11.2 Define genotype and phenotype and give an example each (*hint*: search for the disease Osteogenesis Imperfecta in OMIM).

11.3 Define "edgetic perturbations" and give an example of a causal mechanism.

11.4 Mention two types of biological interactions that do not necessarily include a physical interaction between two molecules.

11.5 Genes A through F participate in a common pathway, and have the following physical interactions: A–B   B–C   B–D   C–E   D–E   E–F
Mutations in genes A, B, E, and F are known to cause familial colorectal cancer. In contrast, no variant of either C or D is known to cause the same phenotype. What could be the reason for that?

11.6 Name two genotype examples. Explain how each genotype leads to a specific phenotype.

11.7 Draw a schematic network and mark the nodes and the edges. What do nodes and edges represent in the network? Name three possible options.

11.8 What is the added value of network-based approaches for investigating disease?

11.9 What is the importance of incorporating context into network? Give an example.

11.10 The gene DAG1 (dystroglycan1) is expressed in most, if not all, tissues. however, it is considered to be causal for muscular dystrophy, and does not affect other tissues. According to this chapter, how can this be explained?

11.11 What are the main approaches for construction of tissue specific interactome?

11.12 What is a disease module?

Note: Solutions are available to instructors at www.cambridge.org/bionetworks.

11.7 Acknowledgments

This research was supported by the United States – Israel Binational Science Foundation (BSF) grant #2011296 and the Joint Broad-ISF Research Grant 2435/16.

References

[1] Weatherall DJ. Phenotype-genotype relationships in monogenic disease: Lessons from the thalassaemias. *Nature Reviews Genetics*, 2001;2:245–255.

[2] Sovio U, Bennett AJ, Millwood LY, et al. Genetic determinants of height growth assessed longitudinally from infancy to adulthood in the northern Finland birth cohort 1966. *PLoS Genetics*, 2009;5:e1000409.

[3] Mengel-From J, Wong TH, Morling N, Rees JL, Jackson IJ. Genetic determinants of hair and eye colours in the Scottish and Danish populations. *BMC Genetics*, 10:88.

[4] Piel FB, Steinberg MH, Rees DC. Sickle cell disease. *New England Journal of Medicine*, 2017;376:1561–1573.

[5] Wjst M, Sargurupremraj M, Arnold M. Genome-wide association studies in asthma: what they really told us about pathogenesis. *Current Opinion in Allergy and Clinical Immunolology*, 2013;13:112–118.

[6] Lee JC, Parkes M. Genome-wide association studies and Crohn's disease. *Briefings in Functional Genomics*, 2011;10:71–76.

[7] Steward CA, Parker APJ, Minassian BA, et al. Genome annotation for clinical genomic diagnostics: strengths and weaknesses. *Genome Medicine*, 2017;9:49.

[8] 1000 Genomes Project Consortium, Auton A, Brooks LD, et al. A global reference for human genetic variation. *Nature*, 2015;526:68–74.

[9] Lyon GJ, Wang K. (2012) Identifying disease mutations in genomic medicine settings: current challenges and how to accelerate progress. *Genome Medicine*, 2012;4:58.

[10] Lage K. (2014) Protein-protein interactions and genetic diseases: The interactome. *B. B. A Molecular Basis of Disease*, 2014;1842:1971–1980.

[11] Barabasi AL, Gulbahce N, Loscalzo J. Network medicine: A network-based approach to human disease. *Nature Reviews Genetics*, 2011:12:56–68.

[12] Karnoub AE, Weinberg RA. Ras oncogenes: Split personalities. *Nature Reviews Molecular Cell Biology*, 2008;9:517–531.

[13] Rolland T, Tasan M, Charloteaux B, et al. A proteome-scale map of the human interactome network. *Cell*, 2014;159:1212–1226.

[14] Hein MY, Hubner NC, Poser I, et al. A human interactome in three quantitative dimensions organized by stoichiometries and abundances. *Cell*, 2015;163:712–723.

[15] Huttlin EL, Ting L, Bruckner RJ, et al. The BioPlex Network: A systematic exploration of the human interactome. *Cell*, 2015;162:425–440.

[16] Wan C, Borgeson B, Phanse S, et al. Panorama of ancient metazoan macromolecular complexes. *Nature*, 2015;525:339–44.

[17] Ecker JR, Bickmore WA, Barroso I, et al. Genomics: ENCODE explained. *Nature*, 2012;489;52–55.

[18] Schwikowski B, Uetz P, Fields S. A network of protein-protein interactions in yeast. *Nature Biotechnology*, 2000;18:1257–1261.

[19] Uetz P, Giot L, Cagney G, *et al*. A comprehensive analysis of protein-protein interactions in Saccharomyces cerevisiae. *Nature*, 2000;403:623–7.

[20] Watts DJ, Strogatz SH. Collective dynamics of "small-world" networks. *Nature*, 1998;393:440–442.

[21] Barabási AL, Albert R. Emergence of scaling in random networks. *Science*, 1999;286:509–512.

[22] Vidal M, Cusick ME, Barabasi AL. (2011) Interactome networks and human disease. *Cell*, 2011;144:986–998.

[23] Jia P, Zhao Z. (2014) VarWalker: Personalized mutation network analysis of putative cancer genes from next-generation sequencing data. *PLoS Computational Biology*, 2014;10:er003460.

[24] Bashashati A, Haffari G, Ding J, et al. DriverNet: Uncovering the impact of somatic driver mutations on transcriptional networks in cancer. *Genome Biology*, 2012;13:R124.

[25] Stuart JM. A gene-coexpression network for global discovery of conserved genetic modules. *Science*, 2003;302:249–255.

[26] Basha O, Tirman S, Eluk A, Yeger-Lotem E. ResponseNet2.0: Revealing signaling and regulatory pathways connecting your proteins and genes–now with human data. *Nucleic Acids Research*, 2013:41W198–W203.

[27] Deberardinis RJ, and Thompson CB. Cellular metabolism and disease: What do metabolic outliers teach us? *Cell*, 2012;148:1132–1144.

[28] Houten SM, Wanders RJA. A general introduction to the biochemistry of mitochondrial fatty acid β-oxidation. *Journal of Inherited Metabolic Disease*, 2010;33:469–477.

[29] Zhong Q , Simonis N, Li QR, et al. (2009) Edgetic perturbation models of human inherited disorders. *Molecular Systems Biology*, 2009;5:321.

[30] Sahni N, Yi S, Zhong Q, et al. Edgotype: A fundamental link between genotype and phenotype. *Current Opinion Genetics Development*, 2013;23:649–657.

[31] Wei X, Das J, Fragoza R, et al. A massively parallel pipeline to clone DNA variants and examine molecular phenotypes of human disease mutations. *PLoS Genetics*, 2014;10:e1004819.

[32] Lambert JP, Ivosev G, Couzens AL, et al. Mapping differential interactomes by affinity purification coupled with data-independent mass spectrometry acquisition. *Nature Methods*, 2013;10:1239–1245.

[33] Wang X, Wei X, Thijssen B, et al. Three-dimensional reconstruction of protein networks provides insight into human genetic disease. *Nature Biotechnology*, 2012;30:159–164.

[34] Meyer M, Beltrán JF, Liang S, et al. Interactome INSIDER: A multi-scale structural interactome browser for genomic studies. bioRxiv preprint available at https://www.biorxiv.org/content/early/201704/12/126862. 2017.

[35] Rolland T, Taşan M, Charloteaux B, et al. A proteome-scale map of the human interactome network. *Cell*, 2014;159:1212–1226.

[36] Sahni N, Yi S, Taipale M, et al. Widespread macromolecular interaction perturbations in human genetic disorders. *Cell*, 2015;161:647–660.

[37] Fuxman Bass JI, Sahni N, Shrestha S, et al. Human gene-centered transcription factor networks for enhancers and disease variants. *Cell*, 2015;161:661–673.

[38] GeneCards Database: GeneCards The Human Gene Database available at www.genecards.org/?vm=r.

[39] Ashburner M, Ball CA, Blake JA, et al. Gene ontology: Tool for the unification of biology. The Gene Ontology Consortium. *Nature Genetics*, 2000;25:25–29.

[40] Kaltenbach LS, Romero E, Becklin RR, et al. Huntingtin interacting proteins are genetic modifiers of neurodegeneration. *PLoS Genetics*, 2007;3:689–708.

[41] Goehler H, Lalowski M, Stelzl U, et al. A protein interaction network links GIT1, an enhancer of huntingtin aggregation, to Huntington's disease. *Molecular Cell*, 2004;15:853–865.

[42] Lim J, Hao T, Shaw C, et al. A protein-protein interaction network for human inherited ataxias and disorders of purkinje cell degeneration. *Cell*, 2006;125:801–814.

[43] Pankow S, Bamberger C, Calzolari D, et al. ΔF508 CFTR interactome remodelling promotes rescue of cystic fibrosis. *Nature*, 2015;528:1–18.

[44] Yi S, Lin S, Li Y, et al. Functional variomics and network perturbation: connecting genotype to phenotype in cancer. *Nature Reviews Genetics*, 2017;18:395–410.

[45] Pujana MA, Han JDJ, Starita LM, et al. Network modeling links breast cancer susceptibility and centrosome dysfunction. *Nature Genetics*, 2007;39:1338–1349.

[46] Goh K, Cusick ME, Valle D, Childs B, Vidal M. The human disease network. *Proceedings of the Natlional Academy of Sciences*, 2007;104:8685–8690.

[47] Sharan R, Ideker T. (2006) Modeling cellular machinery through biological network comparison. *Nature Biotechnology*, 2006;24:427–433.

[48] Sharma A, Menche J, Huang CC, et al. A disease module in the interactome explains disease heterogeneity, drug response and captures novel pathways and genes in asthma. *Human Molecular Genetics*, 2015;24:3005–3020.

[49] Khurana V, Peng J, Chung CY, et al. Genome-scale networks link neurodegenerative disease genes to α-synuclein through specific molecular pathways. *Cell Systems*, 2017;4:157–170.e14.

[50] Lage K, Hansen NT, Karlberg EO, et al. A large-scale analysis of tissue-specific pathology and gene expression of human disease genes and complexes. *Proceedings of the National Academy of Sciences USA*, 2008;105:20870–20875.

[51] Barshir R, Shwartz O, Smoly IY, Yeger-Lotem E. Comparative analysis of human tissue interactomes reveals factors leading to tissue-specific manifestation of hereditary diseases. *PLoS Computational Biology*, 2014;10:e1003632.

[52] Basha O, Barshir R, Sharon M, et al. The TissueNet v.2 database: A quantitative view of protein-protein interactions across human tissues. *Nucleic Acids Research*, 2017;45:D427–D431.

[53] Yeger-Lotem E, Sharan R. Human protein interaction networks across tissues and diseases. *Frontiers in Genetics*, 2015;6:257.

[54] Lan A, Ziv-Ukelson M, Yeger-Lotem E. A context-sensitive framework for the analysis of human signalling pathways in molecular interaction networks. *Bioinformatics*, 2013;29:i210-i216.

[55] Lukk M, Kapushesky M, Nikkilä J, et al. A global map of human gene expression. *Nature Biotechnology*, 2010;28:322–324.

[56] Su AI, Wiltshire T, Batalov S, et al. A gene atlas of the mouse and human protein-encoding transcriptomes. *Proceedings of the National Academy of Sciences of the USA*, 2004;101:6062–6067.

[57] Jongeneel CV, Delorenzi M, Iseli C, et al. An atlas of human gene expression from massively parallel signature sequencing (MPSS). *Genome Research*, 2005;15:1007–1014.

[58] Hillier LD, Lennon G, Becker M, et al. Generation and analysis of 280,000 human expressed sequence tags. *Genome Research*, 1996;6:807–828.

[59] Ramskold D, Wang ET, Burge CB, Sandberg R. An abundance of ubiquitously expressed genes revealed by tissue transcriptome sequence data. *PLoS Computational Biology*, 2009;5:e1000598.

[60] Uhlen M, Fagerberg L, Hallstrom BM, et al. Proteomics: Tissue-based map of the human proteome. *Science*, 2015;347:1260419.

[61] Bradley RK, Merkin J, Lambert NJ, Burge CB. (2012) Alternative splicing of RNA triplets is often regulated and accelerates proteome evolution. *PLoS Biology*, 2012;10:e1001229.

[62] Lizio M, Harshbarger J, Shimoji H, et al. Gateways to the FANTOM5 promoter level mammalian expression atlas. *Genome Biology*, 2015;16:22.

[63] Mele M, Ferreira PG, Reverter F, et al. The human transcriptome across tissues and individuals. *Science*, 2015;348:660–665.

[64] Consortium GTEx. The Genotype-Tissue Expression (GTEx) pilot analysis: Multitissue gene regulation in humans. *Science*, 2015;348:648–660.

[65] Kim MS, Pinto SM, Getnet D, et al. A draft map of the human proteome. *Nature*, 2014;509:575–581.

[66] Wilhelm M, Schlegl J, Hahne H, et al. Mass-spectrometry-based draft of the human proteome. *Nature*, 2014;509:582–587.

[67] de Rie D, Abugessaisa I, Alam T, et al. An integrated expression atlas of miRNAs and their promoters in human and mouse. *Nature Biotechnology*, 2017;35(9):872–878. DOI 10.1038/nbt.3947.

[68] Consortium GTEx. The Genotype-Tissue Expression (GTEx) project. *Nature Genetics*, 2013;45:580–585.

[69] Liu X, Yu X, Zack DJ, Zhu H, Qian J. TiGER: A database for tissue-specific gene expression and regulation. *BMC Bioinformatics*, 2008;9:271.

[70] Bossi A, Lehner B. Tissue specificity and the human protein interaction network. *Molecular Systems Biology*, 2009;5:260.

[71] Emig D, Albrecht M. Tissue-specific proteins and functional implications. *J. Proteome Research*, 2011;10:1893–1903.

[72] Barshir R, Basha O, Eluk A, et al. The TissueNet database of human tissue protein-protein interactions. *Nucleic Acids Research*, 2013;41: D841–D844.

[73] Lopes TJS, Schaefer M, Shoemaker J, et al. Tissue-specific subnetworks and characteristics of publicly available human protein interaction databases. *Bioinformatics*, 2011;27:2414–2421.

[74] Song J, Wang Z, and Ewing RM. Integrated analysis of the Wnt responsive proteome in human cells reveals diverse and cell-type specific networks. *Molecular Biosystems*, 2014;10:45–53.

[75] Liu W, Wang J, Wang T, Xie H. Construction and analyses of human large-scale tissue specific networks. *PLoS One*, 2014;9:e115074.

[76] Magger O, Waldman YY, Ruppin E, Sharan R. Enhancing the prioritization of disease-causing genes through tissue specific protein interaction networks. *PLoS Computational Biology*, 2012;8:e1002690.

[77] Greene CS, Krishnan A, Wong AK, et al. Understanding multicellular function and disease with human tissue-specific networks. *Nature Genetics*, 2015;47:569–576.

[78] Basha O, Flom D, Barshir R, et al. MyProteinNet: Build up-to-date protein interaction networks for organisms, tissues and user-defined contexts. *Nucleic Acids Researcg*, 2015;43:W258–263.

[79] Cumbo F, Paci P, Santoni D, Di Paola L Giuliani A. GIANT: A cytoscape plugin for modular networks. *PLoS One*, 2014;9:(10)e105001.

[80] Micale G, Ferro A, Pulvirenti A, Giugno R. SPECTRA: An integrated knowledge base for comparing tissue and tumor-specific PPI networks in human. *Frontiers in Bioengineering Biotechnology*, 2015;3:1–15.

[81] Schaefer MH, Fontaine JF, Vinayagam A, et al. HIPPIE: Integrating protein interaction networks with experiment based quality scores. *PLoS One*, 2012;7:e31826.

[82] Kotlyar M, Pastrello C, Sheahan N, Jurisica I. Integrated interactions database: Tissue-specific view of the human and model organism interactomes. *Nucleic Acids Research*, 2016;44:D536–D541.

[83] Guan Y, Gorenshteyn D, Burmeister M, et al. Tissue-specific functional networks for prioritizing phenotype and disease genes. *PLoS Computational Biology*, 2012;8:e1002694.

[84] Li M, Zhang J, Liu Q, Wang J, Wu FX. Prediction of disease-related genes based on weighted tissue-specific networks by using DNA methylation. *BMC Medical Genomics*, 2014;7:S4.

[85] Ju W, Greene CS, Eichinger F, et al. Defining cell-type specificity at the transcriptional level in human disease. *Genome Research*, 2013;23:1862–1873.

[86] Cancer Genome Atlas Research Network, Weinstein JN, Collisson EA, et al. The Cancer Genome Atlas Pan-Cancer analysis project. *Nature Genetics*, 2013;45:1113–1120.

[87] Drake JM, Paull EO, Graham NA, et al. Phosphoproteome integration reveals patient-specific networks in prostate cancer. *Cell*, 2016;166:1041–1054.

[88] Geva-Zatorsky N, Dekel E, Cohen AA, et al. Protein dynamics in drug combinations: A linear superposition of individual drug responses. *Cell*, 2010;140:643–651.

[89] Cohen AA, Geva-Zatorsky N, Eden E, et al. Dynamic proteomics of individual cancer cells in response to a drug. *Science*, 2008;322:1511–1516.

[90] Klein AM, Mazutis L, Akartuna I, et al. Droplet barcoding for single-cell transcriptomics applied to embryonic stem cells. *Cell*, 2015;161:1187–1201.

[91] Macosko EZ, Basu A, Satija R, et al. Highly parallel genome-wide expression profiling of individual cells using nanoliter droplets. *Cell*, 2015;161:1202–1214.

[92] Guan Y, Ackert-Bicknell CL, Kell B, Troyanskaya OG, Hibbs MA. Functional genomics complements quantitative genetics in identifying disease-gene associations. *PLoS Computational Biology*, 2010;6:e1000991.

[93] Jerby L, Shlomi T, Ruppin E. Computational reconstruction of tissue-specific metabolic models: Application to human liver metabolism. *Molecular Systems Biology*, 2010;6:401.

[94] Ideker T, Krogan NJ. Differential network biology. *Mol. Syst. Biol.*, **8**.

[95] Bandyopadhyay S, Mehta M, Kuo D, et al. Rewiring of genetic networks in response to DNA damage. *Science*, 2010;330:1385–1389.

[96] Gill R, Datta S, Datta S. A statistical framework for differential network analysis from microarray data. *BMC Bioinformatics*, 2010;11:95.

[97] Gambardella G, Moretti MN, De Cegli R, et al. Differential network analysis for the identification of condition-specific pathway activity and regulation. *Bioinformatics*, 2013;29:1776–1785.

[98] Goenawan IH, Bryan K, Lynn DJ. DyNet: Visualization and analysis of dynamic molecular interaction networks. *Bioinformatics*. 2016;32:2713–2715.

[99] Ha MJ, Baladandayuthapani V, Do KA. DINGO: Differential network analysis in genomics. *Bioinformatics*, 2014;31:3413–3420.

[100] Kitsak M, Sharma A, Menche J, Guney E, Ghiassian SD, Loscalzo J, Barabási AL. Tissue Specificity of Human Disease Module. *Scientific Reports*, 2016, 6:35241.

12 Network Neuroscience

Alberto Cacciola, Alessandro Muscoloni, and Carlo Vittorio Cannistraci

12.1 Introduction

The brain is one of the most fascinating complex systems known to humans, and its network representations built considering different types of brain features are named *connectomes*, which represent indispensable tools to investigate such complexity. The connectomes can emerge at different scales, from the microscopic of neurons and synapses to the macroscopic of the whole brain areas. The features at different scales are captured by means of various imaging technologies, which in turn generate diverse types of signals from which to extract the connectomes. However, the first and main peculiarity of a connectome is that it can represent either the structural wiring (structural connectome) or the functional interaction (functional connectome) between different brain units, which can be defined in relation to the considered scale and the technology adopted to investigate that scale. The interplay between scale, technology, and signal is a triad behind each connectome and determines the way in which the brain information and its paradigm of complexity are processed to form a network.

In the last decades, the focus of the debate about classic cortical localizationist theory [1] has shifted to the idea of a *holistic* and *associationist* brain [2]. The *holistic* theory, almost abandoned, suggested that all the brain regions were reciprocally linked by homogeneous associative pathways forming a wide network, in which cognitive functions resulted from the simultaneous firing of all the regions acting as a whole through the association pathways [2, 3]. On the other hand, according to the more recent *associationist* theory, the brain might consist of several, segregated and parallel networks distributed around critical and participating cortical epicenters [2, 4, 5].

In this relatively new scenario, the field of connectomics has significantly contributed not only to neuroscience but also to precision medicine, by mapping the brain from structure to function and by identifying specific quantitative biomarkers for the assessment of brain disease severity. This has been leading to a better comprehension of the structural neural substrates causing dysfunctions.

To mention some historical remarks, the term connectome, originally coined by Olaf Sporns, Giulio Tononi and Rolf Kotter [6], and independently in a PhD dissertation by Patric Hagmann [7], refers to a matrix representing all the possible pairwise connections between neural-based elements of the brain [6, 7]. To be more precise, the human connectome is more than just a large collection of data. Instead, as we clarified above, it is a comprehensive structural description of the network of elements and connections forming the human brain, ranging over multiple levels of organization (from the microscale to the macroscale), thus reflecting the multiscale nature of brain connectivity and complexity [8].

Neuronal connectivity patterns can be exploited at different levels of scale: Microscale, mesoscale, and macroscale. The microscale allows to study single neurones (each of which is represented as a network node) and synaptic connections linking two or more individual neuronal cells, providing a detailed anatomical description of the basic substrates of the cerebral microcircuits. The mesoscale deals with the activity of small cohorts of neurons (which altogether form a network node) forming a unique unit that is source of signal. This brain connectivity scale can handle for instance structural anatomical levels such as columns and minicolumns. In general, the structural mesoscale connectivity can be investigated by microscopy coupled with invasive tract-tracing approaches [9], whereas the functional one by local field potentials (LFPs) [10, 11] or electrocorticography (ECoG) recordings [12]. Finally, macroscale connectivity points out large-scale anatomical connectivity patterns focusing on the inter-regional white matter pathways connecting distinct neuronal populations. Concepts such as synaptic connections [13] and macroscale connectivity are well known in the broader scientific community, however, other terms such as *columns* and *minicolumns* might need further clarification since they are specific to the neocortex (aka. cortex). A column (aka. macrocolumn) is "an elementary unit of organization in the cortex made up of a vertical group of cells extending through all the cellular layers" [14, 15], and a minicolumn is the basic elementary unit of the cortex constituted by radially oriented pool of neurons in which information integration between afferent and efferent impulses is accomplished [15, 16]. From such assumptions, the reader might be confused on which is the truly basic unit of the cortex, thus this issue has been solved by linking together the two structures and defining a column as "formed by many minicolumns bound together by short-range horizontal connections" [17]. For additional details on the topic the reader can refer to [15, 18]. On the other hand, by placing appropriate electrodes in the brain, LFPs record electric activity not from individual cells, but from the extracellular space as result of the electric current generated by multiple nearby neurons [10, 19]. ECoG, instead, directly records LFPs from the neural tissue via the subdural implantation of either a strip or grid of electrodes on the cortical surface [12].

At the macroscale level, network science has offered a powerful and useful approach to investigate the topological architecture of cerebral networks of living humans both in healthy conditions [20], and in several neurological and neuropsychiatric disorders [21]. *Network neuroscience* is primarily concerned with the topological architecture of the brain network – that is a general set of brain elements endowed with some relationship among them, not with its already known anatomy [22] – and it represents a promising tool for clear understanding of brain network organization cutting across spatial and temporal scales. In brief, brain network topologies can emerge from neural-based connectivity (adjacency) matrices, where each row or column represents different brain units/elements (nodes), and each element of the matrix represents the value of the structural, functional, or effective pairwise relation between two nodes (edges).

Complex network measures can be used to study connectivity relationships in single subjects or between subject groups, characterizing one or more aspects of global and local brain connectivity. Measures of single network elements typically quantify connectivity profiles associated with these elements, thus reflecting the way in which these elements are embedded in the network. Measurement values of all individual elements comprise a distribution (usually "banally" approximated considering its mean), which provides a more global description of the network.

One of the basic and characteristic organization rules on which the brain topology relies on is the tendency of brain networks to cluster into modules with high clustering and short path-length, thus reflecting an intrinsic small-world architecture, functionally segregated (local clustering) and integrated (global efficiency) [23, 24, 25]. Such structural/functional modules involve elements sharing common input and output projections with similar physiological response, thus establishing coherent functional systems. This allows the information to travel rapidly and efficiently between distant brain structures (which functionally contribute to a common and integrated set of tasks or responses) while still preventing the uncontrolled spread of information across the whole network [26].

More recently there has been the emerging idea to investigate brain network features not only related to the topology but also to the latent geometry [27], by means of network embedding in a geometrical space, thus considering the geometry associated to the hidden rules of organization and from which the network topology emerges. In this framework, it is a natural consequence to think that the intrinsic geometry of brain networks could be a useful feature to further depict, investigate, and understand brain connectivity in health and disease.

In this chapter, we will provide a brief but comprehensive overview of the current key approaches to generate structural and functional connectomes at macroscopic scale. We will concentrate on macroscopic connectomes because they are at the moment the most relevant, not only for neuroscience but also for precision medicine. We will also focus on and examine in detail the most common tools and topological measures for the study of complex networks, and how they can be useful for the analysis of brain network organization. Finally, we will provide novel and promising insights in the field of network geometry, which has the potential to promote a significant improvement in revealing and understanding the latent geometry from which emerges a complex network structure in general, and the one of the brain complex system in particular.

12.2 Structural Brain Networks

Structural connectivity is typically assessed by using non-invasive neuroimaging techniques such as magnetic resonance imaging (MRI) that allows for a comprehensive in-vivo connectivity mapping across the entire brain. On the one hand, this method can only provide information at the scale of millimeters and centimeters, thus reducing the precision for nodes and edges definition, on the other hand, it has the advantage to be safe, repeatable, and well tolerable also by patients. Therefore, this permits the study of brain connectivity across the life span and in several brain disorders, and it can also be easily employed for diagnostic purposes. As a general principle, mapping a structural brain connectome typically requires the acquisition of both T1-weighted (T1w) MRI (Figure 12.1 (a)), and diffusion-weighted images (DWIs, see Figure 12.1 (b)), which usually have respectively voxel resolutions of $1 \times 1 \times 1$ mm^3 and of $2 \times 2 \times 2$ mm^3.

MRI uses protons' natural magnetic properties. Protons behave like a small magnet spinning on its randomly aligned axis. An MRI machine creates a magnetic field that forces all the protons axes in the body to align with it. When the protons are aligned with the magnetic field, a radio frequency (RF) is impulsed perpendicularly to the magnetic field through the subject, thus forcing the protons to a spin out equilibrium against the pool of the magnetic field. As a consequence the also protons align themselves perpendicularly to the magnetic field. After the RF impulses finish, the protons realign with the magnetic field (proton relaxation). During this realignment, the absorbed energy is emitted in form of radio waves that can be detected by sensors and transformed into a signal. The amount of released energy and the time of relaxation in general depend on the environment (for instance, the type of tissues) and nature of the molecules. A fast proton realignment creates a bright image. Therefore, magnetic properties can be useful to assess differences between different types of tissues. T1-relaxation is the time to restore the proton axis from the alignment perpendicular to the magnetic field (induced by the RF-impulses perturbation) to the same alignment of the magnetic field. Protons in different tissues relax with different grades, thus tissues have their own specific T1-relaxation time. A typical example – which finds relevance in the clinics - is offered by the T1w images, where the cerebral spinal fluid (CSF) is dark, and blood and fat are bright. Hence, an ischemic injury (for instance, the outcome of a stroke) to the tissue located in a certain brain area appears generally as a visible dark spot in the T1w image (see example in Figure 12.1 (c)).

Diffusion MRI quantifies water molecules diffusion within the brain. When considering an unconstrained medium, water molecules show the same probability of diffusing in any direction (isotropic diffusion), while physical and biological barriers hamper water diffusivity parallel to the barriers (anisotropic diffusion). In nervous tissue, axons act as barriers leading molecular water diffusivity to be preferentially higher along the axis parallel to the fiber trajectory (Figure 12.2). An exhaustive description of diffusion MRI is not the aim of this chapter, however, for a better understanding of this topic we refer to the following books and reviews [28, 29, 30, 31, 32, 33]. In this section, we will discuss some of the critical processing issues and the related pipeline for the construction of a structural connectome starting from the MRI raw data.

During an MRI scan, subjects inevitably undergo various degrees of head movements, which may break the spatial correspondence of the brain across volumes. For this reason, both DWI and T1 images need to be realigned and reoriented to the anterior commissure so that each part of the brain in all volumes is in the same position. Subject motion within the scanner during acquisition may also compromise analysis of DWIs, thus the first processing step is to correct DWIs for motion as well as for susceptibility to distortion artefacts.

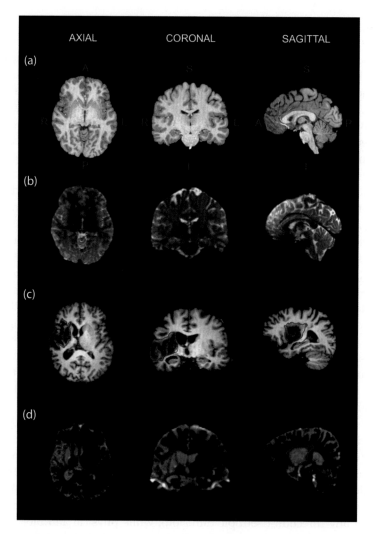

Figure 12.1: Axial, Coronal and Sagittal sections of T1-weighted and diffusion-weighted MRI. In the first row, the red letters A, P, L, R, S and I placed in each section indicate respectively the anterior, posterior, left, right, superior and inferior parts of the brain. T1w images (a) provide a high spatial resolution that allows to easily distinguish the different brain structures and to identify the gray-matter/white-matter borders. Considering the basic physic principles of MRI (i.e., the time of relaxation of the protons) and the peculiar magnetic properties of different tissues, in a T1w image, the CSF appears dark while the white matter, that contains myelinated axons, is bright. Diffusion-MRI (b) instead quantifies the water molecules diffusion within the brain. Higher water diffusion results in greater signal attenuation and therefore appears darker on the image. Indeed, the white matter appears in dark-gray in diffusion-MRI (b) and in bright-grey in T1w (a), while an opposite trend can be appreciated for the CSF. In the clinical scenario, an ischemic event leads to a decreased water diffusion due to reduced blood flow in given brain areas that will appear brighter (hyperintense) on diffusion-MRI images (d) and darker on T1w ones (c). It is worth to note that the sagittal plane in (a) and (b) "cuts" the brain at the level of the midline, whereas in (c) and (d) it is placed at the level of the ischemic injury in order to make it visible. The area of the lesion is pointed out by the red contour (C and D).

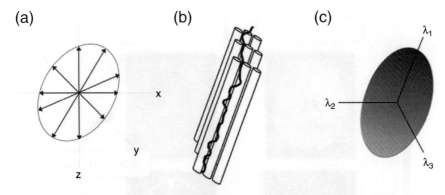

Figure 12.2: Isotropic (a), anisotropic diffusion (b) and tensor representation (c). In an unconstrained medium (a), water molecules show the same probability of diffusing in any direction (isotropic diffusion), therefore a principal diffusion direction cannot be detected. Physical and biological barriers hamper water diffusivity parallel to the barriers, thus leading to an anisotropic diffusion. In the nervous tissue, axons act as barriers forcing molecular water diffusivity to be preferentially higher along the axis parallel to the fiber trajectory (b). Diffusion tensor imaging fits the water diffusion to an ellipsoid that is represented as a tensor. Therefore, the three eigenvectors of the tensor represent the orthogonal basis for the three directions of water molecule diffusivity. The three eigenvalues $(\lambda_1, \lambda_2, \lambda_3)$ instead represent the diffusion water amount in the three orthogonal directions, with the preferential diffusion direction associated with λ_1 that is the largest eigenvalue.

A second processing step in brain connectomics is to reach a reliable anatomical correspondence between T1w images and DWIs, otherwise the nodes may result in being slightly displaced and affected by the CSF signal. Briefly, in order to address this issue, typically the CSF is segmented out from the b0 diffusion image (a type of image without diffusion weighting, whose intensity corresponds to the signal intensity without gradients) and T1w scans. The CSF coming from the T1w image is then linearly and non-linearly warped to match the CSF coming from the b0-image [34]. Such warp generates a deformation field which measures the spatial transformation to match two different brain data and it consists of a 3D displacement vector in each voxel. Finally, the deformation field is applied to the T1w volume, thus allowing for a final anatomical correspondence between the structural (brain areas) information of T1w images and the diffusion (brain connectivity) information of DWIs (see Figure 12.3 for a schematic representation of the co-registration processing).

The third processing step is the diffusion signal modeling. Here, the diffusion signal is modeled in order to fit the probability distribution of the preferential direction of water diffusivity at each voxel. Indeed, in white matter voxels (which contain a wide range of biological structures such as axons, glia, and astrocytes), the water molecule diffusion is hindered in the extra-axonal space and restricted in the intra-axonal space. The main water diffusion directionality in each voxel is inferred by using a diffusion signal modeling, in the simplest case, the diffusion tensor imaging (DTI), which fits the measurements along the different directions to a 3D ellipsoid called tensor. It is worthy to note that DTI suffers from several limitations such as large reconstruction biases and poor reliability for fibers with complex configuration. This is

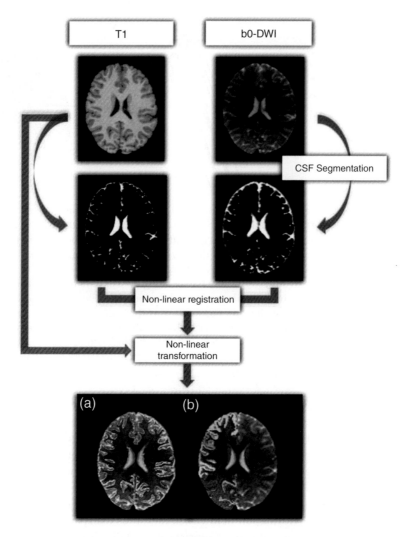

Figure 12.3: Co-registration between T1-weighted and diffusion-weighted images. A fundamental processing step in brain connectomics is to achieve a reliable anatomical correspondence between T1w images and diffusion-weighted images. If this goal is not achieved, the nodes of the connectome may result slightly displaced and affected by the cerebral spinal fluid (CSF) signal. In the proposed figure, we display the axial view of a brain according to T1 and b0-DWI. The pipeline shows the critical steps to reach a good co-registration between T1w and diffusion-weighted images. The CSF is segmented out from both the T1w scans and the b0 unweighted diffusion image and then the CSF coming from the T1w image is non-linearly warped to overlap the CSF coming from the b0-image. In the last step, the non-linear warping field previously obtained was applied to co-register the T1w and diffusion images. The good cortical correspondence between the inner boundaries of the gray-matter obtained by the T1w image (yellow surface cut) and the diffusion image on the background is shown (a). The brain parcellation and segmentation derived from the co-registered T1 is superimposed on the diffusion image (b); for visualization purpose only the two caudate nuclei (the areas in violet at the center adjacent to the later ventricles, which are the hyperintense regions at the center of the brain) and half-brain parcellation are shown.

of major importance considering that more than 90% of white matter voxels contains either crossing, fanning, or kinking fibers [35]. Therefore, several novel sequences and related signal modeling called high angular resolution diffusion imaging (HARDI) have been recently developed and employed for exploring brain connectivity, such as: Constrained spherical deconvolution (CSD; [36]), Q-ball imaging (QBI; [37]), and diffusion spectrum imaging (DSI; [38]). In particular, CSD permits accurate reconstructions with datasets acquired in standard clinical settings if compared with QBI and DSI, which instead require longer acquisition time [39]. Such approaches estimate the fiber orientation distribution function (fODF), which describes the directionality of diffusion in voxels with complex fiber configuration [40].

The fourth step is the fiber tractography. In the latest decades, the development of algorithms together with proper data acquisition, has allowed the development of fiber tractography, which is able to provide a fascinating reconstruction of fiber orientation and trajectories (streamlines) (see Figure 12.4). In the simplest case, called deterministic tractography, only one streamlining process is initiated from a given seed voxel and ended when some termination criteria are reached [41]. Probabilistic tractography instead estimates at each voxel the probability distribution of the preferred water diffusion direction, following different reconstruction routes on the basis of the uncertainty modelled by the diffusion data [41].

In the fifth step, one of the diffusion signal modeling approach is finally combined with the fiber tractography in order to reconstruct white matter pathways and connectivity maps throughout the whole brain. Mapping the connectivity pathway between different brain regions can be easily performed by seeding these areas either directly from the gray matter (although water diffusion is poorly constrained) or preferably from the gray-matter/white-matter interface, thus allowing a more accurate streamlines estimation [42, 43, 44, 45, 46, 47, 48, 49]. Another mechanism is to create a whole brain tractogram by generating at least one million streamlines throughout the entire brain and then to extract the connectivity patterns between the different nodes provided by a parcellation image [50, 51]. Tractography reconstruction using gray-matter seed may shift the overall weight of the resulting connectivity matrix towards short-range connections, whereas tracking from the deep white matter may lead to more prominent connections between far brain areas linked by long-range pathways [52]. For a comprehensive discussion on the pros and cons of the different seeding approaches the reader can refer to [52, 53].

Although the association between each pair of nodes is usually measured by the number of streamlines via which they are interconnected [54, 55], in some cases, some measures of water diffusion (not necessarily and strictly related to the density of axonal connectivity) can be used [56, 57]. It is worthy to note that the number of streamlines does not reflect the number of axons – although it is proportional to that – and it indicates the connection probability between two points in the brain [58]. Regarding water diffusivity measures, the most common are the fractional anisotropy (FA), which quantifies the degree of anisotropy within a voxel, and the mean diffusivity (MD), which instead provides information on the average local diffusivity and water magnitude. Both FA and MD values can be averaged across all the voxels of a given pathway thus providing an indirect index of axonal integrity of that pathway [59].

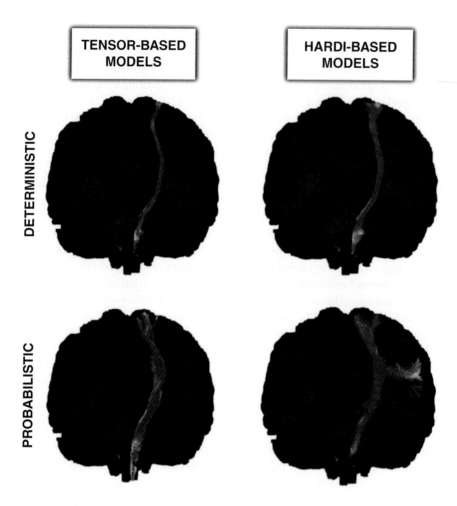

Figure 12.4: Corticospinal tract reconstruction as provided by tensor-based and HARDI-based (high angular resolution diffusion imaging) diffusion signal models combined with deterministic and probabilistic tractography. The tensor model assumes that the water diffusion within a voxel can be approximated by a three-dimensional Gaussian process, which geometrically corresponds to an ellipsoid. Such a single tensor model cannot, however, disentangle the complex white matter architecture consisting of twisting, bending, crossing, and kissing fibers. Even if it has been proposed to fit multiple tensors on the same voxel, correctly estimating the number of tensors to be included is trivial and error prone. Therefore, alternative HARDI "model-free" approaches have been developed in the last decades, such as diffusion spectrum imaging, Q-ball imaging and constrained spherical deconvolution. The figure shows a coronal brain section in 3D view. Diffusion tensor imaging (DTI)-based deterministic tractography, DTI-based probabilistic tractography and HARDI-based deterministic tractography produce only a narrow subset of fibers to the medial part of the sensorimotor. On the other hand, the HARDI-based (in this case constrained spherical deconvolution) diffusion signal modeling, combined with a probabilistic tractography algorithm, is able to reconstruct the fan-shaped configuration of fibers expected from the known anatomy, with tracks reaching also the lateral aspects of the sensorimotor cortex.

Therefore, considering each brain region as a node, the structural connectome could be generated considering as edges one of these possible measures (but not limited to): The inter-nodal number of streamlines, the streamlines length, the tract volume density, tensor-based parameters (i.e., the FA values, the MD values), and the fODF peak. The result is a weighted network – also represented as a $n \times n$ adjacency matrix $C = [c_{ij}]$ – that is not fully connected because not all the inter-nodal connections are anatomically present. The connectome based on the number of streamlines is the most employed and useful for exploring the topological architecture of the human brain inter-region wiring. Figure 12.5 shows the flowchart from data acquisition up to connectivity matrices construction using diffusion MRI and tractography.

Finally, an alternative but less adopted method for exploring the structural brain connectivity is to evaluate the interregional correlation of specific morphological features, such as cortical thickness, surface area, and gray-matter volumes [60]. Briefly, these measures are computed for each node considering the T1w images of each subject, and correlated across individuals in order to create a correlation matrix, which can be considered an indirect estimation of structural brain connectivity, unlike the previous ones (i.e., number of streamlines, FA or MD) that were directly evaluated considering the tractography. It is worth to note that it is common to correct the above mentioned morphometric measures (i.e., gray-matter volumes and thickness) taking into account the whole brain volume of each individual, thus reducing differences related to inter-individual head size.

12.3 Functional Brain Networks

Macroscale functional connectivity networks can be evaluated using functional MRI (fMRI), electroencephalography (EEG), or magnetoencephalography (MEG). The former typically offers higher spatial resolution compared with EEG and MEG, at the cost of a lower temporal resolution ranging generally between 0.5 and 1.5 Hz. fMRI evaluates the fluctuation of the blood oxygenation level dependent (BOLD) signal thus estimating the cerebral blood oxygenation changes in different brain areas. The basic principle on which fMRI relies is that neuronal activity demands high level of oxygen that causes a decrease of deoxygenated blood (paramagnetic), therefore inducing magnetic properties changes in that same brain region. Such fluctuations of oxygenated (diamagnetic) and deoxygenated (paramagnetic) blood are therefore quantified by the MRI scanner – both in resting state (rs-fMRI) and during different tasks – and are considered as a hemodynamic index of local brain activity. Resting state networks (RSNs) activity can be outlined as an automatic condition in which the subject is in a state of absolute rest and does not carry out any physical or mental activities, either with open or closed eyes; definitely, it is not occupied in a particular task. The idea of RSNs was developed after analyzing the functional implication among brain regions demonstrating spontaneous dynamic neural activity recorded in task-free states. The strongest power of such hemodynamic signal in resting conditions has been found at low frequency (<0.1 Hz). However, understanding the functional role and the neurophysiological basis of such spontaneous activity in the resting state has been a matter of debate. Although simultaneous fMRI–EEG recordings

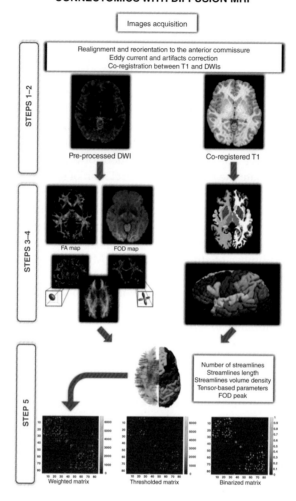

Figure 12.5: Diffusion-MRI structural connectomic workflow. The figure summarizes the fundamental steps to construct a structural connectivity matrix from diffusion-weighted images based on T1-weighted brain parcellation. An axial brain section is displayed both for the diffusion and T1 weighted images. As pre-processing steps both T1 and diffusion images have to be realigned and reoriented to the anterior commissure. Eddy current, motion, and distortion artefacts should be corrected before to accurately co-register the T1 volumes with the diffusion images (STEPS 1–2). Subsequently, the co-registered T1 can be used to parcellate the brain in different brain regions (that will be the nodes of the network), while the pre-processed diffusion-weighted images will be processed to model the diffusion signal (either with tensor-based or HARDI-based models) and to perform whole-brain or seed-based tractography reconstruction. The FA map is a tensor-derived map that displays the amount of water diffusion asymmetry within a voxel and the brighter areas are more anisotropic (i.e., white matter); the FOD map instead is a HARDI-derived map displaying fibre orientation distributions estimated from diffusion data using spherical deconvolution (STEPS 3–4). Finally, the pairwise structural connectivity patterns between the brain regions are estimated and fitted in to a weighted connectivity matrix, that can be thresholded and binarized (STEP 5).

have demonstrated a significant correlation between BOLD signal fluctuations within specific brain networks and the alpha [61] and beta power in the EEG [62], from a theoretical point of view, a single neural rhythm is not likely to be associated with one specific brain functional network. Indeed, combined neurophysiological and MRI studies showed that the spontaneous fluctuations of the BOLD signal are spread across multiple time scales and are topographically and temporally organized in robust distributed networks within the entire brain. In particular, fMRI showed the existence of six main functional brain networks fluctuating simultaneously at low BOLD frequencies (<0.1 Hz), which could be associated with a coalescence of rhythms rather than with one kind of electric oscillation [63]. For the reader interested in the key principles, analyzis and applications of fMRI, see [64, 65].

As for diffusion MRI, fMRI requires some critical pre-processing operations in order to remove physiological noise (due to hearth rate and respiration) as well as to reduce artefacts due to excessive head motion during data acquisition. As first steps, motion and slice-timing correction should be performed. The former is essential considering that, during images acquisition, the subject's movement can cause voxel distortion and can induce a strong variation of the signal. In addition, MRI scanner may acquire the BOLD signal of different slices of the brain at different time points and not in one single moment. However, regarding slice-timing correction there is no consensus on how and if performing it, since it requires interpolation, interacts in unpredictable ways with motion correction and it may not help that much. Instead, it is preferable to introduce a temporal derivative as a regressor when estimating the BOLD response, in order to improve the fitting of responses that are shifted in time [66]. Spatial and temporal filtering are the second step to take into account. The spatial filtering (smoothing) is a fundamental denoising step [67, 68] and consists in averaging the values of one voxel with its neighbors and it can increase the signal to noise ratio by decreasing the variance. The temporal filtering, instead, reduces the temporal noise–removing frequencies within the raw signal that are not of interest or are due to the noise of the scanner drifts as well as to that of the cardiac and respiratory cycles. In addition, resting state images need to be co-registered to the anatomical T1 images and, in turn, both need to be normalized in a template space such as the Montreal Neurological Institute (MNI) template space.

Regardless of the approach used for nodes definition (which will be treated in the next section), the edges of a fMRI network are usually defined by extracting the mean time series BOLD signals for each brain area and computing their pairwise correlations (Figure 12.6).

Functional brain networks can be reconstructed also starting by EEG and MEG signals, which are characterized by high temporal resolution. EEG measures the electric brain activity via electrodes placed on the scalp surface, whereas MEG analyses the weak magnetic fields created by neuronal current by using an array of superconducting quantum interference devices placed around the head.

Herein, we will briefly describe how to use EEG recordings to construct a connectivity matrix. As first pre-processing steps, the raw data need to be band-pass filtered (in order to remove noise and not brain activity-related signals), cleaned, removing artefacts, and re-referenced (although there is no consensus for the "best" common reference site). A further step to improve artefactual EEG portions detection is to

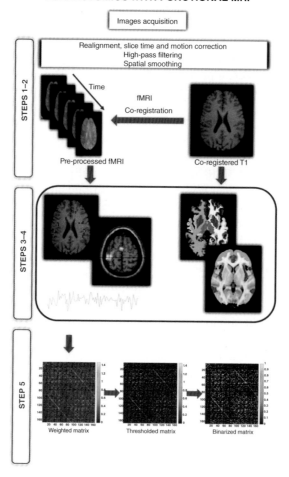

Figure 12.6: Functional MRI connectomic workflow. The figure summarizes the fundamental steps to construct a functional connectivity matrix from functional MRI time series. Axial brain sections are displayed. Fundamental pre-processing steps are represented by: realignment of the volumes (motion correction); slice timing correction (since the slices are scanned at a slightly different time); spatial and temporal filtering. Spatial filtering (smoothing) averages the data voxels with their neighbors, whereas temporal filtering basically removes high frequencies, low frequencies or both, but preserves the signals of interest. This is performed because time series contains scanner-related and physiological signals (that can have both low and high frequency components) as well as high frequency noise (STEPS 1–2). The next steps involve the regional parcellation of the brain from the T1 images and the functional connectivity estimation, which is commonly measured by two families of analytical methods: "seed-based" and "model-free." The seed-based approach estimates the relationship between a "seed" region and the other voxels in the brain, while an alternative model-free approach is the independent component analysis that decomposes the whole-brain BOLD signal into volumetric spatial maps and their related time-courses (STEPS 3–4). Finally, the pairwise functional connectivity patterns (usually measured by the Pearson correlation coefficient) between the brain regions are estimated and fitted into a weighted connectivity matrix, that can be thresholded and binarized (STEP 5).

employ the independent component analysis (ICA) approach, which decomposes the channel data in maximally temporally independent signals. Independent components (ICs) that are not directly related with brain activity are therefore discarded in order to further improve the signal cleaning, whereas good ICs can be retained and can eventually be used to create a ICs-based connectivity matrix.

Once the pre-processing of the raw data is performed, the processed data can be further analyzed to compute functional connectivity measures that can be used as edges of the networks. One of the most common indices used to quantify functional connectivity is the phase lag index (PLI), which evaluates the asymmetry of the distribution of instantaneous phase differences between pairs of electrodes [69]. Spectral coherence instead measures the synchrony of cortical neuronal assemblies, representing the level of synchronization between pairs of EEG electrodes in a given frequency band [70]. In addition, the wavelet correlation has been employed to estimate frequency-dependent time-varying functional connectivity between couples of electrodes [71]. An alternative approach is to compute cortical current density from the channel recordings in order to estimate the location of electric neural generators with optimal localization properties as well as to compute the linear or non-linear brain functional connectivity based on the estimated cortical current density signals [72]. A schematic representation of a simple EEG connectomic workflow is depicted in Figure 12.7.

Regardless of the methods used to construct the functional connectivity matrix (i.e., fMRI, EEG or MEG), the functional brain networks show a complex topology across different frequency bands.

12.4 Node Definition in Structural and Functional Networks

Both structural and functional networks require accurate nodes definition, which can be achieved using a wide range of approaches. One possibility is to define nodes by co-registering the anatomical image to a validated parcellation image. Several parcellations and atlases exist, based on different criteria, such as cyto- and myelo-architetonic features as well as sulcal and gyral boundaries. The former suffer from the limitation that they do not take into account the inter-subjects variability of the cytoarchitetonic boundaries and they do not cover the entire brain. By contrast, looking at the delineation of sulci and gyri led to the construction of probabilistic atlases in which each voxel of the brain has a given probability of belonging to a specific anatomical area.

These templates such as the automated anatomical labeling (AAL) atlas [73], the Harvard–Oxford cortical and subcortical structural atlases [74, 75, 76], the Desikan–Killiany atlas [77] or the Destrieux atlas (provided by the Freesurfer software package) [78, 79] are generated by segmenting individual brain regions either in a single individual (i.e., the AAL) or in a sample of subjects. Considering that such atlases exist in a template space (i.e., the MNI space), they should be warped into the single subject brain. Nodes' definition can be also achieved by clustering together voxels showing similar connectivity profile to other areas during a particular task or in resting state condition [80].

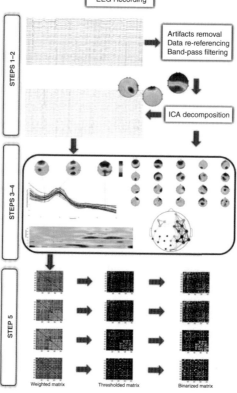

Figure 12.7: EEG connectomic workflow. The figure summarizes the fundamental steps to construct a functional connectivity matrix from EEG recordings. As fundamental pre-processing steps, the raw data needs to be cleaned, removing artefacts, re-referenced and band-pass filtered (in order to remove noise and not brain activity-related signals). Further artefactual EEG portions can be removed using independent component analysis (ICA) algorithm and rejecting bad ICA components that are related to non-brain activity (STEPS 1–2). Functional connectivity can be estimated with different connectivity measures (i.e., spectral power) and using ICA components. On the upper-left corner (STEPS 3–4), the plot shows the channel spectra across different frequencies, with the representation of three scalp maps displaying the distribution of power at three different frequencies. On the upper-right corner (STEPS 3–4), Independent Components containing brain activities that can be used to create the connectivity matrix. On the bottom-left corner (STEP 3–4), the plot shows a time (x axis) – frequency (y axis) decomposition, characterizing the changes of the spectral power considered as a sinusoidal wavelet; the wavelet correlation can be employed to estimate frequency-dependent time-varying functional connectivity between couples of electrodes. Finally, on the bottom-right corner (STEPS 3-4), a representative example of some connectivity patterns between sensors are displayed over a scalp. The nodes of the network are usually represented by the EEG electrodes. However, an alternative approach to estimate EEG functional connectivity is to compute the 3D distribution of the current density on the cerebral cortex with exact localization. Finally (STEP 5), the functional connectivity (in the different frequency bands) between the network nodes is fitted into a weighted connectivity matrix, that can be thresholded and binarized.

When considering EEG and MEG recordings, the nodes of the network can be represented by the single sensors or by pooling together a group of sensors as individual nodes. An alternative approach to define the connectome nodes is the source localization which computes the 3D distribution of the current density on the cerebral cortex with exact localization [81].

Finally, multivariate decomposition of fMRI, EEG and MEG signals using independent component analysis (ICA) can be used to generate nodes including functionally specialized areas [82].

12.5 Tools for Brain Networks Analysis

In the previous sections, we provided a simple but comprehensive overview on the main types of signals from which a connectome can be derived. Once the connectivity matrices are constructed, network science can represent a powerful tool to analyze the topological organization of the brain networks. This has led to an increasing interest for the study of human brain networks during the last decade and consequently to development and implementation of a wide range of freely available toolboxes for network-based analysis, such as: CONN [83], graph-analysis toolbox (GAT; [84]), PANDA [85], BrainNet Viewer [86], GraphVar [87] and the Brain Connectivity Toolbox (BCT; [88]).

These toolboxes present different individual characteristics that include MRI preprocessing, network construction, network analysis/statistics, and network visualization. However, they typically cover only a few of these functions, thus a complete toolbox for brain networks analysis is still lacking. It should be noted that a network analysis toolbox named GRaph thEoreTical Network Analysis (GRETNA; [89]) for imaging connectomics has been recently developed and it can perform fMRI data preprocessing, brain functional networks construction, and computation of the most commonly used global and nodal topological measures with parallel computing ability, thus representing a promising tool in network neuroscience [89].

However, although network analysis is a valuable method to exploit brain topological architecture, it has also some limitations. To date, one of the major issues in connectome analysis is the choice of the right threshold to generate the network topology. Indeed, the threshold chosen is often arbitrary and although it can suppress the false positives adequately, on the other hand, it can increase the chance of false negatives in the network and at the same time may reduce the network to its core components, thus removing group-distinctive network measures. Hence, reliable network analysis requires either a careful choice of the thresholding approach, since it affects network topology in a way that may yield erroneous conclusions [90], or, in alternative, the evaluation of network measures across different threshold levels, as we will discuss later.

In the last years, several approaches have been developed to reduce the randomness of threshold selection. Some authors choose simply a threshold that is considered statistically significant [91]; others have used as criterion for thresholding: (1) the minimal level of network degree, which guarantees the presence of a single connected component [92]; from a theoretical standpoint a random network starts to fragment when its connections density (computed as the ratio between the number of edges and the maximum possible number of edges) is higher than $1/N$, with $N =$ number of the network nodes; (2) the average node degree, which should be maintained in order to ensure that at least the 99% of the nodes are connected in order to investigate the global brain dynamics with biological plausible connections [71]; (3) a minimum mean node degree preserved across all groups [93, 94, 95]; (iv) the false discovery rate (FDR) method to maintain the false positive rate below 5% [71]. For details on each of these specific thresholding strategies please examine the cited references.

Based on this assumption, it appears clear that a bias of node degree properties may affect the network structure and consequently the network measures' comparisons in different groups of connectomes. In particular, the choice of different criteria for different groups should be avoided, because it may also lead to biases, increasing the risk of erroneous inference of significant group differences.

To overcome some of these issues, several authors compute the network measures across a range of thresholds [84, 92], providing a wider framework of network differences, but sometimes leading to more ambiguous results too. Moreover, it has been proposed to summarize the network measures across a range of thresholds, computing the area under the curve for that considered measure across the thresholds [84, 96]. However, the inclusion of low and high range thresholds is likely to lead respectively to false positives and highly disconnected networks, in addition to be unable to detect group differences which can be manifest in a limited range of cut-offs. To overcome these issues, the use of permutation tests has been recently introduced on network metrics [84, 97, 98] and across thresholds [97]. Drakesmith et al. [99], proposed the multi-threshold permutation correction (MTPC) procedure which computes a network measure for all networks thresholded on a very large range of thresholds, and then it tests for group differences using permutation testing, in which only the group assignments are permuted [99]. Finally, network-based statistics (NBS) is a powerful and complementary tool which allows multiple hypothesis tests at the level of interconnected subnetworks, controling the family-wise error when performing analyses associated with a particular effect or contrast of interest [100]. NBS, in fact, overcomes some of the limitations of the generic procedure (such as the FDR) which computes statistical tests and corresponding p-value independently for each link and considering exclusively the strength of that link. On the other hand, a lack is that the subnetworks, nodes, or edges revealed by NBS and FDR approaches, do represent significant between-groups differences, but they are not specific to any network measure, hence they cannot offer information related with a particular property of the topology that differs between the groups.

Although the reader has been already introduced to the concept of topological measures in Chapter 4, in the next section, for completeness, we will outline and discuss the major network measures associated to topological properties that can differ between the connectomes of groups of individuals.

12.6 Topological Measures in Complex Networks

The brain is anatomically and functionally segregated at multiple levels of organization and it consists of local collectives of strongly interconnected cells sharing inputs, outputs and response properties [101]. Specific anatomical areas for dedicated stimuli and responses are also further segregated for more specific stimuli, e.g., visual cortex responds to the visual stimulus in a functionally segregated manner for different aspects of vision such as color, motion, form etc. From the perspective of complex networks analysis, functional brain segregation can relate to the local-derived concept of average clustering coefficient of the functional connectome.

Despite of such spatial segregation, the brain demonstrates global functional integration in various aspects, for instance combining specialized information from different regions to provide a unitary behavioral output, which reflects a coherent response to the integration and combination of multiple local processes. Functional integration can relate to the global-derived concept of average path length in the functional connectome. Specifically, in the structural connectomes, path length means combination of nodes and links resulting in physical information flow; whereas in the functional connectomes, path length means a sequentially coherent statistical relation between subsequent regions, and might not be always supported by physical information flow through anatomical connections.

In complex network analysis, a topological network measure quantifies by means of a unique numerical value the extent to which a certain mechanism of organization (or topological feature) influences the network connectivity. For instance, many network measures take into account mechanisms of segregation or integration of the different parts of a complex network. Network measures can be *stochastic* or *deterministic*. Stochastic measures involve random procedures during their computation, for example the generation of randomized networks (a null model of a given network) based on some topological characteristics preserved from the original network. These randomized networks are used to evaluate the prevalence that a certain topological mechanism of organization shows in the original network in respect to the randomized model of the same network. The stochastic process according to which these null models are created induces the stochasticity in the output of the measure, a good practice is therefore to perform the computation multiple times and to analyze the average behavior and standard error of the measure over the different repetitions. The stochastic measures that will be presented are six: small-worldness (two different types), power-lawness, modularity, structural consistency, and rich-clubness. Deterministic measures, instead, are based on the direct quantification of a considered network topological feature (or rule of organization), e.g., node degree. Some of these measures are even calculated on the basis of other deterministic measures. The randomized networks, which represent the null models, are not required to evaluate deterministic measures: hence, the numerical value associated with the measure computed for a given network is always the same. This often implies that the *computational time* of deterministic measures is in general shorter than the one required by stochastic measures. The deterministic measures that will be presented are nine: average node degree, characteristic path length, average clustering coefficient, efficiency, closeness centrality, node betweenness centrality, edge betweenness centrality, radiality, and local-community-paradigm correlation. A detailed description of both stochastic and determinist measures is provided below.

In addition, a topological measure can be either *local* or *global*. It is local if it makes a statistic evaluation of local topological information in the neighborhood of a node or a link. It is global if it makes a statistical evaluation of global topological information that emerges from nodes or links that are not in a neighborhood. Note that for neighborhood we intend the ensemble of nodes that are first-neighbors of a given node or edge. We will specify whether a measure is local or global in detailed description below.

The Matlab code to compute the measures proposed in the next sections is available at the online repository: https://github.com/biomedical-cybernetics/topological_measures_wide_analysis.

12.6.1 Stochastic Measures

The *small-worldness* [102, 103] was proposed for the characterization of a given network as small-world, meaning that it exhibits a high average clustering coefficient and a low characteristic path length [104]. It relies on comparing a given network with an equivalent random network and lattice network on the basis of the average clustering coefficient, a local measure, and the characteristic path length, a global measure. In 2006, a coefficient called "σ" for characterizing small-world networks was introduced by Humphries and Gurney [102]. To calculate this measure, the average clustering coefficient C and characteristic path length L of the network are compared to C_{rand} and L_{rand} of an equivalent random network (with similar node degree distribution), obtaining the small-world coefficient:

$$\sigma = \frac{C/C_{rand}}{L/L_{rand}}. \tag{12.1}$$

A condition for a network to exhibit small-worldness is that the characteristic path length should be close to that of an equivalent random network, $L \approx L_{rand}$. At the same time, the average clustering coefficient should be close to that of an equivalent lattice network, which also implies that C should be much higher than that of equivalent random network, $C \gg C_{rand}$. These boundary conditions, if met, restrict the value of $\sigma > 1$ for small-world networks. The problem with this coefficient is that even small variations in the already low value of the average clustering coefficient for random networks, C_{rand}, significantly influence the value of the ratio C/C_{rand}. To overcome this problem, a new robust measure was introduced by Telesford et al. [103] which is called "ω." The characteristic path length L is compared to L_{rand} of an equivalent random network and the average clustering coefficient C is compared to C_{latt} of an equivalent lattice network, obtaining the small-world coefficient:

$$\omega = \frac{L_{rand}}{L} - \frac{C}{C_{latt}}. \tag{12.2}$$

Note that C_{rand} is not considered, therefore this measure neglects its fluctuations. Since the boundary conditions for small-worldness are $L \approx L_{rand}$ and $C \approx C_{latt}$, the values of ω are expected to be close to 0 in small-world networks. The equation suggests that the typical range for the coefficient is $\omega \in [-1, 1]$, with positive values representing a network closer to a random one ($L \approx L_{rand}$ and $C \ll C_{latt}$), and negative values representing a network closer to a lattice ($L \gg L_{rand}$ and $C \approx C_{latt}$). Brain connectomes are generally small-world [25, 105].

The *power-lawness* [106] is a local measure that indicates whether the node degree distribution of a particular network follows a power-law behavior or not. If a network presents this property, it can be considered a scale-free network [107]. In order to have a quantitative evaluation, a procedure to compute a *p*-value for considering the power-lawness as a plausible hypothesis has been introduced by Clauset et al. [106]. If the resulting *p*-value is greater than 0.1, the power-lawness hypothesis is accepted, otherwise it is rejected. The statistical test requires the generation of a null distribution, which leads to the stochasticity in the measure. In normal conditions, brain connectomes do not show scale-free distribution [108].

The *modularity* [109, 110] is a global measure that indicates the possible presence of segregated modules or communities in a network. In networks with high modularity, the modules tend to interact densely within themselves but sparsely or not at all between each other. The modularity index Q measures the quality of the best possible partition of nodes, which maximizes the intra-module connectivity and minimizes the inter-modules connectivity. It has been introduced according to the following formula [109]:

$$Q = \sum_{u \in M} \left[e_{uu} - \left(\sum_{v \in M} e_{uv} \right)^2 \right]. \tag{12.3}$$

Where M is a partition of the nodes (whose elements u are called modules) and e_{uv} is the proportion of links in module u connecting to module v. If the number of intra-module edges is no better than random, Q is close to 0, whereas values approaching to the maximum $Q = 1$ indicates strong community structure, although values for such networks typically fall in the range [0.3, 0.7] [109]. Several methods for finding the optimal modularity of a network have been proposed [109, 110, 111] and they rely on different heuristics for sampling the partition space [111]. The randomness in the sampling procedure leads to the stochasticity in the measure.

The *structural consistency* [112], is a global measure that quantifies the link predictability of a complex network. The link predictability characterizes the inherent difficulty to predict the missing or non-observed links of a network regardless of the specific algorithm used for the prediction. Its evaluation relies on a random perturbation (which is origin of stochasticity) and first-order approximation of the adjacency matrix. It is based on the hypothesis that a group of links is highly predictable if their addition does not cause huge structural changes, therefore a network is highly predictable if the removal or addition of a set of randomly selected links does not significantly change the network's structural features [112]. The measure assumes values in the interval [0,1], where 0 indicates absence of link predictability and 1 indicates full link predictability.

The *rich-clubness* [113] defines the extent to which a network is characterized by the presence of a cohort of nodes with a large number of links (rich nodes) that tend to be well connected between each other, creating a tight group (club). The first method introduced for studying the rich-club property is the rich-club coefficient [113], defined for each degree k as the density of the subnetwork composed of the nodes whose degree is larger than k:

$$\varphi(k) = \frac{2E_{>k}}{N_{>k}(N_{>k} - 1)}, \tag{12.4}$$

where $E_{>k}$ is the number of edges between the $N_{>k}$ nodes whose degree is larger than k.

Being the rich-club coefficient a monotonically increasing function even for uncorrelated networks, Colizza et al. [114] suggested a normalization according to the formula:

$$\rho(k) = \frac{\varphi(k)}{\varphi_{rand}(k)}, \quad (12.5)$$

where $\varphi_{rand}(k)$ is the rich-club coefficient computed in an appropriate null-model with the same degree distribution as the original network, usually generated according to an iterative rewiring procedure proposed by Maslov and Sneppen for protein networks [115]. In following years, Jiang and Zhou [116], argued the need of a statistical test for the assessment: a population of random networks is generated according to the null-model, the rich-club coefficient is computed for all of them and for the original network, lastly a one-sided p-value is assigned to each k as the percentage of $\varphi_{rand}(k)$ that are greater or equal than $\varphi(k)$. The same procedure has been applied in further studies while studying the rich-clubness in brain networks [117, 118, 119], using also corrections for multiple comparisons over the range of degree k [118, 119].

Recently, a new statistical test has been proposed, which attributes a unique p-value to indicate the presence of a *significant* rich-club in a given network [120]: (1) A population of random networks are generated using the Cannistraci–Muscoloni (CM) null-model [120], characterized by a lower rich-club coefficient with respect to the Maslov–Sneppen null-model. (2) The rich-club coefficient of the network under investigation is computed for each degree and then normalized (using the difference rather than the ratio, for a fair adjustment over the degree-range) by the mean coefficient of all the random networks. (3) The rich-club coefficient of every random network is also computed for each degree and normalized by the mean coefficient of all the random networks. (4) The maximum value of the normalized rich-club coefficient (peak of deviation from the mean null-model) is computed both for the considered network (observed peak) and for the population of random networks (null distribution of peaks). In practice, the ensemble of all the random peaks (one for each random network generated by the CM model) creates a null distribution of random peaks. (5) A one-sided p-value is computed as the percentage of random peaks greater or equal than the observed peak.

12.6.2 Deterministic Measures

The *average node degree (D)* is a local measure defined as:

$$D = \frac{1}{n} \sum_{i \in N} d_i \quad (12.6)$$

Where N is the set of all the network nodes, d_i is the degree of each node $i \in N$ and n is the size of the set N. The node degree is equal to the number of edges incident to the node and it is the simplest indicator of node centrality in the network.

The *characteristic path length (L)* [103, 104] is a global measure and describes the average of the shortest path lengths between all the pairs of vertices. It is defined as:

$$L = \frac{2}{n(n-1)} \sum_{i<j} sp_{ij}. \tag{12.7}$$

Where sp_{ij} is the shortest path length between a pair of nodes $(i,j) \in N$ and $\frac{2}{n(n-1)}$ is the number of possible node pairs in an undirected network. A small value of characteristic path length in a connectome means that the information flow between the nodes across the network is facilitated, and that the nodes are able to exchange messages between each other easily. In other words, the nodes across connectomes are functionally convergent.

The *average efficiency (E)* [92, 121] is a global measure that quantifies how efficiently the information is exchanged within the network. It is inversely proportional to the L, therefore a network with low characteristic path length is highly efficient.

$$E = \frac{2}{n(n-1)} \sum_{i<j} \frac{1}{sp_{ij}}. \tag{12.8}$$

The *average clustering coefficient (C)* [104], is a local measure and offers an average evaluation of the cross-interaction density between the first neighbors of each node in the network.

$$C = \frac{1}{n} \sum_i \frac{2t_i}{d_i(d_i-1)}. \tag{12.9}$$

Where t_i is the number of cross-interactions that occur between the first neighbors of the node $i \in N$ and $\frac{2}{d_i(d_i-1)}$ is the total number of possible cross-interactions that could occur between them. It assumes values in the range [0,1], large values indicate that the nodes in the network tend to have highly connected neighbors.

The *average closeness centrality (CL)* [122, 123] is a global measure based on the closeness centrality, an indicator of node centrality that evaluates how close a node is from all the others in the network. The closer it is to the others, the faster it can spread information to them. The measure for a single node can be computed as:

$$CL_i = \frac{n-1}{\sum_{j \neq i} sp_{ij}}, \tag{12.10}$$

and CL is obtained as average over the nodes $i \in N$.

When the network is disconnected in multiple connected components and therefore infinite shortest paths are present, the sum of the reciprocal of distances can be adopted instead of the reciprocal of the sum of distances [124].

In brain connectomics, if the average closeness centrality of the network is low then the activity of each node would be functionally more relevant to the other nodes.

The *average node betweenness centrality (NBC)* [125] is a global measure based on the node betweenness centrality, an indicator of node centrality that evaluates how crucial a particular node is in maintaining a path of optimum information flow between any other pair of nodes. The average measure calculates the average stress of information burden on the network nodes. For a single node it is defined as:

$$NBC_i = \sum_{j \neq i \neq k} \frac{\sigma_{jk}(i)}{\sigma_{jk}}, \qquad (12.11)$$

where $i, j, k \in N$, σ_{jk} is the total number of shortest paths between j and k and $\sigma_{jk}(i)$ is the number of those paths passing through i. NBC is obtained as average over the nodes $i \in N$.

The *average edge betweenness centrality (EBC)* [125], is a global measure based on the edge betweenness centrality, an indicator of edge centrality that evaluates how crucial a particular edge is in maintaining a path of optimum information flow between any pair of nodes. The average measure calculates the average stress of information burden on the network edges. For a single edge it is defined as:

$$EBC_{ij} = \sum_{s,t} \frac{\sigma_{st}(e_{ij})}{\sigma_{st}}, \qquad (12.12)$$

where $i, j, s, t \in N$, σ_{st} is the total number of shortest paths between s and t and $\sigma_{st}(e_{ij})$ is the number of those paths passing through the edge e_{ij}. EBC is obtained as average over the edges.

The *average radiality (R)* [126] is a global measure based on the radiality, an indicator of node centrality that evaluates the level of reachability of a node via different shortest paths of the network (i.e., the closer to the rest of nodes, the easier it is to reach). For a single node it is defined as:

$$R_i = \frac{\sum_{j \neq i}(D + 1 - s_{ij})}{n - 1}, \qquad (12.13)$$

where D is the diameter of the network and indicates the length of the maximum shortest path. R is obtained as average over the nodes $i \in N$.

In contrast to the existing node-neighborhood-based local measures, a new strategic shift has been introduced recently in which the focus is no longer only on groups of nodes and their common neighbours, but also on the organization of the links between them [127]. This new idea inspired a theory, which is known as the local community paradigm (LCP theory) and is valid both in monopartite [127], and in bipartite [128, 129] undirected unweighted networks. The LCP theory was proposed to mechanistically and deterministically model local-topology-dependent link-growth in complex networks, and holds that for modeling link prediction in complex networks, the information content related with the common neighbor nodes (CNs) of a given link should be complemented with the topological information emerging from the interactions between them. The cohort of CNs and their cross-interactions – which are called local community links (LCLs) – form what is called a local community. This first part of the theory inspired the Cannistraci variation of the classical CN-based similarity indices for link prediction, named also LCP-based link predictors, for details refer to [127, 128, 129].

Furthermore, the LCP theory holds that in many complex network topologies, the number of CNs of each link in the network is positively correlated with the respective number of LCLs. This second part of the LCP theory motivated a new network measure called *local-community-paradigm correlation (LCP-corr)* [127, 128, 129], which is a local measure that represents an exception with respect to the majority of the previous ones, for two main reasons. First, it is not related with only the node neighborhood but with the node/link neighborhood. Second, the general statistic used to obtain a unique value is not the average but the Pearson correlation. The formula for computing the LCP-corr is:

$$LCPcorr = \frac{\text{cov}(CN, LCL)}{\sigma_{CN} \cdot \sigma_{LCL}}, \text{ when } CN > 0, \qquad (12.14)$$

where cov indicates the covariance operator and σ the standard deviation. This formula is clearly a Pearson correlation between the CN and LCL variables. CN indicates a one-dimensional array. Its length is equal to the number of links in the network that have at least one common neighbor, and it reports the number of common neighbors for each of them. LCL indicates a one-dimensional array of the same size as CN, and it reports the number of local community links between the common neighbors. Mathematically the value of LCP-corr would be in the interval $[-1,1]$. However, extensive tests on many artificial and real complex networks demonstrate that an inverse correlation between CN and LCL is unlikely, therefore the interval is in general between $[0,1]$. In particular, it was revealed that LCP networks (with high LCP-corr, i.e. > 0.7) are very frequent to occur, and they are related to dynamic and heterogeneous systems that are characterized by weak interactions (relatively expensive or relatively strong) that in turn facilitate network evolution and remodeling. These are typical features of social and biological systems, where the LCP architecture facilitates not only the rapid delivery of information across the various network modules, but also the local processing. In contrast, non-LCP networks (with low LCP-corr, i.e. < 0.4) are less frequent to occur and characterize steady and homogeneous systems that are assembled through strong (often quite expensive) interactions, difficult to erase. This non-LCP architecture is more useful for processes where: (1) the storage of information (or energy) is at least as important as its delivery; (2) the cost of creating new interactions is excessive; (3) the creation of a redundant and densely connected system is strategically inadvisable. An emblematic example is the road networks, for which the money and time costs of creating additional roads are very high, and in which a community of strongly connected and crowded links resembles an impractical labyrinth.

In normal conditions, brain connectomes follow LCP organization [127], therefore they are characterized by high LCP-corr, which is in general higher than 0.8.

12.7 Network Neuroscience in the Latent Geometric Space

As previously discussed at the beginning of the chapter, studying the brain as a network of interconnected nodes and the recent developments of network theory contributed to unveil the key structural principles underlying the topology of the healthy human brain [130, 131]. One of the peculiar rules on which brain topology relies is the tendency of the network nodes to cluster into modules with high efficiency and short characteristic path length, thus reflecting an intrinsic small-world architecture, functionally segregated (local clustering) and integrated (global efficiency) [23, 24, 25]. Indeed, structural MRI studies based on DTI have demonstrated that the human brain shows a modularity structure, consisting of around five to six modules, corresponding to known functional subsystems [24, 132]. In addition to the existence of a structural core [24], it is likely that the brain exhibits a rich-club organization, with highly connected and central nodes (hubs) having a strong tendency to be mutually interconnected, thus constituting a focal point for whole-brain communication [117].

Such topological patterns of connectivity are often intricately related to the physical distances between elements in brain networks. Indeed, brain regions that are spatially close have a relatively high probability of being interconnected, while longer white matter projections are more expensive in terms of their material and energy costs, thus making connections between spatially far brain structures less likely [133]. This same concept is also the basic hypothesis on which is founded network geometry, a very active field of network science in recent years. It assumes that the network nodes reside in an underlying hidden metric space (latent geometry), which plays a role in shaping the observed network topology: nodes that are closer to each other in the metric space are more likely to be connected [134]. Hence, understanding whether the latent geometry of the brain connectomes relates to the neuroanatomy is an interesting field of research.

A first attempt to reveal the latent geometry of the brain connectome has been made by Ye et al. [135]. Using dimensionality reduction techniques, one linear (multidimensional scaling) and one nonlinear (Isomap), Ye et al. [135] introduced a new mathematical framework to represent the 3D hidden geometry of the human brain connectome in the Euclidean space and concluded that "this intrinsic geometry only minimally relates to neuroanatomy" [135]. However, since the brain networks are physically expensive systems and it is likely that several features of real brain anatomy have been structured to control such wiring costs [133], the conclusion of Ye et al., [135], could be ascribed to embedding limitations of the algorithms employed.

A more recent study has been performed by Cacciola et al. [27], using the coalescent embedding techniques developed by Muscoloni et al. [136], a class of topological-based unsupervised nonlinear dimension reduction methods [137, 138] able to efficiently map networks in the 2D or 3D hyperbolic space [136]. The hyperbolic space preserves many of the fundamental structural properties observed in real topologies [139], in particular a low characteristic path length, high clustering and a power-law degree distribution, offering a promising universal space of representation for real complex networks. As a first investigation, Cacciola et al. [27], adopted different datasets of healthy individuals, characterized by different types of network weights (e.g., number of streamlines and streamline distance), and employed the coalescent embedding techniques to map an average MR-DTI structural connectivity network in the 2D hyperbolic disk. The first rule of organization of brain networks that emerges in the geometrical space is their structural segregation in two hemispheres (Figure 12.8) [27], which is a simple concept yet quite neglected in previous studies on brain connectomics. Furthermore, their analysis highlights that not only the left–right organization

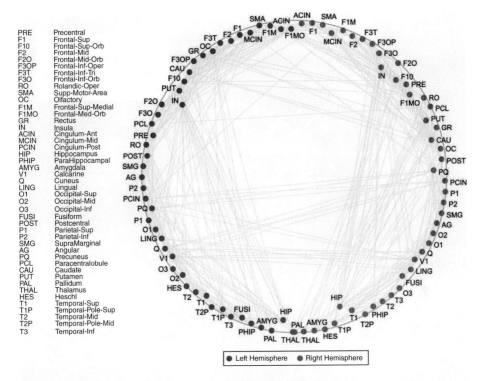

Figure 12.8: 2D coalescent embedding discloses the left–right anatomical arrangement of the brain. Coalescent embedding in the hyperbolic disk (using the ncMCE-EA technique) of an average structural connectivity matrix of healthy controls. The first rule of organization of brain networks that emerges in the geometrical space is their structural segregation in two hemispheres. Indeed, nodes belonging to the left and right hemispheres are perfectly separated over the disk, with all the nodes of the same hemisphere arranged in sequence without any interruption. Interestingly, the angular ordering of the nodes respects the correct anatomic sequence of the brain areas.

of the brain is correctly allocated and segregated in the 2D hyperbolic disk, but also the anterior–central–posterior organization and the well-known brain lobes organization [27]. Since the brain lies on a three-dimensional physical space, Cacciola et al. [27] performed also the coalescent embedding of the MR-DTI structural connectivity networks in the 3D hyperbolic sphere, obtaining a geometric mapping that well resembles the original arrangement of the brain lobes (Figure 12.9). In contrast to the conclusion of Ye et al. [135], this result underlines a close relation between the latent geometry of the brain connectome and the neuroanatomy [27].

As a second investigation, Cacciola et al. [27] tested whether the latent geometry of brain networks in different conditions presents significant differences [27]. Two scenarios have been analyzed: age and pathology. In the first scenario two groups of healthy subjects with different ages were considered, in the second scenario de novo drug naive Parkinson's disease (PD) patients were compared to healthy controls. The structural connectome of each subject was mapped in the hyperbolic space using the coalescent embedding techniques and simple geometric measures (average hyperbolic

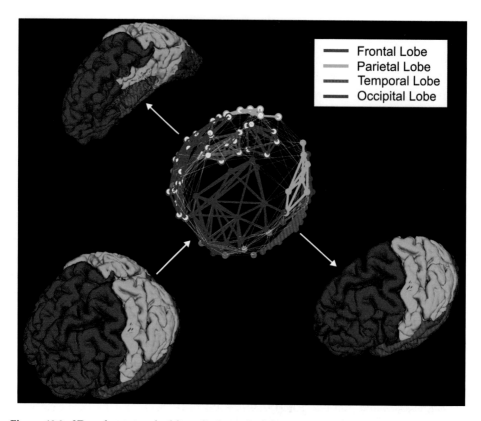

Figure 12.9: 3D coalescent embedding discloses the lobes anatomical arrangement of the brain. Coalescent embedding in the hyperbolic sphere (using the ISO technique) of an average structural connectivity matrix of healthy controls. The figure shows, in a superior-anterior-lateral view, the 3D latent geometry of the brain emerging from the embedding in the hyperbolic sphere. The color-filled circles represent the nodes of the left hemisphere, whereas the white-filled ones represent the nodes of the right hemisphere. Each node is labeled according to its real anatomical localization in the different brain lobes. Nodes in gray color are not assigned to any lobe, since they represent gray-matter structures placed in the deep white matter. The figure highlights a close relation between the reconstructed latent geometry of the brain connectome and the neuroanatomy.

distance or average hyperbolic path between the nodes) were computed as markers. The results indicate that the geometrical markers are able to significantly discriminate between the two groups, and therefore that both age and pathology can affect the latent geometry of the brain networks. By contrast, using the original average weight of the network as marker does not allow us to detect the difference between the two groups [27].

The study of Cacciola et al. suggests that, although the human structural brain networks are weakly hyperbolic, the coalescent embedding algorithms still offer a powerful tool for revealing the latent brain geometry, and pave the way for a network latent-geometry marker characterization of brain diseases, which could be used for diagnostic and prognostic purpose and for therapeutic treatment evaluation [27].

12.8 Brain Network Disorders

Connectomics and network science offer a fascinating and powerful analytic scenario for evaluating network-topology-based disease signatures/markers and for monitoring disease spreading.

As previously outlined in the Introduction section, the human brain is a cost-efficient small-world network, functionally segregated and integrated, thus allowing strong local clustering and efficient long-range connectivity patterns [133]. Brain networks also show a hierarchical modularity architecture, corresponding to known major functional subsystems such as sensorimotor and multimodal associative systems [140, 141]. Furthermore, the human brain is a complex network that seems to exhibit a preponderance of highly connected and central nodes (hubs) with a strong tendency to be interconnected, that together establish a "rich club" [117]. However, the rich club organization of the brain is still a topic under discussion, because the design of a valid test for rich-clubness is an open problem [120]. Finally, it has been recently demonstrated that the topology of the human brain connectome shows a tendency to organize in multiple local communities. This phenomenon has been called local-community-paradigm (LCP) and describes a heterogeneous and dynamic system characterized by weak interactions in an adaptive network structure that allows a global information transmission and processing via multiple local modules [127].

There is little doubt that such optimal architecture emerges during the development and that it is affected by the ageing [142, 143], as well as by the gender (probably due to a possible influence of sex hormones) [144, 145]. These peculiar topological characteristics of the normal brain organization have made possible to exploit the topo-pathological rewiring in several neurological and psychiatric disorders, challenging the classical idea of "local" or "global" brain diseases. Herein, we discuss network studies with a particular focus on neurodegenerative diseases aiming at assessing the way network science is challenging the concept of brain disorders.

Several studies have used network science for connectomic analysis of patients with Alzheimer's disease (AD) and PD, two of the most common neurodegenerative diseases. AD is considered to be a "disconnection syndrome" characterized by a loss of neurons and connections due to a progressive deposition of beta-amyloid in many regions of the neocortex, such as the posterior cingulate cortex, the inferior parietal lobule, the precuneus, the medial frontal lobe as well as the hippocampus [146, 147]. These regions are classic hub nodes of the brain, mediating long-range connectivity patterns between brain modules, thus being crucial for supporting and mediating integrative processes and adaptive behaviours [148]. A growing number of findings demonstrated that these critical and vulnerable regions show a topological disruption in AD supporting the hypothesis that brain hubs are fundamental to understand the clinical anatomy of such neurodegenerative disorder [149]. The involvement of these highly central hubs is one of the most consistent properties of network changes in AD, likely suggesting a relationship between amyloid deposition and hub vulnerability. Besides the involvement of brain hubs, AD is characterized by an altered connectivity both at the local and global level, reflected by lower local efficiency (that captures local connectivity), [94, 150, 151] increased characteristic path length and decreased global efficiency (which mirrors long-range connections) [50, 152, 153]. On the one hand, the loss of such structural and functional long-distance connectivity patterns has been shown in many neuroimaging studies, on the other hand, surprisingly, decreased values of characteristic path length have also been reported, suggesting that the brain puts in action a compensation strategy to contain the loss of long-distance connectivity. Both MRI and magnetoencephalography findings indicate a loss of the small-world network measures in AD, as suggested by reduced global and local clustering in both hippocampi [154], as well as decreased normalized path length, indicating a more randomized topology in AD [155], which leads to the risk of a less efficient and uncontrolled information flow within the whole brain network. Such opposite findings may be related to the imaging techniques (EEG, fMRI, tractography, morphological metrics correlation) and nodes selection employed for constructing the connectomes as well as the use of different variations of topological measures (i.e., path length vs normalized path length). Taken together these results stress the need to develop new tools and network markers which can characterize the topology of brain disorders in an unbiased way. In this regard, the global clustering has been proposed as a useful diagnostic network marker able to discriminate between patients with AD and healthy elderly individuals with high specificity (78%) and sensitivity (72%) [154].

Analogously, network analysis has been recently applied to PD aimed at improving our understanding of the pathophysiological mechanisms, monitoring disease progression, and developing novel biomarkers for diagnosis and treatment efficacy. Several neurobiological studies demonstrated that the pathogenesis of PD begins in the basal ganglia and in turn spreads to numerous extranigral structures [156, 157]. Although the aetiology and pathogenesis of PD is still not completely clear, this disorder is characterized by the accumulation and aggregation of alfa-synuclein in the substantia nigra in the central nervous system [158]. The death of dopamine-generating cells in the substantia nigra leads to the classical motor symptoms such as bradykinesia, resting tremor, rigidity, and late postural instability [159]. However, PD patients may exhibit many non-motor symptoms such as olfactory impairment, constipation and sleep disorders especially at the early stage of the disease [160]. Such neurobiological findings, together with the heterogeneous motor and non-motor symptoms experienced by patients with PD, suggest that PD is a brain multisystemic disorder involving many neural networks within the entire brain. Therefore, the last years have been characterized by a growing interest in investigating the brain connectomes of patients with PD by combining neuroimaging techniques with network science. In particular, many fMRI studies have demonstrated that PD patients show decreased global [161], as well as nodal efficiency especially in the cortico-basal ganglia motor circuit [162]. In line with these findings, the structural neural networks of patients with early PD present an aberrant global topology reflected by lower global clustering coefficient, lower global efficiency and a decreased mean of connections' strength, paralleled by decreased clustering coefficient and strength in crucial brain structures such as amygdala, olfactory cortex, putamen, and pallidum [44]. A common characteristic of PD and AD is the impairment of long-range connections and the increased tendency to reinforce a local topological organization as demonstrated by increased small-worldness, modularity, as well as an altered reorganization of the hub nodes [163], especially in PD patients with mild cognitive impairment. These findings suggest a sort of plastic compensatory mechanism in response to the long-range disconnectivity, underlying the major role of white matter degeneration and neurite dysfunction in the pathophysiology of cognitive impairment in PD [164, 165, 166, 167]. In addition, both non depressed-PD and depressed-PD patients showed increased characteristic path length in specific brain frequencies as compared to healthy controls, indicating an abnormal reorganization of the entire brain leading to a reduced efficiency [168]. By contrast, applying complex network analysis to MEG data, Olde Dubbelink et al. [169] demonstrated that brain networks of early-stage non-medicated PD patients were characterized by preserved characteristic path length in the delta frequency; however, longitudinal analysis showed a reduced normalized path length in the alpha2 frequency band, correlating with worsening motor function [169].

We recall that the characteristic path length is defined as mean of shortest paths between all possible nodes pairs in the network, resulting in a measure for global integration of the network. Theoretically, a shorter path length is not necessarily an advantage in complex networks, since it is the overall structure that must be an effective balance between local specialization and global integration. In addition, when considering the normalized path length, its reduction only indicates that the brain network topology is more similar to that of a random network and does not directly imply a shorter absolute characteristic path length. The loss of structure as expressed by altered values of characteristic path length in PD seems to support the concept of PD as a structural disconnection syndrome too (such as in the case of AD), together with slowing of brain activity and loss of functional connectivity in PD. It is worth to note that, as previously said for AD, these controversial results may be related due to the imaging modality employed and the different way to compute the topological measure.

The application of network science to the study of connectomic disorders is a relatively novel but rapidly developing framework, with a wide range of challenges and issues deserving particular attention. However, such increasing interest in the field of brain network science will certainly pave the way for a step forward in precision medicine, leading to new tools for diagnosis, treatment, and disease spread prediction. In this regard, a promising approach could be investigating the extent to which the connectivity evolution of a network might be predicted [127, 128, 170] by mere topological measures, thus permitting the analysis of network patterns hidden in biomedical data, signals and images.

12.9 Exercises

The exercises 12.1–12.4 are aimed at learning the basic topological network measures, computing them in toy networks. These exercises can be performed on paper. Exercise 12.5 is aimed at analyzing structural and functional connectomes and requires basic programming skills. The material required for Exercise 12.5 can be found at the online repository: https://github.com/biomedical-cybernetics/network_neuroscience_exercises. Exercises 12.6–12.13 are theoretical exercises.

12.1 Considering the networks A and B in Figure 12.10, compute the average node degree, the average clustering coefficient and the characteristic path length. Which topological measure is expected to differ more between the two networks?

12.2 Considering the network C in Figure 12.10, compute the average closeness centrality and the average edge betweenness centrality. According to these two

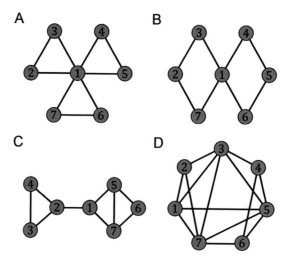

Figure 12.10 Toy-networks for exercises

topological measures, which are the most central node and the most central edge in the network?

12.3 Considering the network B in Figure 12.10, compute the LCP-correlation. Is the network following LCP organization?

12.4 Considering the network D in Figure 12.10, compute the LCP-correlation. Is the network following LCP organization?

12.5 Considering the structural and functional adjacency matrices provided at the online repository, write a script in order to compute for each matrix and for each threshold the average node degree, the average clustering coefficient and the characteristic path length. What is the meaning of the characteristic path length respectively in the structural and functional brain networks?

12.6 Starting from the MRI acquisition, describe the critical processing issues and the related pipeline for mapping a structural connectome.

12.7 Starting from the MRI acquisition, describe the critical processing issues and the related pipeline for mapping a functional connectome.

12.8 Starting from the EEG recording, describe the critical processing issues and the related pipeline for mapping an EEG-based functional connectome.

12.9 Explain the differences between stochastic and deterministic measures. Which group of measures is the most computationally expensive?

12.10 Explain the differences between local and global topological measures. List respectively all the local and global topological measures you know.

12.11 Which are the peculiar characteristics of the brain networks topology?

12.12 What is the "latent geometry" of a network? Why the latent geometry of the brain structural network is expected to be related to the brain neuroanatomy?

12.13 When studying brain network disorders, one might expect to find an increased characteristic path length reflecting an altered global integration of the network. What could be the meaning of a reduced characteristic path length in brain network disorders?

Note: Solutions for Exercises 12.1–12.5 are available to instructors at www.cambridge.org/bionetworks.

References

[1] Catani M, Jones D, ffytche DH. Perisylvian Language networks of the human brain. *Annals of Neurology*, 2005;57(1): 8–16. DOI:10.1002/ana.20319.

[2] Catani M, Dell'Acqua F, Bizzi A, et al. Beyond cortical localization in clinico-anatomical correlation. *Cortex*,2012;48(10):1262–1287. DOI:10.1016/j.cortex.2012.07.001.

[3] Catani M, ffytche DH. "On The study of the nervous system and behaviour." *Cortex*, 2010;46(1):106–109. DOI:10.1016/j.cortex.2009.03.012.

[4] Mesulam MM. Behavioral neuroanatomy: Largescale networks, association cortex, frontal syndromes, the limbic system, and hemispheric specialization. In *Principles of Behavioral and Cognitive Neurology*. Oxford University Press, 2000, 1–120.

[5] Zappalà G, Thiebaut de Schotten M, Eslinger PJ. Traumatic brain injury and the frontal lobes: What can we gain with diffusion tensor imaging? *Cortex*, 2012;48(2):156–165. DOI:10.1016/j.cortex.2011.06.020.

[6] Sporns O, Tononi G, Kötter R. The human connectome: A structural description of the human brain. *PLoS Computational Biology*, 2005;1(4):e42. doi:10.1371/journal.pcbi.0010042.

[7] Hagmann P. From diffusion MRI to brain connectomics. PhD thesis, EPFL. doi:10.5075/epfl-thesis-3230.

[8] Sporns O. The human connectome: A complex network. *Annals of the New York Academy of Sciences*, 2011;1224(1):109–125. doi:10.1111/j.1749-6632.2010.05888.x.

[9] Oh SW, Harris JA, Ng L, et al. A mesoscale connectome of the mouse brain. *Nature*, 2014;508(7495):207–214. doi:10.1038/nature13186.

[10] Narula V, Zippo AG, Muscoloni A, Can local-community-paradigm and epitopological learning enhance our understanding of how local brain connectivity is able to process, learn and memorize chronic pain? *Applied Network Science*, 2017;2(1):28. doi:10.1007/s41109-017-0048-x.

[11] Zippo AG, Storchi R, Nencini S, et al. Neuronal functional connection graphs among multiple areas of the rat somatosensory system during spontaneous and evoked activities. *PLoS Computational Biology*, 2013;9(6):e1003104. DOI:10.1371/journal.pcbi.1003104.

[12] Bastos AM, Vezoli CA, Bosman et al. Visual areas exert feedforward and feedback influences through distinct frequency channels. *Neuron*,2015;85(2):390–401. DOI:10.1016/j.neuron.2014.12.018.

[13] Corti V, Sanchez-Ruiz Y, Piccoli G, et al. Protein fingerprints of cultured CA3-CA1 hippocampal neurons: Comparative analysis of the distribution of synaptosomal and cytosolic proteins. *BMC Neuroscience*, 2008;9(1):36. DOI:10.1186/1471-2202-9-36.

[14] Mountcastle VB. Modality and topographic properties of single neurons of cat's somatic sensory cortex. *Journal of Neurophysiology*, 1957;20(4):408–34. DOI:10.1146/annurev.ph.20.030158.002351.

[15] Buxhoeveden DP, Casanova MF. The minicolumn hypothesis in neuroscience. *Brain*, 2002; 125(5):935–951. DOI:10.1093/brain/awf110.

[16] Rakic P. Confusing cortical columns. *Proceedings of the National Academy of Sciences*, 2008;105(34):12099–12100. DOI:10.1073/pnas.0807271105.

[17] Mountcastle VB. The columnar organization of the neocortex. *Brain*, 1997;120:701–722. DOI:10.1093/brain/120.4.701.

[18] Horton JC, Adams DL. The cortical column: A structure without a function. *Philosophical Transactions of the Royal Society B: Biological Sciences*, 360(1456):837–362. DOI:10.1098/rstb.2005.1623.

[19] Einevoll GT, Kayser C, Logothetis NK, Panzeri S. Modelling and analysis of local field potentials for studying the function of cortical circuits. *Nature Reviews Neuroscience*, 2013;14(11):770–785. DOI:10.1038/nrn3599.

[20] Ingalhalikar M, Smith A, Parker D, et al. Sex differences in the structural connectome of the human brain. *Proceedings of the National Academy of Sciences of the United States of America*, 2014;111(2):823–828. DOI:10.1073/pnas.1316909110.

[21] Fornito A, Zalesky A, Breakspear M. The connectomics of brain disorders. *Nature Reviews Neuroscience*, 2015;16(3):159–172. DOI:10.1038/nrn3901.

[22] Papo D, Buldu JM, Boccaletti S, Bullmore ET. Complex network theory and the brain. *Philosophical Transactions of the Royal Society B*, 2014;369(1653):20130520. DOI:10.1098/rstb.2013.0520.

[23] Gong G, He Y, Concha L, Lebel C, et al. Mapping anatomical connectivity patterns of human cerebral cortex using in vivo diffusion tensor imaging tractography. *Cerebral Cortex*, 2009;19(3):524–536. DOI:10.1093/cercor/bhn102.

[24] Hagmann P, Cammoun L, Gigandet X, et al. Mapping the structural core of human cerebral cortex. *PLoS Biology*, 2008;6(7): 1479–1493. DOI:10.1371/journal.pbio.0060159.

[25] Sporns O, Zwi D. The small world of the cerebral cortex. *Neuroinformatics*, 2004;2(2):145–162. DOI:10.1385/NI:2:2:145.

[26] Hilgetag CC, Kaiser M. Clustered organization of cortical connectivity. *Neuroinformatics*, 2004;2(3):353–360. DOI:10.1385/NI:2:3:353.

[27] Cacciola A, Muscoloni A, Narula V, et al. Coalescent embedding in the hyperbolic space unsupervisedly discloses the hidden geometry of the brain. arXiv preprint available at arXiv:1705.04192 [Q-bio.NC]. https://arxiv.org/abs/1705.04192.2017.

[28] Le Bihan D Looking at the functional architechture of the brain with diffusion MRI. *Nature Reviews Neuroscience*, 2003;4(6):469–80. DOI:http://dx.doi.org/10.1038/nrn1119.

[29] Le Bihan D, Poupon C, Amadon A, Lethimonnier F. Artifacts and pitfalls in diffusion MRI. *Journal of Magnetic Resonance Imaging*, 2006;24(3):478–488. DOI:10.1002/jmri.20683.

[30] Hagmann P, Jonasson L, Maeder P, et al. Understanding diffusion MR imaging techniques: From scalar diffusion-weighted imaging to diffusion tensor imaging and beyond. *RadioGraphics*, 2006;26(suppl_1):S205–223. DOI:10.1148/rg.26si065510.

[31] Beaulieu C. The basis of anisotropic water diffusion in the nervous system: A technical review. *NMR in Biomedicine*, 2002;15;(7–8):435–455. DOI:10.1002/nbm.782.

[32] Basser PJ, Jones DK. Diffusion-tensor MRI: Theory, experimental design and data analysis – a technical review." *NMR in Biomedicine*, 2002;15(7-8):456–467. DOI:10.1002/nbm.783.

[33] Johansen-Berg H, Behrens TEJ. *Diffusion MRI: From Quantitative Measurement to In Vivo Neuroanatomy*. Elsevier/Academic Press. 2013. DOI:10.1016/C2011-0-07047-3.

[34] Besson P, Dinkelacker V, Valabregue R, et al. Structural connectivity differences in left and right temporal lobe epilepsy. *NeuroImage*,2014;100:135–144. DOI:10.1016/j.neuroimage.2014.04.071.

[35] Jeurissen B, Leemans A, Tournier JD, Jones DK, Sijbers J. Investigating the prevalence of complex fiber configurations in white matter tissue with diffusion magnetic resonance imaging. *Human Brain Mapping*, 2013;34(11):2747–2766. DOI:10.1002/hbm.22099.

[36] Tournier JD, Calamante F, Connelly A. Robust determination of the fibre orientation distribution in diffusion MRI: Non-negativity constrained super-resolved spherical deconvolution. *NeuroImage*, 2007;35(4):1459–1472. DOI:10.1016/j.neuroimage.2007.02.016.

[37] Tuch DS. Q-Ball Imaging. *Magnetic Resonance in Medicine*, 2004;52(6):1358–1472. DOI:10.1002/mrm.20279.

[38] Wedeen VJ, Wang RP, Schmahmann JD, et al. Diffusion spectrum magnetic resonance imaging (DSI) tractography of crossing fibers. *NeuroImage*, 2008;41(4):1267–1277. DOI:10.1016/j.neuroimage.2008.03.036.

[39] Tournier JD, Yeh CH, Calamante F, et al. Resolving crossing fibres using constrained spherical deconvolution: Validation using diffusion-weighted imaging phantom data. *NeuroImage*, 2008;42(2):617–625. DOI:10.1016/j.neuroimage.2008.05.002.

[40] Tuch D, Reese TG, Wiegell MR, Wedeen VJ. Diffusion MRI of complex neural architecture. *Neuron*, 2003;40(5):885–895. DOI:10.1016/S0896-6273(03)00758-X.

[41] Descoteaux M, Deriche R, Knösche T, Anwander A. Deterministic and probabilistic tractography based on complex fiber orientation distributions. *IEEE Transactions on Medical Imaging*, 2009;28(2):269–286. DOI:10.1109/TMI.2008.2004424.

[42] Smith RE, Tournier JD, Calamante F, Connelly A. Anatomically-constrained tractography: Improved diffusion MRI streamlines tractography through effective use of anatomical unformation. *NeuroImage*, 2012;62(3):1924–1938. DOI:10.1016/j.neuroimage.2012.06.005.

[43] Grèzes J, Valabrègue R, Gholipour B, Chevallier C. A direct amygdala-motor pathway for emotional displays to influence action: A diffusion tensor imaging study. *Human Brain Mapping*, 2014;35(12):5974–5983. DOI:10.1002/hbm.22598.

[44] Nigro S, Riccelli R, Passamonti L, et al. Characterizing structural neural networks in de novo Parkinson disease patients using diffusion tensor

imaging. *Human Brain Mapping*, 2016;37(12):4500–4510. DOI:10.1002/hbm.23324.

[45] Cacciola A, Calamuneri A, Milardi D, A connectomic analysis of the human basal ganglia network. *Frontiers in Neuroanatomy*, 2017;11:85. DOI:10.3389/fnana.2017.00085.

[46] Cacciola A, Milardi D, Anastasi GP, et al. A direct cortico-nigral pathway as revealed by constrained spherical deconvolution tractography in humans. *Frontiers in Human Neuroscience*, 2016;10:374. DOI:10.3389/fnhum.2016.00374.

[47] Milardi D, Cacciola A, Calamuneri A, et al. The olfactory system revealed: Non-invasive mapping by using constrained spherical deconvolution tractography in healthy humans. *Frontiers in Neuroanatomy*, 2017;11:1–11. DOI:10.3389/fnana.2017.00032.

[48] Milardi D, Cacciola A, Cutroneo G, et al. Red nucleus connectivity as revealed by constrained spherical deconvolution tractography. *Neuroscience Letters*, 2016;626:68–73. DOI:10.1016/j.neulet.2016.05.009.

[49] Cacciola A, Milardi D, Calamuneri A, et al. Constrained spherical deconvolution tractography reveals cerebello-mammillary connections in humans. *Cerebellum*, 2017;16(2):483–495. doi:10.1007/s12311-016-0830-9.

[50] Lo CY, Wang PN, Chou KH, et al. Diffusion tensor tractography reveals abnormal topological organization in structural cortical networks in Alzheimer's disease. *Journal of Neuroscience*, 2010;30(50):16876–6885. DOI:10.1523/JNEUROSCI.4136-10.2010.

[51] Yao Z, Zhang Y, Lin L, et al. Abnormal cortical networks in mild cognitive impairment and Alzheimer's disease. *PLoS Computational Biology*, 2010;6(11):e1001006. DOI:10.1371/journal.pcbi.1001006.

[52] Li L, Rilling JK, Preuss TM, Glasser MF, Hu X. The effects of connection reconstruction method on the interregional connectivity of brain networks via diffusion tractography. *Human Brain Mapping*, 2012;33(8):1894–1913. doi:10.1002/hbm.21332.

[53] Reveley C, Seth AK, Pierpaoli C, et al. Superficial white matter fiber systems impede detection of long-range cortical connections in diffusion MR tractography. *Proceedings of the National Academy of Sciences*, 2015;112(21):E2820–E2828. DOI:10.1073/pnas.1418198112.

[54] Hagmann P, Kurant M, Gigandet X, et al. Mapping human whole-brain structural networks with diffusion MRI. *PLoS ONE*, 2007;2(7):e597. DOI:10.1371/journal.pone.0000597.

[55] Iturria-Medina Y, Sotero RC, Canales-Rodríguez, EJ et al. Studying the human brain anatomical network via diffusion-weighted MRI and graph theory. *NeuroImage*, 2008;40(3):1064–1076. DOI:10.1016/j.neuroimage.2007.10.060.

[56] Robinson EC, Hammers A, Ericsson A, Edwards, AD, Rueckert D. Identifying population differences in whole-brain structural networks: A machine learning approach. *NeuroImage*, 2010;50(3):910–919. DOI:10.1016/j.neuroimage.2010.01.019.

[57] Wen W, Zhu W, He Y, et al. Discrete neuroanatomical networks are associated with specific cognitive abilities in old age. *Journal of Neuroscience*, 2011;31(4):1204–1212. DOI:10.1523/JNEUROSCI.4085-10.2011.

[58] Jones, DK, Knösche TR, Turner R. White matter integrity, fiber count, and other fallacies: The do's and don'ts of diffusion MRI. *NeuroImage*, 2012;73:239–254. DOI:10.1016/j.neuroimage.2012.06.081.

[59] Alexander AL, Hurley SA, Samsonov AA, et al. Characterization of cerebral white matter properties using quantitative magnetic resonance imaging stains. *Brain Connectivity*,2011;1(6):423–446. DOI:10.1089/brain.2011.0071.

[60] Alexander-Bloch A, Giedd JN, Bullmore. Imaging Structural co-variance between human brain regions. *Nature Reviews Neuroscience*,2015;14(5):322–336. DOI:10.1038/nrn3465.

[61] Laufs H, Holt JL, Elfont R, et al. Where the BOLD signal goes when alpha EEG leaves. *NeuroImage*, 2006;31(4):1408–1418. DOI:10.1016/j.neuroimage.2006.02.002.

[62] Buzsaki G, Draguhn A. Neuronal oscillations in cortical networks. *Science*, 2004;304(5679):1926–1929. DOI:10.1126/science.1099745.

[63] Mantini D, Perrucci MG, Del Gratta C, Romani GL, Corbetta M. Electrophysiological signatures of resting state networks in the human brain. *Proceedings of the National Academy of Sciences of the United States of America*, 2007;104(32):13170–13175. DOI:10.1073/pnas.0700668104.

[64] Huettel SA, Song AW, McCarthy G. *Functional Magnetic Resonance Imaging*. Sinauer Associates, Inc.; 2004.

[65] Logothetis NK. What we can do and what we cannot do with fMRI." *Nature*, 2008;453(7197):869–878. DOI:10.1038/nature06976.

[66] Lindquist MA, Loh M, Atlas LY, Wager TD. Modeling the hemodynamic response function in fMRI: Efficiency, bias and mis-modeling. *NeuroImage*, 2009;45(1 Suppl):S187–S198. DOI:10.1016/j.neuroimage.2008.10.065.

[67] Cannistraci CV, Montevecchi FM, Alessio M. 2009. Median-modified wiener filter provides efficient denoising, preserving spot Edge and morphology in 2-DE image processing. *Proteomics*, 2009;9(21):4908–4919. DOI:10.1002/pmic.200800538.

[68] Cannistraci CV, Abbas A, Gao X. Median modified Wiener filter for nonlinear adaptive spatial denoising of protein NMR multidimensional spectra. *Scientific Reports*, 2015;5:8017. DOI:10.1038/srep08017.

[69] Stam CJ, Nolte G, Daffertshofer A. Phase lag index: Assessment of functional connectivity from multi channel EEG and MEG with diminished bias from common sources. *Human Brain Mapping*, 2007;28(11):1178–1193. DOI:10.1002/hbm.20346.

[70] Andres FG, Gerloff, C Coherence of sequential movements and motor learning. *Journal Clinical Neurophysics*, 1999;16(6):520–527. http://www.ncbi.nlm.nih.gov/pubmed/10600020.

[71] Bassett D.S, Meyer-Lindenberg A, Achard S, Duke Bullmore E. Adaptive reconfiguration of fractal small-world human brain functional networks. *Proceedings of the National Academy of Sciences*, 2006;103(51):19518–19523. DOI:10.1073/pnas.0606005103.

[72] Pascual-Marqui RD, Lehmann D, Koukkou M, Kochi K, et al. Assessing interactions in the brain with exact low-resolution electromagnetic tomography. *Philosophical Transactions of the Royal Society of London A: Mathematical, Physical and Engineering Sciences*, 2011;369(1952):3768–3784. DOI:10.1098/rsta.2011.0081.

[73] Tzourio-Mazoyer N, Landeau B, Papathanassiou D, et al. Automated anatomical labeling of activations in SPM using a macroscopic anatomical parcellation of the MNI MRI single-subject brain. *NeuroImage*, 2002;15(1):273–289. DOI:10.1006/nimg.2001.0978.

[74] Makris N, Goldstein JM, Kennedy D, et al. Decreased volume of left and total anterior insular lobule in schizophrenia. *Schizophrenia Research*, 2006;83(2–3):155–171. DOI:10.1016/j.schres.2005.11.020.

[75] Goldstein JM, Seidman LJ, Makris N, et al. Hypothalamic abnormalities in schizophrenia: Sex effects and genetic bulnerability. *Biological Psychiatry*, 2007;61(8):935–45. DOI:10.1016/j.biopsych.2006.06.027.

[76] Frazier JA, Chiu S, Breeze JL, et al. Structural brain magnetic resonance imaging of limbic and thalamic volumes in pediatric bipolar disorder." *American Journal of Psychiatry*, 2005;162(7):1256–1265. DOI:10.1176/appi.ajp.162.7.1256.

[77] Desikan RS, Ségonne F, Fischl B, et al. 2006. An automated labeling system for subdividing the human cerebral cortex on MRI scans into gyral based regions of interest. *NeuroImage*, 2006;31(3):968–980. DOI:10.1016/j.neuroimage.2006.01.021.

[78] Destrieux C, Fischl B, Dale A, Halgren E. Automatic parcellation of human cortical gyri and sulci using standard anatomical nomenclature. *NeuroImage*, 2010;53(1):1–15. DOI:10.1016/j.neuroimage.2010.06.010.

[79] Fischl B, Van Der Kouwe A, Destrieux C, et al. Automatically parcellating the human cerebral cortex. *Cerebral Cortex*, 2004;14(1):11–22. DOI:10.1093/cercor/bhg087.

[80] Dwyer, DB, Harrison BJ, Yucel M. et al. Large-scale brain network dynamics supporting adolescent cognitive control. *Journal of Neuroscience*, 2014;34(42):14096–14107. DOI:10.1523/JNEUROSCI.1634-14.2014.

[81] Lopes da Silva F. Functional localization of brain sources using EEG and/or MEG data: Volume conductor and source models. *Magnetic Resonance Imaging*, 2004;22:1533–538. DOI:10.1016/j.mri.2004.10.010.

[82] Kiviniemi V, Starck T, Remes Y, et al. Functional segmentation of the brain cortex using high model order group PICA. *Human Brain Mapping*, 2009;30(12):3865–86. DOI:10.1002/hbm.20813.

[83] Whitfield-Gabrieli S, Nieto-Castanon A. Conn: A functional connectivity toolbox for correlated and anticorrelated brain networks. *Brain Connectivity*, 2012;2(3):125–141. DOI:10.1089/brain.2012.0073.

[84] Hadi Hosseini SM, Hoeft F, Kesler SR. GAT: A graph-theoretical analysis toolbox for analyzing between-group differences in large-scale structural and functional brain networks. *PLoS ONE*, 2012;7(7):e40709. doi:10.1371/journal.pone.0040709.

[85] Cui Z, Zhong P, Xu Y, He G, Gong. PANDA: A pipeline toolbox for analyzing brain diffusion images. *Frontiers in Human Neuroscience*, 2013;7:42. DOI:10.3389/fnhum.2013.00042.

[86] Xia M, Wang J, He Y. BrainNet viewer: A network visualization tool for human brain connectomics. *PLoS ONE*, 2013;8(7):e68910. DOI:10.1371/journal.pone.0068910.

[87] Kruschwitz JD, List D, Waller L. Rubinov M, Walter H. GraphVar: A user-friendly toolbox for comprehensive graph analyses of functional brain connectivity. *Journal of Neuroscience Methods*, 2015;245:107–115. DOI:10.1016/j.jneumeth.2015.02.021.

[88] Rubinov M, Sporns O. Complex network measures of brain connectivity: Uses and interpretations. *NeuroImage*, 2010;52(3):1059–1069. DOI:10.1016/j.neuroimage.2009.10.003.

[89] Wang J, Wang X, Xia M, et al. GRETNA: A graph theoretical network analysis toolbox for imaging connectomics. *Frontiers in Human Neuroscience*, 2015;9:1–16. DOI:10.3389/fnhum.2015.00458.

[90] Wijk BCM van, Stam CJ, Daffertshofer A. Comparing brain networks of different size and connectivity density using graph theory. *PLoS ONE*, 2010;5(10):e13701. DOI:10.1371/journal.pone.0013701.

[91] Ferrarini L, Veer IM, Baerends E, et al. Hierarchical functional modularity in the resting-state human brain. *Human Brain Mapping*, 2009; 30(7):2220–2231. DOI:10.1002/hbm.20663.

[92] Heuvel MP van den, Stam CJ, Kahn RS, Hulshoff Pol HE. Efficiency of functional brain networks and intellectual performance. *The Journal of Neuroscience*, 2009;29(23):7619–7624. DOI:10.1523/JNEUROSCI.1443-09.2009.

[93] Dimitriadis SI, Laskaris NA, Del Rio-Portilla Y, Koudounis GC. Characterizing dynamic functional connectivity across sleep stages from EEG. *Brain Topography*, 2009;22(2):119–133. DOI:10.1007/s10548-008-0071-4.

[94] Haan W de, Pijnenburg YA, Strijers RLM, et al. Functional neural network analysis in frontotemporal dementia and Alzheimer's disease using EEG and graph theory. *BMC Neuroscience*, 2009;10:101. DOI:10.1186/1471-2202-10-101.

[95] Bernhardt BC, Chen Z, He Y, Evans AC Bernasconi N. Graph-theoretical analysis reveals disrupted small-world organization of cortical thickness correlation networks in temporal lobe epilepsy. *Cerebral Cortex*, 2011;21(9):2147–2157. DOI:10.1093/cercor/bhq291.

[96] Zhang J, Wang J, Wu Q, et al. Disrupted brain connectivity networks in drug-naive, first-episode major depressive disorder. *Biological Psychiatry*, 2011;70(4):334–342. DOI:10.1016/j.biopsych.2011.05.018.

[97] Langer N, Pedroni A, Jäncke L. The problem of thresholding in small-world network analysis. *PLoS ONE*, 2013;8(1):e53199. doi:10.1371/journal.pone.0053199.

[98] Simpson SL, Lyday RG, Hayasaka, Marsh S, Laurienti PJ. A permutation testing framework to compare groups of brain networks. *Frontiers in Computational Neuroscience*, 2013;7171: DOI:10.3389/fncom.2013.00171.

[99] Drakesmith M, Caeyenberghs K, Dutt A, et al. Overcoming the effects of false positives and threshold bias in graph theoretical analyses of neuroimaging data. *NeuroImage*, 2015;118:313–333. DOI:10.1016/j.neuroimage.2015.05.011.

[100] Zalesky A, Fornito A, Bullmore ET. Network-based statistic: identifying differences in brain networks. *NeuroImage*, 2010;53(4):1197–1207. doi:10.1016/j.neuroimage.2010.06.041.

[101] Tononi GO, Sporns O, Edelman GM. A Measure for brain complexity: Relating functional segregation and integration in the nervous system. *Proceedings of the National Academy of Sciences*, 1994;91(11):5033–5037. doi:10.1073/pnas.91.11.5033.

[102] Humphries MD, Gurney K. Network 'small-world-ness': A quantitative method for determining canonical network equivalence. *PLoS ONE*, 2008;3(4):e0002051. DOI:10.1371/journal.pone.0002051.

[103] Telesford QK, Joyce KE, Hayasaka S, Burdette JH, Laurienti PJ. The ubiquity of small-world networks. *Brain Connectivity*, 1(5):367–375. doi:10.1089/brain.2011.0038.

[104] Watts DJ, Strogatz SH. Collective dynamics of 'small-world' networks. *Nature*, 1998;393(6684):440–442. DOI:10.1038/30918.

[105] Bassett DS, Bullmore E. Small-World brain networks. *The Neuroscientist*, 2006;12(6):512–523. DOI:10.1177/1073858406293182.

[106] Clauset A, Rohilla Shalizi Newman MEJ. 2009. Power-Law distributions in empirical data. *SIAM Review*, 2009;51(4):661–703. DOI:10.1214/13-AOAS710.

[107] Barabasi AL. Scale-free networks: A decade and beyond. *Science*, 2009;325(5939): 412–413. DOI:10.1126/science.1173299.

[108] Gastner MT, Ódor G. The topology of large open connectome networks for the human brain. *Scientific Reports*, 2016;6(May): 27249. DOI:10.1038/srep27249.

[109] Newman M, Girvan M. Finding and evaluating community structure in networks. *Physical Review E*, 2004;69(2):1–16. DOI:10.1103/PhysRevE.69.026113.

[110] Newman MEJ. 2006. Modularity and community structure in networks. *Proceedings of the National Academy of Sciences of the United States of America*, 2006;103(23): 8577–8582. DOI:10.1073/pnas.0601602103.

[111] Good BH, De Montjoye YA Clauset A. Performance of modularity maximization in practical contexts. *Physical Review E: Statistical, Nonlinear, and Soft Matter Physics*, 2010;81(4):046106. DOI:10.1103/PhysRevE.81.046106.

[112] Lü L, Pan L, Zhou T, Zhang YC, Stanley HE. Toward link predictability of complex networks. *Proceedings of the National Academy of Sciences*, 2015;112(8):2325–2330. DOI:10.1073/pnas.1424644112.

[113] Zhou S, Mondragón RJ. The rich-club phenomenon in the internet topology. *IEEE Communications Letters*, 2004;8(3):180–182. DOI:10.1109/LCOMM.2004.823426.

[114] Colizza V, Flammini A, Serrano MA, Vespignani A. Detecting rich-club ordering in complex networks. Nature, 2006:2:110–115.

[115] Maslov S, Sneppen K. Specificity and stability in topology of protein networks. *Science*, 2002;296(5569):910–913. DOI:10.1126/science.1065103.

[116] Jiang ZQ, Zhou WX. 2008. Statistical significance of the rich-club phenomenon in complex networks. *New Journal of Physics*, 2008;10. doi:10.1088/1367-2630/10/4/043002.

[117] Heuvel MP van den, Sporns O. Rich-club organization of the human connectome. *The Journal of Neuroscience*, 2016;31(44):15775–1586. DOI:10.1523/JNEUROSCI.3539-11.2011.

[118] Heuvel MP, van den, Kahn RS, Goñi J, et al. High-cost, high-capacity backbone for global brain communication. *Proceedings of the National Academy of Sciences of the United States of America*, 2012l109:11372–1177. DOI:10.1073/pnas.1203593109/-/DCSupplemental.www.pnas.org/cgi/doi/10.1073/pnas.1203593109.

[119] Harriger L, van den Heuvel MP, Sporns O. Rich club organization of macaque cerebral cortex and its role in network communication. *PLoS ONE*, 2012;7(9):e0046497. doi:10.1371/journal.pone.0046497.

[120] Muscoloni A, Cannistraci CV. Rich-clubness test: how to determine whether a complex network has or doesn't have a rich-club? arXive preprint available at arXiv:1704.03526v1 [Physics.soc-Ph]. http://arxiv.org/abs/1704.03526.

[121] Latora V, Marchiori M. Efficient behavior of small-world networks. *Physical Review Letters*, 2001;87(19):198701. dOI:10.1103/PhysRevLett.87.198701.

[122] Bavelas A. Communication patterns in task-oriented groups. *The Journal of the Acoustical Society of America*, 1950;22(6):725–30. DOI:10.1121/1.1906679.

[123] Sabidussi G. The centrality index of a graph. *Psychometrika*, 1966;31(4):581–603. DOI:10.1007/BF02289527.

[124] Opsahl T, Agneessens F, Skvoretz J. Node centrality in weighted networks: Generalizing degree and shortest paths. *Social Networks*, 2010;32(3): 245–251. DOI:10.1016/j.socnet.2010.03.006.

[125] Brandes U. A Faster Algorithm for betweenness centrality. *The Journal of Mathematical Sociology*, 2001;25(2):163–177. DOI:10.1080/0022250X.2001.9990249.

[126] Isik , Baldow C, Cannistraci CV, Schroeder M. Drug target prioritization by perturbed gene expression and network information." *Scientific Reports*, 2015;5:17417. DOI:10.1038/srep17417.

[127] Cannistraci CV, Alanis-Lobato G, Ravasi T. From link-prediction in brain connectomes and protein interactomes to the local-community-paradigm in complex networks. *Scientific Reports*, 2013;3(1613):1–13. DOI:10.1038/srep01613.

[128] Daminelli S, Thomas JM, Durán C, Cannistraci CV. Common neighbours and the local-community-paradigm for topological link prediction in bipartite networks. *New Journal of Physics*, 2015;17(11). DOI:10.1088/1367-2630/17/11/113037.

[129] Durán C, Daminelli S, Thomas JM, et al. Pioneering topological methods for network-based drug–target prediction by exploiting a brain-network self-organization theory. *Briefings in Bioinformatics*, 2017;8(W1):3–62. DOI:10.1093/bib/bbx041.

[130] Bullmore E, Sporns O. Complex brain networks: Graph theoretical analysis of structural and functional systems. *Nature Reviews Neuroscience*, 2009;10(3):186–198. DOI:10.1038/nrn2575.

[131] Bassett DS, Sporns O. Network neuroscience." *Nature Neuroscience*, 2017;20(3):353. DOI:10.1038/nn.4502.

[132] He Y, Evans A. Graph theoretical modeling of brain connectivity. *Current Opinion in Neurology*, 2010;23(4):341–350. DOI:10.1097/WCO.0b013e32833aa567.

[133] Bullmore E, Sporns O. The economy of brain network organization. *Nature Reviews Neuroscience*, 13(5):336–349. DOI:10.1038/nrn3214.

[134] Serrano MÁ, Krioukov D, Boguñá M. Self-similarity of complex networks and hidden metric spaces. *Physical Review Letters*, 2008;100(7):1–4. DOI:10.1103/PhysRevLett.100.078701.

[135] Ye AQ, Ajilore OA, Conte G, et al. The intrinsic geometry of the human brain connectome. *Brain Informatics*, 2015;2(4):197–210. DOi:10.1007/s40708-015-0022-2.

[136] Muscoloni A, Thomas S, Ciucci G, Bianconi G, Cannistraci CV. Machine learning Meets complex networks via coalescent embedding in the hyperbolic space. *Nature Communications*, 2017;8(1):1615. DOI:10.1038/s41467-017-01825-5.

[137] Cannistraci CV, Ravasi FM, Montevecchi Ideker T, Alessio M. Nonlinear dimension reduction and clustering by minimum curvilinearity unfold neuropathic pain and tissue embryological classes." *Bioinformatics*, 2011;27:i531–i539. DOI:10.1093/bioinformatics/btq376.

[138] Cannistraci CV, Alanis-Lobato G, Ravasi T. Minimum curvilinearity to enhance topological prediction of protein interactions by network embedding. *Bioinformatics*, 2013;29:199–209. DOI:10.1093/bioinformatics/btt208.

[139] Krioukov D, Papadopoulos F, Kitsak M, Vahdat A, Boguñá M. Hyperbolic geometry of complex networks. *Physical Review E: Statistical, Nonlinear, and Soft Matter Physics*, 2010;82(3):36106. DOI:10.1103/PhysRevE.82.036106.

[140] Meunier D, Lambiotte R, Bullmore ET. Modular and hierarchically modular organization of brain networks. *Frontiers in Neuroscience*, 2010;4:200. DOI:10.3389/fnins.2010.00200.

[141] Meunier D, Lambiotte R, Fornito A, Ersche K, Bullmore E. Hierarchical modularity in human brain functional networks. *Frontiers in Neuroinformatics*, 2009;3:1–12. doi:10.3389/neuro.11.037.2009.

[142] Meunier D, Stamatakis EA, Tyler LK. Age-related functional reorganization, structural changes, and preserved cognition. *Neurobiology of Aging*, 2014;35(1):42–54. DOI:10.1016/j.neurobiolaging.2013.07.003.

[143] Zhao T, Cao M, Niu H, et al. Age-related changes in the topological organization of the white matter structural connectome across the human

lifespan. *Human Brain Mapping*, 2015;36(10):3777–3792. DOI:10.1002/hbm.22877.

[144] Boersma M, Smit DJA, De Bie HMA, et al. Network analysis of resting state EEG in the developing young brain: Structure comes with maturation. *Human Brain Mapping*,2011;32(3):413–425. DOI:10.1002/hbm.21030.

[145] Schoonheim MM, Hulst HE, Landi D, et al. Gender-related differences in functional connectivity in multiple sclerosis. *Multiple Sclerosis*, 2012;18(2):164–173. doi:10.1177/1352458511422245.

[146] Serrano-Pozo A, Frosch MP, Masliah E, Hyman BT. Neuropathological alterations in Alzheimer disease. *Cold Spring Harbor Perspectives in Medicine*, 2011;1(1):a006189. doi:10.1101/cshperspect.a006189.

[147] Braak H, Braak E. Neuropathological stageing of Alzheimer-related Changes. *Acta Neuropathologica*, 1991;82(4):239–259. doi:10.1007/BF00308809.

[148] Buckner RL, Sepulcre L, Talukdar T, et al. Cortical hubs revealed by intrinsic functional connectivity: Mapping, assessment of stability, and relation to Alzheimer's disease. *Journal of Neuroscience*, 2009;29(6):1860–1873. DOI:10.1523/JNEUROSCI.5062-08.2009.

[149] Crossley, N, Mechelli A, Scott J, et al. The hubs of the human connectome are generally implicated in the anatomy of brain disorders. *Brain*, 2014;137(8):2382–2395. DOI:10.1093/brain/awu132.

[150] Heringa SM., Reijmer YD, Leemans A, et al. Multiple microbleeds are related to cerebral network disruptions in patients with early Alzheimer's disease. *Journal of Alzheimer's Disease*, 2014;38(1):211–221. doi:10.3233/JAD-130542.

[151] Reijmer YD, Leemans A, Caeyenberghs K, et al. Disruption of cerebral networks and cognitive impairment in Alzheimer disease. *Neurology*, 2013;80(15):1370–1377. DOI:10.1212/WNL.0b013e31828c2ee5.

[152] Stam CJ, Jones BF, Nolte G, Breakspear M, Scheltens P. Small-world networks and functional connectivity in Alzheimer's disease. *Cerebral Cortex*, 2007;17(1):92–99. doi:10.1093/cercor/bhj127.

[153] He Y, Chen Z, Evans A. Structural insights into aberrant topological patterns of large-scale cortical networks in Alzheimer's disease." *Journal of Neuroscience*, 2008;28(18):4756–66. doi:10.1523/JNEUROSCI.0141-08.2008.

[154] Supekar K, Menon V, Rubin D, Musen M, Greicius MD. Network analysis of intrinsic functional brain connectivity in Alzheimer's disease. *PLoS Computational Biology*, 2008;4(6):e1000100. doi:10.1371/journal.pcbi.1000100.

[155] Sanz-Arigita EJ, Schoonheim MM, Damoiseaux JS, et al. Loss of 'small-world' networks in Alzheimer's disease: Graph analysis of FMRI resting-state functional connectivity. *PLoS ONE*, 2010;5(11):e13788. DOI:10.1371/journal.pone.0013788.

[156] Dauer W, Przedborski S. Parkinson's disease: Mechanisms and models. *Neuron*, 2003;39(6):889–909. doi:10.1016/S0896-6273(03)00568-3.

[157] Drui G, Carnicella S, Carcenac C, et al. Loss of dopaminergic nigrostriatal neurons accounts for the motivational and affective deficits in Parkinson's disease. *Molecular Psychiatry*, 2014;19(3):358–367. DOI:10.1038/mp.2013.3.

[158] Braak H, Del Tredici K, Rüb U, et al. Staging of brain pathology related to sporadic Parkinson's disease. *Neurobiology of Aging*, 2003;24(2):197–211. doi:10.1016/S0197-4580(02)00065-9.

[159] Jankovic J. Parkinson's disease: Clinical features and diagnosis. *Journal of Neurology, Neurosurgery and Psychiatry*, 2008;79(4):368–376. DOI:10.1136/jnnp.2007.131045.

[160] Schapira AHV, Chaudhuri KR, Jenner P. Non-motor features of Parkinson disease. *Nature Reviews Neuroscience*, 2017;18(7):435–450. DOI:10.1038/nrn.2017.62.

[161] Skidmore FD, Korenkevych D, Liu Y, et al. Connectivity brain networks based on wavelet correlation analysis in Parkinson fMRI data. *Neuroscience Letters*, 2011;499(1):47–51. DOI:10.1016/j.neulet.2011.05.030.

[162] Wei L, Zhang J, Long Z, et al. Reduced topological efficiency in cortical-basal ganglia motor network of Parkinson's disease: A resting state fMRI study. *PLoS ONE*, 2014;9(10). doi:10.1371/journal.pone.0108124.

[163] Baggio HC, Sala-Llonch R, Segura B, et al. Functional brain networks and cognitive deficits in Parkinson's disease. *Human Brain Mapping*, 2014;35(9):4620–4634. DOI:10.1002/hbm.22499.

[164] Agosta F, Canu, E, Stojković et al. The topography of brain damage at different stages of Parkinson's disease." *Human Brain Mapping*, 2013;34(11): 2798–2807. DOI:10.1002/hbm.22101.

[165] Agosta F, Canu E, Stefanova et al. Mild cognitive impairment in Parkinson's disease is associated with a distributed pattern of brain white matter damage. *Human Brain Mapping*, 2014;35(5):1921–1929. DOI:10.1002/hbm.22302.

[166] Baggio HC, B. Segura N, Ibarretxe-Bilbao et al. Structural correlates of facial emotion recognition deficits in Parkinson's disease patients. *Neuropsychologia*, 2012;50(8):2121–228. DOI:10.1016/j.neuropsychologia.2012.05.020.

[167] Hattori T, Orimo S, Aoki S, et al. Cognitive status correlates with white matter alteration in Parkinson's disease. *Human Brain Mapping*, 2012;33(3):727–739. doi:10.1002/hbm.21245.

[168] Qian L, Zhang Y, Zheng L, et al. 2017. Frequency specific brain networks in Parkinson's disease and comorbid depression. *Brain Imaging and Behavior*, 2017;11(1):224–239. doi:10.1007/s11682-016-9514-9.

[169] Olde Dubbelink KTE, Hillebrand A, Stoffers D, et al. Disrupted brain network topology in Parkinson's disease: A longitudinal magnetoencephalography study. *Brain*, 2014;137(1):197–207. doi:10.1093/brain/awt316.

[170] Muscoloni A, Cannistraci CV. Local-ring network automata and the impact of hyperbolic geometry in complex network link-prediction. arXiv preprint available at arXiv:1707.09496 [Physics.soc-Ph]. https://arxiv.org/abs/1707.09496.

13 Cytoscape: A Tool for Analyzing and Visualizing Network Data

John H. Morris

13.1 Introduction

The previous chapters have presented the foundations for the analysis of biological networks, from the specifics of disease prediction and public databases for biological networks to a more general introduction to graph and network theory and various alignment and clustering techniques. This chapter is a little bit of a departure. Rather than discussing specific techniques, this chapter presents a tool, Cytoscape [1, 2], that supports the visualization and analysis of network data, specifically biological networks. One of the key features of this tool is its ability to visualize very complex networks and map a wide variety of data onto that visualization.

The techniques and approaches discussed in previous chapters provide a critical foundation for understanding biological networks, but the capability to visualize the results is equally critical for intuition and understanding as well as scientific communication. The importance of computer-based visualization systems in biology should be unquestioned. We use visualization tools to look at features in genome browsers, images from various microscopy techniques, 3D representations of biological macromolecules, and to create heatmaps, dotplots, Manhattan plots and a wide variety of other visual representations of our data. Interactive visualization tools provide even more capabilities by allow us to explore our data by changing representations, moving things around to improve the clarity, or performing "what-if" analyses that might lead to new directions of inquiry. For networks, as we'll see below, these capabilities allow us to change our representations, map additional data onto our visualization, and change the visual relationships between entities, be they genes, proteins, cells, or organisms.

Cytoscape provides a key set of features to allow the user to explore their network data including the ability to easily augment the network with additional relevant data that can be incorporated into the visualization.

	A	B	C	D	E	F	G	H	I	J	K
1	source	target	interaction	directed	COMMON	gal1RGexp	gal4RGexp	gal80Rexp	gal1RGsig	gal4RGsig	gal80Rsig
2	YGR218W	YGL097W	pp	TRUE	CRM1	-0.018	-0.001	-0.018	6.14E-01	9.79E-01	8.10E-01
3	YGL115W	YGL208W	pp	TRUE	SNF4	-0.111	0.112	-0.221	5.85E-03	1.12E-02	2.75E-03
4	YGR074W	YBR043C	pp	TRUE	SMD1	-0.074	-0.133	0.029	1.19E-01	8.22E-02	8.08E-01
5	YPR113W	YMR043W	pd	TRUE	PIS1	-0.495	0.025	-0.89	1.05E-09	5.46E-01	5.62E-18
6	YLR197W	YDL014W	pp	TRUE	SIK1	0.02	-0.521	0.49	7.02E-01	1.08E-05	3.38E-08
7	YLR197W	YOR310C	pp	TRUE	SIK1	0.02	-0.521	0.49	7.02E-01	1.08E-05	3.38E-08
8	YPR041W	YOR361C	pp	TRUE	TIF5	-0.059	-0.243	-0.177	1.12E-01	6.20E-03	1.17E-02
9	YPR041W	YMR309C	pp	TRUE	TIF5	-0.059	-0.243	-0.177	1.12E-01	6.20E-03	1.17E-02
10	YBR170C	YGR048W	pp	TRUE	NPL4	0.128	-0.134	0.429	7.83E-03	1.35E-01	1.62E-03
11	YML074C	YJL190C	pp	TRUE	NPI46	0.118	-0.38	0.389	2.50E-02	5.30E-05	3.32E-05
12	YER110C	YML007W	pp	TRUE	KAP123	0.05	-0.233	0.43	2.61E-01	2.96E-04	3.63E-07
13	YER111C	YMR043W	pd	TRUE	SWI4	0.195	-0.105	0.16	5.61E-04	3.32E-01	1.56E-01
14	YPR119W	YMR043W	pd	TRUE	CLB2	-0.234	-0.279	-0.342	3.92E-06	6.26E-04	1.64E-05
15	YER112W	YOR167C	pp	TRUE	LSM4	0.193	-0.181	0.151	7.05E-05	3.76E-03	1.37E-01
16	YPR048W	YOR355W	pp	TRUE	TAH18	0.113	-0.191	0.289	1.29E-01	5.46E-01	6.30E-02
17	YPR048W	YDL215C	pp	TRUE	TAH18	0.113	-0.191	0.289	1.29E-01	5.46E-01	6.30E-02
18	YER040W	YPR035W	pd	TRUE	GLN3	0.098	-0.513	0.537	3.84E-01	1.03E-02	1.87E-03
19	YER040W	YGR019W	pd	TRUE	GLN3	0.098	-0.513	0.537	3.84E-01	1.03E-02	1.87E-03
20	YDL030W	YDL013W	pp	TRUE	PRP9	0.244	-0.119	0.339	2.77E-04	2.56E-01	1.18E-02
21	YDL030W	YMR005W	pp	TRUE	PRP9	0.244	-0.119	0.339	2.77E-04	2.56E-01	1.18E-02
22	YCR084C	YBR112C	pp	TRUE	TUP1	0.044	0.704	-0.091	2.86E-01	1.83E-04	6.94E-01
23	YCR084C	YCL067C	pp	TRUE	TUP1	0.044	0.704	-0.091	2.86E-01	1.83E-04	6.94E-01
24	YJL203W	YOL136C	pp	TRUE	PRP21	0.083	-0.46	0.401	1.38E-01	6.54E-04	4.69E-03
25	YER116C	YDL013W	pp	TRUE	YER116C	0.029	-0.11	0.182	5.61E-01	1.24E-01	1.39E-01
26	YOR355W	YNL091W	pp	TRUE	GDS1	-0.176	-0.044	0.43	1.66E-04	4.58E-01	6.27E-03
27	YCR086W	YOR264W	pp	TRUE	YCR086W	-0.081	-0.397	0.381	3.21E-01	8.05E-02	1.72E-01
28	YEL015W	YML064C	pp	TRUE	YEL015W	0.222	-0.171	0.275	3.07E-05	5.29E-03	3.28E-02
29	YLL021W	YLR362W	pp	TRUE	SPA2	-0.155	0.05	-0.036	3.40E-04	2.30E-01	6.22E-01
30	YBL079W	YDL088C	pp	TRUE	NUP170	-0.186	-0.032	-0.42	2.57E-04	4.57E-01	2.95E-09
31	YDR354W	YEL009C	pd	TRUE	TRP4	-0.122	-0.202	-0.253	9.03E-03	2.82E-03	1.21E-03
32	YGR009C	YAL030W	pp	TRUE	SEC9	0.236	-0.185	0.302	2.96E-05	1.74E-02	4.41E-04
33	YGR009C	YOR327C	pp	TRUE	SEC9	0.236	-0.185	0.302	2.96E-05	1.74E-02	4.41E-04

Figure 13.1: Excel spreadsheet showing results from an experiment that combines expression data with a yeast two-hybrid experiment.

Consider two representations of the same network. In Figure 13.1, we show a small portion of an Excel spreadsheet that combines the results from expression data and yeast two-hybrid experiments [3]. Each line contains information about a particular gene, its interacting gene, and information about that gene. In Figure 13.2, we show the exact same dataset rendered as a network, with the expression results for one condition shown as the node fill colors and the node size scaled by the number of neighbors. It is easy to see how the representation shown in Figure 13.2 is much more intuitive to understand, with hub regions and large expression fold changes jumping out at the viewer. This simple example represents the power of network visualization.

Cytoscape is an open-source environment for visualizing and analyzing complex networks and augmenting those networks through the integration of tabular data. Cytoscape's core capabilities include network visualization and the ability to integrate additional data. It also includes a limited set of tools for network analysis. Additional functionality is provided through extensions known as apps. The Cytoscape app store currently contains over 200 apps that provide new analyses, visualization extensions, and integration with additional resources. Cytoscape is a desktop application and is developed by a consortium of university labs. A related project is Cytoscape.js, which is a Web-based visualization tool for rendering networks. Cytoscape provides tools to export networks from the desktop application suitable for visualization on the Web.

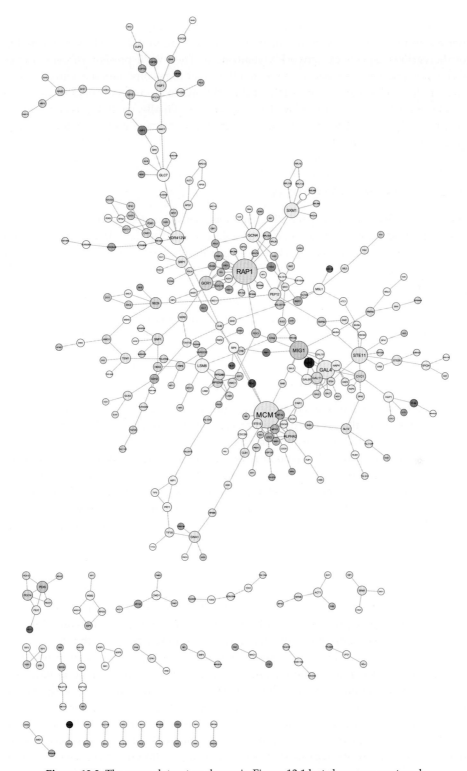

Figure 13.2 The same dataset as shown in Figure 13.1 but shown as a network.

The rest of this chapter will provide a brief taxonomy of networks useful for understanding some aspects of appropriate visualization, followed by a discussion of the various aspects of network visualization. Then we'll present an overview of Cytoscape and some simple exercises to allow you to become familiar with the software, followed by two example biological workflows to more deeply explore the capabilities of the Cytoscape core as well as several apps. Finally, we'll discuss a couple of advanced topics, such as scripting and integrating Cytoscape with other applications such as R [4], or Python [5].

13.2 Network Taxonomy

As you have learned in the preceding chapters, networks are useful ways to express and, combined with graph theoretical techniques, analyze the relationships between items, whether those items are proteins, cells, or individuals. When we start looking at how to visualize networks, it's valuable to think about the *type* of network since different types of networks might have different visualizations that are most appropriate. To help provide a way to talk about these differences, this section will offer three types of networks commonly seen in biological applications: interaction networks, pathways, and similarity networks.

13.2.1 Interaction Networks

Most of the networks discussed in the previous chapters are interaction networks. In biology, these typically represent the interaction between biological macromolecules where interaction might be some directly (e.g., physical) interaction or an indirect (e.g., genetic) interaction. Interaction networks aren't necessarily limited to macromolecules. Cell–cell interaction networks represent a new area of interest, and there are published images where the interactions are between sections of the brain, for example (see Chapter 8). It is still the case, however, that the most common interaction networks we think of are protein–protein interaction (PPI) networks. These networks represent the physical (i.e., binding) interactions between proteins or, in some cases, the genetic interactions.

Typically, in an interaction network such as the yeast protein–protein interaction network shown in Figure 13.3 [5], shows proteins as nodes and the interactions between them as edges. Depending on the nature of the experimental technique, the edges might have weights. For example, in genetic interaction networks, the edges are weighted by the strength of the interaction (as measured by cell growth rate).

There are many examples of interaction networks. Arguably, a map of all of the Internet connections within a company or university is an interaction network. Another very common example is the networks of friends or contacts on social media (see Figure 13.4). This network shows a portion of the Twitter network for Caleb Jones [6], who writes a regular blog on graphs and networks. In this case, the nodes are Twitter handles, which could be individuals, companies, or interest groups. Note, that just as in the protein–protein interaction network shown above, areas of dense connections are readily visible.

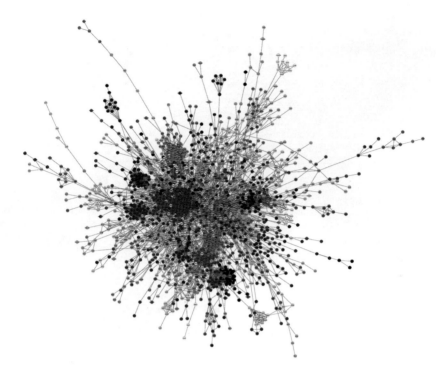

Figure 13.3: Protein–protein interaction network. Nodes represent proteins and edges represent interactions between those proteins.

13.2.2 Pathways

The second type of biological network are similar (and sometimes confused) with the interaction networks described above. Figure 13.5, from Gao et al. 2014 [7], shows some of the genes in cancer-related pathways. There are several differences between Figure 13.5 and either Figure 13.2 or Figure 13.3. Note that Figure 13.5 is much more stylized, and as biological networks go, less dense. If we look at just one of the proteins in this pathway, EGFR, it shows that it interacts with GRB2. However, looking at the STRING ([8, 9]; http://string-db.org) interaction network for EGFR (Figure 13.6) we can see that it interacts with a large number of proteins, including HRAS, which is a downstream target for GRB2. Including all of this detail in the diagram in Figure 13.5 would have obscured the main point the authors were trying to make, which was to highlight the genes that were differentially expressed and those that had a differentially methylated region.

Pathway diagrams are just that – diagrams. They intentionally ignore the complexity of all possible interactions to focus on those that are relevant to a particular cellular process. The primary use of these diagrams is didactic, but they do serve an important purpose in the analysis of many experiments. A common experiment might look at the impact of perturbation on a known pathway, perhaps measuring changes

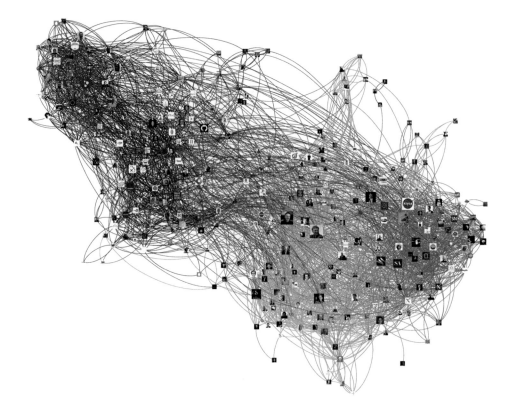

Figure 13.4: Twitter network for Caleb Jones, showing interaction between Twitter handles (http://allthingsgraphed.com/2014/11/02/twitter-friends-network/).

in expression as a result. The fold change expression can be mapped onto a pathway diagram to show the impact on the pathway as a result of the perturbation.

One of the best pathway diagram analogies is the famous London underground map (Figure 13.7). This map is extremely useful for navigating the London underground, but is an abstraction that doesn't attempt to be geographically correct, nor does it attempt to show all transit options in London.

13.2.3 Similarity Networks

The final type of networks that we'll discuss are similarity networks. These networks are somewhat different than interaction networks in that the edges don't represent interactions, but the degree of similarity (or difference). As an example, in a sequence similarity network (SSN; [10]) as shown in Figure 13.8, the edges represent the BLAST [11] similarity between the nodes which are the individual proteins. The colors in the figure show the protein families, which the gray nodes representing proteins that do not yet have any functional annotation. This particular type of similarity network has been gaining increasing use to suggest functional annotations and a number of resources are available to generate them.

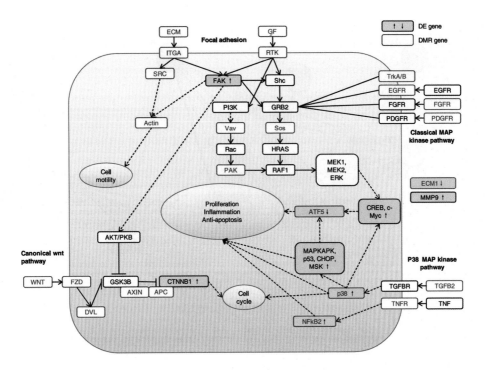

Figure 13.5: Pathway diagram showing genes in selected cancer-related pathways. (Copyright © Gao et al. [7], licensed under CC BY 4.0)

Sequence similarity networks aren't the only biological use case for similarity networks, however. Another common use case are correlation networks, in which the edges represent the correlation between attributes of the nodes. For example, the edges could represent the expression correlation between two genes [12, 13], or as another example, the network might contain only significantly differing genes in group comparison study [14]. This provides an alternative view to the traditional heat map visualization for expression data. Other use cases include chemical similarity networks, which measure the similarity between two compounds, usually based on a Jaccard [15], or Tanimoto [16] coefficient (Figure 13.9).

The most common non-biological example for similarity networks are probably the various recommender systems such as Pandora that attempt to recommend music based on a measure of the "similarity" between two artists or tracks. Similar systems are used to recommend books, movies, or used to group similar pictures together in a personal library.

13.3 Visualizing Networks

There are several ways to visualize networks. It's important to think about the purpose of the visualization before settling on the right style. The right visualization for data exploration might not be right for communicating experimental results. In this section,

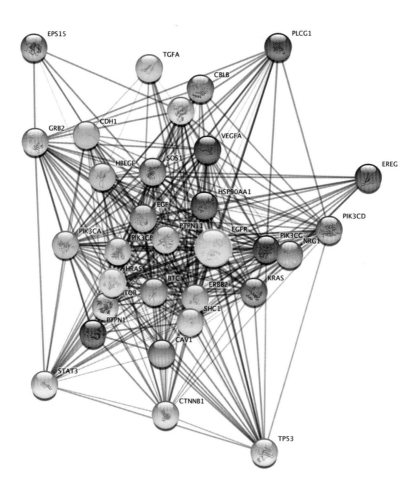

Figure 13.6: STRING network created in Cytoscape from the stringApp. EGFR (slightly larger circle) was the query term and the number of additional proteins was set to 30. The two proteins in yellow, EGFR and HRAS are directly connected by the highlighted (magenta) edge.

we'll talk about a couple of them, and how they might be used, but the focus will be on the traditional node-link visualization. We'll then discuss some of the visual properties of node-link visualizations, types of mappings to map node and edge annotations onto visual properties, layouts, animation, and finally what you might be able to achieve by layering visualizations.

13.3.1 Visual Representations of Networks

The most common representation for a network in biology today is shown in Figure 13.10, which is the source data from Collins et al. [5]. It's clear that there are many

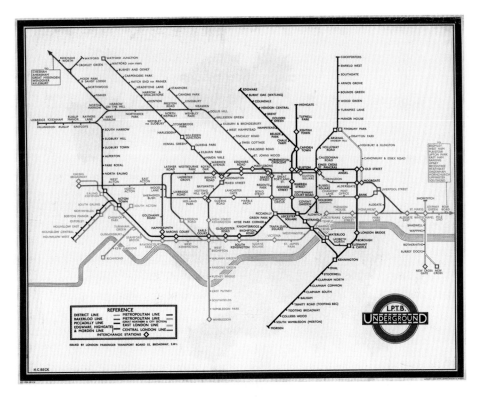

Figure 13.7: Harry Beck's London underground map from 1933. Beck's design is credited as the first diagrammatic (as opposed to geographic) Tube map. Image from https://www.pinterest.co.uk/pin/331436853806951661/.

unsatisfactory aspects to this visualization, including the inability to see distant interactions, clusters, etc. On the other hand, spreadsheets are tools many biologists are comfortable with and often resort to. One of the goals of this chapter is to provide alternatives to this "visualization of last resort."

Mathematically, a graph or network can be expressed as a matrix. An undirected graph is a matrix with only one section below or above the diagonal filled in, and a directed graph has both filled in (note that this is violated when the graph is a hypergraph or a multigraph). Looked at in this way, it's easy to imagine a number of ways this could be visualized. One obvious approach would be to simply visualize the matrix itself as a heat map. If you cluster the heat map using hierarchical clustering, you can get an image of a network that looks like Figure 13.11. There are many variations on this approach, but generally these visualizations are referred to as adjacency matrix visualizations. In some cases, the actual values are put in the cells, but usually some sort of color gradient is used to show areas of network density.

Another format useful for particularly dense networks is BioFabric [17], so called because of its woven appearance. BioFabric organizes entities (e.g., genes or proteins) as lines and the relationships between them are crossings of those lines. The BioFabric representation of the network shown above is shown below as Figure 13.12.

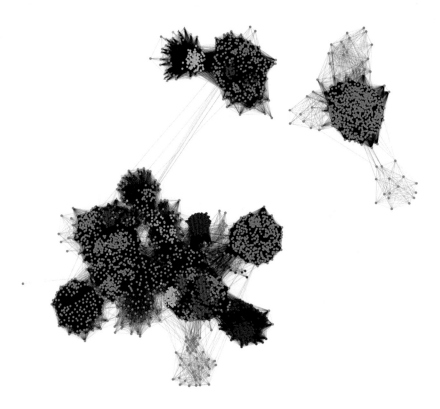

Figure 13.8: Sequence similarity network showing a portion of the amidohydrolase superfamily. Nodes represent proteins and edges represent the BLAST similarity between the protein sequences. Nodes are colored according to families with known function.

The most common representation (other than large Excel spreadsheets) are collectively referred to as node-link diagrams. In these representations, the nodes represent the biological entities and the edges between them represent some sort of relationship (e.g., similarity or interaction). The rest of this section will talk about node-link diagrams and ways to map additional data to enrich those diagrams.

13.3.2 Node-Link Diagrams

Node-link diagrams are what most people think of when they think of networks. As you've learned in the preceding chapters, a graph is defined by a set of vertices (nodes) and edges:

$$G = (V, E). \tag{13.1}$$

Typically, graphs are defined such that an edge connects exactly two nodes, and there can be at most one edge between two nodes. There are two extensions to graphs

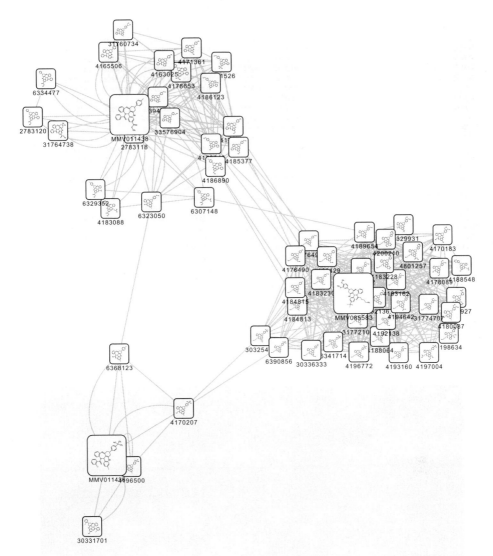

Figure 13.9: Chemical similarity network showing part of a malaria drug screen. Nodes are compounds and edges represent the Tanimoto similarity. The larger nodes with the red outline are compounds that had positive assay results.

however that are useful. In a *hypergraph*, an edge can connect more than two nodes (Figure 13.13). In a *multigraph*, two nodes may be connected by more than one edge (Figure 13.14). Hypergraphs are useful to represent complex relationships such as reactions, which involve multiple inputs (reactants), multiple outputs (products), and possibly modifiers (catalysts). Multigraphs are useful to represent graphs where there are different types of relationships between two nodes. For example, two nodes might represent proteins that are part of the same complex (member relationship) and one protein might also serve to activate the second complex (activates relationship).

# Feature1 Sys. Name	Feature1 Std. Name	Feature2 Sys	Feature2 Std	PE Score	Conf. Score	Direct Score	Indirect Score
Q0010		YAR073W	IMD1	0.368459	0.004676	-0.309804	1.046722
Q0010		YAR075W		0.684409	0.013221	0	1.368818
Q0010		YBL075C	SSA3	0.288681	0.003986	-0.309804	0.887167
Q0010		YBL093C	ROX3	0.430655	0.005211	-0.309804	1.171113
Q0010		YBR011C	IPP1	0.254088	0.003745	-0.619608	1.127784
Q0010		YBR229C	ROT2	0.476909	0.00545	-0.309804	1.263621
Q0010		YBR237W	PRP5	0.677037	0.012592	0	1.354073
Q0010		YCR035C	RRP43	0.404932	0.004991	-0.309804	1.119668
Q0010		YCR063W	BUD31	0.700005	0.014586	0	1.40001
Q0010		YDL192W	ARF1	0.745137	0.019039	0	1.490275
Q0010		YDL209C	CWC2	0.64373	0.009828	0	1.287461
Q0010		YDR458C		0.471205	0.005427	-0.309804	1.252213
Q0010		YEL034W	HYP2	0.510525	0.005574	0	1.021051
Q0010		YEL042W	GDA1	0.716889	0.016139	0	1.433778
Q0010		YER001W	MNN1	0.669922	0.011968	0	1.339845
Q0010		YER066W		0.708273	0.015348	0	1.416547
Q0010		YFR015C	GSY1	0.413135	0.005063	-0.309804	1.136073
Q0010		YGL008C	PMA1	0.590193	0.005998	0	1.180386
Q0010		YGL103W	RPL28	0.348391	0.004511	0	0.696783
Q0010		YGL116W	CDC20	0.64373	0.009828	0	1.287461
Q0010		YGR089W	NNF2	0.677037	0.012592	0	1.354073
Q0010		YGR224W	AZR1	0.745137	0.019039	0	1.490275
Q0010		YHL001W	RPL14B	0.396616	0.004917	0	0.793232
Q0010		YHR098C	SFB3	0.716889	0.016139	0	1.433778
Q0010		YHR158C	KEL1	0.335078	0.004395	-0.309804	0.979959
Q0010		YHR174W	ENO2	0.048993	0.002803	-0.309804	0.407789
Q0010		YHR216W	IMD2	0.663049	0.011426	0	1.326098
Q0010		YIL129C	TAO3	0.551964	0.005692	0	1.103929
Q0010		YJL052W	TDH1	0.324283	0.004301	-0.309804	0.958369

Figure 13.10: Portion of a spreadsheet showing data from a combination of yeast interactome data (Collins et al. [5]).

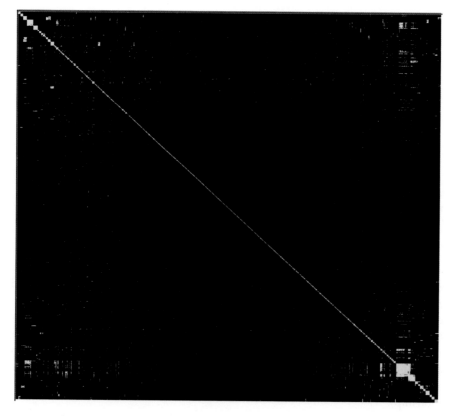

Figure 13.11: Clustered adjacency matrix representation of the Collins yeast interactome data [5].

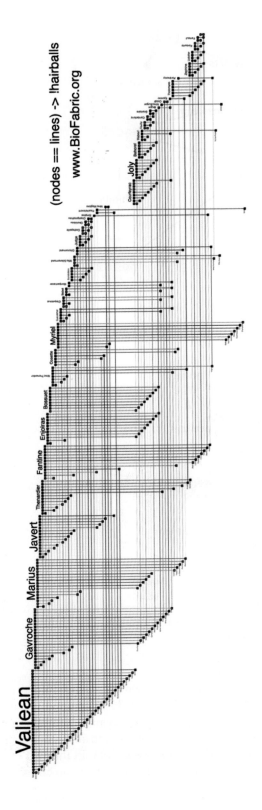

Figure 13.12: An example BioFabric network from www.BioFabric.org. Note that entities (nodes) in this network are represented by horizontal lines and relationships (edges) are represented by vertical lines that terminate on the related entities.

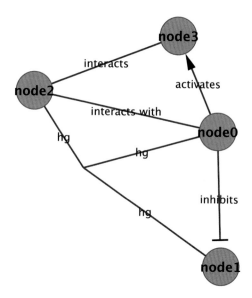

Figure 13.13: An example of a hypergraph. Note that node1, node2, and node0 are all connected by a single edge.

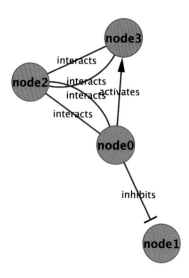

Figure 13.14: An example of a multigraph. Note that there are multiple edges between node0 and node2 and between node2 and node3.

In a node-link diagram, the nodes are typically represented by some form of shape and are connected by one or more lines. Figures 13.3, 13.4, 13.6, 13.8, and 13.9 are all node-link diagrams. As you can see by those figures, by representing graphs this way, we can enhance the visual representation in a number of ways to augment the graph to highlight specific features or visualize additional related data. In the next several sections, we'll discuss some of the ways these visual representations can be used.

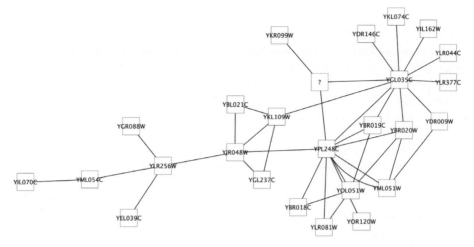

Figure 13.15 Portion of a yeast protein–protein interaction network.

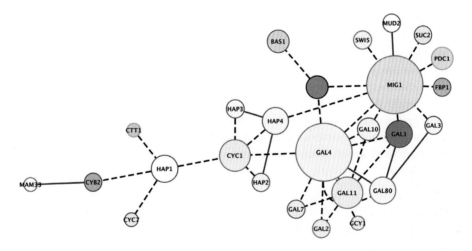

Figure 13.16: Same network as shown in Figure 13.15 after applying visual styles to reflect network attributes and additional experimental data.

13.3.2.1 Visual Properties

There are many aspects of a graph or network visualization that we can manipulate to highlight the graph topology or integrate additional data. Some visual attributes include:

- node fill color, border color, border width, size, shape, opacity, label;
- edge type, color, width, ending type, ending size, ending color.

Figures 13.15 and 13.16 show the same graph (a portion of the network shown in Figure 13.2). In Figure 13.16, we've used the node size to convey the number of neighbors a node has, the color of node to show the fold change of one expression value and the border color to show the fold change of another expression value. The edge shape (dashed vs. solid) and color show the type of the interaction.

While it's certainly the case that one can easily find the hubs (high degree nodes) in Figure 13.15, by changing the node size in Figure 13.16, we are able to significantly enhance that aspect of the visualization, drawing the viewer's attention to the hubs as an important feature. Note that by bringing in additional data, we are able to map several different pieces of biologically relevant information onto our graph, noting for example that gal10, gal1, and gal7 all exhibit large expression fold change values in opposite directions in each of the conditions.

13.3.2.2 Visual Mappings

Some of these visual attributes lend themselves well to continuous values such as the number of neighbors a node has or the number of mutations seen in various cases of a particular cancer. Some are more amenable to discrete values such as the type of protein or where the protein is localized in the cell. We typically think of three types of visual mappings:

- **Passthrough**: the value is directly mapped onto the visual attribute. For example, we could directly map degree onto the size of the node, however, passthrough mappings are typically used for labels, where the text value is utilized as the node or edge label.
- **Discrete**: each value is mapped to a specific visual representation. For example, if we have a value that contains the type of molecule (small molecule, protein, RNA, DNA, etc.), we could map each of those types to a specific shape. Similarly, if we had a value that contains the cellular compartment of the protein, we could map each compartment to a color.
- **Continuous**: a numeric value is mapped onto a visual attribute by providing a ramp or gradient. Mapping expression fold change values into a blue-red gradient, for example.

When choosing mappings, it's important to think about whether the specific mapping is appropriate for the data. For example, a value with 200 discrete values (e.g., GO terms), may be mapped into discrete colors, but it's very difficult for humans to discern differences between that many colors. For an excellent reference on appropriate mappings for visualization, see Tamara Munzner's excellent book: *Visualization Analysis and Design* [18].

Given the number of visual attributes discussed in the section above, and the number of ways values can be mapped into them, it's easy to see that it's possible to integrate a very large number of data values into a single visualization. Easy, but definitely not recommended! Visualizations that provide too much information become cluttered and any attempt to inform the viewer of the important aspects of the biology can be lost. Often it's better to think about providing multiple views that highlight a smaller number of meaningful differences than trying to fit everything into a single visualization.

13.3.2.3 Layouts

A layout is a positioning of the nodes (and possibly edge routing) in a graph visualization. There are many types of layouts and the choice of an appropriate layout algorithm and appropriate parameters can have a significant impact on the clarity

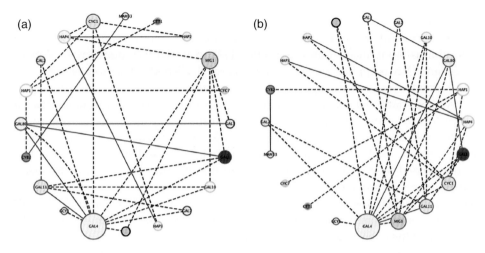

Figure 13.17: Two examples of circle (or radial) layouts: (a) Circular layout with nodes order somewhat random. (b) Circular layout with nodes ordered by degree.

and perceived meaning of the network visualization. Several layout algorithms are discussed below, but this is definitely *not* an exhaustive list, and new algorithms are constantly being developed.

Simple Layout Algorithms The simplest layout algorithm is probably a grid layout, which just organizes all of the nodes in a grid – typically randomly. The major advantage of a grid layout is that it is computationally very inexpensive. Grid layouts should only really be used when there is no relationships (e.g., edges) between nodes that are of importance and the relative positioning of the nodes doesn't have any particular meaning.

An obvious extension to the simple grid layout is to layout all of the nodes around a circle (Figure 13.17 (a)). This can be useful in relatively sparse networks to view the edges between the nodes but doesn't scale well for dense networks. One useful extension to circle layouts is the ability to sort the nodes in some order based on an associated topological measure or data value associated with the biological entity represented by the node (Figure 13.17 (b)).

Another simple layout is a tree or hierarchical layout, which is obviously useful for hierarchical data that is generally acyclic (has no loops), but should be avoided when the data isn't hierarchical, even if the graph suggests a tree-like structure. Remember that the purpose of the layout is to suggest to the viewer the relationship (or lack of one) between the nodes. If the layout algorithm chosen suggests a hierarchical relationship, but there isn't one, this could mislead the viewer.

Force Directed Layouts The most common layouts for node-link diagrams tend to be force-directed layouts. There are several force-directed algorithms, but in general they are all based on the same principle: The nodes are modeled as objects and the edges are modeled as sprints. For example, in the Fruchterman–Reingold algorithm [19], edges have both attractive forces when they are larger than some value and repulsive forces when they are smaller than some value. The value can be based on a set value or vary depending on a weight associated with the edge. Another, similar well-known algorithm is the Kamada–Kawai algorithm [20], in which the edges have a default length and force is applied when the edge exceeds that length. Other implementations add a gravity force that attempts to avoid node overlaps.

An important issue often overlooked by users of network visualization software is that for edge-weighted networks in Euclidean space, any force-directed algorithm is a 2D projection from a possibly larger dimensional space. This can be seen by imagining how one might connect four nodes with edges of fixed, but different lengths. In the general four node case, one can show that this can only be done by using three dimensions space (the number of dimensions is the number of nodes-1). This means that similarly weighted edges won't necessarily have similar lengths. Overall it should approximately average out, but it can be misleading to viewers who might assume that the edge lengths directly imply edge weights. When this is important, it is recommended that some other visual attribute be applied, such as an edge color gradient or edge width.

Network Embedding Another approach to visualizing networks is to take advantage of graph embedding. Simply put, graph embedding transforms the graph, generally to reduce the dimensionality. Examples of graph embedding techniques include PCA, MDS [21], t-SNE [22], Isomap [22], and coalescent embedding [22].

Combining Layouts When creating a visualization and deciding on the positioning of the nodes and edges, it's often the case that you may want to use different layouts for different parts of the network. One example is Figure 13.18 from Yu et al. [23], figure 6, which shows the reconstructed protein interactions and networks from a urinary pellet sample of a patient with a urinary tract infection. In this case, for aesthetic purposes, some of the groups have been laid out using a circular layout and some have used a grid.

Choosing the Layout The purpose of a layout is to convey the relationships (or lack thereof) between the nodes and do so in such a way as to highlight or enhance other visual or topological properties. Two aspects of a layout should be considered when choosing the layout for a network:

- Does the layout highlight the appropriate relationship between the nodes?
- Is the layout aesthetically pleasing?

The importance of the first question is obvious, but the second question is also important. A layout should not appear cluttered (to the extent possible) and should provide a useful global overview of the node relationships. As was discussed at the beginning of this chapter, part of the power of network visualization is the viewer's intuition that can be achieved by looking at the network. A cluttered image, or one in which the most meaningful relationships are obscured can hamper that intuition. Most layout algorithms have several parameters that may be tuned to enhance the positioning of the nodes and/or the routing of the edges. Taking a little time to tune the layout can significantly enhance the resulting visualization.

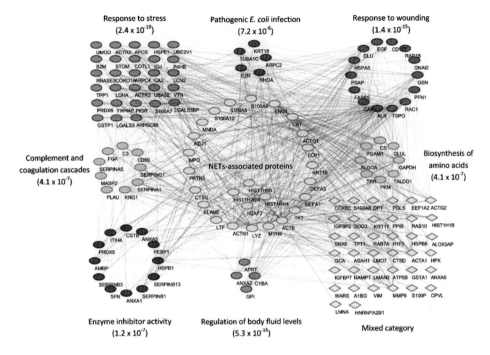

Figure 13.18: Figure showing effective combination of layouts. In this case, this layout uses both circular and grid layouts to orient the nodes. (From Yu et al. [23], figure 6. Used in accordance with the terms of use (http://ivyspring.com/terms).

Finally, it's important to keep in mind that there is no *correct* layout. Trying different layouts and parameters can be useful to help understand the inherent relationships and decide which algorithm highlights the important relationships. Also, as mentioned above, it's also possible to combine layouts by applying a different layout to specific regions of the network.

13.3.2.4 Animation
Many biological networks represent a snapshot in time, or encode several different states depending on various conditions. One way to show the changes in a network over time, or step through different states is by creating an animation that interpolates between the different network. Such interpolations can include fading nodes and edges in and out as network topology changes, moving nodes, interpolating node size and colors, and basically transitioning pretty much any node and edge visual property. Such animations can be used to provide viewers with intuition about changes between network states and to highlight specific dynamic processes.

13.3.2.5 Layering Visualizations
The last aspect of network visualization that we'll mention is the ability to layer other visualizations on top of a network visualization. Typically, this means adding small charts or images onto nodes that show additional information about that node. For example, STRING [8, 9], shows protein structure images on proteins and STITCH

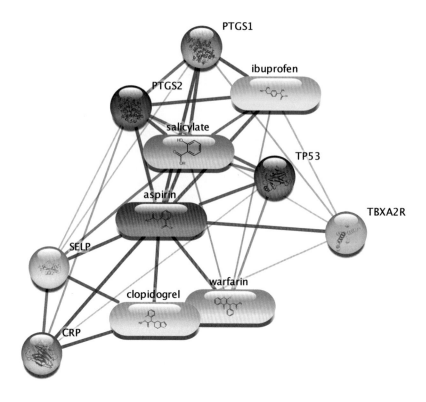

Figure 13.19: STITCH network showing interactions for aspirin. Protein nodes have images depicting protein structures, and small molecules show a 2D depiction of the structure.

[24, 25] shows compound diagrams on small molecules (Figure 13.19). Other examples include showing a bar chart or a small heatmap of expression fold changes on nodes, or even microscope images of cells.

13.4 Getting Started with Cytoscape

13.4.1 Cytoscape in a Nutshell

Cytoscape is an open-source desktop application for the analysis and visualization of networks, and the integration of information into those networks. Cytoscape development is managed by a consortium of institutions, including the University of California at San Diego, the University of California at San Francisco, the Gladstone Institutes, the University of Toronto, the Institute for Systems Biology, the Institute Pasteur, and the National Resource for Network Biology. Cytoscape is a cross-platform application that runs on MacOS, Windows PCs, and Linux desktops. At this point, it does not run on Android or iOS devices. While the focus of Cytoscape is on biological networks and related information, nothing in the core application limits the use of Cytoscape for other network-based visualization and analysis uses.

Cytoscape provides two different data models: A network model and a table model. The network model supports multigraphs, but not hypergraphs, and networks may be hierarchical (i.e., a node may either *contain* or *point to* a network). In general terms, a network just contains information about the topology of the nodes and edges.

The second data model is the table model. The easiest way to think of Cytoscape tables is as a set of spreadsheets. The tables typically contain information about the network, nodes, and edges, but may be unattached to any network objects. For attached tables, the rows of the table are indexed by the internally generated unique identifier of the network, node, or edge that the information refers to. This allows Cytoscape to quickly associate the network objects with associated information and to associate that information with those network objects.

One of the central strengths of Cytoscape is that these two models are kept relatively separate. Cytoscape puts no a priori requirements on the tables to support the network topology or visualization. By default, there are several columns that are created, but additional columns may be imported and merged in from a variety of external data sources.

For network visualization, Cytoscape provides an extensible set of *visual styles* that are used to set default visual properties and to map from data in the tables into the network visual properties. Cytoscape's visual styles include the ability to have all three of the mappings discussed above: Passthrough, discrete, and continuous. The separation between the network model and the table model, and the use of visual styles to control the presentation has been one of the core concepts of Cytoscape since its early inception. This allows other data types to be imported as the underlying science changes, or the use case falls outside the realm of Cytoscape's typical target audience.

Another core concept of Cytoscape is the division of functionality into *core* and *apps*. Cytoscape's core functionality includes the ability to render networks, perform certain layouts, import networks and tables, and perform some limited analysis. However, the Cytoscape App Store [26], provides hundreds of apps that extend Cytoscape by adding new visualization, data import, or analysis options. Users can access the app store via the web (http://apps.cytoscape.org) and if Cytoscape is running, can browse for new apps and install them directly into the running Cytoscape. The app store is also available from within Cytoscape through the Apps→App Manager menu.

13.4.2 The Cytoscape User Interface

As mentioned in the previous section, Cytoscape is a desktop application. The user interface (UI) for Cytoscape (Figure 13.20) is divided up into distinct panels: the **Control Panel** (Figure 13.20 (a)), the **Network Panel** (13.20 (b)), the **Results Panel** (13.20 (c)), and the **Table Panel** (Figure 13.20 (d)). Except for the Network Panel, all of the other panels use a tabbed interface. The **Control Panel** and the **Table Panel** both have a default set of tabs that will appear whenever a network is loaded. The **Results Panel** is hidden by default and will only appear when an App displays data in the panel. In Figure 13.20, the stringApp is showing information about TP53, which is highlighted in the **Network Panel**. As with most desktop applications, there is a menu bar at the top and a small tool bar underneath that. At the bottom of the window are buttons

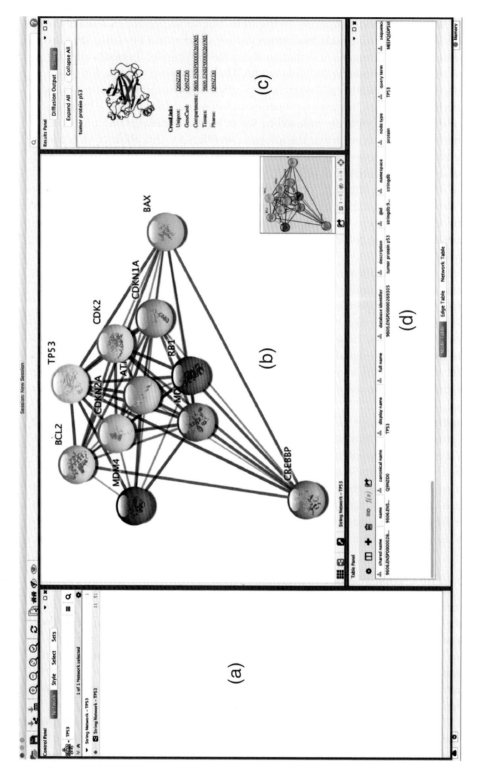

Figure 13.20: Basic Cytoscape user interface showing the four main panels – (a) Control Panel, (b) Network Panel, (c) Results Panel, and (d) Table Panel.

that activate three features: At the far right, the memory button allows users to see how much memory is in use and activate the Java garbage collection feature; at the far left is the jobs button to see what remote jobs are currently running; and next to that is the log button to see the task logger.

13.4.2.1 Control Panel

The **Control Panel** by default has three tabs: **Network**, **Style**, and **Select**. The **Network** tab allows users to search for networks in, or create networks from public databases (top Search bar); see information about networks (middle); and create networks from files by dropping an Excel, csv, sif, or xgmml file onto it.

The **Style** tab is where users can manage the visual style of the network, including the ability to define a default for a visual property, construct mappings from columns in the data panel, or set a visual property for a node or set of nodes. The list of available styles, including a thumbnail of the style, is available in the pull down at the top of the tab. Next to pulldown for available styles is a menu that supports deleting, duplicating, or creating new styles. Underneath the styles pulldown is a **Properties** pulldown, which contains the full list of visual properties that may be added to the style. Note that by default, only a limited set of visual properties are displayed. Finally, the rest of the space is occupied by the visual properties themselves. Note that this tab supports tabs within it for the visual properties of nodes, edges, and networks.

The final default tab is the **Select** tab, which is used to construct filters that will select nodes and edges based on values in associated table columns, node degree, or topology. Filters can be combined using AND and OR operators to create complex selections based on a combination of columns as well as other filter types. For example, Figure 13.21 shows a filter set that selected nodes where there is significant overexpression ($p < 0.05$, fold > 1) or significant underexpression ($p < 0.05$, fold < -1) in two different conditions.

Apps are also able to add tabs to the Control panel. In Figure 13.20, there is a **Sets** tab that is provided by the setsApp [25], and allows users to manipulate sets of nodes and edges.

13.4.2.2 Network Panel

The **Network Panel** is where Cytoscape displays network views, and optionally, an overview pane that always shows the entire network (see Figure 13.22 (a)). The networks can be zoomed using the mouse wheel or the plus (Q) and minus (Q) buttons in the tool bar. To show the entire network, you can use the Q button and to focus the view on only the selected nodes or edges, use the Q button. The network view may be repositioned using the mouse by pressing down on mouse button 1 (usually the left button) and dragging. The view may also be repositioned by dragging the light blue rectangle in the overview window.

Nodes and edges can be selected by clicking on a node or edge with the mouse, or by holding down the shift key and dragging the mouse using mouse button 1. To deselect all nodes and edges, simply click on an empty area in the network.

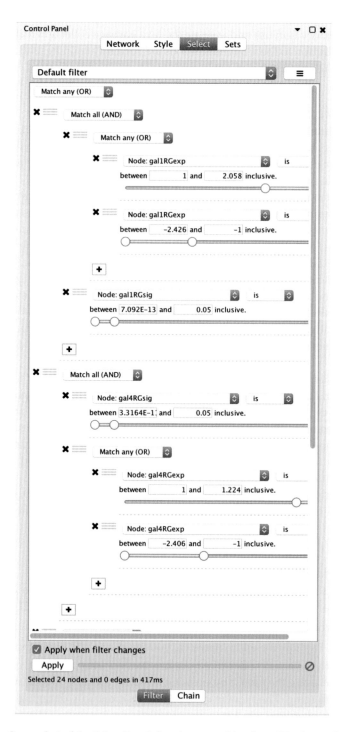

Figure 13.21: Screenshot of the Select Panel showing a combination of Boolean selection parameters.

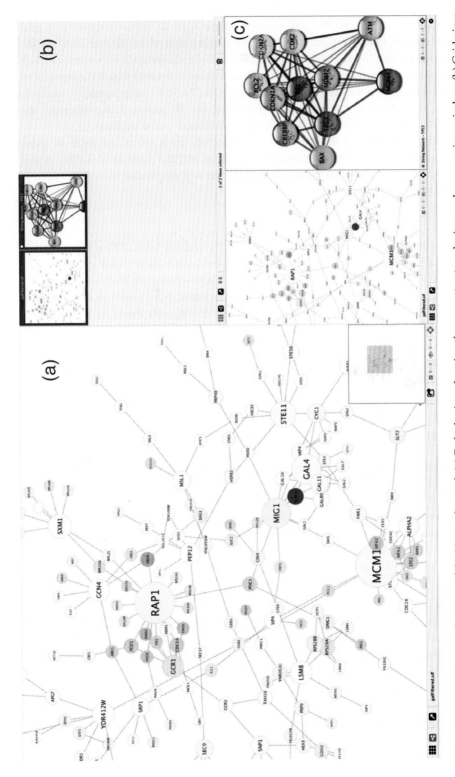

Figure 13.22: Three example views of the Network panel. (a) Default view showing the current network view and an overview window. (b) Grid view showing all of the network views. (c) Comparison view showing multiple network views.

At the bottom of the network panel are a series of controls for network view features. The grid button (▦) will show thumbnails of all network views in a grid in the **Network Panel** (Figure 13.22 (b)). To display that view, double click on the thumbnail or select the show view button (◳). If multiple network views are selected, and the show view icon is selected, a comparison view is created which displays all of the network views in the **Network Panel** (Figure 13.22 (c)). To support use on multiple displays, the user may also detach a view using the ◰ button. This will create the views in independent windows that can be repositioned on the desktop.

On the right side of the network view controls are a set of information icons that show the number of selected nodes and edges and the number of hidden nodes and edges. Finally, at the far right is a button that will display or hide the overview window (✧).

13.4.2.3 Results Panel

The **Results Panel** is almost exclusively used by various apps. Often this will display the results from various calculations, or more information about specific nodes or edges. For example, the Network Analyzer (Tools→NetworkAnalyzer→Network Analysis) will show all of its results in the **Results Panel**, but by default it's detached. Pressing the thumbtack icon will put it back into the **Results Panel**. In fact, this is a feature that works for all of the panels except for the **Network Panel**, which has its own detach mechanism. Other examples of apps that use the results panel include stringApp, chemViz2, and CyBrowser.

13.4.2.4 Table Panel

The final panel is the **Table Panel**, which provides access to all of the tabular data. As with all of the other panels (except the **Network Panel**), the **Table Panel** uses a tabbed interface. By default, the Node Table, Edge Table, and Network Table are shown. Additional tables are shown in separate tabs, and apps will also add information to this panel. The default tables may all be sorted by clicking on a column header, and a toolbar at the top of the panel allows users to change the table mode (how the rows interaction with network selection), change the columns that are shown or hidden, add new columns, and delete columns. Context menus on the default tables (right-click) allow the user to select nodes or edges from the selected rows and apply values to all rows or just selected nodes.

13.4.2.5 View Menu

One of the menus provided in the main Cytoscape toolbar is the **View** menu. This menu provides a number of different options, including the ability to show or hide each of the panels (except the **Network Panel**). It also provides an option to hide all of the panels and maximize the **Network Panel** to fill the screen.

One additional key feature of the **View** menu is the ability to control the level of graphics details shown in the **Network Panel**. To improve interactive performance, the level of detail scales with the number of nodes and edges shown in the current window. As fewer nodes and edges are shown, more detail is included. A very dense network might show nodes as rectangles with no border or labels and edges as straight lines. A zoomed in view of the same network will show labels, add transparency, shapes, edge types, arrows, etc. To force full level of detail, the **View** menu provides the **Show Graphics Details/Hide Graphics Details** toggle.

13.4.3 Getting Data into Cytoscape

One of Cytoscape's strengths is the ability to integrate data from a variety of sources. In some cases, the goal is to integrate experimental data with existing pathway or protein–protein interaction data; in others the network is the experimental data and the goal is to enrich it with additional PPI data, additional experimental data, or various analyses. To achieve these goals, Cytoscape provides tools to import data from various public databases and directly import data from files in a number of formats.

13.4.3.1 Importing Networks from Public Databases

Cytoscape provides an interface to PSICQUIC [27], to import protein–protein interaction networks from one or more of the databases that it exposes. In addition, Cytoscape app provide access to several other public databases including STRING [8, 9], with the stringApp, Reactome [28], with the Reactome FI App, Pathway Commons [29], with CPath2, and WikiPathways [30], with the WikiPathways app. Often importing networks from apps will provide features and/or data not available from PSICQUIC, so if an app exists for your public database of choice, it is recommended that you use that. Typically, an App that provides access to a public database will add an entry to the **Import Network from Public Databases** dialog (Figure 13.23), which is accessible from the **File→Import→Network→Public Databases** menu. Generally, when using the **Import Network from Public Databases** dialog you will select the Data Source, and then enter an appropriate query. The type of query will differ depending on the type of resource. For example, for WikiPathways, you might enter a pathway name, but for the STRING protein query you would enter a list of protein identifiers. The dialog often provides options appropriate to the data source (e.g., species, cutoff values, etc.). The network shown in Figure 13.20 was imported from the STRING database by entering TP53 and setting the option for additional proteins to 10.

As of Cytoscape 3.6, a new interface is provided to provide quicker access to public databases. This new network search field is found in the **Network** tab of the **Control Panel** in Cytoscape. If you enter a comma-separated list of protein identifiers, the stringApp will load that network and display it. Currently, PSICQUIC, STRING, and GeneMania are available via the network search interface, but other resources will quickly be made available.

13.4.3.2 Importing Networks from Files

If your use case requires you to import networks directly from Excel or CSV files, you can either just drag the file and drop it on the network list area of the **Network** tab of the **Control Panel** or use the **File→Import→Network→File...** menu option. In either case, the dialog that results will depend on the type of file. If the file type is a network file format that Cytoscape understands (XGMML, SIF, BioPax, etc.) the file will be

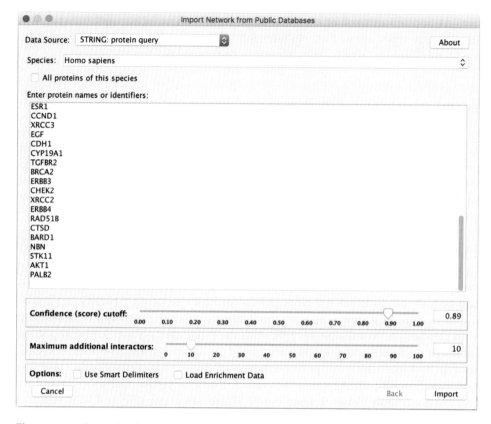

Figure 13.23: Example of an App that has added an entry to the **Import Network from Public Databases** dialog. In this case, the String app has been used to search for a number of human genes related to breast cancer.

opened and the network shown. If, on the other hand, the file is an Excel spreadsheet or CSV file, the **Import Network From Table** dialog will be shown (Figure 13.24). By clicking on the column header an interface provides the ability to select whether the column represents the identifier for the source node, the target node, edge data, or additional information associated with either the source or target nodes. Figure 13.24 shows the interface that results from dropping supplementary table 2 from Jäger et al. [31]. Typically, your data file will need at least a source node column and a target node column, although it is certainly possible to import a "network" with only a source node column but the resulting network will have no edges. The table shown includes columns for the source node identifier (**Bait**), the target node identifier (**Prey**), a number of edge columns, and a column that provides a human readable name for the Prey (**Name**).

Other features of the file importer that can be seen in Figure 13.24 include the ability to change the name of the column, override the datatype that Cytoscape guessed at, and set the list delimiter. The **Advanced Options** button allows the user to change the row to begin importing from (so you can skip over header information), indicate that the first line provides column names, and set a comment character.

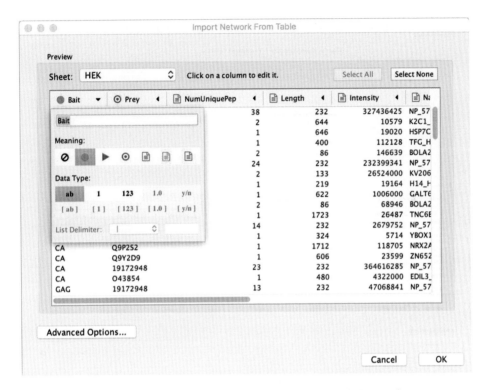

Figure 13.24: Screenshot showing the **Import Network from Table** dialog. In this screenshot, the user has clicked on the header for the Bait column to being up the options for that column.

13.4.3.3 Importing Tables from Files

Once a network has been created, a common need is to augment the nodes or edges in the network with additional data from public sources or experiments. For example, adding expression data from an experiment to a network derived from STRING to be able to calculate active modules. Similar to the way in which networks are imported, data files may be imported into Cytoscape either by dropping them on the **Table Panel** or by using the **File→Import→Table→File...** menu. In either case, the **Import Columns From Table** dialog is displayed (Figure 13.25). This dialog is similar to the network dialog but has some different options that relate specifically to augmenting existing table data. The dialog has two main parts:

- **Target Table Data**, which allows the user to choose to import this as part of an existing network collection (shared data), to only certain networks (local data), or as an unattached table. Depending on this selection, the following options are available:

    – **Select a Network Collection**: If the user chooses "To a Network Collection," this panel allows them to select which collection to import into, whether to import the data as Node Table Columns or Edge Table Columns, and what column in the network to choose as the matching key.

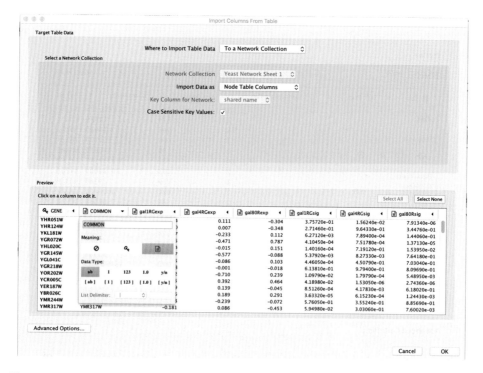

Figure 13.25: Screenshot showing the **Import Columns From Table** dialog. In this case, the user has clicked on the header for the COMMON column to display the options for importing that column.

- **Select Networks**: If the user chooses "To selected networks only," this panel allows them to choose the networks to import into, and as above whether to import the data as Node Table Columns or Edge Table Columns, and what column in the network to choose as the matching key.
- **Set New Table Name**: If the user chooses "To an unassigned table," this panel allows them to name the table.

• **Preview**, which contains a preview of the data and supports the ability to rename columns, choose which column to use as the matching key, and whether or not to import the column. As with the network import dialog, the data types can also be set if Cytoscape guesses wrong.

Clicking OK should add the new columns to the network. If none of the columns have data after the import, it usually means that the key column in the network and the key column in the table don't correctly match. One thing to check if this happens is that the two key columns of the same type, don't have extraneous spaces or quotes, etc.

13.4.3.4 Calculated Data

Another way that data is added to Cytoscape is through the results of calculations or actions of Apps. For example, the results of utilizing Network Analyzer (see below) are stored as columns in the Node and Edge tables. Other apps similarly take advantage of the flexibility of the table model to store results in various tables. These results will be saved as part of the session and restored when the session is restored.

13.4.4 Visualizing Data

In Section 13.3 on Visualizing Networks, we talked about visual attributes and mappings. In this section, we'll discuss how to use visual mappings to visualize the data associated with our networks and discuss some ways to achieve some more sophisticated visualizations.

13.4.4.1 Visual Styles

The **Style** tab in the **Control Panel** is where all of the visual attributes and mappings can be set. The **Style** tab is divided into roughly three sections (see Figure 13.26). The top section (13.26 (a)) provides a list of existing styles that can be applied to the network and a menu that provides items to manage styles (create, delete, copy, etc.). The bottom section (13.26 (c)) provides tabs to select Node, Edge, or Network visual properties. The center section shows the styles themselves. At the top is the **Properties** menu (Figure 13.27) of all visual properties for this type (Node, Edge, or Network). Since many visual properties are rarely used, only the most common

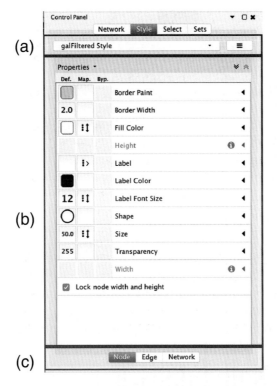

Figure 13.26: Screenshot of the **Style** panel showing: (a) the style selection menu, (b) the list of visual properties and settings for this style, (c) the type of properties being shown.

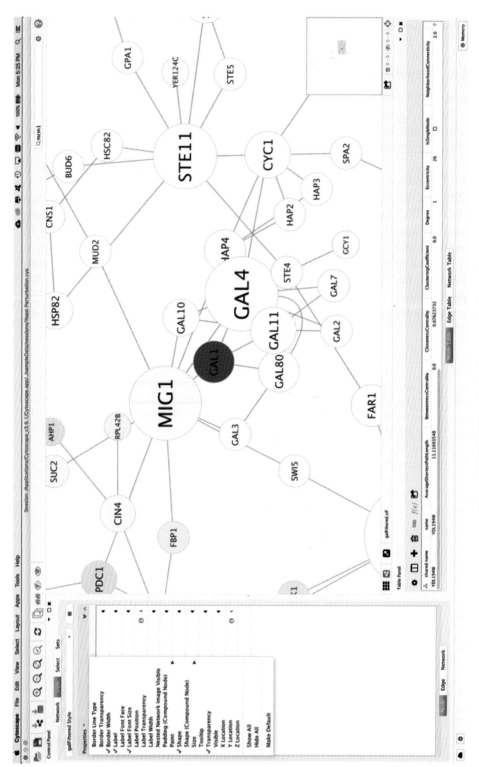

Figure 13.27 Full list of visual properties for nodes.

properties are shown in the panel by default. Underneath the **Properties** menu is a table with four columns and one row for each visual property. The first three columns show the current settings for the property and the fourth column shows the name of the property. The three columns are:

1. **Def.**: This shows the current default value for this property (if any). Clicking on a cell in this column provides an appropriate interface to change the property default. The default value will be used when there is no mapping and no bypass.
2. **Map.**: This shows whether or not there is currently a mapping for this property. There are three mapping types:
   - **Passthrough Mapping**: the attribute value is passed directly to the visual property. The typical example is mapping a text column to the Label visual property.
   - **Discrete Mapping**: each attribute value is mapped to a user-defined visual property value. Discrete mapping is appropriate to map categorical data (e.g. cellular location or major GO category) to color or shape or some other discrete data.
   - **Continuous Mapping**: each attribute value is mapped using a continuous function (e.g., a color gradient or size ramp). Continuous mapping is most useful to map numerical data to a continuous value such as color or size.

   Clicking on a cell in this column will open up a pane to allow the user to set the **Column** to get the attribute values from, the **Mapping Type** and then the **Current Mapping** shown using an appropriate editor. For example, in Figure 13.28, the mapping for *Fill Color* is shown, which is a mapping from the *gal4RGexp* column using a Continuous mapping with a gradient from blue to white to yellow.
3. **Byp.**: This shows a bypass value for the currently selected nodes, edges, or networks. Clicking in this cell is similar to clicking in the **Def.** cell except that the settings only apply to the selected items. Also note that the settings in the **Byp.** column will override any mappings, for the specific nodes, edges, and networks.

13.4.4.2 Visual Property Editors

Cytoscape provides a number of graphical editors that provide appropriate interfaces for creating the mappings between table data and visual properties. There are too many different editors to discuss here, but it's worthwhile discussing four of the most commonly used ones as examples: Color gradient editor, ramp editor, discrete color editors, and shape editors.

The color gradient editor (Figure 13.29) is used to map numerical data values to node fill colors, node border colors, label colors, and edge colors. As can be seen in the figure, the color gradient allows the user to map values to specific colors. New pivot points (the triangles on the top) can be added by clicking on the **Add** button.

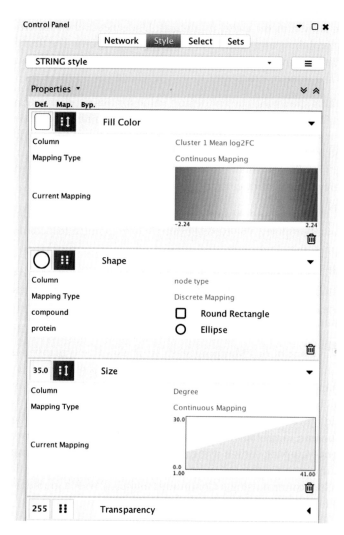

Figure 13.28: Portion of the style panel showing a continuous mapping for fill color (top), a discrete mapping for shape (middle), and a continuous mapping for size (bottom).

The colors of the pivots can be changed by double-clicking on them, which brings up a color editor. The values may be adjusted by dragging the pivots left or right. The arrows at the far left and far right are used to set colors for values that fall outside of the specified range. The min and max values can be adjusted by clicking on the **Set Min and Max...** button. Figure 13.29 shows the mapping of expression fold change values in the gal4Rexp column to the node fill color using a blue/yellow gradient.

The ramp editor (Figure 13.30) maps numerical values to numerical values. Figure 13.30 shows the mapping of the Degree of the nodes from the Degree column to the node size. As with the color gradient editor, the triangles are used to adjust the Degree values to map, but in this case, to change the corresponding node size, the user would drag the squares to the mapped value. **Set Min and Max...** and **Add** behave in a similar manner to the color gradient editor.

CYTOSCAPE: ANALYZING AND VISUALIZING NETWORK DATA 567

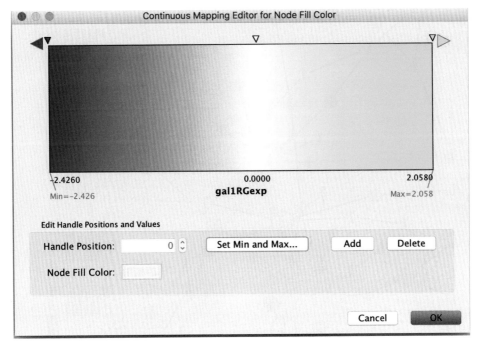

Figure 13.29 Color gradient editor for node fill color.

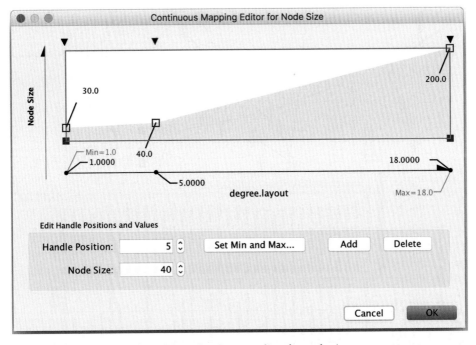

Figure 13.30 Continuous editor for node size.

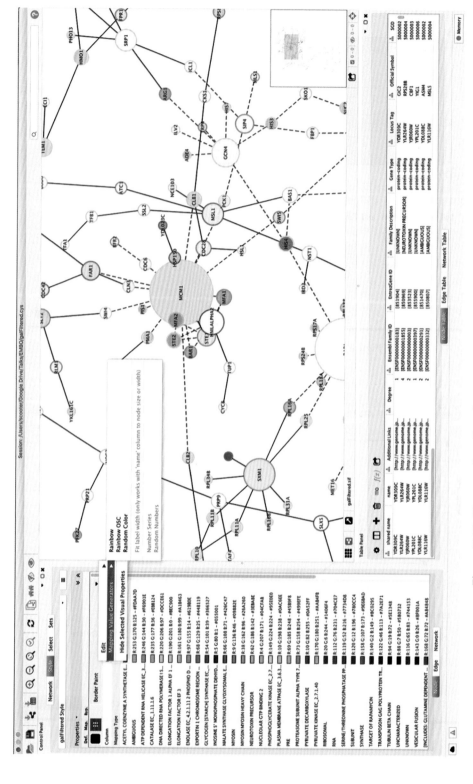

Figure 13.31 Discrete color editor showing options to automatically generate colors.

The discrete color editor in its simplest form provides the ability to assign colors to each individual value one at a time. Because this can be extremely tedious, Cytoscape provides a way to assign a series of colors to values by right-clicking and selecting **Mapping Value Generators**, which provide the option to assign colors using either a Rainbow, Rainbow OSC, or Random model (see Figure 13.31).

The final example is the shape discrete editor, which provides the ability to map discrete values to node shapes. In the middle field in Figure 13.28, the node type column, which contains the values *compound*, or *protein* are mapped to a round rectangle or ellipse.

13.4.4.3 Charts and Graphs

In addition to all of the mappers and editors discussed above, Cytoscape provides an additional tool for mapping from data to visual properties. To accomplish this, Cytoscape provides a set of special visual properties: **Image/Chart 1** through **Image/Chart 9**. These can be added to the style by selecting (for example) **Properties→Paint→Custom Paint 1→Image/Chart 1**. To add the chart or graph you can either click on the **Def.** cell or, if you have a particular set of nodes the you want to add a chart or graph to, you can select those nodes and click on the **Byp.** cell. In either case, the dialog shown in Figure 13.32 appears. This interface allows the user to select from a variety of charts, including bar, box, heat map, line, pie, and ring (or donut) plots. Bar charts have four different options for the bar type: Grouped, Stacked, Heat Strips, and Up-Down. Once the plot type is selected, the data columns can be moved from the *Available Columns* area to the *Selected Columns*. These columns will provide the data for the chart. In the example in Figure 13.32, we are creating a bar chart with heat strips where the three bars will show the data from the columns gal1RGexp, gal4RGexp, and gal80Rexp, respectively. The **Options** tab provides options to change the colors control the labeling and a number of other options appropriate to the type of chart. Figure 13.33 shows the result of creating a bar chart with yellow–orange–red

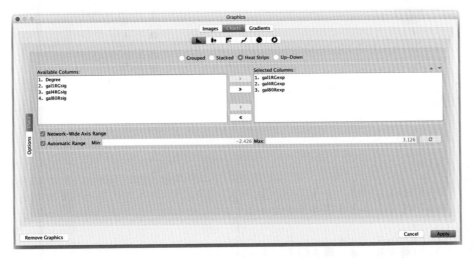

Figure 13.32 Chart editor showing options to create heat strips on nodes.

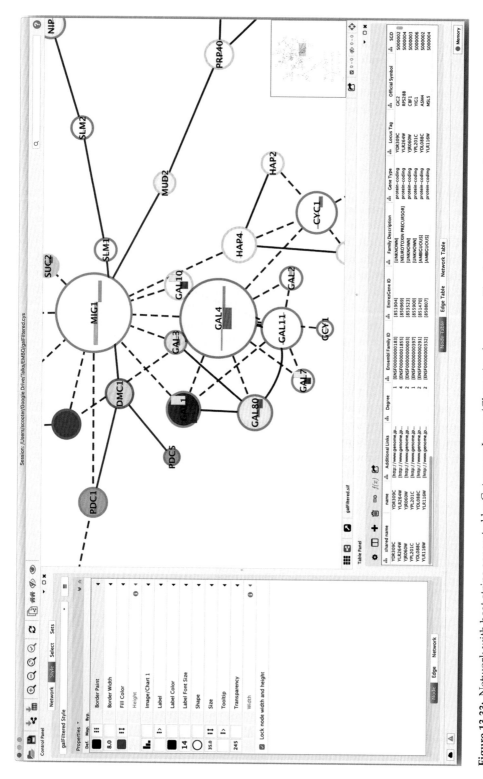

Figure 13.33: Network with heat strips generated by Cytoscape Image/Chart visual property. The charts were created as heat strips with a color palette of YlOrRd chosen in the **Options** tab of the Chart editor dialog to provide a contrast with the node fill color scheme.

(YlOrRd) heat strips. Once a chart is defined for a node, a small icon is displayed in either the **Def.** or **Byp.** cell to show that a chart has been defined for this network (or node).

13.4.4.4 Layouts

We already discussed the possible variety of layouts in a previous section. Cytoscape has support for most of these by default (e.g., Grid, Circular, Force Directed, etc.), and more are available through Apps (e.g., setsApp adds two new layout algorithms). In this section, we'll focus a little on some of the specific features of Cytoscape layouts and how to use them. This is not meant to be an exhaustive list of included and app-based layouts, nor an attempt to explore all of the layout settings that are possible.

Layout Menu Cytoscape's **Layout** menu supports four different, albeit related capabilities: Edge bundling, Node layout tools, Layout settings, and menu items for each of the available layouts.

Edge Bundling The Edge bundling tools in Cytoscape provide a relatively simple mechanism for pulling parallel edges together to form a "bundle." The visual effect is that the perceived complexity is reduced, and major patterns or groupings are emphasized. In Figure 13.34 (a), the STRING network for genes associated with breast cancer was manually separated into two groups. Bundling the edges (Figure 13.34 (b)) simplifies the network and emphasizes the connection between the two groups at the expense of seeing some of the details of the connections. Cytoscape accomplishes this by adding edge "bends" to shape the edges, but it's worth pointing out that the bundling algorithm isn't dynamic – that is, it does recalculate the edge bends if you move nodes around. After changing the position of any nodes, the edge bundling can be recalculated by first clearing the previous calculation (**Clear All Edge Bends**) and then recalculating the bundling with **Bundle Edges**.

Node Layout Tools Typically, the layouts provided by Cytoscape are adequate for visualizing networks, but there are times that more manual approaches are helpful. To support this, Cytoscape provides a set of **Node Layout Tools** from the **Layouts** menu (Figure 13.35). When selected, these tools will appear in the **Network** tab of the **Control Panel**. As shown at the right, there are three basic tools:

1. **Scale:** To expand the distance between the nodes
2. **Align/Distribute/Stack:** To position nodes relative to each other and
3. **Rotate:** To rotate the nodes.

Note that **Scale** and **Rotate** have a checkbox for *Selected Only*. If this is checked and nothing is selected, it will appear that you are scaling or rotating, but nothing will happen. To scale or rotate the entire network, make sure the checkbox is not selected. The **Align/Distribute/Stack** buttons will only be enabled when something is selected.

Layout Settings The **Settings...** dialog from the **Layout** menu provides the user access to the layout properties for each of the algorithms (with the exception of the **yFiles** layouts, which due to license restrictions are not available). Figure 13.36 shows

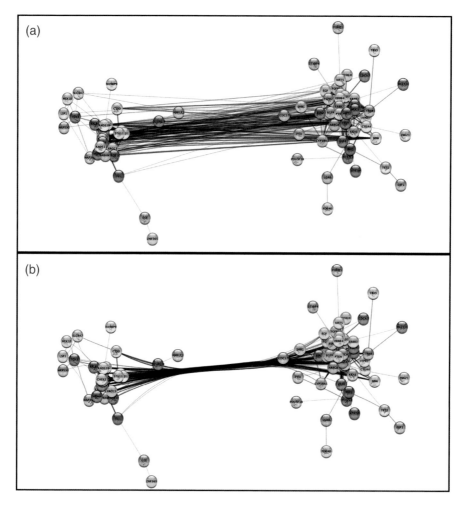

Figure 13.34: Two images of a STRING disease network for breast cancer. The two groups were manually separated for illustrative purposes. (A) shows the network with no edge bundling, and (B) shows the network after edge bundling has been applied.

the properties for the *Prefuse Force Directed Layout*. For algorithms the support edge weights, the first two sections are standard: The column that contains the weights, and then various parameters to help the user select how to interpret the weights. The next section typically contains layouts-specific parameters such as the string length and the string strength. Finally, two common checkboxes are whether two partition the graph (split it into connected components) before performing the layout and whether or not to only apply the layout to the currently selected nodes (if supported). Not all layouts support all options – for example the Grid layout – doesn't support edge weights, graph partitioning or selected only.

Layout Menus Finally, the layout menus themselves. Layout algorithms can advertise support for four different capabilities: Edge weighting, node weighting, the ability

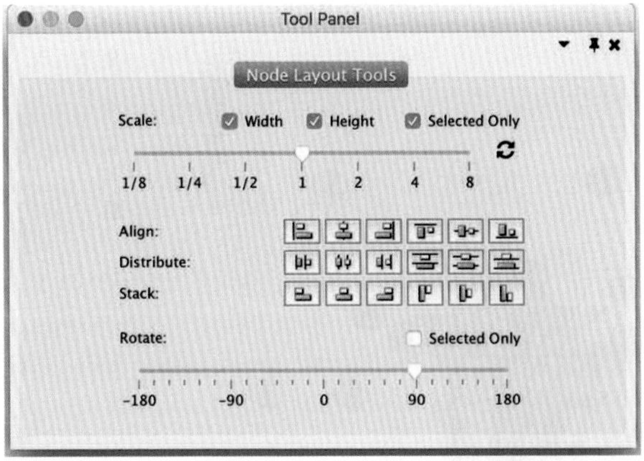

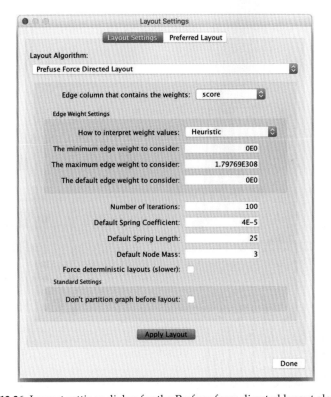

Figure 13.35: Node layout tools dialog. When selected, this dialog will show in the **Network** panel.

Figure 13.36 Layout settings dialog for the Prefuse force directed layout algorithm.

to layout only selected nodes, and support for partitioning the graph before performing the layout. Support for weighting and using only selected nodes is exposed in the menu structure (see Figure 13.37). If the algorithm supports only laying out selected nodes, and some nodes are selected, then the first level will allow the user to determine whether the layout should apply to all nodes or just the selected nodes. If edge

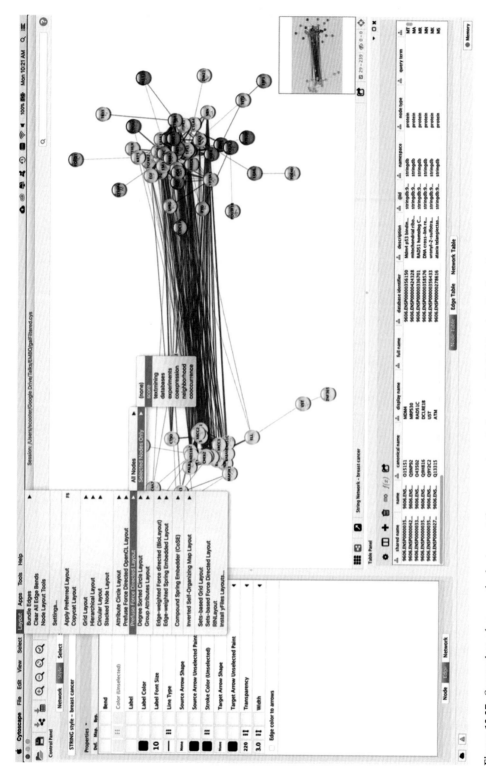

Figure 13.37: Screenshot of menus for laying out the same network shown in 13.34 (a) using Prefuse force directed layout, using only selected nodes, and using the "score" edge attribute to influence the edge length.

weighting or node weighting is supported then the list of columns containing the acceptable weights are shown. Note that if an algorithm supports both edge weights and node weights only the edge weights will be shown and typically, the user will be able to provide the node weights in the Settings dialog.

13.4.5 Network Analysis

The previous 12 chapters have covered a number of techniques for analyzing networks which won't be repeated here. However, in addition to a number of apps which perform various analyses, Cytoscape provides a built-in tool for performing simple network analysis: **NetworkAnalyzer** [32], which is available under the **Tools** menu. To perform basic network analysis, select **Tools→NetworkAnalyzer→Analyze Network**. Then choose whether the analysis should assume whether the network is directed or undirected. This will display a panel that displays various simple parameters such as the network radius, diameter, centralization, number of shortest paths, etc. In addition, various distributions are calculated including node degree, topological coefficients, and betweenness, closeness, and stress centralities. One of the more useful aspects of **NetworkAnalyzer** is that these parameters are saved as columns in the node and edge tables so that they can be utilized for visual mappings and any potential further analyses.

13.4.6 Apps

13.4.6.1 The Cytoscape App Store

In the same way that most modern phone operating systems can be extended by downloading "apps," Cytoscape provides an extension mechanism and an "app store" for searching for apps and adding them to Cytoscape. At the time of this writing, the Cytoscape app store has over 300 apps available that extend Cytoscape by adding new analyses, visualizations, providing access to additional public data repositories, or by providing useful utilities that improve Cytoscape's operation in some way. The easiest way to download or search for apps is to launch Cytoscape, then use a web browser and go to the Cytoscape app store (apps.cytoscape.org). This will allow the app store to install or update apps directly into your running Cytoscape.

13.4.6.2 App Manager

Cytoscape also provides access to the app store from within Cytoscape itself using the app manager, which is available under the menu **Apps→App Manager....** This will bring up an interface that will provide a list of all apps in the app store, show the currently installed apps, and allow apps to be installed, uninstalled or disabled. The information about apps is not nearly as complete as what's available from the web interface to the app store, but the ability to uninstall or disable apps is only available from the app manager.

13.4.6.3 Example Apps

There are far too many apps to discuss in any detail in this chapter, but it's worthwhile to cover a number of sample apps to provide a sense of the breadth of apps that are available. To get started with Cytoscape, it's recommended that you look at the various app "collections," which are a set of apps that have been used together as part of various known biological workflows. Below are a small set of apps that have proven to be very useful.

stringApp The stringApp provides a direct interface to the STRING databases, including STRING [8, 9], STITCH [24, 25], DISEASES [33], and the ability to find interactions from PubMed. This app was written in close collaboration with the STRING team to provide a platform to analyzing larger networks than are typically feasible on the STRING website. The stringApp also downloads additional information about each compound or protein that are stored in node and edge columns and which may be used for further analyses and visualizations.

clusterMaker2 Clustering (unsupervised classification) is a common network analysis technique. clusterMaker2 [34], is an app that provides a number of cluster algorithms including hierarchical clustering, *k*-means, PAM, MCL, Affinity Propagation, and many others. In addition, it provides a small set of dimensionality reduction techniques (PCA, tSNE, and PCoA). Various visualizations are provided such as heatmaps, tree views, and scatter plots that are linked with the network views to support interactive exploration of the data.

Diffusion Diffusion techniques are more sophisticated forms of "guilt by association" approaches. The diffusion app utilizes a random walk with restart (RWR) approach to diffuse values through the network so nodes of low value (based on whatever column is selected) surrounded by high value nodes have their value increased based on the association with the high value nodes. This is a common technique for understanding biological networks as sometimes proteins are critically involved in processes that do not particularly change expression values, for example.

BiNGO Another important, common analysis technique is overrepresentation or enrichment analysis. The basic idea is that given a set of nodes, chosen by some analysis technique, are there terms (e.g., pathways, gene ontology terms, etc.) that more prevalent than would be expected based on the total background set of nodes. BiNGO [35], calculates the enrichment specifically for gene ontology terms and provides a network visualization of the resulting enriched terms, where nodes are terms and the edges are the links between those terms. The node fill color represents how enriched that term is. For more sophisticated enrichment analysis, another app ClueGO [36], is also very popular.

EnrichmentMap EnrichmentMap [37], is similar to BiNGO except that it provides only the visualization component, relying on other tools to calculate the actual enrichment. EnrichmentMap produces a very effective network visualization where nodes are the enriched terms and edges represent the nodes in the original network that are shared between those terms. This allows for a nice grouping of enriched terms.

13.5 Functional Enrichment Workflow

Cytoscape is used for a wide variety of workflows focused on the analysis and visualization of biological data. One of the earliest and still most common uses is the analysis of gene expression data to find subnetworks or modules of genes that are enriched for particular biological functions (pathways or gene ontology terms). For this example workflow, we'll be looking at glioblastoma multiforme using RNASeq data from the cancer genome atlas ([38]; TCGA), protein–protein interaction data from STRING, and several Cytoscape apps.

13.5.1 Apps

First, you'll need to load the following Cytoscape apps from the App store:

- stringApp [9]
- Diffusion [39]
- clusterMaker2 [34]

13.5.2 Data

To save some time, we've provided a data file containing differential expression for 528 glioblastoma multiforme patients downloaded using the firehose_get command downloaded from https://confluence.broadinstitute.org/display/GDAC/Download. The download requested the latest analyses for the GBM cohort (firehose_get analyses latest GBM). The resulting file was post-processed to create mean log2FC (fold change) for all samples, and subsets of samples in each GBM cluster (as identified by the firehose analysis). In addition, mutation counts were added into the spreadsheet. The resulting spreadsheet (GBM-TP.all.tsv available at www.cambridge.org/bionetworks) contains all of the data necessary for the workflow below.

13.5.3 Step-by-Step Workflow

1. **Launch Cytoscape.** If this is your first time launching Cytoscape, refer the manual (http://manual.cytoscape.org) for a basic introduction.
2. **Load apps**. Go to the app store to install the Functional Enrichment Collection (http://apps.cytoscape.org/apps/functionalenrichmentcollection).
3. **Load TCGA data.** We'll actually do this in two parts.

    a. **Create the network.** Start by downloading GBM-TP.all.tsv, then open the file to create a network using **File→Import→Network→File**. This should bring up the "Import Network From Table" dialog. Select **"Select None"** to disable all columns. Then click only on the GeneName column and set this column as the "Source Node" column (green circle). Then click OK. You'll see a warning about no edges, but that's OK. This will create a grid of 1,500 unconnected nodes, where each node represents a gene.

    b. **Add data to our nodes.** To add the rest of the data to our nodes, open the file again, but now use **File→Import→Table→File**. This will bring up the

"Import Columns From Table" dialog. It turns out that all of the defaults are correct for just importing the data, so click on OK. This will import all of the data in the spreadsheet and associate each row with the corresponding node. You should be able to see this in the Table Panel.

4. **Find significant expression changes**. We'll use the "Select" tab in the Control Panel to find the significant overall expression changes.

   a. Open up the Select tab and click on the + button to add a new condition. In this case we're going to add a Column Filter. Select the "Mean log2FC" column and set the values to be between -5 and -1. This will select all genes that are significantly underexpressed on average, across all samples (only one gene should be selected).
   b. Repeat the same process as above, but set the values to be between 1 and 5. You'll also need to change the type of match (at the top of the panel) to "Match any (OR)."

The resulting Select Panel should match Figure 13.38. Only one gene shows significant average expression changes across all samples, RPS4Y1, a ribosomal protein that has been shown to be linked to glioblastoma, but it's

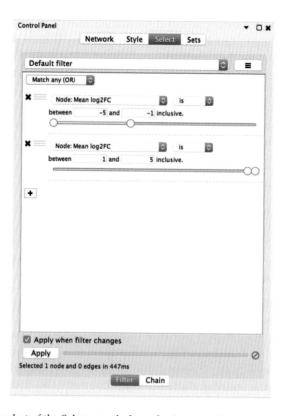

Figure 13.38: Screenshot of the Select panel after selecting significant expression changes (less than -1 or greater than 1).

still a relatively unsatisfying result that only a single protein changes expression in glioblastoma patients. A closer examination might be in order...

5. **Cluster to see expression patterns**. Let's cluster all of the data to see if there are any significant patterns that have been averaged out by looking at all samples in the aggregate. *clusterMaker2* is a Cytoscape app that provides a number of clustering algorithms including hierarchical and k-means.

    a. **Perform hierarchical clustering**. In Cytoscape, select **Apps→clusterMaker→ Hierarchical cluster**. This will bring the settings for hierarchical clustering. Select all of the samples (TCGA-), but none of the other columns in the "Node attributes for cluster" list. The checkboxes "Cluster attributes as well as nodes," "Ignore nodes/edges with no data," and "Show TreeView when complete" should all be checked. Then click OK. The resulting TreeView should look similar to Figure 13.39. Looking at the result, is seems like there are definite bands of expression data that look similar across a group of patients, but differ between groups. In fact, this reaffirms that there are known to be four different subtypes of glioblastoma. Let's take that known data and see if we can subset the data into four subtypes.

    b. **Perform k-means clustering**. In Cytoscape, select **Apps→clusterMaker→ K-Means cluster**. In the resulting settings dialog, set the number of cluster to four, which is the number of known subtypes of glioblastoma, as with

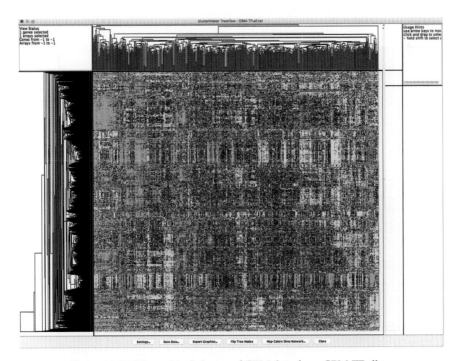

Figure 13.39 Hierarchical cluster of GBM data from GBM-TP.all.tsv

the hierarchical cluster, select all of the samples (TCGA-), check "Cluster attributes as well as nodes" and "Show HeatMap when complete." This will result in a heat map similar to the one shown in Figure 13.40. Note that there are four distinct vertical bands (your results might vary in the ordering in either the vertical or horizontal dimensions). In fact, the firehose analyses identified the same four groupings of patients. In the spreadsheet loaded originally, we've calculated the mean Log2(fold change) for each of the four subtypes.

6. **Find significant expression changes in Cluster 1**. Repeat step 4 above with "Cluster 1 Mean log2FC" as the column. This represents the mean expression across the samples in cluster 1 (the second cluster in Figure 13.40). The result should contain 171 nodes.
7. **Download the PPI from STRING**. From the above selection filter, we've identified 171 nodes that show an average Log2FC greater than 1 or less than −1, and selected those nodes. Now, we can create a network that includes all of those nodes by loading the protein-protein interaction data from STRING. Note that only the selected nodes are shown in the Table Panel.
    a. **Select gene names from Table Panel**. Select everything in the "name" column by clicking into the first cell and then dragging down until you get to the bottom. Then, do a copy (Control-C or Apple-C).
    b. **Paste gene names into STRING network search**. In the Network tab of the Control Panel at the top should be a text field with an icon at the left.

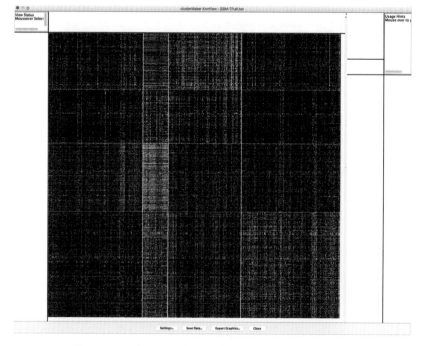

Figure 13.40 *k*-Means cluster of data from GBM-TP.all.tsv

Click on that icon and select "STRING protein query." (If you don't see any STRING options, the stringApp hasn't been loaded.) Then click into the text field and paste the list of genes.
c. **Set STRING search parameters**. Next to the text field is a menu with a list of options. Change the "Confidence (score) cutoff" to 0.8 and the "Maximum additional interactors" to 30. This will get only high quality results (80% confidence) and add 30 extra proteins to the network.
d. **Create the network**. Click on the search (magnifying glass) icon to load the network. The network should appear similar to the Figure 13.41.

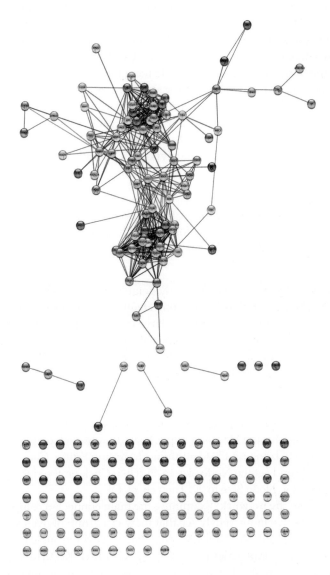

Figure 13.41: STRING network created by querying for the overexpressed and underexpressed genes.

8. **Style the network to show differential expression.** In this step, we'll change the style of the network to highlight the differentially expressed genes.

    a. **Re-import expression data.** First, we need to re-import the expression data for the new network created by the *stringApp*. Similar to step 3b, start by doing **File→ Import→Table→File**. Again, select the GBM-TP.all.tsv file however, now we need to use a different network column to match our names. Change the "Key Column for Network": from "share name" to "query term." STRING uses Ensembl identifiers, but retains the original query term so we can match data against. Now select "OK."
    b. **Disable structure image.** STRING provides some nice images of the 3D structure of the proteins, but we need to disable those to be able to see our expression values clearly. Disable the images by going to: **Apps→STRING→Don't show structure images**.
    c. **Create color gradient for expression data.** To show the expression data, to go the Style tab of the Control Panel and click on the middle square (Map.) of the three Fill Color controls. Set the Column to "Cluster 1 Mean log2FC" and set the Mapping Type to "Continuous Mapping." This will create a gradient using the Brewer palette from blue to red. This is a good starting choice, but feel free to customize it to your liking.

9. **Add styling to show mutations.** In addition to expression fold change, the imported data also includes information about mutation, both across all patients and subdivided by cluster. It might be informative to add that information to our visualization.

    a. **Lock node width and height.** By default, the *stringApp* provides separate values for Node Width and Node Height. In our case, we just want to lock them to be the same so we only have to modify the Node Size. The "Lock node width and height" is a checkbox at the bottom of the Node tab in the Style tab of the Control Panel. Make sure that it's checked.
    b. **Set the default node size.** Click on the leftmost (Def.) box next to Size. Set the default size to 45.0.
    c. **Map mutation to node size.** Click on the middle (Map.) box next to Size. Choose "Cluster 1 Mutations" for the Column and "Continuous Mapping" for the Mapping Type. Then double-click on the ramp that appears to bring up the Continuous Mapping Editor. Click on the leftmost triangle and set the Node Size to 45. Then Click on the rightmost triangle and set the Node Size to 90. Then check OK.
    d. **Examine the network.** Examining the main connected component of the network, we immediately see that there is a group of overexpressed genes (COL6A3, COL1A1, COL1A2, etc.) that also exhibit some mutations. There is also a group of genes (AGT, CXCL2, GABBR1, KCNJ16) that show some mixed expression changes. Finally, there is a group of genes, including

NCAN, BCAN SDC3, and CSPG5 in between these groups that show predominantly underexpression.

At this point in our workflow, we have some options. One typical path might be to find all of the overexpressing genes, use the Diffusion app to perform heat diffusion through the network to grow the selection and then perform functional enrichment analysis on that set of genes and then repeat for underexpressing genes (see step 10). Another approach is suggested by the topology of the network, which strongly suggests groupings in the interactions. We can use clustering based on the network topology and to perform functional enrichment on each cluster (see step 12). We'll take each of those in turn.

10. **Smooth the network using Diffusion**. When we created the STRING network, we added nodes to make sure that well connected nodes that might be silent (i.e. not exhibit changes in expression) could be included. Well connected genes might play a significant role in the biology even if their expression doesn't change due to changes in the expression of their interaction partners. The Diffusion app supports the ability to find those genes by simulating the diffusion of heat through network connections.

    a. **Create subnetwork of the main connected component**. Select all of the nodes and edges in the main connected component by holding mouse button 1 down and sweeping over the nodes. Alternatively, you could select a single node and then repeatedly hitting Control-6 (or ⌘ on the Mac) to continually select the first neighbors of selected nodes until all of the nodes in the connected component are selected. Once the entire component is selected, create a new network by doing **File→New→Network→From selected nodes, all edges**.
    b. **Select overexpressed genes**. Go back to the Select tab and disable (click on the X) the filter for underexpressed genes, then click Apply.
    c. **Diffuse the selection**. In the diffusion app we can diffuse either based on the heat (column value) or just based on the selection. We're going to diffuse based on our expression values, so select **Tools→Diffuse→Selected Nodes with Options** and in the dialog select "Cluster 1 Mean log2FC" as the Heat Column and set a short time of 0.001 seconds. Then click OK.
    d. **Create subnetwork**. Change the Node Rank to about 40 and create a subnetwork by selecting **File→New→Network→From selected nodes, all edges**. Once the network is created, execute an unweighted Prefuse Force Directed Layout (**Layout→Prefuse Force Directed Layout→(none)**.
    e. **Repeat for underexpressed genes**. For underexpressed genes, repeat step c. above except choose all genes with a "Cluster 1 Mean log2FC" less than −1 and in the diffusion app set the time to 0.01 seconds. Unfortunately the Diffusion Output assumes positive diffusion values, so to select the

diffused nodes of interest, you'll need to go to the Table Panel and sort the diffusion_output_heat column, select the negative values less than −0.5, right-click on the column and select "Select nodes from selected rows." Then you can create the subnetwork.

11. **Perform enrichment analysis on the diffused subnetworks**. There are many tools available in Cytoscape to perform functional enrichment. For simplicity we're going to use the built-in functional enrichment available from the *stringApp*. The description below walks through calculating function enrichment of the overexpressed subnetwork, but the same approach will work for the underexpressed network.

    a. **Set network as a STRING network**. In order to retrieve the function enrichment scores the *stringApp* needs to be told that this is a STRING network. A menu item is available for that purpose: **Apps→STRING→ Set as STRING network**.

    b. **Retrieve functional enrichment of a component**. In the subnetwork, there are two connected components of any size. Select all of the nodes in the largest one and retrieve the functional enrichment through the **Apps→STRING→Retrieve functional enrichment**. This will bring up a new Table Panel tab for "STRING Enrichment." There is a pull-down menu that switches the ontology or pathway displayed. In the case of the largest component, we see that there is a significant overexpression of genes involved in the GO Biological Processes: Extracellular matrix organization, collagen catabolic process, and other development processes; they are also associated with the KEGG pathways protein digestion and absorption and ECM-receptor interaction.

    c. **Repeat for other components**.

12. **Topologically cluster the network**. Looking at the protein-protein interaction network (Figure 13.41), we see that there appear to be strong connections between groups of nodes (see discussion in step 9d above). One approach to explore the functional interactions of the network is to use a topological clustering algorithm to partition the network based on the interactions. The *clusterMaker2* app provides a set of topological (network) cluster algorithms. One commonly used one is MCL [40], which performs a Markov-based flow simulation to determine clusters.

    a. **Cluster using MCL**. Make sure the network is connected, then go to **Apps→ clusterMaker→MCL Cluster**. This will bring up the MCL settings dialog. There are a number of options, but most of these don't require any changes for our uses. We're going to treat the network as unweighted so we don't need to provide any Array Sources. We do want to check the "Create new clustered network," however, so we can visualize the results. Then click OK.

13. **Determine functional enrichment of the clusters**. The results of the clustering should look something like Figure 13.42, which is a close-up of the

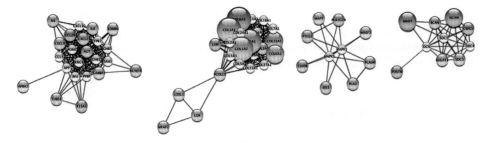

Figure 13.42: The four largest components from the MCL clustering of the network shown in Figure 13.41.

largest components. Of particular interest are the largest four components, which we'll perform functional enrichment.

a. **Set network as a STRING network.** The first step is to inform the *stringApp* that this is a STRING-derived network. A menu item is available for that purpose: **Apps→STRING→Set as STRING network**.

b. **Retrieve functional enrichment of a component.** In the subnetwork, there are four connected components of any size. Select all of the nodes in the second largest one (the one with a number of overexpressed genes and the bulk of the mutations) and retrieve the functional enrichment through the **Apps→STRING→Retrieve functional enrichment**. This will bring up a new Table Panel tab for "STRING Enrichment." There is a pull-down menu that switches the ontology or pathway displayed. Looking at the results we see that there is a significant overexpression of genes involved in the GO Biological Processes: Extracellular matrix organization, collagen catabolic process, and other development processes; they are also associated with the KEGG pathways Protein digestion and absorption and ECM-receptor interaction.

c. **Repeat for other components.**

13.6 Scripting Cytoscape

As can be seen by the above workflow, Cytoscape and the various available apps provide an excellent environment for integrating, analyzing, and visualizing biological data. Once a particular workflow is established, however, there are circumstances where being to automate that workflow becomes extremely important for both documentation and reproducibility. Furthermore, there are many existing workflows that start outside of Cytoscape but include a component of network analysis or visualization that Cytoscape and its apps are well suited for. To support these use cases, Cytoscape provides two scripting options that can work together to support the automation of Cytoscape workflow or the integration of Cytoscape into existing workflows in a straightforward manner. Cytoscape provides a relatively simple textual command syntax that allows commands to be listed in a file and executed as a script, it also provides a REST interface that may be used by external programs such as R or Python. Each of these will be discussed briefly below, but detailed documentation for CyREST [41] in particular is widely available (see https://github.com/cytoscape/cytoscape-automation).

13.7 Command Language

Cytoscape's command language is meant to be a very simple way to execute sets of Cytoscape commands without having to go through the same series of mouse clicks and text entry. In general, a command consists of three parts: **Namespace, command,** and **arguments**. The namespace may refer to some core functionality (e.g., *network*) or an app (e.g., *cluster*). A namespace is always a single word. Commands on the other hand can be multiple words. Arguments are always of the form: **argument=value**. So, an example command to load a network using the string app is:

```
string protein query cutoff=0.8 limit=50 species="homo sapiens"
query="EGFR,BRCA1,BRCA2,TP53"
```

Where *string* is the namespace, *protein query* is the command and the rest of the string are all of the arguments. Note that arguments with spaces need to be quoted.

To experiment with command, use the Cytoscape Command Line Dialog (**Tools→ Command Line Dialog**). This tool allows you to enter and execute commands. In addition the **help** command in the dialog allows you to return all of the available namespaces, their available commands, and possible arguments (see Figure 13.43).

13.7.1 CyREST

CyREST [41], is the REST interface to Cytoscape and it is meant to be used by external programming environments such as R or Python, although it has been used to integrate Cytoscape into other desktop application (e.g., ChimeraX [42], uses CyREST to integrate with Cytoscape for network data). CyREST provides RESTful access to Cytoscape objects such as networks, nodes, edges, tables, rows, and columns. For example, to get a list of networks, you would execute the query: "GET http://localhost:1234/v1/networks." This assumes that Cytoscape is configured to listen on port 1234, which is the default. Similarly, by using POST, PUT, and DELETE HTTP commands, Cytoscape objects can be created, updated, and deleted. The best way to learn about the various REST paths available in Cytoscape is by using the Swagger documentation generated from the **Help→Automation→CyREST API**. This will generate a Web page that supports browsing the various paths, reading the documentation, and executing the query.

In addition to supporting traditional REST interface, CyREST also exposes all of the Commands that are available in the command line. This is one way that apps can expose functionality to callers via REST. A separate Swagger documentation is available by going to **Help→Automation→CyREST Command API**.

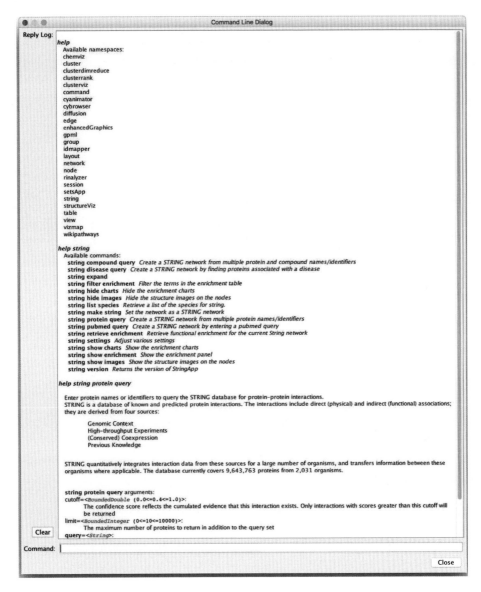

Figure 13.43: Screenshot of the Cytoscape Command Line Dialog showing namespaces, commands for the *string* namespace, and arguments for the *string protein query* command.

Documentation on CyREST, including wrapper methods for R and Python is available on GitHub at https://github.com/cytoscape/cytoscape-automation.

13.8 Exercises

Note: The datasets for the following exercises will be available online at www.cambridge.org/bionetworks.

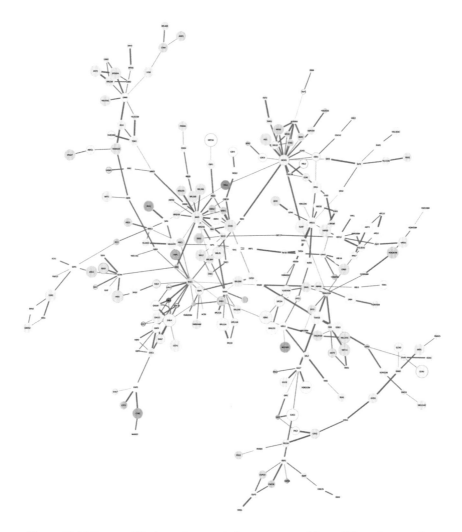

Figure 13.44 Image of the largest connected component of the galFiltered network.

13.1 Cytoscape is delivered with a number of sample files. Using the files galFiltered.xls, galFiltered.nodeAttrTable.xls, galExpData.csv, and galFiltered.edgeAttr Table.xls create a network similar to that shown in Figure 13.44. Figure 13.45 is a zoomed region around MCM1.

13.2 In the above workflow, we analyzed the differential expression results for one of the glioblastoma subtypes (cluster 1). Repeat the analysis for the other three subtypes shown in clusters 2, 3, and 4.

13.3 Looking at the results from the k-means clustering, k-means comes up with a different set of samples in the four categories. Choose one of the categories from k-means (the one with a significant down-regulation of a series of genes would be a good choice) and redo the workflow above. To avoid having to spend a

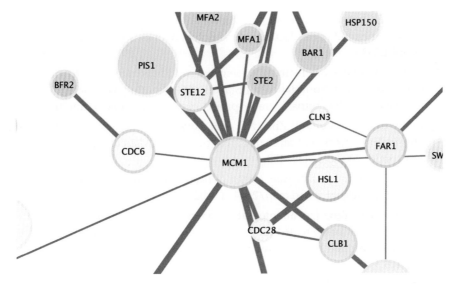

Figure 13.45 Zoomed image of the galFiltered network focused round MCM1.

significant amount of time manipulating spreadsheets, we've provided the four clusters as GBM-TP.clust1.cm.tsv, GBM-TP.clust2.cm.tsv, GBM-TP.clust3.cm.tsv, and GBM-TP.clust4.cm.tsv.

13.9 Acknowledgments

Cytoscape is a team effort and I want to acknowledge the entire team, both past and present (see www.cytoscape.org/development_team.html). Their ongoing effort to improve Cytoscape is one of the reasons it remains such a widely used and successful tool. I also want to acknowledge the Cytoscape leadership: Trey Ideker, Gary Bader, and Alex Pico who manage to keep things moving forward even though we are a very diverse group. I also want to thank Nataša Pržulj for inviting me to be a part of this effort and being patient with me as I tried to figure out how to describe an interactive program in static images and prose. Finally, none of my time on this book would have been possible without the generous support of our funding agency, the National Institute of General Medical Sciences (NIGMS grant P41 GM103504).

References

[1] Cline MS, Smoot M, Cerami E, et al. Integration of biological networks and gene expression data using cytoscape. *Nature Protocols*, 2007;2(10):2366–2382.

[2] Shannon P, Markiel A, Ozier O, et al. Cytoscape: A software environment for integrated models of biomolecular interaction networks. Genome Research, 2003;13(11):2498–2504.

[3] Ideker T, Ozier O, Schwikowski B, Siegel AF. Discovering regulatory and signalling circuits in molecular interaction networks. *Bioinformatics*, 2002;18 Suppl 1:233.

[4] R Development Core Team. *R: A Language and Environment for Statstical Computing*. The R Foundation for Statistical Computing, version 3.5.1;2011. Available online at https://cran.r-project.org/doc/manuals/fullrefman.pdf.

[5] Collins SR, Kemmeren P, Zhao XC, et al. Toward a comprehensive atlas of the physical interactome of saccharomyces cerevisiae. *Molecular Cellular* Proteomics 2007;6(3):439–450.

[6] Jones C. *How to visualize your Twitter Network* (Blog post). Available from http://allthingsgraphed.com/2014/11/02/twitter-friends-network/.

[7] Gao F, Xia Y, Wang J, et al. Integrated analyses of DNA methylation and hydroxymethylation reveal tumor suppressive roles of ECM1, ATF5, and EOMES in human hepatocellular carcinoma. *Genome Biology* 2014; 15(12):9.

[8] Snel B, Lehmann G, Bork P, Huynen MA. STRING: A web-server to retrieve and display the repeatedly occurring neighbourhood of a gene. *Nucleic Acids Research*, 2000 15,;28(18):3442–3444.

[9] Szklarczyk D, Morris JH, Cook H, et al. The STRING database in 2017: Quality-controlled protein-protein association networks, made broadly accessible. *Nucleic Acids Research*, 2017;45(D1):D368.

[10] Atkinson HJ, Morris JH, Ferrin TE, Babbitt PC. Using sequence similarity networks for visualization of relationships across diverse protein superfamilies. *PLoS One*, 2009;4(2):e4345.

[11] Altschul SF, Gish W, Miller W, Myers EW, Lipman DJ. Basic local alignment search tool. *Journal of Molecular Biology*, 1990;215(3):403–410.

[12] Stuart JM, Segal E, Koller D, Kim SK. A gene-coexpression network for global discovery of conserved genetic modules. *Science*, 2003;302(5643): 249–255.

[13] Zhang B, Horvath S. A general framework for weighted gene co-expression network analysis. *Statistical Applications in Genetics and Molecular Biology*, 2005;4:Article 17.

[14] Ciucci S, Ge Y, Duran C, et al. Enlightening discriminative network functional modules behind principal component analysis separation in differential-omic science studies. *Scientific Reports*, 2017;7:43946.

[15] Jaccard P. Distribution de la flore alpine dans le bassin des dranses et dans quelques régions voisines. *Bulletin De La Société Vaudoise Des Sciences Naturelles*, 1901;37:241–272.

[16] Tanimoto TT. *An Elementary Mathematical Theory of Classification and Prediction*. International Business Machines Corporation; 1958.

[17] Longabaugh WJ. Combing the hairball with BioFabric: A new approach for visualization of large networks. *BMC Bioinformatics*, 2012;13:275.

[18] Munzner T, Maguire E. *Visualization Analysis and Design*. CRC Press; 2015.

[19] Fruchterman TMJ, Reingold EM. Graph drawing by force-directed placement. *Software Practice and Experience*, 1991;21(11):1129–1164.

[20] Kamada T, Kawai S. An algorithm for drawing general undirected graphs. *Information Processing Letters*, 1989;31(1):7–15.

[21] Mardia KV. *Multivariate Analysis*, Kent JT, Bibby JM. Academic Press; 1979.

[22] Muscoloni A, Thomas JM, Ciucci S, Bianconi G, Cannistraci CV. Machine learning meets complex networks via coalescent embedding in the hyperbolic space. *Nature Communications*, 2017;8(1):5.

[23] Yu Y, Sikorski P, Smith M, et al. Comprehensive metaproteomic analyses of urine in the presence and absence of neutrophil-associated inflammation in the urinary tract. *Theranostics*, 2016;7(2):238–252.

[24] Kuhn M, von Mering C, Campillos M, Jensen LJ, Bork P. STITCH: Interaction networks of chemicals and proteins. *Nucleic Acids Research*, 2008;36(Database issue):684.

[25] Szklarczyk D, Santos A, von Mering C, Jensen LJ, Bork P, Kuhn M. STITCH 5: Augmenting protein-chemical interaction networks with tissue and affinity data. *Nucleic Acids Research*, 2016;44(D1):380–384.

[26] Lotia S, Montojo J, Dong Y, Bader GD, Pico AR. Cytoscape app store. *Bioinformatics*, 2013;29(10):1350–1351.

[27] del-Toro N, Dumousseau M, Orchard S, et al. A new reference implementation of the PSICQUIC web service. *Nucleic Acids Research*, 2013;41(Web Server issue):601–606.

[28] Fabregat A, Sidiropoulos K, Garapati P, et al. The reactome pathway knowledgebase. *Nucleic Acids Research*, 2016;44(D1):481.

[29] Cerami EG, Gross BE, Demir E, et al. Pathway commons, a web resource for biological pathway data. *Nucleic Acids Research*, 2011;39(Database issue):685.

[30] Kutmon M, Riutta A, Nunes N, et al. WikiPathways: Capturing the full diversity of pathway knowledge. *Nucleic Acids Research*, 2016;44(D1):488.

[31] Jager S, Cimermancic P, Gulbahce N, et al. Global landscape of HIV-human protein complexes. *Nature*, 2011;481(7381):365–370.

[32] Assenov Y, Ramírez F, Schelhorn S, Lengauer T, Albrecht M. Computing topological parameters of biological networks. *Bioinformatics*, 2008;24(2):282–284.

[33] Pletscher-Frankild S, Pallejà A, Tsafou K, Binder JX, Jensen LJ. DISEASES: Text mining and data integration of disease–gene associations. *Methods*, 2015;74:83–89.

[34] Morris JH, Apeltsin L, Newman AM, Baumbach J, Wittkop T, Su G, Bader GD, Ferrin TE. clusterMaker: A multi-algorithm clustering plugin for cytoscape. *BMC Bioinformatics* 2011;12:436.

[35] Maere S, Heymans K, Kuiper M. BiNGO: A cytoscape plugin to assess overrepresentation of gene ontology categories in biological networks. *Bioinformatics*, 2005;21(16):3448–3449.

[36] Bindea G, Mlecnik B, Hackl H, et al. ClueGO: A cytoscape plug-in to decipher functionally grouped gene ontology and pathway annotation networks. *Bioinformatics*, 2009;25(8):1091–1093.

[37] Isserlin R, Merico D, Voisin V, Bader GD. Enrichment map: a cytoscape app to visualize and explore OMICs pathway enrichment results. *F1000Research*, 2014;3:141.

[38] Cancer Genome Atlas Research Network, Weinstein JN, Collisson EA, et al. The cancer genome atlas pan-cancer analysis project. *Nature Genetics*, 2013;45(10):1113–1120.

[39] Carlin DE, Demchak B, Pratt D, Sage E, Ideker T. Network propagation in the cytoscape cyberinfrastructure. *PLoS Computational Biology*, 2017;13(10):e1005598.

[40] van Dongen S, Abreu-Goodger C. Using MCL to extract clusters from networks. In van Helden J, Toussaint A, Thieffry D, eds., *Bacterial Molecular Networks. Methods in Molecular Biology (Methods and Protocols)*, vol. 804. Springer;2012, pp. 281–295.

[41] Ono K, Muetze T, Kolishovski G, Shannon P, Demchak B. CyREST: Turbocharging cytoscape access for external tools via a RESTful API. *F1000Research* 2015;4:478.

[42] Goddard TD, Huang CC, Meng EC, et al. UCSF ChimeraX: Meeting modern challenges in visualization and analysis. *Protein Science*, 2017;14–25.

14 Analysis of the Signatures of Cancer Stem Cells in Malignant Tumors Using Protein Interactomes and the STRING Database

Krešimir Pavelić, Marko Klobučar, Dolores Kuzelj, Nataša Pržulj, Sandra Kraljević Pavelić

14.1 Outline

In this chapter we present and discuss current issues in tumor metastases treatment in the context of the widely accepted cancer stem cells theory. We propose to intensify development of new treatment strategies for malignancies that target tumor "stemness" biomarkers as to advance cancer treatment in general. Analysis and understanding of "stemness" pattern is a prerequisite for such approach and herein we present an example for analysis of different tumors' "stemness" biomarkers by use of the STRING database-generated interactomes.

14.2 Introduction

Recent advances in medicine have proven important in many fields. Examples include stem cell therapy and the development of precision medicine, including personalized oncology. Within the field of personalized oncology, targeted therapy is a particularly important area of both research and clinical application. This approach to tumor treatment has been exploited for several decades and is based on the concept of targeting specific biomarkers, including receptors or kinases, that are involved in tumor cell proliferation, invasion, and metastases. The terms "cancer" or "malignant tumour" are broad, applicable to over 100 different types of disease, which share the common features of abnormal cell growth and spreading from an initial site to distant regions. Depending on the outcome, two types of tumor exist: Benign tumors, which do not spread to other parts of the body, and malignant tumors that usually end as metastases if they are not discovered and treated in a timely manner. The most common varieties of malignant tumors are lung cancer, prostate cancer, breast cancer, colorectal cancer, cervical cancer, gastric cancer, and skin cancer. Numerous factors can influence tumor development and expansion, including chemical agents, diet, infective agents, genetic aspects, physical factors, hormones, and autoimmune diseases. Moreover, tumors are characterized by dysregulation of tissue growth, and genes that regulate these processes bear an important role in tumor development. Such genes are called either *oncogenes*, if they promote the growth of cells, or *tumor suppressor genes*, if they are involved in apoptosis or cell death. Usually mutations in these genes, or other genetic aberrations, are necessary in order to initiate the carcinogenesis process. Types of genetic aberrations associated with different tumors are numerous and have been found to include substitution mutations, insertions, deletions, duplications, rearranging or loss of chromosomes, and gene amplifications [1, 2]. In addition to genetic factors, epigenetic events also contribute to the occurrence and development of tumors. These are functionally relevant modifications of the genome which do not alter the gene sequence itself [3, 4] and include DNA-methylation (either hypermethylation or hypomethylation), histone modification [5], and alterations in chromosomes architecture caused by irregular protein expression [6] (detailed in Chapter 3). Such epigenetic changes can persist during cell division, and in some instances even be inherited by offspring. Epigenetic changes are commonly present in tumors, as illustrated by colon cancer, in which 147 genes have been seen to be hypermethylated and 27 hypomethylated. Of these hypermethylated genes, 10 were hypermethylated in 100% of colon cancer cases examined, while the remainder were heavily methylated in 50% of cases [7]. Finally, tumors are heterogeneous and different tumor types vary substantially, including in the major genetic alterations involved. Considering the biological complexity of tumors and the factors that contribute to them, it is not a surprise that tumors remain among the most difficult pathologies to treat and cure. In recent years, however, growing evidence in the scientific literature suggests a shift in tumor research that may allow the discovery of more effective cures. Indeed, data in the literature demonstrates the existence of two distinct processes occurring during tumor development and distinguishes at least two varieties of malignant cell: Transformed primary tumor cells and metastases. Therefore, novel therapeutic approaches should be designed with the aim of being applicable towards at least these two different types of cells, in spite of them possessing distinct biological features. There is therefore a need for deeper insight into the mechanisms underlying metastatic process and its occurrence.

14.2.1 Metastases and Cancer Stem Cells

Metastases remain incurable, and malignant tumors are still among the main causes of death worldwide. Despite the progress achieved in the past few decades, the exact mechanisms of tumor development and expansion remain elusive [8]. The majority of patients who do not recover from cancer develop systematic metastases and it is these that are the principle cause of death in patients with tumors [9]. It appears that for the process of metastasis to begin, a small niche in tumors, known as cancer stem cells, have to be both present and activated [10]. Indeed, the majority of tumor cells have a limited capacity of self-renewal, but a subset of them, the tumor stem cells, can self-renew indefinitely. They represent 1% of all cancer cells and contribute to increased proliferation and tumor formation. It is plausible that these cells are also the principle cause for both the observed resistance of many tumors to therapy, and also for the occurrence of metastasis [8]. Following their appearance, these cells enter the circulation in the body through the blood system, and can then initiate the process of tumor generation in other distant organs and tissues [9]. The process of metastasis is biologically very complex and presents a series of discrete processes by which cancer cells are transferred from their primary site to distant areas of the body. Tumor cells are thought to invade tissues from their primary sites by moving either into the lymphatic system, or directly to the blood system. Once in the blood, these cells have to survive the attack by the immune system. Under such circumstances, tumor cells require a change in their phenotype, through increased motility or invasive ability. Growth factors encourage these processes to occur, and the mechanisms involved differ substantially from the normal process of mitogenesis.

In addition to survival of the cells in the blood, another key mechanism for metastases is their subsequent colonization. Metastasis colonization is the process by which micrometastases form, followed by their progressive growth and vascularization in distant organs or tissues. Successful colonization depends on the interaction with the microenvironment and the degree of receptivity of the distant tissue. An expansion in blood supply, through vascularization, is essential for the growth of a metastasis [11]. An understanding of the malignant transformation processes and tumor development is especially important for precision medicine, often also known as personalized medicine [12, 13]. These approaches rely heavily on the application of both high-throughput ("-omics") methods and bioinformatics for the diagnosis of the disease, prognosis of its outcome and for the development and evaluation of more effective treatments and drugs. An improved understanding of the patient's personal state of health can also aid in selection of effective and secure individualized therapies, in which side effects are minimized and therapy costs reduced. As an example, in both chemotherapy and radiotherapy used for tumor treatment, a complete cure may not be achieved because some of the cells acquire stem cell-like traits. During radiotherapy, half of tumor cells are usually killed upon treatment but in some tumors, notably breast cancer, residual tumor cells have been found to have transformed into cancer stem cells during the therapy; such cell are known as *induced cancer stem cells* [14]. Moreover, radiotherapy is thought to promote metastases in some cases through the stimulation of signaling molecules, including growth factors and cytokines [15]. Similarly, chemotherapy treatment may also be problematic, as the cytostatic drugs used in reducing the mass of the tumor may also promote the expansion of a small niche of cancer stem cells with metastatic potential [16]. This process may be correlated with the increase in the expression of stem cell markers seen in many tumors, and potentially also with instances of resistance to chemotherapy. To date, some markers of cancer stem cells have been identified, including Nanog and Polycomb Group RING Finger Protein 4 (also known as BMI1), both of which are normally expressed in embryonic cells [17]. Therefore, new treatment strategies are required to target these biomarkers for "stemness," and their respective activities in the development of malignancy.

A targeted therapy directed against those particular cells would be a significant breakthrough in curing patients with metastases. One of the possible approaches might rely on identification of distinct signaling pathways of cancer stem cells, such as those usually expressed during embryonic growth (examples of which include the Hedgehog, Wnt, and Notch signaling pathways) [18]. Hedgehog signaling, for instance, was already seen as being active in some malignancies, and it has been implicated in the development of malignant tumors of the brain, lung, breast, and skin. Some substances are already known to target the Hedgehog processes, including cyclopamine, a steroid alkaloid which blocks Hedgehog signaling [19]. Unfortunately, cyclopamine has a low potency and bioavailability, which has prompted scientists to develop more potent inhibitors such as Vismodegib, Saridegib, and Sonidegib/Erismodegib [20]. Similarly, HhAntag (N-[4-chloro-3-[6-(dimethylamino)-1H-benzimidazol-2-yl]phenyl]-3,5-dimethoxy benzamide) is effective at inhibiting Hedgehog signaling in almost all tumors [21]. Moreover, Wnt signaling plays a role in both benign and malignant tumor development and metastasis, probably through activation of epithelial to mesenchymal transition (EMT). This pathway can be inhibited in hepatocellular tumors using a novel compound, FH-535, in combination with sorafenib [22]. Similarly, Notch signaling is often irregular in malignant tumors and metastasis, and is also known to modulate both EMT and angiogenesis. Its inhibition can be studied using γ-secretase inhibitors, and promising clinical phase I results have already been obtained in patients with advanced-stage solid tumors [23].

A second approach to the treatment of metastases would involve targeting immunomodulatory molecules, such as chemokines and their respective receptors. These molecules are involved in regulation of the chemotaxis of tumor cells. During this process, rearrangement of the extracellular matrix occurs through the activity of enzymes, such as metalloproteases, allowing the cells to more easily cross the basement membrane and enter the blood circulation [24]. For example, the C-X-C Motif Chemokine Receptor 4 (CXCR4), which has Stromal Cell-Derived Factor-1 (SDF-1) as a ligand, regulates chemotaxis in T-cells and has been found to play a role in the breast cancer metastatic processes. A number of possible downstream signaling events are known for this CXCR4-SDF-1 pathway, inlcuding PI3-kinase, FAK, RAFTK/Pyk2, the adaptor-like ubiquitin ligase protein Cbl and SH2-containing tyrosine phosphatase SHP2, all of which are highly relevant to the process of malignancy [25, 26]. Interestingly, interplay of Notch and Wnt signaling with chemokine activation during metastatic processes has been also suggested [27, 28]. Novel targeted therapies might therefore be relevant not only for solid tumor treatment, but also in the treatment of metastasis. However, a deeper understanding is still needed of the molecular events, biomarkers, and signaling processes involved, if substantial advances are to be made in this field. In particular, if cancer stem cells and the stemness phenotype are indeed driving metastasis formation, then we must learn more about the specific molecular features of these cells with regard to each tumor type.

Cancer-like stem cells are difficult both to study and to isolate as such cells derived from solid or hematopoietic tumors retain some characteristics of normal stem cells, notably the ability to generate progeny of all cell types. It is assumed that these cells persist in tumors as a distinct cell population or niche, which can in turn lead to the relapse or metastasis of tumors. Cancer stem cells possess both a long-term capability for self-renewal and also for differentiation into progenitors, which cannot themselves induce a tumor but can contribute to tumor development [29]. This means that only certain stem cell subpopulations have the ability to contribute to tumor progression, implying that they possess distinct intrinsic characteristics, which could theoretically be recognized and then targeted without the need to destroy the entire tumor [30, 31].

The presence of cancer stem cells is not merely theoretical, as a growing body of substantiate evidence has been discovered for the importance of these cells in the clinical outcome of various malignancies, including human leukemia and ovarian cancer [32, 33]. These outcomes can be explained by the existence of cancer stem cells, as described here, however, this does not exclude the possibility that standard clonal evolution of the tumor could also be of simultaneous relevance for the metastatic process. Evidence supporting the cancer stem cells model includes the fact that some metastases indeed, do not share the same properties as primary tumor cells. Furthermore, multiple lines of clinical evidence show that greater genetic complexity (variation in genetic aberrations) exists in metastases compared to primary tumors. This implies that metastases can evolve from small tumors and, even if they are not visible (*cancer of unknown origin*), metastases can still appear 5–7 years before diagnosis of the primary tumor occurs. Additionally, tumor cell dissemination can begin at the pre-invasive stage of tumor progression. Finally, the genotype of metastatic tumour cells is unrelated to the physical size of the tumor.

These findings do not, however, exclude the co-existence of the clonal theory and the dependence of cancer stem cells on proliferation factors for survival and proliferation. Cancer stem cells may give rise to a specific clone that is resistant to therapeutic interventions, as is the case with human chronic myeloid leukemia stem cells that can become insensitive to imatinib, despite inhibition of BCR-ABL activity [33]. Current treatment developments for curing patients with tumors are based on destroying the primary tumor cells, however, metastatic cells may remain unharmed and continue to drive the pathological process. Elimination of the cancer would therefore be impossible without also eliminating the cancer stem cells. To achieve this difficult goal, development of novel strategies is paramount, both for investigation and analysis of cancer stem cell properties and novel stemness biomarkers as potential therapeutic targets. Here, we describe the potential applicability of bioinformatics tools and protein interactome studies in the elucidation of novel potential cancer stem cell features and biomarkers, including stemness patterns and processes.

14.3 From Proteins to Interactomes in Cancer

Proteins are dynamic biomolecules that are heavily regulated in an organism by both post-transcriptional mechanisms and post-translational modifications. These molecules are involved in almost every aspect of cellular function, including biochemical reactions, regulation of cellular metabolism, cell–cell communication, and signaling [34, 35]. Furthermore, the most of these cellular processes are regulated by the formation of protein interaction complexes, involving direct protein interactions at the transcriptional level, and through regulation of cellular signalling pathways [36]. Proteins are therefore responsible for the maintenance of normal cellular homeostasis and reflect almost every functional change at all the levels of organization of living organisms including cells, tissues and organs [37, 38]. Proteins are also involved in the development and maintenance of malignant processes or cancer. Cancer, is a complex multifactorial disease that is associated not only with aberrant protein expression, but also with structural changes in different protein species. These lead to the formation of cancer-related proteins and associated protein complexes, which in turn play an important role in connecting various cellular signaling pathways involved in disease pathogenesis [39, 40].

Since disease processes often manifest, and therapeutic treatments act, at the protein level, the field of proteomics and its associated methodologies have become an invaluable tool for use in many aspects of cancer research, including cancer biomarker discovery. The field of proteomics encompasses the systematic study of the proteome (the complete set of proteins in a living system at a specific moment, and under defined state or set of physiological conditions), including their structure, function, or expression of the proteins, that are subject of structural, functional, and expression proteomics fields, respectively [41, 42, 43, 44]. In particular, expression proteomics occupies an important niche in the field of cancer research, because it provides a comprehensive insight into the cellular mechanisms that underlie disease pathogenesis, and has a role in the identification of novel diagnostic and prognostic cancer biomarkers and processes [45].

Several proteomic strategies have been developed for identification, characterisation and expression analysis of several hundred proteins which are expressed variously in cells, tissues, or biological fluids [46]. These rely mostly on the use of sophisticated high-throughput techniques, based on liquid chromatography coupled with mass spectrometry (LC/MS). These techniques enable the simultaneous expression analysis of a large number of proteins extracted from complex protein mixtures, taken from various biological sources, in a single experiment with high reproducibility, including protein identification and relative or absolute quantification [47, 48]. Quantification can be based on protein labeling or else be label-free. Various stable isotopes labels are used for chemical [49, 50], enzymatic [51], or metabolic [52] labeling of protein samples, while additional labeling methods such as Isobaric tags (iTRAQ), have been developed for relative and absolute protein quantification by LC/MS [53].

In expression proteomics studies, the major difficulty arises from the complexity of the samples, and therefore a number of enrichment and sample pre-fractionation strategies for proteins and peptides have been developed in order to facilitate in-depth analysis of the sub-proteome, such as phosphorylated, glycosylated, or organelle-specific proteins [54]. In general, there are two main expression proteomic approaches which are used in the field of cancer research: "shotgun" proteomics and "targeted" proteomics. In the shotgun proteomic approaches, the extracted proteins from complex biological samples are subjected to enzymatic digestion, usually by trypsin, prior to mass spectrometric analysis, with LC/MS analysis typically employed for quantitation of proteins. This approach relies on the use of integrated chromatographic signal intensity of tryptic peptides as a measure of a protein abundance and subsequent identification of eluted tryptic peptides by use of tandem mass spectrometry (MS/MS) and protein database searching [55, 56]. In contrast, the targeted proteomics approaches rely on single reaction monitoring (SRM) methodology, in which only pre-selected peptides are used for relative or absolute quantitation and identification [57]. Many such processes can occur in parallel (multiple reaction monitoring, MRM) and used for monitoring of hundreds of different protein species derived from a single complex biological sample. The targeted proteomic approach does not suffer from the same limitations in reproducibility and sensitivity caused by high sample complexity that is often encountered in the shotgun proteomics experiments [57].

Modern mass spectrometry platforms are characterized by their substantially increased analytical performance compared to older systems, in particular due to their increased scan speed and resolution. They rely on large, comprehensive proteomic databases. Therefore, modern expression proteomics experiments in the field of cancer research usually deliver results in the form of long and complicated lists of differentially expressed proteins, potentially several hundred or thousand over a wide dynamic range. Such datasets may be difficult to interpret systemically, as cancer is a complex pathology associated with complex interplay of different cellular signaling pathways and biological processes. Therefore, better insight into the systemic features of carcinoma are required, which may be partially achieved through filtering LC/MS results, by which we mean the grouping of proteins relevant for cancer progression with corresponding biological processes, cellular signaling pathways, or analysis of their interaction partners using various bioinformatics tools [58].

Bioinformatics processing of the results obtained through high-throughput proteomic methodologies is the most important step during the post-proteomic analysis, as it may be used in the subsequent generation of hypotheses [59]. Bioinformatics encompasses a wide spectrum of research areas and is considered an interdisciplinary field, combining computer sciences, mathematics, and statistics, in order to classify and analyze large amounts of biological data (detailed in previous chapters). Bioinformatics is therefore expected to yield major contributions in the elucidation of open questions regarding the molecular mechanisms of disease [60]. For example, bioinformatics can be used to perform functional annotation of the large numbers of identified proteins through grouping of differentially expressed proteins according to their molecular function, subcellular localisation, or other biologically relevant features. One of the most widely used bioinformatics tools for functional classification of proteins is the geneontology (GO) database (www.geneontology.org/). GO provides a standardized vocabulary of defined terms with which the properties of gene products can be represented. It therefore enables categorization of genes and their products according to their molecular function, subcellular localization and biological processes.

Another important step in functional analysis of proteomic results is the classification of proteins based on their associated cellular signalling pathways. The most popular pathway databases used for this purpose are Kyoto Encyclopedia of Genes and Genomes (KEGG; www.genome.jp/kegg/) and PANTHER (Protein ANalysis THrough Evolutionary Relationships; www.pantherdb.org/). KEGG is a database resource for the understanding of high-level functions and utilities of the biological system from molecular-level information. It is particularly useful for analysis of "large-scale molecular datasets generated by high-throughput technologies." It forms an integral part of the KEGG PATHWAY database which enables automatic construction of molecular pathways based on a database of manually drawn molecular pathway diagrams (reference maps) [60]. The PANTHER classification system, meanwhile, provides an option to statistically enrich large protein datasets based on the cellular signaling pathways in which these proteins are individually involved [61, 62].

Functional annotations using these tools are often then followed by further, deeper, analysis into the identified proteins' biological roles. For example, protein–protein interactions (PPI) within a given proteomics dataset could be analyzed. PPI are crucial for the correct regulation of almost all biological processes. Conversely, dysregulated biological PPI networks occur in pathological conditions, including cancer and other complex diseases [63]. A comprehensive analysis of PPIs and interaction patterns on a systemic level can therefore provide insight into disease states in a hypothesis-free manner, facilitating the generation of novel hypothesis-driven ideas for therapeutic interventions (detailed in previous chapters)[64]. To date, multiple experimental methods have been used to investigate PPIs. One is based on the use of yeast two-hybrid screening. Another involve the use of powerful mass spectrometry-based methods such as tandem affinity purification. Other experimental methods for PPI detection include protein microarray technology (PMA), surface plasmon resonance (SPR), or the phage–display approach, all of which are described in detail elsewhere [63, 65, 66]. Interestingly, *in silico* computational approaches have greatly contributed to our understanding of PPI, as they facilitate the bypassing of labor-intensive and expensive experimental techniques. *In silico* approaches include methods based on theoretical data, including but not limited to: Predicted interactions from gene fusing events, the mutual occurrence or absence of protein pairs or protein sequences.

In parallel, novel integrative *in silico* computational strategies have been developed based on machine learning algorithms. As an example, one such algorithm utilizes text mining (TM) technology in order to extract and analyze PPI data from collections of full length research papers, including their titles, abstracts, main text, and figures [67]. However, in spite of technological improvements in high-throughput methodologies and *in silico* computational methods for investigating PPI, a large proportion of PPI within humans remains to be explored, and we understand only a fraction of the cellular interaction network [63].

Within bioinformatics, the discipline of *interactomics* therefore aims to investigate interactions between biological macromolecules, including both protein–protein interactions and their biological consequences, in order to elucidate complex biological interaction networks, known as *interactomes*. As we saw in previous chapters, an interactome represents the complete set of known interactions within a specific cell, in particular the interactions of physiologically relevant macromolecules, such as protein–protein interactions. However, understanding of PPI interactions must be combined with knowledge of the proteomes of the relevant type of cell, tissue, or organism in order to derive biologically relevant conclusions. An overview of existing known interactomes is available via Agile Protein Interactomes DataServer (APID; apid.dep.usal.es), which is an interactive bioinformatics web tool that provides a unified generation and delivery of protein interactomes, mapped to their respective proteomes [68].

As described in previous chapters, in the most basic form, PPIs can be mathematically described as a graph G $(V;E)$ that contains nodes, or vertices (each node representing a certain protein) and edges, which together describe physical interactions between two populations of protein species [69]. Such a graphical representation is not, however, an exact replica of the in vivo PPI network, as it does not model the function of the obtained PPI network. For this reason, the theoretical graphical model of a PPI network is usually a labeled graph, with its nodes and edges described by known attributes. For instance, the node attributes can be a protein symbol and protein expression level represented by color, while the edge attributes can indicate the protein's mode of interaction (e.g., physical interaction, inactivation, or repression). It has been shown that biological interaction networks have a scale-free and modular organization (see previous chapters). In scale-free networks, a small number of nodes (hubs) are highly interconnected, and these normally represent proteins with critical roles in biological systems modeled by these networks [68]. PPI networks have modular organization with interconnected subnetworks, or modules, that contain several nodes, serving distinct biological functions [69]. Previously determined PPI interactions are deposited in numerous PPI interaction databases, including the Molecular INTeraction database MINT (MINT) [70], the Biological General Repository for Interaction Datasets (BioGrid) [71], IntAct [72], HomoMint [73], and the Biomolecular Interaction Network Database BIND (BIND) [74]. The main difference between existing interaction databases is in the type of evidence in their PPIs used for the interactome construction. For instance, experimental and prediction-based data may both be used to link proteins into a PPI interactome. Once the PPI interactome is constructed, visualization can be performed by using various resources, with the most widely used PPI analytical bioinformatics platforms being Cytoscape [75] (detailed in Chapter 13) and Search Tool for the Retrieval of Interacting Genes/Proteins (STRING) [76]. In this chapter, we focus on potential applications of the STRING database, while Cytoscape is extensively described in Chapter 13.

STRING is a database of known and predicted protein-protein interactions, which at the time of writing (Version 10.5, current: since May 14, 2017) includes 9,643,763 distinct proteins from 2,031 organisms. The direct and indirect PPI interactions curated within the STRING database are derived principally from experimental data and from computational PPI prediction methods. In addition, the STRING database is connected to external sources and can perform automated text mining of public collections of research articles, such as PubMed [77]. As a consequence of the integration of multiple data sources, STRING can only take into account an individual protein–gene locus or representative protein sequence. This means that alternatively spliced or post-translationally modified forms of proteins are not considered spatially by the STRING database during its analysis [78]. Furthermore, the STRING web server has the capability to visualize protein–protein interaction networks in either a static or interactive mode, all based on the provided gene list and interactions contained in the STRING

Table 14.1: Databases that may be used for the construction and study of PPIs.

Database	Description	Web page
DIP	Database of experimentally determined interactions between proteins.	http://dip.doe-mbi.ucla.edu/
MINT	Database of experimentally verified PPIs, mined from the scientific literature by expert curators.	http://mint.bio.uniroma2.it/mint/
SCOPPI	Database of all domain–domain interactions and their interfaces which can be derived from PDB structure files and SCOP domain definitions.	http://scoppi.biotec.tu-dresden.de/scoppi/
BioGRID	Interaction repository with data on proteins, genetic interactions, chemical associations, and post-translational modifications, from major model organism species.	www.thebiogrid.org/
IntAct	Database system and analysis tools for molecular interaction data derived from curation of the literature or direct user submissions.	www.ebi.ac.uk/intact/

database [79]. PPI networks can be searched for based on protein name, accession number, or amino acid sequence. The most basic interaction unit in STRING database is the functional association of two proteins, which may or may not contribute to a common biological purpose [79]. For each protein–protein associations determined in this manner, a confidence score, or "edge weight" ranging from 0 to 1, is provided in the context of the PPI network. These scores indicate estimated likelihoods for interactions between proteins and may be assumed as biologically relevant and specific, with respect to the two proteins jointly contributing to a shared biological function. The confidence scores are supported by several lines of evidence, which are divided into seven channels, including experimental, database, text mining, co-expression, neighborhood, fusion, and co-occurrence [78]. Additionally, the STRING database has a function allowing for statistical enrichment by Fisher's exact test, followed by correction for multiple testing, of PPI networks on the basis of their biological process and molecular functions, as determined through integration of various functional classification systems (for example, geneontology and KEGG [79]). Other databases that may be of use in the construction and study of PPIs in cancer research are listed in Table 14.1.

14.4 Cancer Stem Cells Biomarkers within a Cell's Protein Interaction Network

Using the currently available protein databases and tools for creation analysis and visualization of PPI networks, we have analyzed PPIs in several pathological states, including a fibromatosis known as Dupuytren's disease, pediatric idiopathic nephrotic syndrome (INS), and breast cancer [80, 81, 82]. In these studies, we successfully identified interesting biomarkers which underlie either disease pathogenesis, or processes that were previously unstudied in the context of these conditions. For instance, we identified an increased expression of receptors ERBB-2 and IGF-1R, as well as activation of Akt signalization in Dupuytren's disease patients palmar nodes by analyzing expression proteomics data through constructed PPI networks and further immunohistochemical validation of the identified targets [80]. Also, we suggested for the first time that many molecular species identified in the study of the idiopathic nephrotic syndrome through PPI construction are involved in oxidative stress, which presents novel opportunities for therapeutic interventions in that group of patients [82].

With the same rationale in mind, we constructed PPIs of biomarkers, selected based on a comprehensive review of the literature regarding stem cell markers in tumors. The search was performed on PubMed database (www.ncbi.nlm.nih.gov/pubmed) with the following keywords: Cancer-like stem cells, tumor stem cells, mesenchymal stem cell marker, tumor marker, stem cell marker, metastatic markers, metastases, prostate cancer, ovarian cancer, glioblastoma, breast cancer, and colorectal cancer. The specific goal of this was to identify common markers of stemness or biological processes that are shared by various types of malignant tumor. This approach was expected to shed new light onto the tumor pathogenesis and metastatic processes. First, we created a comprehensive list of stem cells markers in different tumour types. Second, we used the online database STRING to create the interactomes for prostate and breast cancers, based on this generated list of stem cells markers (Table 14.2). In these two interactomes, we tested the utility of the created stem cell marker PPIs to explain general biological features of the corresponding tumors or its druggable targets. The final goal was the identification of the common stemness elements, or "stemness signature," shared by all types of the analyzed tumors. In generating the interactomes, the "multiple proteins names" search option was used and the ensuing protein interaction networks are presented using the "evidence mode" setting. Two alternative display modes for PPI networks, "molecular action" and "confidence mode," were also available in the database menu of STING, under Settings. Numerical values for basic descriptive network properties of the constructed PPIs networks are given in the description. These are: The average node degree or average binding coefficient per protein as an average number of interactions that a protein forms within the network, the average local clustering coefficient as a measure of node clustering within the network and the PPI functional enrichment p-value. Highly connected PPI networks have high average node degrees and clustering coefficients values. A low PPI enrichment p-value indicates a level of functional organization of nodes (non-randomness) and we observed that anumber of edges (interactions) present in the network were significant. Finally, each of the generated cancer stem cell biomarker PPI were manually inspected for the presence of physiologically relevant interaction hubs, based on the highest number of PPI for individual proteins, including interactions derived from both experimental and predicted data. Some of the most interesting interactions determined to be between certain proteins indicate a potential role in disease pathogenesis, and should therefore be considered for further exploration by experimental methods.

Table 14.2: List of cancer stem cell biomarkers in four types of highly malignant tumors generated by searching PubMed database (www.ncbi.nlm.nih.gov/pubmed).

Prostate cancer	Ovarian cancer	Glioblastoma	Breast cancer	Colorectal cancer
ACVR1B	PBX1	COL5A3	CXCR7	MCC
ACVR1C	STAT3	STAT6	CD44	CD44
STAT3	NANOG	CHI3L1	ALDH1	ALCAM
POU5F1	SOX2	KLRC3	OCT4	FAM49B
SOX2	ROR1	PRUNE2	NANOG	CLU
KLF4	PROM1	FOXD1	RAD51	Lgr5
BRCA1	CCL5	COX7A2L	CDK7	APC
BRCA2	ALDH3A1	GPR133	CCNH	KRAS
KLK3	FOXP1	PROM1	MNAT1	TP53
AR	ABCG2	Notch1	SALL4	DCLK1
CD44	KRAS	HEY1	PRKD1	PROM1
PROM1	BRAF	HEY2	CASP3	ALDH1A1
ABCG2	PTEN	KLF9	FADD	DPP4
CASP3	CTNNB1	SNAI2	SOX2	NANOG
ANXA1	ERBB2	ROR1	IL-6	POU5F1
NANOG	PIK3CA	CXCR4	IL-8	SOX2
SNAI1	TP53	SLC2A1	BCL6	SMAD4
ALDH7A1	NRAS	SerpinB9	TNFRSF10B	KLF5
CK14	USP9X	VEGFA	TNFRSF10D	BMI1
HIF1a	EIF1AX	HIF1a	Lgr5	CD274
VIM	BMP4	EPAS1	TP53	DCLK1
GH1	RPSA	EGFR	CXCR4	BCL2L1
TIMP2	IL17A	Olig2	CXCL1	DLL4
FGF-2	CCR1	POU5F1	HMGCS1	HEY1
MYC	CCR3	NANOG	ABCG2	TFF2)
PSCA	CCR5	SOX2	ERBB2	EpCAM
ESR1	CXCR4	FoxG1	KLF4	ALDH3A1
TNFRSF8	POU5F1	FoxM1	PON1	SNAI1
ICAM1	ALDH1A1	CD9	ANTXR1	IL8
CD34	CD44	ZFX	EGF	KLF4
Flt-1	MyD88	THY1	FGF2	AKT1
KITLG	NES	HBA1	FoxM1	Foxo3a
EGF	CTNNB1	BMI1	ALDH3A1	TNFSF10
ITGA2	Notch1	MELK	STAT3	OXA1L
ITGA6	Notch4	PSPH	PDCD2L	
SOX9	OXA1L	HOXD9	IGF1R	
SFRP1		FUT4	Rab2A	
TGFb1		HDGF	KAT7	
CXCL12		PINK1	BRCA1	
NES		RICTOR	CDH3	
CD28		VTCN1	LBH	
ITGB1		NES	WISP1	

Table 14.2: Cont.

Prostate cancer	Ovarian cancer	Glioblastoma	Breast cancer	Colorectal cancer
BMI1		IGF1	TCF4	
Twist1			CXCR1	
CDH1			CXCR2	
CDH2			B4GALNT1	
FN1			NGFR	
			ITGA6	
			SH2D4A	
			YES1	
			Tbx3	
			PROM1	
			OXA1L	

Cancer stem cells biomarkers listed in Table 14.2 were initially used for construction of cancer stem cell PPI network in prostate cancer (Figure 14.1). This PPI network consists of 48 distinct proteins, that participate in 416 protein–protein interactions. As displayed in Figure 14.1, this interaction network of prostate cancer stem cell biomarkers represents a dense PPI network, with an average binding coefficient per protein of 17.3 and an average local clustering coefficient of 0.742. These figures are indicative of highly confident interactions between the proteins of the network. The PPI enrichment value of this interaction network is 0, demonstrating that the proteins present in the network are at least partially biologically functionally connected. Based on proteins with the highest number of established protein–protein interactions, both experimental and predictive, several major mutually interconnected protein hubs for prostate stem cell biomarkers PPI network were selected as potentially relevant for disease pathogenesis. These include estrogen receptor 1 (ESR1), signal transducer and activator of transcription 3 (STAT3), tissue metallopeptidase inhibitor 2 (MMP2), fibroblast growth factor 2 (FGF2), and stromal cell-derived factor 1 (CXCL12). Interestingly the CD28 protein is the only prostate cancer stem cell biomarker that does not show any known interactions with other proteins present in this PPI network. STAT3 has a central position in the protein interaction network, which is unsurprising given that it is considered to be an oncogene associated with increased proliferation, invasiveness, and immune tolerance in many types of tumor [83].

In addition, there are several lines of evidence to suggest that STAT3 interaction with the above mentioned protein hubs within this network is relevant for prostate cancer progression. For example, interaction between STAT3 protein and estrogen receptor 1 (also known as estrogen receptor α, ERα) might underlie prostate cancer pathogenesis. Indeed, up-regulation of the estrogen receptor expression induced by STAT3 was reported in lung adenocarcinoma cells [84]. Conversely, expression of ERα was seen to stimulate leptin-induced STAT3 activity in breast cancer cells [85]. More importantly, ERα is already established as a proliferative and pro-inflammatory factor in prostate cancer, which activates mitogen-activated protein kinases (MAPK) and phosphoinositide 3-kinase (PI3K), and indeed blockers of ERα are already consid-

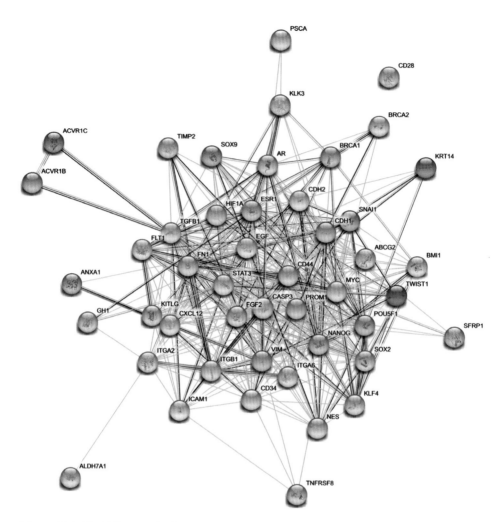

Figure 14.1: The PPI network of prostate cancer stem cell biomarkers generated by using the STRING database network evidence mode (https://string-db.org/).

ered as therapeutic options for prostate cancer [86]. On the other hand STAT3 has been established as transactivator of steroid receptors, including androgen (AR) and estrogen (ER) receptors, in prostate cancer cells [87, 88]. Moreover, it is well known that aberrant activity and expression of growth factor receptors and their ligands is associated with malignant progression in various types of tumor [89]. A further interaction of the STAT3 protein within the network is with basic fibroblast growth factor 2 (FGF2), further indicating its potential biological relevance for prostate cancer, as previously suggested in the scientific literature [90]. Finally, amongst the major hubs observed in the constructed prostate cancer PPI network is the vascular endothelial growth factor receptor FLT-1, which is involved in prostate carcinoma growth and metastasis [91]. In addition to supplying supporting evidence for well-established prostate cancer pathogenesis biomarkers, this newly generated PPI network suggests the existence of putative novel prostate cancer candidates, the details of which remain to be determined. One example would be POU class 5 homeobox 1 (POU5F1), a major regulator during embryonic development that is important in stem cell pluripotency. Loss of function of this protein has already been associated with several types of malignancy [92, 93]

Unlike in prostate cancer, ovarian cancer stem cell PPI network consists of 36 distinct proteins that together form only 196 interactions (Figure 14.2). The average binding coefficient per protein in this network is 10.9, while the average local network clustering coefficient is 0.673. In addition, the PPI enrichment value of the presented protein interaction network is 0. Several network hubs were designated as potentially of physiological relevance, based on the highest number of established interactions. These hubs are interconnected with each other, and include the proteins STAT3, β-catenin 1 (CTNNB1), tumor protein p53 (TP53), phosphatidylinositol-4,5-bisphosphate 3-kinase (PIK3CA), and Notch 1 (NOTCH1) proteins. Organic cationic transporter 3 (OXA1L) and aldehyde dehydrogenase (ALDH3A1) are the only two biomarkers not determined to form connections within this protein interaction network. β-catenin is responsible for regulation of cellular adhesion and whose overexpression is associated with progression of tumors [94]. Furthermore, β-catenin is involved in multiple oncogenic pathways, notably including Wnt signaling, and β-catenin activity is positively regulated by STAT3 signaling in colorectal cancer cell

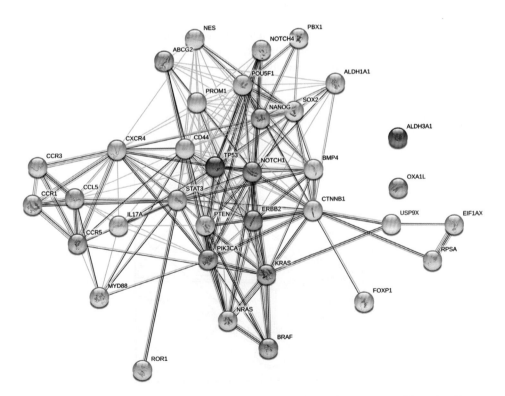

Figure 14.2: The PPI network of ovarian cancer stem cell biomarkers generated by using the STRING database network evidence mode (https://string-db.org/).

lines [95]. In addition, down-regulation of β-catenin reduces STAT3 protein expression in ALK-positive anaplastic large cell lymphoma cell lines [96].

Interaction of β-catenin with STAT3 was similarly observed in the ovarian cancer PPI network described here, pointing to a possible role in pathogenesis of the condition. Indeed, a recent study provides data on the role of STAT3 and Wnt/β-catenin signaling in the maintenance of the stemness phenotype in ovarian cancer, through miR-92a/DKK1–regulatory pathways [97]. A growing body of evidence confirms the role of STAT3 in the development of tumors, and in particular through its ability to inhibit expression of p53 tumor suppressor protein [98]. As might therefore have been expected, an interaction of STAT3 with p53 interaction is also observed in the ovarian cancer PPI network, implying a possible repression of p53 (Figure 14.2). Another interesting observation in this PPI network is the Notch1 and phosphatidylinositol 3-kinase catalytic (PI3KCA) biomarkers hubs. Both of these proteins contribute to tumor stemness and metastasis and are dysregulated in many types of malignant disease [99, 100]. Moreover, Notch 1 upregulates PI3K/Akt signaling, which contributes to anti-apoptotic processes in various tumors, including ovarian cancer [101, 102]. These targets are already being exploited in the development of ovarian cancer treatment options [103].

Finally a large cluster in the ovarian cancer PPI network described here is formed by multiple *chemokines* (CCR1, CCR3, CCR5, and CCL5; Figure 14.3) as well as the inflammatory molecules interleukin 17 and Myd88, that underlie the importance of an active inflammatory environment for ovarian cancer progression [104, 103]. Chemokines are small, secreted proteins with a role in mediating immune cell trafficking. In the tumor microenvironment, chemokines can also be expressed either by tumor or host cells, and they contribute to tumor invasiveness and stemness [105]. Furthermore, ovarian cancer stem cell biomarker chemokine CCL5 may bind to three chemokine receptors present in the network, CCR1, CCR3, and CCR5, and therefore cause accumulation of immunosuppressive myeloid cell in tumor microenvironment.

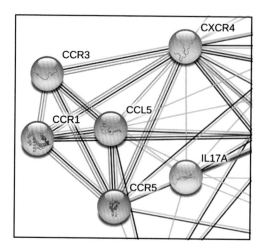

Figure 14.3: Interactions of chemokine receptors and their ligands with the STAT3 protein in the ovarian cancer stem cell biomarker interaction network (https://string-db.org/).

Additionally, it has been suggested that CXCR4 positive tumor cells may have stem-like properties and a high metastatic potential [106]. Moreover, CXCR4 mediated signaling is responsible for the activation of the STAT3 protein in multiple cancers, and has been identified as a possible target for anticancer therapy [107, 108]. Another important hub in this PPI network is KRAS, mutations in which were previously reported in ovarian cancer [109]. The genetic status of this biomarker has also been recently reported as being an important predictor of sensitivity of ovarian cancer cells to decitabine in a xenograft model [110]

The prostate and ovarian cancer PPI networks generated and described here therefore clearly provided a comprehensive insight into significant and biologically relevant processes underlying these malignancies. Therefore, we proceeded to attempt identify common processes (a "stemness pattern") across all five of the types of tumor analyzed, by using all proteins listed in Table 14.2. The obtained "stemness PPI network" (Figure 14.4) generated in this manner comprised 156 nodes, with an

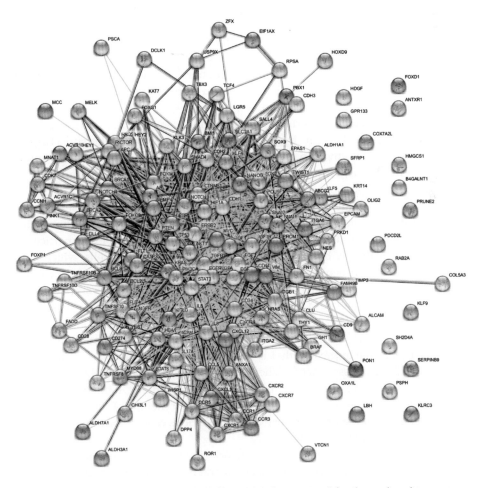

Figure 14.4: The hypothetical "stemness PPI network signature" for the analyzed tumors: Prostate cancer, ovarian cancer, glioblastoma, breast cancer, colorectal cancer.

average node degree of 24 and the average local clustering coefficient of 0.618. The number of predicated interactions that would be predicted using this number of randomly selected proteins from the genome would be 585, but unexpectedly the total number of observed interactions was 1,870. This implies that the collection of proteins analyzed possess considerably more interactions among themselves than would be expected by chance. Such an enrichment is a strong indication that the proteins are biologically connected as a group, and could therefore be hypothetically treated as a putative "stemness PPI signature" for the set of the cancers analyzed. As anticipated, general biological processes observed in the PPI network included the positive regulation of macromolecule metabolic process, regulation of cell proliferation, positive regulation of cellular metabolic process, positive regulation of metabolic process and positive regulation of gene expression. Interestingly, possible common pathways implicated for these malignancies included: Pathways in cancer, cytokine–cytokine receptor interaction, proteoglycans in cancer and the PI3K-Akt signaling pathway (KEGG). Even though many of these processes are well studied in various tumor types, the suggestion here that proteoglycans may play a role in cancer generally, and cancer stem cells specifically, represents an innovative approach arising from this work. In particular, the extracellular matrix (ECM) around cancer cells differs substantially from the ECM of normal cells and tissues [111], and could potentially be exploited as a target for treating tumors [112, 113]. Proteoglycans (PG) are proteins that covalently attach to unbranched carbohydrate chains of repeating disaccharide units, known as glycosaminoglycans (GAGs), and include a variety of molecules such as aggrecan, perlecan, biglycan, lumican, decorin, heparan sulfate PG, and fibromodulin. PGs are hydrophilic and are involved in the spreading of growth factors and the activation of cellular signaling events of importance for tumor proliferation. In contrast to normal cells, cancer cells are negatively charged due to the presence of phosphatidyl-serine, which is exposed on the outer side and can be targeted by certain positively charged cytotoxic peptides. Notable examples include arginine-glycine-aspartic acid (RGD) or asparagine-glycine-arginine (NGR) peptides used to target tumor vasculature [114], or cell-penetrating peptides [115]. In particular, a role for PGs has already been demonstrated in many progenitor cells, which may imply a similar role for them in the maintenance of the stemness signature in cancer. Specific examples include melanoma chondroitin sulphate PG in epidermal stem cells [116], or chondroitin sulphate PG neuronal/glial 2 in the central nervous system [117]. Moreover, modulation of PG expression has been shown to reduce tumor growth, both in vitro and in vivo [118, 119]. PGs were also shown to be potential biomarkers for the personalized medicine approach, that is in the process of prostate cancer diagnosis and therapy [120] and assessment of therapeutic response in cancers [121]. Based on these accumulated data from the literature and scientific evidence, it therefore appears these molecules have the potential to be crucial targets for therapy of cancer stem cells. Further elucidation of their role and expression in various cancers may therefore yield novel opportunities for treatment.

Finally, the common "stemness PPI network signature" for the analyzed tumor types contains several important stem cells markers. This is important for understanding the metastatic processes. Metastases may indeed be partially explained as a consequence of a stem cell-like phenotype responsible for tumor invasion and dissemination to distant sites. Here, we have identified major drivers of this stem-cell phenotype to be the transcriptional regulators POU5F1, NANOG, and SOX2, and the cell surface antigens CD34, and CD44. Together these clearly show stem-like properties in the five cancers analyzed. Targeting of these molecules, and their corresponding biological processes, may therefore interfere with the maintenance of the stem cell-like phenotype of cancer stem cells, and potentially diminish their survival and metastasis.

14.5 Exercises

14.1 Explain the role of proteomics in the field of cancer research.
14.2 Define the concepts of functional, structural, and expression proteomics.
14.3 List and explain the basic strategies of mass spectrometry-based proteomics in the field of cancer research.
14.4 Explain the difference between the *shotgun* and *targeted* proteomic approaches.
14.5 Explain the role of bioinformatics in post-proteomic analysis.
14.6 Describe how results of proteomic research can be processed using bioinformatics tools.
14.7 List several major databases that are commonly used for the functional classification of proteomic research results.
14.8 Explain why it is important to understand PPI when investigating complex diseases such as tumors.
14.9 Explain the importance of *in silico* computational methods in the investigation of PPI.
14.10 Explain how PPI networks can be mathematically described in the simplest way.
14.11 Specify the types of organisation of biological interaction networks.
14.12 Give examples of the various bioinformatics resources that are most commonly used for visualization of PPI.
14.13 Explain the meaning of the confidence score which is provided for description of each protein–protein interaction within the STRING database.
14.14 Construct a protein interaction network in the evidence mode of the STRING database (https://string-db.org/) by using the list of lung cancer stem cells biomarkers generated by PubMed searching (www.ncbi.nlm.nih.gov/pubmed/) and provided in Table 14.3. Describe the protein interaction network generated by selecting the most significant biomarker hubs. Calculate the network *total connectivity* (C) by using the given equation $C = E/N(N-1)$ where where E is the number of edges and N the total number of nodes. Discuss the results obtained.

Note: Solutions are available to instructors at www.cambridge.org/bionetworks.

14.6 Acknowledgments

We greatly appreciate access to equipment belonging to the University of Rijeka, as part of the project "Research Infrastructure for Campus-based Laboratories at University of Rijeka," co-financed by the European Regional Development Fund (ERDF).

Table 14.3:

Lung cancer stem cells biomarker list
OCT4
Nanog
CTNNBIP1
Sema3a
ACVR1B
ACVR1C
PROM1
CD44
ALDH3A1
ABCG2
EPAS1
EpCAM
CD38
Sox2
EGFR
Akt
AKT1
Bcl-2
VMP1
PTEN
SNX1
SGPP1
CAV1
p53
CD24

We acknowledge funding from the Croatian Science Foundation (project no. IP-2013-11-5709 "Perspectives of maintaining the social state: towards the transformation of social security systems for individuals in personalized medicine") and the University of Rijeka (grants 13.11.1.1.11 and 13.11.1.2.01). This work was supported by the European Research Council (ERC) Starting Independent Researcher Grant 278212, the European Research Council (ERC) Consolidator Grant 770827, the Serbian Ministry of Education and Science Project III44006, the Slovenian Research Agency project J1-8155, and the awards to establish the Farr Institute of Health Informatics Research, London, from the Medical Research Council, Arthritis Research UK, British Heart Foundation, Cancer Research UK, Chief Scientist Office, Economic and Social Research Council, Engineering and Physical Sciences Research Council, National Institute for Health Research, National Institute for Social Care and Health Research, and Wellcome Trust (grant MR/K006584/1).

References

[1] Stephens PJ, McBride DJ, Lin MJ, et al. Complex landscapes of somatic rearrangement in human breast cancer genomes. *Nature*, 2009;462(7276):1005–1010.

[2] Berger MF, Lawrence MF, Demichelis F, et al. The genomic complexity of primary human prostate cancer. *Nature*, 2011;470(7333):214–220.

[3] Baylin SB, Ohm JE. Epigenetic gene silencing in cancer: A mechanism for early oncogenic pathway addiction? *Nature Reviews Cancer*, 2006;6(2):107–116.

[4] De S, Michor F. DNA secondary structures and epigenetic determinants of cancer genome evolution. *Nature Structural & Molecular Biology*, 2011;18(8):950–955.

[5] Kanwal R, Gupta S. Epigenetic modifications in cancer. *Clinical Genetics*, 2012;81(4):303–311.

[6] Baldassarre G, Battista S, Belletti B, et al. Negative regulation of BRCA1 gene expression by HMGA1 proteins accounts for the reduced BRCA1 protein levels in sporadic breast carcinoma. *Molecular & Cellular Biology*, 2003;23(7):2225–2238.

[7] Schnekenburger M, Diederich M. Epigenetics offer new horizons for colorectal cancer prevention. *Current Colorectal Cancer Reports*, 2012;8(1):66–81.

[8] Foffi G, Pastore A, Piazza F, Temuss PA. Macromolecular crowding: Chemistry and physics meet biology (Ascona, Switzerland, 10–14 June 2012). *Physical Biology*, 2013;10(4):040301.

[9] Gkountela S, Aceto N. Stem-like features of cancer cells on their way to metastasis. *Biology Direct*, 2016;11:33.

[10] Mimeault M, Batra SK. Molecular biomarkers of cancer stem/progenitor cells associated with progression, metastases, and treatment resistance of aggressive cancers. *Cancer Epidemiology Biomarkers & Prevention*, 2014;23(2):234–254.

[11] Steeg PS. Tumor metastasis: Mechanistic insights and clinical challenges. *Nature Medicine*, 2006;12(8):895–904.

[12] Wells RC. A new president, a new congress and the path to personalized medicine. *Personalized Medicine*, 2009;6(3):235–239.

[13] de Pagter MS, Kloosterman WP. The diverse effects of complex chromosome rearrangements and chromothripsis in cancer development. In Ghadimi B, Ried T, eds., *Chromosomal Instability in Cancer Cells. Recent Results in Cancer Research*, vol. 200. Springer;2015, pp. 165–193.

[14] Chen XW, Liao R, Li D, Sun J. Induced cancer stem cells generated by radiochemotherapy and their therapeutic implications. *Oncotarget*, 2017;8(10):17301–17312.

[15] Wyatt AW, Collins CC. In brief: Chromothripsis and cancer. *Journal of Pathology*, 2013;231(1):1–3.

[16] Korbel JO, Campbell PJ. Criteria for inference of chromothripsis in cancer genomes. *Cell*, 2013;152(6):1226–1236.

[17] Warzecha J, Dinges D, Kaszap B, et al. Effect of the hedgehog-inhibitor cyclopamine on mice with osteosarcoma pulmonary metastases. *International Journal of Molecular Medicine*, 2012;29(3):423–427.

[18] Alam, MM, Sohoni S, Kalainayakan SP, Garrossian M, Zhanget L. Cyclopamine tartrate, an inhibitor of Hedgehog signaling, strongly interferes with mitochondrial function and suppresses aerobic respiration in lung cancer cells. *BMC Cancer*, 2016;16:150.

[19] Justilien V, Fields AP. Molecular pathways: Novel approaches for improved therapeutic targeting of Hedgehog signaling in cancer stem cells. *Clinical Cancer Research*, 2015;21(3):505–513.

[20] Mahindroo N, Connelly MC, Punchihewa C, Yang L, Yan B, Fujii N. Amide conjugates of ketoprofen and indole as inhibitors of Gli1-mediated transcription in the Hedgehog pathway. *Bioorganic & Medicinal Chemistry*, 2010;18(13):4801–4811.

[21] Lindsey Sn Langhans SA. Crosstalk of oncogenic signaling pathways during epithelial-mesenchymal transition. *Frontiers in Oncology*, 2014;4:358.

[22] Messersmith WA, Shapiro GI, Cleary JM, et al. A Phase I, dose-finding study in patients with advanced solid malignancies of the oral gamma-secretase inhibitor PF-03084014. *Clinical Cancer Research*, 2015;21(1):60–67.

[23] Sato, T, Sakai T, Noguchi Y, Takita M, Hirakawa S, Ito A. Tumor-stromal cell contact promotes invasion of human uterine cervical carcinoma cells by augmenting the expression and activation of stromal matrix metalloproteinases. *Gynecologic Oncology*, 2004;92(1):47–56.

[24] Fernandis, AZ, Prasad A, Band H, Klosel R, Ganju RK. Regulation of CXCR4-mediated chemotaxis and chemoinvasion of breast cancer cells. Oncogene, 2004;23(1):157–167.

[25] Gelmini S, Mangoni M, Serio M, Romagnani P, Lazzeri E. The critical role of SDF-1/CXCR4 axis in cancer and cancer stem cells metastasis. *Journal of Endocrinological Investigation*, 2008;31(9):809–819.

[26] Li, XQ, Ma Q, Xu Q, et al. SDF-1/CXCR4 signaling induces pancreatic cancer cell invasion and epithelial-mesenchymal transition in vitro through non-canonical activation of Hedgehog pathway. *Cancer Letters*, 2012;322(2):169–176.

[27] Terzoudi GI, Karakosta M, Pantelias A, Hatzi VI, Karachristou I, Pantelias G. Stress induced by premature chromatin condensation triggers chromosome shattering and chromothripsis at DNA sites still replicating in micronuclei or multinucleate cells when primary nuclei enter mitosis. *Mutation Research/Genetic Toxicology and Environmental Mutagenesis*, 2015;793: 185–198.

[28] Al-Hajj M, Wicha MS, Benito-Hernandez A, Morrison SJ, Clarke MF. Prospective identification of tumorigenic breast cancer cells. *Proceedings of the National Academy of Sciences of the United States of America*, 2003;100(7):3983–3988. Erratum: 2003;100(7):6890.

[29] Chen J, Li Y, Yuet T-S, et al. A restricted cell population propagates glioblastoma growth after chemotherapy. *Nature*, 2012;488(7412): 522–526.

[30] Shackleton M, Quintana E, Fearon ER, Morrison SJ. Heterogeneity in cancer: Cancer stem cells versus clonal evolution. *Cell*, 2009;138(5):822–829.

[31] Eppert K, Takenaka K, Lechman, E, et al. Stem cell gene expression programs influence clinical outcome in human leukemia. *Nature Medicine*, 2011;17(9):1086–1091.

[32] Liu SY, Liu C , Min X, et al. Prognostic value of cancer stem cell marker aldehyde dehydrogenase in ovarian cancer: A meta-analysis. *Plos One*, 2013;8(11):e81050.

[33] Corbin AS, Agarwal A, Loriaux M, Cortes J, Deininger MW, Druker BJ. Human chronic myeloid leukemia stem cells are insensitive to imatinib despite inhibition of BCR-ABL activity. *Journal of Clinical Investigation*, 2011;121(1):396–409.

[34] Altelaar AF, Munoz J, Heck AJ. Next-generation proteomics: Towards an integrative view of proteome dynamics. *Nature Reviews Genetics*, 2013;14(1):35–48.

[35] Ooi HS, Schneider G, Chan YL, Lim TT, Eisenhaber B, Eisenhaber F. Databases of protein–protein interactions and complexes. Databases of protein–protein interactions and complexes. In Carugo O, Eisenhaber F. eds., *Data Mining Techniques for the Life Sciences. Methods in Molecular Biology (Methods and Protocols)*, vol. 609. Humana Press;2010, pp. 145–159.

[36] Buszczak M, Signer RA, Morrison SJ. Cellular differences in protein synthesis regulate tissue homeostasis. *Cell*, 2014;159(2):242–251.

[37] Hubner B. Strickfaden H, Muller S, Cremer M, Cremer T. Chromosome shattering: a mitotic catastrophe due to chromosome condensation failure. *European Biophysics Journal with Biophysics Letters*, 2009;38(6):729–747.

[38] Cremer C, Cremer T, Hens L, Baumann H, Cornelis JJ, Nakanishi K. UV micro-irradiation of the Chinese hamster cell nucleus and caffeine post-treatment. Immunocytochemical localization of DNA photolesions in cells with partial and generalized chromosome shattering. *Mutatation Research*, 1983;107(2):465–76.

[39] Sanz-Pamplona R, Berenguer A, Sole X. Tools for protein-protein interaction network analysis in cancer research. *Clinical & Translational Oncology*, 2012;14(1):3–14.

[40] Patterson SD, Aebersold RH. Proteomics: the first decade and beyond. *Nature Genetics*, 2003;33:311–323.

[41] Graves PR, Haystead TAJ. Molecular biologist's guide to proteomics. *Microbiology and Molecular Biology Reviews*, 2002;66(1):39–63.

[42] Kocher T, Superti-Furga G. Mass spectrometry-based functional proteomics: From molecular machines to protein networks. *Nature Methods*, 2007;4(10):807–815.

[43] Manjasetty BA, Büssow K, Panjikar S, Turnbull AP. Current methods in structural proteomics and its applications in biological sciences. *3 Biotech*, 2012;2(2):89–113.

[44] Shruthi BS, Vinodhkumar P, Selvamani M. Proteomics: A new perspective for cancer. *Advanced Biomedical Research*, 2016;5:67.

[45] Barbosa EB, Vidotto A, Mussi Polachini g, Henrique T, Trovó de Marqui AB, Tajara EH. Proteomics: Methodologies and applications to the study of human diseases. *Revista Associacao Medica Brasileira*, 2012;58(3):366–375.

[46] Pan S, Aebersold R, Quantitative proteomics by stable isotope labeling and mass spectrometry. *Methods in Molecular Biolology*, 2007;367:209–218.

[47] Wong JW, Cagney G, An overview of label-free quantitation methods in proteomics by mass spectrometry. *Methods in Molecular Biology*, 2010;604:273–283.

[48] Shiio Y, and Aebersold R, Quantitative proteome analysis using isotope-coded affinity tags and mass spectrometry. *Nature Protocols*, 2006;1(1):139–145.

[49] Boersema PJ, Raijmakers R, Lemeer S, Mohammed S, Heck AJ. Multiplex peptide stable isotope dimethyl labeling for quantitative proteomics. *Nature Protocols*, 2009;4(4):484–494.

[50] Ye X, Luke B, Anderesson T, Blonder J. 18O stable isotope labeling in MS-based proteomics. *Briefings in Functional Genomics*, 2009;8(2):136-44.

[51] Ong SE, Blagoev B, Kratchmarova I, Kristensen DB, Steen H, Pandey A, Mann M. Stable isotope labeling by amino acids in cell culture, SILAC, as a simple and accurate approach to expression proteomics. *Molecular & Cellular Proteomics*, 2002;1(5):376–386.

[52] Zieske LR, A perspective on the use of iTRAQ (TM) reagent technology for protein complex and profiling studies. *Journal of Experimental Botany*, 2006;57(7):1501–1508.

[53] Zhao YM, Jensen ON. Modification-specific proteomics: Strategies for characterization of post-translational modifications using enrichment techniques. *Proteomics*, 2009;9(20):4632–4641.

[54] Chen EI, Yates JR. Cancer proteomics by quantitative shotgun proteomics. *Molecular Oncology*, 2007;1(2):144–159.

[55] Sallam, RM. Proteomics in cancer biomarkers discovery: Challenges and applications. *Disease Markers*, 2015;1–12.

[56] Boja ES, Rodriguez H. Mass spectrometry-based targeted quantitative proteomics: Achieving sensitive and reproducible detection of proteins. *Proteomics*, 2012;12(8):1093–1110.

[57] Faria SS, Morris C, Sliva A, et al., A timely shift from shotgun to targeted proteomics and how It can be groundbreaking for cancer research. *Frontiers in Oncology*, 2017;7:Article 13.

[58] Morgan WF, Wolff S. Effect of bromodeoxyuridine on induced sister chromatid exchanges. *Basic Life Sciences*, 1984;29 Pt A:281–292.

[59] Rothberg J, Merriman B, Higgs G. Bioinformatics: Introduction. *Yale Journal of Biology and Medicine*, 2012;85(3):305–308.

[60] Du J, Yuan Z, Ma Z, andSong J, Xiea X, Chen Y. KEGG-PATH: Kyoto encyclopedia of genes and genomes-based pathway analysis using a path analysis model. *Molecular BioSystems*, 2014;10(9):2441–2447.

[61] Mi H, Lazareva-Ulitsky B, Loo R, et al. The PANTHER database of protein families, subfamilies, functions and pathways. *Nucleic Acids Research*, 2005;33(Database issue):D284–288.

[62] Yeger-Lotem E, Sharan R. Human protein interaction networks across tissues and diseases. *Frontiers in Genetics*, 2015;6:257.

[63] Bensimon A, Heck AJ, Aebersold R. Mass spectrometry-based proteomics and network biology. *Annual Review of Biochemistry*, 2012;81: 379–405.

[64] Rao VS, Srinivas K, Sujini GN, Sunand Kumar GN. Protein–protein interaction detection: Methods and analysis. *International Journal of Proteomics*, 2014;2014:147648.

[65] Berggard T, Linse S, James P. Methods for the detection and analysis of protein–protein interactions. *Proteomics*, 2007;7(16):2833–2842.

[66] Zahiri J, Bozorgmehr JH, Masoudi-Nejad A. Computational prediction of protein-protein interaction networks: Algorithms and resources. *Current Genomics*, 2013;14(6):397–414.

[67] Alonso-Lopez D, Gutiérrez MA, Lopes KP, Prieto C, Santamaría R, de Las Rivas J. APID interactomes: providing proteome-based interactomes with controlled quality for multiple species and derived networks. *Nucleic Acids Research*, 2016;44(W1):W529–535.

[68] Fung DCY, Li SS, Goel A, Hong SH, Wilkins MR. Visualization of the interactome: What are we looking at? *Proteomics*, 2012;12(10):1669–1686.

[69] Luo F, Yang Y, Chen CF, Chang R, Zhou J, Scheuermann RH. Modular organization of protein interaction networks. *Bioinformatics*, 2007;23(2):207–214.

[70] Licata L, Briganti L, Peluso D, et al. MINT, the molecular interaction database: 2012 update. *Nucleic Acids Research*, 2012;40(D1):D857–D861.

[71] Chatr-aryamontri A, Oughtred R, Boucher L, et al. The BioGRID interaction database: 2017 update. *Nucleic Acids Research*, 2017;45(D1):D369–D379.

[72] Kerrien S, Aranda B, Breuza L, et al. The IntAct molecular interaction database in 2012. *Nucleic Acids Research*, 2012;40(D1):D841–D846.

[73] Persico M, Ceol A, Gavrila C, Hoffmann R, Florio A, Cesareni G. HomoMINT: An inferred human network based on orthology mapping of protein interactions discovered in model organisms. *BMC Bioinformatics*, 2005;6(4):S21.

[74] Bader GD, Betel D, Hogue CWV. BIND: The Biomolecular Interaction Network Database. *Nucleic Acids Research*, 2003;31(1):248–250.

[75] Shannon P, Markiel A, Ozier O, et al. Cytoscape: A software environment for integrated models of biomolecular interaction networks. *Genome Research*, 2003;13(11):2498–2504.

[76] von Mering C, Jensen LJ, Snel B, et al. STRING: known and predicted protein–protein associations, integrated and transferred across organisms. *Nucleic Acids Research*, 2005;33:D433–D437.

[77] Franceschini A, Szklarczyk D, Frankild S, et al. STRING v9.1: Protein–protein interaction networks, with increased coverage and integration. *Nucleic Acids Research*, 2013;41(D1):D808–D815.

[78] Szklarczyk D, Franceschini A, Wyder S, et al. STRING v10: protein–protein interaction networks, integrated over the tree of life. *Nucleic Acids Research*, 2015;43(D1):D447–D452.

[79] Szklarczyk D, Morris JH, Cook H, et al. The STRING database in 2017: Quality-controlled protein–protein association networks, made broadly accessible. *Nucleic Acids Research*, 2017;45(D1):D362–D368.

[80] Kraljevic Pavelic S, Sedic, M, Hock K, et al. An integrated proteomics approach for studying the molecular pathogenesis of Dupuytren's disease. *Journal of Pathology*, 2009;217(4): 524–533.

[81] Sedic M, Gethings LA, Vissers JPC, et al. Label-free mass spectrometric profiling of urinary proteins and metabolites from paediatric idiopathic nephrotic syndrome. *Biochemical and Biophysical Research Communications*, 2014;452(1):21–26.

[82] Ratkaj I, Stajduhar E, Vucinic S, et al. Integrated gene networks in breast cancer development. *Functional & Integrative Genomics*, 2010;(1)10:11–19. Erratum:2011;11(2):381.

[83] Yu H, Lee H, Herrmann A, Buettner R, Jove R. Revisiting STAT3 signaling in cancer: New and unexpected biological functions. *Nature Reviews Cancer*, 2014;14(11):736–746.

[84] Wang HC, Yeh HH, Huang WL, et al. Activation of the signal transducer and activator of transcription 3 pathway up-regulates estrogen receptor-beta expression in lung adenocarcinoma cells. *Molecular Endocrinology*, 2011;25(7):1145–1158.

[85] Binai NA, Damert A, Carra G, et al. Expression of estrogen receptor alpha increases leptin-induced STAT3 activity in breast cancer cells. *International Journal of Cancer*, 2010;127(1):55–66.

[86] Kowalska K, Piastowska-Ciesielska AW. Oestrogens and oestrogen receptors in prostate cancer. *Springerplus*, 2016;5:522.

[87] De Miguel F, Ok Lee S, Onate SA, Gao AC. Stat3 enhances transactivation of steroid hormone receptors. *Nuclear Receptor*, 2003;1(1):3.

[88] Chen TS, Wang LH, Farrar WL. Interleukin 6 activates androgen receptor mediated gene expression through a signal transducer and activator of transcription 3-dependent pathway in LNCaP prostate cancer cells. *Cancer Research*, 2000;60(8):2132–2135.

[89] Venere M, Lathia JD, Rich JN. Growth factor receptors define cancer hierarchies. *Cancer Cell*, 2013;23(2):135–137.

[90] Kwabi-Addo B, Ozen M, Ittmann M. The role of fibroblast growth factors and their receptors in prostate cancer. *Endocrine-Related Cancer*, 2004;11(4):709–724.

[91] Kaliberov SA, Kaliberova LN, Stockard CR, Grizzle WE, Buchsbaum DJ. Adenovirus-mediated FLT1-targeted proapoptotic gene therapy of human prostate cancer. *Molecular Therapy*, 2004;10(6):1059–1070.

[92] Sedaghat S, Gheytanchi E, Asgari M. Expression of cancer stem cell markers OCT4 and CD133 in transitional cell carcinomas. *Applied Immunohistochemistry & Molecular Morphology*, 2017;25(3):196–202.

[93] Petersen JK, Jensen P, Sørensen MD, Kristensen BW. Expression and prognostic value of Oct-4 in astrocytic brain tumors. *PLoS ONE*, 2016;11(12):e0169129.

[94] Morin PJ. Beta-catenin signaling and cancer. *Bioessays*, 1999;21(12): 1021–1030.

[95] Ibrahem S, Al-Ghamdi S, Baloch K, et al. STAT3 paradoxically stimulates beta-catenin expression but inhibits beta-catenin function. *International Journal of Experimental Pathology*, 2014;95(6):392–400.

[96] Anand M, Lai R, Gelebart P. β-catenin is constitutively active and increases STAT3 expression/activation in anaplastic lymphoma kinase-positive anaplastic large cell lymphoma. *Haematologica*, 2011; 96(2):253–561.

[97] Chen MW, Yang ST, Chien MH, et al. The STAT3-miRNA-92-wnt signaling pathway regulates spheroid formation and malignant progression in ovarian cancer. *Cancer Research*, 2017;77(8):1955–1967.

[98] Niu GL, Wright KL, Ma Y, et al. Role of Stat3 in regulating p53 expression and function. *Molecular and Cellular Biology*, 2005;25(17):7432–7440.

[99] Hu, YY, et al. Notch signaling pathway and cancer metastasis. *Notch Signaling in Embryology and Cancer*, 2012;727:186–198.

[100] Guo SC, ML Liu, RR Gonzalez-Perez. Role of Notch and its oncogenic signaling crosstalk in breast cancer. *Biochimica Et Biophysica Acta-Reviews on Cancer*, 2011;1815(2):197–213.

[101] Lu KH, Patterson AP, Wang L, et al. Selection of potential markers for epithelial ovarian cancer with gene expression arrays and recursive descent partition analysis. *Clinical Cancer Research*, 2004;10(10):3291–3300.

[102] Cheaib B, Auguste A, Leary A. The PI3K/Akt/mTOR pathway in ovarian cancer: Therapeutic opportunities and challenges. *Chinese Journal of Cancer*, 2015;34(1):4–16.

[103] Zou W, Wicha MS. Chemokines and cellular plasticity of ovarian cancer stem cells. *Oncoscience*, 2015;2(7):615–616.

[104] Li Z, Block MS, Vierkant RA, et al. The inflammatory microenvironment in epithelial ovarian cancer: A role for TLR4 and MyD88 and related proteins. *Tumor Biology*, 2016;37(10):13279–13286.

[105] Lukaszewicz-Zajac M, Mroczko B, Szmitkowski M. Chemokines and their receptors in esophageal cancer: The systematic review and future perspectives. *Tumour Biology*, 2015;36(8):5707–5714.

[106] Nagarsheth N, Wicha MS, Zou WP. Chemokines in the cancer microenvironment and their relevance in cancer immunotherapy. *Nature Reviews Immunology*, 2017;17(9):559–572.

[107] Xu DS, Li R, Wu J, Jiang L, Zhong HA. Drug design targeting the CXCR4/CXCR7/CXCL12 pathway. *Current Topics in Medicinal Chemistry*, 2016;16(13):1441–1451.

[108] Liu XJ, Xiao Q, Bai X, et al. Activation of STAT3 is involved in malignancy mediated by CXCL12-CXCR4 signaling in human breast cancer. *Oncology Reports*, 2014;32(6):2760–2768.

[109] Dobrzycka B, Terlikowski SJ, Kowalczuk O, et al. Mutations in the KRAS gene in ovarian tumors. *Folia Histochemica Et Cytobiologica*, 2009;47(2):221–224.

[110] Stewart ML, Tamayo P, Wilson AJ, et al. KRAS genomic status predicts the sensitivity of ovarian cancer cells to decitabine. *Cancer Research*, 2015;75(14):2897–2906.

[111] Iozzo RV, Sanderson RD. Proteoglycans in cancer biology, tumor microenvironment and angiogenesis. *Journal of Cellular and Molecular Medicine*, 2011;15(5):1013–1031.

[112] Harisi R, Jeney A. Extracellular matrix as target for antitumor therapy. *Oncotargets and Therapy*, 2015;8:1387–1398.

[113] Yip GW, Smollich M, Gotte M. Therapeutic value of glycosaminoglycans in cancer. *Molecular Cancer Therapeutics*, 2006;5(9):2139–2148.

[114] Boohaker, RJ, et al. The use of therapeutic peptides to target and to kill cancer cells. *Current Medicinal Chemistry*, 2012;19(22):3794–3804.

[115] Regberg J, Srimanee A, Langel U. Applications of cell-penetrating peptides for tumor targeting and future cancer therapies. *Pharmaceuticals*, 2012;5(9):991–1007.

[116] Legg J, Jensen UB, Broad S, Leigh I, Watt FM. Role of melanoma chondroitin sulphate proteoglycan in patterning stem cells in human interfollicular epidermis. *Development*, 2003;130(24):6049–6063.

[117] Pilkington GJ. Cancer stem cells in the mammalian central nervous system. *Cell Proliferation*, 2005;38(6):423–433.

[118] Bruel A, Touhami-Carrier M, Thomaidis A, Legrand C. Thrombospondin-1 (TSP-1) and TSP-1-derived heparin-binding peptides induce promyelocytic leukemia cell differentiation and apoptosis. *Anticancer Res*, 2005;25(2A):757–764.

[119] Seidler DG, Goldoni S, Agnew C, et al. Decorin protein core inhibits in vivo cancer growth and metabolism by hindering epidermal growth factor receptor function and triggering apoptosis via caspase-3 activation. *Journal of Biological Chemistry*, 2006;281(36):26408–26418.

[120] Suhovskih AV, Mostovich LA, Kunin IS, et al. Proteoglycan expression in normal human prostate tissue and prostate cancer. *ISRN Oncology*, 2013;2013:680136.

[121] Harisi R, Dudas J, Nagy-Olah J, et al. Extracellular matrix induces doxorubicin-resistance in human osteosarcoma cells by suppression of p53 function. *Cancer Biology and Therapy*, 2007;6(8):1240–1246.

Index

Locators in **bold** refer to tables; those in *italic* to figures

adjacency
    lists, *130*, 130
    matrices, *129*, 129–130, 291, 294, *295*, 491–492, 499
    matrix visualizations, 541, *544*
Affymetrix SNP microarrays, *10*, 12–14, **15**
aging, 220–223
    PPI network analysis, 194, 199, 212, 220, *221–223*
algorithms; *see also* network alignment
    alignment, 23, 24
    clustering, 242, 256–270, 549, 585
    force-directed, 549–550, *574*, *575*
    FUSE, 394–397, *395*, 397
    gene prioritization, 437
    genotypes, *10*, *16*, 14–18, 21–27
    graph search, 130–131
    hierarchical clustering, 257, 264
    Hungarian, 128, 375
    Isorank, 200, 217, 382–384, *383*
    layouts, 549
    machine learning, 33–39, 601
    mapping, 372
    optimization, 374
    orbit counting, **227**, 228
    pattern mining, 329, 330, 340, 361
    permutation testing, 347
AlignMCL, pairwise network alignment, 382
alignment
    algorithms, 23, 24
    graphs, 374
    network *see* network alignment
    strategy, 216–217
alignment scoring schemes, 375
    agreements and trade-offs, *378*, 378–379
    F-score, 376
    global network alignments, 376–378
    local network alignments, 375–376
    multiple network alignments, 392
    symmetric sub-structure score, 377
AlignNemo, pairwise network alignment, 381–382
Alzheimer's disease, 517–518
animations, node-link diagrams, 551

annotations
    databases, **166**, 165–167, 170
    PPI networks, *171*, *370*, 371
apriori property, pattern mining, 331
articulation points, 168, 177
assortative networks, 135
asthma, *441*, *445*, *447*, 474
asymmetric interactions, 113, 118
automorphism orbits, 203, *384*, 384, *385*
    edge, 204
    node, *205*
average closeness centrality, 511
average clustering coefficient, *135*, 511
average edge betweenness centrality, 512
average efficiency, 511
average node betweenness centrality, 511–512
average node degree, 510
average radiality, 512

Bayesian networks, 39, 41–42, *292*, 291–293
betweenness centrality, 136, 168, *178*
biases
    microarrays, 75, 97
    PPI datasets, 158–159, 438
BioFabric, 542, *545*
bioinformatics
    Hi-C analysis, 82
    lncRNAs analysis, 88–89
    protein-DNA interactions, 600
biological heterogeneity, 159–160, 288–290
biological interpretation, disease modules, 445–448
biological networks (BN), 111, 167, 193–194, 256, 271–272; *see also* molecular networks; visualizing networks
    data integration, 229
    metabolic networks, 113, 418
biomarkers
    cancer precision medicine, 287
    cancer stem cells, **605**, **606**, 603–612
    covariate factors, 349–359
    discovery, 313–314, 359–361
    exercises, 362–364
    ovarian cancer, *608*, *609*, 608–610
    pattern mining, 315–328

621

biomarkers (cont.)
    prostate cancer, 606–608
    statistical redundancy, 341–349
    Tarone's method for discovery, 329–341
bipartite graphs, 113, *123*, 123, *375*, 374–375
bisulfite based arrays, DNA methylation, 72–73
bisulfite conversion, 72–73
BLUEPRINT epigenome, 79, 96
Bonferroni correction, 326–327
Boolean variables, 254–255
brain
    anatomy, *515, 516*
    connectomes, 490–491, 499, *500*, 514
    geometry, 492
    MRI scanning, 492–493
    topology, 492
brain networks
    disorders, 517–519
    functional, 499–503
    structural, 492–499
    tools for analysis, 505–506
BRCA1 gene, 476, 480
breadth first search (BFS), 130, 131
breast cancer, 79, 96, 473–474, *560, 572*

cancer; *see also* tumors
    BRCA1 gene, 476, 480
    breast, 79, 96, 473–474, *560, 572*
    gene mutations, 175–179, **177**, 300, 462, 548–549
    genome atlas, 22, 95, 462, 479–480, 577
    precision medicine, 287
    prostate, 606–608
    proteins, 223–224
    stem cells, 595–598, 603–612, **605, 606**
CART (classification and regression tree classifiers), 37
categorical values, proximity calculation, 256
causal variants, 466–470, 473
cell signaling networks, **112**, 113
centrality
    average closeness, 511
    average edge betweenness, 512
    average node betweenness, 511–512
    betweenness, 136, 168, *178*
    closeness, *135*, 136, 139
    eccentricity, *135*, 137
    eigenvector, *135*, 137
    GDV, 207, 221, *225*
    subgraphs, 137
characteristic path length, 511
chemokines, 597, *609*

ChIA-PET technology, chromatin conformation, 81, 83
ChIP *see* chromatin immunoprecipitation
chromatin, 77
    conformation, 67, 81–83
    higher order organization, 80–87
    modifications, *70*, 315
    topological associated domains, **86**, 86–87
chromatin immunoprecipitation (ChIP), 69, 77–78, 418–419
    data analysis, **79**, 78–79
    differential binding, **80**
chromosome territories, *70*
cis-acting, 67
classification and regression tree classifiers (CARTs), 37
cliques (complete subgraphs), 120, 170, *210*, 265, 269
clonal theory, 598
closeness centrality, *135*, 136, 139
cluster analysis, 241–243, 277
    definitions, 243–245
    exercises, 277–280
    preprocessing, 246–251
    proximity calculation, 252–256
    workflow, *245*, 245–246
cluster evaluation
    external, 271–272
    internal, 272–274
    optimization strategies, 275–277
    validity indices, 270, 271
clustering
    algorithms, 242, 256–270, 549, 585
    coefficients, 170
    data formats, 244–245
    networks, 209, *210*, 209–210
    partitional, 243
    types, 243
clusters, number of, 275, *276*
Cochran-Mantel-Haenszel (CMH) test, 351–354
    minimum attainable P-value, 354–355, *356, 358*
    pruning condition, 355–359
co-expression networks, 419–420
colored graphlets, *195*, 200–202
columns, cortex, 491
comparative genomics, 156
complex diseases, network approaches, 473–474
computational biology workflows, 175–179
computational complexity, 117–118, 128, 197, 200, 226, 297, 338
computational methods, PPI, 222–223, 601

conditional probabilities, 41–42, 291–293, 320
confounding effect, pattern mining, 350, *351*
connectedness
    graph theory, *122*, 122
    subgraphs, *119*
connectomes, brain, 490–491, 499, *500*, 514
contagious diseases spread, 423–424, 426
    transportation networks, 424–425
context-sensitive interactomes, 479–480
contingency table analysis, 31–32
continues variables, proximity calculation, 252–254
correlation networks, 539
correlation, continues variables, 253, *254*
covariate factors, pattern mining, 349–359
CpG islands, 67, 70
curated databases, **161**, *162*, 161–162, 179
cystic fibrosis (CF), 473, 480
Cytoscape, 177, 179, 533–536, 553
    apps, 574–577
    command language, *587*, 586–589
    control panel, 555, *556*
    example workflow, 577–585
    exercises, 589
    hierarchical clustering, 541, *579*, 580
    importing data files, *561*, *563*
    importing from public databases, *560*
    importing networks, *560*
    integration of data, 559–562
    k-means clustering, *580*, 580
    network analysis, 574
    network panel, 557, *558*
    results panel, 557
    scripting, 586
    STRING network, *540*, *560*, *572*, *576*, *579*, *581*
    table panel, 557
    user interface, *554*, 555–559
    view menu, 559
    visualizing data, *565*, 562–573, *574*, *575*, *586*
    visualizing networks, 534, 540–552

DAG1 gene, 476
data integration, 159–160, 288–290
    Bayesian approaches, *292*, 291–293
    biological networks, 229
    early, 289
    heterogeneous, 289, 300–306
    homogeneous, 289, 294–300
    intermediate, 289
    kernel-based methods, *293*, 293–294
    late, 289
    network-based, *291*, 290–291

precision medicine, 287–288, 290–294
protein-protein interactions, 159–160
training methods, *290*, 289–290
databases
    annotation, **166**, 165–167, 170
    curated, **161**, *162*, 161–162, 179
    epigenetic, **75**, 93–96, **97**
    integrated, 160, **164**, **165**, 163–179
    interaction, 160–167, *372*, **463**, 602
    interactome, 439
    lncRNA, **91**
    molecular interaction, **463**
    prediction, *163*, **164**, 162–165, 179
    protein-protein interactions, **112**, 160–167, 602, **603**
    public, 2, 22, 72, 74, 427, 559–562
Davies-Bouldin index, 273
DBSCAN, *267*, 266–268
degree distribution, *133*, 133, *178*, 428
degree of vertices, *119*, 119, 168
de-noising networks, 212, *213*
density, 133, *134*
density based clustering algorithms, 257, 266–268
depth first search (DFS), 130–131
deterministic measures, network topology, 507, 510–513
differential binding, 79, **80**
differential methylation CpGs (DMC), **76**, 76–77
differential methylation regions (DMR), **76**, 76–77
differential network analysis, 481
diffusion tensor imaging (DTI), 495, *498*, *500*
diffusion-based methods, disease identification, 173
diffusion-weighted MRI scanning, *495*, *496*
direct to consumer services, 3, 9, **13**, 45
    predictive genetic risk models, 39–44
directed acyclic graphs (DAG), 39
directed graphlets, 197–198
directed graphs, 113, *118*, 118
directed networks, *402*, 402
disassortative networks, 135
discriminative pattern mining *see* pattern mining
disease gene prioritization, 440, 442
    connectivity-based methods, 440–443
    diffusion-based methods, 443–444
    path-based methods, 443
diseases, 437, 474
    analysis, *438*
    biological interpretation, 445–448
    contagious diseases spread, 423–426

diseases (cont.)
  enrichment, *445*, 444–445
  epigenetic mechanisms, 65–66
  gene prioritization, 440–444
  identification, PPI, 152, 173–174
  interactome analysis, 430
  interactome construction, 427, 438
  molecular basis, *474*, 470–475
  network approaches, 422–423, 437, *438*, 440, 442, 473–474
  resources, 439
  seed clusters, 438–440
  treatment, PPI, 166
  validation, 444
DNA methylation, 66–68, 479; *see also* epigenetics
  bisulfite based arrays, 72–73
  experimental strategies, 69–71
  microarrays, 73–77
  role in genomic profiles, 69
DNA modifications, **68**, *70*
DNA sequencing, next generation, 18–27
domains, proteins, 156
dominant models, 32
dominating set (DS), *211*, 210–211
drug repurposing, 174, 290, 294, 423
  precision medicine, 287
drug targeting, 174
DTC services *see* direct-to-consumer services
Dunn index, 273
dynamic graphlets, *199*, 198–200
dynamic networks, *199*, 198–200, *221*, *225*

eccentricity centrality, *135*, 137
edge conservation, 217, 375
edge correctness, alignment scoring, 377
edge-colored graphlets, 200, 201
edges, 463, 464
edgetic perturbations, 466, 470
edgotype prediction tools, **468**
edgotype scenarios, *467*, 467
effect sizes, SNPs, 5
eigenvector centrality, *135*, 137
electroencephalography (EEG), 501–503, *504*
Encyclopedia of DNA elements (ENCODE), 93
enrichment, disease modules, *445*, 444–445
epigenetic databases, **75**, 93–96, **97**
  BLUEPRINT Epigenome, 79, 96
  Encyclopedia of DNA elements, 93
  Functional Annotation of the Mammalian Genome, 95
  International Human Epigenome Consortium, 96

  Roadmap Epigenomics Project, 95
epigenetics, 65–66, 66, 98; *see also* DNA methylation
  changes in tumors, 595
  disease mechanisms, 65–66
  exercises, 98–100
  higher order chromatin organization, 80–87
  histone modifications, 77–79
  long non-coding RNAs, 87–93
  mapping mechanisms, *72*, *74*
epigenomics, 67
Erdos–Renyi (ER) random graphs, 138–139
error rates, PPI datasets, 156–158
euchromatin, 67
Euclidean distances, 252
Eulerian circuits, 126–127
events, dynamic networks, 198
exact cluster ratio, 392
expectation–maximization (EM) algorithm, 16

false discovery rate (FDR) control, 361
family-wise error rate (FWER), 326, 361
  Bonferroni correction, 326–327
    empirical approximations, 344–346
    Tarone's improved Bonferroni correction, 327–328
FANTOM (Functional Annotation of the Mammalian Genome), 95, **477**
feature selection, cluster analysis, 247–248
Fischer's exact test, 323–325, *336*, *337*
5C technology, chromatin conformation, 81, 83
force-directed algorithms, 549–550, *574*, *575*
4C technology, chromatin conformation, 81, 83
frequent pattern mining *see* apriori property, pattern mining
Fruchterman-Reingold algorithm, 549
F-score, alignment scoring, 376
Functional Annotation of the Mammalian Genome (FANTOM), 95, **477**
functional annotations, PPI networks, 174–175
functional brain networks, 499–503
functional consistency, alignment scoring, 376
functional MRI (fMRI), 499, *502*, 518
FUSE, multiple network alignment method, *395*, 394–397
fuzzy clustering, 243–244
fuzzy C-means (FCM) clustering, 262
FWER *see* family-wise error rate

gain-of-interactions, 466
gap statistic, 276–277
GATK (Genome Analysis Toolkit), 22, 27
GCD (graphlet correlation distance), 215

GCM (graphlet correlation matrix), *207*, 208–209
GDD (graphlet degree distributions), *207*, 208
GDDA (graphlet degree distribution agreement), 214
GDV *see* graphlet degree vectors
GDV-centrality, 207, 221, *225*
GDV-matrices, *207*, 208–209
GDV-similarity, 205–207, 210, 218, *219*
gene co-expression networks, 113
gene duplication and divergence (SF-GD), 139
gene expression analysis, 241
gene mutations, 152, 459, 461
    BRCA1 gene, 476, 480
    cancer, **177**, 175–179, 300, 462, 548–549
    cystic fibrosis, 480
    DAG1 gene, 476
    edgetic perturbations, 466
    KRAS gene, 610
    loss-of-function, 465, 466
    monogenic diseases, 470–473
    oncogenes, 595
    Parkinson's disease, 473, 480
    personalized genetic tests, 3, 7, **13**, 45
    RAS genes, 462
    sickle cell disease, 461–462, 479–480
    tumor suppressor genes, 595
gene ontology (GO) annotation set, 370, *371*
gene prioritization
    algorithms, 437
    disease analysis, 440–444
gene regulatory networks, 418–419
gene signature improvements, PPI networks, 174
generalized random graph models (ER-DD), 138
genetic data, risk prediction, 1–6, 45–47
    exercises, 47–50
    glossary, 4–6
    SNP-disease association, 31–44
    SNPs identification, 9, 30
    tests in healthcare, 6–9
genetic interactions, 113–114, 420–422, 463, 481, 536
genetic tests, healthcare, 6–9
Genome Analysis Toolkit (GATK), 22, 27
genome atlas, cancer, 22, 95, 462, 479–480, 577
genome-wide association studies (GWAS), 2, 473, 479
genotype-phenotype relationships, 460–461, 482
    definitions, 460–461
    exercises, 482–483
    molecular networks, 459, 461–464
    network approaches, diseases, 464–475
    network-based tools, **471, 472**

tissue interactomes, 480–482
tissue-sensitive molecular interaction networks, 475–480
genotypes
    algorithms, 14–18
    calling algorithms, *10*, *16*, 21–27
    definition, 460, 461
geometric graph with gene duplications and mutations (GEO-GD), 136, 139
geometric networks, 139–140
Gini importance, 40
global pairwise network alignment methods
    GRAAL, 384–387
    IsoRank, 382–384
    other, 387–390
GRAAL, global network aligner, 384–387
GRAFENE, 215–216, 220
graph(s), 167
    alignment, 374
    bipartite, 113, 123, *375*, 374–375
    density, **169**, 169
    kernels, 216
    regularization, *300*, 299–300
    types, 122–126
    weighted, 113, *123*, 123
graph based clustering algorithms, 257, 268–270
graph search algorithms, 130–131
graph theory, 111–114, 140
    classic problems, 126–128
    computational complexity, 117–118
    data structures, 128–130
    definitions, 118–119
    degree and neighborhood, 119–120
    exercises, 140–142
    mathematical basis, 114–116
    network measures, 132–140
    search algorithms, 130–131
    spectral graph theory, 131–132
    subgraphs and connectedness, 120–122
    trees, *124*, 124
GraphCrunch, 226, **227**, 227
graphlet correlation distance (GCD), 215
graphlet correlation matrix (GCM), *207*, 208–209
graphlet counting, 196, 197, **227**, 226–229
    orbit-aware, 206, 226
    orbit-unaware, 199, 226
graphlet degree distribution agreement (GDDA), 214
graphlet degree distributions (GDD), *207*, 208
graphlet degree vectors, 203–205, *384*, *385*, 384–387; *see also* GDV-centrality; GDV-matrices; GDV-similarity

graphlet degree vectors (cont.)
    edge, 204
    node-pair, 205
    non-edge, 205
graphlet frequency vector (GFV), 208
graphlet kernel, 293
graphlet-based alignment-free network
        approach (GRAFENE), 215–216, 220
graphlets, *195*, 193–196, *205*
    biological applications, 218–226
    colored, *195*, 200–202
    computational approaches, 209–218
    directed, 197–198
    dynamic, *199*, 198–200
    edge-colored, 200, 201
    exercises, 230–234
    heterogeneous, *195*, 200–202
    homogeneous, *195*, 196, 201–202
    network topology, 196–209
    node-colored, 200–201
    orbits, *203*, 203, *204*, 205
    ordered, *202*, 202
    software tools, **230**, 226–230
    static, *195*, 196, 202
    undirected, 196, 202
    unordered, 196, 202
graph-structured samples, 318
GWAS *see* genome-wide association studies

Hall's theorem, 128
Hamiltonian paths, 127–128
hedgehog signaling pathway, 182, 597
heterogeneity
    biological, 288–290
    cancers, 287
    condition specific, 160
    data integration, 289, 300–306
    experimental, 159
    graphlets, *195*, 200–202
    molecular, 159
    nomenclature, 160
Hi-C analysis, bioinformatics, 82
Hi-C technology, chromatin conformation, 67, 77, 82
    mapping and filtering, 82
    normalization, 82, **84**
    statistical analysis, 84
    tools, **85**
    topological associated domains, **86**, 86–87
    visualization, 84, **86**
hierarchical clustering, 244, 262, *263*, 265–266; *see also* linkage functions

algorithms, 257, 264
Cytoscape, 541, *579*, 580
high angular resolution diffusion imaging
        (HARDI), 497, *498*
high-throughput methods (HT), 154
histone modifications, 77–79
homogeneity
    data integration, 289, 294–300
    graphlets, *195*, 196, 201–202
Hotelling's T2 statistic, 32–33
human tissues, mapping, 463, **477**
Hungarian algorithm, 128, 375
Huntington's disease, 470–473
hypergraphs, *125*, 124–125, 543, *546*
hyper-networks, *403*, 403

IHEC (International Human Epigenome
        Consortium), 96
Illumina NGS Platform, 19–21
Illumina SNP BeadChips, 14, 72, 74
inborn error of metabolism (IEM), 465, 466
induced cancer stem cells, 596
induced conserved sub-structure score (ICS),
        alignment scoring, 377
integrated databases, 160, **164**, **165**, 163–179
interaction databases, 160–167, *372*, **463**, 602
interaction networks, 536–537, *540*
interactome analysis, 151, *163*, 162–165, 427–437, 464
    basic properties, *428*, 427–429
    biological function, 429–430
    construction, 427, 438
    context-sensitive, 479–480
    databases, 439
    diseases, 430
    network localization, 430–432
    randomization, 431–437
interactomics, 601
intermediate data integration, 289
International Human Epigenome Consortium
        (IHEC), 96
inter-organismal networks, 474–475
isomorphic graphs, 120, *121*, 373
isomorphism, 194, 199, 203, 212, 220
IsoRank, global network aligner, 200, 217, *383*, 382–384

Jacard index, 256, 271–272
joint probabilities, 320

k-correctness, alignment scoring, 375
k-coverage, 392

k-means clustering, 258–262, *259, 260, 267*, 580, *580*
k-partite matching, 397
Kamada-Kawai algorithm, 549
kernel functions, 38
kernel-based methods, data integration, *293*, 293–294
KRAS gene, 606

Lance-Williams recurrence formula, 265, **266**
large scale clustering algorithms, 258
largest connected component, alignment scoring, 378
late data integration, 289
layouts, *550, 551*, 570–573
    algorithms, 549
    force directed, 549–550
    node-link diagrams, 549, *550*, 550–551
leukemia, cancer stem cells, 598
linear algebra, 114
link prediction, *213*, 212–213, 511–513
linkage disequilibrium, 5, 31, 34
linkage functions, 257, *263*, 264–265, *267*
    average linkage, 265
    complete linkage, 265
    Lance-Williams recurrence formula, 265, **266**
    single linkage, 265
linkage methods, disease identification, 173
lncRNA *see* long non-coding RNAs
local community paradigm (LCP) theory, 512–513, 517
local network alignments, 375–376
localization
    network, 430–432, *435*, 440, 505
    subcellular, 154, 166, 600
logistic regression models, multi-SNP, 3, 34, 36–37
long non-coding RNAs (lncRNA), 87–88
    algorithms, *92, 94*
    bioinformatic tools, 88–89
    databases, **91**
    epigenetics, 87–93
    precision medicine, 88
    previously annotated, 89–93
    unannotated, 93
long-read sequencing, 71
low-throughput methods (LT), 154

machine learning
    algorithms, 33–39, 601
    data integration, 290–294
    non-negative matrix factorization, 294–295

pattern mining, 313–314, 359–361
precision medicine, 287, 306
macroscale, neuronal connectivity, 491
magnetic resonance imaging *see* MRI
mapping algorithms, 372
mapping mechanisms, epigenetics, *72, 74*
marginal probabilities, 320
matching
    bipartite, *375*, 374–375
    graph theory, 128
    index, 137
    k-partite, 397
matrices
    adjacency, *129*, 129–130, 291, 294, *295*, 491–492, 499
    GDV, *207*, 208–209
    operations, 115
    special, 115–116
    spectral decomposition, 116
MaWish, pairwise network alignment, 381
mean normalized entropy (MNE), 392
medicine, 414, 448
    disease module analysis, 437–448
    disease networks, 422–423
    exercises, 449
    interactome analysis, 427–437
    molecular networks, 415–422
    social networks, 423–426
    types of network, 415
mesoscale, neuronal connectivity, 491
metabolic networks, 113, 418, 419
metastases, 595–598
microarrays, 1, *11*, 9–18, 70
    biases, 75, 97
    DNA methylation, 73–77
    genotyping algorithms, **15**, *16*
    limitations, 71
    vs. next generation sequencing, 26–28
    normalization, 75–76
microscale, neuronal connectivity, 491
minicolumns, cortex, 491
minimum attainable *P*-value, 328, 332–337, *336, 337*, 346, 354–355, *356, 358*
Minkowski distances, 252–253
mixed graphs, 122–123
mixture models, 16
model based clustering algorithms, 257–258
modularity, 509
module-based methods, disease identification, 173
molecular networks, 415, 464
    causal variants, 466–470

molecular networks (cont.)
  co-expression networks, 419–420
  databases, **463**
  diseases, 470–475
  genetic interactions, 420–422
  genotype-phenotype relationships, 459, 461–464
  metabolic networks, 113, 418, 419
  protein-protein interactions, 415–417
  regulatory networks, 418–419
  tissue-sensitive, 475–480
monogenic diseases, 470–473
motifs, network, 138, 197
MRI (magnetic resonance imaging)
  brain structure, 492–493
  diffusion tensor imaging, 495, *498*, *500*
  diffusion-weighted, *495*, *496*
  high angular resolution diffusion imaging, *497*, *498*
  T1-weighted, *68*, *494*, *496*
multi-graphs, 122, *123*, 543, *546*
multilayer networks, *125*, 125–126, *401*, *402*, 401–402
multiple network alignment, *391*
  definitions, 390–391
  FUSE example method, 394–397
  other methods, 397–399
  scoring alignments, 392
  SMETANA example method, 392–394
multiple testing correction, pattern mining, 325–326
multi-SNP association studies, **35**, 33–39
multi-threshold permutation correction (MTPC), 506
mutations *see* gene mutations
myoglobin, *370*

neighborhood of vertices, *119*, 119
NetAligner, pairwise network alignment, 381
netdis, 215
network(s)
  alignment-based comparison, 216–218
  alignment-free comparison, 214–216
  brains *see* brain networks
  clustering, *209*, *210*, 209–210
  comparison, 213–214
  construction, 167–168
  definition, 167
  de-noising, 212, *213*
  edges, 463, 464
  genetic interactions, 113–114, 463, 481, 536
  genotype-phenotype relationships, 464–475
  geometry, 492, *515*, *516*, 513–516
  inter-organismal, 474–475
  localization, 430–432, *435*, 440, 505
  measures, 132–140
  molecular *see* molecular networks
  molecular basis of diseases, *474*, 470–475
  motifs, 138, 197
  neuroscience, 491–492
  nodes, 194, 199, 212, 220, 435–436, 463, 464
  pathways, 537–538, *539*, *541*
  properties, *135*, 132–138
  protein-protein interactions, 111, **112**, *153*, 167–170
  similarity, 538–539, *542*, *543*
  taxonomy, 536–539
  theory *see* graph theory
  topology, 196–209, 492, 506–508
network alignment (NA), 223, 369–373, *374*, 403
  alternative formalisms, 399
  directed networks, 402
  exercises, 404–407
  hyper-networks, *403*, 403, 403
  methods, 400
  multilayer networks, 125–126, *401*, *402*, 401–402
  multiple *see* multiple network alignment
  pairwise *see* pairwise network alignment
  probabilistic networks, *400*, 399–401
  protein-protein interactions, *372*, 371–373
  search-based method, 217–218
  two-stage method, 216–218
network analysis
  Cytoscape, 574
  differential, 481
  protein-protein interactions, 173–175, *176*, 194, 199, 212, 220, *221–223*
network-based data integration, *291*, 290–291
network-based disease modules, 437–438, 440, 442
network-based statistics (NBS), 506
network-based tools, **471, 472**
network models, 138, **462**, 463
  Erdos–Renyi Random Graphs, 138–139
  geometric networks, 139
  scale-free networks, 139
NetworkBLAST, pairwise network alignment, 381
neuroscience, 490–492
  brain network analysis tools, 505–506
  brain network disorders, 517–519
  exercises, 519–520
  functional brain networks, 499–503

network geometry, 513–516
network topology, 506–513
nodes, 503–505
structural brain networks, 492–499
next generation sequencing, 2, 18–27, 70, 477
    vs. microarrays, 26–28
node(s)
    brain networks, 503–505
    conservation, 216–217, 389
    correctness/coverage, alignment scoring, 377
    network, 194, 199, 212, 220, 435–436, 463, 464
    removals, 466
node-colored graphlets, 200–201
node-link diagrams, 542–547
    animations, 551
    combining layouts, 550, *551*
    force directed layouts, 549–550
    layering visualizations, *552*, 552
    layouts, 549, *550*, 550–551
    network embedding, 550
    simple layout algorithms, 549
    visual mappings, 548–549
    visualizing networks, *547*, 547–548
non-negative matrix factorization (NMF), *295*, 294–295
    homogeneous data integration, 298–300
    precision medicine, 294–300
    solutions, 295–298
non-negative matrix tri-factorization (NMTF), *301*, 300–301
    FUSE, 394–397
    heterogeneous data integration, 305
    precision medicine, 300–306
    solutions, 301–305
normalization, cluster analysis, 246–247
Notch signaling pathway, 597

odds ratio, 5, 43
oncogenes, 595
one-mode data format, 244
optimization algorithms, 374
orbit aware graphlet counting, 206, 226
orbit aware quad census (Oaqc), **227**, 228
orbit counting algorithm (Orca), **227**, 228
orbit unaware graphlet counting, 199, 226
orbit weights, *386*
orbits, graphlets, *203*, 203, *204, 205*
ordered graphlets, *202*, 202
ovarian cancer, 305, 476
    biomarkers, *608, 609*, 608–610
    cancer stem cells, 598
overlapping clustering, 243

pairwise network alignment, *391*
    definitions, 373–375
    example method, PathBlast, 379–381
    global methods, 382–390
    other methods, 381–382
    scoring alignments, 375–379
parallel parameterized graphlet decomposition (PGD) library, **227**, 229
Parkinson's disease, 34, 473–475, 480, 518–519
partitional clustering, 243
partitioning around medoids (PAM), 261–262
path lengths, 169–170
PathBlast, pairwise network alignment, *380*, 379–381
pathogenicity, graphlet-based approach, 224
pathway enrichment analysis, 179, **180**
pathways, networks, 537–538, *539, 541*
patient subtyping, cancer precision medicine, 287
pattern enumeration tree, 330, *331*
pattern mining, 315–328
    algorithms, 329, 330, 340, 361
    confounding effect, 350, *351*
    covariate factors, 349–359
    machine learning, 313–314, 359–361
    permutation testing, 346–349
    software tools, **360**
    statistical redundancy, 341–349
    Tarone's method, 329–341
pattern occurrence indicator, 315
peak calling, **79**, 79
Pearson's chi-squared test, 323–325, *336*
permutation importance, 40
permutation testing
    algorithms, 347
    pattern mining, 346–349
personalized genetic tests (PGT), 3, 7, **13**, 45
personalized medicine *see* precision medicine
personalized oncology, 287–288, 594
phenotypes, definition, 460, 461; *see also* genotype-phenotype relationships
power-lawness, 509
PPI *see* protein-protein interactions
precision medicine, 286, *288*, 306, 596
    cancer, 287
    data integration, 287–288
    data integration methods, 290–294
    data integration types, 288–290
    drug repurposing, 287
    exercises, 306–308
    long non-coding RNAs, 88
    machine learning, 287, 306

precision medicine (cont.)
    non-negative matrix factorization, 294–300
    non-negative matrix tri-factorization, 300–306
prediction databases, *163*, **164**, 162–165, 179
predictive genetic risk models, 39–44
preferential attachment, 194, 199, 212, *213*, 220
preprocessing, cluster analysis, 242, 246–251
principal component analysis (PCA), 206, 215–216, 226, *251*, 248–251
probabilistic networks, *400*, 399–401
prostate cancer, biomarkers, 606–608
protein complex detection, 241
protein function prediction, 218–220
protein homology detection, 241–242
protein structure network (PSN), 193
protein-DNA interactions, 600
protein-protein interactions (PPI), 151–154, 179–181, 193, 536, *537*
    annotations, *171*, *370*, 371
    biases, 158–159, 438
    computational biology workflows, 175–179
    computational method types, 156
    computational prediction, 156–159
    data integration, 159–160
    databases, **112**, 160–167, 602, **603**
    dominating set, 211
    exercises, 181
    experimental detection, **155**, 154–156
    high-throughput methods (HT), 154
    human aging, 220–223
    interaction types, 155
    limitations, detection methods, 154, 155
    low-throughput methods (LT), 154
    molecular networks, 415–417
    network alignment, *372*, 371–373
    network analysis, 173–175, *176*, 194, 199, 212, 220, *221*–223
    network visualizations, 170–172
    networks, 111, **112**, *153*, 167–170
    stem cell therapy, 598–603, 603
    tissue interactomes, *479*
proteins
    cancer, 223–224, 598–603
    domains, 156
    functions, *370*, 369–371
    sequencing, 156
proteomics, 599–601
prototype based clustering algorithms, 257
proximity calculation, cluster analysis, 252–256
pruning condition, 338, *339*
    Cochran-Mantel-Haenszel test, 355–359
public databases, 2, 22, 72, 74, 427, 559–562

qualitative annotations, 172
quantitative annotations, 172

Rand index, 271
random forest methods, 37–40
random walks, 132
randomization, network properties, 431–433, *435*, 436–437
    nodes, 435–436
    topology, 433–435
rapid graphlet enumerator (RAGE), **227**, 228
RAS genes, mutations, 462
recessive models, 32
redundancy, pattern mining, *342*, 341–349
regulatory networks, 418–419
relative graphlet frequency distance (RGFD), 214
relative risk, 43
repression, transcriptional, 491
resting state networks (RSN), 499
rich-clubness, 509–510, 517
risk indicators, 43
Roadmap Epigenomics Project, 95

SAND/SAND-3D subgraph tools, **227**, 229
scale-free networks, 139
schema, 160
scoring schemes *see* alignment scoring schemes
search algorithms, graph theory, 130–131
Search Tool for the Retrieval of Interacting Genes/Proteins (STRING), *606–608*
search-based network alignment, 217–218
seed clusters, disease module analysis, 438–440
semantic similarity, alignment scoring, 376
sequencing
    alignment, 369
    by synthesis, 71
    proteins, 156
SF-GD (scale-free gene duplication and divergence), 139
short-read sequencing, 71
sickle cell disease, 461–462, 479–480
significant itemset mining, *317*, 315–317
significant pattern mining *see* pattern mining
significant subgraph mining, 317–318, *319*
silhouette values, 273, *274*
similarity networks, 538–539, *542*, *543*
simultaneous decomposition, 298, *299*
single nucleotide polymorphisms (SNP), 1
    calling and genotyping, 21–27
    definition, 6
    disease association, *10*, *41*, 31–47
    effect sizes, 5

identification, 9, 30
  significant itemset mining, 315
single-SNP association studies, 31–33
small-worldness, 508
SMETANA, multiple network alignment method, 392–394
social networks, *225*, 224–226, 423–426, *538*
  contagious diseases spread, 423–424
  social contagion, 425
  transportation, 424–425
software tools, pattern mining, **360**
spectral clustering, 132
spectral graph theory, 131–132
spinocerebellar ataxia type 1 (SCA1), 473
spreadsheets, visualizing networks, 541, *544*
standardization, cluster analysis, 246–247
static graphlets, *195*, 196, 202
statistical association testing, pattern mining, 318
statistical redundancy, pattern mining, 341–349
stem cell therapy, 594–595
  cancer stem cells, 595–598, **605**, **606**, 603–612
  exercises, 612
  protein interactions, 598–603
  tumor stemness biomarkers, 596–597, **605**, **606**, *610*, 603–612
stochastic measures, network topology, 507–510
STRING (Search Tool for the Retrieval of Interacting Genes/Proteins), *606–608*
structural brain networks, 492–499
structural consistency, 509
subcellular localization, 154, 166, 600
subgraphs, *120*, 120
  centrality, 137
  connectedness, *119*
  isomorphism problem, 373
subset/superset relationships, pattern mining, 341, *343*
support vector machines, 34–38
suppressor genes, tumors, 595
symmetric sub-structure scores, alignment scoring, 377

T1-weighted MRI scanning, *68*, *494*, *496*
targeted therapy, 594
targeting, drugs, 174
Tarone's improved Bonferroni correction, 327–328
Tarone's method, pattern mining, 329–341
ten-eleven-translocation (TET) proteins, 69

tertiary structure, 156–158
testability, pattern mining, 329, *337*, 359
1000 genomes project, 1, 18, 28–29
3C technology, chromatin conformation, 81, 83
time-respecting path, 199, 226
tissue annotation, 166
tissue interactomes, 478–482
  differential network analysis, 481
  genome-wide association studies, 479
  meta-analysis, 481–482
  PPI networks, *479*
  tools, **478**
tissue profiles, **477**
tissue-sensitive molecular networks, 475–480
tissue-specific interactions, 476
topological associated domains (TAD), chromatin conformation, **86**, 86–87
topology, networks, 196–209, 492, 506–508
training methods, data integration, *290*, 289–290
transcriptional regulation networks, 113
transcriptional, repression, 491
transivity clustering, 268–270
transportation networks, contagious diseases, 424–425
trees, graph theory, *124*, 124
tumors; *see also* cancer
  metastases, 595–598
  stemness biomarkers, 596–597, 603–612, **605**, **606**, *610*
  suppressor genes, 595
Turing machines, 117
two-mode data format, 244
two-stage network alignment, 216–218

undirected graphlets, 196, 202
undirected graphs, 118
unordered graphlets, 196, 202

validity indices, cluster evaluation, 270–271
variable importance measures (VIMs), 37
variant calling algorithms, 25–26
variety, databases, 161
variety/velocity/veracity, databases, 160
vector spaces, 116
velocity, databases, 161
veracity, databases, 161
VIM (variable importance measures), 37
visual mappings
  continuous, 548
  discrete, 548

visual mappings (cont.)
    node-link diagrams, 548–549
    passthrough, 548
visualizing networks, 170–172, 533, *534, 535,*
    540–552
    adjacency matrices, 541, *544*
    BioFabric, 542, *545*
    Cytoscape, 534
    node-link diagrams, *547*, 542–548
    spreadsheets, 541, *544*

walks, graph theory, *121*, 121
weighted graphs, 113, *123*, 123
weighted transitive graph projection
    problem (WTGPP), *269*,
    268–269
Wnt signaling pathway, 597
writer enzymes, **68**

yeast two-hybrid (Y2H) assays, 155,
    417, 427